E. D. P. De Robertis, M.D.

Professor of Cytology and
Director of the Institute of Cell Biology
University of Buenos Aires, Argentina

Francisco A. Saez, Ph.D.

Head of the Department of Cytogenetics
Institute of Investigation of Biological Sciences
Montevideo, Uruguay

E. M. F. De Robertis, Jr., M.D., Ph.D.

Senior Instructor of Biochemistry and Research Associate,
Campomar Foundation and Faculty of Sciences
University of Buenos Aires, Argentina

CELL BIOLOGY

SIXTH EDITION

W. B. SAUNDERS COMPANY
Philadelphia, London, Toronto

W. B. Saunders Company: West Washington Square
Philadelphia, PA 19105

1 St. Anne's Road
Eastbourne, East Sussex BN21 3UN, England

1 Goldthorne Avenue
Toronto, Ontario M8Z 5T9, Canada

Listed here is the latest translated edition of this book
together with the language of the translation and
the publisher.

Portuguese (*1st Edition*)—El Ateneo, Buenos Aires,
                         Argentina

French (*1st Edition*)—University of Laval, Quebec,
                      Canada

Hungarian (*2nd Edition*)—Akadeniai Kiodo,
                         Budapest, Hungary

Italian (*2nd Edition*)—Editore Nicola Zannichelli,
                       Bologna, Italy

Japanese (*2nd Edition*)—Asakura, Tokyo, Japan

Polish (*2nd Edition*)—Panstwowe Wydawnictwo,
                      Warsaw, Poland

There is also a Russian translation, *2nd Edition.*

Library of Congress Cataloging in Publication Data

Robertis, Eduardo D   P   de.

Cell biology.

First published in 1946 under title: Citología general.

Includes bibliographies and index.

1. Cytology.   I. Saez, Francisco Alberto, 1898–
   joint author.   II. Robertis, E. M. F. de, joint author.
   III. Title. [DNLM: 1. Cells. QH581 D437c]

QH581.2.R613 1975   574.8'7   74–9431

ISBN 0–7216–3043–X

Cell Biology   ISBN 0-7216-3043-X

Last digit is the print number: 9   8   7   6   5   4

PREFACE AND DEDICATION TO THE SIXTH EDITION

Dedicated to Professor W. W. Nowinski

Under the title *General Cytology* this book was first published in Spanish in 1946 by El Ateneo of Buenos Aires. The original authorship of De Robertis, Nowinski, and Saez continued until the fifth English edition, published in 1970.

It is most unfortunate that Prof. W. W. Nowinski, Ph.D., passed away just before we were to undertake the present revision and could not fulfill his task. We would like to dedicate this edition to his memory, thereby rendering tribute to a colleague and friend who was not only a scientist of foremost excellence but also a scholar of remarkably extensive interests.

To write the sixth edition it was absolutely essential that a young scientist and teacher join the long-lasting effort of the senior authors. The place of Prof. W. W. Nowinski has now been taken by E. M. F. De Robertis, Jr., M.D., Ph.D., who has had the responsibility of thoroughly revising the chapters in the section Molecular Components and Metabolism of the Cell, as well as those included under Molecular Biology. We think that this goal has been achieved and that the readers will find these chapters and the rest of the book much improved over the previous edition.

In the preface of the first edition we stated:

"This book originally arose from the need for a synthesis in the Spanish language of the most important aspects of modern cytology.

"In recent years this branch of biology has shown rapid progress and has become fundamental to the study of the structure and function of living organisms. The cell can be regarded as the vital unit of organisms and the anatomic and physiologic substrate of biologic phenomena. In its morphologic aspect, modern cytology has gone beyond simple description of structures visible to the light microscope; by the application of new methods, an analysis has been begun of submicroscopic organization — the architectural arrangement of the molecules and micelles composing living matter. In this functional aspect, it has transcended the stage of pure description of physiologic changes and seeks an explanation of them in the intimate physicochemical and metabolic processes of protoplasm. Finally, modern cytology, based on the nuclear structures, has tried to interpret and explain the phenomena of heredity, sex, variation, mutation and evolution of living organisms."

Through the three decades that have elapsed since that first edition, these postulates have continued to be valid, but progress has been so rapid and revolutionary that we have revised each edition extensively; in this new edition we have even more thoroughly revised the book. The present title, *Cell Biology,* not only stresses the profound changes that have been introduced but also emphasizes the cell as a fundamental unit in biology.

Although in recent years we have observed the extraordinary development of molecular biology, which stresses the fundamental role of macromolecules such as the proteins and nucleic acids, it is again evident that these advances should be integrated within the framework of the cell as the true structural and functional unit of living matter.

In this book the cell is analyzed at all levels of organization as seen with the various optical instruments (e.g., light and electron microscopes, x-ray diffraction) that are able to reveal its subcellular, macromolecular, and molecular architecture. At the same time, the chemical composition and metabolism of the cell are considered in terms of cytochemistry and function by an analysis of the most important manifestations of cellular activity, such as contractility, excitability, permeability, nutrition, and secretion. This integration is further stressed in the study of the macromolecules that carry biological information, the chromosomes, cell division, and the cytological and molecular bases of genetics.

This book is intended primarily as a textbook for college courses in cell biology and for students who wish to gain a general view of modern cytology for purposes of teaching or investigation in other fields of biology, such as medicine, genetics, physiology, cytogenetics, general biology, general zoology, general botany, agronomy, or veterinary medicine.

The content of the book has been organized in a manner that is most useful to the student, proceeding from simple to more complex matters. Thus, chapters that review the chemistry of the cell, the enzymes, and metabolism are at the beginning of the book, and the study of elementary macromolecular structures and membrane models introduces a consideration of the structural aspects of the cell. To keep the book within reasonable limits of size we have incorporated the new material at the expense of older, less essential material. Most figures are new, and numerous tables and diagrams serving as teaching aids have been added. The material is now divided into eight parts and 25 chapters. Each part has a short introduction that highlights the content. The titles are: Introduction to Cell Biology, Molecular Components and Metabolism of the Cell, Methods for the Study of the Cell, Units of Structure and the Plasma Membrane, The Cytoplasm and Cytoplasmic Organelles, Cellular Basis of Cytogenetics, Molecular Biology, and Cell Physiology.

All the chapters have been brought up to date, and most of them have undergone extensive changes in keeping with the rapid progress in this field of the Life Sciences. The two introductory chapters give an elementary overview of the structure of the cell and define many of the concepts that are dealt with in the realm of Cell Biology. Substantial changes have been introduced in the two chapters dealing with the chemical organization of the cell, as well as in those dealing with the plasma membrane, the cytoplasm and vacuolar system, mitochondria, and the plant cell and the chloroplast. Under the heading Cellular Basis of Cytogenetics there are now five chapters: The Interphase Nucleus and the Chromosomes, Mitosis, Meiosis, Cytogenetics, and Sex Determination and Human Cytogenetics. All these chapters have undergone extensive revisions. The four chapters that now are included under Molecular Biology have undergone so many important changes that they should be considered as completely new; emphasis has been given to the molecular biology of the eukaryotic cell and to the problem of cell regulation and differentiation.

The five chapters concerned with cell physiology have also been changed considerably, and some parts dealing with the mechanism of contraction of undifferentiated cells, the microtubules and microfilaments, and the cell and molecular neurobiology have been entirely rewritten.

One of the new features of this edition is the introduction of short summaries at the end of each section or chapter. Through this device the essential concepts that have been dealt with are emphasized for the benefit of the student.

Because of the elementary nature of this book only a few of the recent and most important references are mentioned by number in the text. Included under Additional Reading are books and general review articles that may be used as a guide to more specific literature.

We were stimulated in our task by the good reception that the previous editions of this book have received in its Spanish, Portuguese, Italian, Japanese, Hungarian, Polish, Russian, and the recent French translations. We have received suggestions and criticisms from colleagues of many countries, and all of them have contributed to the improvement of the book. We want to thank Professors George E. Palade, Elof Carlson, Bernard Strauss, and Lewis J. Kleinsmith for critically reading the former editions.

For the present edition we have been helped by the criticisms and valuable suggestions of Professors Andrew Bajer, Gerard A. O'Donovan, and other reviewers who wish to remain anonymous. We would like to thank the many colleagues around the world who have contributed original figures and tables, which have increased the value of this book.

In the preparation of the present edition we have been greatly helped by our colleagues Drs. A. Pellegrino de Iraldi, D. Zambrano, C. Tandler, Máximo Drets, César Vásquez, Ruben Adler, and Georgina Rodríguez de Lores Arnaiz.

Most of the new diagrams in this edition were ably drawn by Mrs. María Elvira Aued de Rau. The manuscript was typewritten by Mrs. Alicia Fernández de Candame, Julia Elena Connaughton de Wertheim, and Ana Cristina Marazzi de De Robertis.

The W. B. Saunders Company, through Mr. Richard Lampert, has helped immensely with advice on editorial matters and has improved the appearance of the new edition considerably.

A book that attempts to translate into elementary terms the extraordinary advances of modern cytology is possible only with the unselfish collaboration of all who contribute to the progress of this challenging field of biological knowledge.

E. De Robertis

F. A. Saez

E. M. F. De Robertis, Jr.

CONTENTS

Part One
INTRODUCTION TO CELL BIOLOGY

Part Two
MOLECULAR COMPONENTS AND METABOLISM OF THE CELL

Chapter Three
CHEMICAL COMPONENTS OF THE CELL 37

Chapter Four
ENZYMES, BIOENERGETICS, AND CELL RESPIRATION 58

Part Three
METHODS FOR THE STUDY OF THE CELL

Chapter Five
INSTRUMENTAL ANALYSIS OF BIOLOGICAL STRUCTURES 81

Chapter Six
METHODS FOR CYTOLOGIC AND CYTOCHEMICAL ANALYSIS 101

Part Four
UNITS OF STRUCTURE AND THE PLASMA MEMBRANE

Chapter Seven
ELEMENTARY UNITS OF STRUCTURE IN BIOLOGICAL SYSTEMS.............. 133

Chapter Eight
THE PLASMA MEMBRANE.. 145

Part Five
THE CYTOPLASM AND CYTOPLASMIC ORGANELLES

Chapter Nine
THE CYTOPLASM AND VACUOLAR SYSTEM.. 171

Chapter Ten

MITOCHONDRIA 200

Chapter Eleven

THE PLANT CELL AND THE CHLOROPLAST 231

Part Six

CELLULAR BASIS OF CYTOGENETICS

Chapter Twelve

THE INTERPHASE NUCLEUS AND THE CHROMOSOMES 257

Chapter Thirteen

MITOSIS **274**

Chapter Fourteen

MEIOSIS **292**

Chapter Fifteen

CYTOGENETICS. CHROMOSOMAL BASES OF GENETICS **312**

Part Seven
MOLECULAR BIOLOGY

Chapter Twenty
CELL DIFFERENTIATION AND CELLULAR INTERACTION 441

Part Eight
CELL PHYSIOLOGY

Chapter Twenty-one
CELL PERMEABILITY, ENDOCYTOSIS, LYSOSOMES, AND
PEROXISOMES 471

Chapter Twenty-two
PRIMITIVE CELL MOVEMENTS. CILIA, CENTRIOLES,
MICROTUBULES, AND MICROFILAMENTS 497

Chapter Twenty-three

MOLECULAR BIOLOGY OF MUSCLE **523**

Chapter Twenty-four

CELLULAR AND MOLECULAR NEUROBIOLOGY **541**

Chapter Twenty-five

CELL SECRETION **575**

INDEX **595**

Part One

INTRODUCTION TO CELL BIOLOGY

The first two chapters of this book may be considered an elementary introduction to the study of the cell as a biological unit. In living matter there is an integration of different levels of organization, from which the manifestations of life originate. From the morphologic viewpoint, these levels of organization are related to those aspects which can be resolved with the different means of observation: the human eye (anatomy), the various types of microscopes (histology and cytology), and the other methods which facilitate a deeper probe into molecular biology and the ultrastructure of the cell.

Chapter 1 contains a brief account of the history of cell biology, with particular emphasis on the *Cell Theory* and the correlations of cell biology with genetics, cell physiology, and biochemistry. The modern aspects of ultrastructure and molecular biology are also presented. For the benefit of the student literary sources for further reading in cell biology are provided.

In the second chapter the main characteristics of prokaryotic and eukaryotic organisms are emphasized. The bacterium *Escherichia coli* is described as a prokaryotic cell; it is certainly the best known example from the molecular and genetic points of view. It is important that, from the beginning, the reader recognize the similarities and differences between these two types of organisms, although these points will be taken up again in other chapters.

Chapter 2 also deals with the general structure of cells in the living state and after fixation. The main components of the nucleus and cytoplasm are mentioned. It introduces the concept of the life cycle of the cell in direct relation to the processes of mitotic and meiotic division. The concept that the chromosomes are entities able to duplicate themselves

1

and to maintain their morphology and function through successive divisions is introduced. The study of the morphologic constants of the chromosomes, e.g., their number, shape, size, primary, and secondary constrictions, satellites and so forth, which, in general, characterize the karyotype of a species is also included. Chapter 2 is prerequisite to the fifth part of this book where the cellular bases of cytogenetics will be studied. It is important, from the very beginning of these studies, that the reader be exposed to these general concepts to understand better the modern developments in cell biology.

one

INTRODUCTION. HISTORY AND GENERAL CONCEPTS OF CELL BIOLOGY

Ancient philosophers and naturalists, particularly Aristotle in Antiquity and Paracelsus in the Renaissance, arrived at the conclusion that "All animals and plants, however complicated, are constituted of a few elements which are repeated in each of them." They were referring to the macroscopic structures of an organism, such as roots, leaves, and flowers common to different plants, or segments and organs that are repeated in the animal kingdom. Many centuries later, owing to the invention of magnifying lenses, the world of microscopic dimensions was discovered. It was found that a single cell can constitute an entire organism as in Protozoa, or it can be one of many cells that are grouped and differentiated into tissues and organs, to form a multicellular organism.

The cell is thus a fundamental structural and functional unit of living organisms, just as the atom is the fundamental unit in chemical structures. The cell can be considered to be an organism in itself, often very specialized and composed of many elements, the sum of which not only constitutes the cellular unit, but has particular significance to the organism as a whole. If by mechanical or other means cellular organization is destroyed, cellular function is likewise altered. Although some vital functions may persist (such as enzymatic activity), the cell becomes disorganized and dies.

The development and refinement of microscopic techniques made it possible to obtain further knowledge of cellular structure, not only as it appears in the cell killed by fixation, but also as seen in the living state. Biochemical studies have demonstrated that the products of living matter, and even the living matter itself, are composed of the same elements that constitute the inorganic world. Biochemists have isolated from the complex mixture of cell constituents not only inorganic components, but much more complex molecules, such as proteins, fats, polysaccharides, and nucleic acids. Such biochemical studies have also demonstrated the "oneness" of the entire living world. Today it is known that from bacterium to man the chemical machinery is essentially the same, in both its structure and its function (see Monod, 1971).

LEVELS OF ORGANIZATION IN BIOLOGY

The advance of knowledge of the cell has produced a fundamental change in the interpretation of cellular structures. For example, it has been demonstrated that beyond the organization visible with the light microscope are a number of more elementary structures at the macromolecular level that constitute the "ultrastructure" of the cell. We are living in the era of *molecular biology*, that is, the study of the shape, aggregation, and orientation of the molecules that compose the cellular system as a unit.

Modern studies on living matter demonstrate that there is a combination of *levels of organization* which are integrated, and that

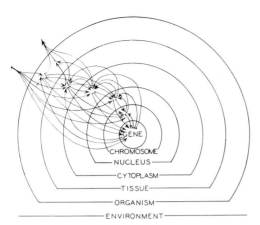

Figure 1–1. The levels of biological organization represented by concentric, interacting shells. (From Weiss, P., *Canad. Cancer Conf.,* 5:243, 1963. © Academic Press, Inc.)

this integration results in the vital manifestations of the organism. The concept of levels of organization as developed by Needham[1] and others implies that in the entire universe—in both the nonliving and living worlds—there are such various levels of different complexity that "The laws or rules that are encountered at one level may not appear at lower levels." This concept can be applied to the different structural constituents of a cell or to the association of numerous cells in a tissue.

In Figure 1–1 the various levels of biological organization are represented by concentric interacting shells—each shell being the environment for the next nearest inner one. The intricacy of the interrelations between the different levels is indicated by the network of arrows interconnecting them, thus giving an idea of the complexities involved in living matter.

Although both inorganic and living matter are composed of the same atoms, there are fundamental differences between them. According to present concepts, in the nonliving world there is a continuous tendency toward reaching a thermodynamic equilibrium with a random distribution of matter and energy, whereas in the living organism there is a high degree of structure and function that is maintained by energy transformations based on continuous input and output of matter and energy.[2]

LIMITS AND DIMENSIONS IN BIOLOGY

Table 1–1 shows the limits that separate the study of biological systems at various dimension levels. The boundaries between levels of organization are imposed artificially by the resolving power of the instruments employed. Note that a great deal of overlapping exists. The human eye cannot resolve (discriminate between) two points separated by less than 0.1 mm (100 μm). Most cells, in general, are much smaller and must be studied under the full resolving power of the light microscope (0.2 μm). However, most cellular components are even smaller and require the resolution of the electron microscope.

From a morphologic point of view, all these fields of biology fall within the discipline of *anatomy* (Gr., *temnein,* to cut; *ana,* up), which is the separation of the different components in such a way as to identify and study them as both isolated parts and integrated parts of the whole organism. Bennett,[3] in a clear interpretation of these concepts, says that "the operational approaches to all branches of anatomy have essential features in common." Whether working in the field of gross, microscopic, or molecular anatomy,

TABLE 1–1. Various Fields of Biology

Dimension	*Field*	*Structures*	*Method*
0.1 mm (100 μm) and larger	anatomy	organs	eye and simple lenses
100 μm to 10 μm	histology	tissues ⎫	various types of light microscopes,
10 μm to 0.2 μm (200 nm)	cytology	cells, bacteria ⎬	x-ray microscopy
200 nm to 1 nm	submicroscopic morphology ultrastructure molecular biology	cell components, viruses	polarization microscopy, electron microscopy
smaller than 1 nm	molecular and atomic structure	arrangement of atoms	x-ray diffraction

one generally proceeds by cutting apart, in some way, the objects of interest. The methodological approach is the same whether a scalpel is used to dissect the cadaver, sections are made for the light or electron microscope, or subcellular components are separated by homogenization and centrifugation. The resolution of different structures into their molecular or atomic elements by means of optical instruments using different electromagnetic waves may be considered still another form of "anatomy." (Fig. 1–2).

To construct a mental image of the molecular organization of a biological system, one should start with knowledge of the main constituent molecules, particularly those of high molecular weight, such as nucleic acids, proteins and polysaccharides. Lipids, although of smaller molecular size, also play an important role as structural components of the cells.

These components must be studied from the point of view of their size, shape, charge, stereochemical characteristics, and main reacting groups. Such studies are difficult when the molecules are isolated or are distributed at random. Frequently, however, the molecules arrange themselves into repeating periodic structures, which can be analyzed with crystallographic techniques. In recent

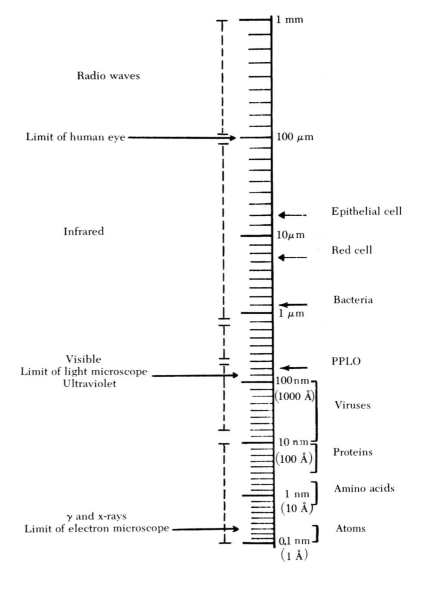

Figure 1-2. Logarithmic scale of microscopic dimensions. Each main division represents a size ten times smaller than the division above. **To the left,** the position of the different wavelengths of the electromagnetic spectrum and the limits of the human eye, the light microscope and the electron microscope. **To the right,** the sizes of different cells, bacteria, PPLO (the smallest known living organism), viruses, molecules and atoms. (Modified from M. Bessis.)

TABLE 1–2. Relationships Between Linear Dimensions and Weights in Cytochemistry*

Linear Dimension	Weight	Terminology
1 cm	1 gm	conventional biochemistry
1 mm	1 mg or 10^{-3} gm	microchemistry
100 μm	1 μg or 10^{-6} gm	histochemistry } ultramicrochemistry
1 μm	1 $\mu\mu$g (or 1 picogram or 10^{-12} gm)	cytochemistry }

*From Engström, A., and Finean, J. B., *Biological Ultrastructure,* New York, Academic Press, Inc., 1958. Copyright 1958, Academic Press, Inc.

years great advances have been made in the detailed x-ray diffraction analysis of the molecular configuration of proteins, nucleic acids and of larger molecular complexes, such as certain viruses. This important field is within the realm of *molecular biology* (Table 1–1).

At a cytologic level, *ultrastructure* or *submicroscopic morphology* is concerned with the larger repeating units that can be analyzed with microscopic techniques. The first technique to be applied, about a century ago, was *polarization microscopy.* German workers, beginning with Nägeli, first recognized ordered structures within biological systems. Later these studies became quantitative and were extended considerably by the work of W. J. Schmidt. This technique makes use of the effect that anisotropic structures have on polarized light (see Chap. 5).

Finally, the most important tool in the study of submicroscopic morphology is the *electron microscope.* With this instrument, direct information can be obtained about structures ranging from 0.4 to 200 nm or more, thus bridging the gap between observations with the light microscope and the world of macromolecules. Results obtained by application of electron microscopy have changed the field of cytology so much that a large part of the present book is devoted to discussions of the achievements obtained by this technique.

In Figure 1–2 the sizes of different cells, bacteria, viruses, and molecules are indicated on a logarithmic scale and compared with the wavelengths of various radiations, as well as the limits of resolution of the eye, the light microscope and the electron microscope. Note that the light microscope (limit of resolution 200 nm) introduces a 500-fold increase in resolution over the eye (10^5 nm), and the electron microscope (0.4 nm) provides a 500-fold increase over the light microscope.

Some cytologic structures, such as mitochondria, centrioles, chromosomes, and nucleoli, can be resolved with the optical microscope; but many more, such as ribosomes, the plasma membrane, myofilaments, chromosomal microfibrils, microtubules, and synaptic vesicles can be resolved only with the electron microscope.

In several chapters of this book examples are given of the different levels of organization of biological structures through the use of different magnifying instruments. From the very start, however, the reader must become aware of the importance of these concepts and be able to visualize the proper level of organization that is being considered, i.e., anatomic, histologic, cytologic, ultrastructural, or molecular (Table 1–1).

Table 1–2 shows the general relationships between some of the linear dimensions used in cytology and the weight of material used in different fields of chemical analysis of living matter. Familiarity with these relationships is essential to the study of cell and molecular biology. The weight of the important components of the cell is expressed in picograms (1 pg = 1 $\mu\mu$g or 10^{-12} gm), or in *daltons.* The dalton is the unit of molecular weight (MW); one dalton equals the weight of a hydrogen atom. For example, a water molecule weighs 18 daltons and a molecule of hemoglobin weighs 64,500 daltons.

The cell is the fundamental structural and functional unit of living organisms. In living matter there are levels of organization that are interrelated in a complex manner and are maintained by energy transformations. The boundaries between these levels of organization are imposed by the resolving power of the eye (0.1 mm), the light microscope (0.2 μm) and the electron microscope. These limits bring about the differentiation between the fields of anatomy, cytology, ultrastructure, and molecular biology. The electron microscope provides direct information about structures ranging in size from 0.4 to 200 nm. It is important to correlate the boundaries with the spectrum of electromagnetic waves and the actual dimensions of cells, bacteria, viruses, and proteins (Fig. 1–2). The student should have a firm understanding of the relationships between linear dimensions and the weights used in cytochemistry. For example, the weight of cell components, such as nucleic acids, is expressed in picograms (10^{-12} gm).

SUMMARY:
Levels of Organization

HISTORY OF CELL BIOLOGY

Cell biology (or as it was formerly called, *cytology*) is one of the youngest branches of the life sciences. It was recognized as a separate discipline by the end of the last century. The early history is intimately bound to the development of optical lenses and to the combination of lenses in the construction of the compound microscope (Gr. *mikros* small + *skopein* to see, to look).

The term *cell* (Gr., *kytos*, cell; L., *cella*, hollow space) was first used by Robert Hooke (1665) to describe his investigations on "the texture of cork by means of magnifying lenses." In these observations, repeated by Grew and Malpighi on different plants, only the cavities ("utricles" or "vesicles") of the cellulose wall were recognized. In the same century and at the beginning of the next, Leeuwenhoek (1674) discovered free cells, as opposed to the "walled in" cells, of Hooke and Grew. Leeuwenhoek observed some organization within cells, particularly the nucleus in some erythrocytes. For more than a century afterward, the observations of these men were all that was known about the cell.

Cell Theory

More directly related to the origin of cell biology was the establishment of the *cell theory,* one of the broadest and most fundamental of all biological generalizations. It states in its present form that all living beings—animals, plants, or protozoa—are composed of cells and cell products. This theory resulted from numerous investigations that started at the beginning of the 19th century (Mirbel, 1802; Oken, 1805; Lamarck, 1809; Dutrochet, 1824; Turpin, 1826), and finally led to the studies of the botanist Schleiden (1838) and the zoologist Schwann (1839), who established the theory in a definite form.

The cell theory has had wide-ranging effects in all the fields of biological research. It was established immediately that every cell is formed by the division of another cell. Much later, with the progress of biochemistry, it was shown that there are fundamental similarities in the chemical composition and metabolic activities of all cells. The function of the organism as a whole was also recognized to be a result of the sum of the activities and interactions of the cell units.[4]

The cell theory was soon applied to pathology by Virchow (1858). Kölliker extended it to embryology after it was demonstrated that the organism develops from the fusion of two cells—the spermatozoon and the ovum.

A more general conclusion was reached at the same time by investigators such as Brown (1831), who established that the nucleus is a fundamental and constant component of the cell. Others (Dujardin, Schultze, Purkinje, von Mohl) concentrated on the description of the cell contents, termed the *protoplasm.*

Thus the primitive idea of *cell* was transformed into the more sophisticated concept of a mass of protoplasm limited in space by a cell membrane and possessing a nucleus. The protoplasm surrounding the nucleus became known as the *cytoplasm* to distinguish it from the *karyoplasm,* the protoplasm of the nucleus.

Once these fundamental theories and concepts were established, the progress of cytologic knowledge was extremely rapid. The extraordinary changes produced in the nucleus at each cell division attracted the attention of a great number of investigators. For example, the phenomena of *amitosis,* or direct division (Remak), and of indirect division were discovered by Flemming (in animals) and by Strasburger (in plants). Indirect division was also called *karyokinesis* (Schleicher, 1878) or *mitosis* (Flemming, 1880). It was proved that fundamental to mitosis is the formation of nuclear filaments, or *chromosomes* (Waldeyer, 1890), and their equal division between the nuclei (daughter cells). Other discoveries of importance were the fertilization of the ovum and the fusion of the two pronuclei (O. Hertwig, 1875). In the cytoplasm the aster (van Beneden, Boveri), the mitochondria (Altmann, Benda) and the reticular apparatus (Golgi) were discovered.

Even while studying tissues as cellular aggregates, biologists concentrated more and more on the cell as a fundamental unit of life. In 1892 O. Hertwig published his monograph *Die Zelle und das Gewebe* in which he attempted to achieve a general synthesis of biological phenomena, based on the characteristics of the cell, its structure and function. He thus created cytology as a separate branch of biology.[4]

If one follows the development of cell biology in the present century it is evident that there were two main reasons for the advance of cytologic knowledge: (1) the increased resolving power of instrumental analysis, essentially the introduction of electron microscopy and x-ray diffraction techniques, and (2) the convergence with other fields of biological research, especially genetics, physiology, and biochemistry.

Cytology and Genetics: Cytogenetics

By the middle of the 19th century the universality of cell division as the central phenomenon in the reproduction of organisms was established, and Virchow expressed it in the famous aphorism *"Omnis cellula e cellula."* From this time on, the study of cells and of heredity and evolution converged, as was well stated by Wilson: "Heredity appears as a consequence of the genetic continuity of the cells by division."

Observations on the germ cells made by van Beneden, Flemming, Strasburger, Boveri, and others gave support to the theory of the continuity of germ plasm proposed by Weissmann in 1883. This theory stated that the transference of hereditary factors from one generation to the next takes place through the continuity of what he called *germ plasm,* located in the sex elements (spermatozoon and ovum), and not through somatic cells.

The discovery of fertilization in animals, foreseen by O. Hertwig but observed directly by H. Fol (1879), and in plants, by Strasburger, led to the theory that the cell nucleus is the bearer of the physical basis of heredity. Furthermore, Roux postulated that chromatin, the substance of the nucleus that constitutes the chromosomes, must have a linear organization. Weissmann stated that the hereditary units are disposed along the chromosomes in an orderly manner.

The fundamental laws of heredity were discovered by Gregor Mendel in 1865, but at that time the cytologic changes produced in the sex cells were not sufficiently known to permit an interpretation of the independent segregation of hereditary characters (see Chapter 15). For this and other reasons, little attention was paid to Mendel's work until the botanists Correns, Tschermack and De Vries in 1901 independently rediscovered Mendel's laws. At that time cytology was advanced enough so that the mechanism of distribution of the hereditary units postulated by Mendel could be understood and explained. It was known that the somatic cells have a double, or *diploid,* hereditary constitution, whereas in the reproductive cells or gametes this constitution is single, or *haploid.* In addition, cytologists had observed that the cycle the chromosomes undergo in *meiosis* of germ cells was related to hereditary phenomena.

In direct accord with these findings, McClung (1901–1902) suggested that sex determination was related to some special chromosomes; this theory was later corroborated by Stevens and Wilson (1905). The experimental demonstration of the chromosome theory of heredity was finally established by Boveri and Baltzer, but it was Morgan and his collaborators, Sturtevant and Bridges, who assigned to the *genes* (Johannsen), or hereditary units, definite loci within the chromosomes. Thereafter experimental research on heredity and evolution became a separate branch of biology, which Bateson in 1906 called *genetics*. Almost from the beginning, however, the science of genetics maintained a close relationship with cytology. From the convergence of cytology and genetics *cytogenetics* originated (see Chapter 15). In the past decade the study of genetics has become linked to biochemistry and has reached the molecular level. Thus the new fields of *biochemical* and *molecular genetics* were established.

Cytology and Physiology: Cell Physiology

Most early cytologic knowledge was based on observations of fixed and stained cells and tissues; this led to the formation of various theories regarding the physicochemical structure of protoplasm. By 1899 interest shifted toward the study of living cells, principally as a result of the work of Fischer and Hardy, who showed that several of the structures observed in fixed cells could be reproduced by the action of fixatives on colloidal models. Various types of movement, such as cyclosis (cytoplasmic streaming), ameboid motion, and ciliary, flagellar, and muscular contraction, were studied at a cellular level.

At the end of the 19th century, Overton advanced the theory that the cell membrane was a lipoidal film. Michaelis made membrane models to study the passage of substances and did the first vital staining of mitochondria. However, the actual technique of vital staining (methylene blue) was introduced by Ehrlich in 1881. The basic concepts of cell irritability and nerve function were established by the middle of the 19th century by Du Bois-Reymond; physiological techniques were then developed to aid in the study of these cells and to measure action potentials and nerve currents.

An important avenue to the study of the living cell was opened in 1909 by Harrison, who demonstrated that nerve cells from an embryo could grow and differentiate in vitro. This gave rise to the technique of *tissue culture,* which, together with the work of Carrel, had a great impact on cytology. With the isolation of pure strains of cells, tissue culture became an ideal technique for the study of structure and behavior of living cells. This analysis was greatly improved by the introduction of phase contrast microscopy and by the use of microcinematography (Lewis and Lewis, Pomerat).

Another method for the study of the physicochemical properties of the living cell—*microsurgery*—came from the field of bacteriology. At the beginning of the twentieth century, Schouten and Barber used fine micropipets, moved by precision instruments, to isolate and culture single bacteria. In 1911 Kite adapted this method to cytology. Levi, Peterfi, Chambers and others have perfected techniques of intracellular operations and have obtained data on the viscosity, hydrogen concentration, and other physicochemical properties of the cell.

Among the important phenomena studied in cell physiology are: the nature of the cell membrane and of active transport across membranes, the reaction of cells to changes in environment, the mechanisms of cell excitability and contraction, cell nutrition, growth, secretion, and other manifestations of cellular activity.

Cytology and Biochemistry: Cytochemistry

Another modern branch of cell biology is *cytochemistry,* the result of the convergence of methods and sciences devoted to the chemical and physicochemical analysis of living matter. Among many outstanding biochemical studies were those of Fischer and Hofmeister in 1902, who independently recognized that the protein molecule consists of a small number of amino acids united by a peptide bond. Of similar importance to cell biology were the earlier investigations of Miescher (1869) and Kossel (1891), who, by the analysis of pus cells, spermatozoa, hemolyzed erythrocytes of birds, and other cell types, isolated the nucleic acids, whose basic role in heredity and protein synthesis has been recognized only recently.

Another great advance was the introduction into biological thinking by Ostwald of the concept of catalytic activity and the discovery that enzymes are the molecular entities used by the cell to produce the various types of energy transformations necessary for maintenance of living activities. The main types of cellular oxidations were discovered by Wieland (1903) and by Warburg (1908), but the final mechanism was discovered much later by Keilin (1934).

Because of the emphasis on morphology, cytologists were at first very slow in grasping the importance of the biochemical approach; at the same time, biochemists, because of the emphasis on organic chemistry, had no interest in cell structure and were concerned principally with isolating chemical components and studying elementary enzyme reactions. The point of convergence of cytology and biochemistry can be traced back to 1934, when Bensley and Hoerr isolated mitochondria from cells in large enough quantities to permit analysis by chemical and physicochemical methods.

This direction was followed with great success by Claude, Hogeboom, and others, and led to the conclusion that mitochondria are centers of cellular oxidations.

Advances in cell fractionation have been of the greatest importance to biology and biochemistry, especially with the development of radioactive tracer techniques which permit a dynamic approach to the study of cell metabolism. Another great advance was the use of the electron microscope for the observation and characterization of cell fractions. These advances have led to the isolation from many different cells not only of mitochondria, but also of chloroplasts, nucleoli, nerve endings, the Golgi complex, nuclei, chromosomes, ribosomes, the mitotic apparatus, and other cell components described in different chapters of this book.

Modern cytochemistry has also developed along the lines of microchemical and ultramicrochemical analysis by means of techniques for assay of minute quantities of material and the isolation of single cells and even parts of cells (see Table 1–2). Chemical analysis can be carried out by cytophotometry, which facilitates the study of the localization of nucleic acids and proteins within parts of a single cell.

Another important branch of cytochemistry arose from the application of numerous enzyme reactions which could be observed under the light and electron microscopes. This last approach is of particular interest since it combines cytochemistry and ultrastructure and thus permits study of the localization of enzymes at the level of resolution of the electron microscope. Of similar importance have been the autoradiographic studies on the localization of radioactive tracers in different cellular structures (see Chapter 6).

Ultrastructure and Molecular Biology

In studying the limits and dimensions in biology (see Tables 1–1 and 1–2), the impact of instrumental analysis was mentioned and the modern fields of ultrastructure and molecular biology were delineated. These are the most advanced branches of biology in which the merging of cytology with biochemistry, physicochemistry, and especially macromolecular and colloidal chemistry becomes increasingly complex. Knowledge of the submicroscopic organization or ultrastructure of the cell is of fundamental importance, because practically all the functional and physicochemical transformations take place within the molecular architecture of the cell and at a molecular level.

The following advances are having an extraordinary impact on biology: the discovery that in the protein molecule the exact sequence of amino acids and the three-dimensional arrangement of the polypeptide chain have a direct relationship to definite biological properties; the studies of active groups in different enzymes; the molecular model of DNA suggested by Watson and Crick in 1953; as well as all the recent knowledge on the stereochemistry of macromolecules. Molecular biology is thus providing valuable information to the fields of genetics (through molecular genetics), biochemistry, and even pathology—the latter with the establishment of molecular diseases. The student is referred to stimulating reviews of the major achievements of molecular and cell biology in articles by Stent (1968) and Herrmann (1968) and to the book by Monod (1971).

Both in ultrastructure and molecular biology the integration between morphology and physiology becomes so intimate that it is impossible to separate them; the concepts of *form* and *function* fuse into inseparable unity.

In summary, it can be said that modern cell biology approaches the problems of the cell at all levels of organization—from molecular structure on. Cell biology is therefore the modern science in which genetics, physiology, and biochemistry converge. Modern cell biologists, without losing sight of the cell as a morphologic and functional unit within the organism, must be prepared to use all the methods, techniques, and concepts of the other sciences and to study biological phenomena at all levels. This is a great challenge, but there is no other way to proceed if the life of the cell and of the organism is to be interpreted mechanistically, i.e., on the basis of combinations and associations of atoms and molecules.

LITERARY SOURCES IN CELL BIOLOGY

The preceding considerations of the present scope of cell biology demonstrate why the sources of literature are wide and multidisciplinary. Current studies are presented at scientific meetings and are published in specialized periodicals. Of the long list of literary sources that could be made, only a few of the more specific ones are mentioned: *Journal of Cell Biology, Experimental Cell Research, Journal of Molecular Biology, Journal of Ultrastructure Research, Zeitschrift für Zellforschung, Journal de Microscopie, Journal of Cell Science, Journal of Cellular and Comparative Physiology, Journal of General Physiology, Chromosoma, Cytogenetics, Heredity,* and *Hereditas.* Papers on cell biology are frequently published in more general periodicals such as *Nature, Science, Proceedings of the Royal Society, Proceedings of the National Academy of Sciences* (Wash.), *Comptes Rendus de l'Académie des Sciences, Naturwissenschaften* and *Experientia,* or even in such specialized publications as *Biochimica et Biophysica Acta, Biochemical Journal* or *Journal of Biological Chemistry.*

Reviews of recent advances are found in the *International Review of Cytology, Advances in Cell and Molecular Biology, Quarterly Review of Biology, Physiological Reviews, Biological Reviews, Advances in Genetics, Plant Physiology* and others.

Very useful in the compilation of a bibliography are special journals which give titles of papers or abstracts of the literature. These journals include: *Biological Abstracts, Index Medicus, Excerpta Medica, Chemical Abstracts* and *Berichte über die Wissenschaftliche Biologie.* Journals such as *Current Contents* or *Bulletin Signalétique du Conseil des Recherches* publish titles of all papers which appear in various journals.

In addition there are many monographs, compendia, and textbooks, that will be mentioned in later chapters of this book, that cover various specialized subjects of cell biology. The most recent, and of widest coverage, are *The Cell,* in six volumes, edited by Brachet and Mirsky, and *Handbook of Molecular Cytology,* edited by A. Lima-de-Faría (1969).

REFERENCES

1. Needham, J. (1936) *Order and Life.* Yale University Press, New Haven, Conn.
2. Bertalanffy, L. von. (1952) *Problems of Life.* John Wiley & Sons, New York.
3. Bennett, H. S. (1956) *Anat. Rec., 125*:2.
4. Hughes, A. (1959) *History of Cytology.* Abelard-Schuman, London and New York.

ADDITIONAL READING

Alexander, J. (1948) *Life: Its Nature and Origin.* Reinhold Publishing Corp., New York.
Baker, J. R. Five articles evaluating the cell theory. *Quart. J. Micr. Sci.,* 1948, *89*:103; 1949, *90*:87; 1952, *93*:157; 1953, *94*:407; 1955, *96*:449.

Bernal, J. D., and Synge, A. (1973) *The origin of life*. In: *Readings in Genetics and Evolution*. Oxford University Press, London.

Brachet, J., and Mirsky, A. E. (1959–1961) *The Cell*. 6 volumes. Academic Press, New York.

Cairns, J., Stent, G. S., and Watson, J. D., eds. (1966) Phage and the origins of molecular biology. *Cold Spring Harbor Laboratory of Quantitative Biology,* New York.

Caullery, M., and Leroy, J. F. (1966) Cytology and histology. In: *Science in the Nineteenth Century*. (Taton, Rene, ed.) Thames and Hudson, London and New York.

Dowben, R. M. (1971) *Cell Biology*. Harper and Row Pub. Co., New York.

Heilbrunn, L. V., and Weber, F., eds. (1953–1959) *Protoplasmatologia, Handbuch der Protoplasmaforschung*. Springer, Vienna.

Herrmann, H. (1968) This is the cell biology that is. *Bull. Inst. of Cell. Biol.* (Univ. of Conn., Storrs, Conn.) *10*:1.

Hughes, A. (1959) *History of Cytology*. Abelard-Schuman, London and New York.

Lima-de-Faría, A., ed. (1969) *Handbook of Molecular Cytology*. North-Holland Pub. Co., Amsterdam.

Loewy, A. G., and Siekevitz, P. (1970) *Cell Structure and Function*. 2nd ed., Holt, Rinehart and Winston, Inc., New York.

Monod, J. (1971) *Chance and Necessity*. Random House, Inc., New York.

Nordenskiöld, E. (1966) *The History of Biology*. Tudor Pub. Co., New York.

Olby, R. C. (1966) *Origin of Mendelism*. Constable, London.

Reinert, J., and Ursprung, H. eds. (1971) *Origin and Continuity of Cell Organelles*. Springer-Verlag, Berlin.

Singer, C. (1950) *A History of Biology*. Henry Schuman, New York.

Stent, G. S. (1968) That was the molecular biology that was. *Science, 160*: 390–395.

Watson, J. D. (1970) *Molecular Biology of the Gene*. W. A. Benjamin, Inc., New York.

Weiss, P. (1963) Cell interactions. *Canad. Cancer Conf., 5*:241–276. Academic Press, Inc., New York.

Wilson, E. B. (1937) *The Cell in Development and Heredity*. The Macmillan Co., New York.

Wolfe, S. L. (1972) *Biology of the Cell*. Wadsworth Publishing Co., Inc., Belmont, Calif.

two

GENERAL STRUCTURE
OF THE CELL

PROKARYOTIC CELLS

The typical cell, with the nucleus and cytoplasm and all the cellular organelles, which is described in this book, is not the smallest mass of living matter or protoplasm (Gr., *protos*, first + *plasma*, formation); simpler or more primitive units of life exist. Thus, unlike the higher types of cells which have a true nucleus (eukaryotic cells), prokaryotic cells (Gr., *karyon*, nucleus), which comprise bacteria and blue-green algae, lack a nuclear envelope, and contain a nuclear substance that is mixed or is in direct contact with the rest of the protoplasm.

The discovery of the *viruses* at the end of the 19th century shed new light on knowledge of these organisms. Recognized at first by the property of being able to pass through the pores of porcelain filters and by the pathologic changes they produced in cells, all viruses can now be visualized with the electron microscope. Viruses can be recognized morphologically and their macromolecular organization can be studied. Although they have properties common to living organisms, such as autoreproduction, heredity, and mutation, viruses are dependent on the host's cells, and are considered obligatory parasites.

The Bacterial Cell

Although this book is dedicated to the more complex eukaryotic cells, it is important to know that most of the present knowledge of molecular biology stems from the study of viruses and bacteria. A bacterial cell such as

Escherichia coli (E. coli) is easily cultured in an aqueous solution containing glucose and some inorganic ions. In this medium, at 37°C, the cell mass doubles and divides in about 60 minutes. This time — the *generation time* — can be reduced to 20 minutes if purines and pyrimidine bases, the precursors of nucleic acids, as well as amino acids, are added to the medium.

As shown in Figure 2–1, one cell of *E. coli* is about 2μm long and 0.8μm thick. It is surrounded by a rigid *cell wall*, 10nm or more thick, containing protein, polysaccharide, and lipid molecules. Inside the cell wall there is the true *cell* or *plasma membrane,* a lipoprotein structure that constitutes a molecular barrier to the surrounding medium. This plasma membrane, by controlling the entrance and exit of small molecules and ions, contributes to the establishment of a special internal milieu for the protoplasm of the bacterium. It is interesting that enzymes involved in the oxidation of metabolites, and which constitute the *respiratory chain,* are associated with this plasma membrane. In eukaryotic cells these enzymes are confined to special organelles in the cytoplasm, the mitochondria. Under the electron microscope (Fig. 2–2) it is possible to recognize light *nuclear* regions or *nucleoids,* where the chromosome of the bacteria, formed by a *single circular molecule of deoxyribonucleic acid (DNA),* is present. It is important to remember that this DNA, which is about 1 mm long (10^6 nm) contains all the genetic information of the organism. In fact, the DNA molecule in this bacterium contains sufficient genetic information to code 2000 to 3000

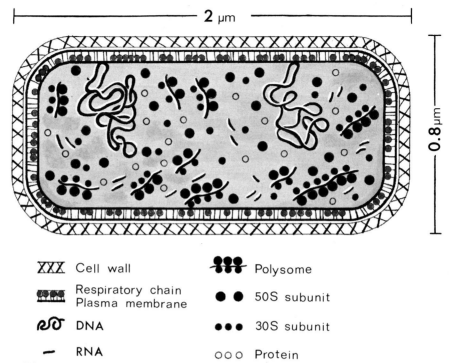

XXX	Cell wall	Polysome
▦	Respiratory chain Plasma membrane	●●	50S subunit
ℛℐ𝒪	DNA	●●●	30S subunit
─	RNA	○○○	Protein

Figure 2–1. Diagram of a single cell of the bacterium *Escherichia coli* containing two chromosomes 1 mm long (10^6 nm) attached to the cell membrane. 50S and 30S refer to ribosomal subunits; see Chapter 18.

different proteins. The DNA molecule is folded and packed within the nuclear region; it lies free in the protoplasm, and is not separated by a nuclear envelope as in the eukaryotic cell. In Figure 2–1 two chromosomes are shown because replication of the DNA has occurred and the cell is ready to divide. An important point shown in this diagram is the attachment of the DNA to the plasma membrane.

Surrounding the DNA, in the dark region of the protoplasm (Fig. 2–2), are 20,000 to 30,000 particles, about 25 nm in diameter, called *ribosomes,* which are composed of *ribonucleic acid (RNA)* and proteins. These particles are the sites of protein synthesis. Ribosomes exist in groups called polyribosomes, or polysomes and are formed of a large and a small subunit.

The remainder of the cell is filled with water, various RNAs, and protein molecules (including enzymes), and various smaller molecules.

Certain motile bacteria have hair-like processes about 10 nm wide and of variable length, called *flagellae,* which are used for locomotion. In contrast with cilia and flagella of eukaryons, which contain several fibrils, each flagellum in bacteria is made of a single fibril.

The Smallest Mass of Living Matter

From what has been said about *E. coli,* it is evident that there must be a minimum size limit for a cell. However, the cell must be large enough (1) to have a plasma membrane, (2) to contain the genetic material necessary to provide a code for the various RNAs involved in protein synthesis, and (3) to contain the biosynthetic machinery where this synthesis takes place.

Among agents that have the smallest living mass, the best suited for study are microbes of the so-called pleuropneumonia-like organism (PPLO) group (see Fig. 1–2), which produce infectious diseases in certain animals and man and which can be cultured in vitro like any bacteria. These agents range in diameter from 0.25 μm to 0.1 μm; thus their size corresponds to that of some of the

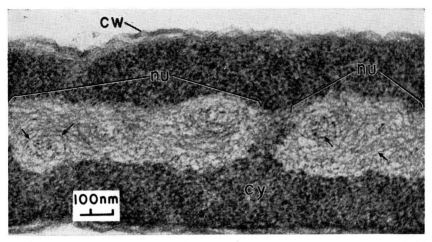

Figure 2–2. Electron micrograph of thin sections of the bacterium *Escherichia coli*. The nucleoid (*nu*) shows the presence of microfibrils of DNA (arrows). Note that the nucleoid lacks a membrane. The cytoplasm (*cy*) is very dense; *cw*, cell wall. ×100,000. (Courtesy of E. Kellenberger.)

large viruses. These microbes are of general biological interest because each is a living organism a thousand times smaller than the average size bacterium and a million times smaller than a eukaryotic cell.[1]

The study of these elementary organisms and of bacteria in general is of paramount importance to cell biology because the organisms represent a simplification of the various patterns of structure and function that are found in higher cells.

EUKARYOTIC CELLS

The eukaryotic cell consists of a small mass of protoplasm, the cytoplasm, containing a nucleus and surrounded by the plasma membrane. The cells of a multicellular organism vary in shape and structure and are differentiated according to their specific function in the various tissues and organs. Because of this functional specialization, cells acquire special characteristics. General characteristics common to all cells invariably do persist, however. In cells that are slightly differentiated, such as the blastomeres or germinative cells, meristematic cells of plants, and others having a relatively simple organization, such as some of the epithelial or connective tissue cells, certain common features may be found. It is these common characteristics that are dealt with in this book.

Shape

Some cells, such as amebae and leukocytes, change their shape frequently. Other cells always have a typical shape, more or less fixed, which is specific for each cell type; e.g., the spermatozoids, infusoria, erythrocytes, epithelial cells, nerve cells, most plant cells, and others.

The shape of cells depends mainly on functional adaptations and partly on the surface tension and viscosity of the protoplasm, the mechanical action exerted by the adjoining cells, and the rigidity of the cell membrane. In chapter 22 it will be shown that certain cell organelles, called *microtubules*, also have an important influence on the shape of the cell.

When isolated in a liquid, many cells become spherical, according to the laws of surface tension. For example, leukocytes in the circulating blood are spherical, but in the extravascular milieu they emit pseudopods (thus exhibiting ameboid movement) and become irregular in shape.

The cells of many plant and animal tissues have a polyhedral shape determined by reciprocal pressures. The original spherical form of these cells has been modified by contact with the other cells, just as each bubble in soap foam is pressed by its neighbors.

Individual cells in a large mass appear to behave like polyhedral solids of minimal sur-

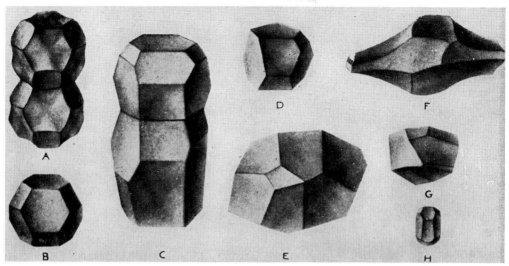

Figure 2–3. Three-dimensional reconstructions of: **A** and **B**, Kelvin's minimal tetrakaidecahedron. **C–H**, wax plate reconstructions of different cell types: **E**, human fat cell; **F, G** and **H**, outer, middle and basal cells of a stratified epithelium of the mouth of a five-month human embryo. Approximate magnifications: C, ×170; D, ×150; E, ×300; F, G and H, ×750. (From F. T. Lewis.)

face which are packed without interstices. Although regular polyhedra of four, six, and twelve sides can be packed without interstices, the fourteen-sided polyhedron (tetrakaidecahedron) satisfies most closely the conditions of minimal surface. The study of bubbles in soap foam by Plateau and Lord Kelvin showed that these conditions of minimal surface exist, and that the average bubble has 14 sides (Fig. 2–3, *A* and *B*).

When observing cells under the microscope, one should always think in terms of three dimensions and observe sections of varied orientations. The best way to learn about the actual shape of cells is by making serial sections of known thickness, drawing all of them and making reconstructions in wax—a procedure similar to that used in anatomic reconstructions.

Figure 2–3, *C* to *H*, shows some reconstructions of different cell types.

Size

The size of different cells ranges within broad limits. Some plant and animal cells are visible to the naked eye. For example, the eggs of certain birds have a diameter of several centimeters and are composed, at least at first, of a single cell. However, this is an exception; the great majority of cells are visible only with the microscope, since they are only a few micrometers in diameter (Fig. 1–2). The smallest animal cells have a diameter of 4 μm.

In tissues of the human body, with the exception of some nerve cells, the volume of cells varies between 200 μm³ and 15,000 μm³. In general, the volume of the cell is fairly constant for a particular cell type and is independent of the size of the organism. For example, kidney or hepatic cells are about the same size in the bull, horse, and mouse; the difference in the total mass of the organ depends on the number, and not the volume, of cells. This is sometimes called the *law of constant volume*.

Structure of the Living Cell

Living cells can be studied only with light microscopes, since they are, in general, too thick to be studied via electron microscopy. In addition, cells must be in a vacuum to be seen with the electron microscope (see Chapter 5). Many animal cells can be observed isolated in an isotonic liquid, such as blood serum, aqueous humor, or physiological salt solutions, or in tissue culture. They appear as irregular, translucent masses of cytoplasm containing a nucleus. In Figure 2–4 most cells are in interphase, the nondividing stage, and show a clear nucleus having one or more nucleoli and separated from the cytoplasm by the nuclear envelope. When cells are about to divide, several refractile bodies, the *chromosomes,* appear in the nucleus.

The cytoplasm appears as an amorphous, homogeneous substance, the *ground cytoplasm,* containing refractile particles of various sizes, among which the mitochondria are the most conspicuous (Fig. 2–4). Frequently the peripheral layer of the cytoplasm, the *ectoplasm,* also called the *cortex,* is relatively more rigid and devoid of granules. The internal cytoplasm, the *endoplasm,* which contains different granules, is less viscous than the ectoplasm.

The living cell can be centrifuged, and the effect on the cell components can be observed in a special centrifuge microscope. For example, if a sea urchin egg is subjected to centrifugation, the various components of the endoplasm become stratified in accordance with their densities (Fig. 2–5). The egg elongates and then becomes constricted in the center. The fat droplets accumulate at

the centripetal pole. Beneath this is a clear, wide zone, the ground cytoplasm, which contains the nucleus. The mitochondria form the next layer, and the yolk bodies, the next. Pigment granules accumulate at the centrifugal pole. The ectoplasm is not displaced by centrifugation, because of its greater viscosity and rigidity; this property appears to depend on the presence of calcium ions, since it liquefies when eggs are treated with substances that bind Ca^{++}, such as oxalate.

Interesting studies on the colloidal properties of cytoplasm and on the physicochemical forces involved have been made (see Chapter 9). By increasing the hydrostatic pressure, the cortex can be liquefied and the cell can no longer change its shape. This effect is reversible within certain limits. The ground cytoplasm behaves, in general, as a reversible sol-gel colloid system. This change can sometimes be produced by mechanical action, a property generally called *thixotropism* (Gr., *thixis,* touch + *trope,* a change).

In addition to mitochondria, other particles observed in the cell, such as highly refractile lipid droplets, yolk bodies, pigment, and secretion granules, are products elabo-

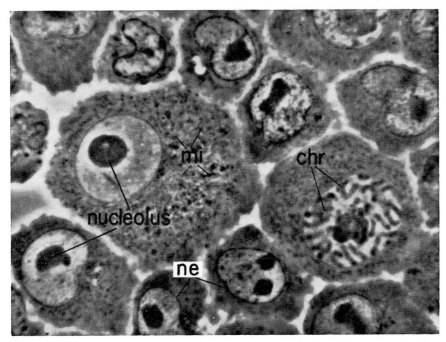

Figure 2–4. Photomicrograph in phase contrast of the living cells from an ascitic tumor. *chr,* chromosomes; *mi,* mitochondria; *ne,* nuclear envelope. (Courtesy of N. Takeda).

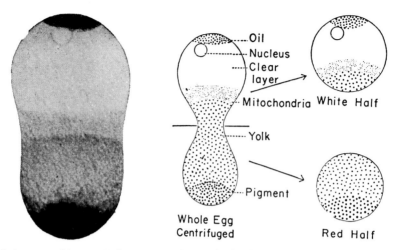

Figure 2–5. **Left,** sea urchin egg *(Arbacia punctulata)* submitted to the action of centrifugal force. The egg has elongated and is being divided into two halves. The cellular materials become stratified (see the description in the text). (Courtesy of Costello.) **Right,** diagram of the stratification of the egg and its division into two halves. (From E. B. Harvey.)

rated by the cell and are found in various amounts. In plant cells, granules called *plastids* can be observed. Among these are the *chloroplasts,* which contain a green pigment—chlorophyll. The function of chlorophyll is *photosynthesis,* a process of immense importance in the biological world (see Chapter 11).

In animal and, more commonly, in plant cells, fluid vacuoles surrounded by a membrane may be found.

Mitochondria and chloroplasts are considered cell *organelles,* because of their general presence and their important function in cells. Other cell organelles, such as the *Golgi* complexes and the *centrioles,* are observed less often in living cells. Some cell structures cannot be resolved with the light microscope, and therefore cannot be seen in living cells.

STRUCTURE OF FIXED CELLS

Examination of the living cell is limited to light microscopy and is based mainly on the differences in refractive index of the different cell components. Sometimes the use of stains that act on the living organism *(vital staining)* facilitates observation of the living cell. However, more important in the morphologic study of the cell are *methods of fixation,* by which cell death results in such a way that

physiologic structure and chemical composition are preserved as much as possible (see Chapter 3).

The complexity of the structural organization in a cell of the higher plants and animals is most impressive. Although there are great differences between the primitive forms of life, such as that illustrated in Figure 2–1, and the higher plant and animal cells, the similarities between primitive and advanced cells are also notable. Eukaryotic cells are characterized by a true nucleus with a *nuclear membrane* or *envelope* which divides the cell into two main compartments: nucleus and cytoplasm. The cytoplasm in turn is limited by the *plasma membrane.* In a plant cell (Fig. 2–6), the plasma membrane is covered and protected on the outside by a thicker cell wall through which there are tunnels, the *plasmodesmata,* by which the cell intercommunicates with neighboring cells via fine cell processes. In animal cells (Fig. 2–7), parts of the plasma membrane are covered by a thin layer of material, which is generally described as the *extraneous coat* of the plasma membrane. The so-called basement membranes shown in Figure 2–7 correspond to this extraneous coat.

The cytoplasmic compartment of the cell has a complex structural organization. When examining the compartment under the electron microscope, one is particularly struck by the prodigious development of membranes. For example, the plasma membrane has

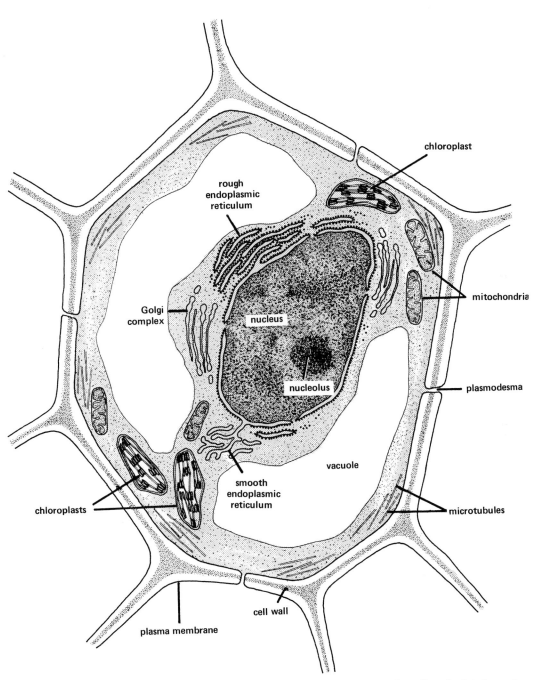

Figure 2-6. Diagram of a plant cell showing the following cell components: the cell wall, the plasmodesmata, chloroplasts, and vacuoles.

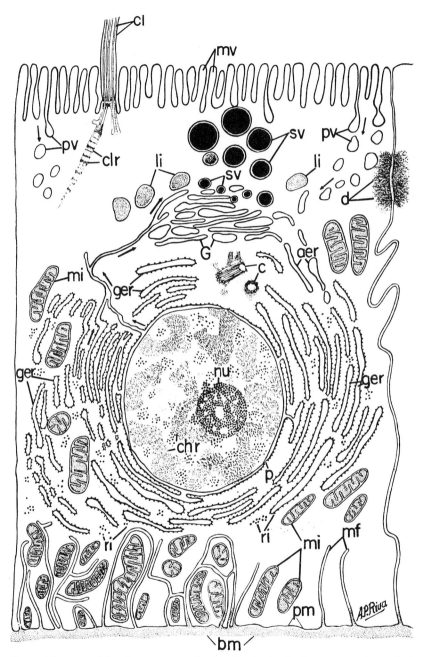

Figure 2–7. General diagram of the ultrastructure of an idealized animal cell. *aer,* agranular endoplasmic reticulum; *bm,* basal membrane; *c,* centriole; *chr,* chromosome; *cl,* cilium; *clr,* cilium root; *d,* desmosome; *G,* Golgi complex; *ger,* granular endoplasmic reticulum; *li,* lysosome; *mf,* membrane fold; *mi,* mitochondria; *mv,* microvilli; *nu,* nucleolus; *p,* pore; *pm,* plasma membrane; *pv,* pinocytic vesicle; *ri,* ribosome; *sv,* secretion vesicle. (From E. De Robertis and A. Pellegrino de Iraldi.)

numerous infoldings and differentiations (Fig. 2–7). In addition, a basic membranous organization is found in cell organelles, such as *lysosomes* (Fig. 2–7, *li*) and *mitochondria* (Fig. 2–7, *mi*), in which one and two membranes separate the interior matrix from the surrounding ground cytoplasm. *Chloroplasts* are organelles with a complex, multilayered organization (see Fig. 2–6). This layered type of structure is also found in certain animal cells, e.g., the myelin sheath and the outer segments of the rods and cones, which are also formed by packed membrane systems. Numerous vesicles, vacuoles, and secretory droplets found in the cytoplasm are also surrounded by membranes.

A complex system of membranes pervades the ground cytoplasm, forming numerous compartments and subcompartments. This system is so polymorphic that it is difficult to describe and to encompass within a single denomination. The term *vacuolar system* seems to be the most general and appropriate description of the fact that the cytoplasm is separated generally into two parts, one contained within the system and the other, the *cytoplasmic matrix* proper, remaining outside. To this vacuolar system belong the *Golgi complex* and the *nuclear envelope;* but the major part is formed by the so-called *endoplasmic reticulum,* which may in turn be differentiated into a *granular* or *rough reticulum,* containing ribosomes, and an *agranular* or *smooth reticulum.* The diagram in Figure 2–7 indicates the possible continuities and functional interconnections of these different portions of the cytoplasmic vacuolar system.

Other cell organelles, the *centrioles,* are involved in cell division. During cell division two centrioles are contained in a clear, gel-like zone, the so-called *aster.* Centrioles are also related to the differentiation of *cilia* and *flagella,* both motile appendixes of the cell (Fig. 2–7, *cl*).

In spite of this complex structural organization, the most important constituents of the cytoplasm are in the *matrix* (ground cytoplasm), which lies outside the vacuolar system. This matrix constitutes the true internal milieu of the cell and contains the following: the *ribosomes,* the principal structures of the machinery for protein synthesis of the cell; glycogen particles; soluble enzymes; microtubules; microfilaments; structural proteins; and all the components found in a primitive organism, excluding DNA. In addition, it is the site of the colloidal changes of the protoplasm and of the production of many cytoplasmic differentiations, such as keratin fibers, and myofilaments.

A detailed study of the vacuolar system, the cytoplasmic organelles and differentiations, and the cytoplasmic matrix is presented in later chapters of the book after the chemical organization of the cell has been discussed.

SUMMARY:
General Cell Structure

Prokaryotic cells such as bacteria and blue-green algae lack a nuclear envelope and contain a single chromosome, formed by a circular DNA molecule, that lies free in the nuclear region (also called nucleoid). A single cell of the bacterium *E. coli* (0.2×0.8 μm) is surrounded by a plasma membrane containing the respiratory enzymes. This membrane is surrounded by a more rigid cell wall. The DNA molecule is 1 mm long and is attached at one point to the plasma membrane. It contains information to code some 3000 protein molecules. The protoplasm contains some 20,000 to 30,000 ribosomes, principally as polysomes.

The smallest mass of living matter is represented by the pleuropneumonia-like organism (PPLO), a small microbe 0.2 to 0.1 μm in length. The minimum living mass should contain (1) a plasma membrane, (2) the genetic material, and (3) the machinery for the synthesis of proteins.

The shape of *eukaryotic* cells is, in general, fixed and

depends on mechanical properties that are intrinsic or depends on the association with other cells. In plant cells the presence of a rigid cell wall is important. Many cells are polyhedral solids and approach the figure of minimal surface area—the *tetrakaidecahedron*. The size of cells varies considerably; the smallest cells may be only a few micrometers in diameter, the largest, several centimeters.

Living cells can be observed only with the light microscope. They show a cytoplasm with an ectoplasm (or cortex) and a more fluid endoplasm, which contains mitochondria, plastids (in plant cells), and various other inclusions. Cell organelles are cell structures that are present in all, or most, cells. Some organelles, such as the Golgi complex or the centrioles, are difficult to see in living cells.

Fixed eukaryotic cells show a complex organization. The cytoplasm is surrounded by the plasma membrane which has many differentiations. The cell wall, in plant cells, and the extraneous coat in animal cells are outside the plasma membrane. The cytoplasm is subdivided into compartments and subcompartments by numerous intracellular membranes that form a large vacuolar system. The main parts of the vacuolar system are the endoplasmic reticulum (rough and smooth), the nuclear envelope with the nuclear pores, and the Golgi complex. Membrane-bound organelles are lysosomes, mitochondria, secretion vesicles, etc. The ground cytoplasm or matrix lies outside the vacuolar system; it contains the ribosomes, microtubules, microfilaments, soluble proteins, etc. It may have also differentiations such as myofilaments and myofibrils, keratin fibers, and others.

THE NUCLEUS AND CHROMOSOMES

The Cell Cycle

The growth and development of every living organism depends on the growth and multiplication of its cells. In unicellular organisms, cell division is the means of reproduction, and by this process two or more new individuals arise from the mother cell. In multicellular organisms, new individuals develop from a single primordial cell, the *zygote;* it is the multiplication of this cell and its descendants that determines the development and growth of the individual.

In many instances cells appear to grow to a certain size before division occurs. This process is repeated in the two daughter cells, so that the total volume eventually becomes four times that of the original cell. The growth of living material progresses rhythmically and according to a geometric progression that has been expressed as follows:

$$\frac{Mn}{Mc}, \frac{2Mn}{2Mc}, \frac{4Mn}{4Mc}, \frac{8Mn}{8Mc}, \text{etc.}$$

where Mn is the *nuclear mass,* and Mc is the *cytoplasmic mass* of the cells. The two masses are in a state of optimum equilibrium, the so-called *nucleoplasmic index* (NP), which is expressed numerically as:

$$NP: \frac{Vn}{Vc - Vn}$$

where Vn is the *nuclear volume,* and Vc is the *cell volume.*

This equilibrium not only refers to a relationship of volumes, but implies a chemical relationship as well.

In general, every cell has essentially two periods in its life cycle: *interphase* (nondivi-

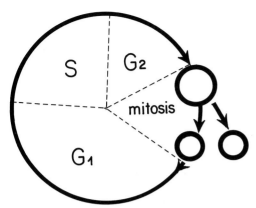

Figure 2–8. Diagram of the life cycle of a cell indicating the mitotic and the interphase periods. Interphase is composed of the G_1, S, and G_2 phases. Duplication of DNA takes place during the synthetic or S phase.

sion) and *division* (which produces two daughter cells). This cycle is repeated at each cell generation, but the length of the cycle varies considerably in different types of cells. Some cells have a short life cycle and cell division takes place frequently, whereas others have an interphase which may be as long as the life of the organism (e.g., nerve cells). During cell division the nucleus undergoes a series of complex but remarkably regular and constant changes in which the nuclear envelope and the nucleolus disappear and the chromatin substance becomes condensed into dark-staining bodies — the *chromosomes* (Gr., *chroma,* color + *soma,* body).

Chromosomes are always present in the nucleus. During interphase they are not generally visible because they are dispersed or hydrated and their macromolecular components are loosely distributed within the nuclear sphere.

In Chapter 17 the cell cycle and the changes occurring in the main chemical components of the cell, i.e., DNA, RNA, and proteins will be considered in detail. Here it will only be mentioned that the duplication of DNA takes place during a special period of the interphase called the *synthetic or S period.* As shown in Figure 2–8 this is preceded and followed by two periods G_1 and G_2 that are intercalated between the S period and mitosis.

Morphology of the Nucleus

The *shape* of the nucleus is sometimes related to that of the cell, but it may be com-

pletely irregular. In spheroid, cuboid, or polyhedral cells, the nucleus is generally a spheroid. In cylindrical, prismatic, or fusiform cells it tends to be an ellipsoid. Examples of irregular nuclei are found in some leukocytes (horseshoe-shaped or multilobate nuclei), certain *Infusoria* (moniliform nuclei), glandular cells of many insects (branched nuclei), spermatozoa (ellipsoid, pyriform and lanceolate nuclei and so forth, according to the species).

By 1905 Boveri had already noted that, in sea urchin larvae, the size of the nucleus was proportional to the chromosome number (ploidy) and increased from haploid to diploid to tetraploid cells. In hepatocytes observed under the light microscope it is easy to recognize a few larger nuclei which correspond to tetraploid or octoploid cells. The size of the nucleus is minimal when most of the chromatin is condensed, as in the small lymphocytes. In ovocytes the nucleus (often called the *germinal vesicle*) is very active and may attain a large volume. In general it may be said that each somatic nucleus has a specific size that depends on the DNA and protein content and is related to its functional activity during interphase.[2]

Almost all cells are *mononucleate,* but *binucleate cells* (some liver and cartilage cells) and *polynucleate* cells also exist. The nuclei of polynucleate cells may be numerous (up to 100 in the polykaryocytes of bone marrow (osteoclasts). In the *syncytia,* which are large protoplasmic masses not subdivided into cellular territories, the nuclei may be extremely numerous. Such is the case with striated muscle fiber and certain algae, which may contain several hundred nuclei.

The *position* of the nucleus is variable, but is generally characteristic for each type of cell. The nucleus of embryonic cells almost always occupies the geometric center, but it commonly becomes displaced as differentiation advances and as specific parts or reserve substances are formed in the cytoplasm. In glandular cells the nucleus is located in the basal cytoplasm.

General Structure of the Interphase Nucleus

In fixed and stained material the structure of the nucleus is distinguished by its complexity and varies according to the type of cell and the fixative used (Fig. 6–7). In general

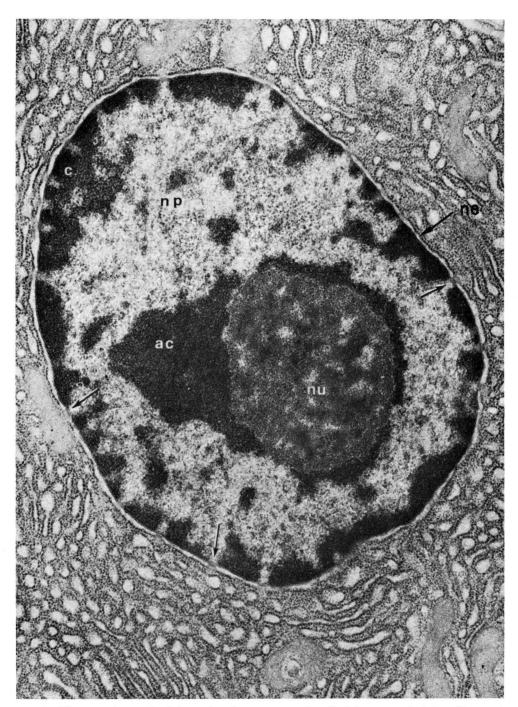

Figure 2–9. Electron micrograph of the nucleus from a pancreatic cell of the mouse. Staining with uranyl acetate enhances mainly the DNA-containing parts of the cell. Arrows in the interchromatin channels point to nuclear pores; *ac*, chromatin associated with the nucleolus (*nu*); *np*, nucleoplasm; *ne*, nuclear envelope. ×24,000. (Courtesy of J. Andrè.)

the following structures are recognized in the interphase nucleus: (1) A *nuclear envelope* that appears as a clear outline on both the cytoplasmic and nuclear sides. Under the electron microscope the nuclear envelope is composed of two membranes and shows the *nuclear pores* (Fig. 2–9). (2) The *nucleoplasm* (or *nuclear sap*) that fills most of the nuclear space. This represents uncondensed regions of *chromatin* (i.e., nucleoproteins) where the chromosomes are largely dispersed into their fine macromolecular components. These regions correspond to the so-called *euchromatin* (Gr., *eu*, true). (3) The *chromocenters or false nucleoli,* that along with twisted filaments of chromatin represent parts of the chromosomes that at interphase remain condensed; i.e., their molecular components are tightly packed. These condensed regions of chromatin, of so-called *heterochromatin,* are frequently found near the nuclear envelope and are also attached to the nucleolus (Fig. 2–9). (4) The *nucleoli,* which are generally spheroid and very large in nerve cells, pancreatic, and other cells and are very active in protein synthesis. They are either single or multiple, usually acidophilic, and contain ribonucleoproteins.

Mitosis and Meiosis

It is important to introduce at this point the essentials of mitosis and meiosis, which are studied in more detail in Chapters 13 and 14.

All organisms that reproduce sexually develop from a single cell, the *zygote,* produced by the union of two cells, the *germ cells* or *gametes* (a *spermatozoon* from the male, and an *ovum* from the female). The union of an egg and a sperm is called *fertilization.* The zygote produced by fertilization develops into a new individual of the same species as the parents.

Every cell of an individual, with the exception of the *gametes,* contains the same number of chromosomes. In the somatic cells of a plant or an animal, chromosomes are paired, one member of each pair originally derived from one parent, the other member, from the other parent. Each member of a pair of chromosomes is called a *homologue,* and commonly pairs of chromosomes (or of homologues) are spoken of when reference is made to the chromosome number of a species. Man

has 46 chromosomes or 23 pairs; the onion has 8 pairs; the toad, 11 pairs; the mosquito, 3 pairs; and so on (see Table 2–1). Homologues of each pair are alike, but the pairs are generally different. The original chromosome number of each cell (diploid number) is preserved during successive nuclear divisions involved in the growth and development of a multicellular organism.

Mitosis

The continuity of the chromosomal set is maintained by *cell division,* which is called *mitosis.* At the time of cell division the nucleus becomes completely reorganized, as illustrated in Figure 2–10. Mitosis takes place in a series of consecutive stages known as prophase, metaphase, anaphase, and telophase. In a somatic cell the nucleus divides by mitosis in such a fashion that each of the two daughter cells receives exactly the same number and kind of chromosomes that the parent cell had.

Figure 2–10 represents two pairs of homologous chromosomes in a diploid nucleus. Each chromosome duplicates some time during *interphase* before the visible mitotic process begins. At this stage and at early *prophase* chromosomes appear as extended and slender threads. At late prophase chromosomes become short, compact rods by the packing of the nucleoprotein fibers. A spindle arises between the two centrioles and the chromosomes line up across the equatorial plane of the spindle at the *metaphase* plate. At *anaphase* each chromosome separates, forming two daughter chromosomes, which go to opposite poles of the cell. Finally, at *telophase* the daughter chromosomes at each pole again become dispersed and two daughter nuclei are formed.

In mitosis the original chromosome number is preserved during the successive nuclear divisions. Since the somatic cells are derived from the zygote by mitosis, they all contain the normal double set, or diploid number *(2n),* of chromosomes.

Meiosis

If the gametes (ovum and spermatozoon) were diploid, the resulting zygote would have twice the diploid chromosome number.

TABLE 2–1.   Diploid (2n) Number of Chromosomes in Some Plants and Animals

Plants Common and Scientific Names	Chromosomes	**Animals** Common and Scientific Names	Chromosomes
Yellow pine, *Pinus ponderosa*	24	Roman snail, *Helix pomatia*	54
Cabbage, *Brassica oleracea*	18	Silkworm, *Bombyx mori*	56
Radish, *Raphanus sativus*	18	Housefly, *Musca domestica*	12
Flax, *Linum usitatissimum*	30, 32	Vinegar fly, *Drosophila melanogaster*	8
Ombu, *Phytolacca dioica*	36	Spanish butterfly, *Lysandra nivescens*	380
Watermelon, *Citrullus vulgaris*	22	Grasshoppers, many Acrididae	24
Cucumber, *Cucumis sativus*	14	Grouse locusts, Tetrigidae	14
Papaya, *Cärica papaya*	18	*Dichroplus silveiraguidoi*	
Upland cotton, *Gossypium hirsutum*	52	(S. American Acrididae)	8
Cherry, *Pronus cerasus*	32	Honeybee, *Apis mellifica*	32, 16
Plum, *Prunus domestica*	48	Mosquito, *Culex pipiens*	6
Pear, *Pyrus communis*	34, 51, 68	Frogs, *Rana* spp.	26
Peanut, *Arachis hypogaea*	40	Tree frogs, *Hyla* spp.	24
Ceibo, *Erythrina cristagalli*	42	Toads, *Bufo* spp.	22
Coffee, *Coffea arabica*	44	Chicken, *Gallus domesticus*	ca. 78
Sunflower, *Helianthus annuus*	34	Turkey, *Meleagris gallipavo*	82
Luzula purpurea	6	Pigeon, *Columba livia*	80
Potato, *Solanum tuberosum*	48	Duck, *Anas platyrhyncha*	80
Tomato, *Lycopersicum solanum*	24	Opossum, *Didelphys virginiana*,	
Tobacco, *Nicotiana tabacum*	48	*D. paraguayensis*	22
Tradescantia virginiana	24	Mouse, *Mus musculus*	40
Banana, *Musa paradisiaca*	22, 44, 55, 77, 88	Rabbit, *Oryctolagus cuniculus*	44
Garden pea, *Pisum sativum*	14	Albino rat, *Rattus norvegicus*	42
Bean, *Phaseolus vulgaris*	22	Common rat, *Rattus rattus*	42
Orange, *Citrus sinensis*	18, 27, 36	Golden hamster, *Mesocricetus Auratus*	44
Apple, *Malus silvestris*	34, 51	Chinese hamster, *Cricetus griseus*	22
Oats, *Avena sativa*	42	Guinea pig, *Cavia cobaya*	64
Indian corn, *Zea mays*	20	Mulita, *Dasypus hybridus* S. America	64
Barley, *Hordeum vulgare*	14	Armadillo, *Dasypus novemcinctus* N. America	64
Summer wheat, *Triticum dicoccum*	28	Dog, *Canis familiaris*	78
Bread wheat, *Triticum vulgare*	42	Cat, *Felis domestica*	38
Rye, *Secale cereale*	14	Horse, *Equus caballus*	64
Rice, *Oryza sativa*	24	Donkey, *Equus asinus*	62
Sorghum spp.	10, 20, 40	Pig, *Sus scrofa*	40
Black sorghum, *Sorghum almum*	40	Sheep, *Ovis aries*	54
Sugar cane, *Saccarum officinarum*	80	Goat, *Capra hircus*	60
Field bean, *Vicia faba*	12	Cattle, *Bos taurus*	60
Onion, *Allium cepa*	16	Rhesus monkey, *Macaca mulatta*	42
Eucalyptus, *Eucalyptus* spp.	22	Gorilla, *Gorilla gorilla*	48
Passion flower, *Passiflora coerulea*	18	Orangutan, *Pongo pygmaeus*	48
		Chimpanzee, *Pan troglodytes*	48
		Man, *Homo sapiens*	46

In order to avoid this, each gamete undergoes a special type of cell division called *meiosis,* which reduces the normal diploid set of chromosomes to a single *(haploid)* set *(n).* Thus, when the ovum and spermatozoon unite during fertilization, the resulting zygote is diploid. The meiotic process is characteristic of all plants and animals that reproduce sexually and it takes place in the course of *gametogenesis* (Fig. 2–10).

Meiosis produces the reduction of the chromosome number by means of two nuclear divisions, the *first* and *second meiotic*

divisions, that involve only a single division of the chromosomes.

The essential aspects of the process are simple. The homologous chromosomes, distinguished by their identical morphologic characteristics, pair longitudinally, forming a bivalent. Each chromosome is composed of two filaments called the *chromatids.* The bivalent thus contains four chromatids and is also called a *tetrad.* In the tetrad each chromatid of the homologue has a single pairing partner. Portions of these paired chromatids may be exchanged from one

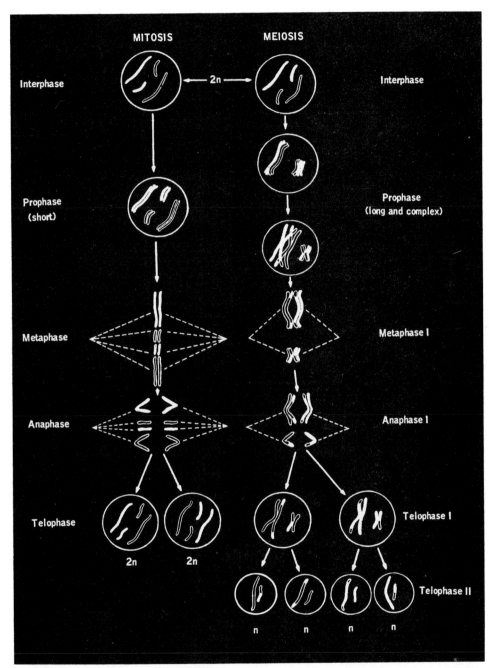

Figure 2–10. Comparative diagram of mitosis and meiosis in idealized cells having four chromosomes (2n). The chromosomes belonging to each progenitor are represented in white and black. In mitosis the division is equational while in meiosis it is reductional, the two divisions giving rise to four cells having only two chromosomes (n). In meiosis there is, in addition, an interchange of black and white segments of the chromosomes.

homologue to the other, giving rise to cross-shaped figures, called *chiasmata*. The chiasma is a cytologic manifestation of an underlying genetic phenomenon called *crossing over* (see Chapter 14).

At metaphase I the bivalents arrange themselves on the spindle, and at anaphase I the homologous chromosomes and their two associated chromatids migrate to opposite poles. Thus, in the first meiotic division the homologous pairs of chromosomes are segregated. After a short interphase the two chromatids of each homologue separate in the second meiotic division, so that the original four chromatids are distributed into each of the four gametes. The result is four nuclei with only a single set (haploid) of chromosomes (Fig. 2–10).

CHROMOSOMES

Of all cellular components observed during mitosis and meiosis, chromosomes have been the most thoroughly investigated. Their presence was demonstrated long before they were named "chromosomes" (Waldeyer, 1888). Forty years earlier, the botanist Hofmeister, while studying the pollen mother cells of *Tradescantia,* drew chromosomes directly from living cells.

A chromosome is considered a nuclear component endowed with a special organization, individuality, and function. It is capable of self-reproduction and of maintaining its morphologic and physiologic properties through successive cell divisions.

Morphology

The morphologic characteristics of chromosomes are best studied during metaphase and anaphase of cell division. Then they appear as cylinders and stain intensely with basic dyes and the Feulgen method (Chap. 6). They are easily observed in vivo by phase microscopy (Fig. 2–4) and they absorb ultraviolet light intensely at 260 nm.

Chromosomes may be studied in tissue sections, but whole preparations obtained by crushing or smearing a small piece of tissue are better suited for microscopic examination. Sex glands, plant meristematic cells, pollen

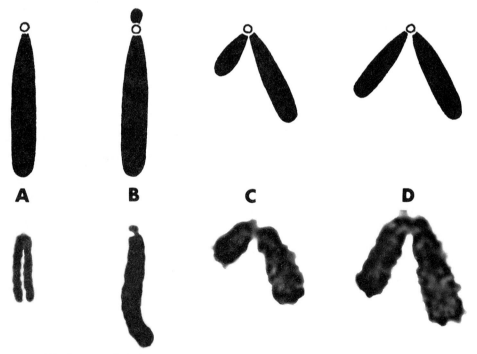

Figure 2–11. The four morphologic types of chromosomes according to the position of the centromere: **A,** telocentric; **B,** acrocentric; **C,** submetacentric; and **D,** metacentric. The upper row shows a diagram of each type of chromosome, whereas the lower row shows a microphotograph of each type.

mother cells, or other tissues can be crushed between a slide and a coverglass and simultaneously fixed and stained with acetic hematoxylin or acetocarmine. The use of hypotonic solutions prior to squashing produces swelling of the nucleus and better separation of the individual chromosomes (see Chapter 16).

Chromosomes are classified into four types by their shape in metaphase or in anaphase, which, in turn, is determined by the position of the centromere (see below) (Fig. 2–11): *telocentric* chromosomes are rodlike and have a centromere situated on the proximal end; *acrocentric* chromosomes are rod-like and have a small, or even imperceptible, arm; *submetacentric* chromosomes have unequal arms and are thus L-shaped; and *metacentric* chromosomes have equal or almost equal arms and thus are V-shaped. The different chromosomal types are constant for each homologous chromosome. The chromosomal type may also be constant throughout a species or even a genus.

Centromere

The shape of chromosomes is determined by the *primary constriction* located at the point where the arms of a chromosome meet (Fig. 2–12). Within the constriction is a clear zone containing a small granule, or spherule. This clear region is the so-called *centromere* or *kinetochore.* It is functionally related to the chromosomal movements that occur during mitosis. For some time the centromere was described as the point of insertion of the spindle fiber. In the chromosomes of *Trillium* the centromere has a diameter of 3 μm and the spherule a diameter of about 0.2 μm. Usually each chromosome has only one centromere (monocentric); however, there may be two (dicentric) or more (polycentric), or the centromere may be diffuse, e.g., as in *Ascaris megalocephala* and *Hemiptera.*

Secondary Constrictions

Other morphologic characteristics are the *secondary constrictions.* Constant in their position and extent, these constrictions are useful in identifying particular chromosomes in a set. Secondary constrictions may be either short or long; they are distributed along the chromosome and are distinguished

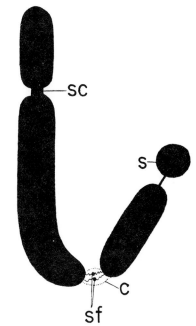

Figure 2–12. Diagram of a metacentric chromosome showing *c,* the centromere; *s,* the satellite; *sc,* secondary constrictions; and *sf,* the spindle fiber.

from the primary constriction by the absence of marked angular deviations of the chromosomal segments (Fig. 2–12).

Telomere

This term applies to each of the extremities of a chromosome. If chromosomes are fractured by x-rays, the resulting segments may fuse again; they will not, however, fuse with the telomere. It appears that the telomere has a polarity that prevents other segments from joining with it.

Satellite

Another morphologic element present in certain chromosomes is the *satellite.* This is a round, elongated body separated from the rest of the chromosome by a delicate chromatin filament, which may be long or short (Fig. 2–12). It is customary to designate as *SAT-chromosomes* those having a satellite. The satellite and the filament are also constant in shape and size for each particular chromosome.

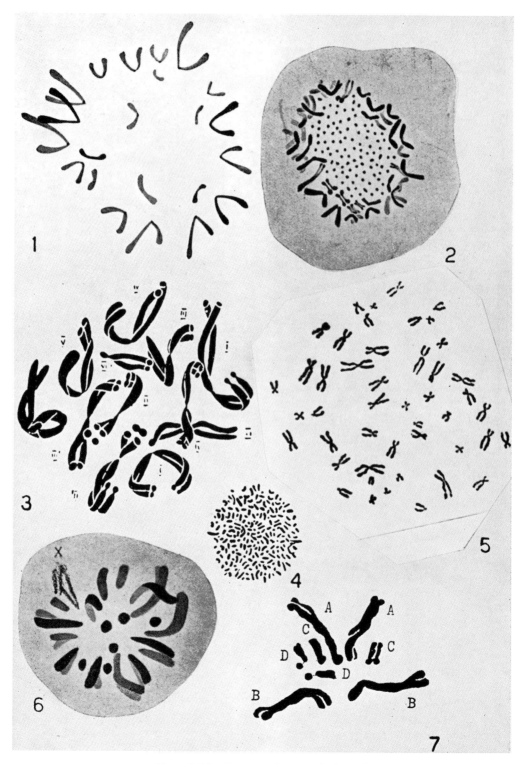

Figure 2–13. *See opposite page for legend.*

Nucleolar Zone

Certain secondary constrictions are intimately associated with the formation of the nucleoli. These specialized regions are the *nucleolar zones (nucleolar organizers)*. Generally there are two chromosomes in each nucleus, called nucleolar chromosomes, that have this special characteristic.

Morphologic Constants in Chromosomes. Karyotype.

The most important characteristics identifying individual chromosomes in mitosis are their number, relative size, structure, behavior, and internal organization. Other characteristics, such as linear contraction and degree of coiling, may be subject to physiologic variations.

Some zoologic groups have typical morphologic characteristics, such as the family *Acrididae* (locusts), which generally have acrocentric chromosomes (Fig. 2–13, *6*), or amphibia (Fig. 2–13, *1*), which have meta-

centric chromosomes. Among plants the form of the chromosomes is more varied; characteristic satellites and constrictions are common (Fig. 2–13, *3* and *7*). The size of the chromosome is relatively constant, thus individualizing a member of a set. The length of a chromosome may vary from 0.2 to 50 μm, the diameter from 0.2 to 2 μm. In humans the most common length of a chromosome is 4 to 6 μm.

In a mitotic nucleus each homologue is not always found near its mate, since the location of each chromosome during this cycle is entirely independent of the others. A single chromosome may occupy any part of the nucleus.

The name *karyotype* is given to the group of characteristics that identifies a particular chromosomal set. The karyotype is characteristic of an individual, species, genus, or larger grouping, and may be represented by a diagram called an *idiogram,* in which the pairs of homologues are ordered in a series of decreasing size (see Chapter 16 for human chromosomes).

Multicellular organisms develop by the division of an initial diploid cell, the **zygote,** which is a product of the union of the haploid gametes (i.e., fertilization). Every cell has a life cycle which is composed of a period of non-divison (interphase), and a period of division (generally, mitosis). The interphase period has a synthetic phase, **S,** in which the DNA duplicates. This **S** phase is preceded and followed by the G_1 and G_2 phase.

SUMMARY:
The Nucleus and Cell Division

Figure 2–13. The characteristics of the somatic chromosomes of some plants and animals. All the figures correspond to metaphase in polar view.

1, the 22 chromosomes of the toad *Bufo arenarum,* showing their morphologic characteristics. In some chromosomes a black point corresponding to the centromere can be seen.

2, the chromosomes of the lizard *Tupinambis teguixin,* showing their 140 elements distributed as microchromosomes in the center and macrochromosomes at the periphery.

3, the complex of chromosomes composed of 12 elements of a member of the Ranunculaceae, *Nigella orientalis,* in which some chromatids can be seen wound about each other, thus forming the relational spiral. The Roman numerals indicate the pairs of homologous chromosomes.

4, the 208 chromosomes of the decapod crustacean *Paralithodes camtschatica,* showing that they are rod-shaped and punctiform.

5, the 46 chromosomes of the human species.

6, the 23 telocentric chromosomes of the locust *Chromacris miles.* The sex chromosome, indicated by X, is found in negative heteropyknosis.

7, the eight chromosomes of the composite plant *Hypochoeris tweedie,* showing a pair of satellites. The letters indicate the pairs of homologous chromosomes. **1,** from Saez, Rojas and De Robertis, 1936; **2,** from Matthey, 1933; **3,** from Lewitsky, 1931; **4,** from Niyama, 1935; **5,** from Saez, 1966; **6,** from Saez, 1930; **7,** from Saez, 1945.)

The size of the interphase nucleus is proportional to the ploidy of chromosomes. By observing a section of the liver, a student may recognize a few larger nuclei corresponding to tetra- and octoploid cells.

The following structures are recognized in the interphase nucleus: (1) the *nuclear envelope*; (2) the *nucleoplasm*, formed by uncondensed regions of chromatin (euchromatin); (3) *chromocenters* and other condensed portions of chromatin (heterochromatin); and (4) nucleoli. The chromosomes are always present in the nucleus, but during interphase they are dispersed, and they are generally not morphologically distinguishable. Each diploid cell has a number of homologous chromosome pairs (23 pairs in human). One chromosome of each pair comes from one parent.

The continuity of the diploid set of chromosomes is maintained by mitosis. Duplication of chromosomes occurs during the *S* phase of interphase; however, by means of *mitosis* (prophase, metaphase, anaphase, and telophase), the daughter chromosomes are distributed equally into the two daughter cells.

Meiosis is a special type of cell division found in germinal cells, by which the number of chromosomes is reduced to the haploid number. Meiosis involves two consecutive divisions without duplication of the chromosomes. The essential feature of meiosis is a long prophase during which the homologous chromosomes pair, forming bivalents. Each bivalent has 4 chromatids (tetrad). Parts of the paired chromatids interchange at cross-points called *chiasmata*, which are manifestations of an underlying genetic recombination (crossing over). After the second meiotic division, the four resulting cells (gametes) are haploid. The morphology of chromosomes can be best studied during metaphase and anaphase. Chromosomes are of four types: (a) telocentric, (b) acrocentric, (c) submetacentric, and (d) metacentric (Fig. 2–11), depending on the position of the centromere or kinetochore, which is at the primary constriction. Other morphological characteristics include the secondary constrictions, the telomeres, the satellites, and the nucleolar zones. The *karyotype* identifies a particular chromosomal set. The *idiogram* is a representation of the chromosome pairs in a series of decreasing size.

SUMMARY:
Prokaryotic and Eukaryotic Cells

It is important, at this point, to summarize some of the concepts presented in this chapter about pro- and eukaryotic cells (see Table 2–2). These two types of cells differ not only in size, but in complexity of structure. The prokaryon nuclear region lacks the nuclear envelope and contains a single DNA molecule that divides by binary fission. The nucleus of the eukaryon has a nuclear envelope composed of two membranes with nuclear pores (Fig. 2–7); the DNA is divided into

TABLE 2–2.   Cell Organization in Prokaryons and Eukaryons

	Prokaryotic Cells	*Eukaryotic Cells*
Nuclear envelope	Absent	Present
DNA	Naked	Combined with proteins
Chromosomes	Single	Multiple
Nucleolus	Absent	Present
Division	Amitosis	Mitosis or meiosis
Ribosomes	$70S(50S + 30S)$*	$80S(60S + 40S)$
Membrane organelles	Absent	Present
Mitochondria	Respiratory and photosynthetic enzymes in the plasma membrane	Present
Chloroplast		Present in plant cells
Locomotion	Single fibril, flagellum	Cilia and flagella

*S refers to Svedberg sedimentation unit, which is a function of molecular size.

several chromosomes, and one or more nucleoli are present; division is carried out by mitosis or meiosis. The prokaryon cytoplasm lacks membrane-bound organelles, and the respiratory or photosynthetic enzymes are located in the plasma membrane. The eukaryon cytoplasm contains many membranous compartments (i.e., endoplasmic reticulum, Golgi complex) and membrane-bound organelles (i.e., mitochondria, lysosomes, chloroplasts, etc.). The plasma membrane of prokaryons is rather simple, whereas that of eukaryons show numerous differentiations (i.e., microvilli, infoldings, desmosomes, etc., (see Fig. 2–7). These and other differences are indicated in Table 2–2.

REFERENCES

1. Morowitz, H. J., and Tourtellotte, M. F. (1962) *Sci. Am., 206*:117.
2. Scheiber, G. (1969) *Internat. Symp. Nuclear Physiology and Differentiation. Genetics, 61*:161.

ADDITIONAL READING

Brachet, J., and Mirsky, A. E., eds. (1959–1961) *The Cell,* 6 Volumes. Academic Press, Inc., New York.
Du Praw, E. J. (1968) *Cell and Molecular Biology.* Academic Press, Inc., New York.
Hamerton, J. L., ed. (1963) *Chromosomes in Medicine.* Medical advisory committee of the National Spastics Society in association with William Heinemann. Little Club Clinics in Developmental Medicine, No. 5.
Head, J. J., ed. (1973) *Readings in Genetics and Evolution.* Oxford University Press, London.
Hughes, A. (1952) *The Mitotic Cycle.* Academic Press, Inc., New York.
John, B., and Lewis, K. R., (1965). *The Meiotic System.* Springer-Verlag, Vienna.
Lewis, K. R., and John, B. (1963) *Chromosome Marker.* J. & A. Churchill, London.
Lima-de-Faría, A., ed. (1969) *Handbook of Molecular Cytology.* North-Holland Pub. Co., Amsterdam.
Margulis, L. (1970) *Origin of Eukaryotic Cells: Evidence and Research.* Yale University Press, New Haven.
Margulis, L. (1971) *Symbiosis and Evolution.* Sci. Am. *225*:48.
Mathey, R. (1949) *Les chromosomes des vertébrés.* L. Rouge, Lausanne.
Mitchison, J. M. (1971) *The Biology of the Cell Cycle.* Cambridge University Press, New York.
White, M. J. D. (1961) *The Chromosomes.* 5th Ed. Methuen & Co., London.
White, M. J. D. (1973) *Animal Cytology and Evolution.* 3rd Ed. Cambridge University Press, London.

Part Two

MOLECULAR COMPONENTS AND METABOLISM OF THE CELL

The general organization of the cell presented in the introductory chapters may best be interpreted on chemical and physicochemical evidence. The reader should review his previous studies of organic chemistry, especially those on proteins, carbohydrates, lipids, and nucleic acids—the main molecular components of the cell. Part two of this book is dedicated to an elementary and abbreviated survey of these molecular components. In Chapter 3 special emphasis is given to some of the stereochemical characteristics of these components, such as the primary, secondary, tertiary, and quaternary structure of proteins, and to the Watson-Crick model of DNA, which explains how the DNA molecule duplicates and functions as the primary storehouse of genetic information. This chapter is also prerequisite to understanding the discussion of cytochemical methods in Chapter 6.

Chapter 4 briefly introduces the concept of enzymes as molecular machines used by the cell to produce all chemical transformations. The notion of the enzyme active site and of its possible interpretation at the molecular level is of particular importance. Enzyme kinetics and the various factors and mechanisms by which an enzymatic reaction may be inhibited or activated are presented. The main metabolic pathways utilized by the cell to obtain chemical energy from different foodstuffs are also introduced. The energy that is produced is then used to synthesize new products which either increase the mass of the cell or are eliminated to the environment as secretions. This suggests the idea of high energy bonds and the concept of bio-energetics, which, in turn, introduces the concept of entropy and its importance in biology. The reader is also introduced to the concepts of anaerobic glycolysis, the Krebs cycle, and oxidative phosphorylation; the last two together embody cell respiration. The ideas presented in Part Two are fundamental to understanding Chapters 10 and 11, which deal with mitochondria and the photosynthetic mechanism of chloroplasts, respectively.

three

CHEMICAL COMPONENTS OF THE CELL

To understand the organization of biological systems, one should first become familiar with the main constituent molecules, particularly those of high molecular weight, such as proteins, nucleic acids, polysaccharides, and lipids.

MOLECULAR POPULATION OF THE CELL

An early approach to the study of the chemical composition of the cell was the biochemical analysis of whole tissues, such as the liver, brain, skin, or plant meristem. This method had limited cytologic value, because the material analyzed was generally a mixture of different cell types and, in addition, contained extracellular material. In recent years the development of cell fractionation methods and various micromethods has led to the isolation of different subcellular particles and thus to more important and precise information about the molecular architecture of the cell (Chapter 6).

The chemical components of the cell can be classified as *inorganic* (water and mineral ions) and *organic* (proteins, carbohydrates, nucleic acids, lipids, and so forth). Some organic components, such as enzymes, coenzymes, and hormones, that have specific activities are mentioned in Chapter 4 and should be studied in detail in biochemistry textbooks.

The protoplasm of a plant or animal cell contains 75 to 85 per cent water, 10 to 20 per cent protein, 2 to 3 per cent lipid, 1 per cent carbohydrates, and 1 per cent inorganic material.

Water, Free and Bound

With few exceptions, such as bone and enamel, water is the most abundant component of cells. It serves as a natural solvent for mineral ions and other substances and also as a dispersion medium of the colloid system of protoplasm. For instance, from microinjection experiments it is known that water is readily miscible with protoplasm. Furthermore, water is indispensable for metabolic activity, since physiologic processes occur exclusively in aqueous media. Water molecules also participate in many enzymatic reactions in the cell and can be formed as a result of metabolic processes.

Water exists in the cell in two forms: *free* and *bound*. *Free water* represents 95 per cent of the total cellular water and is the principal part used as a solvent for solutes and as a dispersion medium of the colloid system of protoplasm. *Bound water*, which represents only 4 to 5 per cent of the total cellular water, is loosely held to the proteins by hydrogen bonds and other forces. It includes the so-called *unmobilized* water contained within the fibrous structure of macromolecules. Because of the asymmetric distribution of charges, a water molecule acts as a *dipole*, as shown in the following diagram.

Because of this polarity, water can bind electrostatically to both positively and negatively charged groups in the protein. Thus, each amino group in a protein molecule is capable of binding 2.6 molecules of water.

Water is also used to eliminate substances from the cell and to absorb heat—by virtue of its high specific heat coefficient—thus preventing drastic temperature changes in the cell.

The water content of an organism is related to the organism's age and metabolic activity. For example, it is highest in the embryo (90 to 95 per cent) and decreases progressively in the adult and in the aged.

Salts and Ions

Salts dissociated into anions (e.g., Cl^-) and cations (e.g., Na^+ and K^+) are important in maintaining *osmotic pressure* and the *acid-base equilibrium* of the cell. Retention of ions produces an increase in osmotic pressure and thus the entrance of water. Some of the inorganic ions, such as magnesium, are indispensable as cofactors in enzymatic activities; others, such as inorganic phosphate, form adenosine triphosphate (ATP), the chief supplier of chemical energy for the living processes of the cell, through oxidative phosphorylation.

The concentration of various ions in the intracellular fluid differs from that in the interstitial fluid (see Table 21–1). For example, the cell has a high concentration of K^+ and Mg^{++}, while Na^+ and Cl^- are localized mainly in the interstitial fluid. The dominant anion in cells is phosphate; some bicarbonate is also present.

Calcium ions are found in the circulating blood and in cells. In bone they combine with phosphate and carbonate ions to form a crystalline arrangement.

Phosphate occurs in the blood and tissue fluids as a free ion, but much of the phosphate of the body is bound in the form of phospholipids, nucleotides, phosphoproteins, and phosphorylated sugars. As primary phosphate ($H_2PO_4^-$) and secondary phosphate (HPO_4^{--}), phosphate contributes to the buffer mechanism, thereby stabilizing the pH of the blood and tissue fluids.

Other ions found in tissues are sulfate, carbonate, bicarbonate, magnesium, and amino acids.

Certain *mineral components* are found in a nonionized form. For example, *iron*, bound by metal-carbon linkages, is found in hemoglobin, ferritin, the cytochromes, and some enzymes (such as catalase and cytochrome oxidase). Traces of *manganese, copper, cobalt, iodine, selenium, nickel, molybdenum,* and *zinc* are indispensable for maintenance of normal cellular activities. For comprehensive reviews, see "The Chemical Elements of Life" (Frieden, 1972) and also Fruton, 1973.

MACROMOLECULES

Structural and other properties of the cell are intimately related to large molecules made of repeating units linked by covalent bonds. These units are called *monomers*, and the resulting macromolecule is called a *polymer*. Molecules having an increasing number of monomers possess widely different characteristics. For example, among the hydrocarbons—methane and ethane are gases, while butane and octane are liquids; further polymerization (20 or more monomers) produces oils, and finally solids, such as paraffins.

The three main examples of polymers in living organisms are as follows: (a) *Nucleic acids* result from the repetition of four different units called *nucleotides*. The repetition of the four nucleotides in the DNA molecule is the primary source of genetic information. (b) *Polysaccharides* can be polymers of monosaccharides, forming starch, cellulose, or glycogen, or may also involve the repetition of other molecules, forming more complex polysaccharides. (c) *Proteins* and *polypeptides* consist of the association in various proportions of some 20 different amino acids linked by peptide bonds. The order in which these 20 monomers can be linked gives rise to an astounding number of combinations in various protein molecules. This can determine not only their specificity, but in certain cases, their biological activity.

Because of their unique structure, nucleic acids and proteins can be considered to be *informational* macromolecules—i.e., carrying biological information. Polysaccharides made of a single type of molecule, on the other hand (e.g., glucose), are *non-informational*.

Amino Acids and Proteins

The building blocks of proteins are the amino acids, which the reader should remem-

TABLE 3–1.   Types of Natural Amino Acids and Abbreviations Used for Them*

Monoamino-monocarboxylic
  Glycine (Gly)
  Alanine (Ala)
  Valine (Val)
  Leucine (Leu)
  Isoleucine (Ile)
Monoamino-dicarboxylic
  Glutamic acid (Glu)
  Aspartic acid (Asp)
Diamino-monocarboxylic
  Arginine (Arg)
  Lysine (Lys)
  Hydroxylysine (Hlys)
Hydroxyl-containing
  Threonine (Thr)
  Serine (Ser)
Sulfur-containing
  Cystine (Cys)
  Methionine (Met)
Aromatic
  Phenylalanine (Phe)
  Tyrosine (Tyr)
Heterocyclic
  Tryptophan (Trp)
  Proline (Pro)
  Hydroxyproline (Hpro)
  Histidine (His)

*From Giese, A. C., *Cell Physiology*, 3rd ed., Philadelphia, W. B. Saunders Co., 1968.

ber from studies of biochemistry (Table 3–1). Essentially, an amino acid is derived from an organic acid in which the hydrogen in the alpha position is replaced by an amino group ($-NH_2$). For example, acetic acid gives *glycine* and propionic acid, *alanine.* Because of the simultaneous presence of acidic carboxyl ($-COOH$) and of basic amino ($-NH_2$) groups, such molecules are called amphoteric.

The free amino acids present in a cell may result from the breakdown of proteins or from absorption from the intercellular fluid. Free amino acids constitute the so-called *amino acid pool,* from which the cell draws its building blocks for the synthesis of new proteins.

The condensation of amino acids to form a protein molecule occurs in such a way that the acidic group of one amino acid combines with the basic group of the adjoining one, with the simultaneous loss of one molecule of water.

The linkage $-NH-CO-$ is known as the *peptide linkage* or *peptide bond.* The formed molecule preserves its amphoteric character, since an acidic group is always at one end and a basic group is at the other, in

addition to the lateral residues (radicals) that can be either basic or acidic (Table 3–1). A combination of two amino acids is a *dipeptide;* of three, a *tripeptide.* When a few amino acids are linked together, the structure is an *oligopeptide.* A *polypeptide* contains a large number of amino acids.

The distance between two peptide links is about 0.35 nm. A protein with a molecular weight of 30,000 consisting of 300 amino acid residues, if fully extended, should have a length of 100 nm, a width of 1.0 nm, and a thickness of 0.46 nm.

Table 3–2 lists the molecular weights of various proteins. The term *protein* (Gr., *proteuo* I, occupying first place) indicates that all basic functions in living organisms depend on specific proteins. They constitute the enzymes and the contractile machinery of the cell, and are present in the blood, and other intercellular fluids. Some long-chain proteins, such as *collagen* and *elastin,* play an important role in the organization of tissues that form the extracellular framework.

For details of the classification of the proteins, the reader is referred to biochemistry textbooks; however, it is important to stress that the properties of proteins vary considerably. For instance, *keratin* and *collagen* are insoluble and fibrous; the *globular proteins,*

TABLE 3–2.   Molecular Weights of Some Proteins

Insulin	12,000
Cytochrome (horse heart)	12,100
Trypsin	23,800
Pepsin	35,500
Ovalbumin	44,000
Serum Albumin (human)	65,000
Hemoglobin (human)	67,000
γ-Globulin (human)	100,000
Catalase	250,000
Collagen	345,000
Thyroglobin (pig)	650,000

e.g., egg albumin and serum proteins, are soluble in water or salt solutions and are spherical rather than threadlike molecules.

The *conjugated proteins* are attached to a non-protein moiety, the so-called *prosthetic group*. To such a group belong the *nucleoproteins* associated with nucleic acids, the *glycoproteins*, the *lipoproteins* (e.g., blood lipoproteins), and the *chromoproteins* that have a pigment as the prosthetic group, such as hemoglobin, hemocyanin, and the cytochromes. Hemoglobin and myoglobin (present in muscle) contain the prosthetic group *heme*, a metal-containing organic compound that combines with oxygen.

Primary Structure of Proteins

The amino acid sequence of the polypeptide chain is known as the *primary structure* of the protein molecule. It is the most important and specific structure and, to a certain extent, determines the so-called *secondary* and *tertiary* structures. Aggregates of protein units containing secondary and tertiary structures constitute the *quaternary* structure.

Determination of the sequence of amino acids has been made possible by the development of a series of methods for the degradation of proteins, methods which finally supplied the first complete analysis of *insulin* (Sanger, 1954). This molecule is composed of two chains: the A-chain consists of 21 amino acids, and the B-chain, 30 amino acids. Both chains are linked by two —S—S— (disulfide) bonds. Figure 3–1 shows the whole sequence in *ribonuclease,* an enzyme that consists of 124 amino acid residues. Other proteins whose structures have been elucidated are hemoglobin, cytochrome *c,* lysozyme, trypsinogen, as well as many others. The longest polypeptide whose primary structure has been determined is glyceraldehyde-3-phosphate dehydrogenase (333 residues) (see Lehninger, 1970).

In the protein molecule, amino acids are arranged like beads on a string (Fig. 3–1), and their sequence is of great biological importance. In the hemoglobin molecule a change in a single amino acid produces profound biological changes (see Table 19–1). A fully extended polypeptide chain, shown with the exact dimensions and bond angles determined by x-ray diffraction, is presented in Figure 3–2.

Secondary Structure of Proteins

In a protein formed by several hundred amino acids, the chain may sometimes be linear, but more frequently it assumes different shapes that constitute what is known as the *secondary structure.* Fibrous proteins are often arranged in an orderly manner that can be analyzed by x-ray diffraction methods (see Chapter 6). This technique has facilitated

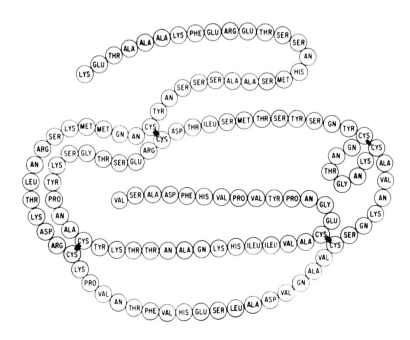

Figure 3–1. The primary structure of bovine pancreatic ribonuclease. Notice the position of the four disulfide bridges between cystine residues. (From C. B. Anfinsen, 1959.)

classification of proteins into three structural types or groups.

The *β-keratin type* has an identity period of about 0.72 nm (Fig. 3–2). The adjacent chains are disposed in a *pleated sheet structure* as shown in Figure 3–3, in which the side-chains of the amino acid residues stick out perpendicular to the plane of the chain. The individual chains are held together by hydrogen bonds, forming a "peptide grid."

The *α-helix structure* found in the *α-keratin type* is produced when the polypeptide chain forms a helical structure, like a spiral winding around an imaginary cylinder, in such a way that hydrogen bonds are established within the molecule and not with an adjacent molecule (Fig. 3–4, *B*). For the *collagen group,* a model made of three helical chains has been proposed (see Fig. 7–1).

Tertiary Structure of Proteins

In the so-called *globular proteins* the polypeptide chain is held together in a definite way to form a compact structure (Fig. 3–4, C). The arrangement in space of such chains is very complex but may be resolved by x-ray diffraction (Fig. 5–10). In globular proteins the chains are folded in a compact way with the polar groups toward the surface, thus leaving little space in the interior for water

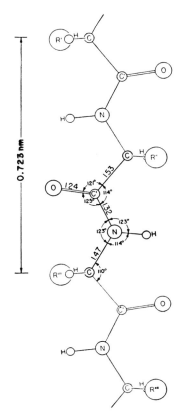

Figure 3–2. Dimensions and angles in a polypeptide chain that is totally extended. (From R. B. Corey and L. Pauling.)

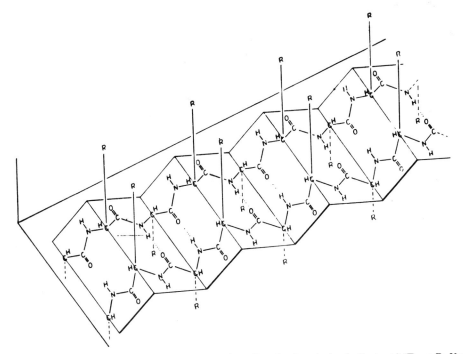

Figure 3–3. Pleated sheet structure of β-protein chains. (See the description in the text.) (From P. Karlson, 1963.)

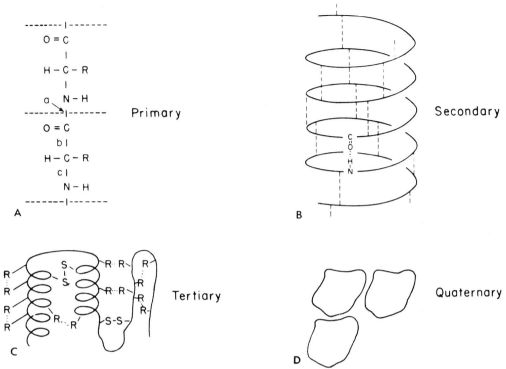

Figure 3–4. Schematic representation of the structural levels in proteins. Amino acid chains are denoted by R, noncovalent interactions by · · ·. (From Nemethy, G., "Proteins (Binding Forces in Secondary and Tertiary Structures)," in *Encyclopedia of Biochemistry*, New York, Van Nostrand Reinhold, 1967.)

molecules. These proteins contain varying amounts of α-helix or β-configurations, and other parts exist as a *random coil,* i.e., in a flexible structure that may change at random.

The spatial arrangement is to some extent predetermined by the sequence of amino acids in the primary structure and by the bonds that can be established among some of the residues. A series of biological properties of proteins, such as enzyme activity and antigenicity, is related to the tertiary structure.

The *denaturation* of a protein is brought about by high temperatures or other nonphysiological conditions, and consists of a disruption of the tertiary structure. This is usually accompanied by loss of biological activity. Sometimes the protein may reassume its natural configuration (*renaturation*) and regain its normal activity.

Quaternary Structure of Proteins

Unlike the primary, secondary, or tertiary structures, which contain a single polypeptide chain, quaternary structure involves two or more chains (Fig. 3–4, *D*). These chains may or may not be identical but in both cases they are linked by weak bonds (noncovalent). For example, the hemoglobin molecule is composed of four polypeptides or subunits, two designated as α and two as β. Separation and association of the subunits may occur spontaneously. Hemoglobin may be broken into two half molecules (two α and two β) by urea. When urea is removed, they reassemble, forming complete, functional molecules. This binding is highly specific and takes place only between the half molecules. This is called the *principle of self-assembly.* This principle also applies to the building up of more complex cellular structures, such as the cell membrane, microtubules, and so forth (see Chapter 7). Many enzymes and other proteins having a molecular weight (MW) above 50,000 daltons probably exist as quaternary structures.

Bonds in the Protein Molecule

Different types of bonds are involved in the structure of proteins. The primary structure (peptide bond) is fully determined by *chemical or covalent bonds*. Between cystine residues, —S—S— bonds of the same nature can be established, as in insulin and ribonuclease (Fig. 3–1). The secondary and tertiary structures are determined by a series of weaker bonds illustrated in Figure 3–5. All these bonds are non-covalent and can be classified as:

Ionic or electrostatic, which bind positive and negative ions that are in close range of 0.2 to 0.3 nm (Fig. 3–5,*a*).

Hydrogen bonds, with a range between 0.25 and 0.32 nm and weaker than ionic bonds. These are essentially electrostatic bonds that form a kind of bridge between two strongly negative atoms such as C, N, or O (Fig. 3–5,*b*).

Weaker bonds, produced by interaction of *non-polar side-chains* and caused by mutual repulsion of the solvent (Fig. 3–5,*c*).

van der Waals forces, produced by interaction between polar side-chains (Fig. 3–5,*d*).

The essential difference between a covalent and a non-covalent bond is in the amount of energy needed to break the bond. For example, a hydrogen bond requires only 4.5 kcal mole^{-1}, as compared with 110 kcal mole^{-1} for the covalent O—H bond in water. In the following chapter, a more detailed discussion will show that covalent bonds are generally broken by the intervention of enzymes, whereas non-covalent bonds are easily dissociated by physicochemical forces.

In a protein, about one half the amino acid residues are hydrophobic, i.e., tending to repel water molecules. Such residues cause the so-called hydrophobic interactions that are important in determining the shape of the globular protein. The interaction between these residues tends to repel the water molecules that surround the protein, thereby causing the globular structure to be more compact.

Electrical Charges of Proteins

In addition to the terminal —NH$_3^+$ and —COO$^-$ charged groups, proteins contain dicarboxylic- and diamino-amino acids (Table 3–1) which dissociate as follows:

1. The acidic groups lose protons and become negatively charged. For example, in aspartic and glutamic acids, the free carboxyl group dissociates into —COO$^-$ + H$^+$.

2. The basic groups, by gaining protons, become positively charged —NH$_2$ + H$^+$ → —NH$_3^+$. This type is found in amino acids with two basic groups, such as lysine or arginine. All these so-called *ion-producing groups* contribute to the acid-base reactions of proteins and to the electrical properties of protein molecules.

The actual charge of a protein molecule is the result of the sum of all single charges. Because dissociation of the different acidic and basic groups takes place at different hydrogen ion concentrations of the medium, pH greatly influences the total charge of the molecule. In an acid medium, amino groups capture hydrogen ions and react as bases (—NH^2 + H$^+$ → —NH$_3^+$); in an alkaline medium the reverse

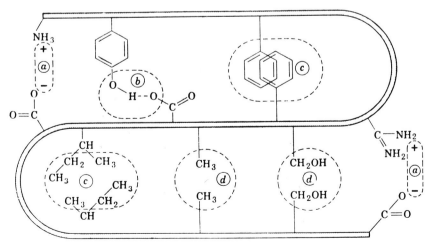

Figure 3–5. Types of noncovalent bonds that stabilize protein structure. (See description in the text.) (From C. B. Anfinsen, 1959.)

takes place and carboxylic groups dissociate ($-COOH \rightarrow COO^- + H^+$). For every protein there is a definite pH at which the sum of positive and negative charges is zero. This pH is called the *isoelectric point* (pI). At the isoelectric point, proteins placed in an electric field do not migrate to either of the poles, whereas at a lower pH they migrate to the cathode and at a higher pH to the anode; this migration is called *electrophoresis*. At the isoelectric point many of the physicochemical properties of the proteins are changed. For instance, viscosity, solubility, hydration, osmotic pressure, and conductivity are at a minimum.

Every protein has a characteristic isoelectric point. For example, in histones and protamines, which are found mainly in the nucleus, the isoélectric point is high (pI 10 to 12). This is because of the presence of numerous diamino-monocarboxylic amino acids. The isoelectric point of gelatin is 4.7 because of the predominance of monoamino-dicarboxylic amino acids.

Carbohydrates

Carbohydrates, composed of carbon, hydrogen, and oxygen, are sources of energy for animal and plant cells; in many plants they also form important constituents of cell walls and serve as supporting elements. Animal tissues have fewer carbohydrates; among the most important are glucose, galactose, glycogen, and amino sugars and their polymers.

Carbohydrates of biological importance are classified as monosaccharides, disaccharides, and polysaccharides. The first two, commonly referred to as *sugars,* are readily soluble in water, can be crystallized, and easily pass through dialyzing membranes. Polysaccharides, on the other hand, neither crystallize nor pass through membranes.

Monosaccharides

Monosaccharides are simple sugars having the empirical formula $C_n(H_2O)_n$. They are classified in accordance with the number of carbon atoms, e.g., trioses and hexoses. The pentoses *ribose* and *deoxyribose* are found in the molecules of nucleic acids, and the pentose *ribulose* is important in photosynthesis (see Chapter 11). Glucose, a hexose, is the primary source of energy for the cell. Other

important hexoses are *galactose* found in the disaccharide lactose, and *fructose* (*levulose*), which forms part of sucrose.

Disaccharides

Disaccharides are sugars formed by the condensation of two monomers of monosaccharides with the loss of one molecule of water. Their empirical formula is therefore $C_{12}H_{22}O_{11}$. The most important of this group are *sucrose* and *maltose* in plants and *lactose* in animals.

Polysaccharides

Polysaccharides result from the condensation of many molecules of monosaccharides, with a corresponding loss of water molecules. Their empirical formula is $(C_6H_{10}O_5)_n$. Upon hydrolysis they yield molecules of simple sugars. The most important polysaccharides in living organisms are *starch* and *glycogen,* which are reserve substances in cells of plants and animals, respectively, and *cellulose,* the most important structural element of the plant cell.

Starch is a combination of two long polymer molecules: *amylose,* which is linear, and *amylopectin,* which is branched (Fig. 3–6). Glycogen may be considered to be the starch of animal cells. It is a polymer composed of many molecules of glucose. It is found in numerous tissues and organs, but the greatest proportion is contained in liver cells and muscle fibers. Cellulose is composed of units of cellobiose ($C_{12}H_{22}O_{11}$). On hydrolysis, cellobiose yields glucose (see Table 11–1).

Complex Polysaccharides and Glycoproteins

In addition to the polysaccharides made of hexose monomers mentioned in the preceding section, there are many more complex long molecules that contain amino nitrogen (e.g., glucosamine) that can, in addition, be acetylated (e.g., acetylglucosamine) or substituted with sulfuric or phosphoric acid. All these polymers are important in molecular organization, particularly as intercellular substances. These polysaccharides may exist either freely or combined with proteins. The most important are:

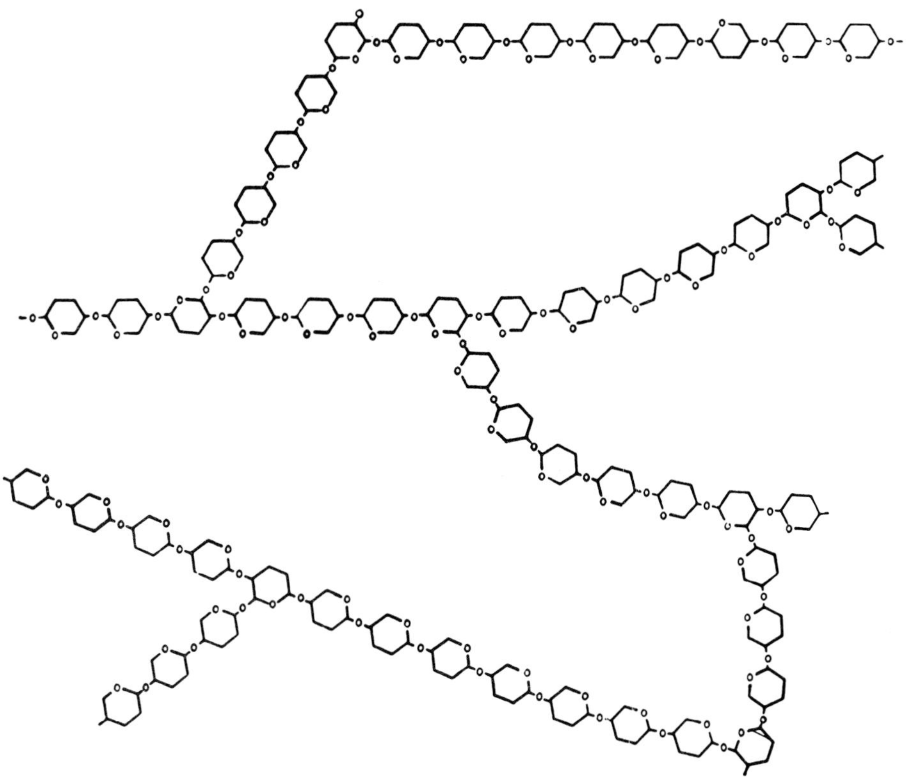

Figure 3–6. Molecular representation of part of a macromolecule of amylopectin. (From Oncley, J. L., *Biophysical Science—A Study Program.* New York, John Wiley & Sons, Inc., 1959.)

Neutral polysaccharides, which contain only acetylglucosamine. The main example is *chitin,* a supporting substance found in insects and crustaceans.

Acidic mucopolysaccharides, which contain sulfuric or other acids in the molecule. These molecules are strongly basophilic. To this group belong *heparin,* an anticoagulant substance; *chondroitin sulfate,* present in the cartilage, skin, cornea, and umbilical cord; and *hyaluronic acid,* in skin and other animal tissues. The latter is hydrolyzed by *hyaluronidase.*

Glycoproteins are complexes composed of a protein and a prosthetic group of carbohydrates. Several monosaccharides such as galactose, mannose, fucose, as well as N-acetyl-D-glucosamine and sialic acid, are found. Glycoproteins can be divided into two major categories: intracellular, and secretory. Intracellular glycoproteins are present in cellular membranes and have important functions in membrane interaction and recognition (see Chapter 8).

There are many glycoproteins secreted by various cells: plasma glycoproteins (i.e., seroalbumins), secreted by the liver; thyroglobulin, produced in the thyroid gland; immunoglobulins, secreted by plasma cells; ovoalbumin, secreted by the hen oviduct; and ribonuclease *B* and deoxyribonuclease, produced by the bovine pancreas. In most glycoproteins the protein is linked to the carbohydrate moiety by a bond between asparagine (Asn) and N-acetyl-D-glycosamine (GlcNAc).

```
——— Asn ———     Protein
         |
      GlcNAc
┌─────────────────────┐
│ Man    —    GlcNAc  │     Core
└─────────────────────┘
  |              |
GlcNAc        GlcNAc
  |              |
 Gal           Gal
  |              |
Fucose       Sialic acid
```

This diagram shows that the prosthetic

group consists of an oligosaccharide core containing only D-mannose and N-acetyl-D-glucosamine. This core may terminate by way of two trisaccharides ending in L-fucose or sialic acid.

In Chapter 9 it will be mentioned that the biosynthesis of glycoproteins is brought about by the successive incorporation of the carbohydrates upon the protein, via a series of specific enzymes called *glycosyltransferases*. Recent evidence obtained by Leloir's group at Buenos Aires indicates that the oligosaccharide core of glycoproteins (formed mainly by N-acetyl-D-glucosamine and mannose) is first assembled on a lipid carrier, and is later transferred to the protein molecule.

Several exocrine glands, such as the salivary glands and the mucous gland of the digestive tract, secrete mucoproteins in which the linkage is between N-acetylglycosamine and serine or threonine in the protein. Sialic acid and L-fucose are the terminal sugars in these compounds.

The functions of glycoproteins are very complex and will be considered in several subsequent chapters, especially those devoted to the cell membrane, the Golgi complex, and secretion. It may be said here that the carbohydrate moiety may be important in determining certain physicochemical properties (the viscosity of mucins may be important in protecting and lubricating the mucous membrane of the digestive tract, for example); and in molecule-membrane interactions (for the determination of blood groups and other immunological properties at the surface of erythrocytes); and in membrane-membrane interactions (see Chapters 8 and 20).

Lipids

This large group of compounds is characterized by their relative insolubility in water and solubility in organic solvents. The cause of this general property of lipids and related compounds is the predominance of long aliphatic hydrocarbon chains or benzene rings. These structures are non-polar and hydrophobic. In many lipids these chains may be attached at one end to a polar group, rendering it capable of binding water by hydrogen bonds.

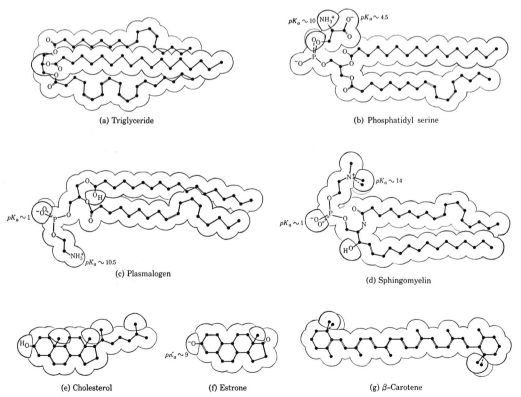

Figure 3–7. Some lipid molecules, showing the three-dimensional array and relative size. (From Oncley, J. L., *Biophysical Science—A Study Program.* New York, John Wiley & Sons, Inc., 1959.)

Following is a classification of lipids:

Simple Lipids

Simple lipids are alcohol esters of fatty acids. Among these are:

Natural fats (glycerides), often called triglycerides (Fig. 3–7,*a*); they are triesters of fatty acids and glycerol.

Waxes, having a higher melting point than natural fats; they are esters of fatty acids with alcohols other than glycerol, such as beeswax.

Steroids

These lipids are characterized by the cyclopentano-perhydro-phenanthrene nucleus (Fig. 3–7,*e*). The steroids include a series of highly important substances in the body, such as sex hormones (Fig. 3–7,*f*), adrenocortical hormones, vitamin D, and bile acids. Steroids that possess an —OH group are called *sterols.* *Cholesterol* is widely distributed and is found in the bile, brain, adrenal glands, and other tissues. It often occurs in ester linkage with fatty acids.

Stereochemically, sterols form complex, rather flattened ring systems. The cholesterol molecule is about 2.0 nm long, 0.7 to 0.75 nm wide and 0.5 nm thick. There is a polar — OH group at one end and a nonpolar hydrocarbon residue at the other (Fig. 3–7,*e*).

Compound (Conjugated) Lipids

Upon hydrolysis these lipids yield other compounds in addition to alcohol and acids. Together with sterols they serve mainly as structural components of the cell, particularly in cell membranes. The following are classified as compound lipids:

Phosphatides (phospholipids), diesters of phosphoric acid that can be esterified with either glycerol, sphingosine, or choline; ethanolamine; serine; or inositol. This group includes the lecithins, cephalins, inositides, and plasmalogens (acetyl phosphatides) (see Figure 3–7 and Table 3–3).

Glycolipids and *sphingolipids* are characterized by the fact that glycerol is replaced by the amino alcohol *sphingosine.* To these groups belong the *sphingomyelins* (Fig. 3–7, *d*), mainly in the myelin sheath of nerves; the *cerebrosides,* which are characterized by the presence of galactose or glucose in the molecule; the *sulfatides,* which contain sulfuric acid esterified to galactose; and the *gangliosides* (Table 3–3).

The gangliosides deserve special mention because of their presence in cell membranes, their possible role as receptors of virus particles, and their influence on ion transport across membranes (see Chapter 21). A ganglioside is a complex molecule containing sphingosine, fatty acids, carbohydrates (lactose + galactosamine), and neuraminic acid. This is a long and highly polar molecule.

TABLE 3–3. Classification of Phosphatides and Glycolipids*

Name	Main Alcohol Component	Other Alcohol Components	P:N Ratio
I. Glycerophosphatides			
1. Phosphatidic acids	Diglyceride (= glycerol diester)		1:0
2. Lecithins	Diglyceride (=glycerol diester)	Choline	1:1
3. Cephalins	Diglyceride (=glycerol diester)	Ethanolamine, serine	1:1
4. Inositides	Diglyceride (=glycerol diester)	Inositol	1:0
5. Plasmalogens ("acetyl phosphatides")	Glycerol ester and enol ether	Ethanolamine, choline	1:1
II. Sphingolipids			
1. Sphingomyelins	N-Acylsphingosine	Choline	1:2
2. Cerebrosides	N-Acylsphingosine	Galactose,† glucose†	0:1
3. Sulfatides	N-Acylsphingosine	Galactose†	(1 H_2SO_4)
4. Gangliosides	N-Acylsphingosine	Hexoses,† hexosamine,† neuraminic acid†	no P

*From Karlson, P., *Introduction to Modern Biochemistry.* 2nd ed., New York, Academic Press, Inc. 1967.

†These components are present as glycosidic linkage, and thus are called glycolipids.

SUMMARY:
Cellular Chemistry

This chapter is an elementary survey of the molecular components of the cell. The protoplasm of a plant or animal cell contains 75 to 85 per cent water, 10 to 20 per cent protein, 2 to 3 per cent lipid and 1 per cent carbohydrate. *Water* is the natural solvent of mineral ions and other small molecules, and is a dispersion medium for the colloid system of the cell. About 95 per cent of the water is *free;* the rest is *bound* or inmobilized within the structure. To each amino group of a protein 2.6 molecules of water are bound. Anions (e.g., Cl^-) and cations (e.g., K^+, Na^+) help to maintain the osmotic pressure and pH of the cell. Inorganic phosphate bound in ATP is the main supplier of chemical energy. The interior of the cell is rich in K^+ and Mg^{++}, while the intercellular fluid contains Na^+ and Cl^-. Iron is bound by metal-carbon linkages in hemoglobin and cytochromes.

Macromolecules are long *polymers* composed of *monomers*. *Nucleic acids* and *proteins* are *informational* macromolecules, whereas simple polysaccharides (e.g., starch, glycogen) are *non-informational.*

Proteins are made of about 20 different monomers—the *amino acids*. These are *amphoteric* molecules because they carry an acidic (—COOH) and a basic group (—NH$_2$). Amino acids are linked by the *peptide bonds,* forming polypeptides. All basic functions of the cell depend on specific proteins. These proteins are *globular* (e.g., egg albumin, serum proteins) or *fibrous*. Conjugated proteins contain a *prosthetic group* (e.g., nucleoproteins, lipoproteins, chromoproteins).

The *primary structure* of the protein is the amino acid sequence; currently, the sequence is known for proteins containing up to 300 amino acids. A change in a single amino acid may produce a profound change in the molecule (e.g., hemoglobins).

According to the secondary structure of proteins, they may be of the β-*keratin,* the α-*helix keratin,* or the *collagen* type. The *tertiary* structure of globular protein is very complex and can be determined by x-ray diffraction. These proteins may contain parts with α-helix and β-configurations, or a random coil. The disruption of the tertiary structure by high temperatures or other agents is called *denaturation.* The *quaternary* structure is characteristic of proteins having more than one subunit. Hemoglobin is a tetramer composed of two α- and two β-chains.

In proteins, non-covalent bonds, i.e., ionic and hydrogen, hydrophobic interactions play an important role in determining the secondary and tertiary structure. The electrical charge of a protein is determined by the *ion-producing groups* (i.e., acidic or basic) and their degree of dissociation at different pHs. At the *isoelectric point* (pI) the net charge is 0. In basic proteins the pI is 10 to 12 (e.g., histones).

Carbohydrates serve as sources of energy or play a structural role (e.g., in cell walls). They are classified as mono-, di-, and polysaccharides. The main monosaccharides are *pentoses* (5 carbons) and *hexoses* (6 carbons). The main polysaccharides, which are composed of glucose monomers, are starch (plants) and glycogen (liver and muscle). Complex polysaccharides contain amino-sugars (i.e., glucosamine), sulfuric acid or phosphoric acid. The acidic mucopolysaccharides are strongly basophilic (i.e., heparin, chondroitin sulfate, hyaluronic acid). *Glycoproteins* are either *intracellular,* present in cell membranes, or *secretory* (e.g., seroalbumins, thyroglobulin, immunoglobulins, ribonuclease, etc.). In most glycoproteins the carbohydrate is linked to the amino acid asparagine by way of N-acetyl-D-glucosamine (GlcNAc). The prosthetic group has an oligosaccharide core of mannose and GlcNAc.

Lipids comprise a large group of different compounds characterized by their solubility in organic solvents. There are *simple* lipids (glycerides), *steroids* (i.e., sex hormones, cholesterol), and *conjugated lipids* (i.e., phosphatides, glycolipids, cerebrosides, sulfatides, and gangliosides). The more important phosphatides are: phosphatidylcholine (lethicin), phosphatidylethanolamine (cephalin), phosphatidyl-inositol, and the acetyl phosphatides.

NUCLEIC ACIDS

Nucleic acids are macromolecules of the utmost biological importance. All living organisms contain nucleic acids in the form of deoxyribonucleic acid (DNA) and ribonucleic acid (RNA). Some viruses may contain only RNA (e.g., tobacco mosaic virus [Fig. 7–1] and poliomyelitis virus), while others have DNA (e.g., bacteriophages [Fig. 5–9], vaccinia, and adenoviruses).

DNA is the major store of genetic information. This information is transmitted by *transcription* into RNA molecules, which are utilized in the synthesis of proteins. In fact, the central dogma of modern biology is:

$$DNA \rightarrow RNA \rightarrow PROTEIN$$

The biological role of nucleic acids is discussed in detail in the chapters dealing with molecular biology (Part Seven). Only those features of their chemical structure relevant to the understanding of their function will be considered in the present chapter.

In higher cells DNA is localized mainly in the nucleus as part of the chromosomes. A small amount of DNA is present in the cytoplasm and contained within mitochondria and chloroplasts. RNA is found both in the nucleus, where it is synthesized, and in the cytoplasm, where the synthesis of proteins occurs (Table 3–4).

Chemical Composition

Nucleic acids are formed of a sugar moiety (pentose), nitrogenous bases (purines and pyrimidines), and phosphoric acid. After a mild hydrolysis the nucleic acids are decomposed into nucleotides.

Nucleotides are the monomeric units of the nucleic acid macromolecule. They result

TABLE 3–4.   Nucleic Acids: Structure, Reactions, and Role in the Cell

	Deoxyribonucleic Acid	*Ribonucleic Acid*
Localization	Primarily in nucleus; also in mitochondria and chloroplasts	In cytoplasm, nucleolus, and chromosomes
Pyrimidine bases	Cytosine Thymine	Cytosine Uracil
Purines	Adenine Guanine	Adenine Guanine
Pentose	Deoxyribose	Ribose
Cytochemical Reaction	Feulgen	Basophilic dyes with ribonuclease treatment
Hydrolyzing enzyme	Deoxyribonuclease (DNase)	Ribonuclease (RNase)
Role in cell	Genetic information	Synthesis of proteins

from the covalent bonding of a phosphate and a heterocyclic base to the pentose (Fig. 3–8). Within the nucleotide, the combination of a base with the pentose constitutes a *nucleoside*. For example, *adenine* is a purine base; *adenosine* (adenine + ribose) is the corresponding nucleoside, and adenosine monophosphate (AMP), the nucleotide (Fig. 3–8).

Nucleic acids are linear polymers in which the nucleotides are linked together by means of phosphate-diester bridges with the pentose moiety. These bonds link the 3′ carbon in one nucleotide to the 5′ carbon in the pentose of the adjacent nucleotide. The backbone of nucleic acids consists, therefore, of alternating phosphates and pentoses. The nitrogenous bases are attached to the sugars of this backbone.

As shown in Figure 3–8 the phosphoric acid uses two of its three acid groups in the 3′,5′ diester links. The remaining negative group confers to the polynucleotide its acid properties and enables the molecule to form ionic bonds with basic proteins. In eukaryotic cells, DNA is associated with *histones* (i.e., basic proteins rich in arginine or lysine), forming a *nucleoprotein*. This anionic group also causes nucleic acids to be highly basophilic, i.e., they stain readily with basic dyes (see Table 6–1).

Pentoses are of two types: *ribose* in RNA, and *deoxyribose* in DNA. The only difference between these two sugars is that the oxygen in the 2′ carbon is lacking in de-oxyribose. *The bases* found in nucleic acids are either *pyrimidines* or *purines*. Pyrimidines have a single heterocyclic ring, whereas purines have two fused rings (Fig. 3–9). In DNA the pyrimidines are *thymine* (T) and *cytosine* (C); the purines are *adenine* (A) and *guanine* (G) (Fig. 3–9). RNA contains *uracil* (U) instead of thymine (Table 3–4). Therefore between RNA and DNA there are two main differences: in the pentose moiety (ribose and deoxyribose, respectively) and in the pyrimidine base (uracil instead of thymine). This explains why radioactive thymidine (i.e., the nucleoside) is used to label DNA and radioactive uridine for RNA in various experiments.

In Chapter 19 it will be shown that in transfer RNA there are methylated bases (e.g., methyladenine, methylguanine, methylcytosine, etc.). Heterocyclic bases absorb ultraviolet light at 260 nm. A cell photographed at this wavelength shows the nucleolus, the chromatin, and the RNA-containing regions of the cytoplasm absorbing intensely (Fig. 3–10).

All the genetic information of a living organism is stored in the linear sequence of the four bases. Therefore, a four letter alphabet (A, T, C, G) must code for the primary structure of all proteins (i.e., composed of 20 amino acids). All the excitement in molecular biology, leading to the unraveling of the genetic code (Chapter 19), began when the structure of DNA was understood.

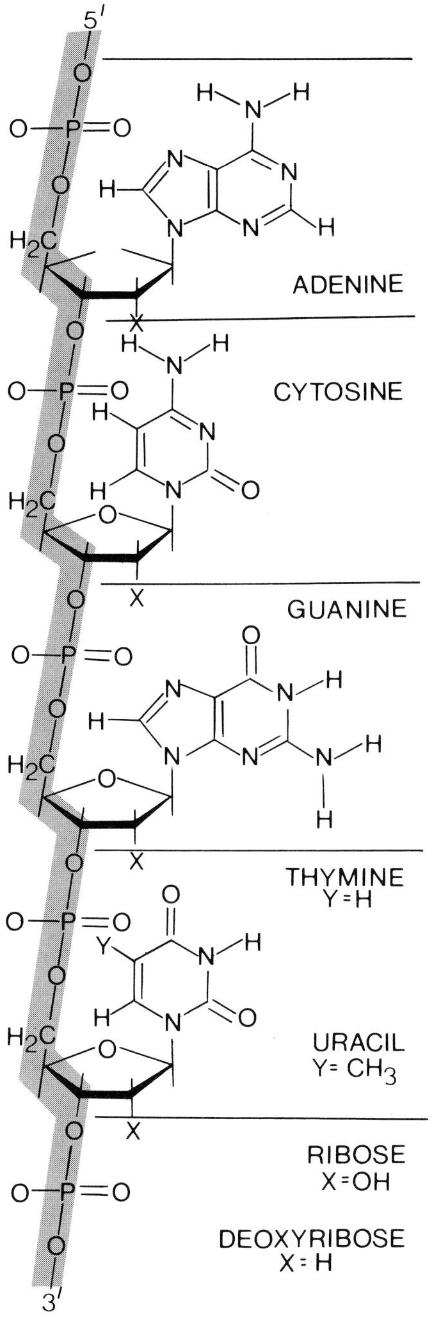

Figure 3–8. A segment of a hypothetical nucleic acid chain showing the nucleotides and their constituent parts.

The Structure of DNA

DNA is present in living organisms as linear molecules of extremely high molecular weight. For example, *E. coli* has a single circular DNA molecule with a molecular weight of about 2.7×10^9 daltons and a total length of 1.4 mm. In higher organisms the amount of DNA may be several thousand times larger; for example, the DNA contained in a single human diploid cell, if fully extended, would have a total length of 1.7 meters.

Between 1949 and 1953 Chargaff et al. studied the base composition of DNA in great detail. It was found that the base composition varies from one species to another, and yet striking regularities were also found. In all cases the amount of adenine was found to be equal to the amount of thymine (i.e., $A = T$). The number of cytosine and guanine bases were also found to be equal (i.e., $G = C$). As a consequence, the total quantity of purines equals the total quantity of pyrimidines (i.e., $A + G = C + T$). On the other hand, there is considerable variation between species regarding the AT/GC ratio. For example, in higher plants and animals AT is in excess of GC, whereas in viruses, bacteria, and lower plants it may be the contrary. For example, in man the AT/GC ratio is 1.52; in *E. coli,* it is 0.93.

The Watson-Crick Model

In 1953, based on the x-ray diffraction data of Wilkins and Franklin, Watson and Crick proposed a model for the DNA structure that provided an explanation for the above-mentioned regularities in base composition and for the biological properties of DNA — particularly its duplication in the cell. The structure of DNA is shown in Figure 3–11. It is composed of two right-handed helical polynucleotide chains that form a *double helix* around the same central axis. The two strands are *antiparallel,* i.e., their 3′,5′ phosphodiester links are in opposite directions. Furthermore, the bases are stacked inside the helix in a plane perpendicular to the helical axis.

The two strands are held together by *hydrogen bonds* established between the pairs of bases. Since there is a fixed distance (i.e., 1.08 nm) between the two sugar moieties in the opposite strands, only certain base pairs can fit into the structure. As may be seen in Figure 3–11, the only two pairs that are possible are AT and CG. Two hydrogen bonds are formed between A and T, and three are formed between C and G. In addition to

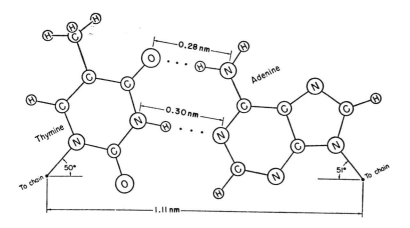

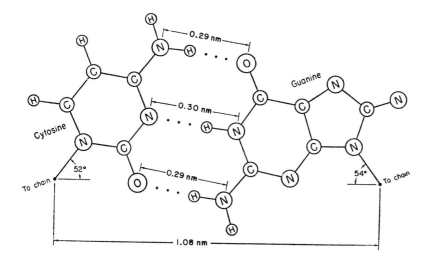

Figure 3–9. The two base pairs in DNA. The complementary bases are thymine and adenine (T—A) and cytosine and guanine (C—G). Observe that between T—A there are two, and between C—G three, hydrogen bonds. The distance between the phosphate sugar chains is about 1.1 nm. (From L. Pauling and R. B. Corey, 1956.)

hydrogen bonds, hydrophobic interactions, established between the stacked bases, are also important in maintaining the double helical structure.

In the Watson and Crick model the distance between the stacked bases is 0.34 nm, which corresponds to the primary period demonstrated by x-ray diffraction. Furthermore, a turn of the double helix is completed in 3.4 nm, a length which corresponds to 10 nucleotide residues (Fig. 3–11). This distance corresponds to a secondary period along the axis. It is evident from observation of a space-filling model of DNA that the double helix has a mean diameter of 2.0 nm; furthermore, two grooves (a major or deep groove, and a minor or more shallow one) are observed (Fig. 3–12).

The *axial sequence* of bases along one polynucleotide chain may vary considerably, but on the other chain the sequence must be *complementary* as in the following example:

First chain: 3′ T, G, C, T, G, T, G, G, T, 5′
Second chain: 5′ A, C, G, A, C, A, C, C, A, 3′

Because of this property, given an order of bases on one chain, the other chain is exactly complementary.

During DNA duplication, the two chains dissociate, and each one serves as a template for the synthesis of two complementary chains. In this way two DNA molecules are produced, each having exactly the same molecular constitution. (The mechanism of DNA duplication will be discussed in Chapter 17.) The varying sequence of the four bases along the DNA chains forms the basis for

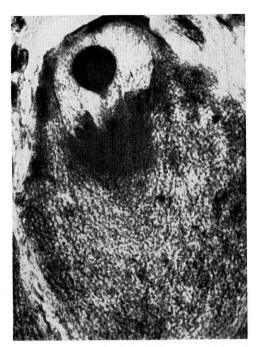

Figure 3-10. Nerve cell photographed with ultraviolet light at 260 nm. The regions in the nucleus and cytoplasm that absorb the ultraviolet light contain nucleic acid. (From H. Hydén.)

genetic information. Four bases can produce thousands of different hereditary characters, because DNA molecules are long polymers along which an immense number of combinations may be produced. (The genetic code will be presented in Chapter 19.)

Denaturation and Renaturation of DNA; Hybridization

Since the structure of the double helix of DNA is preserved by weak interactions (i.e., hydrogen bonds and hydrophobic interactions) it is possible to separate the two strands by heating and other treatments. This separation process, called *melting* or *denaturation of DNA,* can be followed in the spectrophotometer by measuring the absorbance at 260 nm (Fig. 3–13). Since the double helix of DNA absorbs less ultraviolet light than the sum of the two isolated strands, by raising the temperature it may be observed that at a certain point there is a sharp increase in the quantity of ultraviolet light absorbed. The increase in temperature at this point corresponds to the "melting" (or separation) of the two DNA strands and is variable for DNAs of different base composition. Since the temperature to break the GC pair (having three hydrogen bonds) is higher than the one needed to break the AT pair (having two hydrogen bonds), the melting point of a given DNA sample will depend on the AT/GC ratio. If after denaturation the DNA is cooled slowly, the complementary strands will basepair in register and the *native* (double helical) conformation will be restored. This process is called *renaturation* or *annealing.*

Renaturation of DNA is a very useful tool in molecular biology. For example, hy-

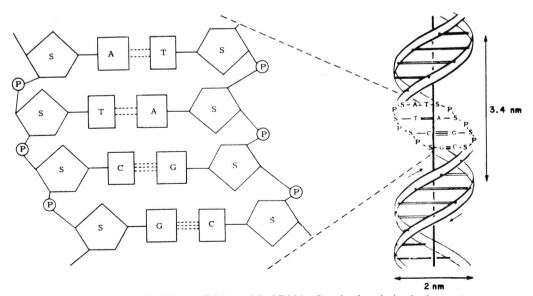

Figure 3-11. The Watson-Crick model of DNA. (See the description in the text.)

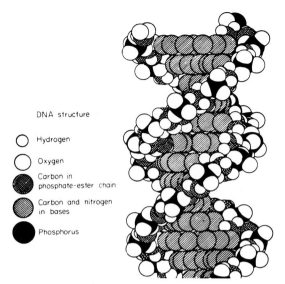

DNA structure

○ Hydrogen

○ Oxygen

◉ Carbon in
  phosphate-ester chain

◐ Carbon and nitrogen
  in bases

● Phosphorus

Figure 3–12. Space-filling model of a segment of DNA molecule showing the double helix with the phosphate-ribose chain and the bases at right angles. In this model it is evident that the surface of the DNA molecule has two grooves—a major and a minor. (See Chapter 17 for a discussion of the probability that histones and protamines fit into the minor groove.) (From Feughelman, et al., 1955.)

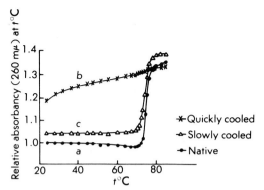

Figure 3–13. Hyperchromic shift in DNA from bacteriophage T2. Observe that the native DNA (*a*), upon reaching a temperature near 80° C, suddenly increases in absorbancy. In *b* and *c*, after heating to 91° C, the DNA was either cooled quickly (*b*) or slowly (*c*). In this last case, the DNA structure is re-formed. (From J. Marmur and P. Doty.)

brid molecules can be formed between DNAs of different species, and the amount of *hybridization* will indicate the degree of genetic similarity. The single strands of DNA may also form hybrids with RNA by complementary base pairing. This property of renaturation has been used to recognize which part of DNA has been used as a template for a given RNA species (see Fig. 18–12). Renaturation studies have led to the discovery of *repeated sequences* of DNA in eukaryotic cells (see Fig. 20–3). When certain sequences are repeated many times (up to 10^6 copies), the rate of renaturation will be much faster than for

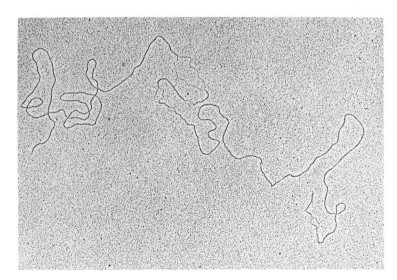

Figure 3–14. Electron micrograph of the lambda phage DNA showing sites in which denaturation (i.e., separation of the strands) has resulted from the action of an alkali medium (pH 11). A map of this partial denaturation is presented below the electron micrograph, showing at scale the two ends of the DNA molecules and indicating by rectangles the position and length of the denatured sites. (Courtesy of R. B. Inman; see Inman, R. B., and Schnös, M. G., *J. Molec. Biol., 49*:93–8, 1970.)

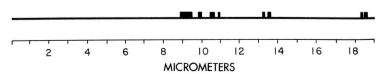

2   4   6   8   10   12   14   16   18

MICROMETERS

TABLE 3-5.   Major Classes of Ribonucleic Acids in E. coli*

Type	Sedimentation Coefficient	Molecular Weight	Number of Nucleotide Residues	Per Cent of Total Cell RNA
mRNA	$6S$ to $25S$	25,000 to 1,000,000	75 to 3,000	~2
tRNA	~$4S$	23,000 to 30,000	75 to 90	16
rRNA	$5S$ $16S$ $23S$	~35,000 ~550,000 ~1,100,000	~100 ~1,500 ~3,100	82

*After A. L. Lehninger, 1970.

those sequences (i.e., genes) represented only once.

Denaturation done under carefully controlled conditions may be used for physical mapping of DNA in a technique known as *partial denaturation mapping*. This technique is based on the fact that the regions rich in AT separate more easily. Under the electron microscope those regions are detected as single-stranded loops. The distance between the loops and the end of the DNA molecule can be measured; this distance is a characteristic of the DNA species (Fig. 3–14).

Structure of RNA

The primary structure of RNA is similar to that of DNA, except for the presence of ribose and uracil in place of deoxyribose and thymine. The base composition of RNA does not follow Chargaff's rules (see above) since RNA molecules are single-stranded. Nevertheless, there is some degree of secondary structure in the different RNA types, because the molecule can form hairpin loops of hydrogen bonded A—U or G—C pairs (see Fig. 19–9).

RNA is synthesized within the nucleus by using only one strand of DNA as template. As shown in Table 3–5 there are three major classes of ribonucleic acid: ribosomal RNA (rRNA), transfer RNA (tRNA), and messenger RNA (mRNA). All of them are involved in protein synthesis. Their structure and function will be studied in detail in the chapters dealing with molecular biology.

SUMMARY: Nucleic Acids

Nucleic Acids are the repositories of genetic information. In general, deoxyribonucleic acid (DNA) is transcribed into ribonucleic acids (RNA), and these are involved in the translation into proteins.

$$DNA \rightarrow RNA \rightarrow Protein$$

DNA is localized mainly in the nucleus (i.e., in the chromosomes); small amounts are in mitochondria and chloroplasts.

Nucleic acids are composed of a pentose, phosphoric acid, purines (adenine, guanine), and pyrimidine bases (thymine, cytosine, uracil). They are linear polymers of *nucleotides* linked by phosphate-diester bonds between the pentoses. Since one acid group is left free, nucleic acids are very acidic and bind to the basic proteins (i.e., protamines and histones).

DNA differs from RNA in the pentose (deoxyribose in DNA; ribose in RNA) and in one of the pyrimidine bases (thymine in DNA; uracil in RNA). Radioactive *thymidine* (i.e., the nucleoside) is used to label DNA; *uridine* is used for RNA.

Genetic information is stored in the linear sequence of bases, the genetic alphabet consists of four letters: A, T, C, and G. The bases of DNA are in certain molar ratios; e.g., A = T, G = C, and A + G = C + T. The ratio AT/GC varies between 1.52 in man and 0.93 in *E. coli.*

Watson and Crick established the structure of the DNA molecule in 1953 on the basis of the x-ray diffraction studies of Wilkins and Franklin. The model is formed of two right-handed helical polynucleotide chains that are *antiparallel*, and in which the bases are stacked perpendicularly. Only two pairs of bases A—T and G—C, having, respectively, two and three hydrogen bonds, are present within the double helix (Fig. 3–11). Other important concepts include the 0.34 nm primary and the 3.4 nm secondary period, the 2.0 nm diameter of the double helix, and the *major* and *minor groove* of the double helix. The axial sequence of bases along one chain is a complement of the other; DNA duplication takes place via a template mechanism, following the unwinding of the two strands (see Chapter 17). The sequence of bases in a long polymer provides an explanation for the immense number of combinations that carry different genetic information.

By an increase in temperature or other treatments, it is possible to denature (i.e., melt) the DNA. The two strands separate, producing a sharp increase in absorbance at 260 nm. More energy is required to break the G—C pair than is necessary to break the A—T pair. By *partial denaturation* the mapping of the regions rich in AT along the DNA molecule can be done and can be studied under the electron microscope. Denaturation may be followed by *renaturation* (i.e., *annealing*) of the complementary strands. Hybrid molecules (*hybridization*) can be made with other DNA or RNA molecules.

The RNA is single-stranded and contains ribose and uracil. In RNA there is some degree of secondary structure (hairpin loops) of hydrogen bonded A—U or G—C pairs. The major classes of RNA are ribosomal RNA, transfer RNA, and messenger RNA.

ADDITIONAL READING

Chargaff, E. E. (1958) Of nucleic acids and nucleoproteins. *Harvey Lect.,* ser. 52 (1956–1957).

Crick, F. H. C. (1957) Nucleic acids. *Sci. Amer., 197*:188.

Davidson, J. N. (1960) *The Biochemistry of the Nucleic Acids.* 4th Ed. John Wiley and Sons, New York.

Dayhoff, M. O., ed. (1969) *Atlas of Protein Sequence and Structure.* Vol. 4, 1969. National Biomedical Research Foundation, Silver Springs, Md.

Frieden, E. (1972) The chemical elements of life. *Sci. Amer., 227*:52.

Fruton, J. S. (1973) Molecules and life. *Nature, 243* :103.

Ingram, V. M. (1966) *The Biosynthesis of Macromolecules.* W. A. Benjamin, Inc., New York.

Inman, R. B., and Schnös, M. G. (1970) Partial denaturation of thymine- and 5-bromouracil-containing λ-DNA in alkali. *J. Molec. Biol., 49*:93.

Karlson, P. (1967) *Introduction to Modern Biochemistry.* 2nd Ed. Academic Press Inc., New York.

Kendrew, J. (1966) *The Thread of Life.* Bell and Sons, London.

Lehninger, A. L. (1970) *Biochemistry.* Worth Publishers, Inc., New York.

Marmur, J., and Doty, P. (1962) Determination of base composition of DNA from its thermal denaturation temperature. *J. Molec. Biol., 5:*109.

Nemethy, G. (1967) Proteins (Binding Forces in Secondary and Tertiary Structures). *Encyclopedia of Biochemistry.* Reinhold Publishing Corp., New York.

Pullman, B., and Weissbluth, M. (1965) *Molecular Biophysics.* Academic Press Inc., New York.

Reithel, J. F. (1967) *Concepts in Biochemistry.* McGraw-Hill Book Co., New York.

Sanger, F. (1965) The structure of insulin. In: *Currents in Biochemical Research.* (Green, D. E., ed.) Interscience Publishers, New York.

White, A., Handler, P., and Smith, E. (1973) *Principles of Biochemistry.* Rev. Ed., McGraw-Hill Book Co., New York.

four

ENZYMES, BIOENERGETICS, AND CELL RESPIRATION

The cell may be compared to a minute laboratory capable of carrying out the synthesis and breakdown of numerous substances. These processes are carried out by enzymes at normal body temperature, low ionic strength, low pressure, and a narrow range of pH.

The enzymes are not randomly distributed within the cell, but are located in various cell compartments. Frequently they are arranged in an orderly fashion within the macromolecular framework of the cell and cell organelles to form what is called a *multi-enzyme system*. Knowledge of the localization and grouping of enzymes within the cell structure is essential and is emphasized throughout several chapters of this book.

Metabolism may be defined as the sum of all chemical transformations in the cell. It comprises both the processes of *catabolism*, by which substances are broken down, and *anabolism*, by which new products are synthesized. Catabolic reactions are mostly *exergonic*, i.e., they liberate energy; anabolic reactions are *endergonic*, i.e., they consume energy. For example, the different substrates taken in by the cell as foodstuffs, such as glucose, amino acids, and lipids, are broken down into smaller molecules, with liberation of energy. This energy may be trapped by substances like adenosine triphosphate (ATP) and, in turn, is utilized by the cell in the synthesis of new and more complex molecules.

The following brief introduction to the study of enzymes is preparatory to the description of their cytochemistry and function in the different cell organelles. Cell metabolism will be discussed to prepare the reader for more specific topics, such as the function of mitochondria and chloroplasts, and active transport. This general introduction should be supplemented by reference to biochemistry and enzymology textbooks.

ENZYMES

Enzymes are the biological catalysts that accelerate chemical reactions inside the cell. They are the largest and most specialized class of protein molecules. To date, over a thousand different enzymes have been identified; many of them have been obtained in pure, and even crystalline, condition. Enzymes represent one of the most important expressions of the genes contained in the DNA molecule. The complex network of chemical reactions which are involved in cell metabolism is directed by enzymes.

Enzymes (E) are proteins with one or more loci called *active sites* to which the *substrate* (i.e., the substance upon which the enzyme acts) attaches. The substrate is chemically modified and converted into one or more products (P). Since this is generally a reversible reaction it may be expressed as follows:

$$E + S \rightleftarrows [ES] \rightleftarrows E + P \qquad (1)$$

where [ES] is an intermediary enzyme-substrate complex. Enzymes accelerate the reaction until an equilibrium is reached. They are so efficient that the reaction may proceed from 10^8 to 10^{11} times faster than in a non-catalyzed condition.

Specificity

A very important feature of enzyme activity is that it is *substrate-specific;* i.e., a particular enzyme will act only on a certain substrate. Some enzymes have nearly *absolute* specificity for a given substrate and will not act on even very closely related molecules, as, for example, stereoisomers of the same molecule. Other enzymes have *relative* specificity, since they will act upon a variety of related compounds. One example of enzyme specificity for proteinases (i.e., proteolytic enzymes) is shown in Figure 4–1. Each enzyme splits the polypeptide chain at different and very precise sites. According to the chemical reactions they perform, enzymes may be classified into: (1) *oxidoreductases* (oxidation-reduction reactions), (2) *transferases* (transfer of groups), (3) *hydrolases* (hydrolytic reactions), (4) *lyases* (addition or removal of groups to, or from, double bonds), (5) *isomerases* (catalytic isomerizations), and (6) *ligases* or *synthetases* (condensation of two molecules by splitting a phosphate bond).

Enzyme Activation; Co-factors

Some enzymes exist in the cell in an inactive form called a *zymogen.* Zymogens are activated by the so-called *kinases.* For example, trypsinogen, produced by pancreatic cells, is activated in the intestine by enterokinase. *Pepsinogen,* secreted by the chief cells of the stomach, is activated by hydrochloric acid (hydrogen ions) secreted by the parietal cells. In the latter case the activation is caused by the splitting off of a small polypeptide, which probably masks the active site of the enzyme (see below). Some enzymes require small non-protein components called *cofactors* for their activity. For example, some enzymes are conjugated proteins having tightly bound *prosthetic groups,* as in the case of the *cytochromes,* which are involved in electron-transfer reactions and have a metalloporphyrin complex.

Other enzymes cannot function without the addition of small molecules called *coenzymes,* which become bound during the reaction. Such inactive enzymes, also called *apoenzymes,* form active *holoenzymes* when joined with a *coenzyme.* For example, *dehydrogenases* utilize either nicotinamide-adenine dinucleotide (NAD^+) or nicotinamide-adenine dinucleotide phosphate ($NADP^+$). These are among the most important coenzymes.

The function of the coenzyme is to transfer the hydrogen nuclei with two electrons from the substrate, thus oxidizing it:

$$Substrate + NAD^+ + ENZYME \rightarrow$$
$$oxidized\ substrate + NADH\ and\ H^+$$

In the reverse direction the substrate is reduced. $NADP^+$ and $NADPH$ behave in a similar way. Both of these coenzymes consist of one mole of adenine, one mole of nicotin-

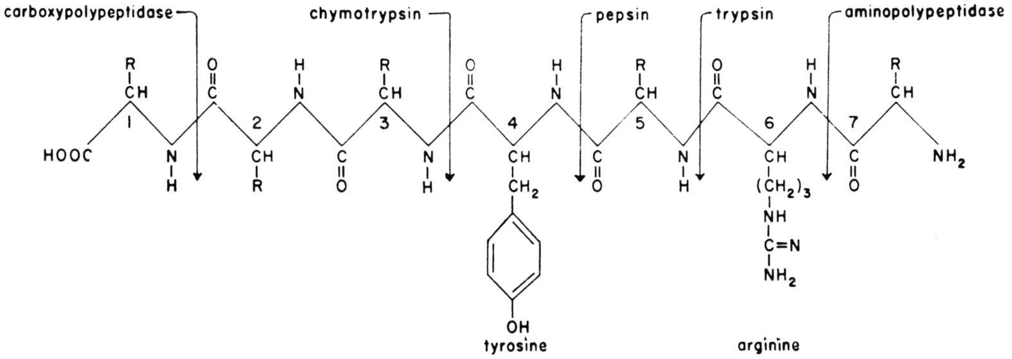

Figure 4–1. A diagram to indicate the specificity of various proteolytic enzymes. The numbers refer to the amino acid residues, of which only two, tyrosine and arginine, are labeled below. The polypeptidases are specific, one to the free carboxyl end (left) of a protein molecule or peptide, the other to the free amino end (right) of such molecules. Pepsin is specific to the amino side of tyrosine (or phenylalanine) residues inside a protein molecule; chymotrypsin is specific to the carboxyl side of such residues; and trypsin is specific to the carboxyl side of arginine or lysine residues. (From Giese, A. C., *Cell Physiology,* 3rd ed., Philadelphia, W. B. Saunders Co., 1968.)

amide, two moles of D-ribose and two or three moles, respectively, of inorganic phosphate. In the cell the energy-producing catabolic processes require NAD^+; the synthetic processes, however, use NADPH. In many coenzymes, as in NAD^+ and $NADP^+$, the essential components are vitamins, particularly those of the B group. Some examples are *pantothenic acid* (vitamin B_5), which forms part of the important coenzyme A; *riboflavin* (vitamin B_2), incorporated into the molecules of flavin-adenine dinucleotide (FAD), and *pyridoxal* (vitamin B_6), a cofactor of transaminases and decarboxylases.

Active Site of the Enzyme

According to the present concept of enzymatic activity, the substrate attaches itself to the protein component of the enzyme, which has on its molecule a place of specific configuration for this purpose. This is called an *active site*. Those parts of the substrate upon which an enzyme acts link themselves to this active site, forming *a lock and key* relationship. This concept was developed to explain the great specificity of enzymes. In fact, a structurally well defined active site will accept only those substrate molecules having

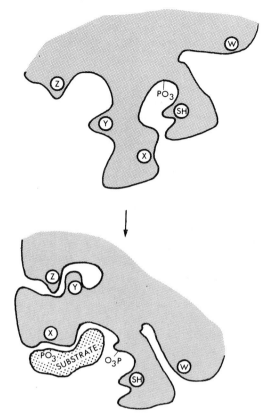

Figure 4–3. Schematic illustration of flexibility in the action of phosphoglucomutase. The **upper part** of the figure represents the enzyme molecule in the absence of substrate. The **lower part** of the figure represents the change in conformation leading to exposure of SH and burying of X, Y, Z and W. (From Koshland, D. E., *Adv. Enzymol., 22*:45, 1960.)

a certain steric configuration and will repell others that differ structurally.

The active site of an enzyme is directly related to the primary structure of the protein. Yet it not only corresponds to a special amino acid sequence, but it also depends on the secondary and tertiary configuration described in Chapter 3 (Fig. 4–2). The classical concept of a rigid lock and key relationship, however, does not explain all aspects of enzyme activity; therefore, a new concept, *the induced fit,* has been developed by Koshland.[1,2] According to this principle, the substrate interacts with the reacting groups of the amino acids within the active site and thereby changes the conformation of the enzyme. In this way, certain chemical groups of the enzyme, essential for catalytic activity, will come in close contact with the substrate (Fig. 4–3).

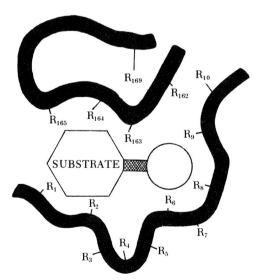

Figure 4–2. Schematic active site of an enzyme. The crosshatched area indicates a bond to be broken in the enzyme action. The R's represent some side chains, and the heavy lines represent the backbone of two segments of the protein chain. (From Koshland, D. E., Jr., *Adv. Enzymol., 22*:45, 1960.)

The binding of the substrate to the active site involves forces of a non-covalent nature (ionic and hydrogen bonds, van der Waals forces), which are of very short range. This explains why the enzyme-substrate complex can be formed only if the enzyme has a site that is exactly complementary to the shape of the substrate. In molecular terms one can explain the function of an enzyme in terms of two steps: (1) the formation of the *specific complex,* and (2) the *catalytic step proper* in which the different mechanisms of catalysis, i.e., hydration, dehydration, transfer of groups, etc., are produced.

The existence of an enzyme-substrate complex (ES) at the active site was postulated by Michaelis and Menten in 1913 on the basis of kinetic evidence. This concept has been of great importance in the understanding of the mechanism of the enzymatic reactions (see below). At present the existence of the [ES] complex has been proven by spectroscopic methods and by the direct isolation of stable covalent derivatives of the complex.

ENZYME KINETICS

The activity of an enzyme depends on a number of external factors such as temperature, hydrogen ion concentration, and so forth; these factors must be kept constant while the kinetics of the enzyme are studied. With external factors kept constant, and in the presence of an excess of substrate, the reaction catalyzed by an enzyme has a velocity that is proportional to the enzyme concentration.

As was mentioned previously, the enzyme-substrate reaction proceeds in two steps (1). The first step can be written as follows:

$$E + S \underset{K_2}{\overset{K_1}{\rightleftharpoons}} [ES] \qquad (2)$$

In the second step the [ES] complex breaks down to form the product and the free enzyme, which will now be available for processing a new substrate molecule:

$$[ES] \underset{K_4}{\overset{K_3}{\rightleftharpoons}} E + P \qquad (3)$$

(K_1, K_2, K_3, and K_4 are rate constants for the reactions.) All steps are reversible, but in general K_4 is negligible, and reaction (3) follows the direction $[ES] \rightarrow E + P$.

As shown in Figure 4–4, the velocity (V) of the reaction depends on the substrate concentration, and the curve describes a hyperbola. At low substrate concentrations, the initial velocity increases rapidly and follows a first order reaction, i.e., the amount of product formed is proportional to the substrate concentration [S]. However, as the [S] increases, the reaction saturates and reaches a point of equilibrium in which the velocity no longer depends on [S]. At this point, because of the great excess of substrate, all the enzyme is in the form of an [ES] complex, and the maximum velocity (Vmax) of the reaction is reached. The equation for this curve is:

$$V = \frac{Vmax\,[S]}{Km + [S]} \qquad (4)$$

This is the Michaelis-Menten equation, which allows for the calculation of the velocity

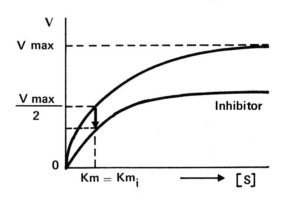

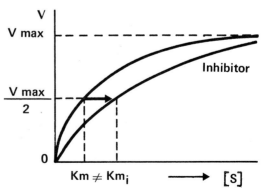

Figure 4–4. Enzyme reaction showing the effect of the substrate concentration [S] on the velocity (V). V max, maximum velocity; Km, Michaelis constant, corresponding to $\frac{V\,max}{2}$. **Upper graph,** the effect of a noncompetitive inhibitor in which Km = Km$_i$. **Lower graph,** the effect of a competitive inhibitor in which Km ≠ Km$_i$ (see the description in the text).

of the reaction for any substrate concentration, provided that Vmax and Km are known. Km is the *Michaelis constant,* which may be defined experimentally as the substrate concentration at which the velocity is half maximal. As shown in Figure 4–4, the Km value can be extrapolated from the point $\frac{Vmax}{2}$ on the ordinate. The Km is expressed in moles of substrate per liter. The smaller the value of Km, the greater the *apparent affinity* of the enzyme for the substrate. In the study of the kinetic characteristics of an enzyme, it is important to know the values of Vmax and Km. From plots such as those of Figure 4–4, however, it is difficult to calculate Vmax since the curve is asymptotic to the abscissa. The calculation is performed more accurately from the double reciprocal, or *Lineweaver-Burk plot,* (i.e., a plot of $\frac{1}{V}$ against $\frac{1}{[S]}$ [Fig. 4–5]). From this plot the value of $\frac{1}{Vmax}$ is obtained at the intercept of the ordinate axis and $-\frac{1}{Km}$ at the intercept of the abscissa.

It was mentioned above that the *hydrogen ion concentration* (pH) plays an important role in the kinetics of an enzyme. By plotting the enzyme activity against increasing values of pH, starting in the acid range, a bell-shaped curve is obtained; the peak of the curve is the optimum pH at which the enzyme exhibits its greatest activity. In extremely alkaline or acid media, the protein may denature, thus causing the enzyme to be irreversibly inactivated.

There are enzymes, however, whose optimum activity lies in a very acid range (e.g., pepsin, pH 2.0), or in a very alkaline range (e.g., alkaline phosphatase, pH 8.5 to 10.0).

Temperature is another factor which influences the kinetics of enzyme reactions. If a low temperature is increased progressively, a level of optimum activity is attained, which then diminishes, and finally stops completely. The activity decreases because of the progressive denaturation of the enzyme.

Enzyme Inhibition

Enzyme inhibition may be *reversible* or *irreversible.* There are two major types of reversible inhibition: *competitive* and *noncompetitive.* Competitive inhibition involves a compound similar in structure to the substrate, which forms a complex with the enzyme:

$$E + I \xrightleftharpoons{K_i} [EI] \qquad (5)$$

where E is the enzyme, I the inhibitor, and K_i the association constant of the enzyme-inhibitor complex. Unlike the [ES] complex (equation 2), the [EI] complex does not break down into the products of reaction and the free enzyme. The inhibition of succinic dehydrogenase by malonic acid, whose molecular structure is very similar to that of succinic acid, serves as an example of competitive inhibition:

COOH COOH
| |
CH_2 CH_2
| |
CH_2 CH_3
|
COOH

Succinic acid　　Malonic acid

In this case, both the substrate (succinate) and the inhibitor (malonate) will compete for the active site, and the enzyme activity will be reduced. This inhibition can be reversed, nevertheless, by increasing the substrate concentration, so that the substrate molecules outnumber those of the inhibitor. Therefore, the *Vmax* is not changed by the competitive inhibition (Fig. 4–4). This fact is even more easily observed in the Lineweaver-Burk plot (Fig. 4–5). Note that in competitive inhibition the Km increases, i.e., the apparent affinity of the enzyme for the substrate decreases.

In non-competitive inhibition the inhibitor and the substrate are not structurally related, and the inhibitor binds to a different site than the substrate. Non-competitive inhibition cannot be reversed by high concentrations of the substrate. As shown in Figure 4–5 the Km remains unchanged, but the Vmax is decreased.

Irreversible inhibition involves either the denaturation of the enzyme (see above), or the formation of a covalent bond with the enzyme. For instance, iodoacetic acid blocks the sulfhydryl groups by alkylation, thus inhibiting certain enzymes. Ferricyanide forms disulfides and produces a similar result. High concentrations of heavy metals may produce

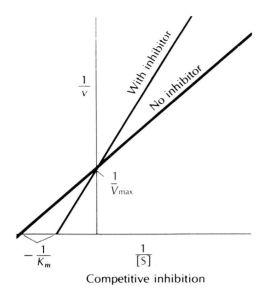

Competitive inhibition

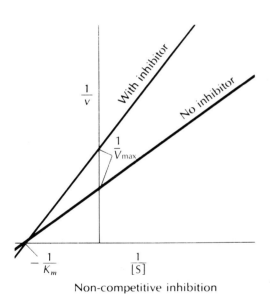

Non-competitive inhibition

Figure 4–5. Lineweaver-Burk plot of the reactions shown in Figure 4–4. Here the plot is between the reciprocal of the velocity $\frac{1}{V}$ and the reciprocal of the substrate concentration $\frac{1}{[S]}$. In the presence of a competitive inhibitor V_{max} is unchanged, while with a non-competitive inhibitor, K_m is unchanged.

irreversible denaturation of the protein moiety of an enzyme; however, low concentrations inhibit the sulfhydryl groups reversibly.

Isoenzymes

It was formerly thought that only one enzyme could act on a given substrate. How-ever, the improvement of preparative techniques, particularly *starch gel electrophoresis,* has aided in the recognition of families of enzymes with identical activity, but with small differences in their molecular structure. It is now well established that these isoenzymes are produced by genetic changes that determine differences in the amino acid sequence.

There are more than 100 enzymes which are known to exist as isoenzymes. One of the best known examples is *lactic dehydrogenase* (LDH), which catalyzes the reaction of pyruvate to lactate. There are five LDH isoenzymes which differ in their electrophoretic mobility and isoelectric point. The relative proportions of these isoenzymes are characteristic for each tissue and for each stage of its differentiation.

In certain cases, an organism may possess a number of enzymes which carry out the same reaction, but which differ considerably in their primary structure and in the genetic locus where they are specified. For instance, *E. coli* has three aspartokinases to catalyze the reaction of aspartate to β-aspartylphosphate. Each of the three enzymes "map" on different parts of the circular chromosome, and they are regulated by different feedback inhibitors (see below).

Allosteric Enzymes

Not all enzymes display the simple hyperbolic kinetics shown in Figure 4–4, and the V vs. [S] curves are sigmoidal (Fig. 4–6). Enzymes having this kinetic behavior are called *regulatory, or allosteric enzymes.* They are *oligomers* containing two (*dimer*), four (*tetramer*), or more subunits which are able to interact with one another. The sigmoidal shape of the curve results from the fact that the binding of the first substrate molecule enhances the affinity for binding the second substrate molecule, and so forth. By observing Figure 4–6 the reader may recognize that there is a region in the curve in which a small increase in [S] causes a very large increase in enzyme activity, i.e., much larger than in the hyperbolic curve.

Allosteric enzymes are of great regulatory value, since large changes in activity can be obtained by small changes in [S].

Two main models have been proposed to explain the complex behavior of allosteric

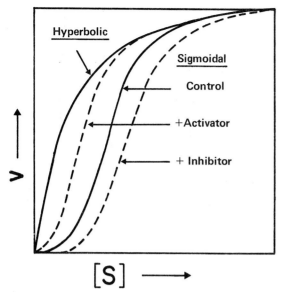

Figure 4-6. Sigmoidal curve characteristic of a regulatory or allosteric enzyme. Observe the difference between a hyperbolic type of curve and the action of an activator and of an inhibitor on the enzyme reaction.

enzymes. In the Monod, Wyman, and Changeux[3] model, the oligomeric enzyme can exist in two different conformational states: R and T.

$$(6)$$

Both states are in equilibrium. In the absence of substrate most of the enzyme molecules will be in the T state. Since the substrate binds preferentially to the R state, the addition of substrate shifts the equilibrium toward the R state. In this way the binding of the first molecules of substrate enhances the binding of those that follow, since more enzyme molecules will be in the R state.

In the Koshland, Nemethy, and Filmer model the binding of a ligand to one of the subunits induces a *change in conformation,* i.e., in its tertiary structure (see Koshland and Neet, 1968). Since the various monomers in-

teract, the change in conformation of one can modify the behavior of the adjacent one in such a way that the binding of the substrate is enhanced. This allosteric change for a tetrameric enzyme can be represented as shown at the bottom of this page.

Regulatory enzymes are sensitive to the so-called *allosteric modifiers* (i.e., modulators) which act as *inhibitors* or *activators.* Such modifiers do not bind to the active site but to a different site—the allosteric one. The binding of the modifier induces a conformational change in the active site that results in an increased or decreased affinity of the enzyme for the substrate. In the case of *aspartate transcarbamylase,* it has been shown that the site which binds the allosteric inhibitor cytidine triphosphate (CTP) is in a special regulatory subunit, different from the catalytic subunit.[4]

Figure 4-6 shows that in the presence of an activator the curve tends toward the hyperbola, while under the influence of an allosteric inhibitor, it is more sigmoidal. As will be shown in the following section, allosteric enzymes are of paramount importance in the regulation of cell metabolism.

REGULATION OF ENZYME ACTIVITY

The living cell seldom wastes energy synthesizing or degrading more material than necessary. Therefore, the thousands of chemical reactions that occur inside the cell must be carefully controlled. The regulation of enzyme activity is produced by two major mechanisms: *genetic control* and *control of catalysis* (Fig. 4-7).

Genetic control implies a change in the total amount of enzyme molecules. Best known examples of this form of regulation are *enzyme induction* and *repression* in microorganisms in which the synthesis of the enzyme is regulated at the gene level by the indirect action of certain metabolites (see Chapter 19). Similar phenomena occur in multicellular organisms with the enzymes related to the endoplasmic reticulum (see Chapter 9).

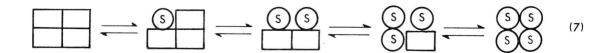

$$(7)$$

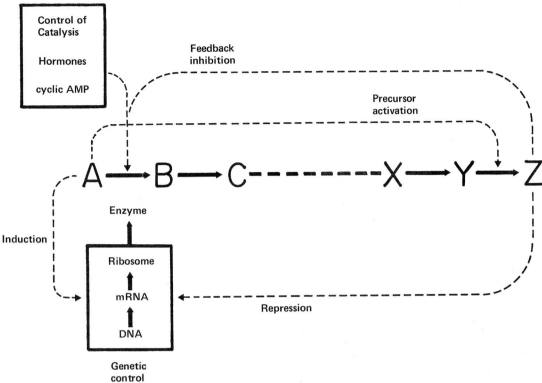

Figure 4–7. Diagram showing that enzyme activity may have a genetic control acting on the synthesis of the enzyme and on control of catalysis—by the action of hormones or cyclic AMP. The mechanisms of *enzyme induction* and *repression* and *feedback inhibition* and *precursor activation* are indicated. (See the description in the text.) (After G. G. Hammes and C. Wu, 1971.)

Control of catalysis involves a change in enzyme activity without a change in the total amount of enzyme synthesized. This is frequently produced in regulatory or *allosteric enzymes* by the action of allosteric activators or inhibitors. *Enzyme interconversion* and *hormonal* control may also act on catalysis. Two important mechanisms of control are: *feedback inhibition* and *precursor activation* (Fig. 4–7). In *feedback inhibition* the end-product of a metabolic pathway acts as an *allosteric inhibitor* of the first enzyme of this metabolic chain. Thus, when enough product is synthesized, the entire chain can be shut off, and useless accumulation of metabolites is avoided. An example is provided by aspartate transcarbamylase, the first enzyme in pyrimidine biosynthesis, which is inhibited by CTP, the end product of this biosynthetic pathway.

In *precursor activation* the first metabolite of a biosynthetic pathway acts as an *allosteric activator* of the last enzyme of the sequence (Fig. 4–7). For example, glycogen synthetase is activated by glucose-6-phosphate, a precursor of glycogen.

Enzyme Interconversions

Some enzymes may exist in two forms of different activity which are interconvertible. Frequently, the mechanism of interconversion consists of *phosphorylation*, i.e., the covalent binding of a phosphate group that is provided by ATP. One of the best known cases is *glycogen phosphorylase*, the enzyme that degrades glycogen into its glucose units, which exists in two forms: *a* and *b*. *Phosphorylase a* is a tetrameric protein having the higher activity; *phosphorylase b* is a less active dimer.

The *b* form can be converted into the *a* form by the covalent binding of four phosphate groups; the reverse phenomenon takes place by *dephosphorylation* (these two processes are shown at the inset of Figure 4–9). By this mechanism the cell regulates the rate of glycogen utilization according to the requirements of the organism.

Cyclic AMP and Hormonal Control

Hormones are molecules that transfer information from one group of cells to another. For example, the cells of the anterior lobe of the hypophysis produce hormones that control the thyroid gland, the gonads, the adrenal gland, and the growth of cartilage. The molecular bases of hormonal action were very poorly understood until 1956, when the late E. W. Sutherland discovered *cyclic adenosine monophosphate* (cAMP) a cyclic nucleotide, that has been found to regulate a large number of metabolic processes.

In 3′5′ cyclic AMP the phosphate group is covalently bound to the 3′ and 5′ carbons of the ribose ring. (In Figure 4–11 the difference from the linear AMP may be deduced.) This cyclic nucleotide is synthesized by adenylate cyclase, an enzyme that is tightly bound to the cell membrane. Many hormones modify the activity of adenylate cyclase and therefore produce a change in the intracellular level of cyclic AMP. In Sutherland's model of hormonal action (Fig. 4–8) the hormone is regarded as a *first messenger* that interacts with specific receptor sites located in the outer surface of the cell membrane. This hormone-receptor interaction results in a change in adenylate cyclase activity whose active site is on the inner surface and which, using ATP as substrate, produces cyclic AMP. This nucleotide is considered as a *second messenger* that carries the information to the metabolic machinery of the cell. The effect of cyclic AMP depends on the target organ; for example, an increase in cyclic AMP will produce glycogen degradation in the liver and steroid production in the adrenal cortex.

Many hormones are known to act by way of specific receptors on adenylate cyclase. Among these hormones are: epinephrine, norepinephrine, glucagon, adrenocorticotropic hormone (ACTH), thyroid-stimulating hormone (TSH), melanocyte-stimulating hormone (MSH), parathyroid hormone, luteinizing hormone (LH), vasopressin, and thyroxine. One of the classic examples of regulation produced by cyclic AMP is the degradation of glycogen. As shown in Figure 4–9 epinephrine and glucagon induce glycogenolysis in the liver. These hormones, acting on the corresponding receptor, stimulate adenylate cyclase and raise the cyclic AMP level, Cyclic AMP activates *protein kinase,* an enzyme that phosphorylates many proteins. Protein kinase phosphorylates *phosphorylase b kinase,* thus converting it into its active form. This kinase is a specific enzyme that will, in turn, phosphorylate *phosphorylase b,* thus

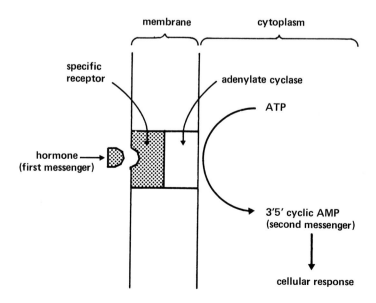

Figure 4–8. Diagram showing the effect of a hormone (first messenger) upon a specific receptor of the cell membrane and its effect on the enzyme adenylate cyclase. (See the description in the text.)

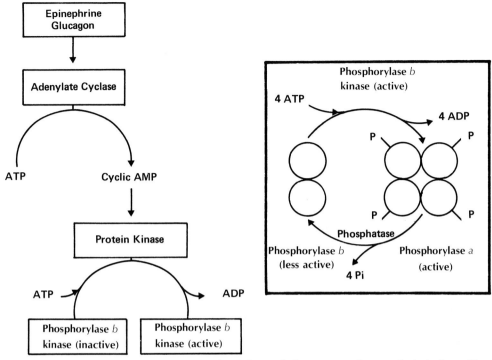

Figure 4–9. Diagram showing the effect of epinephrine and glucagon on glucogenolysis in liver. The figure at right shows that phosphorylase *b*, a less active dimer, is converted into phosphorylase *a*, a more active tetramer, under the action of ATP.

converting it into *phosphorylase a. Phosphorylase a* is the active form of the enzyme that degrades glycogen into its glucose units. In this stepwise manner, the cell amplifies considerably the initial signal given by the hormones at the membrane level (Fig. 4–9).

Cyclic AMP is also involved in certain pathological conditions that are important in medicine. For example, a toxin of *Vibrio cholerae,* the bacterial agent of cholera, activates adenylate cyclase in the intestine. This stimulates salt and water secretion that may lead to a lethal diarrhea. Another aspect that is being actively investigated relates to the area of cancer research. It is known that many cancer cells have low levels of cyclic AMP. Furthermore, it is known that certain abnormal features of cancer cells growing in vitro may be restored to normality by the addition of this cyclic nucleotide to the culture medium.

Knowledge of the localization and grouping of enzymes within the cell structure is essential to the study of cell function. *Metabolism* comprises catabolic reactions (catabolism) that are exergonic and anabolic reactions (anabolism) that are endergonic.

Enzymes are proteins that act as biological catalysts, accelerating chemical reactions. They contain a so-called *active site* to which the substrate attaches, forming a temporary [ES] complex; then the substrate is converted into one or more products and the enzyme becomes free again $(E + S \rightleftarrows [ES] \rightarrow E + P)$. The specificity for the substrate may be absolute, or relative. Some enzymes require *co-factors* for their activity. For example, the prosthetic group in

SUMMARY:
Enzymes in the Cell

cytochromes consists of a metalloporphyrin complex. Other enzymes use small non-protein molecules, i.e., *coenzymes,* which become bound during the reaction to activate the enzyme (apoenzyme + coenzyme → holoenzyme). Important coenzymes are NAD^+ and NADP. Vitamins of the B group, such as nicotinamide, pantothenic acid, riboflavin, and pyridoxal are part of coenzymes, or act as co-factors.

The *lock and key theory* provides one explanation for the specificity of the active site, which is dependent on the primary, secondary, and even the tertiary structure of the protein. According to the *induced-fit theory,* the enzyme-substrate interaction may cause a conformational change in the protein. The binding of the substrate to the active site is by short range, non-covalent forces. After the [ES] complex is formed, the catalytic step, in which the substrate undergoes hydration, dehydration, oxidation, reduction, or transfer of chemical groups (among other processes), proceeds.

In the case of many enzyme-catalyzed reactions, the velocity of the reaction depends on the substrate concentration. The characteristic curve described in these reactions is a hyperbola that reaches a maximum velocity (Vmax) where all the enzyme active sites are saturated. The Km (Michaelis constant) is the substrate concentration at which the velocity is $\frac{Vmax}{2}$ (Fig. 4–4). The smaller Km is, the greater is the affinity for the substrate. The kinetics of the enzyme is greatly influenced by the pH and temperature.

Enzyme inhibition may be reversible or irreversible. Reversible inhibition may be *competitive* or *non-competitive.* In competitive inhibition the inhibitor has a molecular structure similar to that of the substrate. In this case, Vmax is not changed, but the Km increases (Fig. 4–5). In non-competitive inhibition the inhibitor is not structurally related to the substrate; Vmax is decreased and Km remains unchanged (Fig. 4–5).

Isoenzymes are multiple molecular forms of the same enzyme that differ in their electrophoretic mobility. In lactic dehydrogenase, for example, there are five isoenzymes.

Allosteric or regulatory enzymes have a sigmoidal V/[S] curve. They are oligomers composed of two (dimer), four (tetramer), or more protein subunits (i.e., monomers). The binding of a substrate molecule to one subunit enhances the affinity for binding a second S molecule, and so forth. These enzymes have great regulatory value since their activity can be changed by small modifications in the concentration of the substrate [S]. Two models have been suggested to explain this behavior. In one of them a conformational change of the monomers is postulated; in the other, the oligomeric enzyme is thought to exist in two different conformational states.

Regulatory enzymes are sensitive to *modifiers* or *modulators* that bind to an *allosteric site* which, in turn, influences the active site. In the case of aspartate transcarbamylase, the allosteric inhibitor cytidine triphosphate binds to a different subunit than the substrate. The *regulation of enzyme activity* is by two major mechanisms: *genetic control* or *control of the catalytic activity*. In the genetic control there is a change in the amount of enzyme as, for example, in *enzyme induction* and *repression* (see Chapter 19). Control of catalysis is frequent in allosteric enzymes and consists of a change in the activity of the enzyme. This control may be achieved by *feedback inhibition, precursor activation,* and other mechanisms (see Fig. 4–7).

Some enzymes are interconverted from a less active into a more active form. Frequently, this is done by phosphorylation of the enzyme. For example, *phosphorylase b,* a dimer of low activity, is phosphorylated by 4 ATP and produces a more active tetramer by the binding of 4 Pi (Fig. 4–9). A most important mechanism of enzyme regulation is by the action of *hormones (first messenger)* acting on adenylate cyclase. The product, cyclic AMP, is able to carry information inside the cell *(second messenger)*. Many hormones acting on specific *receptors* at the cell membrane may change the cyclic AMP levels, thus producing changes in certain enzymatic pathways. One of the best examples of a hormone acting on a specific receptor is shown in Figure 4–9. Here cyclic AMP activates *protein kinase,* an enzyme that phosphorylates many proteins.

BIOENERGETICS

The energy which the cell has at its disposal exists as chemical energy primarily locked in high energy bonds. The cell uses only part of the *total energy* (H), also called *enthalpy,* contained in a chemical compound. This portion of the total energy, the *free energy* (G), does not dissipate as heat. Expressed as an energy change:

$$\Delta H = \Delta G + T\Delta S \qquad (8)$$

The equation shows that the change in total energy (ΔH) is equal to the change in free or available energy (ΔG) plus the unavailable energy ($T\Delta S$), which dissipates as heat. (In this equation, T is the temperature in degrees Kelvin, and S is the *entropy* of the system.)

Concept of Entropy

It is important to have a clear idea of the role of entropy in biological systems. In equation (8) ΔS is a measure of the irreversibility of a reaction. As the entropy increases, more energy ($T\Delta S$) becomes unavailable, and the process becomes less reversible.

According to the Second Law of Thermodynamics, the entropy of an isolated system of reactions tends to increase to a maximum, at which point an equilibrium is reached and the reaction stops. The concept of entropy is related to the ideas of "order" and "randomness." When there is an orderly arrangement of atoms in a molecule, the entropy is low. The entropy of the system increases when, during a chemical reaction, there is a tendency toward molecular disorder. Thermodynamically, it is well established that the flow of energy proceeds from a higher to a lower level, a phenomenon accompanied by increased entropy. For example, in a cool (low energy) system, the slower moving molecules are better organized in a statistical sense; however, when heat flows from a hotter system and warms up the low energy system, the molecules begin to move faster and the system becomes more disordered. The construction and destruction of a house serves as an illustration of entropy. A great deal of energy (in form of workers' efforts, heat energy, electrical energy, and so forth, is required over a long period of time to build the house, but much less effort is needed to destroy it. The difference between the large amount of energy required to build the house and the much smaller amount needed to destroy it shows that a great deal of energy was lost: this lost energy is entropy.

In any protein molecule (Chapter 3) the sequence of amino acids is very precisely determined. Therefore, the molecule shows a high degree of order and low entropy. The synthesis of such a molecule from the individual amino acids requires considerable amounts of energy or "work" (an *endergonic reaction*). On the other hand, the breakdown of specific proteins into amino acids or into carbon dioxide and water is a highly irreversible process that gives up considerable energy (an *exergonic reaction*).

When energy is forced to flow in a reverse direction (from a lower to a higher level), the entropy decreases. Such processes are thermodynamically unlikely unless they are connected with another system in which the entropy increases accordingly, thus compensating for the decrease. In the plant cell, synthesis of glucose from carbon dioxide and water, simultaneously locking energy derived from the sun into the molecule, is accompanied by a decrease in entropy. On the other hand, in the oxidation of glucose in the animal cell there is a considerable increase in entropy. The interaction of the two systems thus satisfies the second law of thermodynamics, i.e., that entropy must always increase.

These concepts are of great importance in biological systems, since cells are characterized by a high degree of order expressed in their molecular and subcellular structure. When a cell dies, disintegration begins, and entropy increases.

CELL METABOLISM

Energy Cycle

The ultimate source of energy in living organisms comes from the sun. The energy carried by photons of light is trapped by the pigment *chlorophyll*, present in the chloro-

plasts of green plants, and accumulates as chemical energy within the different foodstuffs (see Chapter 11). Without the sun, there would be no life on this planet, but interestingly enough, it has been estimated that all life on earth is driven by only 0.24 per cent of the total energy reaching the earth's surface.

All cells and organisms can be grouped into two main classes, differing in the mechanism of extracting energy for their own metabolism. In the first class, called *autotrophs* (i.e., green plants), CO_2 and H_2O are transformed by the process of *photosynthesis* into the elementary organic molecule of *glucose* from which the more complex molecules are made.

The second class of cells, called *heterotrophs* (i.e., animal cells), obtain energy from the different foodstuffs (i.e., carbohydrates, fats, and proteins) that were synthesized by autotrophic organisms. The energy contained in these organic molecules is released mainly by combustion with O_2 from the atmosphere (i.e., oxidation) in a process called *aerobic respiration*. The release of H_2O and CO_2 by heterotrophic organisms completes this cycle of energy (Fig. 4–10).

There is a small group of bacteria that is able to obtain energy from inorganic molecules. For example, the bacteria of the genus *Nitrobacter* oxidize nitrites to nitrates ($NO_2^- + \frac{1}{2} O_2 \rightarrow NO_3^-$). Other bacteria

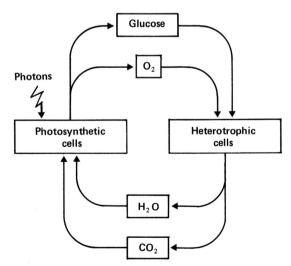

Figure 4–10. Simple diagram of the energy cycle and of the interaction between photosynthetic and heterotrophic cells. (See description in the text.) (After Lehninger, 1970.)

transform ferrous into ferric oxides, and some oxidize SH_2 to sulfate.

Energy Transformation

The *chemical* or *potential* energy of foodstuffs is locked in the covalent bonds between the atoms of a molecule. For example, during hydrolysis of a chemical bond (such as a peptide or an ester bond), about 3000 calories per mole is liberated. In *glucose*, between the atoms of C, H, and O there is an amount of potential energy of about 686,000 calories per mole (i.e., per 180 grams of glucose) which can be liberated by combustion, as in the following reaction:

$$C_6H_{12}O_6 + 6\ O_2 \rightarrow$$
$$6\ H_2O + 6\ CO_2 + 686{,}000\ \text{calories} \quad \textbf{(9)}$$

Within the living cell this enormous amount of energy is not released suddenly, as in combustion by a flame. It proceeds in a stepwise and controlled manner, requiring a great number of enzymes that finally convert the fuel into CO_2 and H_2O.

In the engine of a moving car there are great changes in temperature; within the cell this does not occur. Only a part of the energy liberated from foodstuffs is dissipated as heat; the rest is recovered as new *chemical energy*. The energy liberated in the exergonic reactions resulting from the oxidation of foodstuffs is used in the various cellular functions. Thus, energy may be used: (a) to synthesize new molecules (i.e., proteins, carbohydrates, and lipids) by means of *endergonic* reactions (these molecules can then be used to replace others or for the natural growth of the cell); (b) to perform mechanical work like cell division, cyclosis (cytoplasmic streaming), or muscle contraction; (c) to carry out *active transport* against an osmotic, or ion gradient; (d) to maintain membrane potentials, as in nerve conduction and transmission, or to produce electric discharges (e.g., in electric fish); (e) in cell secretion; or (f) to produce radiant energy, as in bioluminescence. Only in the reactions of group (a) is the energy provided by the foodstuff transformed into chemical bond energy. In all the other reactions chemical energy is transformed into other forms of energy.

High Energy Bonds

Between all these transformations there is a common link, namely, the molecule of *adenosine triphosphate* (ATP). This is a compound found in all cells. Its main characteristic is two terminal bonds with a potential energy much higher than all the other chemical bonds. As shown in Figure 4–11, ATP is composed of the purine *base, adenine,* of *ribose,* and of three molecules of *phosphoric acid.* Adenine plus ribose form the nucleoside *adenosine;* this nucleoside, with the first phosphate, forms adenosine monophosphate, *adenylic acid.* The most important compounds in energy transformation are, however, adenosine diphosphate (ADP) and adenosine triphosphate (ATP). If adenosine is represented by A, and phosphate by P, the simplified formula of ATP and its transformation into ADP is as follows:

$$A{-}P \sim P \sim P \rightleftharpoons A{-}P \sim P + \quad \textbf{(10)}$$
$$Pi + 7300\ \text{calories.}$$

This reaction indicates that the release of the terminal phosphate of ATP produces about 7300 calories, instead of the 3000 calories from common chemical bonds. The reaction ATP $\rightleftharpoons$ ADP plays the central role between the exergonic processes that liberate energy and those that store or transform energy in the various cellular functions.

The high energy $\sim$ P bond enables the cell to accumulate a great quantity of energy in a very small space and to keep it ready for use as soon as it is needed.

Other nucleotides having high energy bonds, such as cytosine triphosphate (CTP), uridine triphosphate (UTP), and guanosine triphosphate (GTP), are involved in biosynthetic reactions. However, the energy source for these nucleoside triphosphates is ultimately derived from ATP, a process aided by a group of enzymes called the *nucleoside diphosphokinases.* The energy obtained by the transfer of the terminal phosphate is channelled into the various synthetic processes by the uridine, guanosine, and cytosine triphosphates. Figure 4–12 indicates the nucleoside triphosphates of the ribose and deoxyribose type (i.e., dATP) that are used as energy sources for the synthesis of important biological compounds. High energy phosphate bonds are also found in phosphocrea-

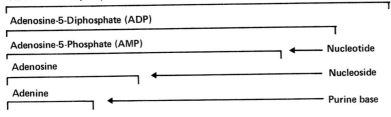

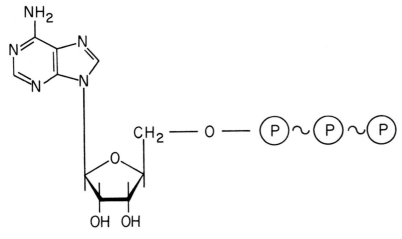

Figure 4-11. Structure of adenosinetriphosphate and its components. Note the presence of two high energy phosphate bonds.

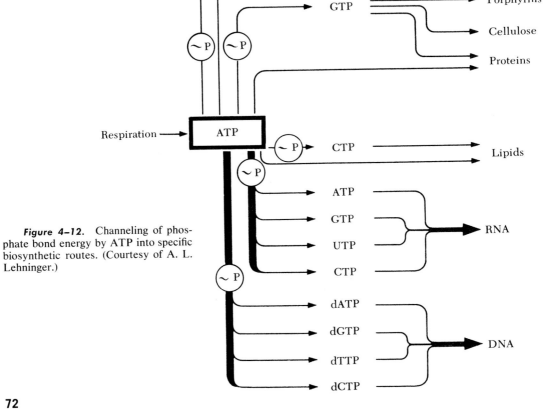

Figure 4-12. Channeling of phosphate bond energy by ATP into specific biosynthetic routes. (Courtesy of A. L. Lehninger.)

tine, acetylphosphate, and phosphoenolpyruvic acid.

CELL RESPIRATION

The mechanisms by which organic substances are degraded and the way in which part of this released energy is stored in ATP as high energy bonds will be considered in the following section. The most common fuel used by the cell is *glucose;* the way in which it is metabolized will depend on the availability of oxygen. Thus, *anaerobic glycolysis* (fermentation) does not require oxygen, but as a result of this process only a small fraction of the chemical energy of glucose is recovered. On the other hand, in the presence of oxygen, by the process of aerobic respiration, glucose is oxidized to CO_2 and H_2O, with a much higher yield in ATP. Some of the main differences between aerobic and anaerobic respiration are summarized in Table 4–1.

Anaerobic Glycolysis

Anaerobic glycolysis will degrade the 6-carbon glucose molecule into two 3-carbon lactic acid molecules:

$$C_6H_{12}O_6 \longrightarrow 2\ C_3H_6O_3 \qquad (11)$$

glucose   lactic acid

This process is achieved in 11 successive steps, each one catalyzed by a different enzyme. Glycolytic enzymes are soluble in the cytoplasmic matrix, and they appear in the last soluble fraction after cell fractionation (Chapter 6). As shown in Figure 4–13, in this chain of reactions the product of one enzyme

serves as substrate for the next reaction. Glucose is first phosphorylated by ATP, with the production of glucose-6-phosphate which is converted into frutose-6-phosphate. After several steps, pyruvate is formed. In Figure 4–13 it may be observed that during this part of the chain reaction 4 ATP molecules are produced. Since 2 ATP were previously used, however, the yield is only two ATP molecules. The general reaction may be written:

$$C_6H_{12}O_6 + 2\ Pi + 2\ ADP \longrightarrow \qquad (12)$$

glucose $2\ C_3H_6O_3 + 2\ ATP + 2\ H_2O$

          lactic acid

The fate of *pyruvate,* a key product in glycolysis, depends on whether oxygen is available. In the case of anaerobic conditions, it will be used as a hydrogen acceptor for the 2 NADH generated during glycolysis, and it is converted into *lactate.* Under aerobic conditions pyruvate is converted into *acetyl coenzyme A* and CO_2 is released. At this point a direct connection with the Krebs cycle is made.

Aerobic Respiration

The term *aerobic respiration* refers to the series of reactions by which organic substances are broken down to CO_2 and H_2O in the presence of molecular oxygen. The general reaction for the degradation of glucose is:

$$C_6H_{12}O_6 + 6\ O_2 \longrightarrow 6\ CO_2 + 6\ H_2O$$

glucose (13)

This process releases 686,000 calories of the chemical energy contained in glucose. During anaerobic glycolysis, however, less

TABLE 4–1.   Some Differences Between Aerobic and Anaerobic Respiration

Aerobic Respiration (Oxidative Phosphorylation)	Anaerobic Respiration (Fermentation)
Uses molecular O_2	Does not use O_2
Degrades glucose to CO_2 and H_2O	Degrades glucose to trioses and other complex organic compounds
Exergonic	Exergonic
Recovers almost 50 per cent of chemical energy	Recovers less chemical energy
Present in most organisms	Present in some microorganisms and important in embryonic and neoplastic cells
Enzymes localized in mitochondria	Enzymes localized in the cytoplasmic matrix
Yields 36 ATP per glucose molecule	Yields 2 ATP per glucose molecule

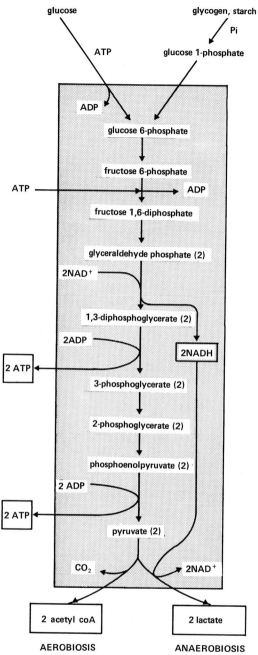

Figure 4–13. Diagram of the degradation of glucose or glycogen by anaerobic glycolysis. (See the description in the text.)

(Fig. 4–14). During a second stage, that takes place in mitochondria, the acetyl groups enter the Krebs cycle, from which CO_2 and hydro-

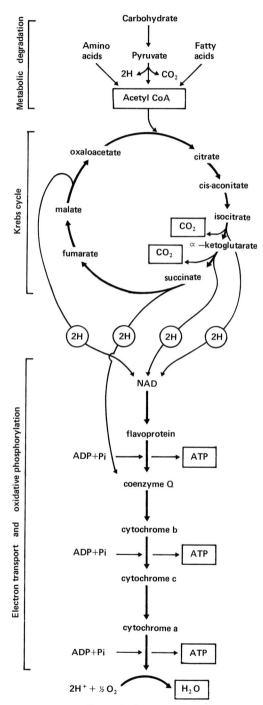

Figure 4–14. General diagram of aerobic respiration showing the Krebs cycle, the respiratory chain, and its coupling with oxidative phosphorylation. (See the description in the text.) (From Lehninger, A. L., *Biochemistry*, New York, Worth Publishers, Inc., 1970.)

than 10 per cent of this amount (i.e., 58,000 calories) is released because its final product, lactic acid, is far more complex than CO_2 and H_2O and, therefore, contains more energy. In the first stage of aerobic respiration all foodstuffs (carbohydrates, amino acids, and fatty acids) are degraded into acetyl groups gen atoms are produced. Finally, in a third stage the energy contained in the hydrogen is taken up and transferred by the *electron transport or respiratory chain* and ATP is produced in the process. The hydrogen atoms ultimately combine with molecular oxygen to yield water.

The Krebs Cycle

The *Krebs cycle,* also called the *tricarboxylic acid cycle,* is the common final pathway of cellular catabolism, in which all "fuel" molecules undergo a final oxidative process. This cycle degrades the acetyl group contained in acetyl coenzyme A to CO_2 and hydrogen atoms:

$$C_2H_4O_2 + 2\ H_2O \longrightarrow 2\ CO_2 + 8\ H \quad (14)$$

As shown in Figure 4–14, the first step of the Krebs cycle consists of the condensation of the acetyl group (2 carbons) with oxaloacetate (4 carbons) to form citrate, a 6-carbon compound. In the following steps of the cycle, two molecules of CO_2 will be released and oxaloacetate is ultimately formed and used again. The eight hydrogen atoms produced will be used to generate 11 ATP molecules by a series of oxidation-reduction reactions in the respiratory chain. This series of reactions illustrates one of the classic examples of nature's economy: the seven enzymes of the Krebs cycle perform their function for *any* type of food which the organism may ingest.

Respiratory Chain

The series of reactions comprising the respiratory chain also occur within the mitochondria and will be described in detail in relation to the molecular structure of the mitochondria in Chapter 10.

In the Krebs cycle, specific *dehydrogenases* transfer hydrogen pairs to nicotine adenine dinucleotide (NAD^+), thus generating $NADH + H^+$, a key coenzyme in catabolic processes. NADH transfers the hydrogen atoms to the *respiratory chain,* where electrons are transported in a series of oxidation-reduction steps to react, ultimately, with molecular oxygen (Fig. 4–14).

In oxidation-reduction reactions there is transfer of electrons from an electron donor to an electron acceptor. *Oxidation* consists of the loss of electrons; *reduction* consists of the gain of electrons. In many biological oxidations electrons are transferred via hydrogen atoms.

Several of the cytochromes of the respiratory chain are iron-containing molecules. During an electron transfer the iron passes from the ferrous to the ferric state, releasing one electron:

$$Fe^{++} \rightleftharpoons Fe^{+++} + e^- \quad (15)$$

This reaction is the basis of all oxidation-reduction processes. When a pair of electrons is transferred from $NADH + H^+$ to molecular oxygen, a large amount of energy is released. In the respiratory chain this transfer is done in a stepwise fashion in which the electron pairs are passed from one acceptor to another, thus delivering energy more gradually.

Oxidative Phosphorylation

Coupled with the respiratory chain is the process of *oxidative phosphorylation.* As shown in Figure 4–14, at three steps in the electron transport there is the formation of one molecule of ATP. The equation for this process can be expressed as follows:

$$NADH + H^+ + 3\ ADP + 3\ Pi + \frac{1}{2} O_2 \longrightarrow$$
$$NAD^+ + 4\ H_2O + 3\ ATP \quad (16)$$

Normally, oxidative phosphorylation is coupled with the respiratory chain; however, these two mechanisms can be separated by the use of *uncoupling agents.* For example, in the presence of 2,4-dinitrophenol, oxygen is consumed, but no ADP is phosphorylated. The molecular mechanisms of oxidative phosphorylation that take place in conjunction with the respiratory chain (in the inner membrane of the mitochondrion) will be described in detail in Chapter 10.

The energy balance of aerobic respiration shows that 36 ATP molecules are produced from each glucose molecule. The overall equation can be written:

$$C_6H_{12}O_6 + 6\ O_2 + 36\ Pi + 36\ ADP \longrightarrow$$
$$6\ CO_2 + 36\ ATP + 42\ H_2O \quad (17)$$

The cell will store 40 per cent of the chemical energy liberated by the combustion of glucose in the form of ATP.

SUMMARY:
Bioenergetics and
Cell Respiration

Bioenergetics deals with the mechanisms by which the cell utilizes chemical energy. A portion of the total energy (H), or *enthalpy,* is dissipated as heat ($T\Delta S$), and another portion is retained as free energy (ΔG). Thus, $\Delta H = \Delta G + [T\Delta S]$ (7). In this equation ΔS is a measure of the *entropy* of the system. When entropy is low, there is relative order in the system; when it increases, the system has a tendency to move toward molecular disorder. Flow of energy is related to an increase in entropy. For the synthesis of a protein, molecular energy is used *(endergonic reaction)* and is then released during the catabolism of the protein *(exergonic reaction).*

The cycle of biochemical energy starts with the *photons* of light trapped by *chlorophyll.* In *autotrophic* organisms (i.e., plants) CO_2 and H_2O are transformed by the process of *photosynthesis* into glucose and other complex molecules. In *heterotrophic* organisms (i.e., animals) the energy is obtained from foodstuffs by combustion with O_2 *(aerobic respiration)* (Fig. 4–10). One mole of glucose (180 grams) contains 686,000 calories. This energy is released in a stepwise fashion and is used for the many functions of the cell. Part of the chemical energy is stored in the ATP molecule, which has two terminal high energy bonds (these bonds contain 7300 calories). The reaction, ATP $\rightleftharpoons$ ADP, releases the energy contained in the $\sim$P bond. Other nucleoside triphosphates (i.e., CTP, UTP, ITP, and GTP) derive their energy from ATP.

Cell respiration is the series of chemical reactions by which organic substances are degraded and energy is released. In *anaerobic glycolysis* the 6-carbon glucose is degraded to two molecules of lactic acid. This process occurs without O_2 via 11 successive steps that take place in the cytoplasmic matrix. The net result of anaerobic glycolysis is the production of two molecules of ATP. The final product under anaerobic conditions, pyruvate, will be transformed into lactate by acting as the hydrogen acceptor for 2 NADH + H^+. In the presence of O_2, pyruvate is converted to CO_2 and acetyl-coenzyme A, which enters the Krebs cycle. Anaerobic glycolysis liberates less than 10 per cent of the energy contained in glucose (58,000 calories).

The *Krebs or tricarboxylic cycle* occurs inside the mitochondrion, and by a complex series of reactions, involving seven enzymes, degrades the acetyl group into 2 CO_2 molecules and H atoms which will be used to generate ATP molecules in the respiratory chain.

The *respiratory chain* or *electron transport system* contains a series of oxidation-reduction systems, involving several cytochromes, in which electrons are transferred according to the reaction $Fe^{++} \rightleftharpoons Fe^{+++} + e^-$. The final cytochrome (cytochrome *a* or cytochrome oxidase) transfers the H atoms to O_2 to produce H_2O. The respiratory chain is coupled with the process of *oxidative phosphorylation* in the inner membrane of the mitochondrion (see Chapter 10).

The final energy balance of aerobic respiration is as follows:

$$C_6H_{12}O_2 + 6\ O_2 + 36\ Pi + 36\ ADP \longrightarrow$$
$$6\ CO_2 + 36\ ATP + 42\ H_2O$$

Thus, in aerobic respiration, 36 ATP molecules are produced from one molecule of glucose. This energy production is equivalent to 40 per cent of the total energy contained in this molecule.

REFERENCES

1. Koshland, D. E., Jr. (1960) *Adv. Enzymol., 22*:45.
2. Yankeelov, J. A., Jr., and Koshland, D. E., Jr. (1965) *J. Biol. Chem., 204*:1593.
3. Monod, J., Wyman, J., and Changeux, J. P. (1965) *J. Molec. Biol., 12*:88.
4. Gerhard, J. C., and Schachman, H. K. (1965) *Biochem., 4*:1054.

ADDITIONAL READING

Atkinson, D. E. (1966) Regulation of enzyme activity. *Ann. Rev. Biochem.; 35*:85.
Baldwin, E. (1967) *Dynamic Aspects of Biochemistry.* 5th Ed. Cambridge University Press, London.
Bernhard, S. (1968) *The Structure and Function of Enzymes.* W. A. Benjamin, Inc., New York.
Boyer, P. D. (1970–1972) *The Enzymes.* 3rd Ed., 8 volumes. Academic Press, Inc., New York.
Bray, H. G., and White, K. (1966) *Kinetics and Thermodynamics in Biochemistry.* 2nd Ed. Academic Press, Inc., New York.
Dixon, M., and Webb, E. C. (1964) *Enzymes.* 2nd Ed. Longmans, Green and Co., London.
Jost, J. P., and Rickenberg, H. V. (1971) Cyclic AMP. *Ann. Rev. Biochem., 40*:741.
Koshland, D. E., Jr., and Neet, K. E. (1968) The catalytic and regulatory properties of enzymes. *Ann. Rev. Biochem., 37*:359.
Krebs, H. A. (1950) The tricarboxylic acid cycle. *Harvey Lect.,* Ser. 44 (1948–49), p. 165.
Latner, A. L., and Skillen, A. W. (1968) *Isoenzymes in Biology and Medicine.* Academic Press, Inc., New York.
Lehninger, A. L. (1965) *Bioenergetics.* W. A. Benjamin, Inc., Menlo Park, Calif.
Lehninger, A. L. (1970) *Biochemistry.* Worth Publishers, Inc., New York.
Monod, J., Changeux, J. P., and Jacob, F. (1963) The catalytic and regulatory properties of enzymes. *J. Mol. Biol., 3*:306.
Pastan, I. (1972) Current directions in research on cyclic AMP. In: *Current Topics in Biochemistry.* (Anfinsen, C. B., Goldberger, R., and Schechter, A., eds.) Academic Press Inc., New York.
Pastan, I. (1972) Cyclic AMP. *Sci. Am. 227*:97.
Robison, A. G., Butcher, W. and Sutherland, E. W. (1971) Cyclic AMP. Academic Press, Inc., New York.
Roodyn, D. B., ed. (1967) *Enzyme Cytology.* Academic Press, Inc., New York.
Stadtman, E. R. (1966) Allosteric regulation of enzyme activity. *Adv. Enzymol., 28*:41.

Part Three

METHODS FOR THE STUDY OF THE CELL

The recent extraordinary progress in cell biology has resulted from the development of new methods for the study of the cell and of its molecular and macromolecular components. In the following chapters the two main groups of techniques employed in cytology are presented.

Chapter 5 includes the methods that employ electromagnetic waves. These may be either visible or ultraviolet radiations, electrons, or x-rays. In studying these methods, it is convenient to review the chapters in physics textbooks that deal with reflection, refraction, interference, and diffraction of electromagnetic waves. This background reading will promote a better understanding of the instruments used for analysis of cellular and subcellular structure. The importance of electron microscopy and x-ray diffraction are emphasized. The images given by the light, phase, interference, polarization, and electron microscopes are discussed in relation to the various physical principles involved. The study of the electron microscope and of the techniques employed to prepare biological material for electron microscopy is of great importance in cell biology.

The discussion of instrumentation is complemented by Chapter 6, which is a brief presentation of the main methods of cytologic and cytochemical analysis used for observation and experimentation. Since the number of techniques by far surpasses the limits of this book, only a few selected examples are mentioned. The methods for study of living cells, the process of fixation for the preparation of cells and tissues, and the mechanism of staining are considered. At present, the cytochemical techniques for the identification and localization of substances within cells are of paramount importance. The examples mentioned deal with the mechanism of cytochemical analysis and the problems that must be overcome.

five

INSTRUMENTAL ANALYSIS
OF BIOLOGICAL
STRUCTURES

Before studying this chapter, the student should be familiar with the limits and dimensions in biology (Chapter 1) and with the optical laws and principles on which the ordinary light microscope is based (Fig. 5–1).

Observation of biological structures is difficult because cells are, in general, very small and are transparent to visible light. The search continues for new instruments designed to provide better definition of cell structure down to the molecular level by an increase in *resolving power* and to counteract the transparency of the cell by an increase of *contrast*.

Resolving Power of the Microscope

In the light microscope, as in any other type of microscope, the resolving power (the capacity of the instrument for showing distinct images of points very close together) depends upon the wavelength (λ) and the numerical aperture (NA) of the objective lens. The *limit of resolution*, defined as the minimum distance between two points that allows for their discrimination as two separate points is:

$$\text{Limit of resolution} = \frac{0.61\lambda}{\text{NA}} \qquad (1)$$

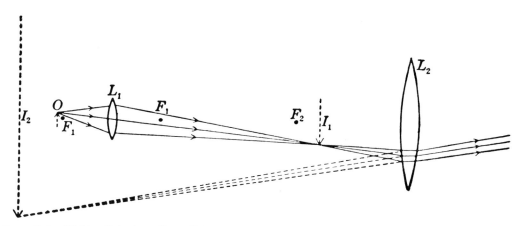

Figure 5–1. Light path in the ordinary light compound microscope. The group of ocular lenses is diagrammatically represented by L_2, the group of objective lenses by L_1. The object (O) on a microscope slide is placed just outside the principal focus of the objective lens (L_1), which has a short focus. This lens produces a real image at I_1, which is formed inside the principal focus of the eyepiece lens (L_2). The eye, looking through the lens L_2, sees a magnified virtual image (I_2) of the image I_1. The eyepiece lens is thus used as a magnifying glass to view the real image (I_1).

The numerical aperture is: $NA = n \times \sin \alpha$. Here, n is the refractive index of the medium and $\sin \alpha$ is the sine of the semiangle of aperture. Remember that the limit of resolution is inversely related to the resolving power; i.e., the higher the resolving power, the smaller the limit of resolution.

Since $\sin \alpha$ cannot exceed 1, and the refractive index of most optical material does not exceed 1.6, the maximal NA of lenses, using oil immersion, is about 1.4. With these parameters it is easy to calculate from formula (1) that the limit of resolution of the light microscope cannot surpass 170 nm (0.17 μm) using monochromatic light of $\lambda = 400$ nm (violet). With white light, the resolving power is about 250 nm (0.25 μm). Since in formula (1) the NA is limited, it is evident that the only way to increase the resolving power is to use shorter wavelengths. In this case, glass lenses are no longer transparent, and other refractive media should be introduced. For example, with ultraviolet radiation of 200 to 300 nm, quartz lenses or reflecting optical instruments should be used, and the resolution is increased only by a factor of two, reaching 100 nm (0.1 μm). By similar reasoning, a microscope using infrared radiation of $\lambda = 800$ nm would have a limit of resolution of 0.4 μm.

METHODS FOR INCREASING CONTRAST

Phase Microscopy

The unaided eye detects variations in wavelength (color) and in intensity of visible light. The majority of cell components are essentially transparent, except for some pigments (more frequent in plant cells) that absorb light at certain wavelengths (colored substances). The low light absorption of the living cell is caused largely by its high water content, but even after drying, cell components show little contrast.

One way of overcoming this limitation is by the use of dyes that selectively stain different cell components and thus introduce contrast by light absorption. In most cases, however, staining techniques cannot be used in the living cell. The tissue must be fixed, dehydrated, embedded, and sectioned prior to staining, and all these procedures may introduce morphologic and chemical changes.

In recent years remarkable advances have been made in the study of living cells by the development of special optical techniques, such as *phase contrast* and *interference microscopy*. These two techniques are based on the fact that although biological structures are highly transparent to visible light, they cause phase changes in transmitted radiations. These phase differences, which result from small differences in the refractive index and thickness of different parts of the object, can now be made more clearly detectable.

Figure 5–2 indicates the effects of a nonabsorbent transparent material (A) and an absorbent transparent material (C) on a light ray. In *A*, the wave impinges on a material that has a refractive index different from that of the medium. In passing through the object, the amplitude of the wave is not affected, but the velocity is changed. If the refractive index of the material is higher than that of the medium, there is a *delay* or *retardation*, also called a *phase change*. After the wave emerges from the object, its original velocity is re-established, but retardation is maintained. This retardation implies a phase change, which can be measured in fractions of a wavelength. The phase change increases in direct proportion to the difference between the refractive indices of the object and the

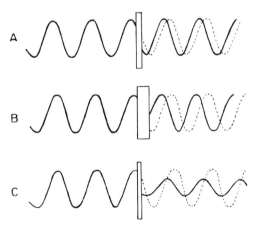

Figure 5–2. Diagram showing: **A,** the effect of a transparent and nonabsorbent material of higher refractive index than the medium, which introduces a phase change (retardation). **B,** the same, but thicker, object. The retardation or phase change is more pronounced. **C,** the effect of a transparent and absorbent object. There is a retardation, but also a decrease, in amplitude (intensity).

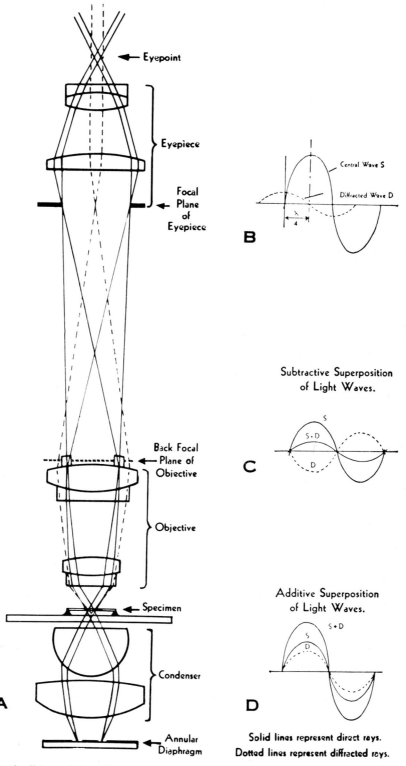

Figure 5–3. **A,** the light path in a phase contrast microscope. **B,** the normal retardation by $\frac{1}{4}$ wavelength of light diffracted by an object, and its difference in phase from the light passing through the surrounding medium. By phase optics the two waves are superimposed to reinforce each other in bright contrast phase as shown in **D** or to subtract from each other as in dark contrast phase shown in **C**. (From the American Optical Company.)

surrounding medium and to the thickness of the object (Fig. 5–2, *B*).

To understand the phase microscope, it is first necessary to analyze the behavior of a ray of light that traverses a thin, transparent particle.

A portion of the light ray that traverses the object (wave *D* in Figure 5–3, *B*) is diffracted and deviates with respect to the

rays that do not traverse the object, or those passing through the exact center (wave *S* in Figure 5–3, *B*).

In biological materials the phase difference between the S and D waves is approximately ¼ wavelength (Fig. 5–3, *B*). These two rays (*S* and *D*) penetrate the objective lens and undergo interference. The resulting ray has a small phase retardation but this is

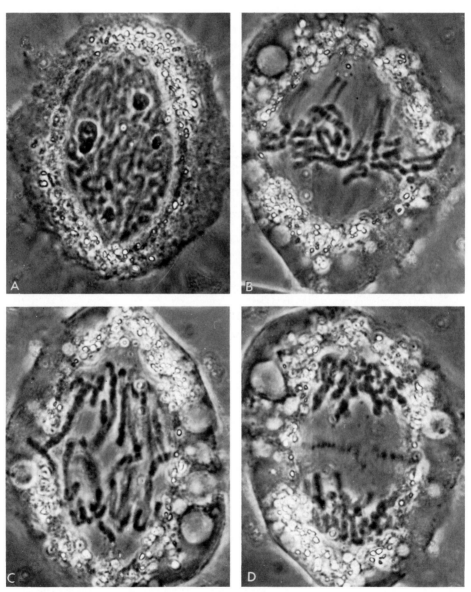

Figure 5–4. Observation by phase contrast microscopy of mitotic cell division in a living cell of endosperm tissue of the plant *Haemanthus*. The same cell has been photographed at the following times: **A**, 10:32 hrs; **B**, 12:48 hrs; **C**, 13:12 hrs and **D**, 13:21 hrs. **A**, late prophase showing the coiled chromosomes and the nucleoli within the nucleus; **B**, metaphase with chromosomes at the equatorial plane; **C**, anaphase; and **D**, telophase showing the chromosomes at the poles and the formation of the phragmoplast at the equatorial plane. ×700. (Courtesy of A. S. Bajer.)

not sufficient to produce a change in amplitude, and thus is not detectable with an ordinary light microscope.

In the phase contrast microscope, the small phase differences are intensified so that they are detected by the eye or the photographic plate. In this type of microscope, the most lateral light passing through the objective of the microscope is advanced, or retarded, by $\frac{1}{4}$ wavelength ($\frac{1}{4}$ λ) with respect to the central light passing the object. An annular phase plate that introduces a $\frac{1}{4}$ wavelength variation is put in the back focal plane of the objective. In addition, an annular diaphragm is placed in the substage condenser (Fig. 5–3). The phase plate is a transparent disk containing an annular groove or elevation of a shape and size that coincide with the direct image of the substage condenser. The phase effect results from the interference between the direct geometric image given by the central part of the objective and the lateral diffracted image, which has been retarded or advanced to $\frac{1}{2}$ wavelength. In *bright,* or *negative, contrast* the two sets of rays are added (Fig. 5–3, *D*) and the object appears brighter than the surroundings; in *dark,* or *positive, contrast* the two sets of rays are subtracted (Fig. 5–3, *C*), making the image of the object darker than the surroundings (Fig. 2–4). Because of this interference, the minute phase changes within the object are amplified and translated into changes of amplitude (intensity). The transparent object thus appears in shades of gray, depending on the product of the thickness and the difference in refractive index of the object with the medium.[1-4]

Phase microscopy is used routinely to observe living cells and tissues, and is particularly valuable for observing cells cultured in vitro during mitosis (Fig. 5–4).

Interference Microscopy

The interference microscope is based on principles similar to those of the phase microscope, but has the advantage of giving quantitative data. Interference microscopy permits detection of small, continuous changes in refractive index, whereas the phase microscope reveals only sharp discontinuities. The variations of phase can be transformed into such vivid color changes that a living cell may resemble a stained preparation.[5-7]

In the interference microscope the light emitted by a single source is split into two beams: one is sent through the object; the other bypasses the object. The two beams are then recombined and interfere with one another, as in the phase microscope. In comparison with the direct beam, the beam that has crossed the object is retarded, which means that it has undergone a phase change. This retardation (Γ) is determined by the thickness of the object (t) and the difference between the refractive indices of the object (n_o) and of the surrounding medium (n_m).

$$n_o - n_m = \frac{\Gamma}{t} \qquad (2)$$

By use of the interference microscope it is possible to measure the dry weight of the object, because this is related to the refractive index. When the object is measured in water, the following relationship applies:

$$C_o = \frac{100\,(n_o - n_w)}{X} \qquad (3)$$

C_o is the percentage concentration of dry material in the object; n_w is the refractive index of water; X is a constant that equals 100 α (α is the specific refractive increment of the material in solution). X is about 0.18 for the major substances of the cell: proteins, lipoproteins, and nucleic acids.[8]

Interference microscopy permits the simultaneous determination of the thickness of the object (t), the concentration of dry matter, and the water content by successive measurements of the optical phase difference in two media of known refractive indices. An example of the application of this method to the study of the sea urchin oöcyte is illustrated in Table 5–1. These measurements show that the percentage concentration of

TABLE 5–1. Measurement of Dry Matter (C_o) in Sea Urchin Oöcytes

	$N_o - N_w$	C_o† *For X = 0.165*
Cytoplasm	0.036	25.5
Nucleus	0.021	16.4

*From Mitchison, J. M., and Swann, M. M., *Quart. J. Micros. Sci., 94*:381, 1953.

†$C_o = \dfrac{100\,(n_o - n_w)}{X}$

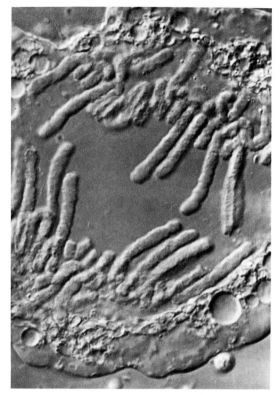

Figure 5–5. Observation of a dividing plant cell at anaphase by interference microscopy with Nomarski optics. The chromosomes separating toward the poles appear three-dimensional; some fine spindle fibers are observed. (Courtesy of A. S. Bajer.)

dry matter is much lower in the nucleus (at this stage of cell development) than in the cytoplasm.

A special variation of the interference microscope is the so-called *Nomarski inter-ference-contrast microscope,*[9] in which a single light beam passes through the object and the objective, but is then divided into two interfering beams via a special birefringent prism. This microscope also includes polar-izer and analyzer filters and a compensating prism above the substage condenser (see Padawer, 1968). The image obtained gives a characteristic relief effect and offers some advantage over ordinary phase contrast optics. It is particularly useful for the study of cells in mitosis, as indicated in Figure 5–5.

Darkfield Microscopy

Darkfield microscopy, also called *ultra-microscopy,* is based on the fact that light is scattered at boundaries between phases having different refractive indexes. The in-strument is a microscope in which the ordinary condenser is replaced by one that illuminates the object obliquely. With this darkfield condenser, no direct light enters the objective; therefore, the object appears bright because of the scattered light, and the back-ground remains dark. In a living cell in a tissue culture, for example, the nucleolus, nuclear membrane, mitochondria, and lipid droplets appear bright, and the background of cytoplasm appears dark.

Under the darkfield microscope objects smaller than those seen with the ordinary light microscope can be detected, but not resolved.

The problem of *detectability* also applies to the other types of light microscopes, par-ticularly those using phase contrast optics. Theoretically a fiber of only 5 nm (i.e., about 40 times smaller than the resolving power) could be detected, provided it had enough contrast in comparison to the background. Several structures of the cell that are too small to be resolved have been detected with the light microscope; e.g., the nuclear en-velope, the endoplasmic reticulum, and bundles of three to eight microtubules (see Bajer and Jensen, 1969).

The detectability of fine structures is important when living cells are observed, as, for example, in following the changes in spindle fibers during the cell cycle (Chapter 13).

Polarization Microscopy

This method is based on the behavior of certain components of cells and tissues when they are observed with polarized light. If the material is *isotropic,* polarized light is propa-gated through it with the same velocity, independent of the impinging direction. Such substances or structures are characterized by having the same *index of refraction* in all directions. On the other hand, in an *aniso-tropic* material the velocity of propagation of polarized light varies. Such material is also called *birefringent* because it presents two dif-ferent indexes of refraction corresponding to the respective different velocities of trans-mission.

Birefringence (B) may be expressed quantitatively as the difference between the two indexes of refraction $(N_e - N_o)$ associated with the fast and slow ray. In practice, the retardation (Γ) of the light polarized in one plane is measured relative to that of light polarized in another perpendicular plane with the polarizing microscope. The retardation depends on the thickness of the specimen (t) in this way:

$$B = N_e - N_o = \frac{\Gamma}{t} \qquad (4)$$

Measurement of the retardation is assisted by a form of compensator introduced into the optical system. The measurement is in nm or in fractions of a wavelength (λ).

The *polarizing* microscope differs from the ordinary one in that two polarizing devices have been added: the *polarizer* and the *analyzer,* both of which can be made from a sheet of polaroid film or with Nicol prisms of calcite. The polarizer is mounted below the substage condenser and the analyzer is placed above the objective lens (Fig. 5–6).

In the crossed position, polarized light is not transmitted. Under this condition, if a birefringent specimen is placed on the stage, the plane of polarization will deviate according to the retardation introduced by the object.

The usual test with the polarizing microscope consists of rotating the specimen to find the points of maximum and minimum brightness. Maximum brightness is obtained when the axis of the object makes a ± 45 degree angle with those of the polarizer and analyzer (Fig. 5–6).

In biological fibers, birefringence is *positive* if the index of refraction is greater along the length of the fiber than in the perpendicular plane, and is *negative* in the opposite case. The sign can be determined by interposing a birefringent material whose slow and fast axes are known. With the improved methods of polarization microscopy now available, retardations of 0.1 nm with a resolution of 0.3 μm can be measured. Since birefringence depends on structural properties that are much smaller than the wavelength of light, polarization microscopy is used for analyzing cell ultrastructure indirectly.

The main types of birefringence are:

Crystalline (Intrinsic) Birefringence. Crystalline birefringence is found in systems in which molecules or ions have a regular asymmetrical arrangement and it is independent of the refractive index of the medium. In structures composed of proteins or lipids, a certain degree of crystalline birefringence

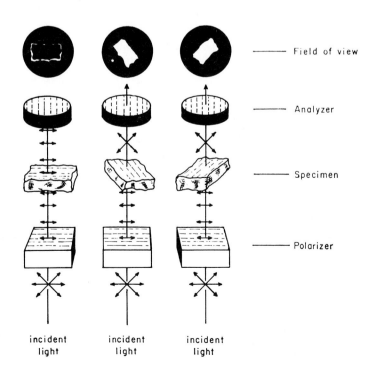

Figure 5–6. Schematic drawing showing variations in darkness and brightness of an anisotropic object when placed between crossed polarizer and analyzer and rotated $\pm$ 45°. (Extracted from Wilson, G. B., and Morrison, J. H., *Cytology,* New York, Reinhold Publishing Corporation, 1961, with the permission of Reinhold Publishing Corporation.)

Field of view

Analyzer

Specimen

Polarizer

incident light   incident light   incident light

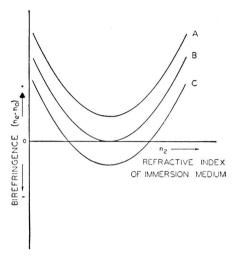

Figure 5–7. Method of determining the sign and relative amount of form and crystalline birefringence by the immersion technique. **A** indicates positive form and positive crystalline birefringence. **B** indicates positive form and no crystalline birefringence. **C** indicates positive form and negative crystalline birefringence. (From F. O. Schmitt.)

may appear, which in both cases is positive. On the other hand, fibers of nucleoprotein have a negative birefringence.

Form Birefringence. This is produced when submicroscopic asymmetrical particles are oriented in a medium of a different refractive index. In this instance, the birefringence is changed when the refractive index of the medium varies. In Figure 5–7 the position of the minimum of the curves indicates whether the form birefringence is pure or is combined with crystalline birefringence (Fig. 5–7).

Strain Birefringence. Certain isotropic structures show strain birefringence when subjected to tension or pressure. It occurs in muscle and in embryonic tissues.

Dichroism. This type of birefringence occurs when the absorption of a given wavelength of polarized light changes with the orientation of the object. In dichroism the changes are in amplitude, that is, in the intensity of the transmitted light. Dichroism can be induced in tissues by some staining procedures. For example, organic dyes, such as congo red or thionine, and colloids, can produce dichroism in certain structures by the special orientation of the molecules.

Electron Microscopy

The electron microscope is the only instrument that permits a direct study of biological ultrastructure. Its resolving power is much greater than that of the light microscope. In the electron microscope streams of electrons are deflected by an electrostatic or electromagnetic field in the same way that a beam of light is refracted when it crosses a lens. If a metal filament is placed in a vacuum tube and heated, it emits electrons that can be accelerated by an electrical potential. Under these conditions, the stream of electrons tends to follow a straight path and has properties similar to those of light. Like light, it has a corpuscular and vibratory character, but the wavelength is much shorter (i.e., $\lambda = 0.005$ nm for electrons and 550 nm for light).

The filament or cathode of the electron microscope emits the stream of electrons. By means of a magnetic coil, which acts as a condenser, electrons are focused on the plane of the object and then are deflected by another magnetic coil, which acts as an objective lens and gives a magnified image of the object. This is received by a third magnetic "lens," which acts as an ocular or projection lens and magnifies the image from the objective. The final image can be visualized on a fluorescent screen or recorded on a photographic plate (Fig. 5–8).

In spite of the apparent similarities shown in Figure 5–8, there are great differences between the light and the electron microscope; one of these is the mechanism of image formation. Whereas in the light microscope image formation depends mainly on the degree of light absorption in different zones of the object, image formation in the electron microscope is due principally to electron scattering. Electrons colliding against atomic nuclei in the object are often dispersed so that they fall outside the aperture of the objective lens. In this case the image on the fluorescent screen results from the absence of those electrons blocked by the aperture. Dispersion may also be the result of multiple collisions, which diminish the energy of the passing electrons.

Electron dispersion, in turn, is a function of the thickness and molecular packing of the object and depends especially on the atomic number of the atoms in the object. The higher the atomic number, the greater is the resultant

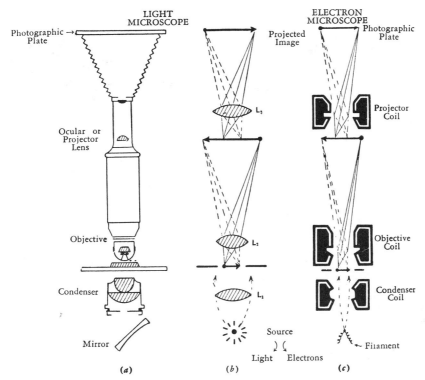

Figure 5–8. Comparison between the optical microscope and the electron microscope. (See the description in the text.) (From G. Thompson.)

dispersion. Most of the atoms that constitute biological structures are of low atomic number and contribute little to the image. For this reason, heavy atoms should be added to the molecular structure.

The greatest advantage of the electron microscope is its high resolving power, which depends on the same variants as does the resolving power of the light microscope. (1).

The wavelength of a stream of electrons is a function of the acceleration voltage to which the electrons are subjected; it can be calculated by the formula of De Broglie:

$$\lambda = \frac{12.2}{\sqrt{V}}\ 0.1\ nm \qquad (5)$$

For example, in one current model of the electron microscope, $V = 50,000$ volts and $\lambda = 0.00535$ nm.

Because of the great aberration of the magnetic lenses, the actual numerical aperture of the electron microscope is small and the limit of resolution is theoretically 0.2 nm (Fig. 1–2). In practice, the limit of resolution for biological specimens is below 1.0 nm.

In the light microscope, magnification is largely determined by the objective, and a maximum magnification of 100 to 120× can be reached. Since the ocular lens can increase this image only 5 to 15 times, a total useful magnification of 500 to 1500× can be achieved.

In the electron microscope the resolving power is so high that the image from the objective can be greatly enlarged. For example, with an initial magnification by the objective of 100×, the image can be magnified 200× with the projector coil, achieving a total magnification of 20,000×.

In the newer instruments a wide range of magnifications can be attained by introducing an intermediate lens. Direct magnifications as high as 160,000× may thus be obtained, and the micrographs may be enlarged photographically to 1,000,000× or more, depending on the resolution achieved.

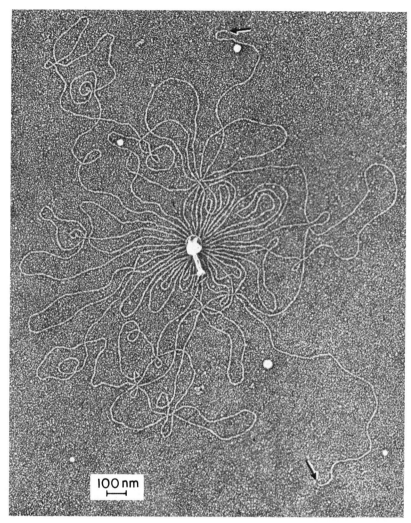

Figure 5–9. Electron micrograph of a bacteriophage (*in the center*) that has undergone an osmotic shock. The DNA molecule that was contained in the "head" of the bacteriophage is now dispersed. Arrows indicate the extremes of the single, unbranched DNA molecule. Preparation shadowcast with platinum. ×76,000. (Courtesy of A. K. Kleinschmidt.)

Preparation of Biological Material for Electron Microscopy

Because of its extraordinary resolving power, the electron microscope seems to be an ideal instrument for the study of cellular ultrastructure. Nevertheless, its usefulness is reduced by a number of technical difficulties and limitations.

One limitation is the low penetration power of electrons. If the specimen is more than 500 nm (0.5 μm) thick, it appears almost totally opaque. The specimen must be deposited on an extremely fine film (7.5 to 15 nm thick) of collodion, carbon, or other substance, to support the specimen, and must be held up by a fine metal grid.

To be observed under the electron microscope, the specimen must first be dehydrated and then placed in a vacuum.

Techniques for preparing specimens vary considerably; several types, which are described below, are often used in biology.

An important method for the study of macromolecules is the so-called *monolayer technique* of Kleinschmidt,[10] in which the macromolecules are extended on an airwater interface before being collected on a film. This method has given excellent results in the demonstration of DNA molecules from various sources (Fig. 5–9).

It has been used similarly for various RNA molecules.[11]

Thick specimens can be disintegrated by mechanical means, such as homogenizers, sonic or supersonic waves, and so forth. The material is thus divided along natural cleavage planes into fragments thin enough to be partially transparent to the electron beam.

Freeze-fracturing and Freeze-etching

The study of the structure of biological membranes has been greatly improved by the use of techniques that involve the freezing and fracturing of specimens.[12] In general, the specimens are submitted to a certain degree of water sublimation in a vacuum, followed by the formation of a replica of the surface by use of carbon and other evaporated elements.

Figure 5–10. Onion root cell which has been submitted to the freeze-etching technique. The upper part of the figure corresponds to the nucleus (*N*) and shows the nuclear pore complexes (*np*) and the nuclear envelope (*ne*). In the cytoplasm, a Golgi complex (*G*) and a large vacuole (*V*) are observed. *cw*, cell wall. ×75,000. (Courtesy of D. Branton.)

The replica, which is detached from the object, reveals a natural-appearing representation of the freeze-etched object (Fig. 5–10). The fracture may disclose either the outer or the inner surface of a membrane, or may even split the membrane lengthwise, thus revealing information about components passing through the cell membrane (see Fig. 8–4, *B*). The effect of etching is to expose and to render more visible the fine surface details. Freeze-etching can be carried out after treatment of the specimen with enzymes, or after labeling with antigens, viruses, or special chemical substances that may yield information about the chemical and biological properties of certain sites in the membrane. With this technique various receptor sites and special chemical groups may be detected (see Moor, 1971; Branton, 1971).

Thin Sectioning

The study of cells and tissues is achieved primarily by the use of *thin sections*.

To satisfy the need for thinner sections hard embedding media have been utilized. Those most often used are acrylic monomers or epoxy resins that impregnate the tissue and then are polymerized by proper catalysts. A water-miscible glycol-methacrylate has been developed for cytochemical studies. In this case the section can be submitted to selective extraction and to the action of various enzymes.[13]

To prepare the extremely thin sections, microtomes with specialized features have been developed. Several microtomes have been designed that have a thermal or a mechanical advance. With both types the thinnest sections that can be made are of the order of 20 nm. The limiting factors seem to be proper embedding and the sharpness of the cutting edge of the microtome. Glass and diamond knives are now in general use. Thin sectioning can be performed at low temperature with simple embedding in gelatin (see Sjöstrand, 1967).

Methods to Increase Contrast

One technique, called *"shadow casting,"* consists of placing the specimen in an evacuated chamber and evaporating, at an angle, a heavy metal such as chromium, palladium, platinum, or uranium from a filament of incandescent tungsten. The material is thus deposited on one side of the surface of the elevated particles; on the other side a shadow forms, the length of which permits determina-

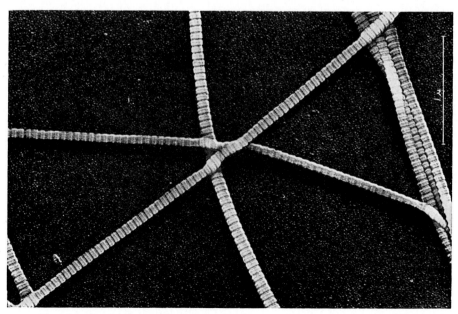

Figure 5–11. Electron micrograph of collagen fibers, shadowed with chromium, from human skin. Bands with a period of 64 nm are shown. ×28,000. (Courtesy of J. Gross.)

tion of the height of the particle. Photomicrographs made of such specimens have a three-dimensional appearance that is not found when other techniques are used (Fig. 5–11).

One of the most important and recent techniques in the study of viruses and macromolecules is *"negative staining."* The specimen is embedded in a droplet of a dense material, such as phosphotungstate, which penetrates into all the empty spaces between the macromolecules.[14] These spaces appear well defined in negative contrast. With this technique the numbers of protein molecules (capsomeres) of different viruses have been determined and interesting observations on cellular structures have been made. Negative staining may be applied to small globular proteins in the range of 10,000 to 40,000 daltons, but only the general shape and size may be determined.[15]

A positive increase in contrast in biological structures has been obtained by the use of substances containing heavy atoms, such as osmium tetroxide, uranyl, and lead ions, which, under certain conditions, act as *"electron stains."* These electron stains are comparable to histologic stains, in that they combine selectively with certain regions of the specimen. Throughout this book several examples of electron staining will be mentioned, some of which are rather selective and may have some cytochemical value.

Use of Tracers

Several biological processes may be studied by the use of appropriate tracers which are revealed by their electron opacity. For example, the uptake of macromolecules into the cells by pinocytosis or phagocytosis (see Chapter 20) or the transport across cellular barriers can be studied with various opaque particles.

The ideal tracer should be: (a) nontoxic and physiologically inert; (b) composed of uniform particles of known size; (c) preserved in situ during the processing of the tissue; and (d) composed of particles of a small size. Several colloidal substances, ranging in size from 1.0 to 10 nm, such as gold, mercuric sulfide, iron oxide, thorium dioxide, and colloidal lanthanum have been used.[16] Ferritin, as well as dextrans of various sizes and other polymers, are now being widely used.[17, 18] A variety of tracers are represented by enzymes that are able to produce a reaction that greatly enhances the contrast of biological specimens. Various *peroxidases* are demonstrated by their reaction with peroxide and 3,'3'-diaminobenzidine, for example (see Chapter 6).[19] One of the smallest tracers of this kind is the so-called *microperoxidase,* which has a molecular weight of only 1900 daltons.[20]

High Voltage Electron Microscopy

While most electron microscopes use accelerating voltages between 50 and 100 kilovolts, there are instruments now in operation that greatly surpass this voltage and reach 500 to 3000 kV. The design of these newer instruments is essentially similar to the older models, but the construction is much more massive to permit higher acceleration, greater magnetic excitation for the lenses, and the shielding needed to protect against x-radiation. While its main application is in metallurgy, the high-voltage electron microscope is being used increasingly in studies of biological material to examine very thick sections (up to 5 μm), with less radiation damage resulting from ionization and temperature effects. With this instrument there is the possibility of examining living cells. The practice of obtaining stereomicrographs by tilting the specimen also gives a greater amount of three-dimensional information.[21–23]

Scanning Electron Microscopy

A surface view of a specimen may be obtained by using the secondary electron emission that is ejected after the primary electron beam has interacted with the surface of a thick specimen (Fig. 5–12). Special preparative techniques, such as freeze-drying or freeze-substitution (Chapter 6) are needed to preserve the surface topography of the sample. In some cases, to increase the scattering power of the surface structures, infiltration with electron-contrasting chemicals or surface coating may be used. Frozen tissues may be examined with the scanning electron microscope, provided the specimen is maintained at low temperature (i.e., −120 to −150 C). The scanning electron microscope usually has a lower resolving power than the transmission electron microscope; however, with special electron emitters (i.e., lanthanum

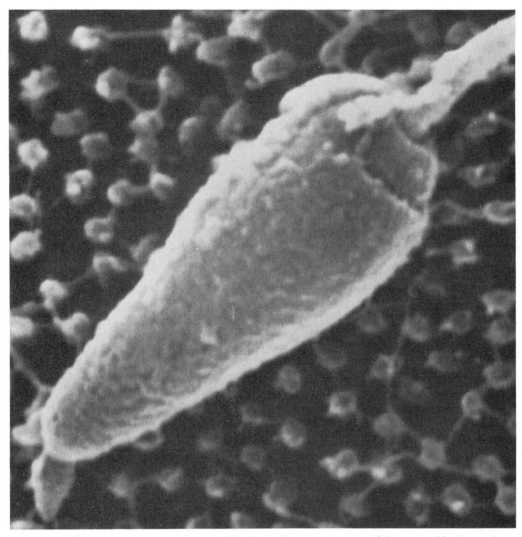

Figure 5–12. Scanning electron micrograph of the head of a spermatozoon of the sea urchin *Strongylocentrotus purpuratus,* attached to the vitelline surface of the egg by the acrosome. Observe that the egg surface has regularly spaced projections representing microvilli. (From Tegner, M., and Epel, D., *Science, 179*:685, 1973, Copyright 1973 by the American Association for the Advancement of Science.)

hexaborite) it is possible to obtain resolutions of the order of 3.0 nm. These brighter and more concentrated electron sources permit x-ray microanalysis of the secondary emission by which qualitative and quantitative estimates of the distribution of certain elements in the specimen are possible (see Crewe, 1971; Catley, 1972; Heywod, 1971; Echlen, 1971; Carr, 1971).

X-ray Diffraction

This technique is based on the diffraction of radiations when they encounter small obstacles. If a ray of white light (wavelength averaging 0.5 μm impinges upon a diffraction grating that has 1000 lines per millimeter (1 μm spacing), it will be diffracted and will show the various bands of the spectrum. If the wavelength of the light is known, the spacing can be calculated from the diffracted angles, and vice versa. This type of grating would be too wide for x-rays and no diffraction would be produced.

Laue suggested that gratings of much smaller dimensions, such as those found in natural crystals, would be necessary for the diffraction of x-rays. The atoms, ions, or molecules in crystals constitute a true lattice of molecular dimensions capable of diffracting radiations of this wavelength (see Figure 1–2). This technique has its widest application in the study of inorganic and organic crystals, in which it is possible to determine the precise spatial relationships between the constituent atoms. An analysis of the structure of complex organic molecules, such as proteins and nucleic acids, is much more difficult because of the great number of atoms involved in a single molecule and the irregularities in three-dimensional architecture that most of these large and complex molecules have (Fig. 5–13). However, as shown in Chapter 3, this configuration of molecules is so vitally important to the understanding of biological function that a great deal of work by Pauling, Perutz, Kendrew, Wilkins, and others has been carried out to elucidate it. The study of molecular structure, such as of hemoglobin, myoglobin, DNA, and collagen, has been of fundamental importance in the development of molecular biology.

In essence, the technique of x-ray diffraction employs a beam of collimated x-rays that traverse the material to be analyzed (e.g., crystal of hemoglobin, DNA, or collagen fiber); a photographic plate is placed beyond this to record the diffraction pattern.

A series of concentric spots or bands, caused by interference between the different diffracted rays, may appear on the plate. The distance between these spots and the center of the pattern depends upon the spaces between the regularly repeating units, or *periods of identity,* in the specimen that produced the diffraction—the smaller the angle of diffraction, the greater the distance between the repeating units; the sharper the spots, the more regular the spacing (Fig. 5–14).

A crystalline structure can be considered a three-dimensional lattice in which the atoms are regularly spaced along the three principal axes. The so-called *unit cell* is a solid parallelepiped that represents the minimal repeating unit within the crystal. In practice, it is simpler to think of the crystal as composed of sets of superimposed lattice planes such as indicated in Figure 5–15. In this diagram, *d* is the spacing of the diffracting planes and θ, the angle of incidence.

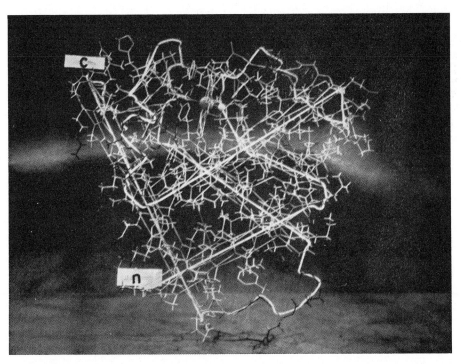

Figure 5–13. Model of the myoglobin molecule. The white cord represents the course of the polypeptide chain. The iron molecule is indicated by a grey sphere. The two terminals of the protein molecule are indicated by *c* and *n*. (From Kendrew, J. C., *Science, 139*:1259–1266, 1963, Copyright 1963 by the American Association for the Advancement of Science.)

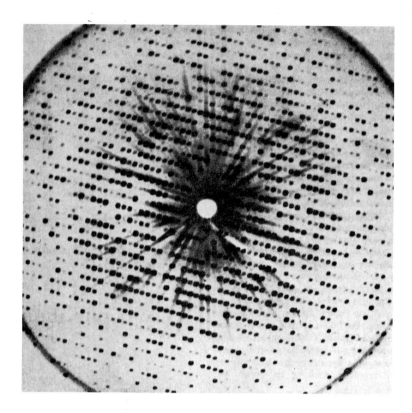

Figure 5-14. X-ray diffraction pattern of a myoglobin crystal. (From Kendrew, J. C., *Science, 139*:1259–1266, 1963, Copyright 1963 by the American Association for the Advancement of Science.)

According to Bragg's law, *d* can be calculated as follows (*n* is an integer corresponding to the diffraction order):

$$n\lambda = 2d \sin \theta \qquad (6)$$

Knowing the wavelength and the angle of incidence of a definite spot in the diffraction pattern, the spacing producing the diffraction can be calculated.

In oriented protein chains that characterize many biological fibers, the equatorial points are thought to indicate the lateral separation between the individual chains; the meridional points, in certain cases, represent the distance between the amino acid residues.

In the more refined methods of x-ray analysis of macromolecules, such as myoglobin, hemoglobin, and DNA, not only is the distance within the unit cell calculated, but also the *scattering power* of the individual atoms. This is similar to electron dispersion in that the scattering power is related to the atomic number. By introducing heavy atoms (such as mercury) into known points of the organic molecule, it is possible to increase their scattering power. The heavy atom serves as a landmark for the reconstruction of the

molecule. This is accomplished by a very complex process that involves plotting the electron density and the Fourier synthesis of all the component waves or diffraction orders. From this mathematical synthesis a three-

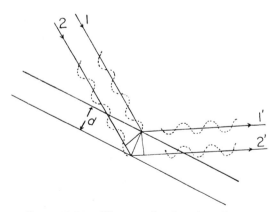

Figure 5-15. Diagram showing the effect of an x-ray beam incident upon two parallel planes in a crystal lattice that are separated by a distance, *d*. Incident rays *1* and *2* form an angle with these planes, and two secondary or diffracted rays are produced (*1'* and *2'*). By simple geometrical considerations, Bragg's law ($n\lambda = 2d \sin \theta$) can be deduced and the distance (*d*) calculated.

dimensional representation of the object can be constructed, and models of the entire molecule made (Fig. 5–13).

X-ray diffraction is one of the most important tools in molecular biology and ultrastructure because it permits the biologist to determine not only the orientation of the molecules, but also the exact distances that separate them and even to recognize their atomic organization (Table 1–1).

SUMMARY:
Microscopy

Observation of biological structures is difficult because of their small size and lack of contrast. Optical instruments are specially designed to overcome both difficulties. In the *light microscope* the limit of resolution: $Lm = \dfrac{0.61\lambda}{NA}$ (1) depends on the wavelength of the light and the numerical aperture. The resolving power is the inverse of the limit of resolution. The limit of resolution is, in general, 0.25 μm; with ultraviolet light it can be reduced to 0.1 μm.

Phase microscopy is used for the study of living cells which are, in general, transparent to light. The principle on which the phase microscope is based is that the light passing through an object undergoes a retardation, or phase change, which normally is not detected. In this instrument, however, the phase difference is advanced or retarded one fourth of the wavelength (λ), and the small variations in phase produced by the various structures are thereby made visible. In phase microscopy the phase changes are translated into changes in light intensity.

Interference microscopy is based on similar principles, but it provides the microscopist with quantitative information regarding the thickness, dry matter, and water content of the object. The *Nomarski interference contrast* microscope, in particular, gives extraordinary images of living cells, with a relief effect.

Darkfield microscopy, also called ultramicroscopy, is based on the scattering of light and uses a darkfield condenser. With this microscope, as well as with phase contrast optics, objects smaller than the wavelength of light can be *detected,* although not *resolved.* Theoretically, a fiber of only 5.0 nm could be detected, provided it had enough contrast.

Polarization microscopy uses polarized light. A birefringent material (also called *anisotropic*) has two different indexes of refraction in two perpendicular directions. Birefringence is related to retardation and to the thickness of the object: $B = N_e - N_o = \dfrac{\Gamma}{t}$ (4). In the polarizing microscope there is a polarizer and an analyzer (prisms or polaroid films) that are placed perpendicular to the specimen. The object is rotated 360 degrees, and if it is birefringent, it shows positions of maximum or minimum brightness. Most biological fibers show birefringence along the axis; a fiber of nucleic acid has negative birefringence. Crystalline, or intrinsic birefringence is independent of the refractive index of the medium, whereas form birefringence changes when the refractive index of the medium varies. In general, both types are present. *Dichroism* is a special type of birefringence in which there is a change in absorption of polarized light with a change in the orientation of the object.

Electron microscopy is the best method for studying biological ultrastructure. Electrons are emitted and accelerated in a vacuum tube, and the electron beam is deflected by electromagnetic coils acting as condenser, objective, and projector lenses. The image formed on the fluorescent screen depends on the dispersion of electrons (electron scattering) by the atomic nuclei present in the object. This dispersion depends on the thickness of the object, the molecular packing, and, in particular, on the atomic number. Most atoms in biological objects do not scatter the electrons; heavy atoms are used as "electron stains" to increase the contrast. The resolving power depends on the wavelength ($\lambda = 0.005$ nm) and numerical aperture, as in the light microscope. The resolution reached is 0.3 to 0.5 nm, and the final magnification can be 10^6 times, or more.

The preparative techniques are of fundamental importance in observation of biological material. Macromolecules such as DNA and RNA can be studied by the monolayer technique. Thick specimens can be studied via the technique of *thin sectioning,* in which the material is embedded in plastic and is cut with glass or diamond knives. The structure of membranes may be observed by freeze-fracturing and freeze-etching. The contrast can be increased by shadow casting, negative staining, or electron staining. Osmium tetroxide is used both as a fixative and an electron stain. Uranyl acetate and lead hydroxide are widely used for staining. Electron opaque substances, such as colloids and certain enzymes that give an opaque reaction, are used as tracers in electron microscopy for the study of several biological processes (e.g., pinocytosis). Thick sections and even living cells may be observed with high *voltage electron microscopy.*

Scanning electron microscopy gives a surface view of structures by secondary electrons reflected by them. Special techniques and microscopes are used for this purpose.

X-ray diffraction is a technique used in molecular biology, especially for the study of nucleic acid and protein structure. The method is based on the fact that a lattice of molecular dimensions (i.e., a crystal) produces a diffraction of x-rays. The most sophisticated techniques permit the determination of the complete three dimensional structure of proteins such as myoglobin, hemoglobin, and others.

REFERENCES

1. Barer, R. (1956) In: *Physical Techniques in Biological Research,* Vol. 3, p. 30. (Oster, G., and Pollister, A. W., eds.) Academic Press, Inc., New York.
2. Bennett, A. H., et al. (1951) *Phase Microscopy. Principles and Application.* John Wiley & Sons, New York.
3. Zernike, F. (1955) *Science, 121*:345.
4. Richards, O. W. (1954) *Science, 120*:631.
5. Engström, A., and Finean, J. B. (1958) *Biological Ultrastructure.* Academic Press, Inc., New York.
6. Hale, A. J. (1958) *The Interference Microscope in Biological Research.* E. & S. Livingstone, Edinburgh.
7. Mellors, R. C., ed. (1959) *Analytical Cytology.* 2nd Ed. McGraw-Hill Book Co., New York.
8. Davies, H. G., and Wilkins, M. H. F. (1952) *Nature (London), 189*:541.
9. Nomarski, G. (1955) *J. Phys. Radium (Paris), 16*:95.
10. Kleinschmidt, A. E., Lang, D., Yacherts, D., and Zahn, R. K. (1962) *Biophys. Biochim. Acta, 61*:857.
11. Granboulan, N. K., Scherrer, F. W., Jr., and Franklin, R. M. (1967) *Second International Symposium on Medical and Applied Virology,* p. 366. (Saunders, M., and Lennette, E. H., eds.) Warren H. Green, St. Louis.
12. Steere, R. L. (1957) J. Biophys. Biochem. Cytol., *3*:45.
13. Leduc, E. H., and Bernhard, W. J. (1967) *Ultrastruct. Res., 19*:196.
14. Brenner, S., and Horne, R. W. (1959) *Biochim. Biophys. Acta, 34*:103.
15. Mellerna, J. E., Van Bruggen, F. J., and Gruber, M. (1968) *J. Molec. Biol., 31*:75.
16. Revel, J. P., and Karnovsky, M. J. (1967) *J. Cell Biol., 33*:C7.
17. Farquhar, M. G., Wissing, S. L., and Palade, G. E. (1961) *J. Exp. Med., 113*:47.
18. Simionescu, N., Simionescu, M., and Palade, G. E. (1972) *J. Cell Biol., 53*:365.
19. Graham, R. C., and Karnovsky, M. J. (1965) *J. Histochem. and Cytochem., 14*:291.
20. Feder, N. J. (1971) *J. Cell Biol., 51*:339.
21. Hama, K., and Porter, K. R. (1969) *J. Microscopie, 8*:149.
22. Cosslett, V. E. (1969) *J. Rev. Biophys., 2*:95.
23. Favard, P., Ovtracht, L., and Carasso, N. (1971) *J. Microscopie 12*:301.
24. Tegner, M., and Epel, D. (1973) *Science, 179*:685.

ADDITIONAL READING

Branton, D. (1971) Freeze-etching studies of membrane structure. *Philos. Trans. R. Soc. London* [Biol. Sci.], *261*:133.
Carr, K. E. (1971) Applications of scanning electronmicroscopy in biology. *Int. Rev. Cytol., 30*:183.
Catley, C. W. (1972) *The Scanning Electron Microscope.* Cambridge University Press, New York.
Crewe, A. V. (1971) High resolution scanning microscopy of biological specimens. *Philos., Trans. R. Soc. London* [Biol. Sci.], *261* :61.
Du Pouy, G. (1973) Three-megavolt electron microscopy. *Endeavour* (London), *32*:66.
Echlen, P. (1971) The applications of scanning electron microscopy to biological research. *Philos. Trans. R. Soc. London* [Biol. Sci.], *261*:51.
Engström, A., and Finean, J. B. (1958) *Biological Ultrastructure.* Academic Press, Inc., New York.
Heywood, V. H. (1971) *Scanning Electron Microscopy: Systematic and Evolutionary Applications.* Systematics Assoc., Special Vol., No. 4. Academic Press, Inc., New York.
Hollenberg, M. J., and Erickson, A. M. (1973) The scanning electron microscope: potential usefulness to biologists. *J. Histochem. Cytochem.* 21:107.
Kendrew, J. C. (1963) Myoglobin and the structure of proteins. *Science, 139*:1259.
Kopac, M. J. (1959) Micrurgical studies on living cells. In: *The Cell,*

Vol. 1, p. 161. (Brachet, J., and Mirsky, A. E., eds.) Academic Press, Inc., New York.

Moor, H., (1971) Recent progress in the freeze-etching technique. *Philos. Trans. R. Soc. London* [Biol. Sci.] *261*:121.

Muhlethaler, K. (1971) Studies on freeze-etching of cell membranes. *Int. Rev. Cytol., 31*:1.

Oster, G. (1956) X-ray diffraction and scattering. In: *Physical Techniques in Biological Research,* Vol. 2, p. 441 (Oster, G., and Pollister, A. W., eds.) Academic Press, Inc., New York.

Padawer, J. (1968) The Nomarski interference-contrast microscope. *J. Roy. Microsc. Soc., 88*:305.

Park, R. B. (1972) Freeze-etching: A classical view. *Ann., New York Acad. Sci., 195*:262.

Pease, D. C. (1964) *Histological Techniques for Electron Microscopy,* 12th Ed. Academic Press, Inc., New York.

Sjöstrand, F. S. (1967) *Electron Microscopy of Cells and Tissues.* Academic Press, Inc., New York.

Weibel, E. R. (1972) The value of stereology in analysing structure and function of cells and organs. *J. Microsc.* (Oxford), *95*:3.

Wischnitzer, S. (1970) *Introduction to Electron Microscopy,* 2nd. Ed. Pergamon Press, Inc., New York.

six

METHODS FOR CYTOLOGIC AND CYTOCHEMICAL ANALYSIS

In cell biology many different types of specimens and techniques are used in the analysis of cytologic and chemical organization. In general, one or a few types of cells are best suited to each particular problem. Sometimes an entire branch of cytology has developed from the choice of a special material or the development of a certain technique. A few examples follow: The chromosomes and their behavior during mitosis and meiosis are best studied in the sex glands of insects, in meristems of roots and stems, or in the pollen mother cells of plants; mitochondrial movements are best studied in tissue cultures; cell permeability, in erythrocytes; pinocytosis, in amebae; and protein synthesis, in reticulocytes. Cytogenetics is best studied in *Drosophila*, Neurospora, and human chromosomes; molecular genetics, in bacteria and viruses.

As mentioned in the previous chapter, the two main procedures of instrumental analysis are: (1) direct observation of living cells, and (2) observation of killed cells that have been subjected to procedures that preserve morphology and composition, i.e., *fixation*.

Examination of Living Cells

Cell Culture

The culture of animal and plant cells outside the organism permits the observation of living cells under favorable conditions (see Fig. 9–3). Furthermore, a cell culture represents a much simpler experimental system than a whole animal, and provides a system that can be studied under carefully controlled conditions.

Since 1912, when Carrel first succeeded in growing tissue explants for many cell generations, considerable progress has been made in the techniques of cell culture. At the present time these techniques represent some of the most powerful methods for the study of fundamental problems in cell biology. In early days of tissue culturing the technique consisted of explanting small portions of different tissues (preferably embryonic) in a medium consisting of blood serum and embryo extract, plus saline solution. This system was extremely complex from the chemical viewpoint, and it was only after 1955 that the first chemically defined culture media became available.[1] At present, the nutritional requirements of eukaryotic cells are well known, and most cells can be grown, with the addition of a small percentage of serum, in synthetic media (see Paul, 1970).

Three main types of cultures can be distinguished: *primary, secondary,* and those using *established cell lines. Primary cultures* are those obtained directly from animal tissue. The organ is aseptically removed, cut into small fragments, and treated with trypsin. This proteolytic enzyme has the property of dissociating the cell aggregates into a suspension of single cells, without affecting viability. Thereafter, the cells are plated in sterile Petri dishes and grown in the appropriate culture medium. This culture can be trypsinized and replated in a fresh medium, resulting in a *secondary culture.*

The other major type of culture utilizes

101

established cell lines which have been adapted to prolonged growth in vitro. Among the best known cell lines are the HeLa cells, obtained from a human carcinoma, the L and 3T3 cells from mouse embryo, the BHK cells from baby hamster kidney, and the CHO cells from Chinese hamster ovary.

Normal mammalian cells do not survive indefinitely in culture and, after a variable time in vitro, they fail to divide and eventually die.[2] Occasionally, some cells will survive and will grow permanently in culture. These established cell lines differ from normal cells in many respects: they grow more tightly packed; they have lower serum requirements; and they are usually *heteroploids,* i.e., their chromosome number varies from one cell to another. Despite these abnormalities, established cell lines are very useful as model systems for the study of cancer (see Tooze, 1973).

One of the major advances in cell culture was the obtaining of *clones,* i.e., a population of cells derived from a single parent cell. After several days of incubation of cells plated at high dilutions, rounded colonies grow and adhere to the Petri dish. All the cells of this clone are derived from a single cell. If this colony is carefully trypsinized and replated, large numbers of cells may be obtained.[3]

Cell culture techniques have wide application in cell biology (see Pollack, 1973). In Chapter 20 some of the problems involved in the control of cell growth in tissue culture, particularly the so-called *contact inhibition* of movement and cell division when normal cells come into contact will be considered. It will be shown that in many cases normal cells are able to "communicate" with one another via special cell contacts (see Chapter 20). Most permanent cell lines do not display contact inhibition, and when they are injected into an animal they develop a cancer-like growth pattern. In several of these cell lines the cells do not communicate with one another (see Fig. 20–12).

Cell cultures are also used as an experimental system for the induction of cancer by virus. For example, normal fibroblasts infected in vitro with polyoma virus become *transformed* into cancer cells, and the growth becomes uncontrollable (see Chapter 20).

Another important field under rapid expansion is concerned with the *genetics of somatic cells* (see Chapter 15). Four main types of mutants have been obtained in cultured animal cells:[4] *drug-resistant mutants; auxotrophic mutants* (which require some non-essential nutrient for growth); *temperature-sensitive mutants;* and *natural mutations,* which may be obtained from patients with a genetically determined disease (see Table 19–2). These mutant cells may be recombined with other cell strains by the use of *cell fusion techniques* (see Harris, 1970). When two cell types are grown together, fusion may occasionally occur and may result in the production of *heterokaryons,* i.e., cells that contain two nuclei—one from each parent cell. The frequency of cell fusion may be increased 100-fold by the addition of the inactivated *Sendai virus* to the culture medium. When the two nuclei of the heterokaryon enter mitosis synchronously, a single metaphase plate is formed, and the daughter cells will contain chromosomes from both parent cells, thus giving rise to hybrid cells which will continue to divide (see Handmaker, 1973). With time, these cell hybrids may lose chromosomes from one or the other parent cell, and at the same time they may lose certain genetic characteristics. In a human mouse hybrid, for example, after several generations the human chromosomes are lost. Sometimes the loss of a certain genetic trait can be correlated with the elimination of a chromosome and this permits the mapping of a mutation in a specific chromosome (see Harris, 1970).

Microsurgery

This is another method that has contributed considerably to the knowledge of the living cell. Instruments such as micropipets, microneedles, microelectrodes, and microthermocouples are introduced into cells with the aid of a special apparatus that controls the movement of these instruments under the field of the microscope. Examples of microsurgical procedures are the dissection and extraction of parts of cells or tissues, the injection of substances, the measurement of electrical variables, and the grafting of parts from cell to cell (see Chapter 20).

Figure 16–3 shows an example of the application of microsurgery to the study of electrical potentials at the plasma and nuclear membranes.

Fixation

Fixation brings about the death of the cell in such a way that the structure of the living cell is preserved with the addition of a minimum number of artifacts. Some fixation methods, at the same time, are useful in maintaining the chemical composition of the cell as intact as possible.

The choice of a suitable fixative is dictated by the type of analysis desired. For example, for studying the nucleus and chromosomes, *acid fixatives* are frequently used. Acetone, formaldehyde, and glutaraldehyde, which produce minimal denaturation and preserve some enzyme systems, are used for the study of enzyme activity.[5-6]

The majority of fixatives act essentially upon the protein portion of the cell. A good fixative should be selected to precipitate protein in the finest way and, if possible, in ultramicroscopic aggregates, so that the appearance of the cell is not modified.

Some fixing agents, such as formaldehyde, glutaraldehyde, dichromate, and mercuric chloride produce strong cross linkages between protein molecules. For example, aldehydes react with the amino, carboxyl, and indole groups of a protein and then produce methylene bridges with other protein molecules. The two-step reaction is shown below. Glutaraldehyde has two aldehyde groups ($HOC—CH,—CH_2—CH_2—COH$) that can react with amino groups in two adjacent protein monomers.

Chromium salts (e.g., potassium dichromate) produce oxidation and chromium linkages between proteins. They also bind the phospholipids. Mercuric chloride acts on sulfhydryl, carboxyl, and amino groups of proteins, producing mercury linkages between molecules.

When a piece of tissue is immersed in a fixing liquid, cellular death does not occur instantaneously, and "postmortem" alterations due to anoxia, changes in the concentration of hydrogen ions, and enzymatic action (autolysis) may occur. The fixative penetrates the tissue by diffusion in such a way that the most external cells are fixed more rapidly and with fewer artifacts than the central cells. For this reason, every fixed tissue has a *gradient of fixation,* which depends upon the *penetrability* of the fixative and its progressive *dilution* with the liquid of the cells. The rate of fixative penetration

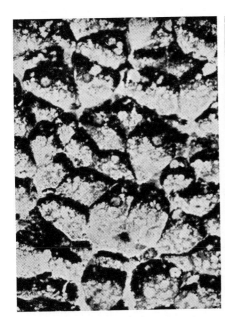 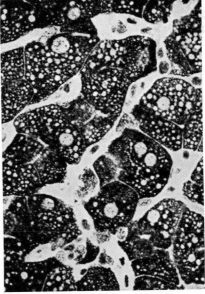

Figure 6-1. **A,** liver cells of *Ambystoma* fixed in Zenker-formol. The diffusion current produced by the chemical fixative (from the lower to the upper part of the figure) displaces the glycogen of the cell. **B,** liver cells of *Ambystoma* fixed by freezing-drying. The glycogen appears to be distributed homogeneously in the cytoplasm. Spheroid nuclei and lipid droplets are distinguishable. Stain: Best's carmine. (Courtesy of I. Gersh.)

$$-\overset{\displaystyle |}{\underset{\displaystyle H}{N}}-H \quad + \quad HCHO \quad \rightarrow \quad ----NH \cdot CH_2OH$$

amino group   formaldehyde   methylol

$$-NH \cdot CH_2OH \quad + \quad -\overset{\displaystyle |}{\underset{\displaystyle H}{N}}-H \quad \rightarrow \quad -NH-CH_2-HN + H_2O$$

methylol   amino group   methylene bridge

depends also on the protein barrier of precipitation produced at the periphery of the tissue. For example, with osmium tetroxide the precipitation is very fine, and a barrier preventing further passage of the fixative is produced. For this reason only very thin pieces (0.5 to 1.0 mm thick) are fixed in osmic liquids.

Diffusion currents that displace the soluble components, such as glycogen, may be observed (Fig. 6–1, A). Fixatives may also extract soluble substances, such as electrolytes, soluble carbohydrates, and even some lipids.

The preservation of a structure by fixation depends, to a great extent, on the degree of organization at the macromolecular level. In a well organized structure, such as a chromosome, a mitochondrion, or a chloroplast, a great number of interacting forces hold the molecules together, and the action of the fixative is insufficient to break structural relationships. However, less organized regions of the cell, such as the cytoplasmic matrix, are more difficult to preserve, and the production of fixation artifacts is more likely to occur.

Osmium Tetroxide

Osmium tetroxide (OsO_4) is one of the most frequently used fixatives for investigation of cell structure under the electron microscope. The reaction that this fixative has with lipids is probably due to double bonds that form unstable osmium esters, which decompose to deposit osmium oxides or hydroxides. The fixative causes proteins to gel initially, presenting a homogeneous structure under the electron microscope. This initial gelation may then be followed by further oxidation and solubilization of some products.

Based on studies on the binding of osmium tetroxide by different substances, it has been found that nucleic acids do not bind OsO_4. Osmium fixation has been improved by introducing buffer solutions at physiologic pH[7] maintaining osmotic pressure, adding calcium ions,[8] and maintaining a temperature of about 0° C.

Freeze-Drying

This method consists of rapid freezing of the tissues, followed by dehydration in a vacuum at a low temperature. The initial freezing is generally accomplished by plunging small pieces of tissue in a bath of liquid nitrogen cooled to a temperature of from −160 to −190° C. Fixation in liquid helium near absolute 0° (Kelvin) has also been used. The tissues are dried in a vacuum at −30 to −40° C. Under these conditions the water in the tissues is changed directly into a gas, and dehydration is achieved.

The advantages of this method are obvious. The tissue does not shrink; fixation is homogeneous throughout; soluble substances are not extracted; the chemical composition is maintained practically without change; and the structure, in general, is preserved with very few modifications produced by the ice crystals (Fig. 6–1, B). In addition, fixation takes place so rapidly that cell function can be arrested at critical moments, such as when kidney cells are excreting colored material.

The freeze-drying technique should be considered as intermediary between the examination of fresh and fixed tissues, since many of the cellular components are preserved in the same soluble form as in the living state. Since some cells can resist rapid freezing, this procedure is commonly used to keep them alive (e.g., spermatozoa).

Freeze-Substitution

In the freeze-substitution method, the tissue is rapidly frozen and then kept frozen at a low temperature (−20 to −60° C.) in a reagent that dissolves the ice crystals (e.g., ethanol, methanol, or acetone). The advantages of this method are somewhat similar to those just mentioned for fixation by freeze-drying.

Embedding and Sectioning

Tissues should be conveniently sectioned before they are observed under the microscope. For this purpose, *freezing microtomes,* cooled with liquid carbon dioxide, are frequently used. Instruments consisting of a microtome enclosed in a chamber at low temperature—the so-called *cryostat*—can make sections of fixed or fresh tissue for cytochemical purposes (Fig. 6–2).

For the most frequently used sectioning techniques the tissue is *embedded* in a material that imparts the proper consistency to the section. For sections to be observed under the light microscope, *paraffin* or *celloidin* is generally used. The fixed tissue is dehydrated and then penetrated by the embedding material. This requires a proper intermediary solvent (e.g., xylene or toluene for paraffin; ethanol-ether for celloidin). (See Chapter 5 for a discussion of embedding methods required for electron microscopy.)

Cytologic Staining

Most cytologic stains are solutions of organic aromatic dyes. Since the time of the pioneer work of Ehrlich, two types of

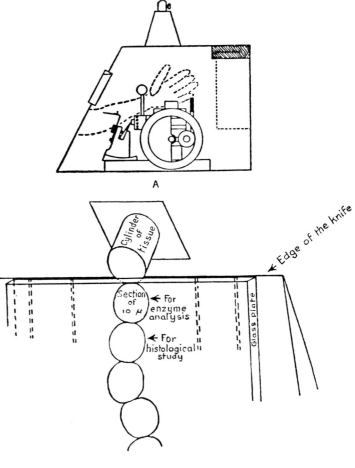

Figure 6–2. **A,** microtome for frozen sectioning. The apparatus is kept in a refrigerated container that keeps the tissue and the sections frozen. **B,** scheme of sectioning with the freezing microtome. The cylinder of tissue is sectioned and the sections are collected so that each alternate piece is used for enzymatic analysis and the others for histologic control. (Modified from Linderstrom-Lang.)

dyes have been recognized: basic and acid. In a basic dye the *chromophoric group,* which imparts the color, is basic (cationic). For example, methylene blue is a chlorhydrate of tetramethylthionine, in which the basic part carries the blue color. Eosin is generally used as potassium eosinate, in which the base is colorless. Sometimes the two components of the salt are chromophoric, e.g., eosinate of methylene blue. The most frequently used chromophores for acid dyes contain nitro ($-NO_2$) and quinoid ($O=\!\!<\!\!\square\!\!>\!\!=O$) groups. Basic chromophores contain azo ($-N\!\!=\!\!N-$) and indamin ($-N\!\!=$) groups. For example, picric acid has three nitro groups (chromophores) and one OH group, also called *auxochrome,* by which the dye combines with the tissue:

$$OH$$
$$NO_2 \quad \quad NO_2$$
$$NO_2$$

Mechanism of Staining

The properties that enable proteins, certain polysaccharides, and nucleic acids to ionize either as bases or acids should be noted (Chapter 3). Acid ionization may be produced by carboxyl ($-COOH$), hydroxyl ($-OH$), sulfuric ($-HSO_4$), or phosphoric ($-H_2PO_4$) groups. Basic ionization results from amino ($-NH_2$) and other basic groups in the protein. At pH values above the isoelectric point, acid groups become ionized; below the isoelectric point, basic groups dissociate (see Chapter 3). Because of this property, at a pH above the isoelectric point, proteins will react with basic dyes (e.g., methylene blue, crystal violet, or basic fuchsin) and below it, with acid dyes (e.g., orange G, eosin, or aniline blue). The intensity of staining with basic or acid dyes depends on the degree of acidity or alkalinity of the medium (Fig. 6–3). By measuring the amount of dye bound as a function of the pH of the medium, curves can be obtained that are typical of various proteins, nucleic acids, and mucopolysaccharides.

The net charge of nucleic acids is determined primarily by the dissociation of the phosphoric acid groups, and the isoelectric point is very low (pH 2 or less). For this reason, staining with basic dyes (e.g., toluidine blue or azure B) at low pH values is selective for nucleic acids. Toluidine blue is frequently used to stain ribonucleic acid, and its specificity can be demonstrated by previous hydrolysis with ribonuclease (Fig. 6–4).

Some histochemical methods based on these staining properties of proteins are now in use. One of the best known is the fast green method for histones.

Metachromasia

Some basic dyes of the thiazine group, particularly thionine, azure A, and toluidine blue, stain certain cell components a different color than the original color of the dye. This property, called *metachromasia,* has interesting histochemical and physicochemical

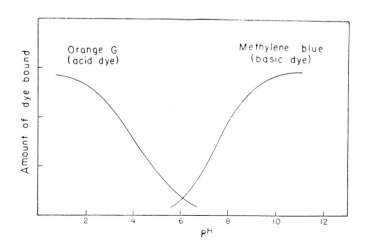

Figure 6–3. Curves indicating the amount of stain fixed by a protein at different pH values. Both the acid and the basic staining show a minimum fixation at the isoelectric point of the protein. (From Singer, M., and Morrison, P. R., *J. Biol. Chem., 175*:1, 133, 1948.)

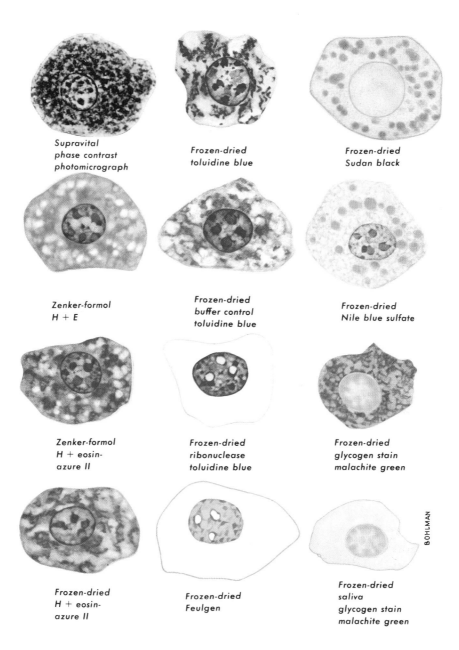

Supravital
phase contrast
photomicrograph

Frozen-dried
toluidine blue

Frozen-dried
Sudan black

Zenker-formol
H + E

Frozen-dried
buffer control
toluidine blue

Frozen-dried
Nile blue sulfate

Zenker-formol
H + eosin-
azure II

Frozen-dried
ribonuclease
toluidine blue

Frozen-dried
glycogen stain
malachite green

Frozen-dried
H + eosin-
azure II

Frozen-dried
Feulgen

Frozen-dried
saliva
glycogen stain
malachite green

BOHLMAN

Figure 6–4. Mouse liver cells fixed and stained by a variety of cytochemical procedures to show distribution of deoxyribonucleic acid, ribonucleic acid, glycogen and lipid droplets. Since these tests were all done on material fixed by freezing and drying, some of the sections are compared with similarly stained sections fixed by Zenker-formol. For further orientation, the fixed cell stained by hematoxylin and eosin and the unfixed cell photographed by phase contrast are also shown. ×1500. (Courtesy of I. Gersh, *in* Bloom, W., and Fawcett, D. W., *Textbook of Histology,* 9th ed., Philadelphia, W. B. Saunders Co., 1968.)

implications.[9] The reaction occurs in muco-polysaccharides and, to a lesser extent, in nucleic acids and some acid lipids. This reaction is strong in cells that contain sulfate groups (such as chondroitin sulfate), e.g., cartilage and connective tissue.

In mucus-secreting cells, basophilic leukocytes, and mast cells, the mucoproteins are not stained the normal color of the dye, but acquire a red-violet tint (metachromatic reaction). Some of the intercellular substances that take a similar stain are the matrix of cartilage, tendons, and cornea, and the gelatinous substance of the umbilical cord.

Some investigators believe that metachromasia depends on the formation of dimeric and polymeric molecular aggregates of dye on these high molecular weight compounds.[10] The same basic dyes do not form polymers when acting upon nucleic acid. It is believed that a distance of about 0.5 nm between the anionic groups is necessary for metachromatic staining.[11]

SUMMARY:
Observation of Living and Fixed Cells

Observation of cells can be made directly on living specimens or after cells have been fixed. *Tissue culture* consists of explanting a piece of tissue in a suitable medium (e.g., plasma, embryonic extract, or synthetic). The cells spread and divide, forming a zone of growth. Pure strains (i.e., clones) of cells may be obtained after isolation of cells by trypsin. Organ as well as tissue cultures can also be grown. *Microsurgery* is a technique that may give information about the physicochemical properties of living cells. Through this method, transplants of subcellular parts—including the nucleus—can be made.

Fixation is brought about by chemicals which preserve cell structure. For nuclei and chromosomes, acid fixatives are preferred. To preserve the activity of certain enzymes, acetone, formaldehyde, and glutaraldehyde are used. Aldehydes react with amino groups of proteins, forming methylene bridges). Glutaraldehyde has two aldehyde groups that can react with adjacent protein monomers. Fixatives penetrate by diffusion and produce a gradient of fixation. The preservation of cell structure depends on the tightness of macromolecular organization. Osmium tetroxide and glutaraldehyde are widely used in fixation for electron microscopy.

Freeze-drying consists of the rapid freezing of tissue in liquid nitrogen (−160 C to −190° C), followed by dehydration at −30 to −40° C. The water is sublimed in a vacuum. In *freeze-substitution* the water of the frozen tissue is dissolved with alcohols or acetone. *Embedding* in paraffin or celloidin is used for light microscopy; special plastics are used for electron microscopy. Microtomes and ultramicrotomes are then used for *sectioning* the embedded tissues.

Cytologic stains may be basic (e.g., methylene blue) or acid (e.g., eosin). Chromophores for acid dyes are nitro and quinoid groups; for basic dyes they are azo and indamin groups. The mechanism of staining is based in the ionization of acid groups (carboxyl, hydroxyl, sulfuric, phosphoric) or basic groups (amino) in proteins, polysaccharides, and nucleic acids. The staining of a protein is minimal at the isoelectric point; it stains with acid dyes below, and basic dyes above, the isoelectric point (Fig. 6–3)). Nucleic acids are stained selectively at low pH because the isoelectric point is at pH 2.

Metachromasia, the property by which a basic dye has a different color in tissue than in solution, is found in mucopolysaccharides, particularly those containing sulfate groups. This particular characteristic of the dye is probably due to dimerization or polymerization.

HISTOCHEMISTRY AND CYTOCHEMISTRY

The immediate goal of *cytochemistry* is the identification and localization of the chemical components of the cell. As R. R. Bensley, one of the founders of modern cytology, once said: The aim of cytochemistry is the outlining "within the exiguous confines of the cell of that elusive and mysterious chemical pattern which is the basis of life." This aim is quantitative as well as qualitative; and once achieved, the next step is to study the dynamic changes in cytochemical organization taking place in different functional stages. In this way it is possible to discover the role of different cellular components in the metabolic processes of the cell.

Cytochemistry is included within the more general subject of *histochemistry,* which deals with the chemical characterization and localization of substances or groups of substances in the cells and intercellular materials of a tissue.[13-14.]

Modern cytochemistry has followed three main methodological approaches. Of these, only one can be considered strictly *microscopic,* because it comprises a series of chemical and physical methods used to detect or measure different chemical components within the cell.

The other two methods are more closely related to biochemistry and microchemistry, since they involve the development of techniques for the assay of small quantities of material (Table 1-2). One method relies on conventional *biochemical* techniques for the isolation and investigation of subcellular fractions. The other applies the techniques of *microchemistry* and *ultramicrochemistry* to the study of minute quantities of material, or even parts of a cell.

Cell Fractionation Methods

These methods involve, essentially, the homogenization or destruction of cell boundaries by different mechanical or chemical procedures, followed by the separation of the subcellular fractions according to mass, surface, and specific gravity. The various cell fractions are then analyzed by biochemical or microchemical methods.

Results of these analytical techniques are of particular interest when based on accurate and precise cytologic analysis of the cell fraction, preferably by electron microscopy.

Many different methods of cell fractionation are in use. Most of them are based on the homogenization of the cell in aqueous media—usually sucrose solutions in various concentrations (Fig. 6–5). Cell fractionation can also be carried out in nonpolar media, as in Behrens' method for the separation of nuclei. This method has the advantage of reducing the loss of soluble substances, such as proteins and certain enzymes.

A standard cell fractionation procedure is shown diagrammatically in Figure 6–5. The liver of an animal is first perfused with an ice-cold saline solution, followed by cold 0.25 M sucrose. The tissue is then forced through a perforated steel disk and homogenized in 0.25 M sucrose. This classical type of cell fractionation is directed toward the subdivision of the cell components into four morphologically distinct fractions (nuclear, mitochondrial, microsomal, and soluble). In some glandular tissue a fifth fraction containing secretory granules may be obtained.

Note carefully that it is necessary to differentiate between *cell fractions* and the parts of the cell (*organelles*) contained in the

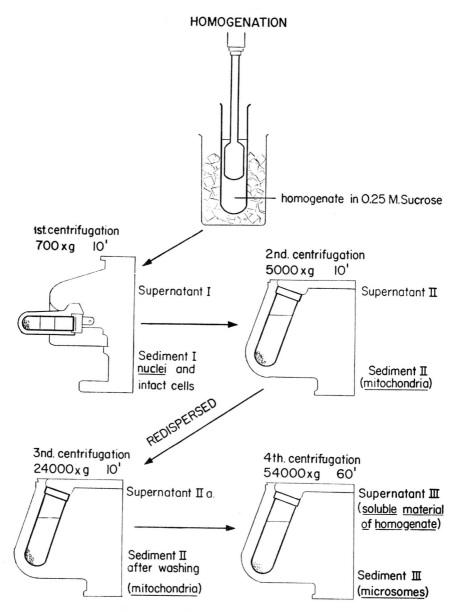

Figure 6–5. Diagram of a cell fractionation.

fraction. For example, the mitochondrial fraction of the liver is composed principally of mitochondria, but the "mitochondrial fraction" of the brain is very heterogeneous and contains nerve endings and myelin in addition to free mitochondria.[15] "Microsomes" do not exist, as such, in the cell; the "microsomal" fraction is composed mainly of broken parts of the endoplasmic reticulum, including the ribosomes, the Golgi complex, and other membranes (see Chapter 9).

Differential Centrifugation

In the example given above, the method used to separate the subcellular particles is called *differential centrifugation*. Depending on the strength of the centrifugal field needed, *standard centrifuges* or *preparative ultracentrifuges* are used. The effect of the centrifugal field on particles of different sizes is indicated in Figure 6–6. Initially, all the particles are distributed homogenously (A); as centrifuga-

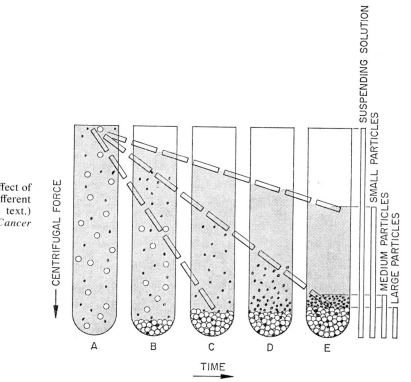

Figure 6–6. Diagram of the effect of a centrifugal field on particles of different sizes. (See the description in the text.) (From Anderson, N. G., *Natl. Cancer Inst. Monogr.* 21, June, 1966.)

tion proceeds, (B) → (E), the particles settle according to their respective sedimentation ratios. Complete sedimentation of the larger particles is achieved in (C), while in (E), the medium-sized particles have settled. (The distribution of the particles at the end of stage (E) is shown by the bars at the right; some cross-contamination of particles of different sizes is observed.) The time required for each particle to come to rest at the bottom of the tube depends on its size and density. A similar principle may be used for separation of much smaller particles, such as viruses or macromolecules (i.e., nucleic acids and proteins), with the *analytical ultracentrifuge*. This device has a transparent window in the tube, and with suitable optical and electronic techniques the moving particle boundaries can be visualized and the *sedimentation coefficient* can be determined. This coefficient, expressed in Svedberg (S) units, is related to the molecular weight of the particle (for example, transfer RNA, with 4S, has a MW of 25,000 daltons).

Gradient Centrifugation

Improvement in the technique of differential centrifugation may be achieved by using a density gradient, which may be either *discontinuous* or *continuous*. If it is discontinuous, the centrifuge tube is loaded with steps of varying densities (for example, sucrose varying in molarity between 1.6 and 0.5 from the bottom). However, mixing, at varying rates, of two concentrations of sucrose produces a continuous gradient. Once the gradient is formed, the material is layered on the top and centrifuged until the particles reach equilibrium with the gradient. For this reason, this type of separation is also called *isopycnic* (equal density) centrifugation[16] (see Fig. 24–17).

Improvements in this type of fractionation technique include the use of heavy water, cesium chloride, and media with different partition coefficients.[17]

To avoid drastic changes in osmotic pressure, macromolecular media such as gly-

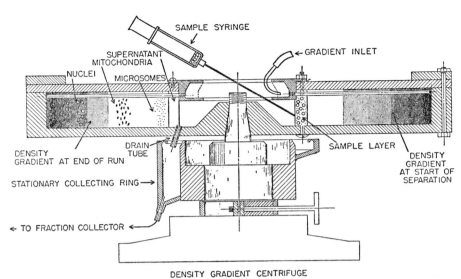

Figure 6-7. Schematic section through a zonal rotor. (See the description in the text.) (From Anderson, N. G., *Natl. Cancer Inst. Monogr. 21*, June, 1966.)

cogen and Ficoll are used. In these cases the gradient is based on differences in viscosity of the medium.

Zonal Centrifugation

An important development has been achieved with the use of the so-called *zonal rotors*. As shown in Figure 6-7, the density gradient is formed while the rotor is spinning; then the sample is layered and centrifuged until the isopycnic zonal layering of the particles is reached. At this moment, an injection of a denser sucrose solution pushes the layers toward the center where they are collected in tubes of a fraction collector (see Anderson, 1966).

Buoyant Density. Isopycnic centrifugation in preparative or zonal rotors permits the determination of the buoyant density of a macromolecule, i.e., the density at which it will reach an equilibrium with the suspending medium. This is important in studies of the molecular biology of nucleic acids (Fig. 6-8). In Chapters 17, 18, and 19 several examples will be given of DNAs with different buoyant densities.

Microchemistry and Ultramicrochemistry

In recent years many ingenious methods have been devised for the quantitative analy-

sis of extremely small quantities of substances.[18-20] For example, if freshly frozen tissue is sectioned in a *cryostat* (Fig. 6-2), some sections can be weighed on a balance made of a fine *quartz fiber*, and enzymatic determinations can be carried out using ultramicropipets and burets and microcolorimetric or microspectrophotometric methods. Of even higher sensitivity are *microfluori-*

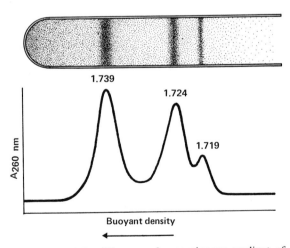

Figure 6-8. Diagram of a continuous gradient of cesium chloride (CsCl) showing the position of three bands with the corresponding buoyant density expressed in grams per cm². Since these bands correspond to nucleic acid molecules the concentration of the molecules is measured by absorbancy at 260 nm ($A_{260\,nm}$).

metric methods, which can be used to determine different enzymes and coenzymes.

Also of considerable interest are *micromanometric* methods. One of these employs the Cartesian diver microrespirometer, which is 1000 times more sensitive than the classic Warburg manometer. With this instrument the oxygen consumption of a single sea urchin egg can be measured during short intervals. Based on similar principles is the *Cartesian diver balance* of Zeuthen,[21] by which a single ameba can be weighed with great accuracy.[22]

By *microchromatographic* and *microelectrophoretic* methods the ribonucleic acid content of a single nerve cell has been determined.[23]

Cytochemical and Histochemical Staining Methods

For the cytochemical and histochemical determination of a substance certain conditions must be fulfilled: (A) The substance must be immobilized at its original location. (B) The substance must be identified by a procedure that is specific for it, or for the chemical group to which it belongs. This identification can be made by: (1) chemical reactions similar to those used in analytical chemistry, but adapted to tissues, (2) reactions that are specific for certain groups of substances, and (3) physical methods.

To demonstrate proteins, nucleic acids, polysaccharides, and lipids, some chromogenic agents that bind selectively to some specific groups of these substances may be used. Only a few cytochemical stainings that are widely used are mentioned here:

Detection of Proteins

Millon Reaction. A nitrous-mercuric reagent applied to the tissue reacts with the tyrosine groups present in the side-chains of the protein, forming a red precipitate.

Diazonium Reaction. The chromogenic agent, a diazonium hydroxide, reacts with tyrosine, tryptophan, and histidine groups, forming a colored complex.

Detection of —SH Groups. Certain reagents bind —SH by a mercaptide covalent linkage. A red sulfhydryl reagent, 1-(4-chloromercuri-phenylazo)-naphthol-2, was first used for this purpose. The —SH content

of the cell can be measured quantitatively by photometric analysis of tissues stained by this technique.[24] Other methods for detection of —SH groups are also widely used.[25–26]

Detection of Arginine. The Sakaguchi test for arginine has been introduced in histochemistry. A reddish color is produced when the tissue sections are treated with an alkaline mixture of α-naphthol and sodium hypochlorite. A high concentration of arginine in a tissue is indicative of basic proteins, such as histones.

Schiff's Reagent for Aldehydes

Deoxyribonucleic acid, certain carbohydrates, and lipids can be demonstrated with a single reagent for aldehyde groups, the so-called *Schiff's reagent*. This reagent is made by treating basic fuchsin, which contains parafuchsin (triaminotriphenyl-methane chloride), with sulfurous acid. Parafuchsin is converted into the colorless compound bis-N aminosulfonic acid (Schiff's reagent), which is then "recolored" by the aldehyde groups present in the tissue (Fig. 6–9).

In the histochemical tests involving Schiff's reagent, three types of aldehydes may be involved: (1) *free aldehydes,* which are naturally present in the tissue, such as those giving the plasmal reaction; (2) *aldehydes produced by selective oxidation* (which give the PAS reaction); and (3) aldehydes produced *by selective hydrolysis* (which give the Feulgen reaction).

Detection of Nucleic Acids

Cytochemical staining methods for nucleic acids depend on the properties of the three components of the nucleotide (phosphoric acid, carbohydrate, and purine and pyrimidine bases [Chap. 3]).

Both DNA and RNA absorb ultraviolet light at 260 nm, because of the presence of nitrogenous bases. The deoxyribose present in DNA is responsible for the Feulgen reaction, which is specific for this type of nucleic acid (Table 6–1).

The phosphoric acid residue is responsible for the basophilic properties of both DNA and RNA. Among the basic stains, azure B gives a specific reaction with DNA and RNA. Another stain for DNA, based on the use of methyl green, also depends on the phosphoric acid residues.

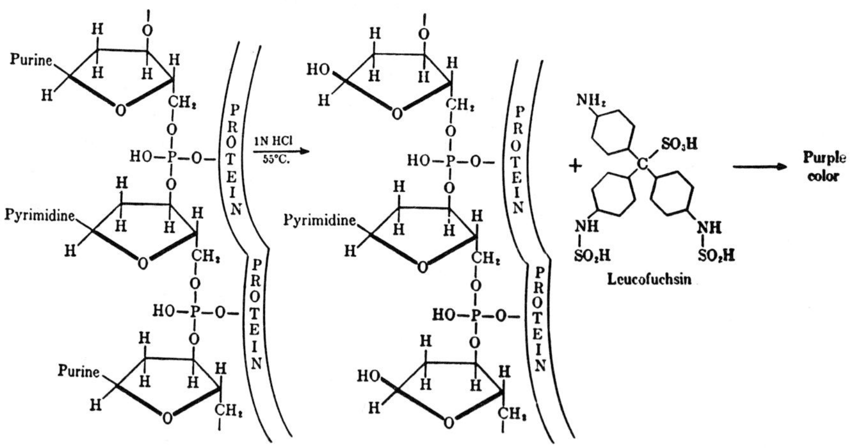

Figure 6–9. Chemistry of the Feulgen reaction. Acid hydrolysis removes the purines and liberates the aldehyde groups, which react with leucofuchsin (Schiff's reagent), resulting in a purple color. In the diagram the size of deoxypentose is greatly exaggerated in relation to the protein. (From Lessler, M. A., *Internat. Rev. Cytol.*, 2:231, 1953.)

Feulgen Reaction. DNA can be studied by means of the *nucleal reaction*, a technique developed in 1924 by Feulgen and Rossenbeck. Sections of fixed tissue are first submitted to a mild acid hydrolysis and then treated with Schiff's aldehyde reagent. This hydrolysis is sufficient to remove RNA, but not DNA. The reaction takes place in the following stages: (1) The acid hydrolysis removes the purines at the level of the purine-deoxyribose glucosidic bond of DNA, thus unmasking the aldehyde groups of deoxyribose. (2) The free aldehyde groups react with Schiff's reagent (Fig. 6–9).[27] The specificity of the reaction can be confirmed by treating the sections with deoxyribonuclease, which removes DNA.[28]

The Feulgen reaction is positive in the nucleus and negative in the cytoplasm (Fig. 6–4). In the nucleus, the masses of condensed chromatin (i.e., heterochromatin) are intensely positive; the nucleolus is Feulgennegative (Fig. 6–10).

Periodic Acid–Schiff (PAS) Reaction

McManus[29] devised a reaction based on the oxidation with periodic acid of the "1,2-glycol" group of polysaccharides with libera-

TABLE 6–1.　Some Specific Reactions Used in Cytophotometric Analysis*

Substance Tested For	Reaction or Test	Maximum Absorption (wavelength in nm)
Total nucleotides	Natural absorption of purines and pyrimidines	260
Soluble nucleotides	Natural absorption of purines and pyrimidines	260
Ribonucleic acid (RNA)	Natural absorption of purines and pyrimidines	260
Deoxyribonucleic acid (DNA)	Natural absorption of purines and pyrimidines	260
Deoxyribonucleic acid (DNA)	Feulgen nucleal reaction for deoxyribose	550 to 575
Deoxyribonucleic acid (DNA)	Methyl green	645
Nucleic acids (phosphoric acid groups)	Azure A	590 to 625
Protein (free basic groups)	Fast green	630
Protein (tyrosine)	Millon reaction	355
Polysaccharides with 1,2-glycol groupings	Periodic acid-Schiff reaction (PAS)	~550

*From Moses, M. J. (1952) *Exp. Cell Res.*, suppl. 2:76.

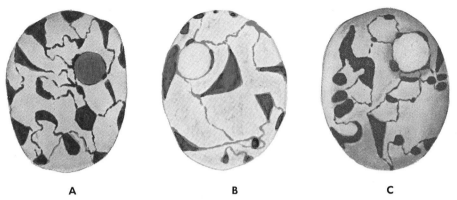

Figure 6–10. Interphase nuclei of pancreatic cells fixed by freeze-drying. **A,** Azan staining: the nucleolus (in red), the chromonemic filaments with their enlarged portions (chromocenters) and the nuclear sap are visible. **B,** Feulgen reaction: the nucleolus gives a negative reaction; in the nuclear sap the reaction is slightly positive. **C,** action of ribonuclease and staining with Azan. The nucleolus does not stain, owing to the digestion of the ribonucleic acid. (From De Robertis, Montes de Oca, and Raffaele, *Rev. Soc. Arg. Anat. Normal Patol.,* 1945.)

tion of aldehyde groups, which give a positive Schiff reaction (Fig. 6–11). This test is done on plant cells for starch, cellulose, hemicellulose, and pectins; and on animal cells for mucin, mucoproteins, hyaluronic acid, and chitin.

Since the PAS reaction is given by a number of substances, different tests can be applied to improve its specificity. For example, enzymes, such as amylase, can be used to remove glycogen (Fig. 6–4), or hyaluronidase can be used to remove hyaluronic acid. Similarly, some extraction or blocking procedures serve this purpose.

Detection of Lipids

Fat droplets can be demonstrated with osmium tetroxide, which stains them black by reacting with unsaturated fatty acids. Staining with Sudan III or Sudan IV (scarlet red) has a greater histochemical value. These stains act by a simple process of diffusion and solubility and accumulate in the interior of the lipid droplets. Sudan black B has the advantage of being dissolved also in phospholipids

and cholesterol, and of producing greater contrast (Fig. 6–4). Nile blue sulfate is used to detect acidic lipids, which include fatty acids and phospholipids.

Plasmal Reaction. Long-chain aliphatic aldehydes occurring in plasmalogens give the so-called *plasmal reaction* upon direct treatment of the tissue with Schiff's reagent. Since the substances giving the plasmal reaction are soluble in organic solvents, the tissue is not embedded in the usual way but is studied in frozen sections. The compounds are free aldehydes such as *palmitaldehyde,* CH_3 $(CH_2)_{14}CHO$, and *stearaldehyde,* CH_3-$(CH_2)_{16}CHO$, corresponding to palmitic and stearic acids, respectively, which together constitute the so-called *plasmal.*

Detection of Enzymes

Because of the inactivating action of most fixatives, special preparation of tissues for enzyme chemistry is necessary. To detect some enzymes, unfixed frozen sections are made in a cryostat; in other cases, the enzyme resists a brief fixation in cold acetone,

Figure 6–11. Chemical diagram of a polysaccharide, showing the action site of periodic acid in the PAS reaction of McManus. The resulting aldehydes react with Schiff's reagent.

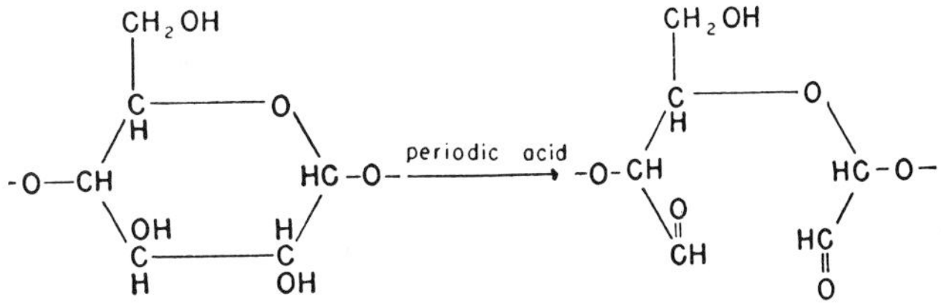

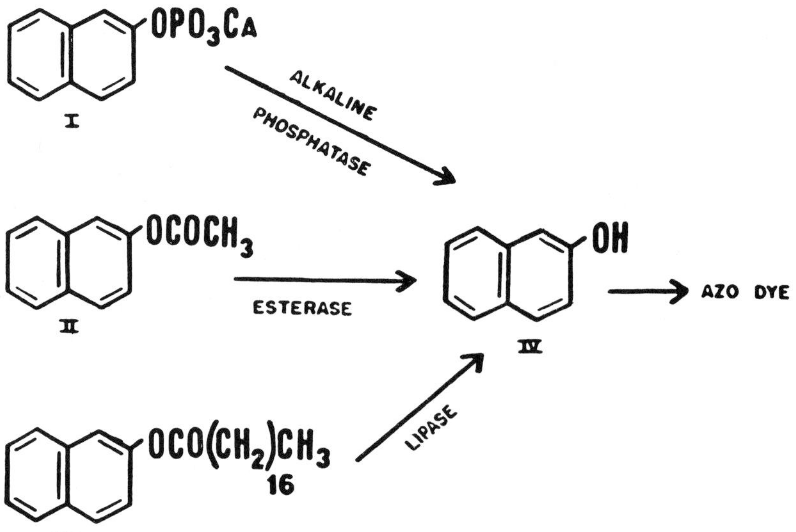

Figure 6-12. Diagram showing the cytochemical steps in the methods used to demonstrate hydrolytic enzymes. (See the description in the text.) (From Nachlas, M. M., and Seligman, A. M., *J. Natl. Cancer Inst.*, 9:415, 1949.)

formaldehyde, glutaraldehyde, and other dialdehydes.[5]

Techniques for identifying and localizing enzymes are based on the incubation of the tissue sections with an appropriate substrate. For example, in the Gomori method for detecting alkaline phosphatase, phosphoric esters of glycerol are used as the substrate.[30] The phosphate ion liberated by hydrolysis is converted into an insoluble metal salt (generally in the presence of Ca^{++}), and the metal, in turn, is visualized by conversion into metallic silver, lead sulfide, cobalt sulfide, or other colored compounds. In another method[31] a phosphoric ester of β-naphthol is used as the substrate. The hydrolysis liberates β-naphthol, which, in the presence of a diazonium salt, couples immediately, giving a colored azo component at the site of enzymatic activity (Fig. 6-12). Other hydrolytic enzymes, such as esterase, lipase, acid phosphatase, sulfatase, and β-glucuronidase, can be detected with this method by changing the conditions and substrate.

Phosphatases. Phosphatases are enzymes that liberate phosphoric acid from many different substrates. A number of phosphatases are known, and they differ with respect to substrate specificity, optimum pH, and the action of inactivators and inhibitors. The best known are the *phosphomono-*

esterases, which hydrolyze simple esters held by P—O bonds, and the *phosphamidases,* which hydrolyze P—N bonds. Table 6-2 indicates some of the most common enzymes studied cytochemically and some of the substrates used. An example of the alkaline phosphatase reaction is shown in Figure 6-13.

Esterases. Esterases are enzymes that

TABLE 6-2. Some Phosphatases Studied Cytochemically

Type	*Substrate*
Phosphomonoesterases	
Alkaline phosphatase	α- or β-glycerophosphate
Acid phosphatase	Naphthylphosphate
Adenosine triphosphatase (ATPase)	Adenosine triphosphate
5-Nucleotidase	5-Adenylic acid
Phosphamidase	Phosphocreatine Naphthyl phosphoric acid diamines
Glucose-6-phosphatase	Glucose-6-phosphate
Thiamine pyrophosphatase	Thiamine pyrophosphate
Pyrophosphatase	Sodium pyrophosphate Dinaphthyl pyrophosphate
Phosphodiesterases	
Ribonuclease	Ribonucleic acid (RNA)
Deoxyribonuclease	Deoxyribonucleic acid (DNA)

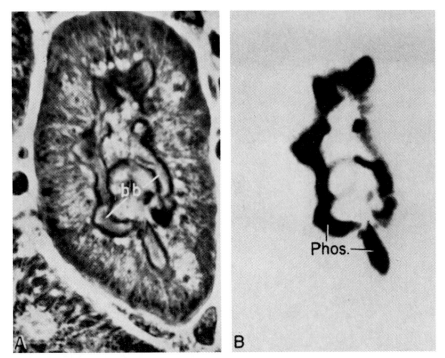

Figure 6–13. Proximal convoluted tubule from a mouse kidney after freezing-substitution. **A,** observation in phase contrast with a medium of refractive index n = 1460; *bb,* brush border. **B,** observation with transmitted light; *Phos.,* alkaline phosphatase reaction. (Courtesy of B. J. Davies, and L. Ornstein.)

catalyze the following reversible reaction:

$$—COOR + HOH \leftrightharpoons R—COOH + R'OH$$

Esterases may be divided into *simple esterases (aliesterases),* which hydrolyze short chain aliphatic esters; *lipases,* which attack esters with long carbon chains; and *cholinesterases,* which act on esters of choline. Cholinesterases are usually subdivided into "true" (specific), which hydrolyze acetylcholine, and "pseudo" (nonspecific), which act on other choline esters.[32]

Other *hydrolytic enzymes* studied cytochemically are β-D-glucuronidase, β-D-galactosidase, aryl sulfatase, and aminopeptidase. Figure 6–14, B shows an example of the reaction of aminopeptidase, a proteolytic enzyme that can attack peptide bonds adjacent to a terminal α-amino group.

Oxidases. Oxidases are enzymes that catalyze the transfer of electrons from a donor substrate to oxygen (Chap. 4). They usually contain iron, e.g., peroxidase and catalase, or copper, e.g., tyrosinase and polyphenol oxidase. Another enzyme in this series

is monoamine oxidase, which is involved in the metabolism of indole and catecholamines.

Colorless substrates, such as benzidine, are used to detect peroxidases. These substrates are transformed into stained dyes by H_2O_2 in the presence of the enzyme. *Cytochrome oxidase* gives the so-called *Nadi* reaction. It oxidizes the Nadi reagent, a mixture of α-naphthol and dimethyl paraphenylene diamine (Fig. 6–15). The reagent 3',3'-diaminobenzidine (DAB) has allowed the study of peroxidase[33] and cytochrome oxidase at the electron microscope level.[34] It is postulated that oxidized DAB is polymerized into a macromolecule that reacts with osmium tetroxide (see Figure 10–10).

Dehydrogenases. Most oxidation reactions that are catalyzed by enzymes are dehydrogenations, i.e., transfer of electrons from the substrate (proton donor) to the oxidizing agent or electron acceptor (see Chapter 4).

The pyridine nucleotide-linked dehydrogenases require the coenzymes NAD^+ or $NADP^+$. Among the best known NAD^+ enzymes are lactic acid dehydrogenase, which

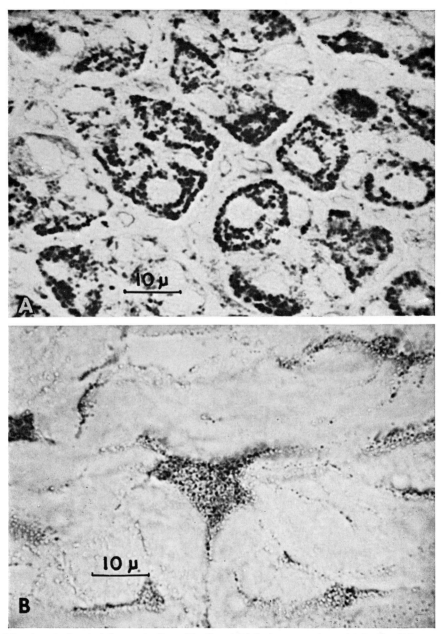

Figure 6–14. **A,** cytochemical demonstration of lactic acid dehydrogenase in parietal cells of the rat stomach. The insoluble formazan produced by the reaction delineates the mitochondria. **B,** amino-peptidase activity in fibroblasts of the rat dermis. ×1500. (Courtesy of B. Monis.)

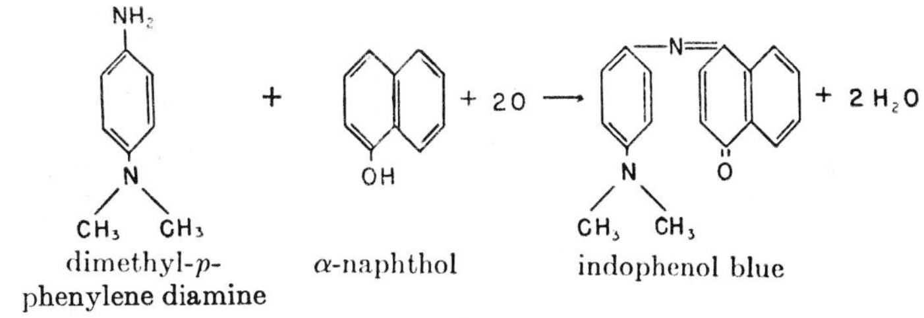

Figure 6-15. Nadi reaction for cytochrome oxidase.

dimethyl-*p*-phenylene diamine   α-naphthol   indophenol blue

converts lactic acid into pyruvic acid, and malic acid dehydrogenase, which converts malic acid into oxaloacetic acid. Among the $NADP^+$ enzymes are isocitric acid dehydrogenase and the malic enzyme (malate → pyruvate + CO_2). Tellurite, triazole, and tetrazolium have been used as electron acceptors, but of these only triazole and tetrazolium, which produce an insoluble chromogenic formazan dye, have been successful from a histochemical standpoint.

Figure 6–16 represents the mechanism of these histochemical reactions, and Figure 6–14 *A* illustrates the detection of lactic acid dehydrogenase in parietal cells of the stomach.

Histochemical Methods Based on Physical Determinations

Cytophotometric Methods

Several cell components display a specificity in the way in which they absorb ultraviolet light. For example, the absorption range of nucleic acids is about 260 nm, whereas that of proteins is 280 nm (Fig. 6-17). Some histochemical staining reactions give specific absorption in the visible spectrum and can be analyzed quantitatively with instruments called *cytophotometers*.

A typical apparatus for absorption cytophotometry is represented in Figure 6–18. By changing the light source and the optical system, this instrument can be used for either the ultraviolet or the visible spectrum. The absorption is measured directly by means of a photomultiplier or by densitometry on calibrated photographic plates. Table 6–1 indicates some of the histochemical reactions that can be analyzed by cytophotometric methods.

The specific ultraviolet absorption of nucleic acids is due to the presence of purine and pyrimidine bases, and, for this reason, is the same in DNA, RNA, and nucleotides (Table 6–1). By ultraviolet cytophotometry the two types of nucleic acids can be localized, but not distinguished (Fig. 6–17). On the other hand, the Feulgen reaction shows the presence of DNA (Fig. 6–4). This reaction can be adapted to quantitative determinations of DNA in tissue sections. Monochromatic light of 550 nm, corresponding to the maximum absorption of this stain, is used for this purpose.[27, 28] The Millon reaction can be used to determine the protein content (Table 6–1).

Figure 6–16. Schematic representation of the transfer of electrons to tetrazolium salt. *FAD*, flavin adenine dinucleotide; *FMN*, flavin mononucleotide; *NAD*+, nicotinamide dinucleotide; *NADP*+, nicotinamide triphosphate.

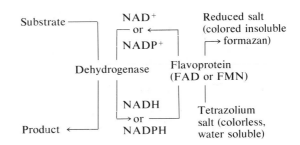

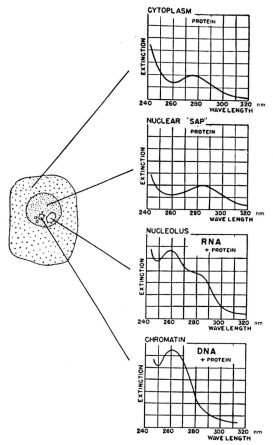

Figure 6–17. Ultraviolet absorption spectrum of the cytoplasm, nuclear "sap," nucleolus and chromatin. The extinction at different wavelengths is indicated. (Courtesy of T. Caspersson.)

Fluorescence Microscopy

In this method tissue sections are examined under ultraviolet light, near the visible spectrum, and the components are recognized by the fluorescence they emit in the visible spectrum. Two types of fluorescence may be studied: natural fluorescence (*autofluorescence*), which is produced by substances normally present in the tissue, and *secondary fluorescence,* which is induced by staining with fluorescent dyes called *fluorochromes.*

Certain proteins can be tagged with fluorescent dyes without denaturing the molecule. These fluorescent proteins may then be injected into the animal and localized in sections within the cell or in the extracellular space.[35]

The most important advantage of fluorescence microscopy is its great sensitivity. This is particularly important for vital studies, since only a low concentration of fluorescent dye is necessary, and thus there is a minimum amount of interference with the normal physiology of the tissue.

Fluorescence often yields specific cytochemical information because some of the normal components of the tissue have a typical fluorescent emission. Thus vitamin A, thiamine, riboflavin, and other substances can be detected. The cytochemical value of the method is increased considerably by spectrographic analysis of the radiation. Sometimes certain substances incorporated in cells, e.g., sulfonamides, can be localized.[36]

The most common pattern of autofluorescence is a weak, diffuse, bluish fluorescence of the cytoplasm, with a yellow and stronger fluorescence of the granules; usually the nucleus is not fluorescent. Mitochondria of the liver and kidney give a strong fluorescence, calcium deposits appear yellow-white, and free porphyrins have a strong red fluorescence. An important application concerns the so-called lipogenic pigments, which are found in a great number of cells and which increase as the cell ages. It is thought that these pigments represent different degrees of oxidation and polymerization of unsaturated fatty acids. With fluorescence two types of pigments, the so-called *lipofuscin* and the *ceroid,* can be determined.[36]

An important development has been the use of paraformaldehyde fixation, which, in freeze-dried tissues, produces condensation with catecholamines and indolamines, emitting a green and a yellow fluorescence, respectively.[37] This fluorescent reaction has also been studied by microspectrographic methods.[38]

Immunocytochemistry

Cytochemical techniques have been developed to localize antigens at the light and electron microscopic levels. Antibodies are produced by plasmocytes against most macromolecules (antigens) and also against small molecules, provided they are bound to larger molecular species. Antibodies are present in the γ-globulin fraction of the serum and generally have a sedimentation constant of 6S. Figure 6–19 shows the general principle of some of the cytochemical techniques involving the use of labeled antibodies. In the

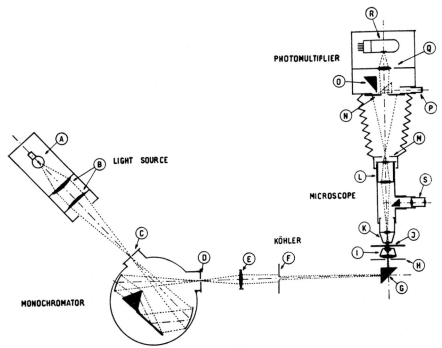

Figure 6–18. Cytophotometer used at the Instituto de Anatomía General y Embriología. *A*, light source of a tungsten light; *B*, condenser lens; *C* and *D*, entrance and exit slits of the monochromator; *E*, lens; *F*, diaphragm; *G*, prism; *H*, diaphragm; *I*, condenser; *J*, slide; *K*, objective; *L*, ocular; *M* and *N*, diaphragms; *O*, prism for observation of the final image; *P*, to displace prism; *Q*, lens; *R*, photomultiplier; *S*, lateral view. (Courtesy of A. O. Pogo and J. Cordero Funes.)

upper portion an antigen is shown binding to an unlabeled antibody to form a complex. A similar *direct reaction* may be produced by coupling the γ-globulin to a fluorescent dye or to a molecule that is opaque to the electron beam. In Figure 6–19 (lower portion) the so-called *indirect method,* which is more widely used, is indicated. Here the unlabeled antigen-antibody complex reacts with an antiglobulin antibody labeled with fluorescein to form a complex visible under the fluorescence microscope[39] or with ferritin for the electron microscope.

Antibodies coupled with ferritin (a macromolecule with a MW of 650,000 daltons, a diameter of 8.5 nm, and an iron content of 23 per cent) are easily detected under the electron microscope because of the four dense points, with a diameter of about 1.5 nm each, exhibited by the ferritin molecule[40] (Fig. 21–7).

Rabbits may be immunized by the injection of enzymes such as alcaline phosphatase[41] and peroxidase[42] (Fig. 6–20). The tissue sections are then exposed to the enzyme which is revealed by the DAB method described previ-

ously under *oxidases.* It was demonstrated that peroxidase antibodies first appear in the perinuclear space of plasmocytes and then accumulate in the cisternae of the endoplasmic reticulum and in some of the lamellar portions of the Golgi complex.[43]

In the widely used Coon's technique, the antibodies are coupled with fluorescein isocyanate. The tissues are frozen and sectioned in a cryostat and then the sections are stained with the coupled antiserum. This method has been widely used to localize viruses and bacterial antigens.

The pituitary hormone ACTH is localized in the basophilic cells of the pituitary gland, and several of the enzymes produced by the pancreas have also been localized by this technique. Figure 6–21 illustrates an example of the degree of localization that can be achieved with Coon's technique. The myofibril shown in the figure has been stained with an antibody against myosin, and it demonstrates that this protein is strictly localized in the A bands. The antibody against hyaluronidase is localized in the acrosome of the spermatozoon (see Chapter 23).

antigen + antibody   complex

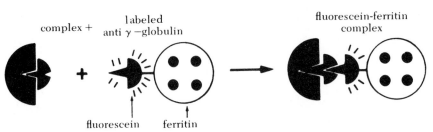

complex +   labeled   fluorescein-ferritin
             anti γ −globulin   complex

             fluorescein   ferritin

Figure 6–19. Diagram of the general principles involved in immunocytochemical methods. **Above,** the antigen-antibody reaction. **Below,** the indirect method by which anti γ-globulin (antibody) is labeled either with fluorescein for the light microscope or with ferritin for the electron microscope. (See the description in the text.) (Courtesy of W. Bernhard.)

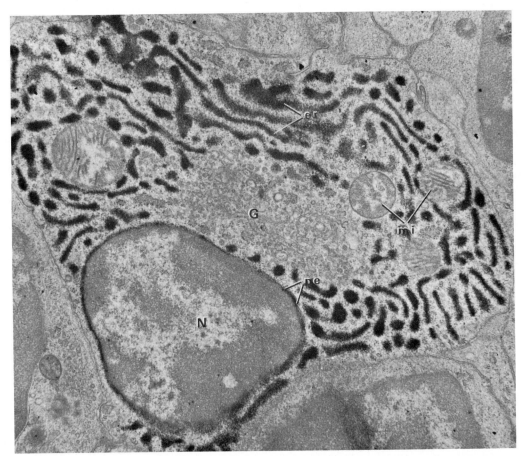

Figure 6–20. Electron micrograph of an immature plasmocyte showing the localization of antiperoxydase antibodies in the cisternae of the endoplasmic reticulum (*er*) and nuclear envelope (*ne*). The Golgi complex (*G*) contains no antibodies at this time; *mi,* mitochondria; *N,* nucleus. ×18,400. (Courtesy of E. H. Leduc.)

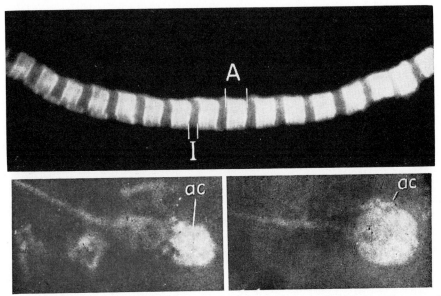

Figure 6-21. **Above,** striated myofibril isolated and stained with a fluorescent antibody against myosin. Notice that the antigen (myosin) is located exclusively in the A-bands. Observation made with the fluorescence microscope. (Courtesy of H. Marshall.) **Below,** bull spermatozoon (*left*) and guinea pig spermatozoon (*right*) incubated with the corresponding hyaluronidase antibody. Notice that in both cases the bright fluorescence is localized at the acrosome (*ac*). ×1250. (Courtesy of R. E. Mancini.)

Radioautography in Cytochemistry

One of the most important modern cytochemical methods is based on the use of substances labeled with radioisotopes. These are incorporated in the cell and are then localized with a photographic emulsion.

Radioautography is based on the capacity of radioisotopes to act on the silver bromide crystals of the emulsion. The tissue section is put in contact with the photographic emulsion for a certain period; then the radioautograph is developed like an ordinary photograph. By comparing the radio-autograph with the cells in the tissues seen under a microscope, the radioisotope can be localized fairly accurately.

Radioisotopes used in radioautography may emit one or more of three types of radiation: α- and β-particles and γ-rays. α-Particles are positively charged helium nuclei that produce straight tracks in the emulsion; they can easily be traced back to the point of origin. In biological work they have limited use because they are produced primarily by heavy metals. β-Particles are electrons, which may have different energy levels. Their

tracks are tortuous and may vary in length from a few microns to a millimeter, depending on their energy. γ-Rays are not important in radioautography. Most of the isotopes used are β-emitters.[45]

In the *stripping film* technique, a 5 μm emulsion on a gelatin base of 10 μm is used. The film is stripped off the glass plate and floated in a water bath. Then the section mounted on a glass slide is immersed beneath the floating film. Upon withdrawal of the specimen, the emulsion covers the preparation tightly. After exposure for different lengths of time according to the isotope used (several weeks or months), the radioautograph is developed. Quantitative results are obtained by determining the density of the particles in the radioautographs by various optical methods, or by counting the grains.

Substances marked with the β-emitter ^{14}C have been widely used in radioautography. With such an emitter, *track radioautography* can also be used[46]—a thick liquid emulsion is applied instead of a photographic film. With this technique the contact is even more intimate than with the film stripping technique, and it is possible to follow and count the

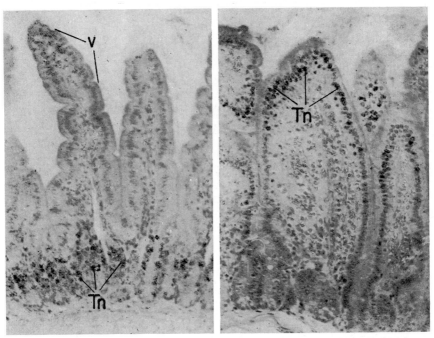

Figure 6–22. Section of intestine of a mouse injected with tritiated thymidine. **Left,** animal killed eight hours after injection; **right,** 36 hours after injection. *Tn,* tagged nuclei; *V,* villus. (See the description in the text.) (Courtesy of C. P. Leblond.)

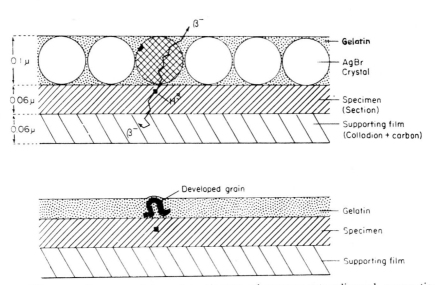

Figure 6–23. Diagrammatic representation of an electron microscope autoradiograph preparation. **Top,** *during exposure*: The silver halide crystals, embedded in a gelatin matrix, cover the section. A beta particle, from a tritium point source in the specimen, has hit a crystal (cross-hatched), causing the appearance of a latent image on the surface (black speck on upper left region of crystal). **Bottom,** *during examination and after processing*: The exposed crystal has been developed into a filament of silver; the nonexposed crystals have been dissolved. The total thickness has decreased because the silver halide occupied approximately half the volume of the emulsion. (Courtesy of L. G. Caro.)

single tracks of β-particles coming out from a definite area, thus providing a quantitative estimation. The resolution of this method is of the order of 1 to 2 μm.

Substances labeled with tritium (^{3}H), a weak β-emitter, are most widely used in radioautography. For the study of deoxyribonucleic acid (DNA) metabolism of the cell, tritiated thymidine is used and is specific for this type of nucleic acid. Important investigations of the mechanism of DNA replication (see Chapter 17) and RNA metabolism have been made with appropriate precursors. In the example shown in Figure 6–22, the nuclei tagged with tritiated thymidine and initially present at the bottom of the intestinal crypts are found after 36 hours near the tip of the villus. This illustrates most graphically the life cycle of the cell—after a division at the bottom of the crypt, the cell ascends along the

epithelium until, after a few hours, it is destroyed at the tip of the villus.

A two-emulsion radioautographic technique has been developed that can distinguish β-particles emitted from ^{14}C and ^{3}H atoms. With this method a tritiated precursor of DNA and a ^{14}C-labeled precursor of RNA or of a protein can be employed simultaneously.[47]

The technique of radioautography with tritiated substances has been used at the level of the electron microscope. Special thin liquid emulsions are applied on thin sections, forming a monolayer of silver halide crystals (Fig. 6-23). After proper development, the silver grains that were hit by the β-emission stand out on the electron microscopic image. A resolution of about 0.1 μm (100 nm) has been achieved (see Fig. 24–5).[48–51]

SUMMARY: Cytochemistry

Cytochemistry consists of the identification and localization of chemical components of the cell. Cytochemical (and histochemical) studies may be based on four main analytical techniques: (a) separation of cell fractions by conventional biochemical techniques; (b) isolation of minute amounts of tissues, and even single cells, by micro- and ultramicromethods; (c) direct detection of cell components in the cell by chemical staining; and (d) use of measurement of physical parameters.

Cell fractionation consists of the homogenization of tissue and the separation of the various components of the cell according to mass and specific gravity. In a standard *differential centrifugation,* by the use of increasing centrifugal fields, four main fractions are obtained: nuclear, mitochondrial, microsomal, and soluble. Standard centrifuges and preparative ultracentrifuges are used. The time required for each particle to settle depends on its size and density. Cell fractionation can be improved through the use of discontinuous or continuous *density gradients.* In this case centrifugation is continued until the various particles reach an equilibrium with the density of the gradient (i.e., isopycnic centrifugation). In *zonal rotors* the density gradient is formed and the sample is layered while the rotor is spinning. Small particles, such as ribosomes or macromolecules (DNA, RNA, proteins), may be studied with the *analytical ultracentrifuge* and at the same time the sedimentation constant, *S,* which is related to the molecular weight, can be determined.

Isopycnic ultracentrifugation in gradients of cesium chloride permits demonstration of the existence of DNAs of different *buoyant densities. Microchemical* and *ultramicro-chemical* methods are used in the quantitative analysis of tissue sections or isolated cells. Micromanometric methods permit the measurement of the oxygen consumption and of the weight of a single microorganism, such as an ameba.

Cytochemical staining is the method most often used in cytochemical studies. The substance to be investigated must be immobilized and then identified by chemical reactions. For the study of *proteins* the Millon, and the diazonium reactions may be used. SH groups in proteins may be determined by certain reagents. Arginine, which is present in high concentration in histones, is detected by the Sakaguchi test. *Schiff's reagent,* which is the leukobase of basic fuchsin, is recolored by *aldehyde* groups. This reagent is used in the *Feulgen reaction* for DNA. A prior mild acid hydrolysis removes RNA and unmasks aldehyde groups from deoxyribose linked to purine bases (Fig. 6–9). The *periodic acid-Schiff (PAS) reaction* occurs in polysaccharides after oxidation of the 1,2-glycol groups by periodic acid. The *plasmal reaction* occurs in free aldehydes present in certain plasmalogens.

Detection of enzymes is accomplished by means of frozen sections made in a cryostat, or after light fixation with aldehydes. In general, the section is incubated with specific substrates, and the product is converted into a metal precipitate or a colored compound. Various phosphatases, esterases, cholinesterases, and other hydrolytic enzymes may be studied cytochemically. *Oxidases* and *dehydrogenases* are detected by special reagents, such as the Nadi reagent for cytochrome oxidase. The DAB reagent (3',3'-diaminobenzidine) is widely used for peroxidases and other oxidases. Tetrazolium salts produce insoluble formazan dyes with some mitochondrial enzymes.

Several cytochemical methods are based on *physical determinations*. Ultraviolet absorption at 260 nm is characteristic of nucleic acids and results from the presence of purine and pyrimidine bases. Cytophotometric methods can be used in the ultraviolet or the visible spectrum to make quantitative determinations in cells and tissues. Thus, the Feulgen reaction permits the determination of the DNA content of the nucleus and even of single chromosomes.

Fluorescence microscopy, with ultraviolet light near the visible spectrum, is used to study the natural fluorescence *(autofluorescence)* of cell components such as vitamin A, thiamine, riboflavin, calcium, and phorphyrins. More important is the use of fluorescent dyes *(secondary fluorescence)* in several cytochemical methods. Catechol and indolamines produce a fluorescent reaction with aldehydes.

Immunocytochemistry uses antibodies for the localization of macromolecules (antigens) in cells. (Antibodies are present in the γ-globulin fraction of serum.) In the direct reaction the antibody may be coupled with a fluorescent dye or an opaque molecule (e.g., ferritin) for the electron microscope. In the *indirect* method the unlabeled antibody-antigen complex reacts with a labeled anti-γ-globulin antibody (Fig. 6–19). This technique permits the study of antibody formation in plasmocytes and the localization of various proteins in different cells.

Radioautography uses substances labeled with radioisotopes. These radioisotopes are then detected using a photographic emulsion that is developed like an ordinary photograph. Most radioisotopes used are β-emitters and contain ^{14}C or tritium. Several techniques, such as the *stripping film*, the *track radioautography* in liquid emulsions, and the thin liquid emulsion for *electron microscope radioautography* are employed. In most cases ^3H-thymidine is used for the study of DNA. ^3H-leucine, or other tritiated amino acids, are used for protein synthesis. Radioautography is one of the most frequently used techniques for following within the cell structure the mechanisms of DNA replication, DNA-RNA transcription, and their translation into protein.

REFERENCES

1. Eagle, H. (1955) *Science, 122*:501.
2. Hayflick, L., and Moorehead, P. (1961) *Exp. Cell Res., 25*:585.
3. Puck, T. T., and Marcus, P. I. (1955) *Proc. Nat. Acad. Sci. USA, 41*:432.
4. Thompson, L. H., and Baker, R. M. (1973) In: *Methods in Cell Biology,* Vol. 6, Chapter 7. (Prescott, D. M., ed.) Academic Press, Inc., New York.
5. Sabatini, D. D., Bensch, K. G., and Barrnett, R. J. (1963) *J. Cell Biol., 17*:19.
6. Gersh, I. (1959) Fixation and staining. In: *The Cell,* Vol. 1, p. 21 (Brachet, J., and Mirsky, A. E., eds.) Academic Press, Inc., New York.
7. Palade, G. E. (1952) *J. Exp. Med., 95*:285.
8. De Robertis, E. (1956) *J. Biophys. Biochem. Cytol., 2*:785.
9. Schubert, M., and Hamermann, D. (1956) *J. Histochem. Cytochem., 4*:158.
10. Michaelis, L., and Granick, S. (1945) *Amer. Chem. Soc., 67*:1212.
11. Sylvén, B. (1954) *Quart. J. Micr. Sci., 95*:327.
12. Lison, L. (1960) *Histochimie et Cytochimie Animale;* 3rd Ed. Gauthier-Villars, Paris.
13. Pearse, A. G. E. (1961) *Histochemistry: Theoretical and Applied,* 3rd Ed. J. & A. Churchill, London.
14. Glick, D. (1959) Quantitative microchemical techniques of histo- and cytochemistry. In: *The Cell,* Vol. 1, p. 139. (Brachet, J., and Mirsky, A. E., eds.) Academic Press, Inc., New York.
15. De Robertis, E., Pellegrino de Iraldi, A., Rodriguez de Lores Arnaiz, G., and Salganicoff, L. (1962) *J. Neurochem., 9*:23.
16. Anderson, N. G. (1956) Techniques for the mass isolation of cellular components. In: *Physical Techniques in Biological Research,* Vol. 3, p. 300 (Oster, G., and Pollister, A. W., eds.) Academic Press, Inc., New York.
17. Albertsson, P. (1960) *Partition of Cell Particles and Macromolecules.* John Wiley & Sons, New York.
18. Kirk, P. L. (1950) *Quantitative Ultramicroanalysis.* John Wiley & Sons, New York.

19. Eränko, Q. (1955) *Quantitative Methods in Histology and Microscopic Histochemistry.* Little, Brown and Co., Boston.
20. Lowry, O. H. (1957) Micromethods for the assay of enzymes. In: *Methods in Enzymology,* Vol. 4. (Colowick, S. P., and Kaplan, N. O., eds.) Academic Press, Inc., New York.
21. Zeuthen, E. (1946) *C. R. Lab. Carlsberg, série chim., 25*:191.
22. Holter, H., and Zeuthen, E. (1948) *C. R. Lab. Carlsberg, série chim., 26*:7, 243.
23. Edström, J. E. (1956) *Biochim. Biophys. Acta, 22*:378.
24. Bennett, H. S., and Watts, R. M. (1958) In: *General Cytochemical Methods,* Vol. 1, p. 317. (Danielli, J. F., ed.) Academic Press, Inc., New York.
25. Barrnett, R. J., and Seligman, A. M. (1955) *J. Histochem. Cytochem., 3*:406.
26. Cafruny, E. J., Di Stefano, H. S., and Farah, A. (1955) *J. Histochem. Cytochem., 3*:354.
27. Lessler, M. A. (1953) *Internat. Rev. Cytol., 2*:231.
28. Brachet, J. (1957) *Biochemical Cytology.* Academic Press, Inc., New York.
29. McManus, F. A. (1946) *Nature* (London), *158*:202.
30. Gomori, G. (1952) *Microscopic Histochemistry.* University of Chicago Press, Chicago.
31. Nachlas, M. M., and Seligman, A. M. (1949) *J. Natl. Cancer Inst., 9*:415.
32. Koelle, G. B., and Friedenwald, J. S. (1949) *Proc. Soc. Exp. Biol. Med., 70*:617.
33. Graham, R. C., and Karnovsky, M. L. (1966) *J. Histochem. Cytochem., 14*:291.
34. Seligman, A. M., Karnovsky, M. J., Wasserknig, H. L., and Hanker, J. S. (1968) *J. Cell Biol., 38*:1.
35. Mancini, R. E. (1963) *Internat. Rev. Cytol., 14*:193.
36. Price, G., and Schwartz, S. (1956). Fluorescence microscopy. In: *Physical Techniques in Biological Research,* Vol. 3, p. 91. (Oster, G., and Pollister, A. W., eds.) Academic Press, Inc., New York.
37. Carlsson, A., Falck, B., and Hillarp, N. (1962) *Acta Physiol. Scand., 56*(suppl.):196.
38. Ritzen, M. (1967) *Exp. Cell Res., 44*:250, and *45*:178.
39. Coons, A. H. (1956) *Internat. Rev. Cytol., 5*:1.
40. Rifkind, R. A., Osserman, E. F., Hsu, K. C., and Morgan, C. (1962) *J. Exp. Med., 116*:423.
41. Scott, G., Avrameas, S., and Bernhad, W. (1968) *C. R. Acad. Sci. (Paris), 266*:746.
42. Leduc, E. H., Avrameas, S., and Bouteille, M. (1968) *J. Exp. Med., 127*:109.
43. Stengerger, L. A. (1967) Electron microscopic inmunocytochemistry. *J. Histochem. Cytochem., 15*:139.
44. Nakane, P. K. (1970) *J. Histochem. Cytochem., 18*:9.
45. Taylor, J. H. (1956) Autoradiography at the cellular level. In: *Physical Techniques in Biological Research,* Vol. 3, p. 546. (Oster, G., and Pollister, A. W., eds.) Academic Press, Inc., New York.
46. Ficq, A. (1955) *Exp. Cell Res., 9*:286.
47. Baserga, R., and Nemeroff, K. (1962) *J. Histochem. Cytochem., 10*:628.
48. Caro, L. G., and van Tubergen, R. P. (1962) *J. Cell Biol., 15*:173.
49. Caro, L. G. (1964) In: *Methods in Cell Physiology,* Vol. 1, p. 327. (Prescott, D. M., ed.) Academic Press, Inc., New York.
50. Saltpeter, M. M. (1966) General area of autoradiography at the electron microscope level. In: *Methods in Cell Physiology,* Vol. 2 (Prescott, D. M., ed.) Academic Press, Inc., New York.
51. Stevens, A. R. (1966) *High Resolution Autoradiography,* Vol. 2, p. 255. (Prescott, D. M., ed.) Academic Press, Inc., New York.

Allfrey, V. G. (1959) The isolation of subcellular components. In: *The Cell*, Vol. 1, p. 193. (Brachet, J., and Mirsky, A. E., eds.) Academic Press, Inc., New York.

Anderson, N. G., ed. (1966) The development of zonal centrifuges. *Natl. Cancer Inst. Monogr., 21.* Bethesda, Md.

Avrameas, S. (1972) Enzyme markers: their linkage with proteins and use in immuno-histochemistry. *Histochem. J.* (London), *4*:321.

Baserga, R., and Malamud, D. (1969) *Autoradiography: Techniques and Application.* Harper and Row, Publishers, New York.

Brachet, J. (1957) *Biochemical Cytology.* Academic Press, Inc., New York.

Burstone, M. S. (1962) *Enzyme Histochemistry and Its Application in the Study of Neoplasms.* Academic Press, Inc., New York.

Coons, A. H. (1956) Histochemistry with labeled antibody. *Internat. Rev. Cytol., 5*:1.

Danielli, J. F. (1953) *Cytochemistry, A Critical Approach.* John Wiley & Sons, New York.

Deane, H. W. (1958) Intracellular lipids: Their detection and significance. In: *Frontiers in Cytology.* (Palay, S. L., ed.) Yale University Press, New Haven, Conn.

De Duve, C. (1971) Tissue fractionation. Past and present. *J. Cell Biol., 50*:20d.

Engström, A. (1956) Historadiography. In: *Physical Techniques in Biological Research,* Vol. 3, p. 489. (Oster, G., and Pollister, A. W., eds.) Academic Press, Inc., New York.

Ficq, A. (1959) Autoradiography. In: *The Cell,* Vol. 1, p. 67. (Brachet, J., and Mirsky, A. E., eds.) Academic Press, Inc., New York.

Giacobini, E. (1968) Chemical studies on individual neurons, Part I. Vertebrate nerves. *Neurosci. Res., 1*:1.

Giacobini, E. (1969) Chemical studies on individual neurons, Part II. Invertebrate nerve cells. *Neurosci. Res., 2*:111.

Glick, D. (1959) Quantitative microchemical techniques of histo- and cytochemistry. In: *The Cell,* Vol. 1, p. 139. (Brachet, J., and Mirsky, A. E., eds.) Academic Press, Inc., New York.

Gomori, G. (1952) *Microscopic Histochemistry.* University of Chicago Press, Chicago.

Hale, A. J. (1957) The histochemistry of polysaccharides. *Internat. Rev. Cytol., 6*:194.

Handmaker, S. D. (1973) *Ann. Rev. Microbiol., 27*:189.

Harris, M. (1964) *Cell Culture and Somatic Variation.* Holt, Rinehart & Winston, Inc., New York.

Heidelberg, M. (1967) Some contributions of immunochemistry to biochemistry and biology. *Ann. Rev. Biochem., 36*: part 1, 1.

Hopwood, D. (1972) Theoretical and practical aspects of glutaraldehyde fixation. *Histochem.* (London), *4*:267.

Lison, L. (1960) *Histochimie et cytochimie animale,* 3rd Ed. Gauthier-Villars, Paris.

Mazurkiewicz, J. E., and Nakane, P. K. (1972) Light and electron microscopic localization of antigens in tissues embedded in polyethylene glycol with a peroxidase-labeled antibody method. *J. Histochem. Cytochem., 20*:969.

Miller, H. R. P. (1972) Fixation and tissue preservation for antibody studies. *Histochemical J., 4*:305.

Pearse, A. G. E. (1968) *Histochemistry, Theoretical and Applied,* 3rd Ed. J. & A. Churchill, London.

Price, G., and Schwartz, S. (1956) Fluorescence microscopy. In: *Physical Techniques in Biological Research,* Vol. 3, p. 91. (Oster, G., and Pollister, A. W., eds.) Academic Press, Inc., New York.

Paul, J. (1970) *Cell and Tissue Culture,* 4th Ed. The Williams & Wilkins Co., Baltimore, Md.

Pollack, R. (1973) *Readings in Mammalian Cell Culture.* Cold Spring Harbor Laboratory, Cold Spring Harbor, New York.

Puck, T. T. (1972) *The Mammalian Cell as a Microorganism.* Holden-Day Inc., San Francisco.

Tooze, J., ed. (1973) *The Molecular Biology of Tumor Viruses,* Chapter 2. Cold Spring Harbor Laboratory, Cold Spring Harbor, New York.

Waymouth, C. H., ed. (1970) *Advances in Tissue Culture.* The Williams & Wilkins Co., Baltimore, Md.

Part Four

UNITS OF STRUCTURE
AND THE PLASMA
MEMBRANE

In the following two chapters an attempt is made to convey how the molecular components presented in Part Two are organized to form more complex structures, which may be analyzed with the various optical instruments. The different levels of organization—from the molecular to the subcellular —are considered.

When elementary structures (i.e., macromolecular, fibrous and membranous) are dealt with it is possible to see how they can be formed by the interaction of different molecules. The use of molecular models has been a great aid in interpreting the images observed with the electron microscope.

The investigation of the cell membrane, since it is of macromolecular dimensions, is related to these elementary structures. In this section it will be shown that a better understanding of the structure and properties of biological membranes may be attained from a study of the models of lipoprotein membranes. The numerous differentiations of the cell membrane in the various cell types will be considered here, but a discussion, in detail, of the physiology of the cell membrane will be reserved for Chapter 21, along with a consideration of cell permeability.

seven

ELEMENTARY UNITS OF STRUCTURE IN BIOLOGICAL SYSTEMS

The molecular constituents of the cell, described in Chapter 3, can interact among themselves and become organized into supramolecular units. These units, in turn, are parts of structures recognizable within the cells by means of the electron microscope.

When the molecules are associated linearly, the elementary units are primarily *unilinear* (fibrous); when the molecules are extended in two dimensions forming thin membranes, the units are two-dimensional; and when they are crystalline or amorphous particles, the units are *three-dimensional.*

Many of the molecular components are polymers in themselves (e.g., proteins are polymers of amino acids) but, in turn, are monomers of larger units and can polymerize end to end or can interact laterally to form the fibrous, membranous, or crystalline structures.

ELEMENTARY STRUCTURES AND THEIR FUNCTION

In certain systems elementary structures may aggregate to form higher types of organization visible under the light microscope and even to the naked eye. In animal and plant tissues there are several series of components with this type of organization. They can be classified into three categories: *subcellular,* which comprise parts of cells, such as membranes, cilia, and chromosomes; *extracellular,* such as collagenous and elastic fibers, membranes of cellulose, or chitin situated outside

the cells; and *supracellular,* which are macroscopic structures, such as hair, bone, and muscle with a more complex supramolecular organization.

The function of these elementary structures in biological systems will be mentioned throughout this book. Several of these molecular systems are involved in *mechanical functions;* e.g., collagen fibers form tendons; fibrin fibers are used in blood clotting to prevent bleeding; and muscle proteins interact to produce shortening during contraction. Several of these supramolecular structures have *enzymatic properties* and may constitute *multienzymatic complexes.* The storing (coding) and transmitting of *genetic information* is one important function of these complexes. Research has shown that most of the fundamental functions of biological systems, such as osmotic work, association of cells, permeability, and oxidations, are intimately related to these basic molecular structures.

Molecular Shape of Proteins

Since proteins are most often involved in the formation of the elementary structures associated with biological systems, it is important to remember the size and shape of these molecules (Table 7–1). In the case of the soluble proteins (see Chapter 3), which are important components of biological tissues, it will be noted that the α-helix content of the polypeptide chain varies from about 100 to 30 per cent. The asymmetry of the molecule is,

TABLE 7-1. Molecular Structure of α-Proteins*

	Helix Content (%)	Mol. Wt.	Model	Approximate Length
Tropomyosin	>90	53,000		40 nm
Light meromyosin fr. 1	>90	135,000		80 nm
Paramyosin	>90	200,000		140 nm
Myosin	65	530,000		140 nm
Heavy meromyosin	50	350,000		40 nm
Fibrinogen	30	340,000		46 nm
Prekeratin	~40	640,000		
Flagellins	~40	20–40,000		3 to 4 nm
GLOBULAR PROTEINS				
Myoglobin	70	17,000		3 nm
Bovine serum albumin	45	68,000		5 nm

*(From Cohen, C., *in* Wolstenholme, G. E. W., and O'Connor, M. (Eds.), *Ciba Foundation Symposium*, p. 101, London, J. & A. Churchill, 1966.)

in general terms, proportional to the α-helix content. For example, the muscle proteins (i.e., tropomyosin, light meromyosin, paramyosin, myosin, and heavy meromyosin) have a helix content above 50 per cent and are elongated molecules. On the other hand, the so-called globular proteins have basically a spherical shape.[1]

Both the globular and the elongated protein molecules may associate to form elementary structures of various degrees of complexity. For example, the contraction of a muscle depends on the formation of complexes in which several of the proteins indicated in Table 7-1 are integrated in an elaborate macromolecular machinery (see Chapter 23).

Principles in the Assembly of Macromolecules

In Chapter 3 the concept of *self-assembly*, by which several protein subunits may form more complex arrangements, was mentioned in conjunction with the quaternary structure of proteins. In hemoglobin, for example, α- and β-chains interact to form the complete molecule.

Other examples include the case of multienzyme complexes, some of which are found free in the cytoplasm (e.g., the huge *pyruvate dehydrogenase* complex of *Escherichia coli*, which contains three groups of enzymes and a total of 88 protein subunits). Other multienzyme complexes are embedded in membranes, such as in the case of the respiratory chain in mitochondria (see Chapter 10).

In self-assembly, the protein subunits contain the necessary information to produce the larger complex by means of physicochemical forces. Complex macromolecules, and even subcellular structures, may be formed in the cell by the principle of self-assembly. In addition to this simple case of self-assembly, in which no other component is involved, there is also the principle of *aided assembly*, in which certain enzymes may prepare the macromolecules for assembly (e.g., fibrin). Finally, there is the case of *directed assembly*, in which a previous structure is needed for the organization of the new macromolecule. Note that in the duplication of DNA and the transcription of RNA (see Chapters 17 and 19) a template is needed to direct the assembly of the other macromolecule.

Assembly of Viruses

Viruses represent a beautiful example of structures in which the principles of macromolecular assembly are in action.

The tobacco mosaic virus (TMV), for example, is a particle 40×10^6 daltons in mass, with the form of a cylinder of 16×300 nm. This cylinder contains a single-stranded molecule of RNA consisting of 6500 nucleotides, forming a helix with a radius of 4.0 nm, and having a cylindrical cavity of 2.0 nm. Associated with this RNA helix and forming the protein coat are 2130 identical protein subunits of 18,000 daltons (Fig. 7–1). It has

Figure 7–1. Diagram of the molecular organization of the tobacco mosaic virus. In the center there is a spiral of RNA which is associated with protein subunits. There is one protein monomer for every three bases in the RNA chain. (From Caspar, D. L., and Klug, A., *Cold Spring Harbor Sympos. Quant. Biol.*, 27:1, 1962.)

been found possible to dissociate the RNA and the protein subunits and later to reassociate them in order to reconstitute active virus particles. The RNA molecule appears to influence the assembly of the protein subunits (Klug, 1972).

Many viruses have been observed to display *icosahedral symmetry*. According to Caspar and Klug (1962), this symmetry depends on the fact that the assembly of the protein subunits (i.e., the *capsomeres*) causes the *capsid* of the virus to be at a state of minimum energy. Such icosahedral symmetry has been found in a virus as small as the ϕX174, which has only 12 capsomeres, and in one as large as the adenovirus, which has 252 capsomeres.

The *bacteriophages* are viruses which attach to bacteria and infect them by injecting their own DNA content. An interesting example of a complex virus is the small bacteriophage ϕ 29 of *Bacillus subtilis*, which has the elaborated macromolecular structure shown in Figure 7–2. This phage contains seven major structural proteins.[2]

The head is a prolate icosahedron composed of two types of protein subunits with a pentameric and hexameric arrangement, and containing a double-stranded DNA molecule 5.7 μm long. The head is covered by thin fibers composed of another protein. Attached to the head of the phage is a neck with three types of subunits and a tail with a single subunit. A total of 172 protein molecules, of which 145 are in the head capsid, are found in a single phage. By the use of ethylenediaminetetraacetate (EDTA) or dimethyl-sulfoxide it has been possible to produce a progressive disruption of the various parts of the phage. Thus, the head fibers, the tail, and the neck pieces can be detached in a sequential manner. DNA can also be removed, thus leaving the empty capsids. These various disrupted structures can then be characterized by electron microscopy (Fig. 7–3), and the proteins can be separated by polyacrylamide gel electrophoresis. These various processes constitute a complete biophysical and biochemical characterization of the structural components of this phage.[3]

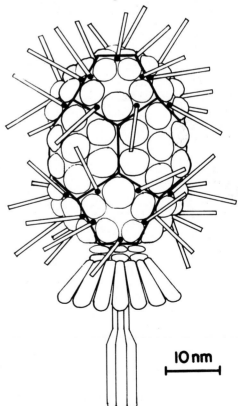

Figure 7–2. Diagram of the small bacteriophage ϕ29 of *B. subtilis* having a total of 172 protein molecules of which 145 integrate the capsid. (Courtesy of C. Vásquez.)

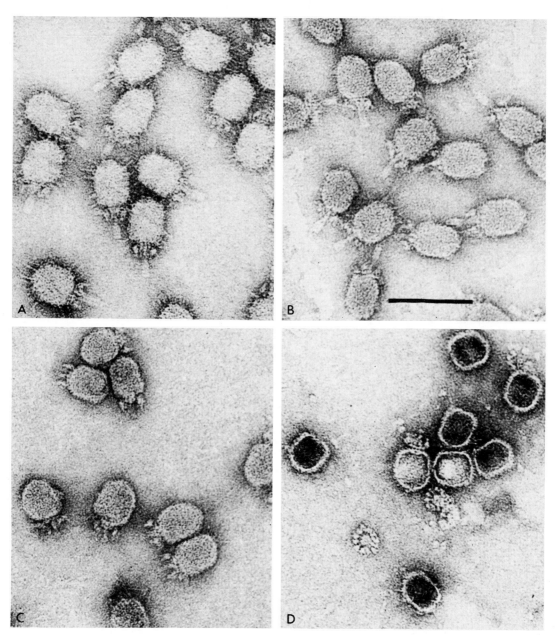

Figure 7–3. Electron micrographs of ϕ29 bacteriophage of *B. subtilis* showing a progressive disruption of the various parts. **A,** normal phages; **B,** phages that have lost the head fibers; **C,** phages that have lost the tail; and **D,** phages that have lost the neck piece. In this last case some capsids have lost their content of DNA. Scale in B = 50 nm. (Courtesy of C. Vásquez.)

The assembly of macromolecules may be brought about by straight *self-assembly,* in which no other component is involved; by *aided assembly,* in which an enzymatic process may aid in the process; or by *directed assembly,* in which a template directs the formation. Self-assembly is found in some multienzyme complexes and in certain virus particles. (At the present time the more complex viruses have not been assembled in vitro.)

In the assembly of macromolecules the size and shape of the protein subunits may play an important role. In some instances, such as with the TMV virus and the ribosomes (see Chapter 18), the interaction between RNA and protein may be fundamental to guiding the assembly of the final structure.

**SUMMARY:
Assembly of
Macromolecules**

Collagen as an Example of a Fibrous Unit

Collagen may be used to illustrate the principles involved in the formation of large molecular complexes. One of the most abundant proteins in the animal kingdom, collagen is synthesized primarily by the fibroblasts and is an important part of major fibrous components of the body, such as skin, tendon, cartilage, and bone. The large aggregates are visible to the naked eye and under the light microscope, but the intimate structure at the molecular level can be studied only by combining electron microscopy, x-ray diffraction, chemical analysis, and other techniques. It has been found, for example, that collagen fibers can be dissociated into smaller and smaller units by the action of acids and then reassembled.

The basic collagen molecule has a molecular weight of 360,000, a length of about 280 nm, and a width of 1.4 nm. It consists of three chains coiled together in a helical fashion as shown schematically in Figure 7–4.[4] It is interesting to recall that collagen has a rather simple amino acid constitution: about one third is glycine, another third is proline and hydroxyproline, and the rest is other amino acids.

The molecular unit of collagen—also called "tropocollagen"—can be considered a macromolecular monomer,[5-7] because it is capable, by interaction, of forming different collagen structures. The tropocollagen molecule is polarized in the sense of having a definite linear sequence of the amino acid residues in the intramolecular strands. In fact, in relation to its interaction, the tropocollagen molecule behaves as if it had a "head" and "tail".

The study of native *collagen fibers* with x-ray diffraction and electron microscopy has shown that they are composed of *fibrils* that have a repeating period of 70 nm, which is reduced to 64 nm after drying (Fig. 5–11). The relationship of the tropocollagen molecule of 280 nm to this period of the fibrils has been

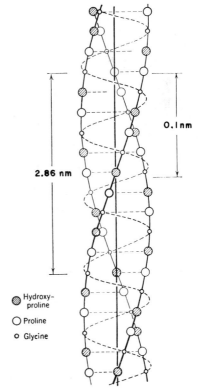

2.86 nm

0.1 nm

⊘ Hydroxy-proline
◯ Proline
○ Glycine

Figure 7–4. Structure of the collagen molecule with the three-stranded helix. (From Rich, A., *Biophysical Science*, New York, John Wiley & Sons, Inc., 1959.)

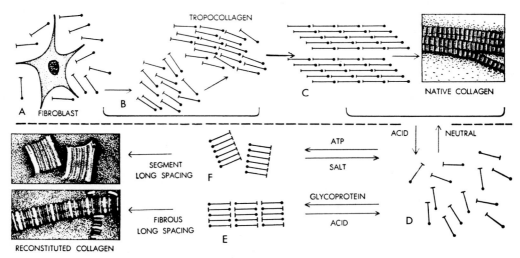

Figure 7–5. Diagram of the formation and reconstitution of collagen. A fibroblast (**A**), manufactures *tropocollagen* molecules (**B**), which form *native collagen* (**C**). Collagen fibrils are solubilized in acid (**D**) and the resulting tropocollagen, in the presence of glycoprotein, produces *fibrous long spacing* (**E**) and, with addition of ATP, *segment long spacing* collagen (**F**). The long spacing of 280 nm results from the lateral aggregation of the tropocollagen molecules without overlapping. The 70 nm spacing of native collagen fibrils is due to the overlapping of the tropocollagen molecule. (Courtesy of J. Gross, 1961.)

clarified by reconstituting collagen in the presence of some glycoproteins or ATP. This results in the formation of two other types of fibers (Fig. 7–5): one is composed of long fibrils having a spacing of 280 nm; the other contains short segments having a similar period, but showing no polymerization. The most probable explanation for these findings is that the collagen fibrils—with a period of 70 nm—result from the lateral association of tropocollagen molecules, which overlap at intervals of one fourth their length. It is assumed that in this instance the molecules are longitudinally associated "heads" with "tails" (Fig. 7–5).

In the case of fibrous collagen with long spacing, there is no lateral overlapping, and the tropocollagen molecules are assembled side by side and randomly linked in a linear direction. In the segments with long spacing it is supposed that the tropocollagen molecules do not overlap laterally, and, because they are all in phase, they cannot link longitudinally (Fig. 7–5).[8]

Blood Clotting

The important process of *blood clotting* involves fibrinogen and thrombin. The fibrinogen molecule is asymmetrical and has a molecular weight of 340,000.

Under the electron microscope it appears to be composed of three beads, each about 6.5 nm, connected by a very thin strand of 1.5 nm (Table 7–1). The total length varies between 23 and 46 nm, depending on the pH.[9]

Under the action of thrombin, which splits off a small peptide from fibrinogen, fibrinogen is activated and starts to interact with other monomers. The end-to-end association forms long fibrin fibrils,[10] but apparently there is also some lateral staggering and cross-linking with other fibers to form a network. As clotting progresses, aided by the blood platelets,[11] fibrin retracts, squeezing out the serum, and the blood clot is completed.

The nature of the physicochemical forces involved in these macromolecular interactions varies considerably. For example, the fact that collagen fibrils are soluble in weak organic acids implies that salt linkages and hydrogen bonds are involved. In blood clotting, the process of binding the molecules is more complex, since, as indicated, the interaction involves enzymatic action.

Stronger bonds, such as —S—S— linkages, are involved in other proteins, such as those forming the various types of keratin fibers. Within the cell, loose and reversible aggregations of corpuscular proteins may occur. These globular-fibrous transformations take place in some processes involving displacement of parts of the cell matrix, such as ameboid motion, cyclosis, or the formation of the mitotic apparatus. The formation of microtubules and microfilaments is generally involved in these transformations (Chap. 22).

Particulate Glycogen

Another interesting example of molecular interaction is observed in the glycogen deposits found in liver cells, muscle, and in many other tissues. The branched structure of the polysaccharides amylopectin and glycogen, is based on 1,6-α-glycosidic bonds, as mentioned in Chapter 3. Electron microscopy has revealed that glycogen particles have three structural levels of organization, each with a characteristic size and morphology.[12,13] The largest units—called α-particles—are spheroid and measure 50 to 200 nm with a mean of 150 nm. These particles have a morular aspect, which indicates that they are composed of smaller units—the β-particles—which are ovoid or polyhedral and measure 30 nm in diameter (Fig. 7–6). Finally, within the β-particles a finer structure—the γ-particles—composed of rods of 3 × 20 nm can be

observed. These three different units can be demonstrated by acidic treatment of particulate glycogen (Fig. 7–6). Glycogen synthesis is achieved in two successive steps which can be followed under the electron microscope.

ELEMENTARY MEMBRANOUS STRUCTURES

Biological membranes are known to result from interaction between lipids and proteins, but the molecular arrangement of these two components is difficult to ascertain. The use of models and artificial monomolecular films has increased our understanding of the natural structures.

Monolayer Films

The structural importance of certain lipids was mentioned in Chapter 3. Fatty acids, phospholipids, cholesterol, and cholesterol esters can be packed in single layers of constant thickness, and the orientation of the lipids within this structure depends on the dipolar constitution of a polar group and a nonpolar hydrocarbon chain (Fig. 7–7, A). These properties of the lipids can be studied by forming films on the surface of water.

The technique of making *monolayer (monomolecular) films* is of considerable biological importance. The *film balance,* devised by Langmuir in 1917, is still the principal instrument used to study these films. Essentially, it is a shallow trough filled with water on which the substance is spread. A bar or barrier can be pushed across the trough to compress the film. The surface pressure exerted by the film is measured by a sensitive floating, suspended balance. For example, if stearic acid is dissolved in a volatile solvent and deposited on the water, the molecules will spread until they reach an equilibrium. Upon evaporation of the solvent, a film one molecule thick is formed. Because the molecule is bipolar, the polar group (—COOH) is attracted by the water molecule, and the nonpolar hydrocarbon chain tends to stand straight on the surface. At first, some molecules are not well aligned because of the ample space, but as the barrier is pushed across the trough and the surface area is reduced, the molecules are compressed until

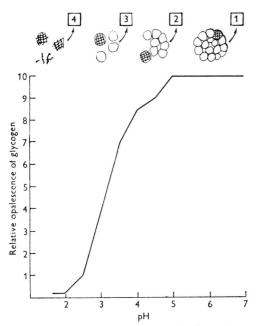

Figure 7–6. The structure of particulate glycogen. At decreasing pH values the glycogen particle dissociates progressively. (See the description in the text.) (From Drochmans, P., *Biochem. Soc. Symp., 23:*127, 1963.)

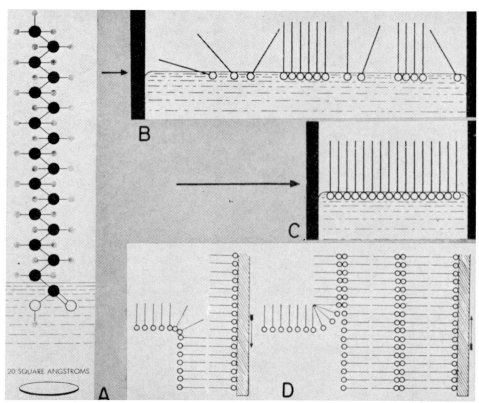

Figure 7–7. Diagram of the technique of making monolayer films. **A,** a molecule of stearic acid with the polar group dipped in water. **B,** at low compression, molecules are oriented at different angles or form packed aggregates. **C,** at high compression, molecules are tightly packed and are vertical. Circles represent polar groups and straight lines the nonpolar hydrocarbon chains. **D,** method of building up molecular films at an air-water interface. *Left,* a glass slide previously coated with a monomolecular film of barium stearate (notice the polar groups attached to the glass surface) is dipped in water that has a monomolecular film at the interface. The second monomolecular layer attaches to the first by the nonpolar ends. *Right,* several bimolecular layers of barium stearate have been deposited on the glass slide by successive dips into the water. (A, B, C, from H. E. Ries, Jr.; D, courtesy of D. Waugh.)

they form a packed film (Fig. 7–7, *B* and *C*). Under these conditions, molecules exert a pressure that can be measured with the film balance. When the number of molecules and the total surface occupied by the film at the maximum compression are known, the average area of each molecule can be calculated. For example, the stearic acid molecule occupies about 0.2 square nanometer units (Fig. 7–7, *A*).

By this method the thickness of the monolayer can also be measured (2.5 nm for stearic acid). Thickness depends on the number of carbons. The monomolecular film can be deposited on the surface of a glass slide dipped into the water. As shown in Figure 7–7, *D*, by successive dippings bimolecular or multimolecular layers are built up. Multi-layered systems can be obtained that give

coherent x-ray diffractions from which the distance or period between the layers can be measured.

Lipid Bilayers

Because of the many experimental applications the so-called lipid bilayers are even of greater interest than the lipid monolayers. In this case an artificial lipid membrane is produced across a small hole separating two aqueous solutions (Fig. 7–8). A droplet of a lipid dissolved in an organic solvent is applied within this hole. After a few minutes, a thin film having no interference colors (i.e., a black film) is produced.[14]

The apparatus indicated in Figure 7–8, or a similar one, permits the study of the electrical properties of the lipid bilayer as well as its fixation and removal for study under the electron microscope.[15] These artificial membranes are 6 to 9 nm in thickness and have a trilaminar structure, suggesting that they are composed of two layers of lipids, with the hydrophilic groups toward the water interphase and a hydrophobic region in the middle (see below). Lipid bilayers have biophysical properties that are comparable to some biological membranes. They differ, however, in their higher electrical resistance and in the fact that they do not show selectivity for the passage of various ions (see Stoeckenius and Engelman, 1969; Henn and Thompson, 1969).

The incorporation of certain polypeptides and proteins (i.e., ionophores) may alter the properties of these membranes considerably. With the addition of these molecules, the membrane may acquire selectivity, electrical excitability, and even chemical receptor properties.[16]

Lipid-Water Systems. Myelin Figures

Bulk systems containing phospholipids and water can also be produced. In this case the molecular association of lipids in water depends on the temperature and concentration of the components and may be studied by x-ray diffraction.[17] If the amount of water is small and the temperature low, the lipid molecules crystallize. At intermediate lipid concentrations, the molecules become associated into micelles, which may have different shapes and become dispersed in water. Two liquid-crystalline phases may be recognized: (1) *Lamellar* (Fig. 7–9, a), formed by alternate layers of lipid and water. Although the thickness of the lipid is always the same in this system, the thickness of the water may vary with the concentration from 1.0 nm to over 6.0 nm. (2) *Hexagonal* (Fig. 7–9, b), having a two-dimensional hexagonal lattice. The interior of the cylinders is filled with water, and the cylinders themselves are embedded in a lipid matrix.[18]

Studies of these systems at different temperatures have shown that they may exhibit special transitions that can be detected by x-ray diffraction, as well as with other physical methods. At these transitions, the so-called *thermotropic mesomorphism* occurs. This phenomenon is apparently due

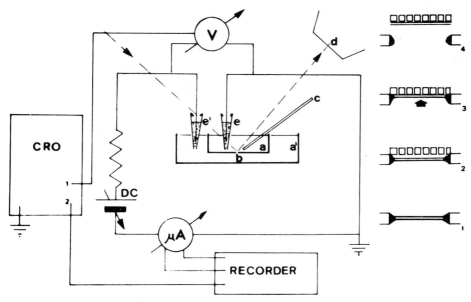

Figure 7–8. **Left,** diagram of the apparatus used to study the electrical properties of artificial membranes. The membrane is formed in a 1 mm hole (*b*) at the bottom of a teflon cup (*a*) immersed in a dish (*a'*). The electrical measurements are done via calomel electrodes (*e, e'*). The membrane is polarized at a constant voltage by means of a DC source. Measurements are made by a microammeter (*μA*) and a voltmeter (*V*). An oscilloscope *(CRO)* and a recorder may be used to register the conductance changes. The pipette (*c*) may be used to study the effect of drugs; (*d*) stereomicroscope. **Right,** diagram of the technique used to remove the artificial membrane and bring it to observation under the electron microscope. *1,* artificial membrane; *2,* placing of the grid; *3,* change in pressure to stick the membrane to the grid; *4,* removal of the grid with the membrane. (From Parisi, Reader, Vásquez, and De Robertis, unpublished.)

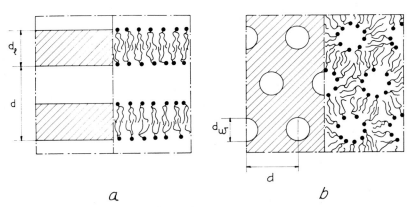

Figure 7–9. The structure of liquid crystalline phases of phospholipid-water systems. **a,** lamellar configurations and **b,** hexagonal configurations; d_l, thickness of lipid layer; d_w, thickness of water layer; d, period. (From Luzzati, V., and Husson, F., The structure of the liquid-crystalline phases of lipid-water systems. *J. Cell. Biol., 12*:215, 1962.)

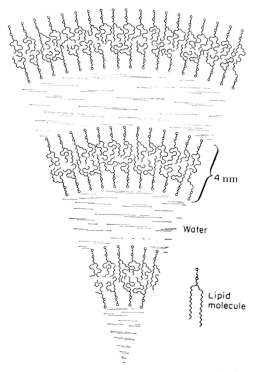

Figure 7–10. Molecular arrangements of lipids and water in a myelin figure. The polar groups are represented by circles. The width of the bimolecular lipid layers is approximately constant while the water layers change in width with the degree of hydration of the specimen. (Courtesy of W. Stoeckenius.)

to the change of the hydrocarbon chain from a liquid to a solid state. With a temperature above the transition point the mobility of the lipid molecules is increased. In biological membranes it is considered that the phospholipids are in a liquid state at body temperature. In Chapter 8 it will be mentioned that the *"fluidity"* of the membrane depends on the composition of the membrane—particularly the chain length—and degree of saturation of the fatty acids.[19-21]

The *lamellar* type of lipid-water phase forms the so-called *myelin figures.* If phospholipids extracted from brain or other tissues are mixed with water, then wormlike, concentric, semiliquid structures, which flow from the lipid phase, appear. These structures have a strong birefringence with a radially oriented axis. The lipid molecules are disposed in bimolecular layers attached by their nonpolar interfaces (Fig. 7–10).

The myelin figures can be studied by x-ray diffraction or fixed with osmium tetroxide and observed under the electron microscope.[22] These models are of considerable interest for the interpretation of the electron microscope image. The micrographs show alternate parallel light and dark bands, which repeat at approximately 4 nm. The dark bands should be attributed to the osmium deposits in this multilamellar structure.

In the *hexagonal* array mentioned above (Fig. 7–9,*b*), the electron microscope shows a similar pattern with the dense dots in the place of the water phase, thus indicating that osmium is bound to the polar groups.

More complex models of membrane structures can be made by incorporating some protein into the myelin figure. A double layer of lipid coated on both sides with protein gives two parallel dense lines of 2.5 to 5.0 nm, separated by a lighter interspace of 2.0 to 2.5 nm.

Several models of lipid-water and lipid-protein systems may be used to clarify our understanding of natural biological membranes. Lipid monolayers and bilayers, as well as myelin figures or crystalline phospholipid-water phases with a hexagonal configuration may be produced and analyzed with polarization microscopy, x-ray diffraction, or electron microscopy. Such studies may give information about the relative position of polar and nonpolar groups and their association with proteins. Some inferences, based on studies with the electron microscope, may be made about the electron density image.

In his *Chance and Necessity* (highly recommended reading) Monod considers, under *"Molecular Ontogenesis"* many of the concepts that have been discussed in this chapter. He explains that the way in which structures of higher and higher order are built in the cell is determined by the genetic information contained in DNA. The information contained in this molecule determines the primary sequence of the polypeptide chain in the protein, which in turn determines the secondary and tertiary structures, and, finally, the formation of the oligomeric complexes. The interaction of different proteins with lipids and nucleic acids, on the other hand, results in the formation of macromolecular complexes and structures of a higher order of complexity.

SUMMARY:
Artificial Membrane Systems

REFERENCES

1. Cohen, C. (1966) *Ciba Foundation Symposium*. p. 101. (Wolstenholme, G. E. W., and O'Connor, M., eds.) J. & A. Churchill, London.
2. Mendez, E., Ramirez, G., Salas, M., and Viñuela, E. (1971) *Virology*, 45:567.
3. Salas, M., Vasquez, C., Mendez, E., and Viñuela, E. (1972) *Virology*, 50:180.
4. Crick, F. H. C., and Rich, A. (1957) In: *Recent Advances in Gelatine and Glue Research*. p. 20. Pergamon Press, London.
5. Schmitt, F. O., Gross, J., and Highberger, J. H. (1955) *Symp. Soc. Exp. Biol.*, 9:148.
6. Schmitt, F. O. (1959) In: *Biophysical Science*, p. 349. (Oncley, J. L., et al., eds.) John Wiley & Sons, New York.
7. Gross, J. (1961) *Sci. Am.*, 204:120.
8. Hodge, A. J., and Schmitt, F. O. (1958) *Proc. Natl. Acad. Sci. USA*, 44:418.
9. Hall, C. E. (1963) *Lab. Invest.*, 12:998.
10. Porter, K. R., and Hawn, C. V. A. (1949) *J. Exp. Med.*, 90:225.
11. De Robertis, E., Paseyro, P., and Reissig, M. (1953) *Blood J. Hemat.*, 8:7.
12. Drochmans, P. (1963) In: Methods of separation of subcellular structural components. *Biochem. Soc. Symp.*, 23:127.
13. Drochmans, P. (1968) *Excerpta Medica Internat. Cong. Ser.*, 166:49.
14. Mueller, P., Rudin, D. O., Ti Tien, H., and Wescott, W. C. (1963) *J. Phys. Chem.*, 67:534.
15. Vasquez, C., Parisi, M., and De Robertis, E. (1971) *J. Membr. Biol.*, 6:353.
16. Parisi, M., Rivas, E., and De Robertis, E. (1971) *Science*, 172:56.
17. Luzzati, V., Reiss-Husson, F., and Saludjian, P. (1966) *Ciba Foundation Symposium*. p. 69. (Wolstenholme, G. E. W., and O'Connor, M., eds.) J. & A. Churchill, London.
18. Luzzati, V., and Reiss-Husson, F. (1962) *J. Cell Biol.*, 12:207.

19. Luzzati, V. (1968) *Biological Membranes* (Chapman Edition), Academic Press, Inc., New York, p. 71.
20. Chapman, D. (1969) *Lipid Res., 4*:251.
21. Engelman, D. (1970) *J. Mol. Biol., 47*:115.
22. Stoeckenius, W. (1962) In: The interpretation of ultrastructure. (Harris, R. J. C., ed.) *Symp. Internat. Soc. Cell Biol.,* Vol. 1, p. 349.

ADDITIONAL READING

Caspar, D. L., and Klug, A. (1962) Physical principles in the construction of regular viruses. *Cold Spring Harbor Sympos. Quant. Biol., 27*:1.

Fenner, F. (1968) *The Biology of Animal Viruses.* Academic Press, Inc., New York.

Fernández-Morán, H. (1959) Fine structure of biological lamellar systems. In: *Biophysical Science.* (Oncley, J. L., et al., eds.) John Wiley & Sons, New York.

Gross, J. (1961) Collagen. *Sci. Am., 204*:120.

Henn, F. A., and Thompson, T. E. (1969) Synthetic lipid bilayer membranes. *Ann. Rev. Biochem., 38*:241.

Hodge, A. J. (1959) Fibrous proteins of muscle. In: *Biophysical Science.* p. 409. (Oncley, J. L., et al., eds.) John Wiley & Sons, New York.

Klug. A. (1972) Assembly of tobacco mosaic virus. *Fed. Proc., 31*:30.

Miller, A., and Parry, D. A. D. (1973) Structure and packing of microfibrils in collagen. *J. Mol. Biol., 75*:441.

Monod, J. (1971) Chance and Necessity. Random House, Inc., New York.

Stoeckenius, W., and Engelman, D. M. (1969) Current models for the structure of biological membranes. *J. Cell Biol., 42*:613.

Wolstenholme, G. E. W., and O'Connor, M., eds. (1966) *Principles of Biomolecular Organization.* Ciba Foundation Symposium. J. & A. Churchill, London.

eight

THE PLASMA MEMBRANE

The cell has a different internal milieu from that of its external environment. For example, the ionic content of animal cells is quite dissimilar from that of the circulating blood. This difference is maintained throughout the life of the cell by the thin surface membrane, the *plasma membrane,* which controls the entrance and exit of molecules and ions. The function of the plasma membrane of regulating this exchange between the cell and the medium—generally called *permeability*—is studied in Chapter 21. The emphasis of this chapter is on the structural aspects, as well as the possible molecular organization, of the plasma membrane.

This membrane is so thin that it cannot be resolved with the light microscope, but in some cells it is covered by thicker protective layers that are within the limits of microscopic resolution. For example, most plant cells have a thick cellulose wall that covers and protects the true plasma membrane (Fig. 2–6). Some animal cells are surrounded by cement-like substances that constitute visible cell walls. Such layers, also called *extraneous coats,* generally play no role in permeability, but do have other important functions.

ISOLATION OF THE PLASMA MEMBRANES

Several methods have been used to isolate plasma membranes from a variety of cells, i.e., liver cells, striated muscle, *Amoeba proteus,* sea urchin eggs, or Erhlich ascitis cells.[7] In most cases the purity of the fraction has been controlled by electron microscopy, enzyme analysis, the study of surface anti-gens, and other criteria (see De Pierre and Karnovsky, 1973). Plasma membranes are more easily obtained from erythrocytes subjected to hemolysis.

To obtain plasma membranes from human red blood cells, the cells are treated with hypotonic solutions that produce swelling and then loss of the hemoglobin content (i.e., *hemolysis*). The resulting membrane is generally called a *red cell ghost.* Two main types of ghosts may be produced: *resealed ghosts* and *white ghosts.* The so-called re-sealed ghosts are produced when hemolysis is milder; the ghosts can be treated with substances that produce restoration of the permeability functions (i.e., resealing). White ghosts are formed if hemolysis is more drastic. There is complete removal of the hemoglobin, and the ghosts can no longer be resealed (i.e., white ghosts). These ghosts can be used for biochemical, but not physiological, studies. Even in the white ghosts, however, there are components that do not belong strictly to the membrane. For example, there are *microfilaments* attached to the inner surface of the membrane that contain a special protein, the so-called *spectrin,* which has a molecular weight of about 250,000. These microfilaments may act as a kind of molecular framework that gives stability to the membrane.[2] The study of these and other types of microfilaments that may play important mechanical roles in cells will be considered in Chapter 22.

Chemical Composition

The plasma membrane is composed mainly of protein, lipid, and a small percentage (1 to 5 per cent) of oligosaccharides that

TABLE 8-1.   Lipid and Protein Ratios in Some Cell Membranes*

Species and Tissue		Protein (%)	Lipid (%)
Human	CNS myelin	20	79
Bovine	PNS myelin	23	76
Rat	Muscle (skeletal)	65	35
Rat	Liver	60	40
Human	Erythrocyte	60	40
Rat	Liver mitochondrion	70	27–29

	Molar ratio			Area ratio
	Amino Acid	Phospholipid	Cholesterol	Protein: Lipid
Myelin	264	111	75	0.43
Erythrocyte	500	31	31	2.0

*Triggle, D. J., *Neurotransmitter–Receptor Interactions,* New York, Academic Press, Inc., 1971, p. 122.

may be attached to either the lipids (glycolipids) or the proteins (glycoproteins). From the data shown in Table 8–1 it is evident that there is a wide variation in the lipid-protein ratio between different cell membranes. Myelin is an exception, in the sense that the lipid predominates; in the other cell membranes there is higher protein/lipid ratio. In Table 8–1 it may be observed that in myelin the area occupied by the protein is insufficient to cover that of the lipids, whereas in a red cell ghost the opposite situation is found.

Lipids

The main lipid components of the plasma membrane are the phospholipids, cholesterol, and galactolipids; their proportion varies in different cell membranes. As shown in Table 8–2 myelin differs, with regard to lipid composition, from other cell membranes found in the brain.[3] The major proportion of membrane phospholipids is represented by phosphatidylcholine, phosphatidylethanolamine, and sphingomyelin, all of which have no net charge at neutral pH (i.e., *neutral phospholipids*) and tend to pack tightly in the bilayer (see Chapter 7). (This property is also shared by cholesterol.) Five to 20 per cent of the phospholipids are acidic, including: phosphatidylinositol, phosphatidylserine, cardiolipin, phosphatidylglycerol, and sulfolipids. *Acidic phospholipids* are negatively charged and in the membrane are associated principally with proteins by way of lipid-protein interactions.[1]

Recent studies demonstrate that there is an asymmetry in the erythrocyte membrane regarding not only the protein, but also the phospholipids. There are more choline phospholipids and glycolipids in the external half of the bilayer and more amino phospholipids in the inner, or cytoplasmic, half of the bilayer. It is assumed that this asymmetry is rather stable and that there is no exchange of lipids across the bilayer (see Bretscher, 1973).

Carbohydrates

In red cell ghosts hexose, hexosamine, fucose, and sialic acid are bound mainly to

TABLE 8-2.   Molar Ratios of Lipids in Some Subcellular Fractions of Cerebral Cortex

	Phospholipids	Cholesterol	Galactolipids	Protein Amino Acids
Myelin	2.4	2.4	1	29
Nerve-ending membranes	5.4	4.2	1	83
Synaptic vesicles	10.8	6.7	1	142
Mitochondria	14.4	4.7	1	415

proteins; the same is true of carbohydrates in liver membranes. Sialic acid is sensitive to neuraminidase and is attached to proteins by N-acetylgalactosamine on the outer surface of the membrane.

Because of the presence of sialic acid residues, as well as carboxyl and phosphate groups, the outer surface of the membrane is negatively charged; consequently, positively charged proteins may be bound by electrostatic interactions to the plasma membrane. Only a small amount of sialic acid exists in the form of *gangliosides* (i.e., glycolipids) in the plasma membrane of liver. However, gangliosides are important constituents of the neuronal surface and are probably involved in ion transfers. The preferential localization of gangliosides in the acetylcholinesterase rich nerve-ending membranes has been demonstrated.[4]

Membrane Proteins

Proteins represent the main component of most biological membranes (Table 8–1). They play an important role, not only in the mechanical structure of the membrane, but also as carriers or channels, serving for transport; they may also be involved in regulatory or ligand-recognition properties. In addition, numerous enzymes, antigens, and various kinds of receptor molecules are present in plasma membranes.

Peripheral and Integral Proteins. Recent studies tend to differentiate membrane proteins into two more or less definite groups (see Singer, 1971): the *peripheral or extrinsic proteins* and the *integral or intrinsic proteins.* Table 8–3 shows some of the criteria that are used to differentiate between these two types of membrane proteins. *Peripheral proteins* are separated by mild treatment, are soluble in aqueous solutions, and are usually free of lipids. Examples include: the above-mentioned *spectrin,* which may be removed from red cell ghosts by chelating agents[5]; cytochrome *c,* found in mitochondria; and acetylcholinesterase, in electroplax membranes, which are easily removed in high salt solutions.[6, 7] *Integral proteins* represent more than 70 per cent of the two protein types and require drastic procedures for isolation. Usually they are insoluble in water solutions and need the presence of detergents to be maintained in a non-aggregated form. The study of integral proteins from different membranes has shown that they are rather heterogeneous in relation to molecular weight. These proteins may be attached to oligosaccharides, thus forming glycoproteins (see Chapter 3) or to special phospholipids, thereby constituting lipoproteins (i.e., proteolipids).

One of the major intrinsic proteins of erythrocyte ghosts has a molecular weight of 95,000 and is highly asymmetrical. Several

TABLE 8–3. Criteria for Distinguishing Peripheral and Integral Membrane Proteins*

Property	Peripheral Protein	Integral Protein
Requirements for dissociation from membrane	Mild treatments sufficient: high ionic strength, metal ion chelating agents	Hydrophobic bond-breaking agents required: detergents, organic solvents, chaotropic agents
Association with lipids when solubilized	Usually soluble-free of lipids	Usually associated with lipids when solubilized
Solubility after dissociation from membrane	Soluble and molecularly dispersed in neutral aqueous buffers	Usually insoluble or aggregated in neutral aqueous buffers
Examples	Cytochrome *c* of mitochondria; Spectrin of erythrocytes	Most membrane-bound enzymes; histocompatibility antigens; drug and hormone receptors

*From Singer, S. J., "The Molecular Organization of Biological Membranes," *in* Rothfield, L. I., (Ed.), *The Structure and Function of Biological Membranes,* New York, Academic Press, Inc., 1971, p. 145.

methods have demonstrated that this protein is related to anion permeability.[8] It is thought that this protein molecule has a roughly globular shape and spans the lipid bilayer. There are some 500,000 molecules of this type per erythrocyte. It has been found that the anion channels are inactivated by reagents that react with external amino groups (see Bretscher, 1973). Using lithium diiodosalicylate a *chaotropic* (i.e., water disordering) agent or phenol (Table 8–3), a glycoprotein, called *glycophorin*, is released from the red cell ghost. This protein has a molecular weight of 55,000, of which 60 per cent is carbohydrate. Near the COOH end of the molecule there is a region that is very hydrophobic and which interacts with the lipids of the membrane. The NH_2 end is more hydrophilic, is exposed to the external environment, and has the attached oligosaccharides that are at the outer surface of the membrane.[9] The COOH end is probably exposed to the interior of the red cell (Fig. 8–1). At the outer surface glycoproteins contain protein-bound antigens for the ABO blood groups and others, such as: the MN groups reacting with rabbit antisera, the influenza virus, phytohemoagglutinin, and wheat germ agglutinin. It has been calculated that there are some 700,000 copies of this protein per human red cell. Glycophorin accounts for 80 per cent of the carbohydrate and 90 per cent of the negatively charged sialic acid present in the cell surface. The model that has been suggested for this major glycoprotein of the erythrocyte is that of a molecule that spans the plane of the membrane.[10] In the diagram of Figure 8–1 the various segments of glycophorin, i.e., the hydrophilic N-terminal, the receptor region with the oligosaccharides, the hydrophobic region across the membrane, and the hydrophilic C-terminal are indicated.[11]

In red cell membranes there is an evident asymmetry of the membrane proteins; most of them are associated with the cytoplasmic surface of the membrane.

Among the most hydrophobic integral proteins of the membrane, the so-called *proteolipids* are characterized by their strong association with lipids and the fact that they are soluble in organic solvents. First isolated by Folch and Lees[12] from myelin, proteolipids are found in practically all cell membranes, and in many of them they represent receptor proteins for synaptic transmitters (see De Robertis, 1971).

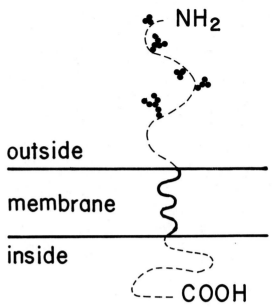

Figure 8–1. Diagram of a glycophorin molecule, a glycoprotein of the red cell ghost which traverses the membrane. The —NH_2 end with the attached oligosaccharides is exposed to the external environment. (See description in the text.) (From Marchesi, S. L., Steers, E., Marchesi, V. T., and Tillack, T. W., *Biochemistry,* 9:50, 1970.)

Enzymes

Some 30 enzymes have been detected in isolated plasma membranes. Those most constantly found are 5'-nucleotidase, Mg^{2+} ATPase, $Na^+ - K^+$ activated - Mg^{2+} ATPase, alkaline phosphatase, adenyl cyclase, acid phosphomonoesterase, and RNAse.

Some enzymes have a preferential localization; for example, alkaline phosphatase and ATPase are more abundant at the bile capillaries, while disaccharidases are present in microvillae of the intestine. A specific localization with a mosaic arrangement has been postulated for some of these enzymes. Disaccharidase forms 5 to 6 nm globular units coating the membrane of the microvillae.[13] The plasma membrane lacks the respiratory chain and glycolytic activity.

Of all the enzymes mentioned, $Na^+ - K^+$ activated - Mg^{2+} ATPase is one of the most important because of its role in ion transfer across the plasma membrane (see Chapter 21). This enzyme is dependent on the presence of lipids and is inactivated when all lipids are extracted.

The chemical composition of the plasma membrane is rather complex. It contains an almost continuous and asymmetrical framework of lipids with neutral phospholipids and cholesterol in the bilayer and acidic phospholipids bound to the proteins. The membrane proteins are the site of numerous membrane functions, which will be studied in Chapter 21. About 70 per cent of the proteins are of the integral or intrinsic type, and the rest are of the peripheral or extrinsic type. *Glycophorin*, the major glycoprotein of the red cell, and *spectrin* are characteristic examples of these two types. Among the most hydrophobic intrinsic proteins, the proteolipids may have important drug receptor properties. Surface antigens are carried out principally by glycoproteins.

SUMMARY:
Chemistry of the
Plasma Membrane

STRUCTURE OF THE PLASMA MEMBRANE

Before the isolation of plasma membranes, theories on the molecular structure of the membrane were generally based on indirect information. Since substances soluble in lipid solvents penetrate the plasma membrane easily, Overton postulated in 1902 that the plasma membrane is composed of a thin layer of lipid. In 1926 Gorter and Grendell found that the lipid content of hemolyzed erythrocytes was sufficient to form a continuous layer 3 to 4 nm thick over the entire surface, and postulated that the plasma membrane is composed of a double layer of lipid molecules. This theory was also supported by electrical measurements that indicated a high impedance at the plasma membrane. The high impedance is due to the fact that it is difficult for ions to penetrate a lipid layer.

More direct information concerning the role of lipids in membranes was provided by *artificial membranes* of phospholipids separating two liquid chambers (see Chapter 7).

Other indirect information about the molecular structure of the plasma membrane came from the study of the interfacial tension of different cells. Tension at a water-oil interface is about 10 to 15 dynes per centimeter, whereas surface tension of cells is almost nil. For example, in sea urchin egg the surface tension of cells has been calculated to be 0.2 dyne per centimeter. It has been postulated that the low tension is due to the presence of protein layers on the lipid components. In

fact, when a very small amount of protein is added to a model lipid-water system, the surface tension is lowered comparably.

To explain all these properties Danielli proposed that the plasma membrane con-

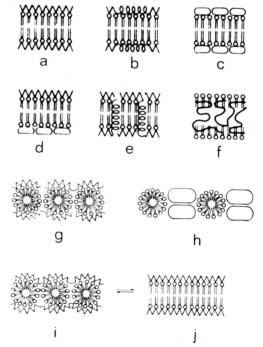

Figure 8-2. A variety of molecular models proposed for the plasma membrane: **a-f**, models based on a lipid bilayer structure; **g-j**, models based on globular arrangements. **a**, protein in β-form; **b**, α-helix; **c**, globular protein; **d**, asymmetry in the protein; **e**, partial penetration with protein channels or pores; **f**, protein within the lipid bilayer; **g**, lipid micelles with β protein; **h**, lipid micelles with globular protein; **i** and **j**, globular-bilayer transformation. (Courtesy of A. L. Lehninger.)

tained a *lipid bilayer,* with protein adhering to both lipid-aqueous interfaces.[14]

Figure 8–2, *a* to *f,* illustrates different membrane models based on the concept of a lipid bilayer, but which differ in the type of protein (i.e., globular, α, or β helix) or the penetration of proteins within the bilayer.[15]

Other models containing globular lipid micelles, globular proteins, or a combination of both have been proposed (Fig. 8–2, *g to i*);

a globular-bilayer transition has also been postulated (Fig. 8–2, *j*). These globular models do not account satisfactorily for the high electrical impedance.

Fine Structure of the Plasma Membrane. The Unit Membrane.

Electron microscopy has thrown some

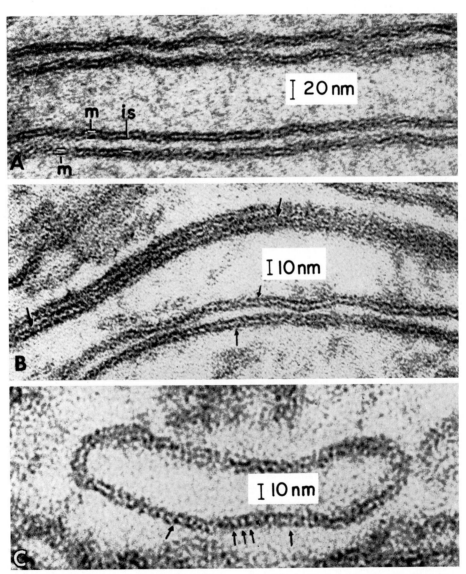

Figure 8–3. **A,** electron micrograph of cell membranes of intestinal cells (*m*), showing the three-layered structure (unit membrane). *is,* intercellular space. ×240,000. **B,** cell membranes in the rat hypothalamus showing the unit membrane structure and, with arrows, some finer details across the membrane. The upper arrows indicate a region in which the two cell membranes are adherent (*tight junction*) and the intercellular space has disappeared. ×360,000. **C,** the same as **B,** showing fine bridges (arrows) across the unit membrane. ×380,000. (From E. De Robertis.)

light on the fine structure of the plasma membrane and has revealed the numerous structural differentiations that this membrane and the underlying cell cytoplasm have in different cell types. To resolve the structure of the plasma membrane, extremely thin sections ($\sim$ 20 nm) must be used, otherwise one would observe different orientations of the plasma membrane with respect to the plane of the section. The membrane appears most thin when it is exactly perpendicular to the plane of the section. Definite plasma membranes of 6 to 10 nm have been observed at the surface of all cells. The plasma membranes of two cells that are in close contact appear as dense lines separated by a space of 11 to 15 nm, which is strikingly uniform and contains a material of low electron density (Fig. 8–3, A). This intercellular component can be considered as a kind of cementing substance.[16] As will be shown later, the plasma membranes of adjacent cells may be totally adherent at certain points, forming the so-called tight junctions (Fig. 8–3, B).

The plasma membrane of most types of cells appears to be three-layered. The two outer dense layers are about 2.0 nm thick and the middle clear layer about 3.5 nm. This structure, called the "unit membrane," is found in most intracellular membranes.[16]

As was mentioned in Chapter 7 the electron microscopic image of the plasma membrane was interpreted in relation to studies of artificial models. The less dense middle layer of the plasma membrane corresponds to the hydrocarbon chains of the lipids. The extraction of the lipids from a membrane fixed in aldehydes does not change the unit membrane structure. This indicates that the protein is the main contributor to the electron microscopic image.

Specializations of the Unit Membrane

Differences in thickness and asymmetry of the layers have been observed in various membrane types.[17] Some finer details have also become apparent, such as small discontinuities at the dense layers and, particularly, bridges across the light central layer, which suggest the intercalation of proteins (Fig. 8–3, C).

The thickness of the unit membrane has been found to be greater in the plasma membrane (10 nm) than in the intracellular membranes of the endoplasmic reticulum or Golgi complex (5 to 7 nm).[18]

The asymmetry of the membrane may be determined by the presence of electron-dense material on the outer or inner surfaces of the unit membrane. For example, the surface coating of mucopolysaccharides, which has been called *glycocalyx*[19] (see below), may be 0.1 to 0.5 μm thick in microvilli,[20] tapering off in the form of a filamentous material. Another differentiation, this time on the inner surface, is observed in some invaginations, 100 to 150 nm deep, of the plasma membrane that are involved in the pinocytosis of fluids and uptake of proteins. These are the so-called *coated vesicles* which have a filamentous material 20 nm long adherent to the plasma membrane.[21]

In invertebrate cells there are special contacts between plasma membranes called *septate desmosomes* and *gap junctions* which permit intercellular communications and electrical coupling. Such junctions have a hexagonal honeycomb structure. Furthermore, globular units 5 nm in diameter have been observed in membranes of the endoplasmic reticulum, and, according to some interpretations, such a globular structure may have a transient existence in other membranes as well.[22, 32]

The observation of thin sections by electron microscopy has led to the concept of a trilayered plasma membrane, also called the "unit membrane." This is interpreted as indicating that the electron-dense outer layers correspond to the protein, and the less dense middle layer corresponds to the hydrocarbon chains of the lipids. The unit membrane concept, however, is certainly an oversimplification; numerous fine details suggest that the molecular organization of the membrane is much more complex.

SUMMARY: Structure of the Plasma Membrane

Important information about the molecular structure of the cell membrane derived from studies of natural multi-layered lipoprotein systems, such as the myelin sheath and the photoreceptors.

The Myelin Sheath

The myelin sheath is a lipoprotein membrane that surrounds the axon, or axis-cylinder, of the nerve fiber. In peripheral nerves this sheath is formed by the Schwann cells. In central nerves the myelin sheath is produced by the activity of the oligodendroglial cells.

The myelin sheath is a very special membrane structure in which the lipids are more abundant than the protein (Table 8–1). In central myelin there are two main protein moieties: the proteolipid and the basic protein. Another difference between this and other membranes is the apparent lack of intercalation of the proteins into the lipid bilayer. The particles observed in erythrocyte membranes and in other freeze-fractured membranes are absent in myelin (see Fig. 8–4 and Branton, 1971). It has been known for over a century that the myelin sheath has a strong birefringence, which indicates a high degree of organization at a submicroscopic level.

Studies with x-ray diffraction have revealed a spacing of 17 nm in amphibian and 18.0 to 18.5 nm in mammalian peripheral nerves. Within this period, the proportion corresponding to the lipid, protein, and water content has been estimated.

In the molecular model shown in Figure 8–5 the existence of lecithin-cholesterol (i.e., glycerophosphatides) and sphingomyelin-cholesterol complexes is postulated. The first type can be accommodated within the thickness of the lipid layer (L), but in the second, the longer sphingomyelin molecules must interdigitate in order to fill the same space.[23] This model also accounts for the localization of protein (HP) and water (HL) and is in accord with the view that each x-ray diffraction period of 18 nm corresponds to two unit membranes, i.e., to two lipid bilayers separated by a watery space (H), that originally corresponds to extracellular space.

Electron microscopic studies have confirmed that myelin has a multilayered membranous structure. In most cases, however, the x-ray diffraction period is reduced to 10 to 12 nm, thereby introducing a great many artifacts. Recently, with a highly polar embedding medium based on polymerized glutaraldehyde-urea, the periodicity obtained was of the same order as by x-ray diffraction. With this technique the lipids are not extracted; this explains the lack of collapse of the structure.[24]

The intraperiod space H in Figure 8–5 appears in the electron micrograph as a relatively thick layer that is darkly stained in Figure 8–6. It is possible that this space corresponds to hydrated carbohydrate representing the "glycocalyx" (see below) that covers the outer leaflet of the two bilayers.

The less dense lines of Figure 8–6 correspond to the hydrocarbon chains of the lipids, and the fine lines (mp) correspond to the main lines of the period.

In nerve conduction the myelin sheath seems to function as an insulator, preventing the dissipation of energy into the surrounding medium. It might act not only as a dielectric (insulating) material but also as a kind of resonant conductor in which the energy waves resonating in the lipid layers between the protein membranes could pass with maximum speed and minimum loss of energy[25] (see Chapter 22).

Photoreceptors

The retinal rods and cones are highly differentiated cells that have at their outermost segment a lipoprotein structure that is specialized for photo-reception. Studies with the polarization microscope suggest a submicroscopic organization consisting of transversely oriented protein layers alternating with lipid molecules arranged longitudinally along the axis of the photoreceptor. This type of layered organization has been demonstrated by electron microscopy in fragmented rod outer segments and in thin sections of the retina.[26] These observations indicate that the rod consists of a stack of superimposed disks (several hundred) along the axis. These disks

(*Text continued on page 156.*)

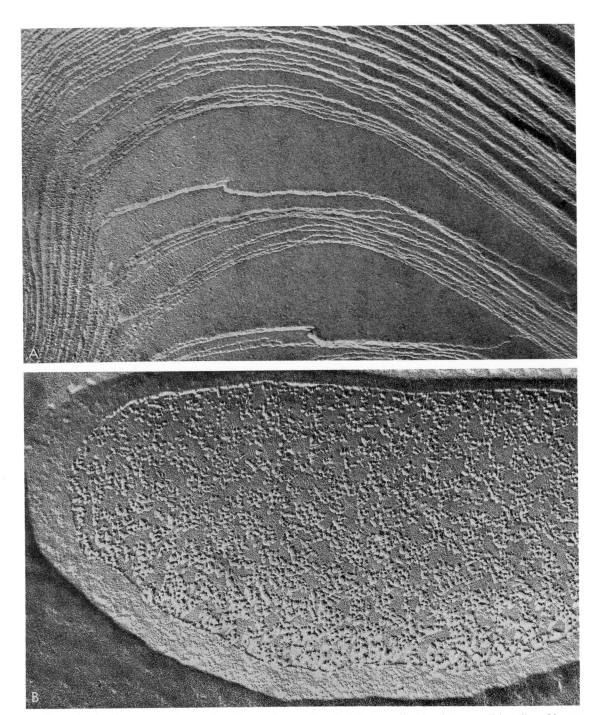

Figure 8–4. **A,** electron micrograph of freeze-fractured and etched myelin showing several lamellae. Observe that the surfaces of the fracture are smooth. ×100,000. **B,** electron micrograph of a freeze-fractured and etched red cell ghost. Most of the surface shows particles that are intercalated in the plane of cleavage of the membrane. ×88,000. (Courtesy of D. Branton.)

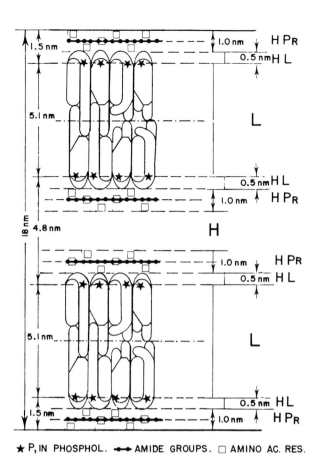

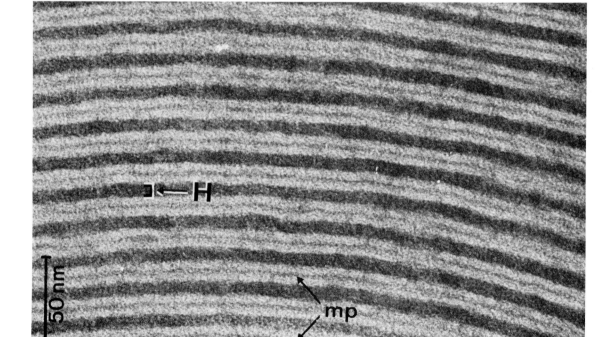

Figure 8–5. Molecular model of the myelin sheath. *HPr*, protein layer represented by a chain backbone; *HL*, water layer; *L*, lipid bilayer made of lecithin-cholesterol and sphingomyelin-cholesterol complexes (these are interdigitating); *H*, intraperiod water space. (See the description in the text.) (From Vandenheuvel, F. A., Structural studies of biological membranes: the structure of myelin. *Ann. N.Y. Acad. Sci.*, *122*:70, 1965. © The New York Academy of Sciences, 1965; reprinted by permission.)

★ P, IN PHOSPHOL.   ⬤—⬤ AMIDE GROUPS.   ☐ AMINO AC. RES.

Figure 8–6. Electron micrograph of a section of the myelin sheath using the glutaraldehyde-urea method of embedding. *H*, the intraperiod space stained with silico-tungstic acid; *mp*, the main lines of the period. (See description in the text.) ×440,000. (Courtesy of R. G. Peterson and D. C. Pease.)

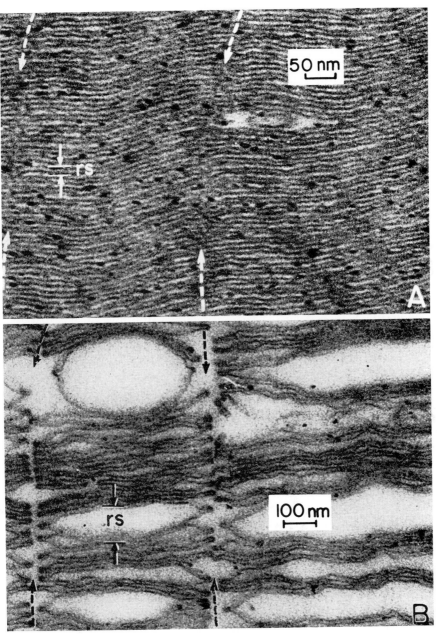

Figure 8–7. Electron micrographs of the outer segment of the retinal rods of a toad. **A,** fixation with isotonic osmium tetroxide maintains the regular organization of the retinal sacs, ×190,000; **B,** fixation with hypotonic osmium tetroxide (Palade's method) produces great swelling of the sacs and separation of the membranes. (From De Robertis, E., and Lasansky, A., "Ultrastructure and chemical organization of photoreceptors," *in* Smelser, G. K. (Ed.), *The Structure of the Eye,* New York, Academic Press, Inc., 1961.)

are really flattened sacs[27] made of two membranes, which surround a thin space of 3 nm and become continuous at the edges (Fig. 8–7). The space between the rod sacs is 5 to 12 nm. The cone outer segments, with minor differences, have a similar structure.[28]

Rod sacs are highly sensitive to osmotic change. In hypotonic solutions they swell considerably, and the inner space between the membranes becomes very large (Fig. 8–7,B).

In the frog the rod outer segments can easily be isolated and be submitted to chemical analysis. The protein composition is simpler than in other membranes. The photopigment protein, *rhodopsin*, represents 40 per cent of the total mass of the rod sacs; another 40 per cent is lipid. Rhodopsin is an integral type of protein (see above), requiring detergents to be extracted. In the plane of the membrane the rhodopsin molecules have a liquid-like distribution.[29] X-ray diffraction studies have led to the interesting conclusion that this molecule may change its position within the lipid layer. In the dark (non-stimulated condition) the molecule is about one third embedded in the lipid and two thirds exposed to the water surface. Upon activation by light (bleached rhodopsin) the molecule becomes more deeply embedded within the lipid bilayer. This change is presumably caused by changes in surface charges of the molecule.[30]

Photoreceptors transform light energy into another type of energy that can be conducted as nerve impulses. This process is based on a cycle of chemical reactions, which involve the visual pigments present in the protein membranes of the rod and cone sacs. This multilayered structure is a very effective system that facilitates the maximum absorption and utilization of light by the chromophoric groups present in the visual pigments (retinenes). The acute sensitivity of the photoreceptors, which can react to a single photon, can be explained by the fact that the possibility of striking a sensitive molecule is increased by a factor of hundreds or thousands by the molecular organization of the photoreceptor.

It is thought that the interaction with photons of light produces liberation of Ca^{2+} ions which, in turn, inhibit the sodium current that is characteristic of the photoreceptor in the dark condition. In this way photoreceptors act as energy transducers, transforming light energy into electrical signals.

THE FLUID MOSAIC MODEL OF MEMBRANE STRUCTURE

Present knowledge about the molecular organization of biological membranes is still rather indirect and comes mainly from an integration of the data on chemical analysis (see above) and of the application of several biophysical techniques, some of which (e.g., polarization microscopy, x-ray diffraction, and electron microscopy with freeze-etching) were discussed in Chapter 6. Other techniques that have yielded important information and will only be mentioned here are: nuclear magnetic resonance, optical rotatory dispersion, circular dichroism, electron spin labeling techniques, and differential thermal analysis. They have contributed to a better understanding of how the main components of the plasma membrane are integrated at the molecular level.

The important concepts that have emerged from the various methods of investigation are summarized in the so-called *fluid mosaic model* of membrane structure. This postulates: (1) that the lipid and integral proteins are disposed in a kind of mosaic arrangement; and (2) that biological membranes are quasi-fluid structures in which both the lipids and the integral proteins are able to perform translational movements within the overall bilayer. The concept of fluidity implies that the main components of the membrane, i.e., lipids, proteins, and oligosaccharides, are held in place only by means of non-covalent interactions.[31] This type of structure is supported by the fact that the components of the cell membrane can be dispersed by solvents, detergents, or denaturing agents that do not involve the breaking of truly chemical bonds. To better understand the molecular organization of the membrane, it is necessary to remember that, not only the lipids, but also many of the intrinsic proteins and glycoproteins of the membrane, are amphipatic molecules. The term *amphipathy*, coined by Hartley in 1936, refers to the presence, within the same molecule, of hydrophilic and hydrophobic groups. As was mentioned in Chapter 7 these amphipatic molecules constitute liquid crystalline aggregates in which the polar groups are directed toward the water phase and the non-polar groups are situated inside the bilayer (Fig. 7–9).

In the fluid mosaic model represented in Figure 8–8 the integral proteins of the membrane are intercalated to a greater or lesser extent into a rather continuous lipid bilayer.[32–33] This arrangement is based on the fact that these integral proteins are also amphipatic, with polar regions protruding from the surface and non-polar regions embedded in the hydrophobic interior of the membrane (see Singer and Nicholson, 1972). This arrangement may explain why different enzymes and antigenic glycoproteins may have their active sites exposed to the outer surface of the membrane. It is conceivable that a protein of appropriate size or a cluster of protein subunits may pass across the entire membrane.[34, 35] Such traversing proteins could be in contact with the aqueous solvent on both sides of the membrane (Fig. 8–1). A similar model has been postulated for the cholinergic receptor (see De Robertis, 1971). This model also reflects the generally accepted view—based on calorimetric and x-ray diffraction studies—that the major proportion of the phospholipids in the membrane are arranged in a bilayer form.[36–38]

One of the major supports for this mosaic model of the membrane, with intercalating proteins, comes from the use of freeze-etching techniques in erythrocytes and other cell membranes (see Chapter 5). The red cell ghosts show a large number of particles, about 8 nm in diameter, that are interpreted as representing proteins embedded within the

plane of cleavage which passes through the middle of the lipid bilayer (Fig. 8–4, B). Some 4200 particles per square micrometer have been counted, and there are more particles attached to the inner half of the bilayer than to the outer half.[39–40] They are randomly distributed and some of them probably correspond to the anionic permeation sites. The fluidity of the plasma membrane is supported by studies on lipid bilayers, using spin labeling techniques, which suggest a rapid lateral diffusion of the lipids.[41, 42] The fact that the integral proteins of the membrane can also undergo translational displacements within the bilayer is strongly supported by the experiments on cell fusion.[43] Two cell types (i.e., mouse L cells and human transformed cells), having different surface antigens, were marked with the corresponding antibodies labeled with two distinct fluorescent dyes. These cells were then induced to fuse under the influence of Sendai viruses. While at the beginning both cell surfaces could be recognized by their differing labels, after 40 minutes considerable intermixing of the antigens had occurred, so that the two labels could no longer be recognized. This intermixing was not prevented by inhibitors of protein synthesis, but was retarded by temperatures below 20° C. More recently, in surface antigens of cultured muscle cells, a diffusion rate of 1×10^{-9} cm^2 per second was calculated.[44] Also the so-called clustering or "capping" effect observed in lymphocytes may result in the displacement of antigenic molecules within the surface membrane.[45] In this case, upon

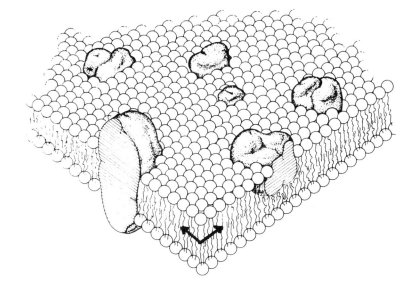

Figure 8–8. The fluid mosaic model of the cell membrane. (See the description in the text.) Observe that the integral proteins may be embedded in the lipid bilayer and may even go across it. The bulk of the lipid is disposed in an interrupted bilayer. The arrows indicate the plane of cleavage in the freeze-etching techniques as used in Figure 8–4, B. (From Singer, S. J., and Nicholson, G. L., *Science, 175:*720, 1972, Copyright 1972 by the American Association for the Advancement of Science.)

treatment with a labeled antibody, the antigens are first randomly distributed, but after some time they become clustered and agglomerate at one pole of the cell (i.e., capping) where pinocytosis of the antigen-antibody complexes takes place. This process is also inhibited by temperatures that produce solidification of the lipid bilayer.

SUMMARY:
The "Fluid Mosaic" Model

One of the most favored models for the plasma membrane is the so-called fluid mosaic structure. According to this model, there is a rather continuous lipid bilayer into which the integral proteins of the membrane are intercalated; both these components are capable of translational diffusion within the overall bilayer. The mosaic nature of the membrane is favored by the results of freeze-etching techniques in which protein particles are shown at the plane of cleavage of the bilayer. The fluidity of the lipids is supported by many indirect studies based on x-ray diffraction, differential thermal analysis, and electron spin techniques. The fluidity of the integral proteins is supported by experiments on cell fusion and on those of clustering and "capping" of surface antigens.

DIFFERENTIATIONS AT THE CELL SURFACE

Regions of the cell surface of certain cells are related to absorption, secretion, fluid transport, and other physiologic processes.[46, 47] Topographically they are referred to as specializations of the cell surface, specializations of contact surfaces between cells, and specializations of the cell base.

Figure 8–9 is a diagram of a columnar epithelial cell in which various types of differentiations of the cell membrane may be observed. The apical surface is projected into slender processes called *microvilli*. On the surface corresponding to the edge of the cell in contact with an adjacent cell there are several differentiations (see below). At the base of the cell the plasma membrane is covered by a thick basement membrane of extracellular material where infoldings of the plasma membrane may be observed.

Microvilli

In the intestinal epithelium microvilli are very prominent and form a compact structure that appears under the light microscope as a *striated border*. These microvilli, which are 0.6 to 0.8 μm long and 0.1 μm in diameter, represent cytoplasmic processes covered by the plasma membrane. Within the cytoplasmic core fine microfilaments are observed which in the subjacent cytoplasm form a terminal web. The outer surface of the microvilli is covered by a coat of filamentous material (fuzzy coat) composed of glycoprotein macromolecules (Fig. 8–10). Microvilli increase the effective surface of absorption. For example, a single cell may have as many as 3000 microvilli, and in a square millimeter of intestine there may be 200,000,000. The narrow spaces between the microvilli form a kind of sieve through which substances must pass during absorption.

Numerous other cells, in addition to intestinal epithelium, have microvilli, although fewer in number. They have been found in mesothelial cells, in the epithelial cells of the gall-bladder, uterus, and yolk sac, in hepatic cells, and so forth.

The *brush border* of the kidney tubule is similar to the striated border, although it is of larger dimensions. An amorphous substance between the microvilli gives a periodic acid–Schiff reaction for polysaccharides. Between the microvilli, at the base, the cell membrane invaginates into the apical cytoplasm. These invaginations are apparently pathways by which large quantities of fluid enter by a process similar to pinocytosis. (Other specializations of the cell surface, such

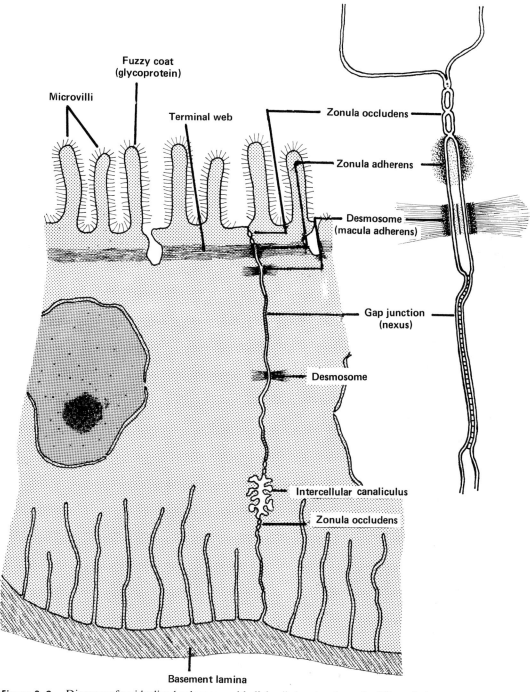

Figure 8–9. Diagram of an idealized columnar epithelial cell showing the main differentiations of the cell membrane. **To the right,** at a higher magnification, the series of differentiations found between two epithelial cells are indicated. (See the description in the text.)

as cilia and flagella, are described in Chapter 21.)

Desmosomes or Macula Adherens

Essentially four types of differentiations are present at the lateral surfaces of epithelial cells: (1) the *macula adherens* or *desmosome;* (2) the *zonula adherens,* also called *intermediary junction or terminal bar;* (3) the *zonula occludens* or tight junction; and (4) the *gap junction or nexus.* The first three types may be observed in the electron micrograph of Figure 8–10. The fourth appears to play a role in intercellular communication (see Fig. 8–9, B).

The so-called *desmosomes,* found in a number of epithelial cells, appear under the light microscope as darkly stained bodies at the midpoint of what was once interpreted as an intercellular bridge. These structures, are formed by a circular area, about 0.5 μm in diameter, of the plasma membranes of two adjacent cells that are separated by a distance of 30 to 50 nm. Under the membrane there is a dense intracellular plaque toward which numerous *tonofilaments* converge. These filaments describe a kind of loop in a wide arc and course back into the cell. Within the intercellular gap a coating material may be observed which sometimes forms a discontinuous middle dense line. This extracellular material contains acid mucopolysaccharides and proteins. In fact, desmosomes are broken by trypsin, collagenase, and hyaluronidase and are sensitive to agents that chelate calcium. The dense intracellular attachment plaques are digested by proteolytic enzymes.[48]

While the tonofilaments provide the intracellular mechanical support, cellular adhesion at the desmosome depends on the extracellular coating material. Frequently there are regions of looser contact between the desmosomes and even intercellular spaces for free circulation of fluids.

Along the basal surface of some epithelial cells *hemidesmosomes* may be observed. These are similar to desmosomes in fine structure, but represent only half of them, the outer side frequently being substituted with collagen fibrils.[49]

Zonula Adherens

The zonula adherens or *terminal bar* also called *intermediary junction* is generally found at the interface between columnar cells just below the free surface. Under the electron microscope the terminal bar appears somewhat similar to the desmosome. The membrane is thickened and the adjacent material is dense, but filaments are generally lacking (Fig. 8–10).

Within the *zonula adherens* there are parts of the two adjacent membranes that come into close contact at certain points, forming a kind of anastomosing network that extends around the apical region of the cell. Such a network is especially visible in freeze-etched preparations.[50]

Tight Junctions

Cell contacts may be specially differentiated to create a barrier or "seal" to diffusion. As shown in Figure 8–10, between the two cells a series of differentiated zones start from the apical region and form a tripartite complex with the following components: the tight junction, the intermediary junction, and the desmosome.[51]

In the *tight junction* (zonula occludens) the adjacent cell membranes have fused, and therefore there is no intercellular space for a variable distance (Fig. 8–10, *1–2*). The tight junction is situated just below the apical border, and at this point the outer leaflets of the unit membranes fuse in a single intermediary line. Experiments have demonstrated the relationship between tight junctions and epithelial permeability. For example, macromolecules put into the lumen cannot penetrate the intercellular space. Tight junctions may also play an important role in brain permeability at the level of the blood-brain barrier and the synaptic barrier (Fig. 8–11).

Gap Junctions and Electrical Coupling

In Chapter 20 the problem of electrical coupling and junctional communications will be discussed, and in Chapter 24 the so-called electrical synapses will be mentioned. At present some of the possible structural bases of these phenomena will be considered.

Electrical coupling seems to be related to the so-called *gap junctions* or *nexus* that are shown in the diagram of Figure 8–9. An excellent material for the study of gap junctions is the myocardial tissue, in which the

(*Text continued on page 164*)

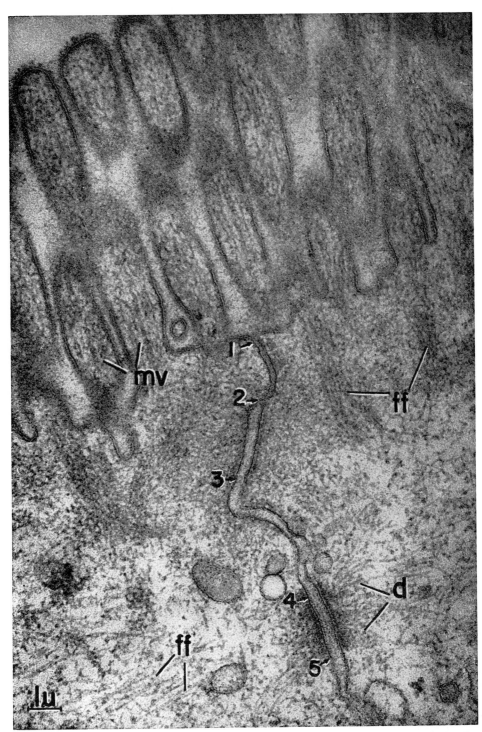

Figure 8–10. Electron micrograph showing the apical region of contact between two intestinal cells. *1-2*, tight junction; *2-3*, intermediary junction; *4-5*, desmosome. (See the description in the text.) *d*, desmosome; *ff*, fine filaments in the matrix; *mv*, microvilli. ×96,000. (From Farquhar, M., and Palade, G. E., *J. Cell Biol., 17*:375, 1963.)

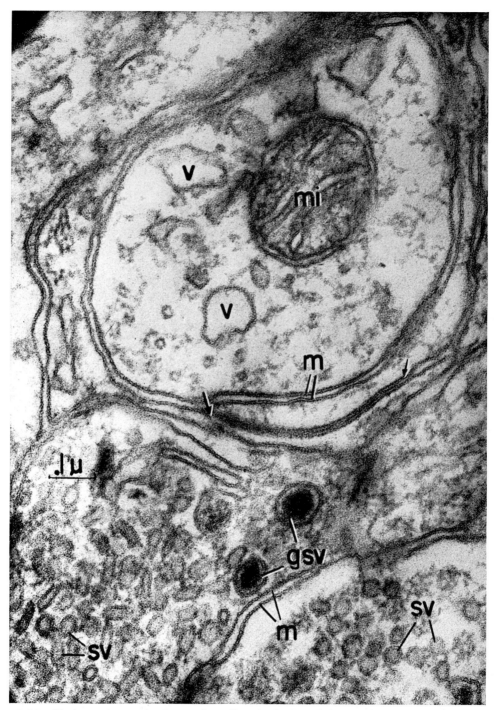

Figure 8–11. Electron micrograph of synaptic endings in the hypothalamus of the rat. *gsv*, granular synaptic vesicle; *m*, cell membrane; *mi*, mitochondrion; *sv*, synaptic vesicle; *v*, vacuoles. Note that the membranes of all these components show the unit structure with a triple layered organization. Notice the tight junction that surrounds a synaptic ending between the arrows. ×135,000. (From E. De Robertis, unpublished.)

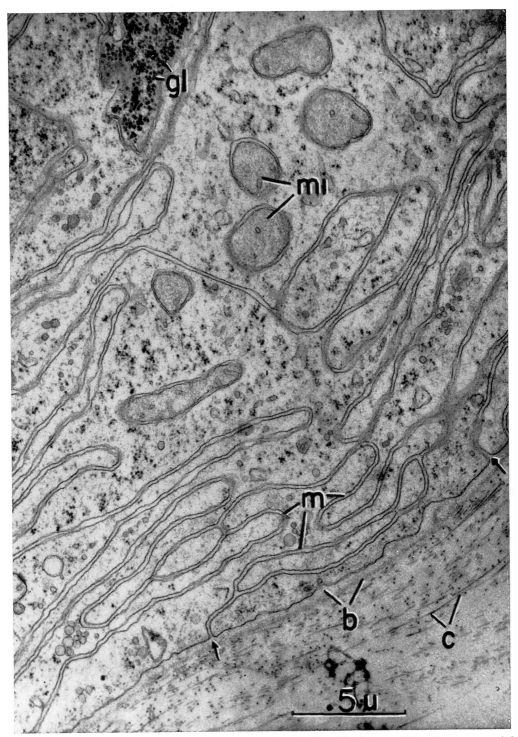

Figure 8–12. Electron micrograph of the basal region of a Müller cell of the toad showing numerous infoldings of the basal cell membrane (*m*). Arrows indicate openings of the intermembranal space at the base. *b*, basement membrane; *c*, collagen fibrils; *gl*, glycogen; *mi*, mitochondria. ×60,000. (Courtesy of A. Lasansky.)

action potential is transmitted from cell to cell by an electrical coupling. In between these cells junctions have been observed which have a minute gap of about 2 nm, which is permeable to some electron opaque substances such as lanthanum salts. When studied in tangential sections, gap junctions show a hexagonal array having a unit size of 9 nm. These hexagons comprise the gap and the outer leaflets of the two opposing membranes. The electron dense material is apparently able to penetrate between the hexagons and into the core of the prisms.[52] The freeze-etching technique (Chapter 6) has demonstrated this hexagonal pattern also in the middle region of the gap junction.[50]

At the *cell base* of certain cells involved in rapid water transport, numerous infoldings of the plasma membrane penetrate deeply into the cell (Fig. 8–12). In a three-dimensional view, these folds form septa that subdivide the basal cytoplasm into narrow compartments containing large mitochondria.

COATS OF THE CELL MEMBRANE

Glycocalyx

At the beginning of this chapter the *coats* surrounding cell membranes were mentioned. These are very conspicuous in eggs of marine animals and in amphibia. A glycoprotein-like substance called *mucin* is the main constituent. Mucins also cover and protect the cell surface lining the gastrointestinal tract. Polysaccharides constitute the *pectin* and *cellulose* of plant cell walls and the *chitin* of

crustacea. (The cell wall of plant cells is discussed in Chapter 12.)

Glycoproteins and polysaccharides, in the form of hyaluronic acid, are found at the base of most epithelial cells, in capillaries, and also in many intercellular spaces. The name *glycocalyx*[19] has been coined to designate the glycoprotein and polysaccharide covering that surrounds many cells.

The glycocalyx contains the oligosaccharide side-chains of the glycolipids and glycoproteins which are exposed to the outer surface of the membrane. In the so-called *greater membrane model* of the plasma membrane shown in Figure 8–13, both the cell coat and the lipoprotein structure are shown. The cell coat has negatively charged sialic acid termini, both on the glycoproteins and gangliosides, which may bind Ca^{2+} and Na^+ ions.

When the membranes are treated with *neuraminidase,* an enzyme that removes sialic acid, there is a reduction in the negative charge of the membrane.

The use of PAS and Alcian blue staining may render the surface of many cell types more visible.[53] Staining with ruthenium red or with lanthanum[54] may reveal the thin cell coat present in biliary capillaries (Fig. 8–14) and in other cells.

In Chapter 10 it will be mentioned that the biosynthesis of the glycoproteins forming the glycocalyx takes place in the ribosomes of the endoplasmic reticulum and that the final assembly with the oligosaccharide moiety is accomplished in the Golgi complex. The cell coat can be considered to be a secretion product of the cell which is incorporated into cell surface and undergoes continuous renewal.

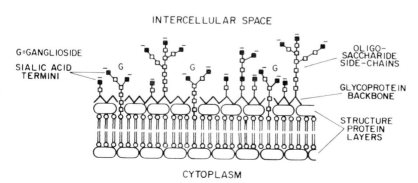

Figure 8–13. A greater membrane model including the cell coat on the outer surface. (From Lehninger, A. L., *Proc. Nat. Acad. Sci. U.S.A.*, 60:1069–80, 1968.)

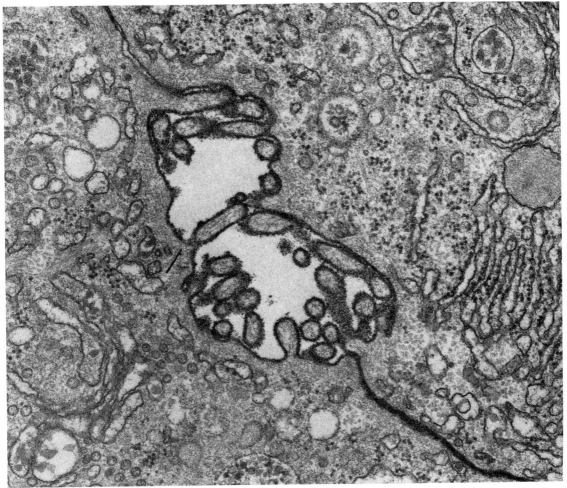

Figure 8–14. Electron micrograph of a biliary capillary stained with lanthanum nitrate to enhance the electron density of the surface coat of cells. Note the dark material between the liver cells and around the microvilli. ×38,400. (Courtesy of D. Ferreira.)

Functions of the Glycocalyx

Although the cell coat is not absolutely necessary for the integrity of the cell and for the permeability of the plasma membrane it, nevertheless, performs functions that are of great significance to studies in cell biology.

Molecular Recognition. The evidence suggesting that molecular entities used by the cell for *specific recognition* are in the cell coat will be discussed in Chapter 20 (see section on cell interactions). The glycoproteins and glycolipids may possibly represent the molecular sites by which different cell types recognize one another, thus contributing to the formation of cell layers or compact tissues. This recognition also involves the important phenomenon of *contact inhibition* that also will be presented in Chapter 20. It is evident that the function of molecular recognition reaches its maximum expression in the nervous system, where a neuron can make synaptic contacts with numerous other neurons, thereby forming specific neuronal circuits of immense complexity. This property of the cell membrane (i.e., molecular recognition) is dependent on the expression of genes located within the nucleus. Certainly, molecular recognition is one of the fields of cell biology and neurobiology which must be strongly developed in the future (see Singer and Rothfield, 1973).

Antigenicity. In the erythrocyte membrane and other cell membranes there are specifically determined antigens for the A, B, and O blood groups and others that have been mentioned previously. Among the antigens are those of histocompatibility that permit the recognition of the cells of one organism and the rejection of other cells that are alien to it (e.g., the rejection of grafts from another organism). This function, essentially, is related to molecular recognition.

Filtration. The extracellular coats that surround many vertebrate capillaries, especially the kidney glomerulus, act as a filter and regulate the passage of molecules according to size. Hyaluronate in connective tissue may control diffusion.

Microenvironment. The glycocalyx may change the concentration of different substances at the surface of the cell, not only functioning as diffusion barriers but also affecting the cationic environment of the cell because of their charge. In this respect they are similar to exchange resins used in chemistry. For example, a muscle cell with its excitable plasma membrane is surrounded by a glycocalyx that can trap sodium ions. Certain components, such as hyaluronate, can drastically change the electrical charge and pH at the cell surface. Because of this, enzymes present at the plasma membrane may change their activity while they are kept in the microenvironment of the cell.

Enzymes. In the intestinal cells the apical coat is remarkably stable and cannot be separated from the underlying striated border made of microvilli.[55] Histochemical techniques have demonstrated alkaline phosphatase in the coat as well as on the surface of the microvilli. When these structures are isolated, practically all the enzymes involved in the terminal digestion of carbohydrates and proteins are found in them.[56, 57]

SUMMARY:
The Cell Surface

A cell coat or glycocalyx made of glycoproteins, glycolipids, and polysaccharides is present in most cell types. The macromolecules forming the glycocalyx are synthesized and secreted by the cell by way of the endoplasmic reticulum and Golgi complex. In addition to serving as a protective cover for the plasma membrane, the glycocalyx has important functions related to molecular recognition and specific interactions between cells. It is the site of antigens such as the A, B, and O blood groups and others that are related to the rejection of cellular grafts. In certain tissues the cell coat may act as a filter to control diffusion of fluids, or may create a special microenvironment for the cell. Certain enzymes may be located in the cell coat.

REFERENCES

1. Maddy, A. H. (1966) *Internat. Rev. Cytol., 20*:1.
2. Nicolson, G. L., Marchesi, V. T., and Singer, S. J. (1971) *J. Cell Biol., 51*:265.
3. Lapetina, E. G., Soto, E. F., and De Robertis, E. (1968) *J. Neurochem., 15*:437.
4. Lapetina, E. G., Soto, E. F., and De Robertis, E. (1967) *Biochim. Biophys. Acta, 35*:33.
5. Marchesi, S. L., Steers, E., Marchesi, V. T., and Tillack, T. W., (1970) *Biochemistry, 9*:50.
6. Silman, H. I., and Karlin, A. (1967) *Proc. Natl. Acad. Sci., 58*:1664.
7. De Robertis, E., and Fiszer de Plazas, S. (1970) *Biochim. Biophys. Acta, 219*:388.

8. Knauf, P. A., and Rothstein, A. (1971) J. Gen. Physiol., 58:221.
9. Marchesi, V. T., Tillack, T. W., Jackson, R. L., Segrest, J. P., and Scott, R. E. (1972) *Proc. Natl. Acad. Sci., 69*:1445.
10. Bretscher, M. S. (1971) *J. Molec. Biol., 59*:351.
11. Segrest, J. P., Kahane, I., Jackson, R. L., and Marchesi, V. T. (1973) *Arch. Biochem. Biophys. 155*:167.
12. Folch, J., and Lees, M. (1951) *J. Biol. Chem., 191*:807.
13. Emmelot, P. (1968) Plasma Membranes. *Excerpta Medica. Internat. Cong. Ser., 166*:16.
14. Danielli, J. F., and Harvey, E. N. (1934) *J. Cell. Comp. Physiol., 5*:483.
15. Lehninger, A. L. (1968) *Proc. Natl. Acad. Sci. USA, 60*:1069.
16. Robertson, J. D. (1959) *Biochem. Soc. Symp., 16*:3.
17. Sjöstrand, F. S. (1963) *Nature, 199*:1262.
18. Yamamoto, T. (1963) *J. Cell Biol., 17*:413.
19. Bennett, H. S. (1963) *J. Histochem. Cytochem., 11*:14.
20. Ito, S. (1965) *J. Cell Biol., 27*:475.
21. Roth, T. F., and Porter, K. R. (1964) *J. Cell Biol., 20*:313.
22. Lucy, J. A. (1964) *J. Theoret. Biol., 7*:360.
23. Vanderheuvel, F. A. (1965) *Ann. N.Y. Acad. Sci., 122*:57.
24. Peterson, R. G., and Pease, D. C. (1972) *J. Ultrastruct. Res., 41*:115.
25. Engström, A., and Finean, J. B. (1958) *Biological Ultrastructure*. Academic Press, New York.
26. Sjöstrand, F. S. (1953) *Experientia, 9*:68.
27. De Robertis, E. (1956) *J. Biophys. Biochem. Cytol., 2*:319.
28. De Robertis, E., and Lasansky, A. (1958) *J. Biophys. Biochem. Cytol., 4*:743.
29. Blasie, J. K., and Worthington, C. R. (1969) *J. Molec. Biol., 39*:417.
30. Blasie, J. K. (1972) *Biophys. J., 12*:191.
31. Gitler, C. (1972) *Ann. Rev. Biophys. Bioenz., 1*:51.
32. Wallach, D. F. H., and Zahler, P. H. (1966) *Proc. Natl., Acad. Sci. USA., 56*:1552.
33. Lenard, J., and Singer, S. J. (1966) *Proc. Nat. Ac. Sci., 56*:1828.
34. Steck, T. L., Fairbanks, G., and Wallach, D. F. H. (1971) *Biochemistry, 10*:2617.
35. Bretscher, M. S. (1971) *J. Mol. Biol., 58*:775.
36. Steim, J. M., Tourtellote, M. E., Reinert, J. C., Mc Elhaney, R. N., and Rader, R. L. (1969) *Proc. Nat. Acad. Sci., 63*:104.
37. Engelman, D. M. (1970) *J. Mol. Biol., 47*:115.
38. Wilkins, M. H. F., Blaurock, A. E., and Engelman, D. M. (1971) *Nature (New Biol.), 230*:72.
39. Pinto da Silva, P., and Branton, D. (1970) *J. Cell Biol. 45*:598.
40. Tillack, T. W., and Marchesi, V. T. (1970) *J. Cell Biol., 45*:649.
41. Hubbell, W. L., and McConnell, H. M. (1971) *J. Am. Chem. Soc., 93*:314.
42. Devaux, P., and McConnell H. M. (1972) *J. Am. Chem. Soc., 94*:4475.
43. Frye, L. D., and Edidin, M. (1970) *J. Cell Sci. 7*:319.
44. Edidin, M., and Frambrough, D. (1973) *J. Cell Biol., 57*:27.
45. Taylor, R. B., Duffus, W. P. H., Raff, M. C., and de Petris, S. (1971) *Nature (New Biol.), 233*:225.
46. Sjöstrand, F. S. (1956) *Internat. Rev. Cytol., 5*:455.
47. Fawcett, D. (1958) Structural specializations of the cell surface. In: *Frontiers in Cytology*. (Palay, S. L., ed.) Yale University Press, New Haven, Conn.
48. Douglas, W. H. J., Ripley, R. C., and Ellis R. A. (1970) *J. Cell Biol. 44*:211.
49. Kelly, D. E. (1966) *J. Cell Biol., 28*:51.
50. Goodenough, D. A., and Revel, J. P. (1970) *J. Cell Biol. 45*:272.
51. Farquhar, M., and Palade, G. E. (1963) *J. Cell Biol., 17*:375.
52. Revel, J. P., and Karnovsky, M. J. (1967) *J. Cell Biol., 33*:C7.
53. Rambourg, A., and Leblond, C. P. (1967) *J. Cell Biol., 32*:27.
54. Shea, S. M. (1971) *J. Cell Biol. 51*:611.
55. Ito, S. (1969) *Fed. Proc., 28*:12.
56. Miller, D., and Crane, R. K. (1961) *Biochim. Biophys. Acta, 52*:293.
57. Ugolev, A. M. (1965) *Physiol. Rev., 45*:555.

ADDITIONAL READING

Branton, D. (1971) Freeze-etching studies of membrane structure. *Philo. Trans. R. Soc. London* [Biol. Sci.] *261*: 133.

Bretscher, M. S. (1973) Membrane structure: some general principles. *Science, 181*:622.

De Pierre, J. W., and Karnovsky, M. L. (1973) Plasma membranes of mammalian cells. *J. Cell Biol., 56*:275.

De Robertis, E. (1971) Molecular biology of synaptic receptors. *Science, 171*:963.

Fawcett, D. W. (1958) Structural specialization of the cell surface. In: *Frontiers in Cytology.* (Palay, S. L., ed.) Yale University Press, New Haven, Conn.

Fox, C. F. (1972) The structure of cell membranes. *Sci. Am., 226*:30.

Hendler, R. W. (1971) Biological membrane ultrastructure. *Physiol. Rev., 51*:66.

Ito, S. (1969) Structure and function of the glycocalyx. *Fed. Proc., 28*:12.

Raff, M. C., and de Petris, S. (1973) Movement of lymphocyte surface antigens and receptors: the fluid nature of the lymphocyte plasma membrane and its immunological significance. *Fed. Proc., 32*:48.

Rambourg, A. (1971) Morphological and histochemical aspects of glycoproteins at the surface of animal cells. *Int. Rev. Cytol., 31*:57.

Robertson, J. D. (1959) The ultrastructure of cell membranes and their derivatives. *Biochem. Soc. Symp., 16*:3.

Singer, S. J. (1971) The molecular organization of biological membranes. *In: Structure and Function of Biological Membranes.* (Rothfield, L. I., ed.) Academic Press, Inc., New York, p. 145.

Singer, S. J., and Nicolson, G. L. (1972) The fluid mosaic model of the structure of cell membranes. *Science, 175*:720.

Singer, S. J. and Rothfield, L. I. (1973) Synthesis and turnover of cell membranes. *Neurosci. Res. Program Bull., 11*:181.

Sjöstrand, F. S. (1967) The structure of cellular membranes. *Protoplasma, 63*:248.

Vanderkooi, G. (1972) Molecular architecture of biological membranes. *Am. New York Acad. Sci., 195*:6.

Winzler, R. J. (1969) In: *Red Cell Membrane* (Jamieson, G. A., and Greenwalt, T. J., eds.) J. B. Lippincott Company, Philadelphia, p. 157.

Part Five

THE CYTOPLASM AND CYTOPLASMIC ORGANELLES

In the following three chapters the structural, biochemical, and physiological characteristics of the cytoplasm of animal and plant cells, as well as their main organelles, are considered. The discussion is based on the latest studies of electron microscopy, cytochemistry, and structural evolution of biological systems.

Chapter 9 is a discussion of the structure and function of the cytoplasm, in particular, the matrix—the true internal milieu of the cell. Cytoplasm is capable of carrying on biosynthesis, glycolysis, and many fundamental functions related to the movement of the cell and to cell differentiation. The ribosomes, which may also be present in the matrix, will be discussed in Chapter 18.

The cytoplasmic vacuolar system, which comprises the nuclear envelope (see Chapter 12), the endoplasmic reticulum, and the Golgi complex, is also described in Chapter 9. Combining electron microscopy, cell fractionation methods, and biochemical analysis has proved to be a most valuable approach to the investigation of the structure and function of these intracellular membranes. The vacuolar system subdivides the cytoplasm into several compartments that may function independently. It is postulated that this system interchanges, circulates, and segregates the products that are absorbed by the cell or that are synthesized on the ribosomes. Some of these products may be prepared for export (i.e., secretion) out of the cell. The recent studies on the isolation and biochemical characterization of the Golgi complex will be mentioned. The general function of the Golgi complex in secretion and its relation to the production of glycoproteins will be emphasized.

In general terms, most of the metabolic and biosynthetic functions of the cell occur in the cytoplasm. The cytoplasm is differentiated by the activity of the genes contained in the nucleus and becomes adapted to the division of cellular work. In different types of cells, therefore, the cytoplasm may be markedly different while the nucleus appears to be comparatively uniform.

Chapter 10 presents the mitochondria as macromolecular machines whose chemical and molecular organization is admirably adapted to their function in cell respiration. Special emphasis will be placed on the ultrastructure and compartmentalization of these organelles and on the studies concerning the separation of these compartments. The coupling of oxidation and phosphorylation will be explained on the basis of the special localization of the enzymes and their peculiar asymmetry on the mitochondrial crests. The function of mitochondria is intimately related to the conformational changes which they undergo during their function. The concept that both the mitochondrion and the chloroplast contain DNA and special ribosomes and are capable of some local protein synthesis is discussed in Chapter 10. This idea is based on the theory that these organelles contain genetic information and a certain degree of autonomy within the cell.

Chapter 11 is dedicated to the plant cells and emphasizes some of the differences between them and animal cells, in particular, the presence of rigid cell walls, a type of cell division peculiar to plants, and the presence of special organelles, generally called plastids, which are related to the special metabolic properties of plants. Most of this chapter concerns the ultrastructure and macromolecular organization of the chloroplasts and their fundamental function in photosynthesis.

nine

THE CYTOPLASM
AND VACUOLAR SYSTEM

The present chapter is, in some ways, an expansion of Chapter 2, in which the general structure of prokaryotic and eukaryotic cells was considered. In spite of the differences between these two types of cells, outlined in Table 2–1, the similarities are also notable. The cytoplasmic matrix of a eukaryotic cell contains the same components as a bacterium: ribosomes, RNA molecules, globular proteins, enzymes, and so forth (Fig. 2–1). The new components that have evolved in higher cells are the many intracellular membranes that constitute the vacuolar system, with its several portions, and the membrane-bound organelles (e.g., mitochondria, chloroplasts, lysosomes, peroxisomes, and vacuoles). As a consequence of the vacuolar system, numerous compartments and subcompartments are formed in the cell. Of all these cellular subdivisions, the nucleus, where the main portion of the DNA is confined, is most important (see Chapter 12). The vacuolar system is little developed in most embryonic plant and animal cells (Fig. 9–1), but it increases in complexity with cell differentiation.

CYTOPLASMIC MATRIX

The most important part of the cell, and that which constitutes the true internal milieu, is the *cytoplasmic matrix.* The colloidal properties of the cell, such as those essential to sol-gel transformations, viscosity changes, intracellular motion (cyclosis), ameboid movement, spindle formation, and cell cleavage, depend, for the most part, on the cytoplasmic matrix. Furthermore, the cytoplasmic matrix is the site of many fibrillar differentiations found in specialized cells, such as keratin fibers, myofibrils, microtubules, and filaments, some of which are described in detail in later chapters.

After the nuclear, mitochondrial, and microsomal fractions have been separated by cell fractionation (Chapter 6), the remaining supernatant, or *soluble fraction* or *cytosol* (Fig. 6–5), contains the soluble proteins and enzymes found in the cytoplasmic matrix. These proteins and enzymes constitute 20 to 25 per cent of the total protein content of the cell. Among the important *soluble enzymes* present in the matrix are those involved in glycolysis and in the activation of amino acids for protein synthesis. The enzymes of many reactions that require ATP are found in the soluble fraction. Soluble (transfer) RNA is also found in this part of the cell.

Ergastoplasm

At the end of the 19th century, it was discovered that portions of the cytoplasm in certain cells have a differential staining property. Because these areas stained with basic dyes, they were called the *basophilic,* or *chromidial, cytoplasm* (Hertwig). The still common name *ergastoplasm* (Gr., *ergazomai,* to elaborate and transform) was coined by Garnier in 1887 to imply that biosynthesis is the fundamental role of this substance.

The ergastoplasm includes basophilic regions of the ground cytoplasm, such as the

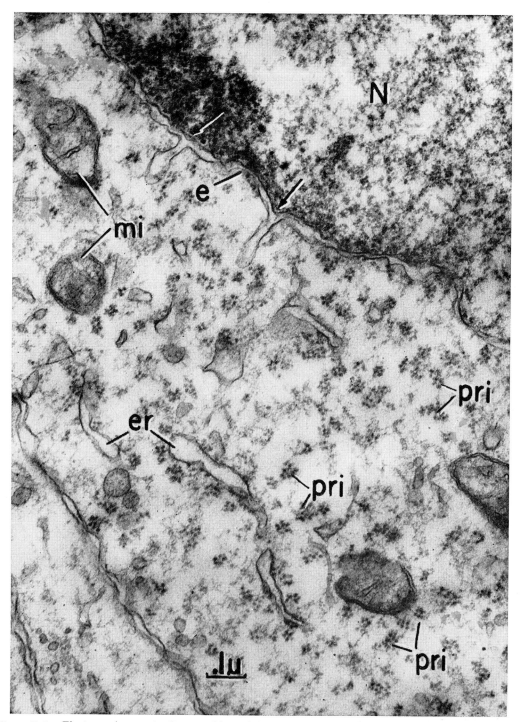

Figure 9-1. Electron micrograph of a neuroblast of the cerebral cortex of a rat embryo, showing the cytoplasm rich in matrix with numerous ribosomes and little development of the vacuolar system. *e*, nuclear envelope sending projections into the cytoplasm (arrows); *er*, endoplasmic reticulum; *mi*, mitochondria; *N*, nucleus; *pri*, polyribosomes (groups of ribosomes). ×45,000. (From E. De Robertis.)

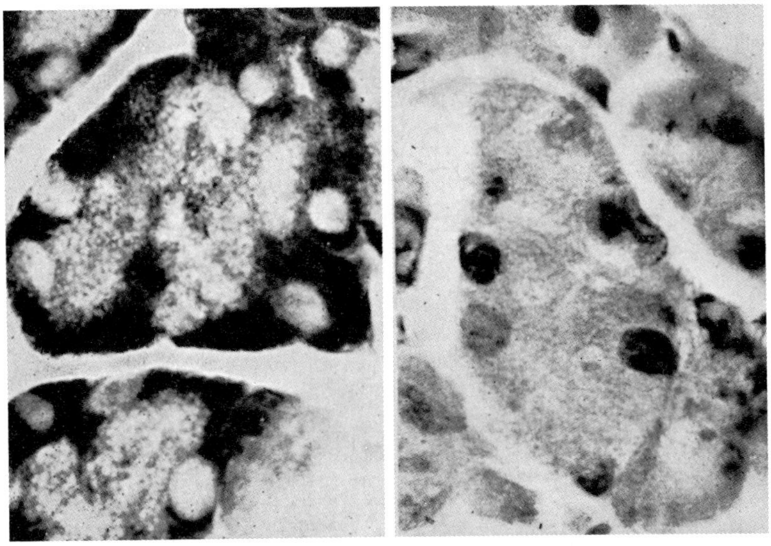

Figure 9–2. **Left,** pancreatic acini frozen and dried and stained with toluidine blue. The basophilic substance appears intensely stained. **Right,** same, but after digestion with ribonuclease; the basophilic substance has disappeared. (From E. De Robertis.)

Nissl bodies of the nerve cells, the basal cytoplasm of serous cells (e.g., secretory cells of the pancreas and the parotid gland and chief cells of the stomach), and the basophilic clumps of liver cells. Caspersson, Brachet, and others demonstrated that the intense basophilic property of the ergastoplasm is due to the presence of ribonucleic acid (Figs. 6–4 and 9–2).

It has been shown that the ergastoplasm loses its staining properties if the cell is treated with ribonuclease, an enzyme that hydrolyzes RNA (Figs. 6–4 and 9–2). RNA is contained principally in the ribosomes, and as a result, a relationship between the ergastoplasm and protein synthesis was postulated.

Properties of the Cytoplasmic Matrix

The cytoplasm along with its related membrane can be considered to be a highly heterogeneous colloid system in which the matrix represents the least compact part and the intracellular membranes, the relatively more dense portions. Some *mechanical properties* such as elasticity, contractility, cohesion, rigidity, and intracellular movements (e.g., cyclosis), are related to the cytoplasmic matrix, but the vacuolar system may also be involved. Some of the properties may be changed by the application of moderate *hydrostatic pressure*, as in the case of cyclosis, ameboid movement, cell division, and migra-

tion of pigment in chromatophores (pigment-carrying cells), which can be inhibited. This inhibition is due to the *solation* induced by pressure on the gelled portions of the cytoplasm.

Viscosity. Environmental or internal factors can change the *viscosity* of different cells. In *amoeba,* viscosity is temperature-dependent, and within certain limits it is reversible. Above 30° C, there is an abrupt increase in viscosity, because of permanent injury by heat[1]. Viscosity is decreased, however, in *anaerobiosis* and in hypotonic solutions. During the mitotic cycle there are continuous changes in viscosity. Table 9–1 shows some viscosity values of various cells.

TABLE 9–1. Viscosity of Various Cells*

Substance	*Temperature (°C)*	*Viscosity (in Centipoises)*
Water	25	0.8937
Sucrose solution		
(20%)	20	1.960
(60%)	20	56.5
Nerve fiber	20	5.5
Amoeba dubia	18	2
Slime molds	20	9 to 18
Chara	20	10
Arbacia egg	20	7
Paramecium	20	50

*From Heilbrunn, L. V., *An Outline of General Physiology,* 3rd ed., Philadelphia, W. B. Saunders Company, 1952.

pH and Oxidation-Reduction. By injecting pH indicators (i.e., substances which change color according to the *hydrogen ion concentration*), the pH of the cytoplasm and other parts of the cell can be determined. In general, the cytoplasmic matrix is slightly acid (pH about 6.8).

Differences of pH have been recorded in at least three regions of the cell. A number of vacuoles surrounded by a membrane can be observed in animal cells and, particularly, in protozoa and plant cells. The content of these vacuoles may be either basic or acidic (pH may be as low as 5.0). Another region that shows a differential pH is the aqueous nucleoplasmic matrix, which, in a variety of plant and animal cells, has a pH of 7.6 to 7.8.

Characteristic of the protoplasm is its buffering capacity. The pH of the cell can be altered by adding acids or alkalis to the medium or by injecting the same into the cell, but the original pH value is rapidly reestablished as long as the vitality of the cell has not been altered.

The *oxidation-reduction potential* (the reducing ability) of the cytoplasm can be determined by introducing into the cell dyes that change color or are decolorized when reduced. This color reaction depends on the partial pressure of oxygen in the medium and the concentration of enzyme systems and metabolites found in the cell. Furthermore, it indicates the process by which chemical energy is used by the cell. For example, in the ameba the oxidation-reduction potential of the cytoplasm is approximately -0.275 volt in anaerobiosis and $+0.070$ volt in aerobiosis.

Microfilaments

In the cytoplasmic matrix of many cells, the electron microscope has revealed fine *microfilaments,* some of which are related to the plasma membranes (see Chapter 8). (Other structures also made visible include some specialized microfilaments and fibrils present in muscle cells, and the keratin tonofilaments of epithelial cells.) The microfilaments in the matrix are more evident when they are organized in a parallel array; they are most prominently revealed after aldehyde fixation. Some of these filaments appear to originate by polymerization in response to environmental stresses or to cell motion and presumably confer the gelled consistency to

parts of the cell matrix (see Wessells et al., 1971).

Microfilaments have a diameter of 4 to 6 nm and are at least of two types: the *L-type* and the *S-type.* The so-called L-type (i.e., from lattice) is made of short segments forming a kind of network that comes near to the plasma membrane and is inserted on its inner surface. In the intestinal cell, for example, the segments form the *terminal web,* just below the apical membrane. These filaments are characterized by their sensitivity to *cytochalasin B,* a drug that inhibits several functions of the cytoplasmic matrix: cytoplasmic streaming, cell motility, phagocytosis, discharge of secretion, cell polarity, and others (see Chapter 22 and Allison, 1973).

The S-type of filament forms a filamentous sheath beneath the plasma membrane. These filaments are elongated and frequently are arranged in parallel. These S filaments are insensitive to cytochalasin *B* and have properties that are similar to the protein *actin* from muscle. Both types of microfilaments and their possible function in the cell will be presented in detail in Chapter 22.

Microtubules

In addition to the microfilaments the cytoplasmic matrix of higher cells contains *microtubules,* first observed in homogenates and the axoplasm of nerves (neurotubules).[2, 3] Microtubules are hollow cylinders of indefinite length and a diameter of about 25 nm that are formed by the association of globular proteins (i.e., monomers) arranged in a helical structure. In the wall of each microtubule 13 subfilaments with globular structures of 4 to 5 nm can be recognized.[4] Since microtubules are the main components of the mitotic spindle and of cilia and flagella, a detailed study of them will be made in the corresponding chapters.

The microtubules present in the cytoplasmic matrix are rather unstable and require the use of aldehyde fixation (especially glutaraldehyde fixation; see Chapter 6) to stabilize their structure. In many cells microtubules constitute a rigid framework of elastic rods, a kind of *cytoskeleton* which may determine the shape of the cell during embryogenesis. For example, the growth of the axon in neuroblasts, the development of the fine processes of certain protozoa (i.e., *axonemes*), and the change in shape of embryonic cells, are dependent on the presence of the microtubules that are arranged along the main axis. Another function of microtubules is the determination of channels in the cytoplasm for the displacement of cytoplasmic material (i.e., granules, vesicles). The protein component of microtubules—the so-called *tubulin* —is a dimer with a total molecular weight of 110,000 daltons; the monomer of tubulin is half this weight. This protein binds specifically the alkaloids *colchicine* and *vinblastine*, but does not interact with cytochalasin *B*. The binding tends to disrupt the microtubules into the monomers and to prevent their reorganization (see Chapters 13 and 22 for more details and the corresponding literature).

SUMMARY: Cytoplasmic Matrix

Several of the mechanical properties of the cytoplasm depend on the fact that this is a polyphasic colloid system in which the matrix is the less dense, and the vacuolar system the more dense, portion. Cyclosis, ameboid movement, cell division, and other mechanical properties of the cytoplasm can be inhibited by hydrostatic pressure. The viscosity of the cell is changed by environmental and internal factors. The pH of the cell varies in different subcellular compartments and is more alkaline in the nucleus than in the cytoplasm. The oxidation-reduction potential of the cytoplasm is -0.275 volt in anaerobiosis and $+0.070$ volt in aerobiosis.

The cytoplasmic matrix, in addition to having an amorphous background structure, contains microfilaments and microtubules. Besides specialized microfilaments (e.g., tonofilaments of keratin, and myofilaments), there are at least two types of microfilaments with a diameter of 4 to 6 nm, that can be distinguished by their morphology and properties. Type-L forms a kind of network and is sensitive to cytochalasin *B*. Type-S forms sheaths beneath the plasma membrane and contains an actin-like protein with a molecular weight of 45,000 daltons. These filaments are probably important in cell locomotion and in *sol-gel transitions* of the cytoplasm. The microtubules are cylindrical structures of 25 nm mean diameter, formed by the polymerization of tubulin. They form a kind of cytoskeleton for the matrix and are important in the formation of the spindle, cilia, and flagella.

CYTOPLASMIC VACUOLAR SYSTEM

As diagrammed in Figures 2–6 and 2–7, which represent typical cells from higher plants and animals, the cytoplasm is traversed by a complex system of membrane-bound tubules, vesicles, and flattened sacs that have many intercommunications. This membrane system should be interpreted in its three-dimensional array as a vast network of closed or open cavities that subdivide the cytoplasm into two main compartments: one enclosed within the membrane; the other situated outside (the cytoplasmic matrix).

The name *cytoplasmic vacuolar system* seems the most descriptive for this intracellular membranous organization. The main components are: the *endoplasmic reticulum*, the *nuclear envelope*, and the *Golgi complex*. Figure 2–7 emphasizes that these different parts of the vacuolar system are made continuous at certain points by permanent or intermittent channels.

The vacuolar system was not discovered until the techniques for electron microscopy of intact cultured cells and thin sections became available. Other advances were made by cell fractionation methods followed by biochemical analysis and the use of cytochemical techniques for the study of specific components—particularly enzymes—at both the light and electron microscopic levels. As in many other areas of cell biology, rapid progress has resulted from the convergence of various technical and scientific approaches.

A cell can no longer be considered as a bag containing enzymes, ribonucleic acid (RNA), deoxyribonucleic acid (DNA), and solutes surrounded by an outer membrane, as in the most primitive bacterium (Fig. 2–7). Numerous membrane-bound compartments are responsible for vital cellular functions, among which are the separation and association of enzyme systems, the creation of diffusion barriers, the regulation of membrane potentials, ionic gradients, and different intracellular pH values, and other manifestations of cellular heterogeneity. Furthermore, there is evidence that enzymes are spatially organized, forming multienzyme systems within the insoluble membranous framework of the cell. For this reason, in studying the cell one is dealing not with individual enzymatic reactions, but with integrated enzyme systems.[5]

General Morphology of the Vacuolar System

In 1945 the first observations of cultured fibroblasts revealed a lacelike reticular component of the cytoplasm[6] (Fig. 9–3). Under the phase contrast microscope the cytoplasm of a living cell, excluding the mitochondria and some inclusions, appears structureless (Fig. 9–3, *A*), while in a similar, but fixed, specimen, the electron microscope reveals a special reticular component. This is a network of membrane-bound cavities that may vary considerably in size and shape. Since this network is more concentrated in the endoplasm of the cell than in the so-called ectoplasm (peripheral region), the name *endoplasmic reticulum* was proposed. In addition to vesicles and tubules, this system may show large flattened sacs (cisternae). An electron micrograph of a culture cell examined *in toto* shows a three-dimensional view of the endoplasmic reticulum. In it, the shape, distribution, and interconnections of the tubules, vesicles, and cisternae throughout the cytoplasm are clearly visible (Fig. 9–3, *B*).

A more detailed analysis of the vacuolar system, by which a great variety of animal and plant cells could be observed, was possible after the introduction of thin sectioning.[7–9] It was found that the vacuolar system varied considerably in different cells and within a single cell in the different cytoplasmic regions.

The recognition of the ribosomes as morphological entities revealed that some of them were free in the cytoplasmic matrix as isolated units or as polysomes, whereas others were attached to some of the membranes of the vacuolar system (Fig. 9–1). Based on the association with ribosomes and on other morphological details, the vacuolar system may be subdivided into its two main portions: the *endoplasmic reticulum* and the *Golgi complex*. The endoplasmic reticulum may be further subdivided to include: (1) the *nuclear envelope* (see Chapter 12), (2) the *rough* (or granular) *endoplasmic reticulum* with attached ribosomes, and (3) the *smooth* (or agranular) *endoplasmic reticulum,* without ribosomes.

In Figure 9–4 the different portions of the vacuolar system are represented three-dimensionally. The arrows indicate the possible relationship between the rough and the smooth endoplasmic reticulum and the Golgi complex.

As already noted, the development of the endoplasmic reticulum varies considerably in various cell types. It is often small and relatively undeveloped in eggs and in embryonic or undifferentiated cells but increases in size and complexity with differentiation. In spermatocytes only a few vacuoles can be observed. A simple smooth endoplasmic reticulum is found in cells engaged in lipid metabolism, such as adipose, brown fat, and adrenocortical cells. In the interstitial cells of the testis of the opossum a considerable amount of smooth endoplasmic reticulum has been observed.[10]

Rough Endoplasmic Reticulum

The rough endoplasmic reticulum is especially well developed in cells actively

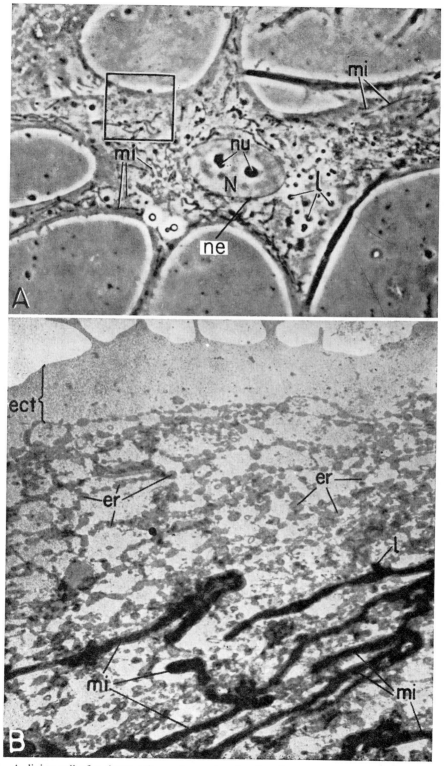

Figure 9–3. **A,** living cell of a tissue culture observed under the phase contrast microscope: *l,* lipid; *mi,* mitochondria; *ne,* nuclear envelope; *nu,* nucleoli. The region indicated in the inset is similar to B. (Courtesy of D. W. Fawcett.) **B,** electron micrograph of the marginal region of a mouse fibrocyte in tissue culture: *er,* endoplasmic reticulum; *mi,* filamentous mitochondria; *l,* lipid. The peripheral region (*ect*) is homogeneous. ×7000. (Courtesy of K. R. Porter.)

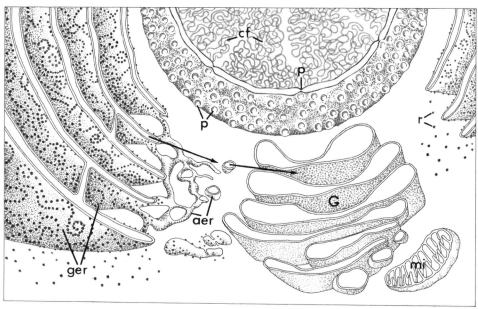

Figure 9–4. Three-dimensional diagram of the vacuolar system of the cell. The nucleus with the chromosomal fibrils (*cf*) show interchromatin channels (arrow) leading to nuclear pores (*p*). Note the double membrane organization of the nuclear envelope. Cisternae of granular endoplasmic reticulum (*ger*) are interconnected and have ribosomes attached to the outer surface. Some of these cisternae are extended by tubules of agranular endoplasmic reticulum (*aer*). *G,* Golgi complex; *mi,* mitochondria; *r,* free ribosomes. The large arrows indicate the probable dynamic relationship of the portions of the vacuolar system.

engaged in protein synthesis, such as the enzyme-producing cells. In general, it occupies the regions of the cytoplasm that appear to be basophilic (corresponding to the *ergastoplasm*) when viewed under the light microscope. This property is due to the presence of the RNA-containing ribosomes; and the basophilic areas of the cytoplasm, in general, coincide with regions in which there is rough endoplasmic reticulum. Some exceptions may be found in embryonic plant and animal cells, which also may be strongly basophilic (i.e., containing many ribosomes) but which have little development of membranes.

In cells of the pancreatic acinus the rough endoplasmic reticulum is highly developed and consists of stacks of large flattened cisternae covered with ribosomes and occupying the base and lateral regions of the cell (Fig. 9–5). In liver cells the rough endoplasmic reticulum is distributed throughout the cytoplasm in groups of cisternae which apparently act as a functional unit. There is evidence, in fact, that the product synthesized within one of these units proceeds radially to the edge of the cisternae, which form a con-

tinuous system with the tubular elements of the smooth endoplasmic reticulum (see Claude, 1969).

The cavity of the rough endoplasmic reticulum is sometimes very narrow, with the two membranes closely apposed; but more frequently there is a true space between the membranes that may be filled with a material of varying opacity. This space is much distended in certain cells actively engaged in protein synthesis, such as the plasma cells and goblet cells. In these cases a dense macromolecular material can be observed inside the cisternae. In the pancreas, intercisternal secretion granules, smaller than the zymogen granules, may be observed.[11]

The membrane of the endoplasmic reticulum is about 5 to 6 nm thick. Although it is thinner than the plasma membrane, it does exhibit a "unit membrane" structure, i.e., two dense layers separated by a lighter one. The total surface of the endoplasmic reticulum contained in 1 ml of liver tissue has been calculated to be of about 11 square meters, two thirds being of the granular, or rough type.[12]

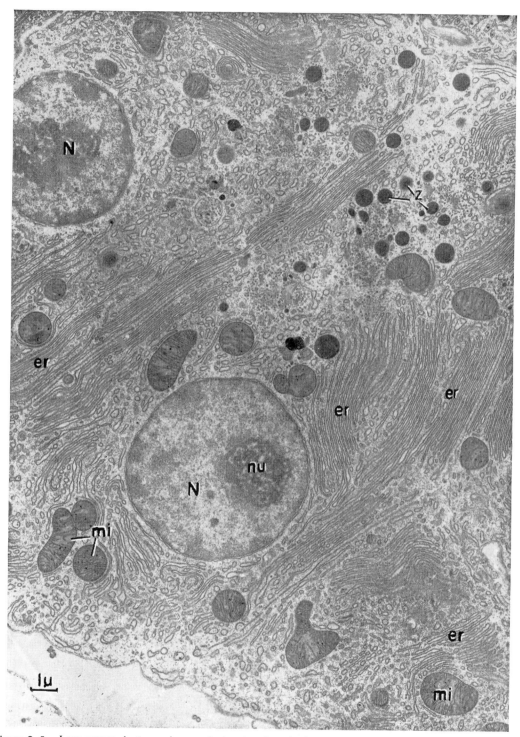

Figure 9–5. Low power electron micrograph showing the submicroscopic organization of a cell of the pancreatic acinus. At the base and lateral portions, the cell is rich in rough endoplasmic reticulum (*er*). In the apex the zymogen granules (*z*) are apparent. *mi*, mitochondria; *N*, nucleus; *nu*, nucleolus. ×8000. (Courtesy of K. R. Porter.)

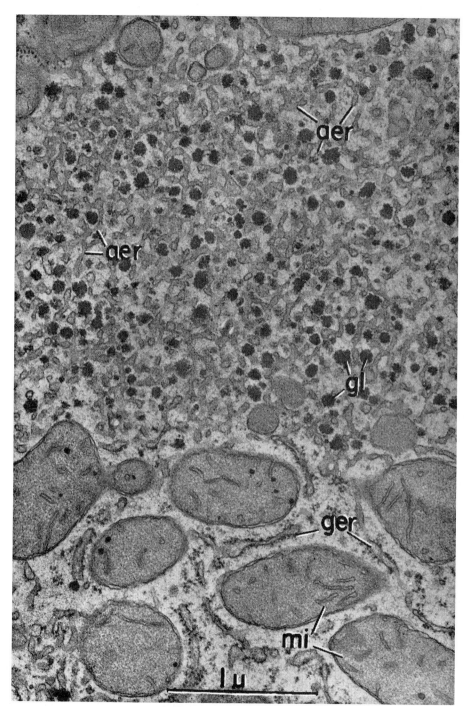

Figure 9–6.   Electron micrograph of the cytoplasm of a liver cell. **At the bottom,** the rough or granular endoplasmic reticulum (*ger*) and mitochondria (*mi*): **at the top,** the smooth or agranular endoplasmic reticulum (*aer*) mixed with glycogen particles (*gl*). ×45,000. (Courtesy of G. E. Palade.)

Smooth Endoplasmic Reticulum

Although the smooth endoplasmic reticulum forms a continuous system with the rough portion, it has a different morphology. In the liver cell it is made of tubular elements that apparently start at the edge of the cisternae and make a tubular network that pervades large regions of the cytoplasmic matrix. These fine tubules, with some fenestrated smooth membrane segments, are present in regions rich in glycogen and can be observed as dense particles in the matrix (Fig. 9–6). Another characteristic observed in liver cells is the relationship of the smooth endoplasmic reticulum to *peroxisomes* (i.e., organelles containing peroxidase, catalase, and other oxidases [see Chapter 21]), and with the Golgi complexes.

**SUMMARY:
Cytoplasmic Vacuolar System**

The cytoplasm of animal and plant cells is traversed by a complex system of membranes that form closed compartments in structural continuity. The main components of the system are the endoplasmic reticulum, with its three portions (the nuclear envelope, the rough and the smooth endoplasmic reticulum, and the Golgi complex). The rough endoplasmic reticulum has ribosomes attached to its outer surface and is particularly well developed in the basophilic regions of the cytoplasm (i.e., the ergastoplasm). The tubular and cisternal cavities of the endoplasmic reticulum may be closed, but more often contain material that has been synthesized on the ribosomes (e.g., protein) or by the enzymes present in the membranes (e.g., lipids, oligosaccharides). The smooth endoplasmic reticulum is devoid of ribosomes. Frequently it forms a tubular network, and in the liver it is related to glycogen deposits and peroxisomes.

Fractionation of the Vacuolar System

Microsomes

In Chapter 6 the concept of the "microsomal fraction" that can be isolated by differential centrifugation following the separation of the nuclear and mitochondrial fractions was introduced (Fig. 6–5). Electron microscopic studies have revealed that such a fraction is rather complex. In addition to fragments of the plasma membrane, microsomes include various parts of the vacuolar system, i.e., the rough and smooth endoplasmic reticulum and the Golgi complex (Fig. 9–7). Thus, the concept of microsome-at-large must not be confused with that of a particular membrane system of the cell. Microsomes, as discrete entities, are not found in the intact cell, but are the result of the fragmentation of most of the cytoplasmic membranous components. The microsomal fraction from the liver is more heterogeneous than that obtained from the pancreas (Fig. 9–8, *A*), and this reflects the different organization of the vacuolar system in both types of cells (see above). Refinements in the methods of cell fractionation (in particular, the use of gradient centrifugation) have led to the separation of microsomes derived mainly from the rough, or the smooth endoplasmic reticulum. Sections to come later will show that the Golgi complex can also be separated in a rather homogeneous fraction. Figure 9–8, B shows that after treatment with desoxycholate, a surface-active agent, microsomal membranes are solubilized, and the ribosomes may be isolated.

Chemical Composition. Microsomes constitute about 15 to 20 per cent of the total mass of the cell. They contain 50 to 60 per cent of the RNA of the cell which is due

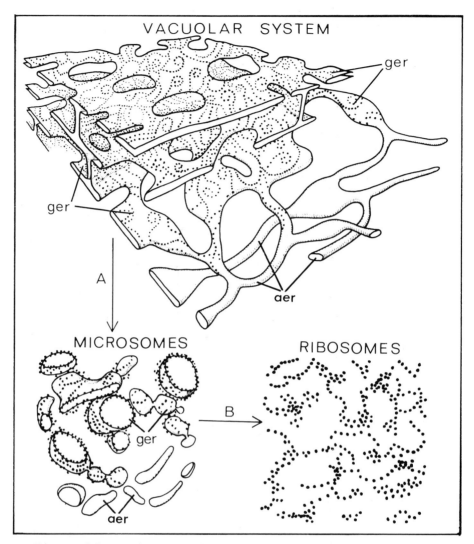

Figure 9–7. Diagram of the vacuolar system with the granular endoplasmic reticulum (*ger*) and agranular endoplasmic reticulum (*aer*). **A**, microsomes, produced by homogenation; **B**, free ribosomes, after membranes are lysed by detergent action.

to the presence of ribosomes (see Chapter 18). Microsomal membranes have a high lipid content which includes: phospholipids, neutral lipids, phosphatidylinositol, plasmalogens, and some gangliosides. There is more lipid in relation to proteins in the smooth endoplasmic reticulum and Golgi membranes than in the rough endoplasmic reticulum. The latter also contains less sphingomyelin and cholesterol.[13] Microsomal membranes also have some special proteins—particularly of the enzymatic type—which serve as markers for the recognition of this particular fraction and to differentiate it from Golgi membranes, secretory granules, or plasma membranes.

The membrane proteins of the endoplasmic reticulum have been studied by polyacrylamide gel electrophoresis with sodium dodecylsulfate. As many as 30 polypeptide bands with moleclar weights ranging from 15,000 to 150,000 daltons have been identified in the rough endoplasmic reticulum of the pancreas. These bands differ from the bands (which are few in number) found in the Golgi membranes and from those corresponding to zymogen granules.[14]

Some of the complex enzymatic activities of microsomes are indicated in Table 9–2. Some important enzyme markers are electron transport systems that may serve for the

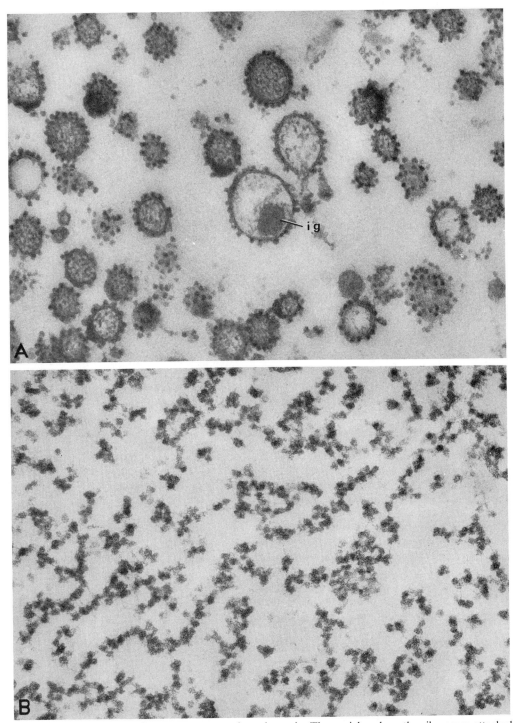

Figure 9-8. **A,** microsomes from the pancreas of a guinea pig. The vesicles show the ribosomes attached to the outer surface. *ig,* intracisternal granule. ×91,200. **B,** ribosomes from the pancreas after solubilization of the membranes. ×136,000. (Courtesy of G. E. Palade.)

TABLE 9–2.   Some Microsomal Enzyme Activities*

Synthesis of glycerides:
   Triglycerides
   Phosphatides
   Glycolipids and plasmalogens
Metabolism of plasmalogens
Fatty acid synthesis
Steroid biosynthesis:
   Cholesterol biosynthesis
   Steroid hydrogenation of unsaturated bonds
$NADPH_2 + O_2$-requiring steroid transformations:
   Aromatization
   Hydroxylation
$NADPH_2 + O_2$-requiring drug detoxification:
   Aromatic hydroxylations
   Side-chain oxidation
   Deamination
   Thio-ether oxidation
   Desulfuration
L-Ascorbic acid synthesis
UDP-uronic acid metabolism
UDP-glucose dephosphorylation
Aryl- and steroid-sulfatase

*Modified from Rothschild, J., The isolation of microsomal membranes. *In The Structure and Function of the Membranes and Surfaces of Cells. Biochem. Soc. Symp. 22*:4, 1963, New York, Cambridge University Press.

oxidation of steroids and drugs: *NADPH-cytochrome-c-reductase* with the *cytochrome b_5* and the *cytochrome P-450*. This last component is characterized by its absorption at 450 nm and is a protein of 50,000 daltons.[15] (Cytochrome b_5 and the reductase have also been isolated.[16]) Other important enzymes are Mg^{2+}-activated ATPase and glucose-6-phosphatase. In the liver this last enzyme has the important function of splitting off the phosphate from glucose-6-phosphate, thus allowing glucose to be liberated into the blood, probably by way of the lumen of the endoplasmic reticulum (see Loewey and Siekevitz, 1969). As shown in Figure 9–9 glucose-6-phosphate originates from the degradation of glycogen in the cytoplasmic matrix and the enzyme, that is embedded in the membrane, probably acts in a vectorial manner sending glucose into the lumen of the endoplasmic reticulum. Glucose could then find its way through the vacuolar system into the blood stream.

Studies done in newly born animals reveal that during differentiation of the liver cell glucose-6-phosphatase and NADPH-cytochrome-C-reductase appear first in the rough, and then in the smooth, portions of the endoplasmic reticulum.[17]

Table 9–2 shows that microsomes are involved in the biosynthesis of triglycerides, phosphatides, and other lipids, including the steroids. They contain enzymes that bring about aromatization, hydroxylation, deamination, and other chemical transformations of different active substrates, thus contributing to their inactivation. In liver, this general function of microsomes is related to *detoxification*. Several of the enzymes that metabolize drugs are more concentrated in the smooth, than in the rough, endoplasmic reticulum.

Synthesis and Turnover of Intracellular Membranes

The origin of the endoplasmic reticulum is not definitely known. Observations such as those shown in Figure 9–1 suggest that it may develop by evagination from the nuclear envelope. At telophase, however, the nuclear envelope is re-formed by vesicles of the endoplasmic reticulum. The close relationship between these two portions of the vacuolar

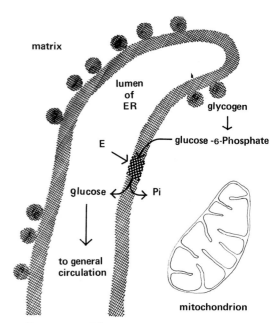

Figure 9–9. Diagram of the intervention of the smooth endoplasmic reticulum in glucogenolysis with the consequent release of glucose. The enzyme (*E*), glucose-6-phosphatase, is present in the membrane and has a vectorial disposition by which it receives the glucose-6-phosphate from the matrix surface. The product—glucose—penetrates the lumen of the endoplasmic reticulum.

system is also suggested by cytochemical studies.

The relationship between rough and smooth endoplasmic reticulum may be studied in differentiating cells. In rat liver cells before birth there is a preferential increase of the rough type, whereas after birth the growth is mainly of the smooth type. Studies using the protein or lipid precursors ^{14}C-leucine and ^{14}C-glycerol have shown that in the period of rapid growth of the endoplasmic reticulum, the incorporation into proteins and lipids is greater in the rough than in the smooth type. This finding suggests that the synthesis of membranes follows the direction rough → smooth endoplasmic reticulum.

In addition to the *structural and functional continuity* between the various membrane compartments of the vacuolar system there is also a *temporal continuity*. This refers to the fact that one cell receives a full set of membranes from its ancestor cell. On the basis of present experimental evidence, there is no *de novo* synthesis of membranes; they grow by expansion of pre-existing membranes.[18, 19]

The studies on subfractionation of microsomes mentioned above suggest that in spite of the many interconnections between the rough and smooth portions of the endoplasmic reticulum, and also with the Golgi complex, these several portions of the vacuolar system apparently do not mix between them. Such a conclusion is derived mainly from the observations of a differential chemical composition regarding the proteins and enzymes. Other interesting results refer to turnover rates of some of the specific proteins. The different half-lives found for cytochrome b$_5$ (60 hours), cytochrome-C-reductase (40 hours)[16] and NADH-glycohydrolase (16 days)[20] strongly suggest that the membranes are not degraded as a unit, but that each of their components is removed and degraded independently.[15]

**SUMMARY:
Microsomal Fraction**

By differential centrifugation most of the components of the vacuolar system are separated out in the so-called *microsomal fraction*. Further separation between rough and smooth endoplasmic reticulum and Golgi complexes may be achieved by gradient centrifugation; the membranes and the ribosomes may also be separated. Numerous proteins are present in the microsomal membranes, some of which are the enzyme markers, NADPH-cytochrome-c-reductase; cytochrome b$_5$ and cytochrome P-450. These are electron transport systems involved in the oxidation of steroids and drugs. Glucose-6-phosphatase is a vectorial enzyme that is involved in the liberation of glucose from the liver. Microsomes also contain enzymes for the biosynthesis of triglycerides, phosphatides, and steroids. The general function of detoxification of the liver is localized in the endoplasmic reticulum.

The fact that, in spite of the many interconnections between the various segments of the vacuolar system, there are striking differences in chemical composition and enzymes suggests that the components of the corresponding membranes probably do not mix together.

Functions of the Endoplasmic Reticulum

The polymorphic aspect of the vacuolar system in a variety of cells and its change with differentiation and cellular activity has led to various functional interpretations. The most reliable are those based on experimental evidence and quantitative data derived from cell fractionation.

Mechanical Support. The vacuolar system, along with the cytoplasmic matrix, participates in many of the mechanical functions of the cell. By dividing the fluid content of the cell into compartments, the vacuolar system provides supplementary mechanical support for the colloidal structure of the cytoplasm.

Exchange. The enormous internal surface provided by the endoplasmic reticulum (about 11 m²/ml in liver cells) provides an idea of the importance of the exchanges taking pace between the matrix and the inner compartment.

It is known that in the cell the vacuolar system has *osmotic properties.* After isolation, microsomes expand or shrink according to the osmotic pressure of the fluid. Diffusion and active transport may take place across the membranes of the vacuolar system. As in the plasma membrane, the presence of *carriers* and *permeases* that are involved in active transport across the membrane has been postulated (see Chapter 21).

Ionic Gradients. The existence of a vacuolar system separating the cytoplasm into two compartments makes possible the existence of ionic gradients and electrical potentials across these intracellular membranes. This concept has been applied especially to the *sarcoplasmic reticulum,* a specialized form of endoplasmic reticulum found in striated muscle fibers, which is considered as an intracellular conducting system.[21] It has been postulated that a portion of the sarcoplasmic reticulum transmits impulses from the surface membrane into the deep regions of the muscle fiber. (A more detailed study of the sarcoplasmic reticulum is presented in Chapter 23.)

Circulation and Membrane Flow. The existence of a directional flow of material with locks at certain points of the system has been postulated. The endoplasmic reticulum may act as a kind of *circulatory system* for intracellular circulation of various substances.[7] Membrane flow may also be an important mechanism for carrying particles, molecules, and ions into and out of the cells by way of the vacuolar system.[22] As shown in Figure 9–10, if there is a region, AA′, in which the plasma membrane is being actively synthesized and another, B, in which it is broken down, membrane flow in the direction of the arrow will occur. By this mechanism, particles attached to the surface of the cell or suspended in the fluid medium can be incorporated into the cytoplasm (Fig. 9–10, 2). A similar mechanism, but working in a reverse direction (*f* to *a*), can effect the transport of a particle from the interior of the cytoplasm to the outer medium (i.e., secretion; see Chapter 25). The continuities observed

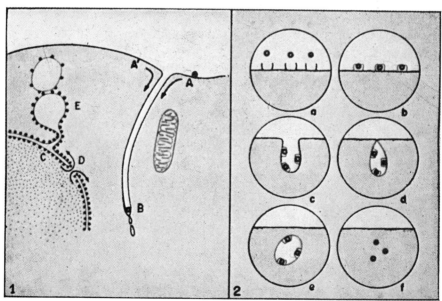

Figure 9–10. **1,** diagram representing the hypothesis of membrane flux. **2,** diagram representing the concept of transportation by vesiculation of the membranes. (See the description in the text.) (From Bennett, S., *J. Biophys. Biochem. Cytol.,* 2:99, 1956.)

in some cases between the endoplasmic reticulum and the nuclear envelope suggest that the membrane flow may also be active at this point. This flow would provide one of the several mechanisms for export of RNA and nucleoproteins from the nucleus to the cytoplasm (see Chapter 17). More will be said later about this possible mechanism of membrane flow, particularly in the chapters concerned with pinocytosis, phagocytosis, and secretion.

Synthesis of Proteins. Protein synthesis is intimately related to the ribosomes (see Chapter 19). When proteins are synthesized to be incorporated or used by the cell, e.g., hemoglobin and fibrous proteins, the membranes of the endoplasmic reticulum are not involved, and the products are stored in the cell matrix. The endoplasmic reticulum is active when the proteins are made for export, e.g., synthesis of tropocollagen, serum proteins and secretion granules. The two-dimensional array of polyribosomes on the surface of the endoplasmic reticulum probably facilitates the activity of messenger RNA and protein synthesis (see Chapter 19). The protein molecules that are discharged from the ribosomes penetrate into the cavity of the endoplasmic reticulum and are stored and segregated for export outside the cell. In the case of some serum proteins (particularly γ-globulin) synthesized within the plasma cells, it has been possible to follow the storage process by means of an antibody conjugated to ferritin or by antibodies produced against enzymes (see Chapter 16).

It is interesting to note that the transport of the material through the cavities of the endoplasmic reticulum may be much faster than the actual synthesis and flow of the membranes which serve as the container for the material. During the transport of these products, three types of membranes (endoplasmic reticulum → Golgi membrane → plasma membrane) should interact, and these three compartments should be connected and disconnected by fusion and fission of the membranes. The kinetics of this multistep transport system will be studied in Chapter 25 (see Jamieson, 1971).

Detoxification. Large amounts of drugs such as phenobarbital administered to an animal result in increased activity of enzymes related to detoxification, as well as other enzymes, and a considerable hypertrophy of the smooth endoplasmic reticulum.[31] This mechanism for detoxification also applies to endogenous or administered steroid hormones. Carcinogens such as 3-methylcholanthrene and 3,4-benzopyrene are among the most potent inducers of drug-metabolizing enzymes.[32]

The effect of these drugs is to produce a true induction of the enzymes of the endoplasmic reticulum. In other words, there is an increased synthesis of the enzymes which can be prevented by inhibitors of protein synthesis, such as puromycin.[32] The NADH-cytochrome-c-reductase and cytochrome P-450 content of a liver cell, for example, may increase 50 to 100 per cent after prolonged phenobarbital administration.[15]

Synthesis of Lipoproteins. While rough endoplasmic reticulum predominates in cells actively synthesizing proteins, the smooth type is abundant in those involved in the synthesis of lipids.[23] An interrelationship of the membranous components of the vacuolar system has been observed during the synthesis of triglycerides and also during the formation of lipoprotein complexes. This interrelationship seems to be associated mainly with the smooth endoplasmic reticulum and the Golgi complex.[24]

Synthesis of Glycogen. In fasted animals it was found that the residual glycogen remained associated with the tubules and vesicles of the endoplasmic reticulum.[25] When feeding was resumed, there was an increase in smooth endoplasmic reticulum which maintained its association with the accumulating glycogen. Also, in plant cells the smooth endoplasmic reticulum develops along the surface where the cellulose walls are being formed.[26]

The enzyme UDPG-glycogen transferase,[27] which is directly involved in the synthesis of glycogen by addition of uridine diphosphate glucose (UDPG) to primer glycogen, is bound to the glycogen particle rather than to the membranous component.[28] This suggests that the reticulum is related to glycogenolysis, but not to glycogenesis.[21]

In prenatal liver cells, just before birth, the amount of glycogen increases and then decreases simultaneously with an increased amount of glucose-6-phosphatase. This depletion of glycogen is accompanied by an increase in smooth endoplasmic reticulum.[29, 30]

It was mentioned that the withdrawal of glucose from the glycogen deposits could be mediated by the smooth endoplasmic reticulum by way of the glucose-6-phosphatase.

SUMMARY:
Endoplasmic Reticulum

Numerous functions are attributed to the endoplasmic reticulum. It contributes to the *mechanical support* of the cytoplasm, has *osmotic properties,* and is involved in intracellular *exchanges* between the matrix and the internal cavity. These exchanges are brought about by diffusion or *active transport,* in which carriers and permeases may be involved. In liver there are about 11 square meters of membrane per ml available for exchange. *Ionic gradients* and electrical potential may generate across these membranes, and *conduction of intracellular impulses* have been postulated in the case of the sarcoplasmic reticulum. The endoplasmic reticulum serves as a *circulatory system* for the transport of various substances. Although slower, the flow of membranes may be effective in various types of intracellular transport. The *synthesis of proteins* for export is one of the main functions of the rough endoplasmic reticulum. The *synthesis of lipids and lipoproteins* is associated with the rough and smooth endoplasmic reticulum. The *synthesis of glycogen* is accomplished in the cytoplasmic matrix, but the smooth endoplasmic reticulum is involved in *glucogenolysis* through the action of glucose-6-phosphatase. Another important function of the smooth endoplasmic reticulum is in the *detoxification* of many endogenous and exogenous compounds. The prolonged administration of certain drugs produces an increase in the smooth endoplasmic reticulum and the induction of the specific enzymes.

GOLGI COMPLEX

In 1898 by means of a silver staining method, Golgi discovered a reticular structure in the cytoplasm. The name Golgi "apparatus," generally given to this structure, is confusing because it suggests a definite relationship with the physiologic processes of the cell. Today it seems more appropriate to use the name "Golgi substance" or "Golgi complex," to refer to this material that has special staining properties (Fig. 9–11). Because its refractive index is similar to that of the matrix, the Golgi complex is difficult to observe in living cells. The use of the electron microscope has provided a distinct image of this component, and its submicroscopic structure has been revealed.

Morphology

Electron microscopy has revealed that the Golgi complex consists mainly of membranes that belong to the vacuolar system of the cell. One of the main characteristics of the complex is the lack of ribosomes. In fact, the Golgi complex appears to be surrounded by a zone from which ribosomes apparently are excluded. For this reason there is no protein synthesis in this organelle. The localization and organization of Golgi membranes and certain cytochemical and biochemical properties clearly differentiate the Golgi complex from the endoplasmic reticulum.

In cells that have a polarized structure the Golgi complex is, in general, a single large structure and occupies a definite position between the nucleus and the pole of the cell where the release or secretion takes place. This is the case, for example, in thyroid cells (Fig. 9–11), in the exocrine pancreas, or in the mucous cells of the intestinal epithelium. In nerve (Fig. 9–11) and liver cells (Fig. 9–12) and in most plant cells the Golgi complexes are multiple and do not show a special polarity. In liver cells a rough evalua-

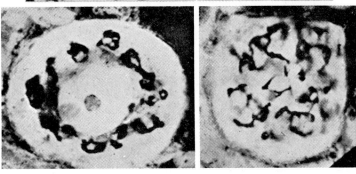

Figure 9-11. **Above,** Golgi complex (Golgi apparatus) in cells of the thyroid gland of the guinea pig, apical position. Osmic impregnation. **Below left,** ganglion cell, perinuclear Golgi apparatus. **Below right,** same structure as at left, optical section tangential with respect to the nucleus. Silver impregnation. (From E. De Robertis).

tion indicates that there are some 50 Golgi complexes (formerly called *dictyosomes*) per cell. These structures represent some 2 per cent of the total cytoplasmic volume.[33] In plant cells and in invertebrate tissues Golgi complexes are dispersed throughout the cytoplasm. Although the localization, size and development of these organelles vary from one cell to another and also with the physiologic state of the cell, they show morphological characteristics that permit their differentiation from the other parts of the vacuolar system.

In general, three membranous components are recognized under the electron microscope: (1) flattened sacs (i.e., cisternae); (2) clusters of tubules and vesicles of about 60 nm; and (3) larger vacuoles filled with an amorphous or granular content. The Golgi cisternae are arranged in parallel and are separated by a space of 20 to 30 nm. Often they are arranged concentrically with a convex and a concave face. There may be from three to seven of these structures in most animal and plant cells. In certain algae, however, there may be as many as 10 or 20 cisternae. A stack of such cisternae generally encloses a region of cytoplasm containing

large vacuoles (Fig. 9–13). On the convex surface of the Golgi complexes there is a system of small vesicles and tubules that is more easily recognized by negative staining in isolated complexes. These tubules converge upon the Golgi complex, forming a kind of fenestrated plate. Most of the vesicles (60 nm in diameter) observed in sections at this side of the Golgi complex, are probably profiles of the tubules. These may be connected with tubular elements of the smooth endoplasmic reticulum. The large vesicles represent regions in which secretory products of the endoplasmic reticulum and Golgi complex are being concentrated. These *concentrating* or *secretory vesicles* may lead to the formation of zymogen granules in the case of many enzyme-producing cells (see Chapter 25).

In general, the stacks of cisternae show a definite polarization. In the Golgi complex a *proximal* or *forming face* closer to the nuclear envelope or to the smooth endoplasmic reticulum and a *distal* or *maturing face,* associated with the formation of the secretory vesicles may be distinguished. These frequently appear as dilatations at the edge of the cisternae. It has been postulated that between the forming and the maturing face

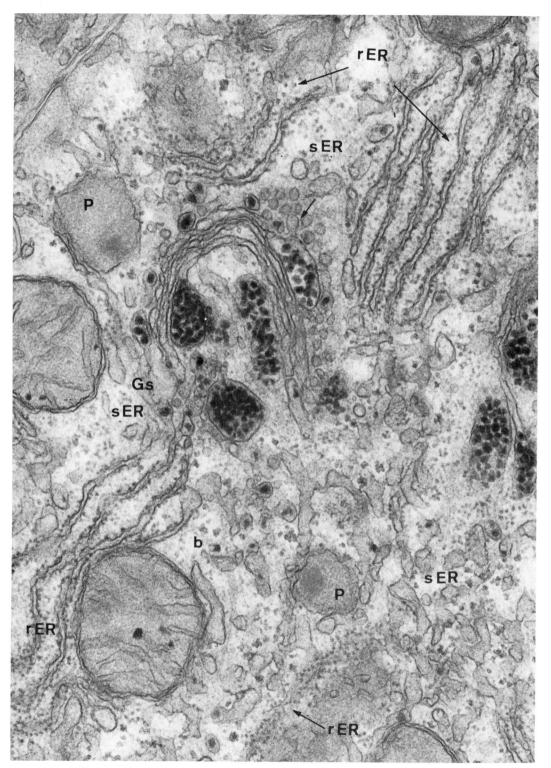

Figure 9-12. Electron micrograph of a liver cell of an animal having a diet rich in fat. The synthesis and transport of the lipoprotein granules is observed. The rough endoplasmic reticulum (*rER*) is observed as two stacks of lamellae converging toward a Golgi complex having Golgi sacs (*Gs*) on its convex or "forming face." There are portions of smooth endoplasmic reticulum (*sER*) connecting both parts of the vacuolar system. Mitochondria and peroxisomes (*P*) are observed. ×56,000. (Courtesy of A. Claude.)

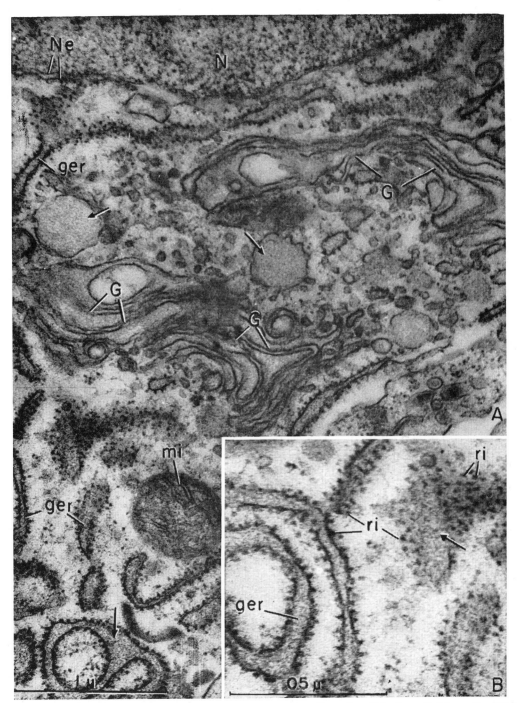

Figure 9–13.   Electron micrograph of a plasma cell showing near the nucleus (*N*) a large Golgi complex (*G*) formed of flat cisternae and small and large vesicles. Some of the large vesicles (arrows) are filled with material. Surrounding the Golgi complex is abundant granular endoplasmic reticulum (*ger*) having cisternae filled with amorphous material (arrows). *mi*, mitochondrion; *Ne*, nuclear envelope; *ri*, ribosomes. ×48,000; inset ×100,000. (From E. De Robertis and A. Pellegrino de Iraldi.)

there is a morphological transition of the membranes. They are thinner and more similar to the endoplasmic reticulum at the forming face and become thicker and more densely stained at the maturing face. Some authors have interpreted these transitions as being suggestive of the theory that the Golgi complex is a site in which the membranes undergo a certain differentiation. They are supposed to become more similar in structure and composition to the plasma membrane and in this way more capable of fusing with it (see Morré et al., 1971).

Golgi Complex and Lipoprotein Particles

The transitions between the various parts of the Golgi complex are more clearly observed in liver cells under certain physiologic conditions. The following experimental conditions make it possible to trace the synthesis and transport of lipoproteins: One week prior to sacrifice, rats are submitted to partial hepatectomy (i.e., removal of 50 per cent of the liver). On the day of the sacrifice, the rats fast for a few hours and then two to three hours before their death are given a diet rich in fat (i.e., 20 per cent butter). Under these conditions it is possible to follow the synthesis and transport of the lipoproteins, which appear as discrete dense granules of about 40 nm. As shown in Figure 9–12 these granules appear first within tubules of the smooth endoplasmic reticulum and then enter the outer fenestrated cisternae of the Golgi. After longer periods they accumulate in the large vacuoles that are formed by dilatation of the edge of the sacs and, finally, they are detached as secretory vesicles. Although not recognizable morphologically, the protein moiety of the lipoprotein is synthesized in the rough endoplasmic reticulum. The granules, however, become visible in the smooth portion because of the addition of triglycerides.[33]

Golgi Complex and Cell Secretion

Electron microscopy has brought new evidence of the morphologic relationship between the *Golgi complex* and *secretion*, which

was postulated by Cajal in 1914 in his study on goblet cells. In this type of cell, the sacs of the Golgi complex are related on one side to the cisternae of the endoplasmic reticulum and on the other to the secretion droplets. In the plasma cell shown in Figure 9–13, the Golgi complex occupies a large region near the nucleus and the granular endoplasmic reticulum surrounds the complex. The smaller vesicles in the center correspond to the centrosphere region. In this and in other cells the Golgi complex is topographically related to the centrioles.

A particularly interesting example of the complex is that present in developing mammalian spermatids, in which the Golgi complex is related to the formation of the acrosome.[34]

As will be mentioned in Chapter 25, the Golgi complex concentrates the products of secretion which come from the endoplasmic reticulum in a more dilute form. This function may be demonstrated by radioautography at the optical and electron microscopic levels.[35]

Also related to the function of concentrating secretion products is the homology that has been suggested between the Golgi complex and the contractile vacuole found in lower animals and protozoa. In these lower forms, vacuoles similar in structure to the Golgi complex are found. Such vacuoles, by their contraction, expel water from the cytoplasm into the medium.[36]

Golgi Complex and Lysosomes

In addition to serving as the site for packaging the secretory products and providing a limiting membrane to zymogen granules, the Golgi membranes are involved in the formation of the *primary lysosomes.* These are now interpreted as a special type of secretion, and the relationship between them and the Golgi complex will be considered in detail in Chapter 21. It has been observed that certain vacuoles in the Golgi complex give the first reaction of acid phosphatase[37] (a reaction characteristic of lysosomes). The Golgi region gives a positive PAS reaction, and it has been suggested that glycoproteins are formed in the Golgi complex and migrate by way of small vesicles to the surface of the cell to constitute the cell coat (see Chapter 8 and below).[38]

The Golgi complex is a differentiated portion of the vacuolar system that is morphologically and functionally related to the endoplasmic reticulum on one side, and to secretory vesicles on the other. It is composed of stacks of curved cisternae that are differentiated between a *forming face* and a *maturing face*. At the forming face, there are tubular and vesicular elements that may form a fenestrated plate. On the maturing face, secretion vacuoles are formed by dilatation and pinching off of the edges of the cisternae. The Golgi complexes are large, and in cells having a polarized structure, they are single. In nerve, liver, plant, and invertebrate cells there are numerous small Golgi complexes, formerly called *dictyosomes*, distributed throughout the cytoplasm. The morphological interrelationship between the various parts of the Golgi complex and both the endoplasmic reticulum and secretion vesicles may be observed—in particular, in the case of the lipoprotein granules secreted by the liver cells. This same relationship is also observed in enzyme-secreting cells (see Chapter 25) and in the production of the acrosome in the spermatid. Cytochemical studies have revealed that the Golgi complex is the site of formation of the primary lysosomes (see Chapter 21).

SUMMARY:
Golgi Complex

Isolation of the Golgi Complex

Biochemical Studies

Although for many years the Golgi complex had only been isolated from cells of the epididymis,[39–41] in recent years the isolation has been achieved in numerous plant and animal cells.[42, 43] The methods used are based on gentle homogenization, which tends to preserve the stacks of cisternae, thus allowing large fragments of Golgi complexes to be obtained by differential and gradient centrifugation (see Chapter 5). The Golgi complexes have a lower specific density than the endoplasmic reticulum or the mitochondria and they are equilibrated in a band having a density of 1.16. Under the electron microscope the stacked cisternae appear to be bordered by an extensive system of tubules and vesicles (the secretory products remain within these vesicles) (Fig. 9–14). Washing the Golgi complexes in distilled water results in further purification with a consequent loss of the secretory components.

Chemical Composition

The Golgi complex isolated from rat liver consists of about 60 per cent protein and 40 per cent lipid. By the use of gel electrophoresis it was found that the Golgi complex and the endoplasmic reticulum contain some proteins in common, but the former has fewer protein bands. In the plasma membrane fraction, however, fewer proteins were observed.[44] The decreasing complexity in protein band patterns from endoplasmic reticulum, through Golgi complex, to plasma membranes, is in some ways consistent with the view that membrane proteins are synthesized by ribosomes of the rough endoplasmic reticulum and then transferred to other portions of the vacuolar system.

The Golgi fraction shows very low levels of RNA, DNA, and polysaccharides. It contains sialic acid and other reactive sugars, as well as glycolipids. Of the phospholipids, phosphatidylcholine is most abundant in the endoplasmic reticulum and least abundant in plasma membrane, with the Golgi membranes in an intermediary position. Sphingomyelin is more concentrated in plasma membranes. The neutral lipid fraction contains principally cholesterol, cholesterol esters, and triglycerides.

Enzyme Markers

In the isolation of subcellular organelles it is of great importance to determine the

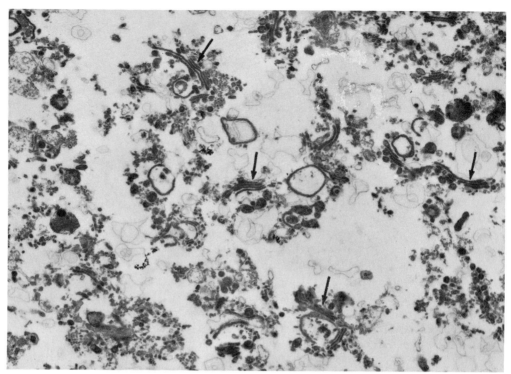

Figure 9–14. Isolated Golgi complexes from liver cells. The complexes which best show the stacks of cisternae are indicated by arrows. (Courtesy of D. J. Morré.)

degree of contamination between the various fractions. Table 9–3 shows the various enzyme markers used for plasma membranes (5'-nucleotidase), for the endoplasmic reticulum (glucose-6-phosphatase), for mitochondria (cytochrome oxidase), and for peroxisomes (uric acid oxidase). The relative proportions in which they are found in the Golgi fraction of the liver are also given. As a consequence of these studies it has been calculated that a fraction purified by washing contains small percentages of contamination: plasma membrane, 6 per cent; endoplasmic reticulum, 2 per cent; and mitochondria, 1 per cent.[45] The enzymes that are concentrated in the Golgi fraction are thiamine-

TABLE 9–3. Distribution of Enzyme Markers in the Corresponding Fraction and in the Golgi Complex from Rat Liver*

Enzyme	Fraction	Specific Activity in Corresponding Fraction	Specific Activity in Golgi Apparatus	Ratio
5'-Nucleotidase	Plasma membrane	41.9	5.8	0.14
Glucose-6-phos-phatase	Endoplasmic reticulum	11.4	1.1	0.10
Cytochrome oxidase	Mitochondria	67.8	1.8	0.03
Uric acid oxidase	Peroxisomes	50.0	0.5	0.01

*From Morré, D. J., Keenan, T. W., and Mollenhauer, H. H., Golgi Apparatus Function in Membrane Transformations and Product Compartmentalization: Studies with Cell Fractions Isolated from Rat Liver in *Advances in Cytopharmacology*, Vol. 1, First International Symposium on Cell Biology and Cytopharmacology. (Clementi, F., and Ceccarelli, B., eds.) New York, Raven Press, 1971, Table 1.

TABLE 9–4. Glycosyl Transferase Activity in Golgi Complex Fraction of Rat Liver*

Glycosyl Transferase	Specific Activity† in		Per Cent of Total Activity in Golgi Complex
	Total Homogenate	Golgi Complex	
Sialyl	50	422	44
Galactosyl	11	128	42
N-acetylglucosaminyl	24	219	43
Galactosyl-N-acetylglucosamine	6	64	40

*Modified from Morré, D. J., Keenan, T. W., and Mollenhauer, H. H., Golgi Apparatus Function in Membrane Transformations and Product Compartmentalization: Studies with Cell Fractions Isolated from Rat Liver *in Advances in Cytopharmacology*, Vol. 1, First International Symposium on Cell Biology and Cytopharmacology. (Clementi, F., and Ceccarelli, B., eds.) New York, Raven Press, 1971. (Unpublished data of H. Schachter.)
†Specific activity of the four glycosyl transferases is expressed in $m\mu$ moles/hrs/mg protein of sugar nucleotide bound to the protein acceptor.

pyrophosphatase[46] and several glycosyl transferases. The Golgi fraction also contains acid phosphatase and other lysosomal enzymes, presumably in relation to primary lysosomes (see Chapter 21). Cytochemical studies have shown that certain activities involving nucleoside diphosphatases are localized in the Golgi region.

The most characteristic enzymes of the Golgi fraction are those related to the transfer of oligosaccharides to proteins (i.e., glycosyl transferases) with the resulting formation of glycoproteins (see below). Table 9–4 shows the specific activity of four glycosyl transferases that are involved in the transfer of CMP-neuraminic acid (CMP-NANA), UDP-galactose, and UDP-acetylglucosamine to protein acceptors. (All of these transferases are highly concentrated in the Golgi fraction of the rat liver.)

Synthesis and Secretion of Glycoproteins

The synthesis and secretion of glycoproteins are related to the endoplasmic reticulum and the Golgi complex. In these processes the coupling of the two parts of the vacuolar system with the extracellular space and the plasma membrane becomes evident and is of great interest in cell biology. A liver cell secretes two types of glycoproteins—one that is liberated into the extracellular space to be used elsewhere, and the other that becomes incorporated into the cell membrane as soon as it is secreted and thereby forms the cell coat (see Chapter 8).

From numerous studies using radioactive precursors (e.g., galactose, fucose, and sialic acid) the following biosynthetic scheme has emerged for the two types of secreted glycoproteins just mentioned. The protein backbone of the glycoprotein is synthesized by membrane-bound ribosomes and then the monosaccharides are added, one by one, as the protein moves through the channels of the endoplasmic reticulum and Golgi complex. In this process the system of glycosyl transferases that are bound to the membranes of the Golgi play an important role. The enzymes adding the oligosaccharide core containing N-acetyl-glucosamine and mannose are, at least in part, assembled in the rough endoplasmic reticulum. The fucosyl- and the sialyl-galactosyl-N-acetyl-glucosamine termini of the prosthetic group, on the other hand, are incorporated in the Golgi complex by the corresponding transferases.[47] The incorporation of the various precursors can be followed in the cellular organelles by the use of cell fractionation methods.

In Table 9–5 the three main patterns of incorporation are indicated. When D-mannose is used as precursor, the rough endoplasmic reticulum is the first to be marked and then the other compartments: Golgi complex → secretory granules → extracellular space or plasma membrane. When L-fucose or D-galactose is used the Golgi complex is the first to be labeled. With D-glucosamine both the rough endoplasmic reticulum and the Golgi complex are simultaneously labeled. A comparative study of the secretion of glycoproteins and the secretion of albumin by the liver cell revealed differences in the time of transport through the different segments of the vacuolar system. While albumin passes

TABLE 9–5. Incorporation of Various Precursors into Glycoproteins*

Pattern Type	Radioactive Precursor	Interpretation of Time Course Experiments
A	Leucine D-mannose	RER → Golgi → SG, EC, or PM ↑ Label

B	Sialic acid
	L-fucose
	D-galactose
		RER → Golgi → SG, EC, or PM
		       ↑
		     Label

C	D-glucosamine	RER → Golgi → SG, EC, or PM ↑ ↑ Label Label

*Abbreviations: *RER,* rough endoplasmic reticulum; *SG,* secretory granule; *EC,* extracellular space; *PM,* plasma membrane. This is a table which presents, in abbreviated form, experiments by various groups of investigators. The experimental work involved the injecting of whole animals, perfused tissues, or tissue slices with the precursors and the fractionating of the tissue at various time intervals. (Modified from Schachter, H., *J. Biol. Chem.,* 248:974–76, 1973.)

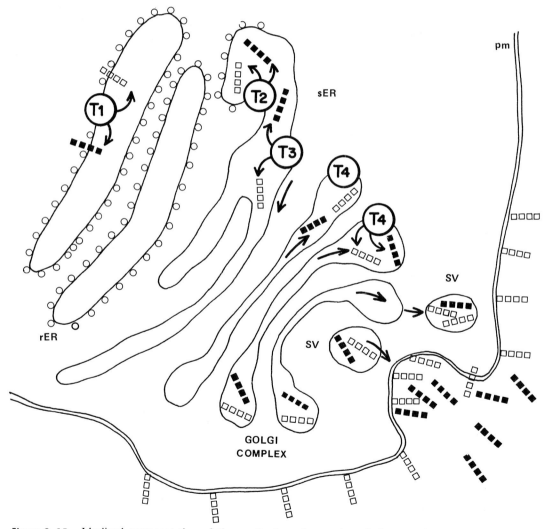

Figure 9–15. Idealized representation of the synthesis and secretion of glycoproteins throughout the various parts of the vacuolar system: rough (granular) endoplasmic reticulum (*rER*), smooth (agranular) endoplasmic reticulum (*sER*), Golgi complex, and secretory vesicles (*SV*). One of the secretory vesicles is shown opening into the plasma membrane (*pm*). T1, T2, T3, and T4 represent various transferases involved in the glycosidation of protein cores synthesized in the rER. Observe that two types of glycoproteins are produced, i.e., those that are incorporated into the plasma membranes, and those that are secreted and carried away from the cell. (Modified from figure in Schachter, H., Jabbal, I., Hudgin, R. L., and Pinteric, L., *J. Biol. Chem.,* 245:1090, 1970.)

directly into the cisternal cavities, glycoproteins remain attached to the membranes, where they are submitted to glycosidation by the glycosyl transferases.[48]

For cytological studies using electron microscope radioautography ^{3}H-fucose is preferred because it does not enter into mucopolysaccharides, and in glycoproteins fucose is present only in the terminal group. In intestinal cells, two minutes after the injection of ^{3}H-fucose the labeling is almost exclusively in the Golgi complex. Then there is migration of the glycoproteins to the sides of the cell and into the apical cell membranes, where it is concentrated four hours after injection. The intracellular migration of the glycoproteins is carried out by small vesicles formed in the Golgi region.[49]

The recent discovery of various lipid-sugar compounds (i.e., polyisoprenoid sugars) in animal tissues has raised the possibility that these substances could serve as intermediates in the synthesis of glycoproteins (see Leloir, 1971). These compounds might be active in the Golgi complex of the cell. In addition, the Golgi appears to be involved in the addition of sulfate to the carbohydrate moiety of the glycoproteins. In cartilage cells, mucopolysaccharides as well as glycoproteins are synthesized in the Golgi complex.[50, 51]

The diagram of Figure 9–15 is an idealized representation of the synthesis and secretion of glycoproteins and the intervention of the various segments of the vacuolar system: rough and smooth endoplasmic reticulum→ Golgi membranes → secretory vesicles and plasma membrane. In the diagram T^1, T^2, T^3, and T^4 represent the various transferases involved in the synthesis of the glycoproteins.

SUMMARY: Isolation of the Golgi Complex

The Golgi complexes of many plant and animal cells can be separated by gentle homogenization, followed by differential and gradient centrifugation. Both the endoplasmic reticulum and mitochondria have a higher specific density than the Golgi complex (density: 1.16). The Golgi from liver contains 60 per cent protein and 40 per cent lipids. (The pattern of protein is less complex than in the endoplasmic reticulum and more complex than in the plasma membrane.)

Various subcellular fractions can now be characterized by enzyme markers, and the degree of contamination of the Golgi fraction by mitochondria, plasma membranes, endoplasmic reticulum, and peroxisomes may be determined. The enzymes that appear most concentrated in the Golgi fraction are thiamine-pyrophosphatase and several glycosyl transferases that transfer oligosaccharides to the glycoproteins.

One of the best examples of functional coupling between the various compartments of the vacuolar system is the synthesis and secretion of glycoproteins that are released into the extracellular space or become incorporated into the cell coat. After the synthesis of the protein backbone in the rough endoplasmic reticulum a series of glycosyl transferases first adds the oligosaccharides of the core and then those of the terminal group.

Reading of Chapter 25, in which the general processes of cell secretion will be considered, should provide more insight into the function of the Golgi complex. Points to be clarified include: (1) the fact that the Golgi compartment is physiologically coupled with the endoplasmic reticulum and the plasma membrane; (2) the fact that the Golgi complex provides membranes for the packing of secretory products which are then delivered to the extracellular space by way of fusion with the plasma membrane; and (3) the existence of special locks in the transport and processes of fusion of membranes.

REFERENCES

1. Heilbrunn, L. V. (1958) The viscosity of protoplasm. In: *Protoplasmatologia,* II, Cl (Heilbrunn, L. V., and Weber, F., eds.) Springer-Verlag, Vienna.
2. De Robertis, E., and Schmitt, F. O. (1948) *J. Cell. Comp. Physiol., 31*:1.
3. De Robertis, E., and Franchi, C. M. (1953) *J. Exp. Med., 98*:269.
4. Porter, K. R. (1961) In: *The Cell,* Vol. 2, p. 621 (Brachet, J., and Mirsky, A. E., eds.) Academic Press, New York.
5. Siekevitz, P. (1959) *Ciba Foundation Symposium on the Regulation of Cell Metabolism,* p. 17. (Wolstenholme, G. E. W., and O'Connor, C. M., eds.) J. & A. Churchill, London.
6. Porter, K. R., Claude, A., Fullman, E. F. (1945) *J. Exp. Med., 81*:233.
7. Palade, G. E. (1956) *J. Biophys. Biochem. Cytol., 2*:85.
8. Sjöstrand, F. S. (1956) *Internat. Rev. Cytol., 5*:456.
9. Haguenau, F. (1958) *Internat. Rev. Cytol., 7*:425.
10. Christensen, A. K., and Fawcett, D. W. (1960) *Anat. Rec., 136*:333.
11. Palade, G. E. (1956) *J. Biophys. Biochem. Cytol., 2*:417.
12. Weibel, E. R., Stäubli, W., Gnägi, R., and Hess, F. A. (1969) *J. Cell Biol., 42*:68.
13. Meldolesi, J., Jamieson, J. D., and Palade, G. E. (1971) *J. Cell Biol., 49*:130.
14. Meldolesi, J., and Cova, D. (1972) *J. Cell Biol., 55*:1.
15. Dehlinger, P. J., and Schimke, R. T. (1971) *J. Biol. Chem., 246*:2574.
16. Kuriyama, Y, Omura, T., Siekevitz, P., and Palade, G. E. (1969) *J. Biol. Chem., 244*:2017.
17. Dallner, G. P., Siekevitz, P., and Palade, G. E. (1966) *J. Cell Biol., 30*:97.
18. Leskes, A., Siekevitz, P., and Palade, G. E. (1971) *J. Cell Biol., 49*:264.
19. Eytan, G., and Ohad, I. (1972) *J. Biol. Chem., 247*:112.
20. Bock, K. W., Siekevitz, P., and Palade, G. E. (1971) *J. Biol. Chem., 246*:188.
21. Porter, K. R. (1961) *J. Biophys. Biochem. Cytol., 10*:219.
22. Bennett, S. (1956) *J. Biophys. Biochem. Cytol., 2*:99.
23. Christensen, A. K., and Fawcett, D. W. (1961) *J. Biophys. Biochem. Cytol., 9*:653.
24. Claude, A. (1968) *J. Cell Biol., 39*:25a.
25. Porter, K. R., and Bruni, C. (1960) *Cancer Res., 19*:997.
26. Porter, K. R., and Machado, R. D. (1960) *J. Biophys. Biochem. Cytol., 7*:167.
27. Leloir, L. F., and Cardini, C. E. (1957) *J. Amer. Chem. Soc., 79*:6340.
28. Luck, D. J. L. (1961) *J. Biophys. Biochem. Cytol., 10*:195.
29. Peters, V., Kelly, G., and Dembitzen, H. (1963) *Ann. N.Y. Acad. Sci., 111*:87.
30. Rosen, S. I. (1964) *J. Cell Biol., 23*:78a.
31. Jones, A. L., and Fawcett, D. W. (1966) *J. Histochem. and Cytochem., 14*:215.
32. Conney, A. H., Schneidman, K., Jacobson, M., and Kuntzman, R. (1965) *Ann. N.Y. Acad. Sci., 123*:98.
33. Claude, A. (1970) *J. Cell Biol., 47*:745.
34. Burgos, M. H., and Fawcett, D. W. (1955) *J. Biophys. Biochem. Cytol., 1*:4.
35. Caro, L. G. (1961) *J. Biophys. Biochem. Cytol., 10*:37.
36. Gantenby, J. B., Dalton, A. J., and Felix, M. D. (1955) *Nature, 176*:301.

37. Novikoff, A. B., Essner, E., and Quintana, N. (1964) *Fed. Proc.,* *23*:1010.
38. Rambourg, G. (1966) *Anat. Rec., 154*:41.
39. Schneider, W. C., Dalton, A. J., Kuff, E. L., and Felix, M. D. (1953) *Nature, 172*:161.
40. Kuff, L., and Dalton, A. J. (1959) In: *Subcellular Particles,* p. 114. (Hayashi, T., ed.) The Ronald Press Co., New York.
41. Schneider, W. C., and Kuff, E. L. (1954) *Amer. J. Anat., 94*:209.
42. Morré, D. J., Cheetham, R., and Yunghans, W. (1968) *J. Cell Biol., 39*:961.
43. Fleischer, B. (1969) *Fed. Proc., 28*:404.
44. Yunghans, W. N., Keenan, T. W. and Morré, D. J. (1970) *Exp. and Mol. Pathol. 12*:36.
45. Morré, J., Keenan, T. W., and Mollenhauer, H. H. (1971) In: *Advances in Cytopharmacology,* Vol. 1, p. 159. (Clementi, F., and Ceccarelli, B., eds.) Raven Press, New York.
46. Novikoff, A. B. In: *The Neuron* pp. 255, 319. (Hyden, H., ed.) Elsevier Publishing Co., Amsterdam.
47. Schachter, H., Jabbal, I., Hudgin, R. L., and Pinteric, L. (1970) *J. Biol. Chem., 245*:1090.
48. Redman, C. M., and Cherian, M. G. (1972) *J. Cell Biol., 52*:231.
49. Bennett, G., and Leblond, C. P. (1971) *J. Cell Biol., 51*:875.
50. Neutra, M., and Leblond, C. P. (1969) *Sci. Am., 220*:100.
51. Young, R. W. (1973) *J. Cell Biol., 57*:175.

ADDITIONAL READING

Allison, A. C. (1973) The role of microfilaments and microtubules in cell movement, endocytosis and exocytosis. In: "Locomotion of Tissue Cells," *Ciba Found. Symp., 14*:109, Elsevier Publishing Co., Amsterdam.
Claude, A. (1969) Microsomes, endoplasmic reticulum and interactions of cytoplasmic membranes. In: *Microsomes and Drug Oxidations,* p. 3, Academic Press, Inc., New York.
Haguenau, F. (1958) The ergastoplasm: its history, ultrastructure, and biochemistry. *Internat. Rev. Cytol., 7*:425.
Jamieson, J. D. (1971) Role of the Golgi complex in the intracellular transport of secretory proteins. In: *Advances in Cytopharmacology,* Vol. I, p. 183. (Clementi, F., and Ceccarelli, B., eds.) Raven Press, New York.
Leloir, L. F. (1971) Two decades of research on the biosynthesis of saccharides. *Science, 172*:1299.
Loewy, A. G., and Siekevitz, P. (1969) Cell Structure and Function, 2nd Ed., Holt, Rinehart and Winston, Inc., New York.
Morré, J., Keenan, T. W., and Mollenhauer, H. H. (1971) Golgi apparatus function in membrane transformations and product compartmentalization. In: *Advances in Cytopharmacology,* Vol. I, p. 159. (Clementi, F., and Ceccarelli, B., eds.) Raven Press, New York.
Neutra, M., and Leblond, C. P. (1969) The Golgi apparatus. *Sci. Amer., 220*:100.
Palay, S. L. (1958) The morphology of secretion. In: *Frontiers in Cytology.* (Paley, S. L., ed.) Yale University Press, New Haven, Conn.
Pollard, T. D., and Korn, E. D. (1972) The contractile proteins of *Acanthamoeba castellani Cold Spring Harbor Symp. Quant. Biol., 37*:573.
Porter, K. R. (1961) The ground substance; observations from electron microscopy. In: *The Cell,* Vol. 2, p. 621. (Brachet, J., and Mirsky, A. E., eds.) Academic Press, Inc., New York.
Wessells, N. K., Spooner, D. S., Ash, J. F., Bradley, M. O., Ludueña, M. A., Taylor, E. L., Wrenn, J. T., and Yamada, K. M. (1971) Microfilaments in cellular and developmental processes. *Science, 171*:135.

ten

MITOCHONDRIA

Mitochondria (Gr., *mito-*, thread + *chondrion,* granule), granular or filamentous organelles present in the cytoplasm of protozoa and animal and plant cells, are characterized by a series of morphologic, biochemical, and functional properties. Among these are their size and shape, visibility in vivo, special staining properties, specific structural organization, lipoprotein composition, and content of a large "battery" of enzymes and coenzymes that interact to produce cellular energy transformations. From the physiological viewpoint, mitochondria are energy-transducing systems that recover the energy contained in foodstuffs (through the Krebs cycle and the respiratory chain) and convert it by phosphorylation into the high energy phosphate bond of adenosine triphosphate (ATP) (see Chapter 14). Thus, mitochondria are the "power plants" that produce the energy necessary for many cellular functions (Fig. 10–1).

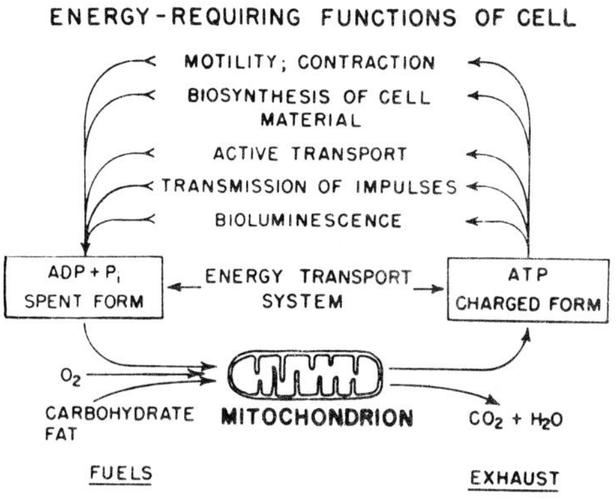

First observed at the end of the nineteenth century and described as "bioblasts" by Altmann (1894), these structures were called "mitochondria" by Benda (1897). Altmann predicted the relationship between mitochondria and cellular oxidation, and Warburg (1913) observed that respiratory enzymes were associated with cytoplasmic particles.

A most important advance was the first isolation of liver mitochondria by Bensley and Hoerr in 1934. This established the possibility of a direct study of the organelles by biochemical methods. The final demonstration that the mitochondrion was indeed the site of cellular respiration was made in 1948 by Hogeboom et al.[1] In recent years important advances in the study of its ultrastructural organization have been made with the aid of the electron microscope. The study of this organelle is particularly thrilling because the mitochondrion is one of the best known examples of structural-functional integration within the cell.[2,3] Lately, more progress has been made, with the demonstration that mitochondria contain a specific type of DNA, that they have their own machinery for protein synthesis, and that they may participate in inheritance and differentiation.

Figure 10–1. Diagram showing that the mitochondrion constitutes the central "power plant" of the cell. The adenosine triphosphate (ATP) produced is used in the various functions that are indicated. (From Lehninger, A. L., *Physiol. Rev., 42*:3, 467, 1962.)

Morphology

In Vivo Examination

Although the examination of mitochondria in living cells is somewhat difficult because of their low refractive index, they can be observed easily in cells cultured in vitro, particularly under darkfield illumination and phase contrast (Fig. 9–3). This examination has been greatly facilitated by coloration with a dilute solution of *Janus green*. The resultant greenish blue stain is due to the action of the cytochrome oxidase system present in mitochondria, which maintains the dye in its oxidized (colored) form. In the surrounding cytoplasm the dye is reduced to a colorless leukobase.

Micromanipulation has demonstrated that mitochondria are relatively stable and can be displaced by the microneedle without alterations. The *specific gravity* is greater than that of the cytoplasm. By ultracentrifugation of living cells at 200,000 to 400,000 g, mitochondria are deposited intact at the centrifugal pole (Fig. 2–5).

Volume-Shape Changes and Motion

Observation in vivo is of particular interest when supplemented by time-lapse cinematography. In cultured fibroblasts, continuous and sometimes rhythmic changes in volume, shape, and distribution of mitochondria can be observed. The two main types of motion are *agitation* and *displacement* from one part of the cell to another. Furthermore, filamentous mitochondria can fragment into granules which may reunite. Some movements are passive and due to cytoplasmic streaming.

Active changes in volume and shape of mitochondria may be caused by chemical, osmotic, and mechanochemical changes. In living cells, low amplitude contraction cycles associated with oxidative phosphorylation have been observed. It is known that cyanide, dinitrophenol, and other oxidative inhibitors produce swelling, and that ATP in excess produces contraction of mitochondria. Swelling and contraction of mitochondria also occur by changing the osmotic pressure of the medium. Inorganic phosphate, reduced glutathione, Ca^{++}, and fatty acids cause swelling, whereas ATP prevents it. The property of swelling and contraction is best studied in isolated mitochondria and will be considered again in the discussion of the physiology of mitochondria.

Fixation, Size, and Shape

Since mitochondria are labile structures that are readily disintegrated by the action of fixatives, fixation should be by methods that stabilize the lipoprotein structure by the prolonged action of oxidizing agents, such as osmium tetroxide, chromic acid, and potassium dichromate. Iron hematoxylin (Regaud) and acid fuchsin (Altmann) are the commonly used stains. In Chapter 6 some of the cytochemical staining methods for mitochondria were mentioned (Fig. 6–16).

The *shape* of mitochondria is variable, but in general these organelles are *filamentous* or *granular* (Fig. 10–2). During certain functional stages, other derived forms may be seen. For example, a long mitochondrion may swell at one end to assume the form of a *club* or be hollowed out to take the form of a *tennis racket*. At other times mitochondria may become *vesicular* by the appearance of a central clear zone. The morphology of mitochondria varies from one cell to another, but it is more or less constant in cells of a similar type or in those performing the same function.

The *size* of mitochondria is also variable; however, in most cells the width is relatively constant (about 0.5 μm), and the length is variable, reaching a maximum of 7 μm. Depending on the functional stage of the cell, however, it is possible to find rods that are either thinner or thicker. The size and shape of the fixed mitochondria depend also on the osmotic pressure and the pH of the fixative. Using buffered fixatives at physiologic pH has been shown to be important.

Table 10–1 indicates some quantitative data regarding the size of mitochondria in liver cells as determined by electron microscopy.[4]

Distribution

Mitochondria are, in general, uniformly distributed throughout the cytoplasm, but there are many exceptions to this rule. In some cases, they accumulate preferentially around the nucleus or in the peripheral cytoplasm. Overloading with inclusions, such as glycogen and fat, displaces these organelles. During mitosis, mitochondria are concentrated near the spindle, and upon division of the cell they are distributed in approximately equal number between the daughter cells.

The distribution of mitochondria within the cytoplasm should be considered in relation to their function as energy suppliers. In some cells they can move freely, carrying ATP

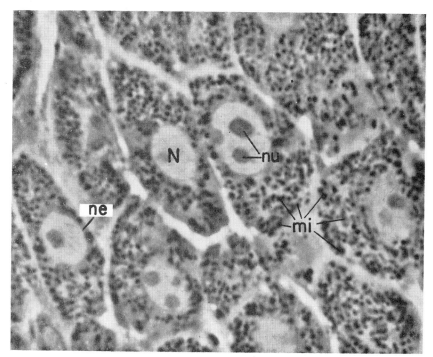

Figure 10-2. Liver cells of the rat fixed at −180° C and dehydrated in acetone at −40° C. Photomicrograph made with phase contrast microscope in a medium of n = 1.460. *mi*, mitochondria; *N*, nucleus; *ne*, nuclear envelope; *nu*, nucleoli. (Courtesy of S. Koulish.)

where needed, but in others they are located permanently near the region of the cell where presumably more energy is needed. For example, in certain muscle cells (e.g., diaphragm), mitochondria are grouped like rings or braces around the I-band of the myofibril. In the rod and cone cells of the retina all mitochondria are located in a portion of the inner segment. The basal mitochondria of the kidney tubule are intimately related to the infoldings of the plasma membrane in this region of the cell. It is assumed that this close relationship with the membrane is related to the supply of energy for the active transport of water and solutes.

Orientation. Mitochondria may have a more or less definite orientation. For example, in cylindrical cells they are generally oriented in the basal-apical direction, parallel to the main axis. In leukocytes, mitochondria are arranged radially with respect to the centrioles. It has been suggested that these orientations depend upon the direction of the diffusion currents within cells and are related to the submicroscopic organization of the cytoplasmic matrix and vacuolar system.

Number. Mitochondria are found in the cytoplasm of all aerobically respiring cells, with the exception of bacteria in which the respiratory enzymes are located in the plasma membrane (Fig. 2-1). The mitochondrial content of a cell is difficult to determine, but, in general, it varies with the cell type and functional stage. It is estimated that in liver, mitochondria constitute 30 to 35 per cent of the total protein content of the cell, and in

TABLE 10-1.   Measurements in Rat Liver Mitochondria*

	Peripheral Cells	Midzonal Cells	Central Cells
Cytoplasmic volume (%)	19.8	19.1	12.9
Number per cell	1060.0	1300.0	1600.0
Diameter (μm)	0.56	0.47	0.32
Length (μm)	3.85	4.32	5.04

*From Loud, A. V., *J. Cell Biol., 37*:27, 1968.

kidney, 20 per cent. In lymphoid tissue the value is much lower. In mouse liver homogenates there are about 8.7×10^{10} mitochondria per gram of fresh tissue.[5] A normal liver cell contains about 1000 to 1600 mitochondria (Table 10–1), but this number diminishes during regeneration and also in cancerous tissue.[6] This last observation may be related to decreased oxidation that accompanies the increase to anaerobic glycolysis in cancer. Another interesting finding is that there is an increase in the number of mitochondria in the muscle after repeated administration of the thyroid hormone, thyroxin. An increased number of mitochondria has also been found in human hyperthyroidism.

Some oöcytes contain as many as 300,-000 mitochondria—the largest number recorded for a cell. There are fewer mitochondria in green plant cells than in animal cells, since some of their functions are taken over by chloroplasts.

SUMMARY:
Nature of Mitochondria

Mitochondria are organelles present in the cytoplasm of all eukaryotic cells. They provide an energy-transducing system by which the chemical energy contained in foodstuffs is converted, by oxidative phosphorylation, into high-energy phosphate bonds (ATP). Mitochondria may be observed in the living cell; visibility is increased by the vital stain, *Janus green*. They display passive and active motion and show changes in volume and shape that are related to their function. Swelling of mitochondria can be induced by Ca^{2+}, various hormones, and certain drugs.

The morphology of mitochondria is best studied after fixation. In general, they are rod-shaped, with a diameter of about 0.5 μm and a variable length that may range up to 7 μm. There are 1000 to 1600 mitochondria in a liver cell and 300,000 in some oöcytes. Green plants contain fewer mitochondria than animal cells. The distribution of mitochondria may be related to their function as suppliers of energy. Their orientation in the cell may be influenced by the organization of the cytoplasmic matrix and vacuolar system.

STRUCTURE

As indicated in Figure 10–3, a mitochondrion consists of two membranes and two compartments, the larger of which contains the *mitochondrial matrix*. An outer limiting membrane, about 6 nm thick, surrounds the mitochondrion. Within this membrane, and separated from it by a space of about 6 to 8 nm, is an inner membrane that projects into the mitochondrial cavity complex infoldings called *mitochondrial crests*. This inner membrane, also about 6 nm thick, divides the mitochondrion into two chambers or compartments: (1) the outer chamber contained between the two membranes and in the core of the crests, and (2) the inner chamber, bound by the inner membrane. This inner chamber is filled with a relatively dense material usually called the *mitochondrial matrix*. This is generally homogeneous, but in some cases it may contain a finely filamentous material, or small, highly dense granules (see Figure 10–21). These granules are now considered to be sites for binding divalent cations, particularly Mg^{2+} and Ca^{2+}.[7] The mitochondrial crests that project from the inner membrane are, in general, incomplete septa or ridges that do not interrupt the continuity of the inner chamber; thus, the matrix is continuous within the mitochondrion.

Further studies have shown that the mitochondrial membranes may be more complex, with two layers of high electron opacity and a less opaque middle layer. This corresponds to the unit membrane structure mentioned in Chapter 8 (Fig. 10–4). As in the case of the plasma membrane (Chapter 8), after

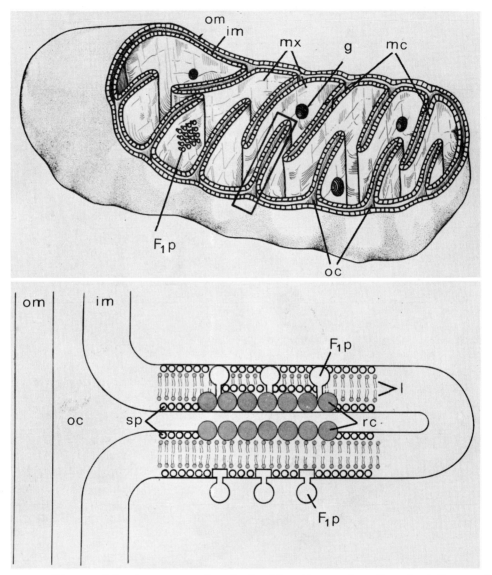

Figure 10–3. Diagram of the ultrastructure of a mitochondrion. **Above,** a three-dimensional diagram of a mitochondrion showing: the outer membrane (*om*), the inner membrane (*im*), the mitochondrial matrix (*mx*), the mitochondrial crests (*mc*), and granules (*g*) present in the matrix and containing calcium and magnesium. The outer chamber (*oc*), between the membranes, and the F_1 particles (F_1p) are also indicated. **Below,** the molecular organization of a mitochondrial crest. (This portion corresponds to the inset in the upper figure.) Notice that the respiratory chains (*rc*) are arranged along the outer edge of the inner membrane. The F_1 particles are probably within the membrane in the intact mitochondrion, but they become exposed (lower part) with osmotic treatment and negative staining (see Figure 10–5); lipid layer (*l*); structural protein (*sp*). (Lower portion of the figure courtesy of A. L. Lehninger.)

most of the lipids are extracted, the unit membrane structure remains intact.[8] The outer and inner membranes and the crests can be considered to be solid molecular films with a compact molecular structure; the matrix is gel-like and contains a high concentration of soluble proteins and smaller molecules. This double (solid-liquid) structure is important in providing an explanation for some of the mechanical properties of mitochondria, e.g., deformation and swelling under physiologic or experimental conditions.

The use of negative staining has permitted recognition of other details of structure. If a mitochondrion is allowed to swell and break in a hypotonic solution and is then im-

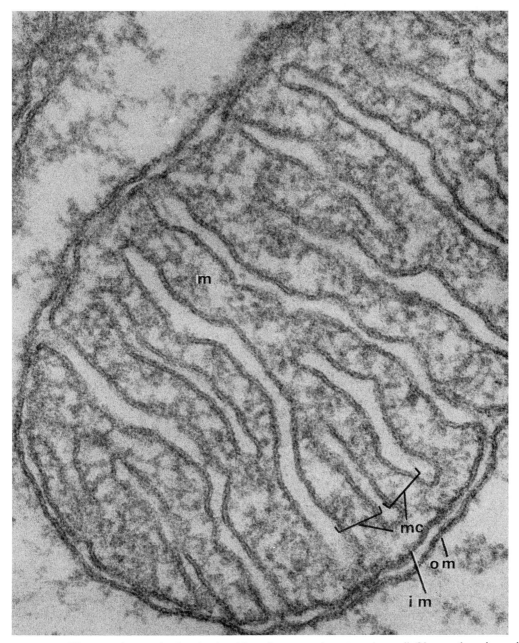

Figure 10–4. Electron micrograph of a mitochondrion of a pancreatic centroacinar cell. Observe the unit membrane structure in the outer membrane (*om*), the inner membrane (*im*), and mitochondrial crests (*mc*). *m*, matrix. ×207,000. (Courtesy of G. E. Palade.)

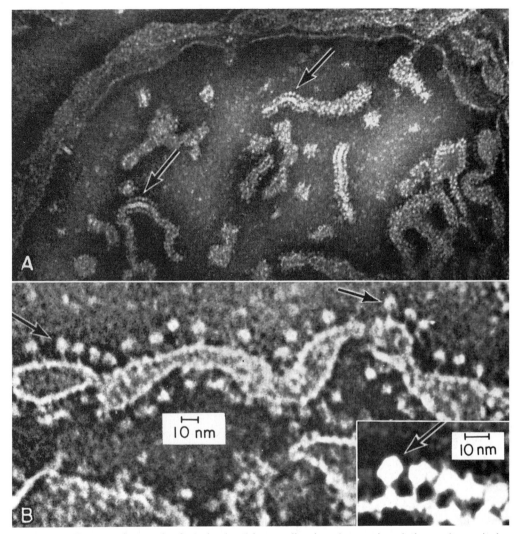

Figure 10–5. Electron micrograph of a mitochondrion swollen in a hypotonic solution and negatively stained with phosphotungstate. **A,** at low power; isolated crests can be observed in the middle of the swollen matrix. Arrows point to some of these crests. **B,** at higher magnification (×500,000), a mitochondrial crest showing the so-called "elementary particles" on the surface adjacent to the matrix. Inset at ×650,000, showing the elementary particles with a polygonal shape and the fine attachment to the crest. (Courtesy of H. Fernández-Morán.)

mersed in phosphotungstate, the inner membrane and the crests appear covered by particles of 8.5 nm that have a stem linking them with the membrane (Fig. 10–5). These so-called "elementary," or "F_1 particles,"[9] are regularly spaced at intervals of 10 nm on the inner surface of these membranes. According to some estimates, there are 10^4 to 10^5 elementary particles per mitochondrion.[14] These particles represent a special ATPase involved in the coupling of oxidation and phosphorylation.[10] F_1 particles are contained within the thickness of the inner membrane and are not

seen in sections; only when mitochondria are opened by hypotonic treatment and negatively stained are they extruded on the inner surface and made visible (Fig. 10–5).

Structural Variations

It is presumed that a common pattern of mitochondrial structure developed at an early stage of evolution and was subsequently transmitted, without considerable modifications, from protozoa to mammals and from algae to flowering plants. Detailed structural

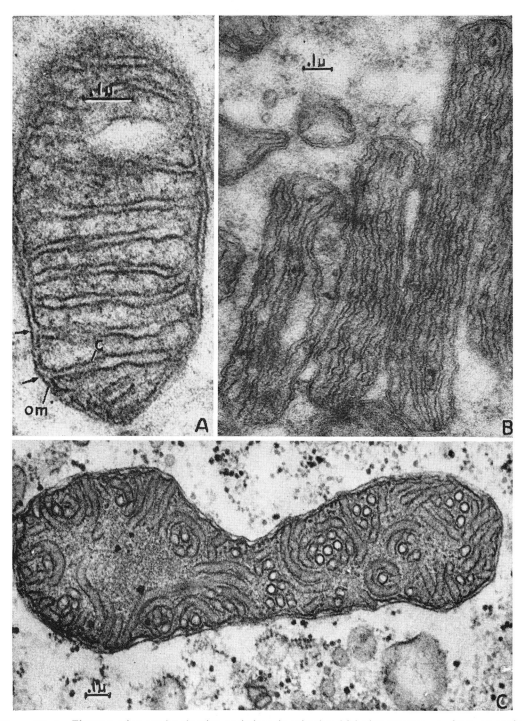

Figure 10–6. Electron micrographs showing variations in mitochondrial ultrastructure. **A,** from rat testicle: *c,* transverse crests; *om,* outer membrane; arrows show origin of crests at the inner membrane. **B,** from ovotestis of *Helix.* Longitudinal crests in mitochondria of spermatocytes. **C,** from *Paramecium,* tubular crests. **A,** ×130,000; **B,** ×68,000; **C,** ×60,000. (Courtesy of J. Andrè.)

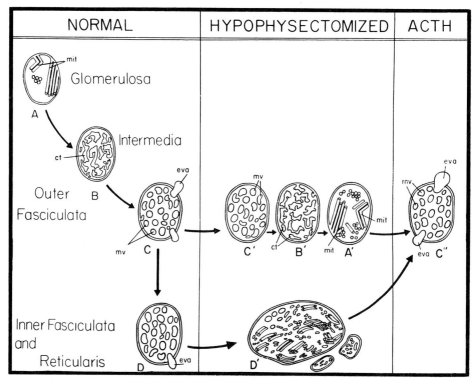

Figure 10–7.   General diagram of the mitochondrial changes found in adrenal cortex after hypophysectomy and restorational therapy. *A, B, C,* and *D,* normal mitochondria at the glomerulosa, intermedia, outer fasciculata, inner fasciculata, and reticularis. *C', B',* and *A',* progressive changes of mitochondria of the outer fasciculata, leading to a pattern similar to that of glomerular cells (*A*). (Notice the lack of extruding vacuoles.) *D',* a chondriosphere of the inner fasciculata and reticularis formed by fusion of altered mitochondria. *C'',* mitochondria of the fasciculata after the injection of ACTH in a hypophysectomized animal. Notice the restoration of the vesicular pattern and the extruding vacuoles. *eva,* extruding vacuoles; *mit,* mitochondrial tubules; *mv,* mitochondrial vesicle. (From Sabatini, D. D., De Robertis, E., and Bleichmar, H. B., *Endocrinology, 70:*390, 1962.)

variations can be observed, however. The crests (Fig. 10–6) may be arranged longitudinally (e.g., in nerve and striated muscle) or they may be simple or branched, forming complex networks. In protozoa, insects, and adrenal cells of the glomerular zone, the infoldings may be tubular instead of lamellar, and the tubules may be packed in a regular fashion (Fig. 10–6).

The number of crests per unit volume of a mitochondrion is also variable. Mitochondria in liver and germinal cells have few crests and an abundant matrix, whereas those in certain muscle cells have numerous crests and little matrix. In some cases the crests are so numerous that they may have a quasi-crystalline disposition.[11] The greatest concentration of crests is found in the flight muscle of insects. In general, there seems to be a correlation between the number of crests and the oxidative activity of the mitochondrion.

A particularly interesting variation in fine structure is observed in cells of the different regions of the adrenal cortex (Fig. 10–7). One characteristic of these mitochondria is the enlargement of the space within the crests or tubules which, according to the diagram in Figure 10–3, corresponds to the outer compartment of the mitochondrion. This is very conspicuous and appears as tubular or vesicular openings, which are much less opaque than the inner matrix. These structures seem to be related to the specific secretory activity of the gland. In this case, mitochondria, in addition to functioning in cell oxidations, are actively engaged in the synthesis of steroid hormones.[12]

Relationship with Lipids.   Various authors

since the time of Altmann have observed that the *disposition of lipids* may be related to mitochondrial activity. In pancreas and liver cells, for example, after a short period of starvation, the mitochondria come into contact with lipid droplets. The relationship may be so tight that only the inner mitochondrial membrane can be seen adjacent to the lipid in some regions. The electron microscopic images suggest that an active process of fat utilization takes place under the action of the fatty acid oxidases present in mitochondria.[13]

Accumulation of Protein and Other Substances. An accumulation of pigment derived from hemoglobin has been observed in mitochondria of amphibians. Ferritin molecules

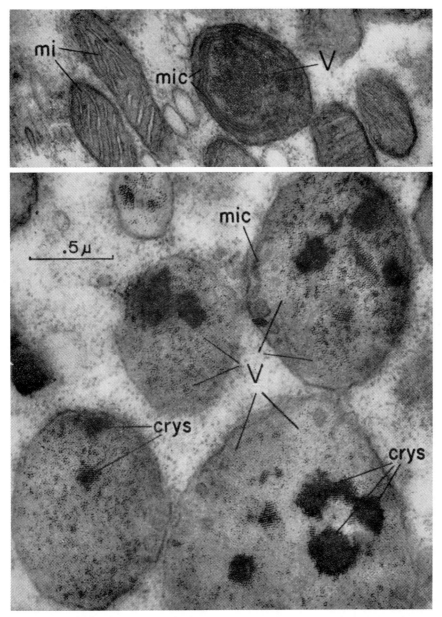

Figure 10–8. Electron micrographs of mitochondria of the egg of *Planorbis*. **Above,** normal mitochondria (*mi*) with crests, one showing an accumulation of protein molecules. *mic,* mitochondrial crests; *V,* yolk. **Below,** same structures as above, but with more development of the yolk. Mitochondria are being transformed into yolk platelets. The protein molecules have a crystalline disposition (*crys*). ×50,000. (Courtesy of P. Favard and N. Carasso.)

accumulate within mitochondria in subjects suffering from Cooley's hereditary anemia. Another example is the transformation of mitochondria into yolk bodies in eggs of the mollusk *Planorbis*.[14] Early oöcytes contain typical mitochondria, in which numerous dense protein molecules appear. In more advanced stages, mitochondria may become transformed into larger yolk bodies. In these, the masses of protein molecules may assume a regular crystalline disposition (Fig. 10–8). In amphibian oöcytes, hexagonal, crystalline yolk bodies also form within mitochondria.[15]

Degeneration of Mitochondria. Mitochondria are one of the most sensitive indicators of injury to the cell. Although mitochondria may be readily altered by the action of various agents, the changes are, within certain limits, reversible. If the alteration reaches a certain critical point, however, it becomes irreversible, and this is generally considered degeneration of the mitochondria. Essentially there are three types of change: (1) fragmentation into granules, followed by lysis and dispersion; (2) intense swelling with transformation into large vacuoles; and (3) a great accumulation of materials with transformation of mitochondria into hyalin granules. This last change is characteristic of the so-called cloudy swelling and hyalin degeneration that frequently results in cellular death.

A relatively frequent observation in an otherwise normal cell is the presence of degenerating mitochondria in foci of autolysis, constituting a type of lysosome (cytolysosome).

Another type of degeneration is the fusion of mitochondria to form large bodies called *chondriospheres*. This degeneration has been found in patients with scurvy and seems to be normal in the adrenal gland of the hamster.[16] In the hamster, in the deepest region of the adrenal cortex there are mitochondria that undergo a process of flattening into thin, multilamellar sheaths. At the same time there is a concentric apposition of several of these mitochondria, a process that results in the formation of large lamellar chondriospheres.

SUMMARY:
Structure of Mitochondria

The mitochondrion contains two compartments — an inner one filled with the mitochondrial matrix and limited by an inner membrane and an outer one located between the inner and outer membranes. Both membranes have a trilaminar (unit membrane) structure; however, by negative staining the outer membrane is smooth but the inner membrane shows, on its inner surface, particles of 8.5 nm linked to the membrane (F_1 particles). It will be shown later that these particles contain a special ATPase. Complex infoldings of the inner membrane, called mitochondrial crests, project into the matrix. The shape and disposition of these crests vary in different cells and their number is related to the oxidative activity of the mitochondrion. Mitochondria may show a close relationship with lipid droplets. They may accumulate iron-containing pigments or protein molecules, thereby forming yolk bodies. Injury to mitochondria may produce degenerative changes consisting of fragmentation, intense swelling, or accumulation of material. Degenerating mitochondria may be found forming cytolysosomes or large chondriospheres.

Isolation of Mitochondria and Mitochondrial Membranes

In Chapter 6 the methods of cell fractionation for the separation of the mitochondrial fraction were described (Fig. 6–5). Morphologically homogeneous fractions of mitochondria may be isolated from liver, skeletal muscle, heart, and other tissues. Since the early investigations of Hogeboom et al.

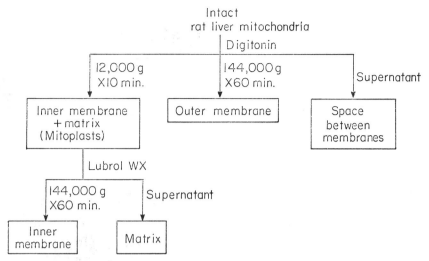

Figure 10–9. Fractionation procedure used to separate the outer and inner membranes of the mitochondria. This method also permits separation of the matrix and provides information about the content of the outer chamber of the mitochondrion. (Courtesy of C. Schnaitman and J. W. Greenawalt.)

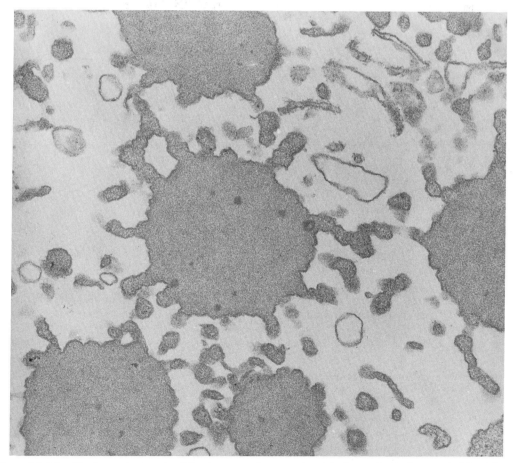

Figure 10–10. Electron micrograph of the inner membrane and matrix (i.e., mitoplast) separated from liver mitochondria. Note the fingerlike processes and the intact appearance of the inner membrane. ×97,500. (Courtesy of C. Schnaitman and J. W. Greenawalt.)

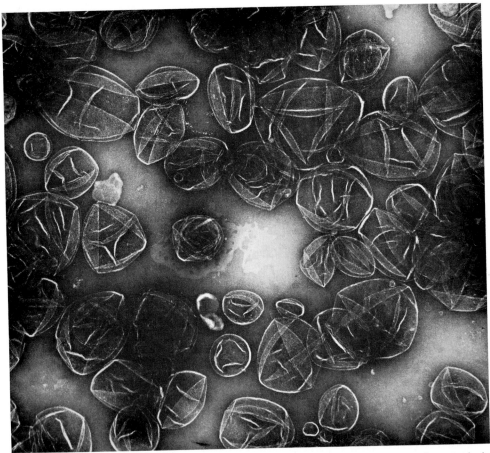

Figure 10-11.   Isolated outer mitochondrial membranes from liver. Whole mounted preparation negatively stained. Notice the folded bag appearance of the membrane and the absence of F_1 particles. ×35,000. (Courtesy of D. F. Parsons.)

(1948)[1] and others it has been demonstrated that the mitochondrion has a lipoprotein composition—65 to 70 per cent protein, and 25 to 30 per cent lipid. Most of the lipid content consists of phosphatides (e.g., lecithin and cephalin); cholesterol and other lipids are present in small amounts. Ribonucleic acid was consistently found in about 0.5 per cent of the dry weight.

In recent years the two mitochondrial membranes and the compartments they limit have been separated by density gradient centrifugation.[17-20] The outer membrane can be separated by causing a swelling, with breakage followed by a contraction of the inner membrane and matrix. Figure 10-9 shows one of the most frequently used procedures in which two detergents, digitonin and lubrol, are used. The outer membrane is

much lighter and needs stronger centrifugal forces or a less dense gradient to be separated.

A so-called *mitoplast*, which includes the inner membrane and matrix both intact, has been produced by separating these two elements from the outer membrane with digitonin (Fig. 10-10). The mitoplast has pseudopodic processes and is able to carry out oxidative phosphorylation. This separation has provided a clearer definition of the exact localization of the mitochondrial enzyme systems and has revealed interesting differences between the two membranes.

The outer membrane fraction has a 40 per cent lipid content (compared to 20 per cent in the inner membrane), contains more cholesterol, and is higher in phosphatidyl inositol; on the other hand, it is lower in cardiolipin. These figures demonstrate that a fundamental difference between the outer and inner membrane is in the lipid/protein ratio (i.e., about 0.8 in the outer membrane and about 0.3 in the inner membrane). This low lipid/protein ratio indicates that in the inner membrane there is a greater degree of intercalation of the protein within the lipid bilayer (see the fluid model of membrane structure in Chapter 8). This fact of structure has been confirmed by electron microscopic observations with freeze-fracture methods.[21]

From the morphologic viewpoint the outer membrane lacks the elementary particles that are prominent, when negatively stained, in the inner membrane (Fig. 10–5). As shown in Figure 10–11, the outer membrane has a characteristic "folded bag" appearance in these preparations.

Compartmentation of Enzymes

Table 10–2 lists the enzymes present in the various mitochondrial fractions. The activities of some enzymes that may be considered as markers of the fractions are presented in Table 10–3.

The *outer membrane* contains an NADH-cytochrome-*c*-reductase system which consists of a flavoprotein and cytochrome b_5. The most specific enzyme system of the outer membrane is *monoamine oxidase,* which may serve as an enzyme marker (Table 10–3).[18] This membrane also contains kynurenine hydroxylase, fatty acid coenzyme A ligase, a phospholipase, and various enzymes of the phospholipid metabolism. The supernatant, obtained after treatment with digitonin (Fig. 10–9), is considered as originating from the outer compartment between the two mem-

TABLE 10–2. Enzyme Distribution in Mitochondria*

Outer membrane
    Monoamine oxidase
    Rotenone-insensitive NADH-cytochrome *c*
      reductase
    Kynurenine hydroxylase
    Fatty acid CoA ligase

Space between outer and inner membranes
    Adenylate kinase
    Nucleoside diphosphokinase

Inner membrane
    Respiratory chain enzymes
    ATP synthetase
    Succinate dehydrogenase
    β-Hydroxybutyrate dehydrogenase
    Carnitine fatty acid acyl transferase

Matrix
    Malate and isocitrate dehydrogenases
    Fumarase and aconitase
    Citrate synthetase
    α-Keto acid dehydrogenases
    β-Oxidation enzymes

*Courtesy of A. L. Lehninger.

branes. In this compartment are found adenylate kinase, nucleoside diphosphokinase, DNAase I, and 5'-endonuclease.

The *mitoplast fraction* (inner membrane plus matrix) contains the components of the respiratory chain and oxidative phosphorylation and the soluble enzymes present in the matrix.[19]

The *inner membrane* carries all the components of the respiratory chain and the oxidative phosphorylation system (see below). It also has glycerol-phosphate dehydrogenase, choline-dehydrogenase, and several *carriers*

TABLE 10–3. Enzyme Activity of Various Mitochondrial Components*

Component	Inner Membrane	Outer Membrane	Matrix	Outer Chamber
Enzyme marker	Cytochrome oxidase	Monoamine oxidase	Malate dehydrogenase	Adenylate kinase
Specific activity of marker in component	9315	551	3895	6690
Specific activity in whole mitochondria	1980	22	2608	421
% Protein in component	21.3	4.0	66.9	6.3

*From Schnaitman, C. A., and Greenawalt, J. W. *J. Cell Biol.,* *38*:158, 1968.

or translocators for the permeation of phosphate, glutamate, aspartate, ADP, and ATP. It is calculated that the respiratory chain represents about 20 per cent and the phosphorylating system about 15 per cent of the total protein content of the membrane. These impressive figures give an idea of the importance of these systems in the inner membrane and mitochondrial crests.

The inner membrane contains a remarkably high concentration of cardiolipin (polyglycerophosphatides), which appear to be important in all systems involving electron transport.[20] The *inner compartment* or *mitochondrial matrix* contains all the soluble enzymes of the citric acid or Krebs cycle and those involved in the oxidation of fatty acids. The matrix also contains DNA, ribosomes, and other RNA species involved in the synthesis of protein (see below).

SUMMARY:
Isolation of Mitochondria

Mitochondria, as well as their membranes and compartments, may be separated by subcellular fractionation. The so-called mitoplast is a mitochondrion from which the outer membrane has been stripped by osmotic action or by digitonin. The outer membrane comes off in a lighter fraction because it has a much higher lipid/protein ratio than the inner membrane. By negative staining it shows a smooth surface and a folded bag appearance.

The mitochondrial enzymes show a definite compartmentation. The outer membrane contains NADH-cytochrome *c* reductase, which consists of a flavoprotein and cytochrome b_5. Monoamine oxidase is the specific enzyme marker of this membrane. The *outer compartment* contains adenylate kinase and other soluble enzymes. The *inner membrane* carries all the components of the respiratory chain and of oxidative phosphorylation. Taken together, these components represent 35 per cent of the protein of the membrane. In addition to other bound enzymes the inner membrane contains several specific carriers or translocation proteins involved in the permeation of metabolites. The *mitochondrial matrix* contains soluble enzymes of the Krebs cycle, DNA, RNA, and other components of the machinery for protein synthesis of the mitochondrion.

MOLECULAR STRUCTURE AND FUNCTION

The various functions of mitochondria are so intimately related to the structure at the molecular level that one cannot be studied without the other. This section requires some knowledge of the enzymatic mechanisms in which mitochondria are involved. The concepts relating to *bioenergetics,* summarized in Chapter 4, should be reviewed.

To express the complexity of the enzyme system in numerical terms, in a mitochondrion more than 70 enzymes and coenzymes and numerous cofactors and metals essential to mitochondrial functions work together in an orderly fashion. Besides oxygen, the only fuel that a mitochondrion needs is phosphate and adenosine diphosphate (ADP); the principal final products are ATP plus CO_2 and H_2O. Figure 4–14 indicates the final common pathway of biological oxidation, which takes place within the mitochondrion. The three major foodstuffs of the cell (carbohydrate, fat, and protein) are ultimately degraded in the cytoplasm to a two-carbon unit that is bound to coenzyme A to form acetyl coenzyme A. When this penetrates the mitochondrion the acetate group enters the *Krebs tricarboxylic (citric) acid cycle* in which, after a complex series of steps involving several enzymes, it is decarboxylated, losing CO_2. At several points in the cycle, pairs of electrons (or their equivalent hydrogen atoms) are removed by dehydrogenases and enter into the *respiratory chain (electron transport system),* at the end of which they combine with molecular oxygen to form water.

Respiratory Chain

The early work of Warburg and Keilin, in which spectrophotometric techniques were used, led to the concept that oxidations are brought about by carriers of electrons arranged in a chain of increasing redox potential. The main components of the respiratory chain then recognized were the NADH-linked dehydrogenases, the flavin-linked dehydrogenases, and the various cytochromes (cytochrome b,c,a,a_3). Because all these components, with the exception of cytochrome c, are firmly bound to the inner membrane, it has been difficult to isolate them

individually and to reconstitute the complete respiratory chain.

In recent years several additional centers able to accept electrons from NADH or succinate have been discovered (see Slater 1972 and 1973). For example, a lipid soluble quinone, the so-called *ubiquinone* (coenzyme-Q), was found in addition to enzyme-bound copper and two classes of iron compounds (one containing iron-sulfur, the other, high-spin iron). According to Slater, there are about 30 centers in mitochondria that are able to accept electrons in the respiratory chain.

Various parts of the respiratory chain may be separated as multimolecular complexes containing considerable amounts of lipid, essential to the activity of the complex.[22, 23] Green and associates have recognized four main complexes which can reconstruct the electron transport chain, if mixed in correct stoichiometric ratios. Three of these four complexes (I, III, and IV) are indicated in Figure 10–12. Complex II corresponds to the succinate-Q-reductase (i.e., succinate dehydrogenase). In Table 10–4 the various components of the respiratory and ATP-synthesizing system of heart mitochondria are mentioned. Complex IV or cytochrome-c-oxidase contains two heme *a*

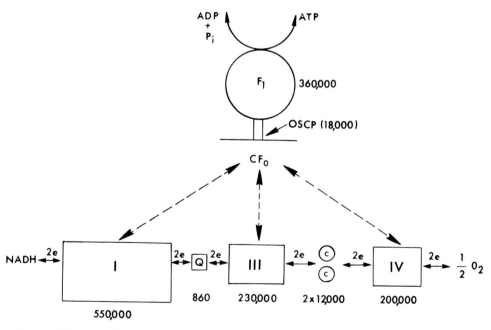

Figure 10–12. Diagram of the electron-transferring complexes (i.e., respiratory chain) and the ATPase complex (F_1) present in the mitochondrial inner membrane; complex I, NADH-Q-reductase; complex III, QH_2-cytochrome-c-reductase; complex IV, cytochrome-c-oxidase; CF_0, F_0 coupling factor; *OSCP,* oligomycin sensitive protein; Q, ubiquinone; c, cytochrome. (See the description in the text.) (Courtesy of E. C. Slater.)

groups and two copper atoms per molecule. Complex III, also called QH_2-cytochrome-c-reductase, contains at least two cytochromes b, probably two iron-sulfur centers, cytochrome c_1, and two or three high-spin iron centers. Complex I is the NADH-Q-reductase. In Figure 10–12 these complexes are drawn to the scale of molecular size, and the relationship with ubiquinone (coenzyme Q), cytochrome c, and oxygen is indicated.

Phosphorylating System

Figure 10–12 also shows the relationship between the respiratory chain and the *factor of Racker* (F_1) or *mitochondrial ATPase*, responsible for the synthesis of ATP. The F_1 particle has a molecular weight of 360,000 (Fig. 10–12) and is represented structurally by the 8.5 nm head of the elementary particle observed with the electron microscope (Fig. 10–5). The F_1 particle is attached to a piece of the membrane.[25] This attachment contains a protein of low molecular weight (OSCP) that binds oligomycin (see Racker, 1970).

The elegant experiments shown in Figure 10–13 have permitted the dissociation of the phosphorylating system from the respiratory chain followed by the reassociation of the entire system. If mitochondria are exposed to sound waves, the inner membrane is disintegrated and forms closed vesicles

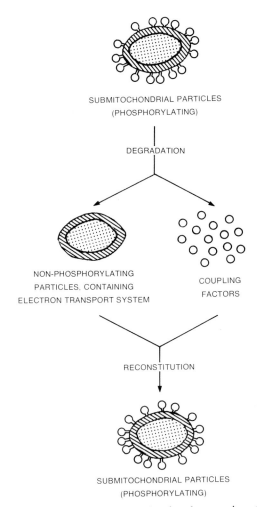

Figure 10–13. Diagram showing the experiment by which the submitochondrial particles, corresponding to the inner mitochondrial membranes, are submitted to urea to remove the coupling factor F_1, thus leaving non-phosphorylating particles. The lower part of the figure shows the subsequent reconstitution of a phosphorylating submitochondrial particle. (From Racker, E., The membrane of the mitochondrion. *Sci. Am., 218*:32, 1968. Copyright © 1968 by Scientific American, Inc. All rights reserved.)

TABLE 10–4. Stoichiometry of Electron-Transfer and ATP-Synthesizing Components in Heart Mitochondria*

Component†	Stoichiometry	Molecular weight $(\times 10^{-3})$
I	1	$1 \times 550 = 550$
III	4	$4 \times 230 = 920$
c	8	$8 \times 12 = 96$
IV	8	$8 \times 200 = 1600$
ATPase	4	$4 \times 360 = 1440$
OSCP	4	$4 \times 18 = 72$
	29	4678
CFo	4	
Q	64	

*Modified from Slater, E. C., Electron transfer and energy conservation. *9th International Congress on Biochemistry*, Stockholm, 1973.

†Components: I, Complex I; III, Complex III; c, cytochrome c; IV, Complex IV; OSCP, oligomycin sensitive protein; CF_0, F_0 coupling factor; and Q ubiquinone or coenzyme Q. (See Figure 10–12 for further information.)

with the elementary particles attached. These vesicles can carry out respiration and oxidative phosphorylation. After treatment with urea, which removes the F_1 coupling factor, the membranes no longer phosphorylate. Reconstitution of the complete phosphorylating system can be achieved by mixing the non-phosphorylating membranes with the F_1 particles. After it is isolated, the F_1 particle causes the hydrolysis of ATP to ADP and phosphate; however, when it is attached to the membrane it functions as a synthesizing enzyme (Fig. 10–12). Other properties of the

F_1 factor are also changed after isolation; for example, it is no longer sensitive to oligomycin and its ATPase activity is lost in the cold. An F_0 coupling factor, which is able to restore oligomycin sensitivity and to reverse the effect of cold, has been isolated.[26]

Topology of the Respiratory Chain and Phosphorylating System

The observations described above and summarized in Figure 10–12 indicate that there is a fine topology of the respiratory chain across the inner membrane of the mitochondrion; in other words, this membrane is highly asymmetrical.

Figure 10–3 illustrates two proposed molecular arrangements of the respiratory chain in a mitochondrial crest and the relationship of the chain to the structural proteins, lipids, and the F_1 particles of Racker. The upper part shows the probable arrangement in an intact mitochondrion, in which the F_1 particles are not protruding on the inner surface of the crests. The lower part corresponds to what is actually observed after osmotic disruption and negative staining of the mitochondria, with the F_1 particles corresponding to the elementary particles of Fernández-Morán (Fig. 10–5). The diagram emphasizes that each respiratory chain is related to three ATPases at which points the three molecules of ATP are produced. An-

other feature of Figure 10–3 is that on the inner membrane of the mitochondrion both the respiratory chain and the F_1 coupling factor show a special asymmetrical arrangement, i.e., a peculiar "sidedness." The sidedness of succinate dehydrogenase and cytochrome c has been demonstrated by specific indicators that produce depositions of electron dense material within the mitochondrial crests.[27] By using both a special tetrazole, which is converted to formazan by succinate dehydrogenase (see Chapter 6), and 3',3'-diaminobenzidine (DAB), which is oxidized by cytochrome c, it has been shown that the dense products accumulate on the outside surface of the mitochondrial crest (Fig. 10–14).

From experimental evidence it can be demonstrated that the electron transport system is accessible to NADH and succinate only from the inner matrix side, while cytochrome c is better reached from the outer surface of the membrane. In addition to this transversal topology there is also a lateral topology of the components of the respiratory chain and oxidative phosphorylation. The current theory is that the compact assemblies of macromolecules are arranged in a mosaic which is regularly spaced by other proteins and lipids.

The number of assemblies varies according to the tissue and the folding of the membrane. A mitochondrion from liver has about 15,000 assemblies, while one from the flight muscle of an insect may have as many as 100,000. Within this assembly or *unit of*

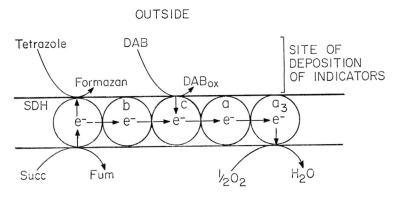

Figure 10–14. Diagram showing the sidedness of the respiratory chain within the mitochondrial crest. The reaction of succinic dehydrogenase (SDH) produces deposits of formazan on the outer edge of the crest. The same is observed with the oxidation products of DAB (3',3'-diaminobenzidine) in the reaction for cytochrome c. **Right above**, a diagram of the electron microscopic view of the reaction. (See the description in the text.) Interpretation of the results of Seligman, A. M., Karnowski, M. J., Wasserkrug, H. L., and Hanker, J. S., *J. Cell Biol.*, *38*:1, 1968. (Courtesy of A. L. Lehninger.)

phosphorylation the main enzymes are in equimolecular quantities (see Table 10–4). This finding has resulted from the use of ultra-rapid spectrophotometric studies[28] and, more recently, from the determination of binding sites of drugs that specifically affect the various components of the system (Fig. 10–12). For example, *aurovertin* binds to the ATPase; *oligomycin* acts at the level of the binding of ATPase to the membrane; *antimycin* is an inhibitor acting on the QH_2 cytochrome-*c*-reductase; *rotenone* acts on NADH-Q-reductase; and *cyanide* has its well known poisoning effect on cytochrome-*c*-oxidase.

The minimum unit of oxidative phosphorylation contains some 30 proteins with a total molecular weight approaching 5×10^6 daltons (Table 10–4). This unit or assembly must, therefore, cover a considerable area of the membrane. From biochemical and electron microscopic studies it is possible to imagine a mosaic organization for the inner membrane of the mitochondrion in which each unit of phosphorylation should cover a surface area of about 20×20 nm.[29]

Link Between the Respiratory Chain and Phosphorylation

The above-mentioned studies have demonstrated that for each pair of electrons received from the Krebs cycle and transported by the electron transfer system there is one molecule of ATP synthesized by the ATPase. The fundamental problem is that of establishing the nature of the link between these two systems (represented by the dotted lines between the respiratory chain and the F_1 coupling factor in Figure 10–12).

According to the *chemiosmotic hypothesis*,[30] the interaction is of an electrical nature. Because of the vectorial arrangement of these systems, an electrical potential across the inner mitochondrial membrane should be created every time electrons flow through the respiratory chain. Based on this hypothesis the ATPase synthesizes ATP under the influence of the electrical field. Since it operates at *long-range* distances across the membrane there is no need for a direct contact between the molecular complexes of the respiratory chain and the ATPase. An alternate theory—the *chemical or conformational* hypothesis[31]—postulates that the two systems are in molecular contact and the information is transmitted by *short-range* interactions which may be, in part, electrostatic in nature (see Slater, 1973). The fact that the proteins involved in the unit of phosphorylation are in stoichiometric relationship favors this hypothesis. As shown in Table 10–4, however, it is evident that there are a greater number of respiratory chains than ATPases. In other words, a single ATPase molecule must receive the message from more than one electron transfer complex (Fig. 10–2). It is postulated that upon acceptance of the two electrons, each electron transfer protein undergoes a conformational change which is then transmitted to the ATPase by short-range interactions.

SUMMARY:
Energy Transduction
in Mitochondria

Mitochondria function as energy-transducing organelles into which the major degradation products of cell metabolism penetrate and are converted into chemical energy (ATP) to be used in the various activities of the cell. This entire process requires the entrance of O_2, ADP, and phosphate, and it brings about the exit of ATP, H_2O, and CO_2. The process is based on three coordinated steps: (a) the *Krebs cycle,* carried out by a series of soluble enzymes present in the mitochondrial matrix, which produces CO_2 by decarboxylation and removes electrons from the metabolites (Fig. 4–14); (b) the *respiratory chain* or *electron transport system,* which captures the pairs of electrons and transfers them through a series of electron carriers, which finally leads by combination with activated oxygen to the formation of H_2O; (c) a *phosphorylating system,* tightly coupled to the respiratory chain, which at three points gives rise to ATP molecules. Most of the components of the respiratory chain, i.e., NADH-Q-reductase and succinate-Q-reductase, cytochromes *b* and *a,* and cytochrome-*c*-oxidase and other electron carriers are tightly bound to the inner membrane. Only cytochrome *c* and ubiquinone (coenzyme Q) can be removed easily. The components of the respiratory chain can be separated into four complexes of enzymes (complexes I to IV).

The phosphorylating system is composed of the F_1 coupling factor of Racker, an ATPase present within the head of the elementary particle. (This enzyme synthesizes ATP from ADP and phosphate.) The F_1 particle is attached to a piece of membrane by a protein binding oligomycin. By proper methods it is possible to separate the F_1 factor from the membranes which no longer phosphorylate. The entire system can then be reconstituted by addition of the two components.

The inner mitochondrial membrane shows a fine topology of these components transversally and also laterally, in a mosaic arrangement. Within each phosphorylation unit the main enzymes are in equimolecular quantities (Table 10–4); however, there are fewer F_1 factors than there are respiratory chains. The entire phosphorylating assembly may reach 5,000,000 daltons and may cover a surface of 20×20 nm. The nature of the link between the respiratory chain and the ATPase is unknown, but there are two favored hypotheses: the *chemiosmotic* and the *chemical or conformational*. According to the chemiosmotic hypothesis the ATPase synthesizes ATP under the influence of the vectorial field generated by the transfer of the electrons through the respiratory chain; this effect operates at long range and direct molecular contact is not needed. The chemical or conformational hypothesis postulates that the transfer of electrons originates a conformational change in the proteins of the respiratory chain which, by short-range interaction, transmits the signal to the ATPase that is needed for the synthesis of ATP.

Permeability of Mitochondria

Since the most important metabolic activities of mitochondria take place within the inner mitochondrial compartment, there should be a rapid and active flow of certain metabolites across the two membranes. The products of extramitochondrial metabolism, for example, must reach the mitochondrial matrix in order to undergo oxidation (Fig. 4–14), and ADP and phosphate must enter to form ATP. Simultaneously, end products such as H_2O, ATP, urea, and ammonia must leave the mitochondrion.

There are important differences in permeability between the mitochondrial membranes. The outer membrane is freely permeable to electrolytes, water, sucrose, and molecules as large as 10,000 daltons. The inner membrane, on the other hand, is normally impermeable to ions, as well as to sucrose.

The inner membrane utilizes specific carriers for the translocation of various substances. These *carriers,* or *translocators,* are thought to be genetically determined proteins present in the inner membrane that facilitate the diffusion of metabolites which normally do not cross membranes. The existence of specific carriers for ATP (or ADP), phosphate, succinate (or malate), isocitrate, glutamate, aspartate, and bicarbonate has been suggested. A carrier mechanism has been postulated for Ca^{2+}, Mn^{2+}, or Sr^{2+},[32] and there are evidences that some carriers could be localized at tight junctions between the outer and inner membrane (see below and Lehninger, 1971).

Conformational Changes in Mitochondria

At the beginning of this chapter it was noted that low amplitude contraction cycles that could be associated with stages in oxidative phosphorylation were observed in living cells. Similar observations have been made in isolated mitochondria by using absorbancy or light scattering to study them.[33, 34] Such changes were interpreted as being the result of small variations in volume, due to an energy linked swelling-contraction phenomenon. The changes in absorbancy and light scattering, however, may also reflect a rearrangement of the internal structure of the mitochondrion without a concomitant modification of the actual size. By means of a quick sampling method, which permits the fixation of isolated mitochondria at different stages of their metabolism, reversible ultrastructural changes were observed.[35]

Mitochondria may alter their internal conformation between the two extreme states shown in Figure 10–15. One is the so-called *orthodox* state that is usually observed in intact tissues. The other corresponds to the *condensed* state, in which there is a dramatic contraction of the inner compartment of the mitochondria accompanied by accumulation of fluid in the outer compartment. In the orthodox conformation the inner membrane shows the characteristic crests; the matrix fills practically the entire volume of the mitochondrion and has a reticular or granular aspect (Fig. 10–15,*A*). In the condensed conformation the inner membrane is folded at random, and the matrix, now more homogeneous, represents only about 50 per cent of the mitochondrial volume (Fig. 10–15,*B*). In this state, at certain points, it is possible to see tight junctions between the inner and outer membrane of the mitochondrion. More than 100 tight junctions per mitochondrion may be found.

The electron transport system is required for the change from condensed to orthodox conformation to take place. Inhibition of the respiratory chain by cyanide, antimycin A, or amytal will impair this transformation.[36] The orthodox state is induced when the external ADP becomes low and there is none left to be phosphorylated. If at this time ADP is added, respiration is rapidly enhanced, and the contraction of the inner membranes takes place. It is thought that during the transition from the orthodox to the condensed stage there is a change both in the inner membrane and in the matrix. The inner membrane is believed to contain the contractile elements or "mechano-enzymes,"[37] but the possible role of the matrix in this contraction should also be considered.

Swelling and Contraction

During swelling the mitochondrial volume may increase three to five times its normal value in the absence of ADP. With the addition of ATP, or the restoration of respiration, the mitochondrion may regain its original size. Studies on the phenomenon of swelling demonstrate that mitochondria have an important function in the uptake and extrusion of intracellular fluid.

Agents known to induce swelling are phosphate, Ca^{2+}, reduced glutathione, and, in particular, the thyroid hormone thyroxine. Thyroxine is the most effective swelling agent and is capable of producing swelling in physiologic concentrations.

Since mitochondria have two membranes and two compartments, they may swell in two general ways (Fig. 10–16). Water, K^+, and Na^+ penetrate both membranes very rapidly and may produce the structural changes shown in Figure 10–16,*C* and *E*, with dilution of the matrix. On the other hand, sucrose may produce an "inflation" of the outer chamber or the space within the crest without dilution of the matrix (Fig. 10–16,*B* and *D*).

Figure 10–17 shows the swelling effect of thyroxine as measured by light absorption and the reversible contraction produced by the addition of ATP. This experiment also shows that the P:O ratio (i.e., the ratio between the passage of inorganic phosphorus to ADP and the O_2 consumed) is lowered by thyroxine and returned to normal range by ATP. The water uptake is associated with the

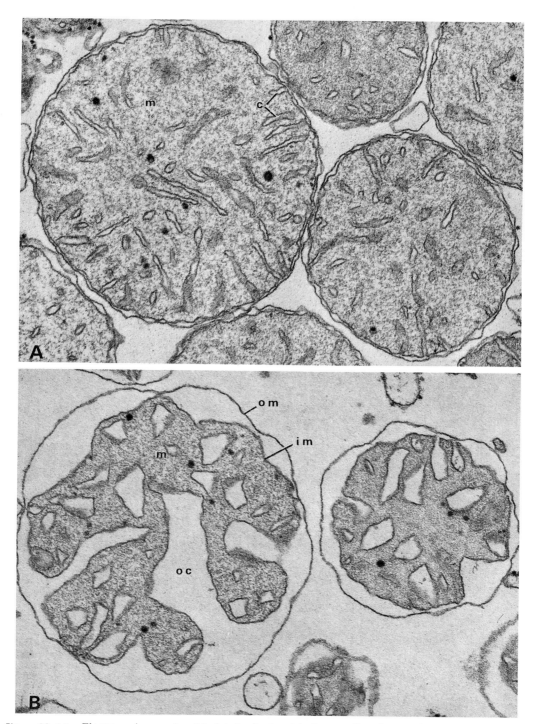

Figure 10–15. Electron micrographs of isolated mitochondria from rat liver in two extreme conformation states. ×110,000. **A,** the *orthodox* conformation. The inner membrane is organized into crests (*c*), and the matrix (*m*) fills the entire mitochondrion. **B,** the *condensed* conformation. Mitochondrial crests are not observed, and the outer chamber (*oc*) represents about 50 per cent of the volume; *om,* outer membrane; *im,* inner membrane. (Courtesy of C. R. Hackenbrock.)

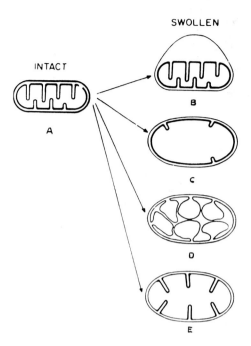

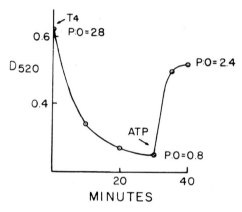

Figure 10–16. Structure of a normal intact mitochondrion (*A*) and of different types of swelling. Entrance of solutes in the outer chamber produces dilation of the intermembranal space (*B*), or of the intracristal spaces (*D*). Penetration in the inner chamber produces dilution of the matrix with (*C*) or without (*E*) unfolding of the crests (From Lehninger, A. L., *Physiol. Rev., 42*:467, 1962.)

Figure 10–17. Swelling of rat liver mitochondria in the presence of thyroxine and later contraction by ATP. The decrease in optical density corresponds to an increase in water content, and vice versa. Note that the P:O ratio declines during swelling, but is restored again during the contraction stage. *P:O*, ratio between the passage of inorganic phosphorus to ADP and the O_2 consumed. (From Lehninger, A. L., *Pediatrics, 26*:466, 1960.)

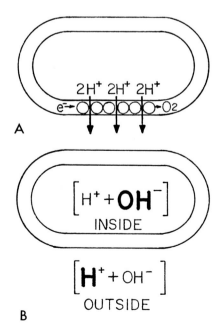

Figure 10–18. Diagram showing: **A,** the electron transport system pumping out H^+. **B,** the final result: an increase in OH^- inside the mitochondrion and the release of H^+. (Courtesy of A. L. Lehninger.)

uncoupling of oxidative phosphorylation, which is restored to normal levels after the water extrusion.[37] Mitochondria can actively squeeze out water and small molecules in the proportion of several hundred per each ATP molecule split.

Cation Accumulation in Mitochondria

Although respiration and oxidative phosphorylation are the most important functions of mitochondria, another related function is the accumulation of cations. Ca^{2+} can be concentrated in isolated mitochondria up to several hundred times normal values.[38] Phosphate enters the mitochondria along with Ca^{2+}. The amounts of Ca^{2+} and phosphate accumulated may be so great that the dry weight may increase by 25 per cent, and microcrystalline, electron-dense deposits may become visible within the mitochondria.[39] Electron microscopy coupled with high resolution microincineration have been used to study these deposits.[40] This process usually occurs in the osteoblasts present in tissues undergoing calcification.

In the presence of Ca^{2+}, mitochondria no longer phosphorylate but, instead, accumulate Ca^{2+} and phosphate. Both oxidative phosphorylation and accumulation of Ca^{2+} depend on the maintenance of cell respiration (i.e., the Krebs cycle). The accumulation of Ca^{2+} and other cations inside the mitochondrion is accompanied by loss of H^+, so as to maintain an electrical equilibrium.

As shown in the upper portion of Figure 10–18, electrons are permitted to pass through the oxidase system in the membrane, but the H^+ ions are restrained on one side of the membrane and the OH^- on the other. The OH^- ions, driven away from phosphate by the action of the F_1 coupling factor, become neutralized inside the mitochondrion at the same time that H^+ is released on the outer surface of the membrane. It is evident, then, that the matrix becomes more alkaline because of OH^- accumulation, while acid is simultaneously released to the outside (Fig. 10–18). Interestingly enough, chloroplasts behave precisely in the reverse direction by pumping out OH^- and accumulating H^+.

Since most metabolic processes in mitochondria occur in the inner compartment, an active flow of metabolites across both membranes takes place. The outer membrane is freely permeable, but the inner one is rather impermeable to ions and metabolites and must use specific carriers or translocators to achieve this goal. Carriers for ATP (or ADP), phosphate, succinate or malate, isocitrate, glutamate, aspartate, bicarbonate, and Ca^{2+} have been postulated. In isolated mitochondria cycles of contractions associated with oxidative phosphorylation have been observed. Physiologically, the mitochondrion passes from a conformational stage called *orthodox* into that called *contraction*. In the orthodox conformation, found at low external ADP, the matrix fills the entire volume of the inner compartment, and the crests may be seen clearly. In this stage the organelle does not phosphorylate. If ADP is added, the contraction stage is brought about, the matrix becomes condensed, and about 50 per cent of the water diffuses into the outer compartment.

Mitochondrial swelling is induced by several agents, such as phosphate, Ca^{2+}, and especially thyroxine. This hormone increases the volume of the mitochondrion by water uptake at the same time that it uncouples phosphorylation. With ATP the swelling is reversed; Ca^{2+} and phosphate are accumulated in large amounts. Mitochondria tend to pump out H^+ and to retain OH^-. Chloroplasts act in a reverse manner and tend to pump out OH^- and retain H^+.

**SUMMARY:
Other Functions
of Mitochondria**

MITOCHONDRIA AS SEMIAUTONOMOUS ORGANELLES

In recent years an entirely new area of knowledge in the study of mitochondria and chloroplasts has been opened by the demonstration that these organelles behave with a certain degree of autonomy within the cells. Mitochondria contain DNA molecules and ribosomes and may synthesize protein. Furthermore, they undergo division and may carry biological information, which represents a type of cytoplasmic inheritance. These recent developments are interesting when viewed in the light of past history. Early cytologists speculated about the possible function of these organelles. In 1890, Altmann and Schimper postulated that mitochondria and chloroplasts might be intracellular parasites that had entered the cytoplasm and established a symbiotic relationship with the cell; bacteria would have originated the mitochondria, and blue-green algae the chloroplasts. The name *bioblasts,* applied to mitochondria by Altmann, emphasized the self-duplicating nature of these structures.

Mitochondrial DNA

Although between 1956 and 1957 Chevremont and colleagues demonstrated that under certain conditions the mitochondria of cultured cells gave a positive Feulgen reaction, until 1960 it was generally agreed that DNA was exclusively localized within the cell nucleus. In 1963 M. and S. Nass[41] observed filaments within mitochondria that were interpreted as DNA molecules. This finding was later fully confirmed, both in sections and in DNA extracted and studied by the surface spreading technique.

A single mitochondrion may contain one or more DNA molecules depending on its size; i.e., the larger the mitochondrion, the more DNA molecules are present. Mitochondrial DNA appears as a highly twisted, double-stranded molecule having a circular shape (Fig. 10–19). In most mammalian cells the DNA is 5.5 μm long, whereas in yeast and *Neurospora,* it is longer (e.g., 26 μm in yeast).[45]

Mitochondrial DNA differs from nuclear DNA in several respects. The guanine-cytosine content is higher in mitochondrial

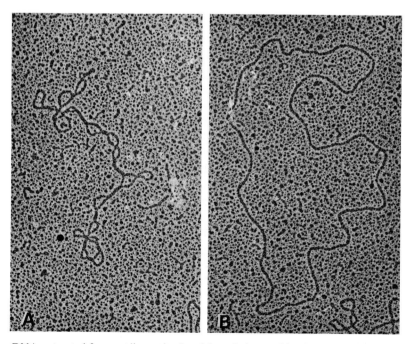

Figure 10–19.   DNA extracted from rat liver mitochondria and observed by the spreading technique. **A,** configuration in twisted circle ("super-coiled"); **B,** configuration in open circle. (Courtesy of B. Stevens.)

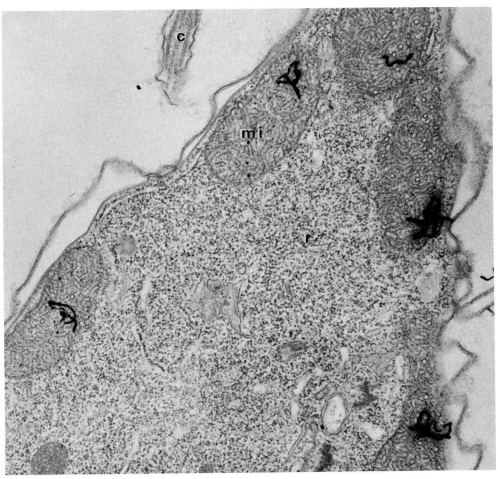

Figure 10–20. Incorporation of ³H-thymidine in mitochondria of *Tetrahymena pyriformis*. Note the localization of the synthesized DNA in mitochondria. ×31,000. (Courtesy of J. Andrè.)

DNA, and consequently the buoyant density is also higher (Rabinowitz, 1968). Another difference is the higher denaturation temperature of mitochondrial DNA and the facility with which it renatures. The amount of genetic information carried by mitochondrial DNA is not sufficient to provide specifications for all the proteins and enzymes present in this organelle. Since mitochondrial DNA provides information for the intrinsic ribosomal and transfer RNAs, about one-third of the nucleotides should be used for this purpose. The remaining two-thirds could possibly code for about 3000 amino acids, which corresponds to a protein of 360,000 or 10 proteins of 36,000 daltons each.[46]

In mammalian mitochondria only 5 to 10 per cent of the protein synthesis is carried out by its own synthetic machinery. Other synthetic processes come from the cytoplasm of the cell, and are under the control of the nuclear DNA. All soluble enzymes of the matrix, the cytochromes, cytochrome oxidase, and ATPase are formed outside the mitochondrion and then enter this organelle to be integrated into the inner membrane (see Dawidowicz and Mahler, 1973; González Cadavid et al., 1973). Apparently, only insoluble proteins of the intrinsic type, present in the inner membrane, are formed by the mitochondrial synthetic machinery.

Mitochondrial DNA may duplicate by the usual mechanism, which will be described in Chapter 17. Incorporation of ³H-thymidine into DNA has been observed in mitochondria of *Tetrahymena* (Fig. 10–20). All mitochondria incorporate ³H-thymidine in a population-doubling time.[47] Furthermore, mitochondria contain DNA polymerase, needed for DNA synthesis,[48] which is different from the nuclear DNA polymerase.[49] In synchronized cultured human cells labeled with ³H-thymidine, it was found that mitochondrial DNA was synthesized during a period extending from the G_2 phase to cytokinesis.[50]

Protein Synthesis and Ribosomes in Mitochondria

Mitochondria contain ribosomes and polyribosomes (see Chapter 19) which can

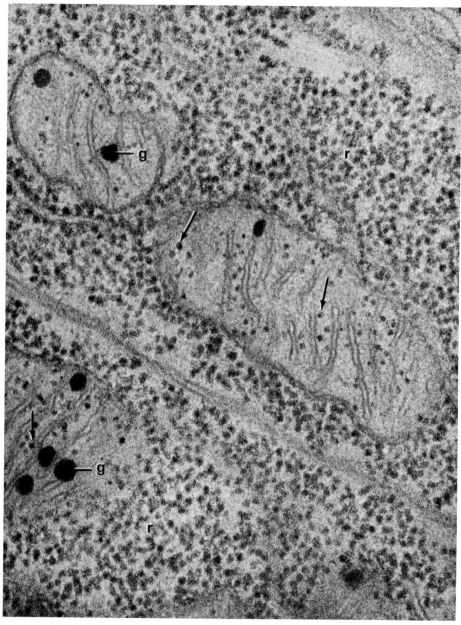

Figure 10–21. Electron micrograph of the intestinal epithelium showing a large accumulation of ribosomes (*r*) in the cytoplasm. In mitochondria, dense granules (*g*) and ribosomes (arrows) are observed. ×95,000. (Courtesy of G. E. Palade.)

be demonstrated, in many cases, by electron microscopy (Fig. 10–21). They have also been clearly observed in chloroplasts.[51]

In yeast and *Neurospora*, ribosomes have been assigned to a 70S class—similar to that of bacteria; in mammalian cells, however, mitochondrial ribosomes are definitely smaller, have a total sedimentation of 55S, and have subunits of 35 and 25S.[52] Ribosomes in mitochondria appear to be tightly associated with the membrane, and their formation seems to be related to the lipid content of the membrane (see Linnane, 1971). In yeast grown under anaerobic conditions, for example, there are fewer unsaturated fatty acids and no ribosomes. Another difference between mitochondrial and cytoplasmic protein synthesis is the action of certain inhibitors. Protein synthesis in mitochondria is inhibited by chloramphenicol (as it occurs with bacteria), while the cytoplasmic protein synthesis is not affected.

Biogenesis of Mitochondria

Two main mechanisms for the biogenesis of mitochondria have been postulated: origination by division from parent mitochondria or origination *de novo* from simpler building blocks. Probably, the truth lies somewhere between. Most of the mitochondrial molecular components are synthesized outside the organelle and are under nuclear control, but some essential parts are produced autonomously and are transmitted from one cell generation to another.

In tissue culture mitochondria are often seen to fragment, as well as to fuse. Mitochondria are distributed between the daughter cells during mitosis, and their number increases during interphase. It has been observed with time-lapse cinematography that mitochondria gradually elongate and then fragment into smaller mitochondria. This observation has been verified in *Neurospora*.[53] After labeling of a choline deficient mutant of *Neurospora* with radioactive choline, the radioactivity was followed in the mitochondria of the second and third generations. By radioautography it was found that all mitochondria of the original progeny were labeled. The mitochondria of each daughter cell were also labeled but contained about half the radioactivity. This seemed to indicate that mitochondria had divided and grown by the addition of new lecithin molecules to the existing mitochondrial framework.

Yeast cells grown anaerobically lack a complete respiratory chain (cytochromes *b* and *a* are absent); under the electron microscope they show no typical mitochondria. Only some of the membranes contain the two primary dehydrogenases of the respiratory chain. When the yeast cells are placed in air, however, these membranes fuse, unfold, and form true mitochondria that contain the cytochromes. This observation is somewhat in agreement with the *de novo* synthesis theory on the biogenesis of mitochondria.

The differences between the outer and the inner membrane should be considered in relation to the origin of mitochondria. The chemical composition of the outer membrane is similar to that of the endoplasmic reticulum (Chapter 8) and very different from that of the inner membrane. Such similarities and differences are also reflected in the mean life of the proteins. In the endoplasmic reticulum and the outer mitochondrial membrane the mean life is about five to six days and in the inner membrane about 12 days.[54] Furthermore, the outer membrane is more active in lipid synthesis than the inner membrane. Continuities between the outer membrane of the mitochondrion and the endoplasmic reticulum,[55] and even with the axon membrane, have also been observed.[56]

Prokaryotic Origin of Mitochondria

The view postulated by early cytologists that mitochondria represent symbiotic organisms living in association with higher cells may now be reinstated in terms of modern cell and molecular biology. The homologies between mitochondria and bacteria are numerous and, considered from an evolutionary viewpoint, they may be more than circumstantial.

There are similarities in the *localization of the respiratory chain*. In bacteria the electron transport system is localized in the plasma membrane, which can be compared to the inner membrane of mitochondria (Fig. 2–1). Certain bacteria have membranous projections extending from the plasma membrane forming the so-called *mesosomes*.[57] Such membranous projections which are comparable to mitochondrial crests have been separated and are shown to contain the respiratory chain.[58] The inner mitochondrial membrane and the matrix may represent the original symbiont; these may be enclosed within a membrane of cellular origin.

The *mitochondrial DNA* is circular, as it is frequently found in chromosomes of prokaryons. It replicates and divides into several circles that may be found in a single mitochondrion. *Mitochondria contain ribosomes* that are smaller than those belonging to the cell-at-large, however, they are generally smaller than those in bacteria.

Protein synthesis in mitochondria and in bacteria is inhibited by *chloramphenicol,* whereas the extramitochondrial protein synthesis of the higher cell is not affected.

In mitochondria there is evidence of a *DNA-dependent RNA synthesis,* which indicates partial autonomy of this organelle. The amount of information carried by the mitochondrial DNA, however, is insufficient for an autonomous biogenesis, thereby causing mitochondria to depend on the nucleus and the cytoplasm of the cell for synthesis of most of their specific constituents.

According to the *symbiont hypothesis* the host cell is conceived of as an anaerobic organism deriving its energy from glycolysis, a process that occurs in the cytoplasmic matrix (see Chapter 4), the parasite contains reactants of the Krebs cycle and the respiratory chain and is able to carry on respiration and oxidative phosphorylation. The symbiont hypothesis is even more plausible in the case of plant cells, since the parasite would be the chloroplast, an autotrophic microorganism able to transform energy from light. (see Margulis, 1971).

REFERENCES

1. Hogeboom, G. H., Schneider, W., and Palade, G. (1948) *J. Biol. Chem., 172:*619.
2. Goodwin, T. W., and Lindberg, O. (1961) Biological structure and function. *Proc. First IUB/IUBS Internat. Symp.,* Vol. 2. Academic Press Inc., New York.
3. Lehninger, A. L. (1962) *Physiol. Rev. 42:*3, 467.
4. Loud, A. V. (1968) *J. Cell Biol., 37:*27.
5. Shelton, E., Schneider, W. C., and Striebich, N. J. (1953) *Exp. Cell Res., 4:*32.
6. Allard, G., de Lamirande, G., and Cantero, A. (1953) *Canad. J. Med. Sci., 30:*543.
7. Peachey, L. D. (1962) In: *Fifth International Congress for Electron Microscopy,* Vol. 2, p. 00-3. (Breese, S. S., Jr., ed.) Academic Press Inc., New York.
8. Fleischer, S., Fleischer, B., and Stoeckenius, W. (1967) *J. Cell Biol., 32:*193.
9. Fernández-Morán, H. (1963) *Science, 140:*381.
10. Racker, F. (1967) *Fed. Proc., 26:*1335.
11. Slautterback, D. (1965) *J. Cell Biol., 24:*1.
12. Sabatini, D., De Robertis, E., and Bleichmar, H. (1962) *Endocrinology, 70:*390.
13. Palade, G. E. (1958) *Anat. Rec., 130:*352.
14. Carasso, N., and Favard, P. (1958) *C. R. Acad. Sci. (Paris), 246:*1594.
15. Ward, R. T. (1962) *J. Cell Biol., 14:*309.
16. De Robertis, E., and Sabatini, D. (1958) *J. Biophys. Biochem. Cytol., 4:*667.
17. Levy, M., Toury, R., and André, J. (1967) *Biochim. Biophys. Acta, 135:*599.
18. Schnaitman, C. V., Erwin, V. G., and Greenawalt, J. W. (1967) *J. Cell Biol., 32:*719.
19. Sottocasa, G. L., Kuylenstierna, B., Ernster, L., and Bergstrand, A. (1967) *J. Cell Biol., 32:*415.
20. Parsons, D. F., Williams, G. R., Thomson, W., and Chance, B. (1967) In: *Mitochondrial Structure and Compartmentation,* p. 5. (Quagliariello, E., et al., eds.) Adriatica (Libreria) dell' Universita, Bari.
21. Melnick, R. L., and Packer, L. (1971) *Biochim. Biophys. Acta, 253:*503.
22. Sottocasa, G. L. (1967) *Biochem. J., 105:*1.
23. Green, D. E., and Goldberger, R. F. (1966) *Molecular Insights into the Living Process.* Academic Press, Inc., New York.

24. Fleischer, S., Fleischer, B., and Stoeckenius, W. (1967) *J. Cell Biol.,* *32*:193.
25. Kagawa, Y., and Racker, E. (1966) *J. Biochem., 241*:2461.
26. Racker, E. (1968) *Sci. Am. 218*:32.
27. Seligman, A. M., Karnowski, M. J., Wasserkrug, H. L., and Hanker, J. S. (1968) *J. Cell Biol., 38*:1.
28. Chance, B., and Williams, G. R. (1956) *Adv. Enzymol., 17*:65.
29. Klinenberg, M. (1967) In: *Mitochondrial Structure and Compartmentation.* (Quagliariello, E., et al., eds.) p. 124 Adriatica (Libreria) dell'Universita, Bari.
30. Mitchell, P. (1967) *Fed. Proc., 26*:137.
31. Boyer, P. D. (1965) In: *Oxidases and Related Redox Systems.* (King T. S., et al., eds.) Vol. 2, p. 994, John Wiley & Sons, Inc., New York.
32. Reynafarje, B., and Lehninger, A. L. (1969) *J. Biol. Chem., 244*:584.
33. Chance, B., and Packer, L. (1958) *Biochem. J., 68*:295.
34. Packer, L. (1963) *J. Cell Biol., 18*:487.
35. Hackenbrock, C. R. (1966) *J. Cell Biol., 30*:269.
36. Hackenbrock, C. R. (1968) *J. Cell Biol., 37*:345.
37. Lehninger, A. L. (1964) *The Mitochondrion.* W. A. Benjamin, Inc., New York.
38. Vasington, F. D., and Murphy, J. V. (1962) *J. Biol. Chem., 237*:2670.
39. Greenawalt, J. W., Rossi, C. S., and Lehninger, A. L. (1964) *J. Cell Biol., 23*:21.
40. Thomas, R. S., and Greenawalt, J. W. (1968) *J. Cell Biol., 39*:55.
41. Nass, M. M. K., and Nass, S. (1963) *J. Cell Biol., 19*:593.
42. Sinclair, J. H., and Stevens, B. (1966) *Proc. Natl. Acad. Sci. USA, 56*:508.
43. Kroon, A. M., Borst, P., Van Bruggen, E. F., and Ruttenberg, G. J. C. M. (1966) *Proc. Natl. Acad. Sci. USA, 56*:1836.
44. Dawid, I. B., and Wolstenholme, D. R. (1967) *J. Molec. Biol., 28*:233.
45. Wolstenholme, D. R., and Dawid, I. B. (1967) *Chromosoma, 20*:445.
46. Freeman, K. B., Haldar, D., and Work, T. S. (1967) *Biochem. J., 105*:947.
47. Parsons, J. A., and Rustad, R. (1968) *J. Cell Biol., 37*:683.
48. Parsons, P., and Simpson, M. (1967) *Science, 155*:91.
49. Meyer, R. R., and Simpson, M. V. (1968) *Proc. Natl. Acad. Sci. USA, 61*:130.
50. Koch, J., and Stokstad, E. L. R. (1967) *European J. Biochem., 3*:1.
51. Stutz, E., and Noll, H. (1967) *Proc. Natl. Acad. Sci. USA, 57*:774.
52. Attardi, B., Attardi, G., and Aloni, Y. (1971) *J. Molec. Biol., 55*:231, 251, and 271.
53. Luck, D. J. L. (1965) *J. Cell Biol., 24*:461.
54. Bucher, T. (1968) *Excerpta Medica Int. Cong. Ser. 166*:5.
55. Robertson, J. D. (1961) Cell membrane and the origin of mitochondria. In: *Regional Neurochemistry* (Kety, S. S., and Elkes, J., eds.) Pergamon Press, New York.
56. De Robertis, E., and Bleichmar, H. B. (1962) *Z. Zellforsch., 57*:572.
57. Fitz-James, P. C. (1960) *J. Biophys. Biochem. Cytol., 8*:507.
58. Salton, M. R. J., and Chapman, J. A. (1962) *J. Ultrastruct. Res., 6*:489.

ADDITIONAL READING

Avron, M., and Chance, B. (1966) Relation of phosphorylation to electron transport in isolated chloroplast. *Brookhaven Symp. Biol., 19*:149.
Baxter, R. (1971) Origin and continuity of mitochondria. In: *Origin and Continuity of Cell Organelles,* p. 46. (Reinert, J. and Ursprunz, H. eds.) Springer Verlag, Berlin.
Borst, P. (1972) Mitochondrial nucleic acids. *Annu. Rev. Biochem., 41*:333.
Chance, B. (1972) The nature of electron transfer and energy coupling reactions. *FEBS.* (Federation of European Biochemical Societies) *Letters, 23*:1.
Dawidowicz, K., and Mahler, H. R. (1973) Synthesis of mitochondrial proteins. In: *Gene Expression and Its Regulation,* p. 503, (Kenney, F. T., et al, eds.) Plenum Publishing Corp., New York.
Gonzalez-Cadavid, N. F., Herrera, F., Guevara, A., and Viera, A. (1973)

Stimulation of the synthesis of inner mitochondrial membrane proteins in rat liver by cuprizone. In: *Gene Expression and its Regulation,* p. 523. (Kenney, F. T., et al eds.) Plenum Publishing Corp., New York.

Hall, D. O., and Palmer, J. M. (1969) Mitochondrial research today. *Nature,* (London) *221*:717.

Lehninger, A. L. (1964) *The Mitochondrion.* W. A. Benjamin, Inc., New York.

Lehninger, A. L. (1965) *Bioenergetics.* W. A. Benjamin, Inc., New York.

Lehninger, A. L. (1971) The molecular organization of mitochondrial membranes. In: *Advances in Cytopharmacology,* Vol. I, p. 199. Raven Press, New York.

Linnane, A. W. (1971) *Acta Cient., Venezolana, 22*: suppl. 2, 51.

Margulis, L. (1971) Cell organelles such as mitochondria may have once been free-living organisms. *Sci. Am., 225*:48.

Mitchell, P. (1967) Proton-translocation phosphorylation in mitochondria, chloroplasts and bacteria: natural fuel cells and solar cells. *Fed. Proc., 26*:1370.

Nass. M. M. K. (1969) Mitochondrial DNA. *Science, 165*:25.

Novikoff, A. B. (1961) Mitochondria (chondriosomes). In: *The Cell,* Vol. 2, p. 299. (Brachet, J., and Mirsky, A. E., eds.) Academic Press Inc., New York.

Rabinowitz, M. (1968) Extracellular DNA. *Bull. Soc. Chim. Biol. (Paris), 50*:311.

Racker, E. (1968) The membrane of the mitochondrion. *Sci. Am., 218*:32.

Racker, E. (1970) *Membranes of Mitochondria and Chloroplasts.* Van Nostrand Reinhold Co., New York.

Slater, E. C. (1972) Mechanism of energy conservation. In: *Mitochondrial Biomembranes.* p. 133. North-Holland Publishing Co., Amsterdam.

Slater, E. C. (1973) Electron transfer and energy conservation. *9th International Congress on Biochemistry,* Stockholm.

Wagner, R. P. (1969) Genetics and phenogenetics of mitochondria. *Science, 163*:1026.

eleven

THE PLANT CELL AND
THE CHLOROPLAST

The emphasis in this chapter is on some special characteristics of plant cells, in particular, the thick *cell wall* outside the plasma membrane and certain organelles—the *plastids*—that are related to the synthesis and accumulation of various substances. Of the plastids, the most important are the *chloroplasts*, which, along with the mitochondria, are biochemical machines that produce energy transformations. In a chloroplast the electromagnetic energy contained in light is trapped and converted into chemical energy by the process of *photosynthesis*.

CHARACTERISTICS OF PLANT CELLS

Cell Walls

Characteristic of the structure of the plant cell are rigid walls that surround and protect the plasma membrane (Fig. 2–6). These walls constitute a framework that provides the mechanical support for plant tissues. The thick walls, which formed a regular pattern, were the first structures to be recognized, microscopically, by Robert Hooke. This arrangement prompted him to name the structures he saw *cells* (Chap. 1). The cell

Figure 11–1. Diagram showing the time sequence of the formation of the various types of cell wall layers in a tracheid. *P.W.*, primary wall; *S.W.*, secondary wall. Notice the arrangement of fibrils and other structures in the different membranes. (From Mühlethaler, K., "Plant Cell Walls," *in* Brachet, J., and Mirsky, A. E. (Eds.), *The Cell*, Vol. 2, New York, Academic Press, Inc., 1961.)

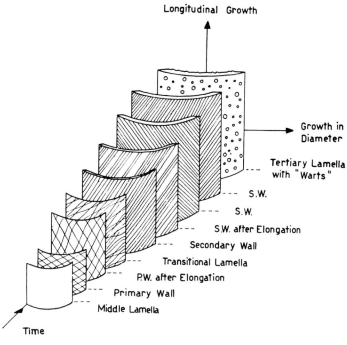

Longitudinal Growth

Growth in Diameter

Tertiary Lamella with "Warts"

S.W.

S.W.

S.W. after Elongation

Secondary Wall

Transitional Lamella

P.W. after Elongation

Primary Wall

Middle Lamella

Time

TABLE 11-1. Plant Cell Wall Substances

Substance	Chemical Unit	Staining Reaction
Cellulose	Glucose	Chlorzinc iodide (stains violet)
Hemicellulose	Arabinose, xylose, mannose, galactose	None specific
Pectin substances	Glucuronic and galacturonic acids	Ruthenium red
Lignin	Coniferyl alcohol	Phloroglucinol hydrochloride (stains rose); chlorzinc iodide (stains yellow)
Cuticular substances	Fatty acids	Sudan III (stains orange)
Mineral deposits	Calcium and magnesium in the form of carbonates or silicates	

walls are composed principally of *cellulose* produced by the cell. Adjacent cell walls are cemented together with *pectin*.

Cell walls are complex and highly differentiated in some tissues and develop in special sequences. In certain cells, primary, secondary, and tertiary walls, deposited in layers during growth and differentiation, have been described. These three types of cell walls can be differentiated by chemical composition and by the special disposition of the *microfibrils*—the building blocks of most cell walls (Fig. 11-1).

Both the *primary* and *secondary walls* are composed mainly of the polysaccharide cellulose, but other substances may be incorporated, especially in secondary walls. Lignin or suberin may be added to the primary wall, and, in epidermal cells, cutin and cutin waxes may produce an impermeable surface coating that reduces water loss. In many fungi and in yeasts the cell wall is composed of *chitin*, a polymer of glucosamine. Table 11-1 lists substances, in addition to cellulose, found in the cell wall, as well as data on the chemical composition and staining reactions of these substances.

In some tissues the *tertiary wall* is deposited at the interior of the secondary wall; it has a special structure as well as

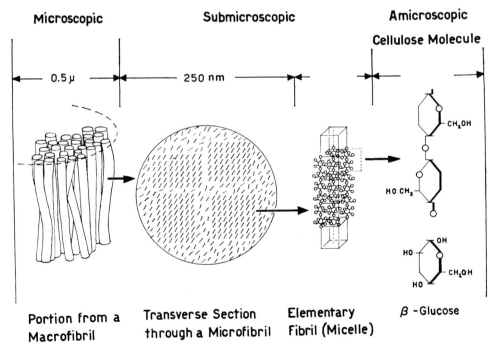

Figure 11-2. Structural elements of cellulose at different levels of organization. (From Mühlethaler, K., "Plant Cell Walls," *in* Brachet, J., and Mirsky, A. E., (Eds.), *The Cell*, Vol. 2, New York, Academic Press, Inc., 1961.)

different chemical and staining properties. The principal component of this wall is *xylan*, rather than cellulose.

The deposition of the various walls, along with cellular growth and differentiation, should be considered in time sequence. During growth the outermost, older parts of the cell wall are severely stretched, which may reorient and even tear the microfibrils. Various processes of tearing—*multinet, tip,* and *mosaic growth*—may thus be produced.[1] To a great extent the cell wall determines the shape of the cell and serves as a criterion for the classification of plant tissue, (i.e., parenchyma, collenchyma, and fibers.)[2] (Consideration of these classifications is beyond the scope of this book.)

The cell wall is a product of the cytoplasm, and its development begins with the formation of the *phragmoplast* and *cell plate* immediately after nuclear division. The primary cell wall is composed essentially of microfibrils of cellulose that may run in all directions within the plane of the wall. These constitute a loose framework that contains large amounts of water and noncellulose substances. It is generally agreed that the growth of both the primary and secondary walls is by apposition. In the secondary wall, however, the microfibrils are parallel and more densely packed than in the primary (Fig. 11–1).[1]

Figure 11–2 indicates schematically the structural elements of cellulose, from the microscopic to the molecular level. Macrofibrils visible with the light microscope are composed of *microfibrils* about 25 nm in diameter, each of which is, in turn, composed of about 2000 cellulose chains. About 100 cellulose chains are held together in an elementary fibril. X-ray diffraction reveals a crystalline pattern in cellulose, with a repeating period of 1.03 nm along the fiber axis. This corresponds to a *cellobiose* unit composed of two β-glucose molecules.

Plasmodesmata: Continuity of Cytoplasm

A characteristic of most plant cells is the presence of bridges of cytoplasmic material that establish a continuity between adjacent cells. These bridges, called *plasmodesmata*, pass through the thickness of the pectocellulose membrane. They have a thin plasma membrane and may contain tubules that establish continuities between the endoplasmic reticulum of both cells. The presence of plasmodesmata permits the free circulation of fluid, which is essential to the maintenance of plant cell tonicity, and probably also allows passage of solutes and even of macromolecules. According to these concepts, cell walls do not represent complete partitions between cells, but constitute a vast syncytium supported by a skeleton formed by the pectocellulose membranes.

The formation of plasmodesmata is related to the formation of the *cell plate,* mentioned earlier, which appears at the equator of dividing cells during telophase.[3] At this time the cell plate is crossed by vesicles and tubules of the endoplasmic reticulum that determine the location of the plasmodesmata. (see Chap. 13).[4, 5]

Cytoplasmic Matrix and the Vacuolar System

In Chapter 2 the basic similarities between the cytoplasm of animal and plant cells were mentioned.[6] In meristematic cells the membranes of the cytoplasmic vacuolar system are relatively scanty and are masked by the numerous ribosomes that fill the cytoplasmic matrix. Indeed, in undifferentiated cells most of these particles are not attached to the membranes, but are free in the matrix.

When meristematic cells are fixed with glutaraldehyde it is possible to observe a system of *microtubules* (Chap. 22) below the plasma membrane and oriented tangentially to the cell. These structures are apparently very labile and cannot be observed with other types of fixatives in plant cells. This component of the cell matrix is made of unbranching tubular structures 25 nm in diameter and several microns in length.[6] A possible role of these microtubules in wall deposition has been postulated. Microtubules are present in the strands of cytoplasm that underlie the points where the secondary wall is being deposited in spiral or reticulate patterns.[7, 8]

The endoplasmic reticulum in plant cells may serve similar functions as those described in animal cells (Chap. 9). For example, the secretion of protein material within dilated cisternae has been observed in radish root cells.[9]

The vacuolar system becomes more and

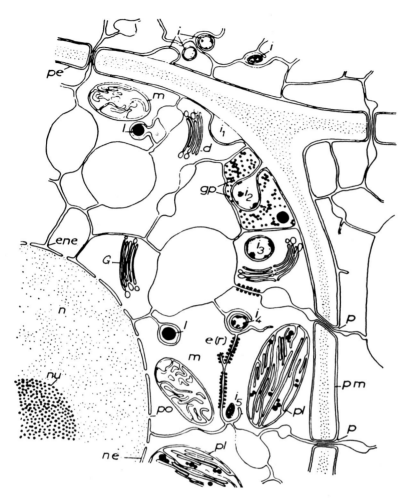

Figure 11–3. Diagram showing the interpretation of plant ultrastructure and its intercellular relationships. *G*, Golgi complex; *ene*, evagination of nuclear envelope, *er*, endoplasmic reticulum; i_1-i_5, steps in pinocytosis; *l*, lipid; *m*, mitochondrion; *ne*, nuclear envelope; *n*, nucleus; *nu*, nucleolus; *p*, plasmodesmata; *pl*, plastid; *pm*, plasma membrane; *po*, pores. (Courtesy of R. Buvat.)

more developed with the differentiation of the cell. In leaf primordia, rough and smooth endoplasmic reticulum have been observed, but the more differentiated cells show fewer ribosomes and a vacuolar system containing large vacuoles filled with fluid.

The diagram in Figure 11–3 indicates the possible connections of the vacuolar system with the nuclear envelope and the plasma membrane. It has been postulated that the plasma membrane of the plant cell invaginates and actively takes in fluid (pinocytosis). The same process also occurs in numerous animal cells.[10]

The great development of the vacuolar system during cell differentiation is related to the intense hydration of the cytoplasm. This process may give rise to huge vacuoles that are filled with liquid and may be confluent. As a result, the cytoplasm may become compressed in a thin layer against the cellulose

membrane and may show cytoplasmic movements, called *cyclosis* (Chap. 22).

Golgi Complex

As in invertebrate material, the Golgi complex in plants appears as discrete bodies (formerly called *dictyosomes*) dispersed throughout the cytoplasm (Fig. 11–4). The Golgi complex has a platelike arched shape and is the size of a mitochondrion or smaller. The fine structure consists of a stack of flattened vesicles (cisternae) that are slightly dilated at the edges. Surrounding the cisternae are other vesicles formed by localized dilatations; these vesicles are probably a product of the activity of this cell structure.

Golgi complexes are dispersed throughout the cytoplasm without definite polarization. At telophase they aggregate at the periphery of the cell plate and form small vesi-

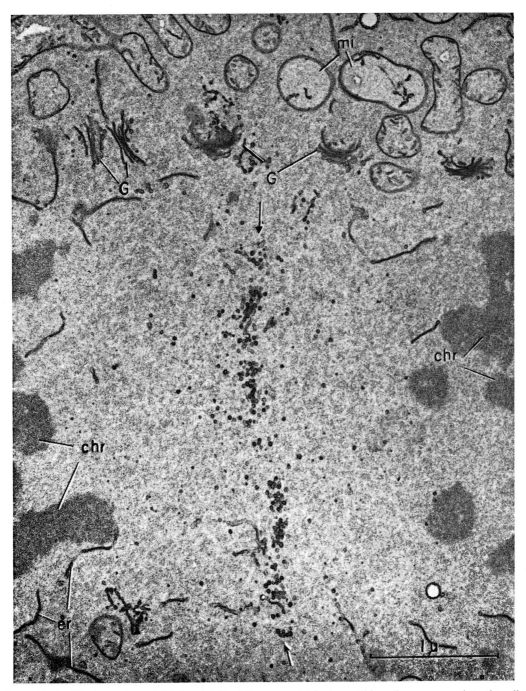

Figure 11–4. Electron micrograph of root cells of *Zea mays* at telophase. This region corresponds to the cell plate. Note at the top the marginal mitochondria (*mi*) and the Golgi complex (*G*) (dictyosomes). Between the arrows the vesicles are aligned to form the first evidence of a cell plate. *chr*, telophase chromosomes in the two daughter cells; *er*, endoplasmic reticulum. ×45,000. (Courtesy of W. Gordon Whaley and H. H. Mollenhauer.)

cles, which fuse to form the plate (Fig. 11–4). As in animal cells, the Golgi complexes of some plant cells (e.g., root cap cells of maize) are directly related to secretion. The Golgi cisternae become filled with secretion products, which are then concentrated and discharged.[11]

Golgi complexes and their associated vesicles are numerous in cells involved in the synthesis of mucilage, and in those of the outer root cap of the bean. The mucilage is produced at the expense of the starch bodies present in plastids.[12] The Golgi complexes of plant cells contain specific enzymes such as thiamine pyrophosphatase and inosinic diphosphatase. Incorporation of labeled glucose is highest during the formation of the cell plate.[13]

Mitochondria

Mitochondria of plant cells have a structure essentially similar to that in animal cells. In meristems, mitochondria have relatively few crests and an abundant matrix. During differentiation this internal structure may vary. In cells engaged in photosynthesis (leaf cells), mitochondria show an increased num-

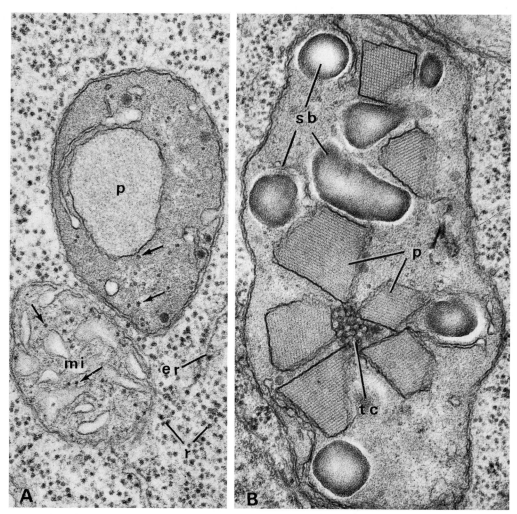

Figure 11–5. Electron micrograph of plastids in bean root tips. **A,** a young plastid and a mitochondrion (*mi*) are observed in a meristematic cell. The plastid contains a protein inclusion (*p*). The stroma is dense and contains granules and ribosomes (arrows). Numerous ribosomes (*r*) are in the cytoplasm. **B,** a plastid containing several crystalline protein inclusions (*p*) and starch bodies (*sb*). Near the center it has a system of tubules called the tubular complex (*tc*). A, ×67,000; B, ×77,000. (Courtesy of E. H. Newcomb.)

ber of crests; in cells containing starch granules (amyloplasts), mitochondria remain undifferentiated, as in meristems (Fig. 11–5, *A*).

One of the important points still under discussion is the relationship between mitochondria and chloroplasts. Guillermond[14] studied this problem in leaf meristems and postulated that in early stages there are two types of organelles. One is typically composed of short mitochondria and the other of long filamentous bodies. These elongated organelles increase in thickness and may give rise to vesicles, starch granules, or chloroplasts, whereas the short mitochondria remain unchanged. According to this view, both types of organelles are independent. The introduction of electron microscopy has, in general, confirmed this viewpoint. Although early plastids, also called proplastids, in some ways resemble mitochondria, they are readily distinguishable because of fewer projections of the inner membrane, their large size, and the presence of dense granules in the matrix (Fig. 11–5, *A*).[15]

The general morphology of plant cells shows many details in common with animal cells; the most characteristic differences are the *cell wall* and the *plastids*.

Cell walls develop in a certain sequence: primary, secondary, and tertiary walls are laid down during cell differentiation. The primary wall, a product of the cytoplasm, appears with the formation of the phragmoplast and cell plate (Chap. 13). The main component of the cell wall is cellulose, which is deposited in the form of microfibrils; pectin is used to cement adjacent walls; and chitin is present in fungi and yeasts. In secondary and tertiary walls xylan, lignin, suberin, and other components may be added to the cellulose. The cytoplasm of plant cells is bridged by plasmodesmata, thereby establishing intercellular communication. Microtubules are present below the plasma membrane underlying the deposition of the cell wall.

The vacuolar system is greatly developed in certain differentiated cells and is related to the formation of vacuoles. The Golgi complex is arranged in the form of cisternae and associated vesicles. This organelle is related to the formation of the cell plate and to secretory processes. Mitochondria in plant cells are similar to those of animal cells and can be differentiated from the early plastids or proplastids.

SUMMARY:
Characteristics of
Plant Cells

PLASTIDS

In 1883 Schimper first used the term *plastid* for cytoplasmic organelles present in eukaryotic plant cells. Animal cells, fungi, bacteria, and blue-green algae lack plastids; however, the electron microscope has revealed some distinct lamellar structures thought to be equivalent to plastids in photosynthetic bacteria and blue-green algae.

Plastids are characterized by the presence of pigments, such as *chlorophyll* and *carotenoids,* and the capacity to synthesize and accumulate reserve substances, such as starch, fats, and proteins.

Leukoplasts—colorless rodlike or spheroid plastids—are found in embryonic and germ cells. During embryonic development leukoplasts in certain differentiated zones of the root produce starch granules, called *amyloplasts* (Fig. 11–5, *B*). They may be seen under the polarization microscope, because of their characteristic birefringence, or they may be distinguished by histochemical reactions for starch. Leukoplasts are also found in meristematic cells and in those regions of the plant not receiving light.

Plastids located in the cotyledon and the primordium of the stem are colorless at first, but eventually become filled with chlorophyll and acquire the characteristic green color of chloroplasts.

In addition to chloroplasts, other colored plastids may be observed. These are grouped under the name *chromoplasts.* Yellow or orange chromoplasts occur in petals, fruits, and roots of certain higher plants. In general, they have a reduced chlorophyll content and are thus less active photosynthetically. The red color of ripe tomatoes is the result of chromoplasts that contain the red pigment *lycopene,* a member of the carotenoid family. Chromoplasts containing various pigments (e.g., *phycoerythrin* and *phycocyanin*) are found in algae.

Leukoplasts have been observed which may accumulate protein (*proteinoplasts*), starch (*amyloplasts*), or fat (*elaioplasts*); some more specialized plastids produce essential oils (see Amelunxen and Gronau, 1969). It was noted in *Phaseolus vulgaris,* that a single leukoplast may store both starch and protein, i.e., it may qualify as both an amyloplast and a proteinoplast (Fig. 11–5, *B*). The starch granules are deposited in the matrix or stroma of the plastid, while the protein, which sometimes shows a crystalline arrangement, is always accumulated within membrane-bound sacs. Also present in the stroma are ribosome-like particles, lamellae, phyto-ferritin granules, and a tubular complex that arises by invagination from the inner plastid membrane.[15]

Chloroplasts

Chloroplasts are the plastids that are most common and of greatest biological importance, since by *photosynthesis* they produce most of the chemical energy used by living organisms. Without chloroplasts there would be no animals, because animals feed on the foodstuffs produced by plants.

Morphology

The *shape, size,* and *distribution* of chloroplasts may vary in different cells within a species, but these organelles are relatively constant within cells of the same tissue. In leaves of higher plants, each cell contains a large number of spheroid, ovoid, or discoid chloroplasts. Some are club-shaped, having a thin middle zone and bulging ends filled with chlorophyll. Chloroplasts are frequently vesicular, with a colorless center. The presence of starch granules is detected by the characteristic blue iodine reaction. Algae often possess a single huge chloroplast that appears as a network, a spiral band, or a stellate plate (Fig. 11–6).

The *number* of chloroplasts is relatively constant in the different plants. In higher plants there are 20 to 40 chloroplasts per cell. It has been calculated that the leaf of *Ricinus communis* contains about 400,000 chloroplasts per square millimeter of surface area. When the number of chloroplasts is insufficient, it is increased by division; when excessive, it is reduced by degeneration.

The *size* of chloroplasts varies considerably. The average diameter in higher plants is 4 to 6 μm. This is constant for a given cell type, but sexual and genetic differences are found. For instance, chloroplasts in polyploid cells are larger than those in the corresponding diploid cells. In general, chloroplasts of plants grown in the shade are larger and contain more chlorophyll than those of plants grown in sunlight.

Chloroplasts are sometimes distributed homogeneously within the cytoplasm, but are frequently packed near the nucleus, or close to the cell wall. The *distribution* and *orientation* of the chloroplasts within the cell may vary with the amount of light energy. Chloroplasts apparently multiply by division — elongation of the plastid and constriction of the central portion. They have a higher density than the cytoplasm and migrate to the centrifugal pole of the cell when submitted to the action of centrifugal force.

Observation of living epidermal cells from leaves of *Iris* and other genera has shown that chloroplasts are displaced and deformed by the action of cytoplasmic streaming (cyclosis). In addition to this passive *motility,* active movements of an ameboid or contractile type, which are sometimes related to the degree of illumination, have been observed.

Changes in shape and volume caused by the presence of light have been observed in chloroplasts isolated from spinach. The volume decreases considerably after the chloro-

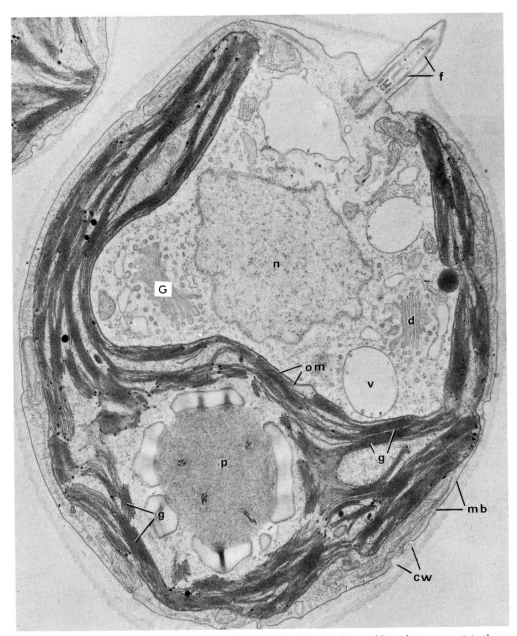

Figure 11–6. Electron micrograph of *Chlamydomonas* showing the huge chloroplast; grana (*g*), the pyrenoid (*p*), Golgi complex (G), flagellum (*f*), membrane (*mb*), cell wall (*cw*), nucleus (*n*), outer membrane of the chloroplast (*om*), vacuole (*v*), × 8000. (Courtesy of G. E. Palade.)

plasts are struck by light and photophosphorylation is initiated; this effect is reversible.[16] In the dark, contraction of chloroplasts may be induced by addition of ATP.[17] Two proteins having contractile properties, which may account for this phenomenon, have been extracted from isolated chloroplasts.[18]

Chloroplasts are distinguished from mitochondria and other plastids by their greater resistance to osmotic changes and fixatives. When placed in distilled water they generally swell and take on a granular appearance. In an isotonic sucrose solution their size and morphologic characteristics remain unchanged. These osmotic properties are due to the membrane surrounding the chloroplast.

Chemical Composition

Chloroplasts are isolated by differential centrifugation after the cell has been homogenized by special procedures. Table 11–2 shows the approximate chemical composition of isolated chloroplasts in higher plants. About 80 per cent of the protein is insoluble and is intimately bound to lipids to form lipoproteins. A structural protein has been isolated which, under certain conditions, may form one-to-one complexes with chlorophyll. This protein accounts for about 40 per cent of the total.[19] An important part of the remaining protein is represented by chloroplast enzymes, which may be soluble or built into the structure of the protein. The lipid fraction comprises neutral fats, steroids, waxes, and phospholipids.

The main component of chloroplasts is the green pigment *chlorophyll,* an asymmetrical molecule having a hydrophilic head made of four pyrrole rings bound to each other and forming a porphyrin (Fig. 11–7). This part of the molecule is similar to some animal pigments, such as hemoglobin and cytochromes. In chlorophyll, however, there is a Mg atom forming a complex with the four rings. In animal pigments the Mg is replaced by Fe. Chlorophyll has a long hydrophobic chain (phytol chain) attached to one of the rings (Fig. 11–7).

In higher plants there are two types of chlorophyll—*a* and *b* (Table 11–2). In chlorophyll *b* there is a —CHO group in place of the —CH$_3$ group, indicated by a circle in Figure 11–7. Three other types of chlorophylls are found in algae, diatoms, dinoflagellates, and in photosynthetic bacteria. Unlike the cytochromes, the chlorophylls are not bound to proteins. The absorption spectrum of chlorophyll is 663 nm in the isolated condition, but in the cell it absorbs at 672 and 683 nm. The absorption spectrum of chlorophyll coincides with that of the photosynthetic efficiency, indicating that this pigment is the main light-trapping molecule. Using refined spectroscopic methods, however, a small amount of pigment absorbing at 700 nm has been found. This pigment is called P700 because it is bleached when it absorbs at that wavelength. (It will be shown later that P700 is related to one of the photosystems in photosynthesis.)

Pigments that belong to the group called *carotenoids* are masked by the green color of chlorophyll. In autumn, the amount of chlorophyll decreases and the other pigments become apparent. These belong to the *carotenes* and *xanthophylls,* which are both related to vitamin A. Carotenes are characterized chemically by the presence of a short chain of unsaturated hydrocarbon, which makes them completely hydrophobic. Xanthophylls, on the contrary, have several hydroxyl groups.

RNA has been found in an average of 3

TABLE 11–2.   Approximate Chemical Analysis of Chloroplasts of Higher Plants*

Constituent	Per Cent of Dry Weight	Components			
Proteins	35–55	About 80% is insoluble			
Lipids	20–30	Fats	50%	Choline	46%
		Sterols	20	Inositol	22
		Waxes	16	Glycerol	22
		Phosphatides	2–7	Ethanolamine	8
				Serine	0.7
Carbohydrates	Variable	Starch, sugar phosphates (3–7%)			
Chlorophyll	9	Chlorophyll a	75%		
		Chlorophyll b	25		
Carotenoids	4.5	Xanthophyll	75		
		Carotene	25		
Nucleic acids					
RNA	2–3				
DNA	<0.02–.01				

*From Granick, S. "The Chloroplasts: Inheritance, structure, and function," *in* Brachet, J., and Mirsky, A. E. (Eds.) *The Cell,* Vol. 2, New York, Academic Press, Inc., 1961.

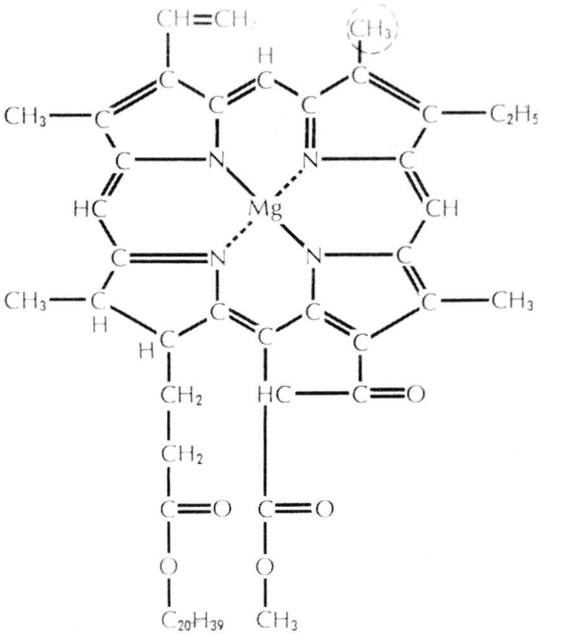

Figure 11-7. Structure of chlorophyll *a*. Observe the four pyrrole rings forming a complex with Mg, and the long hydrophobic phytol chain.

tion of chloroplasts. Chloroplasts and the chlorophyll dissolved in an alcohol or acetone solution show a red *fluorescence*. However, chlorophyll in colloidal suspension in water is not fluorescent. On the other hand, fluorescence persists if chlorophyll is absorbed and forms a monomolecular film. These facts led to the conclusion that chlorophyll within the plastids is disposed in monomolecular layers.

Observation of thin sections under the electron microscope has revealed that the chloroplast has a double limiting membrane. The inner structure varies considerably whether chloroplasts of algae (Fig. 11-6) or of higher plants are considered. In both groups

to 4 per cent of the dry weight of plastids. In *Chlamydomonas* (Fig. 11-6), bodies giving a Feulgen reaction typical of DNA have been observed within the chloroplast.[20] In these and other chloroplasts, DNA has been related to the presence of a special nonchromosomal genetic system (cytoplasmic heredity).

Chloroplasts also contain some cytochromes, vitamins K and E, and metallic atoms, such as Fe, Cu, Mn, and Zn. Some chloroplast enzymes will be considered in relation to the function of chloroplasts later in this chapter.

Ultrastructure and Grana

Many chloroplasts have a heterogeneous structure made up of small granules called *grana*, which are embedded within the stroma, or matrix. The size of the grana varies from 0.3 to 1.7 μm depending on the species. The smallest ones, within the limit of microscopic visibility, are more numerous. They are flat bodies shaped like platelets or disks, which in a lateral view appear as dense bands perpendicular to the chloroplast surface (Fig. 11-8, *A*).

Several indirect methods have been employed in the study of the molecular organiza-

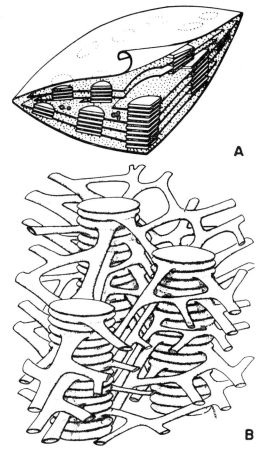

Figure 11-8. **A,** diagram of a chloroplast showing the inner structure with the grana disposed in stacks perpendicular to the surface. (From Erickson, G. A., Kahn, E., Wallis, B., and von Wettstein, D.) **B,** diagram of the ultrastructure of three grana showing the anastomosing tubules that join some of the membranous compartments of the grana. (From Weier, T. E., Stocking, C. R., Thompson, W. W., and Drever, H., *J. Ultrastruct. Res.*, 8:122, 1963.)

thin membranes form the basic structure. In green algae these membranes are more or less continuous, forming flattened sacs or discs without grana.[21, 22]

In higher plants, the entire organelle is enclosed by two concentric membranes, which show the unit membrane structure (Chap. 8). The chloroplast is filled with a matrix or stroma where the grana and the intergranal connecting membranes are encountered. In spinach cells, for example, it has been found that each chloroplast contains from 40 to 60 grana, each about 0.6 μm in diameter.

Grana are cylindrical structures formed by the superimposition of double membranous sacs (Fig. 11–8). In some chloroplasts these sacs appear to be linked by a system of intergranal lamellae.[21] As shown in Figure 11–8, B, grana are formed by stacked closed compartments called *thylakoids*.[23] The number of thylakoids per granum may vary from a few

to 50 or more. In some cases the granum extends, like a cylinder, across the entire width of the plastid. Adjacent grana may be interconnected by a network of flexuous, anastomosing tubules, which join certain compartments, but not others. Under conditions that produce swelling of grana in vivo (e.g., in plants returned to light after continued existence in darkness, or in those with a zinc deficiency), or in vitro (e.g., by osmotic action in isolated plastids), the swelling is confined to the cavity contained within the thylakoid, a situation that is similar to that in retinal rods (Fig. 8–7). After destruction of the intergranal network, grana may separate and become individual entities.

Origin of the Lamellar Structure of the Chloroplast

Figure 11–9 traces the development of a chloroplast during the ontogenesis of a plant.

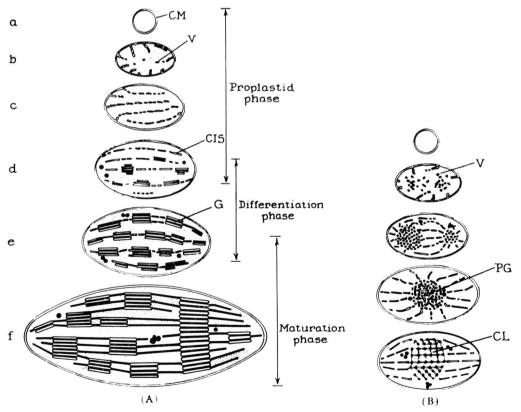

Figure 11–9. **A,** phases in the development of a proplastid into a chloroplast in the presence of light. **B,** same, but in the dark, showing the formation of the primary granum (*PG*), or prolamellar body. *CIS,* flattened cisternae; *CL,* crystal lattice; *CM,* double chloroplast membrane; *G,* granum; *V,* vesicles. (Modified from von Wettstein, D., *J. Ultrastruct. Res., 3*:235, 1959.)

Proplastids are limited by a double membrane. In the presence of light the inner membrane grows and gives off vesicles that arrange themselves to form larger discs. In the granal regions stalks of closely packed lamellar sacs or thylakoids are built. In the mature chloroplasts some compartments of the grana remain connected by intergranal tubules. When plants are grown under low light intensity (*etiolation*), the vesicles formed in the proplastid aggregate, forming one or several *prolamellar bodies*. Sometimes the vesicles form a crystalline pattern consisting of regularly connected tubules[24, 25] (Fig. 11–9, *B*). When these plants are re-exposed to light, the vesicles may fuse into layers and again develop into grana.

As in the case of mitochondria, the chloroplast is constructed of two membranes, each having different functions. The inner membrane, containing the photosynthetic and electron transport systems, forms the grana and the intergranal tubules, which are similar to mitochondrial crests.

The inner membrane and related structures are regulated by genetic factors (DNA) contained both in the nucleus and inside the chloroplast, as well as by external factors such as light, metabolic inhibitors, plant hormones, and minerals.[26]

SUMMARY:
Plastids

Eukaryotic plant cells have specialized organelles—the *plastids*—which contain pigments and may synthesize and accumulate various substances. *Leukoplasts* are colorless plastids; amyloplasts produce starch; *proteinoplasts* accumulate protein; and *elaioplasts* produce fats and essential oils. *Chromoplasts* are colored plastids that contain less chlorophyll than the chloroplasts, but more carotenoid pigments, such as lycopene. Some plastids may store starch and protein at the same time.

Chloroplasts are the most important and most common plastids. Their shape, size, number, and distribution vary in different cells, but are fairly constant for a given tissue. In higher plants they are discoid, 4 to 6 μm in diameter, and number about 20 to 40 per cell. The size and number are genetically controlled; in polyploids, therefore, they are larger. Chloroplasts multiply by division. The quantity of light available causes chloroplasts to undergo changes in shape and volume caused by contraction or swelling.

Chloroplasts are composed of proteins and lipids; about 80 per cent of the protein is insoluble and bound to the lipids, thereby forming lipoproteins. Chlorophyll, the main pigment, is an asymmetrical molecule with a porphyrin head composed of four pyrrole rings and forming a complex with a Mg atom. The molecule also has a long hydrophobic phytol chain. There are several types of chlorophylls (*a,b,c,d,* and *e*). Types *a* and *b* are in higher plants. This green pigment absorbs at 663 nm in the isolated condition; another pigment called P700 is bleached at 700 nm.

The structure of chloroplasts consists of two concentric membranes and an internal system of flattened sacs, the so-called *thylakoids,* that intercommunicate by a system of tubules. A number of piled thylakoids form a granum. A chloroplast may contain 40 to 60 grana of about 0.6 μm in diameter. The membrane of the thylakoids and tubules limit an internal cavity; there is a *stroma* outside the membrane which fills the space of the chloroplast.

The development of chloroplasts begins with a *proplastid* having a double membrane. Under conditions of adequate light the inner membrane invaginates, forming vesicles that will give rise to the tubules and thylakoids of the grana. If the plant develops under light of low intensity (i.e., etiolation) the vesicles aggregate, forming prolamellar bodies. (This process reverts in the presence of light.) As in the case of mitochondria, the inner membrane, grana, and tubules contain the photosynthetic and electron transport system. These membranes, therefore, bear similarities to mitochondrial crests.

PHOTOSYNTHESIS

Photosynthesis is one of the most fundamental biological functions. By means of the chlorophyll contained in the chloroplasts green plants trap the energy of sunlight emitted as photons (i.e., exciton) and transform it into chemical energy. This energy is stored in the chemical bonds that are produced during the synthesis of various foodstuffs.

The previous chapter has emphasized how mitochondria can utilize and transform the energy contained in the foodstuffs by oxidative phosphorylation. Photosynthesis is somewhat the reverse process (Table 11–3). Chloroplasts and mitochondria have many structural and functional similarities, but there are also several differences.

The overall reaction of photosynthesis is:

$$nCO_2 + nH_2O \xrightarrow[\text{chlorophyll}]{\text{light}} (CH_2O)_n + nO_2 \qquad (1)$$

This indicates that, essentially, photosynthesis is the combining of carbon dioxide and water to form various carbohydrates with loss of oxygen.

It has been calculated that each CO_2 molecule from the atmosphere is incorporated into a plant every 200 years, and that all the oxygen in the atmosphere is renewed by plants every 2000 years. Without plants there would be no oxygen in the atmosphere, and life would be almost impossible.

The carbohydrates first formed by photosynthesis are soluble sugars; these can be stored as granules of starch or other polysaccharides inside the chloroplasts or, more usually, inside the amyloplasts. After several steps involving different types of plastids and enzymatic systems, the photosynthesized material is either stored as a reserve product or used as a structural part of the plant (e.g., cellulose).

In early studies it was rightly suggested that in reaction (1) H_2O was the hydrogen donor much in the same way that H_2S is the donor in sulfur bacteria. Thus, reaction (1) can be written as follows:

$$2nH_2O + nCO_2 \longrightarrow nH_2O + nO_2 + (CH_2O)_n \qquad (2)$$

TABLE 11–3. Differences Between Photosynthesis and Oxidative Phosphorylation

Photosynthesis	Oxidative Phosphorylation
Only in presence of light; thus periodical	Independent of light; thus continuous
Uses H_2O and CO_2	Uses molecular O_2
Liberates O_2	Liberates CO_2
Hydrolyzes water	Forms water
Endergonic reaction	Exergonic reaction
$CO_2 + H_2O + energy \rightarrow foodstuff$	$Foodstuff + O_2 \rightarrow CO_2 + H_2O + energy$
In chloroplasts	In mitochondria

Reaction (2) shows that water is the H_2 donor and all the O_2 liberated comes from water. Experiments using water labeled with heavy oxygen ($H_2{}^{18}O$) have confirmed this. In this process water participates primarily as a proton and electron donor. Reaction (2) involves a complex series of steps, of which some take place only in the presence of light and the others take place also in darkness — hence the names *light* and *dark reactions*. In the first, light is absorbed and used by chlorophyll; this is the *photochemical (Hill) reaction*. (In 1939, Robert Hill found that leaves ground in water, to which hydrogen acceptors were added [e.g., quinone], give off O_2 when exposed to light, without synthesizing carbohydrates.) In the second or dark reaction CO_2 is fixed and reduced by thermochemical mechanisms.

Photochemical Reaction

In the study of the photochemical reaction it is necessary to recall the process of oxidative phosphorylation of mitochondria (Chapter 10). In oxidative phosphorylation the flow of electrons is from $NADH_2$ to O_2, following the path of standard oxidation-reduction potential (i.e., from -0.32 to $+0.82$ volts); in photosynthesis, the opposite process takes place. The electrons flow from H_2O to $NADPH_2$ (i.e., from $+0.82$ to -0.32 volts) because the light absorbed causes the boosting of the electrons to higher levels of energy.

In photosynthesis the photoreactive center is functionally and structurally connected to a chain of electron carriers that leads the flow of electrons. This center is generally known as the *photosynthetic electron transport system* and is coupled with the phosphorylation of ADP to ATP.

Photosystems I and II. There is experimental evidence that the efficiency of photosynthesis is a function of the wavelength of the impinging light. For example, if chloroplasts are simultaneously excited at 680 and 600 nm, there is considerable stimulation of the photochemical reaction. The existence of two systems of light absorption have been postulated: Photosystem I is stimulated at the longer wavelength and is not accompanied by the production of O_2, whereas photosystem II is activated at the shorter wavelength and yields O_2 (Fig. 11–10).

Photosystem I is made of units containing about 200 molecules of chlorophyll *a* and 50 molecules of carotenoids. One molecule of pigment P700 is related to this unit and is the one that can trap excitons and release electrons. Photosystem II contains about 200 molecules of chlorophyll *a* and some 200 of chlorophyll *b, c,* or *d,* depending on the species; it may also contain molecules of xanthophyll. The complete set of pigments and lipids, constituting these two photosystems, is thought to be clustered in a structural and functional unit. The two photosystems have been separated by treatment with digitonin and other detergents, followed by density gradient centrifugation.

According to the diagram of the Figure 11–10, both photosystems operate in a sequential and interrelated fashion. When photosystem I is excited, electrons are boosted to a higher energy level, thereby reducing ferredoxin. This compound is then reoxidized by the transfer of electrons to $NADP^+$, forming $NADPH_2$. Two photons of light (i.e., two excitons) appear to be needed to produce one molecule of $NADPH_2$. This reducing component, having high energy electrons, can be used in a variety of biochemical reactions — particularly in the synthesis of carbohydrates.

From Figure 11–10 it is evident that to restore the electrons to photosystem I this process of light absorption must be coordinated with photosystem II. In photosystem II the quanta of light remove electrons from the hydrogen of water and boost them to a higher energy level, thus reducing plastoquinone. From here the electrons are brought back to photosystem I by way of an electron transfer system (i.e., cytochrome *b*559, cytochrome *f,* and plastocyanine). From this diagram it is evident that only the excited photosystem I results in the release of O_2 from water (the Hill reaction).

Electron Transport Systems. As shown in Figure 11–10, *ferredoxin* plays a role in the transport of electrons from photosystem I to $NADP^+$. Ferredoxin is a protein of about 11,600 daltons and contains iron and sulfur, but lacks a porphyrin group. Other components are a *ferredoxin-NADP oxireductase,* a flavoprotein that uses $NADP^+$ as an electron acceptor. Another electron acceptor, not yet identified, is indicated by *Z* in Figure 11–10 (see Lehninger, 1970). In the transfer of electrons from photosystem II to I there are other carriers. For example, there is

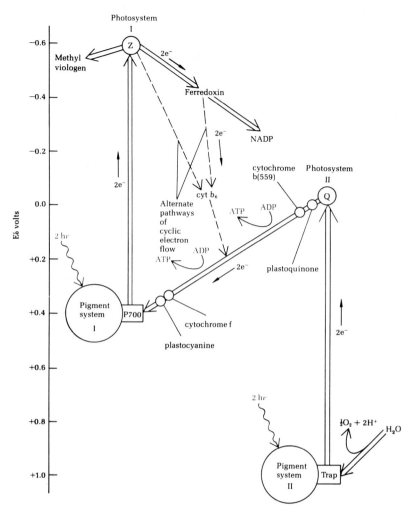

Figure 11-10. Diagram showing the arrangement of Photosystems I and II. These two systems of photosynthesis are connected by an electron transport chain situated between the acceptor Q and P700, a pigment in Photosystem I. Observe that starting from H_2O and with the release of O_2 electrons are boosted from Photosystem II toward Photosystem I, finally reaching NADPH + H^+. On the left the redox potential of the different components of the system is indicated. (See the description in the text.) (From Lehninger, A. L., *Biochemistry: The Molecular Basis of Cell Structure and Function*, New York, Worth Publishers, Inc., 1970.)

cytochrome f (Latin, *frons*, leaf), a hydrophobic protein of 100,000 daltons isolated in nonpolar solvents; other carriers include the *cytochromes* b_6 and b_3 (or *559*), tightly bound to the membrane structure. Chloroplasts also contain a copper-protein called *plastocyanine* and two quinones—*vitamin K_1* and *plastoquinone*—the latter similar to ubiquinone in mitochondria. The sequence in the transfer of electrons from photosystem II to I begins with a non-identified carrier *Q*. This transfer is followed by the sequence: plastoquinone → cytochrome *b559* → cytochrome *f* → plastocyanine.

Photophosphorylation. As in the case of mitochondria, in which the electron transport system is coupled with phosphorylation of ADP, in chloroplasts there is a light-induced phosphorylation that is coupled with the

electron transfer chain of photosystem II (Fig. 11-10). This process can be uncoupled by 2,4-dinitrophenol that also acts in mitochondria. Chloroplasts also contain a coupling factor that has a Ca^{2+}-activated ATPase.

In contrast to the oxidative phosphorylation of mitochondria, O_2 is not used in photophosphorylation of chloroplasts (Table 11-3). Green plants can produce 30 times as much ATP by photophosphorylation as by oxidative phosphorylation in their own mitochondria. In addition, these plants contain many more chloroplasts than mitochondria.

In the overall photosynthetic pathway, for every one molecule of O_2 liberated eight light quanta are needed, and two molecules of $NADPH_2$ are formed; the efficiency is close to 25 per cent. Along the electron transport there is the synthesis of three ATP molecules.

Dark (Thermochemical) Reaction in Photosynthesis

The diagram in Figure 11–10 shows that along with the energy provided by ATP, the reduced $NADPH_2$ can bring about the reduction of atmospheric CO_2 and combine it with the hydrogen to form the various carbohydrates. This process involves many steps, which have been elucidated principally by the use of radioactive CO_2 in a series of brilliant experiments.[27] The reactions involved (Fig. 11–11) are so rapid that they appear one second or less after the addition of $^{14}CO_2$. These reactions occur in complete darkness if the plant were previously exposed to light. For details of the photosynthetic carbon cycle, see Figure 11–11 and refer to biochemistry textbooks.

In cells exposed to $^{14}CO_2$ for five seconds, the dominating compound is 3-phosphoglyceric acid; from this, all the compounds shown in the cycle in Figure 11–11 originate. Two triose phosphate molecules unite to form hexose (fructose) diphosphate, from which glucose phosphate is formed. Then, from glucose phosphate various disaccharides and polysaccharides are formed. As shown in Figure 11–11, the initial enzyme, *carboxydismutase*, is responsible for the formation of phosphoglyceric acid molecules from ribulose diphosphate and CO_2. Under the action of many enzymes different hexoses, heptoses, and pentoses are formed.

Correlation between Structure and Function in Chloroplasts

As in the case of mitochondria, it is now possible to correlate chloroplast structure and function at the molecular level.

Isolated chloroplasts have the biochemical machinery necessary to perform both the light and dark reactions.[28] However, the enzymes involved in the dark reactions are easily soluble in water so that by fractionation of chloroplasts the two systems of reactions can be separated.[29] After sonic disruption a green sediment is obtained which carries out the light reactions (production of O_2, reducing compounds, and ATP) while the supernatant contains the enzymes involved in the dark reactions (fixation of CO_2, etc.). Thus, the

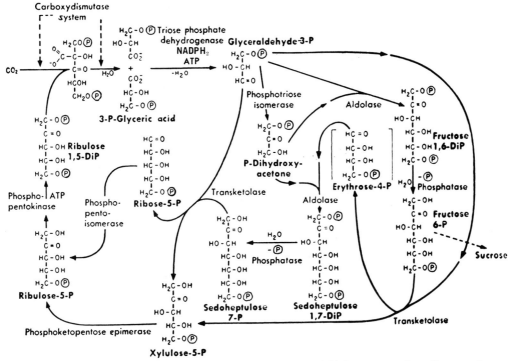

Figure 11–11. Details of the dark reactions in photosynthesis. The initial enzyme *carboxydismutase* is responsible for the formation of glyceraldehyde-3-phosphate into which the CO_2 is added. The various steps in the pentose cycle are indicated. (Courtesy of J. A. Bassham and M. Calvin.)

enzyme carboxydismutase is contained almost exclusively in the supernatant.

The electron microscope has revealed that the green sediment is composed of the lamellar system of the chloroplast forming the grana. Chlorophyll is uniformly distributed within the lamellar structure.[30] Small frag-

ments of these lamellae give the Hill reaction, but fix CO_2 only in the presence of the supernatant containing the stroma proteins.

A recent methodological advancement— the use of a special embedding technique in an aqueous medium—prevents the removal of lipids and chlorophyll. This technique has

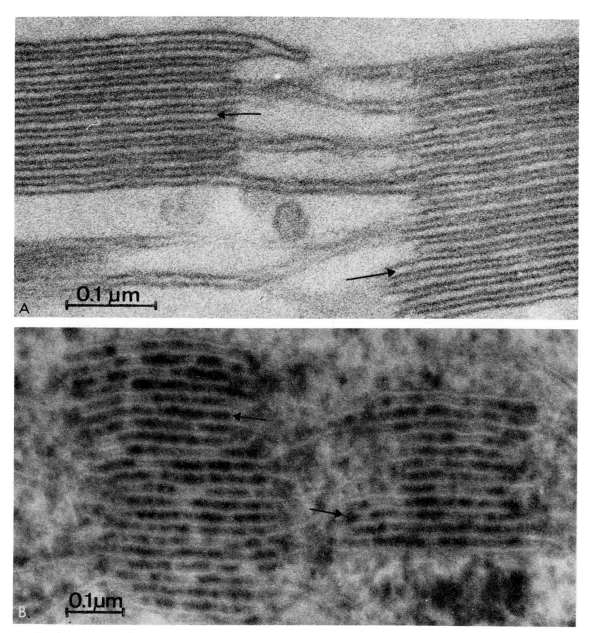

Figure 11–12. **A,** electron micrograph of a section of two chloroplast grana and their connecting tubules (see Fig. 11–8). The arrows indicate the cavity of the membranous compartment of the grana (i.e., thylakoid). ×240,000. **B,** same structure as in *A,* but after a photo-oxidation reaction with 3′,3′-diaminobenzidine. The products are accumulated within the cavity of thylakoids (arrows). (See the description in the text.) ×153,000. (Courtesy of I. Nir and D. C. Pease.)

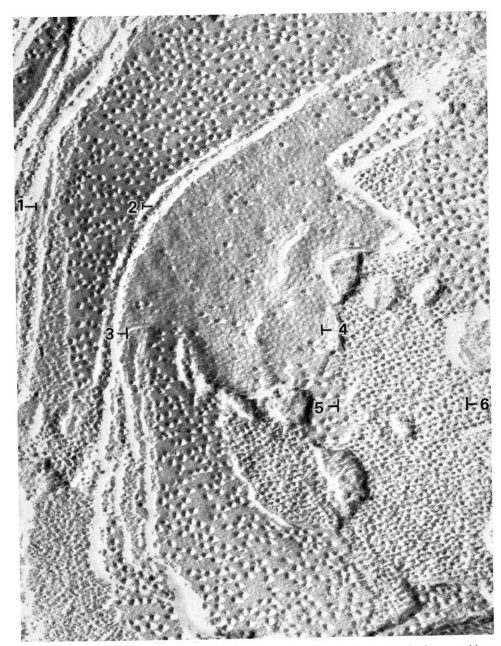

Figure 11–13.   Electron micrograph of the fracture of a chloroplast lamella obtained by the freeze-etching technique. Region *1-2* contains particles of the quantasome type. In region *5-6* particles are smaller (11nm) and tightly packed. In region *3-4* the membranes show few particles. ×90,000. (Courtesy of R. B. Park.)

made it possible to obtain more information about the cytochemical localization of both photosystems. By using 3'3'-diaminobenzidine (DAB) in the presence of light, the products of photo-oxidation accumulate in the cavity of the thylakoid, indicating that *photo-system I* is probably localized at the inner side of the thylakoid membrane (Fig. 11–12). The photoreduction of ferrocyanide, which indicates the localization of *photosystem II*, causes dense deposits in the partitions between thylakoids, in this case indicating a possible localization at the outer surface of the membrane. As in the case of the mitochondria, there is a special functional asymmetry across the membrane of the chloroplast.[31]

Electron microscopic studies of the inner surface of the thylakoid, or compartment forming the grana, has established the existence of a paracrystalline array of particles of 20×10 nm, called *quantasomes*.[32] It was found that as few as 3 to 6 of these particles, forming aggregates of 50×10 nm, still produced the Hill reaction and, in the presence of CO_2 and the supernatant fraction, were able to fix CO_2.[33]

When the fine structure of chloroplasts is studied by the freeze-etching technique (Chap. 5) three types of membrane surfaces are observed: one containing the quantasome particles, another in which smaller particles (11 nm) are more densely packed, and finally a membrane surface with a rough texture and few or no particles.[34, 35] According to this view both the quantasomes and the smaller particles lie within the membrane of the thylakoid (Fig. 11–13). The significance of the quantasomes is still uncertain. In fact the interpretation that they represent photosynthetic units within the thylakoid membrane has been disputed.[36–38]

Conformational Changes and Ion Movements

Chloroplasts undergo striking conformational changes which are dependent on the action of light. In the dark they undergo swelling and separation of the membranes of the thylakoid. This phenomenon is reversible; in the presence of light the membranes again return to close apposition. It was mentioned in Chapter 9 that mitochondria eject H^+ ions during the electron transport and that the interior becomes more alkaline (Fig. 10–17). In the case of chloroplasts the opposite reactions occur; under the action of light H^+ ions are absorbed, and the interior becomes acidified. At the same time K^+ and Mg^{2+} are ejected into the medium. In the dark, H^+ is ejected and K^+ and Mg^{2+} are reabsorbed by the chloroplasts. Thus, the chloroplast membrane has a vectorial organization that is the reverse of the mitochondrial membrane. In physical terms it is understood that dark reactions, being of a thermochemical nature, do not need an ordered structural arrangement (i.e., they are carried out by soluble enzymes). On the contrary, photophosphorylation requires the separation of positive and negative charges into specific pathways of electron flow.

SUMMARY:
Photosynthesis and the Chloroplast

Photosynthesis is the process by which chloroplasts trap light quanta (i.e., photons, excitons) and transform them into chemical energy. Photosynthesis is, in some ways, the reverse of oxidative phosphorylation in mitochondria. The overall reaction of photosynthesis is:

$$2nH_2O + nCO_2 \longrightarrow nH_2O + nO_2 + (CH_2O)n$$

Thus, using H_2O as hydrogen donor and CO_2 from the atmosphere, carbohydrates are synthesized, and O_2 is released.

Photosynthesis consists of a *photochemical reaction* (Hill reaction), which occurs in the presence of light, and a *dark* or *thermochemical reaction*. In the first reaction, O_2 is released when the chloroplasts are exposed to light; in the second, CO_2 is fixed and carbohydrates are formed. In the photochemical reaction electrons flow from H_2O to $NADPH_2$ (i.e., from +0.82 to −0.32 volts of redox potential) because electrons are boosted to high energy levels by the absorbed light. In photosynthesis there are two photosystems (I and II) that are excited at different wavelengths. *Photosystem I* (excited at 680 to 700 nm) comprises 200 molecules of chlorophyll *a*, 50 of carotenoids, and one of P700. *Photosystem II* (excited at 600 nm) contains chlorophyll *a* and *b;* only this second system is associated with the release of O_2 from H_2O. Both photosystems operate in a sequential and interrelated fashion (Fig. 11–10). In photosystem I, two quanta of light, trapped by P700, boost electrons which reduce *ferredoxin,* a protein containing iron and sulfur. In turn, ferredoxin transfers the electrons to $NADP^+$, reducing it to $NADPH_2$. The electrons in photosystem I are restored by the second system. Here the light quanta remove the electrons from the hydrogen of water, releasing O_2. The electrons are transferred to P700 by a transport system which comprises several cytochromes and *plastocyanine* (a copper-containing protein). Two quinones — *vitamin K_1* and *plastoquinone* — are also included in this electron transport system. As in mitochondria, the transfer of electrons is coupled with the phosphorylation of ATP at the level of photosystem II. The dark, or thermochemical, reaction involves many steps which start with the uptake of CO_2 and its reduction by $NADPH_2$ to form the various carbohydrates (Fig. 11–11). In several steps of this complex cycle of reactions, ATP is also used as an energy source. A dominating compound is 3-phosphoglyceric acid which gives rise to glucose phosphate, from which come the various disaccharides and polysaccharides. The formation of phosphoglyceric acid depends on the initial enzyme, *carboxydismutase.* As in mitochondria, there is a strict structural-functional correlation in chloroplasts. All the enzymes of the dark reactions (like those of the Krebs cycle) are soluble and are contained in the stroma of the chloroplast. The photosystems, including electron transport and photophosphorylation, form part of the membranes of the chloroplasts.

Cytochemically, it is possible to demonstrate an asymmetry of the thylakoid membrane: photosystem I is probably localized at the inner surface and photosystem II, at the outer surface. By freeze-fracture techniques large particles (quantasomes) and smaller particles may be observed within the membrane of the thylakoid.

Chloroplasts undergo conformational changes. In the dark, they swell and the cavity of the thylakoid expands; H^+ is ejected and K^+ and Mg^{2+} are reabsorbed. Under the action of light the interior of the chloroplasts becomes more acid by the uptake of H^+. At the same time there is release of K^+ and Mg^{2+} and the cavity of the thylakoid becomes smaller.

Chloroplasts as Semiautonomous Organelles

Chloroplasts, like mitochondria, exhibit a certain degree of functional autonomy within the intracellular environment. In fact, they undergo division and may contain some genetic information (cytoplasmic inheritance).

Chloroplasts contain their own DNA, different from nuclear DNA, special ribosomes, smaller than those present in the cytoplasm, and all the necessary molecular machinery to achieve protein synthesis.

Since the early work of Schimper and Meyer (1883) it has been accepted that plastids multiply by fission. This is easily observed in unicellular algae that contain only one chloroplast (Fig. 11–6). In the alga *Nitella* a division cycle of 18 hours has been recorded cinematographically.[39] The division process is as orderly as chromosomal division. Plastid reproduction by fission implies a growth process of the daughter plastids. (The way in which plastids become differentiated during development and how they change in fine structure under the action of different external factors such as light have been mentioned earlier.)

There are several examples of inheritance that do not follow typical Mendelian segregation, thereby indicating an extrachromosomal type of heredity, that has been shown experimentally. When the unicellular green alga *Euglena* is grown in the dark it contains small, colorless proplastids. If under these conditions the nucleus is irradiated with an ultraviolet microbeam, the proplastids are still capable of developing into chloroplasts.[40] However, if the cytoplasm is irradiated with the nucleus shielded, a considerable number of colorless colonies (lacking chloroplasts) develop. These results suggest that there is a cytoplasmic DNA in this alga. (For details on the genetic information of plastids see Stubbe, 1971.)

It is now generally accepted that a characteristic DNA occurs in chloroplasts of algae and higher plants. DNA regions resembling bacterial nucleoids were identified in thin sections under the electron microscope.[41-43] The quantity of DNA per chloroplast varies slightly with the species and is in the range of 2 to 5×10^{15} gm, or slightly less than in the bacterium *E. coli*. Segments of DNA as long as 150 μm have been separated from chloroplasts.[44] The replication of chloroplast DNA has been followed with ^{3}H-thymidine. The presence of ribosome-like particles within

chloroplasts is easily demonstrated by electron microscopy. These ribosomes are smaller than cytoplasmic ribosomes. Polysomes have also been separated from chloroplasts.[45-47]

The discovery of special ribosomes associated with chloroplasts provided evidence that these organelles contain a specific protein synthesizing system.[48] In the presence of CO_2 as the sole source of carbon, chloroplasts actively incorporate amino acids into proteins. Protein synthesis is preferentially inhibited by chloramphenicol concentrations that do not affect protein synthesis in the cytoplasm.[49]

The involvement of the two types of protein synthesis in the assembly of chloroplasts may be studied by using cycloheximide to inhibit synthesis due to cytoplasmic ribosomes.[50]

It is evident that chloroplasts have sufficient DNA to code for a number of proteins and to contain all the necessary mechanisms for a DNA-RNA directed protein synthesis. It is still not known, however, which proteins are specified by the chloroplast DNA-RNA system. There is genetic evidence that various photosynthetic enzymes are under nuclear control. Like the mitochondrion, the chloroplast might exert a dual control over some of the structural proteins probably being coded by its own DNA.

Chloroplasts have been described as having many of the characteristics of a semi-autonomous or symbiotic organism living within the plant cells. They divide, grow, and differentiate; they contain DNA, ribosomal RNA, messenger RNA, and are able to conduct protein synthesis. It has been suggested that chloroplasts may have resulted from a symbiotic relationship between an autotrophic microorganism, one able to transform energy from light, and a heterotrophic host cell (Chap. 10). Although this hypothesis is highly attractive, still it is evident that the electron transport system in chloroplasts and the enzymes required for making the photosynthetic pigments are controlled by the nuclear genes.[51]

REFERENCES

1. Mühlethaler, K. (1961) Plant cell walls. In: *The Cell*, Vol. 2, p. 85. (Brachet, J., and Mirsky, A. E., eds.) Academic Press, Inc., New York.
2. Erickson, R. O. (1959) Patterns of cell growth and differentiation in plants. In: *The Cell*, Vol. 1, p. 497. (Brachet, J., and Mirsky, A. E., eds.) Academic Press, Inc., New York.
3. Porter, K. R., and Machado, R. D. (1960) *J. Biophys. Biochem. Cytol.*, 7:167.

4. Frey-Wyssling, A., López-Sáez, J. F., and Mühlethaler, K. (1964) *J. Ultrastruct. Res., 10*:422.
5. Hepler, P. K., and Newcomb, E. H., (1967) *J. Ultrastruct. Res., 19*:498.
6. Porter, K. R. (1966) In: *Principles of Biomolecular Organization*, p. 308. *Ciba Foundation Symposium* (Wolstenholme, G. E. W., ed.) J. & A. Churchill, Ltd., London.
7. Hepler, P. K., and Newcomb, E. H. (1964) *J. Cell Biol., 20*:529.
8. Cronshaw, J., and Bouck, G. B. (1965) *J. Cell Biol., 24*:415.
9. Bonnett, H. T., and Newcomb, E. H. (1965) *J. Cell Biol., 27*:423.
10. Buvat, R. (1959) *Ann. Soc. Nat. Bot., 11e série*, p. 121.
11. Mollenhauer, H. H., and Whaley, W. G. (1963) *J. Cell Biol., 17*:222.
12. Northcote, D. H., and Pickett-Heaps, J. D. (1966) *Biochem. J., 98*:159.
13. Dauwalder, M., Whaley, W. G., and Kephart, J. (1969) *J. Cell Sci., 4*:455.
14. Guillermond, A. (1922) *C. R. Acad. Sci. (Paris), 175*:283.
15. Newcomb, E. H. (1967) *J. Cell Biol., 33*:143.
16. Itoh, M., Izawa, S., and Shikata, K. (1963) *Biochim. Biophys. Acta, 66*:319.
17. Packer, L. (1966) In: *Biochemistry of Chloroplasts*, Vol. 1, p. 233. (Goodwin, T. W., ed.) Academic Press, Inc., New York.
18. Ohnishi, T. (1964) *J. Biochem. (Tokyo), 55*:494.
19. Criddle, R. S., and Park, L. (1964) *B. B. Res. Commun., 17*:74.
20. Ris, H., and Plaut, W. (1962) *J. Cell Biol., 13*:383.
21. Weier, T. E., Stocking, C. R., Thompson, W. W., and Drever, H. (1963) *J. Ultrastruct. Res., 8*:122.
22. Sager, R., and Palade, G. E. (1957) *J. Biophys. Biochem. Cytol., 3*:463.
23. Menke, W. (1962) *Ann. Rev. Plant Physiol., 13*:27.
24. Wilsenach, R. (1963) *J. Cell Biol., 18*:419.
25. Wettstein, D. von (1959) *J. Ultrastruct. Res., 3*:235.
26. Park, R. B. (1968) In: *Organizational Biosynthesis*, p. 373 (Vogel, H. J., et al., eds.) Academic Press, Inc., New York.
27. Calvin, M. (1962) *Science, 135*:879.
28. Arnon, D. I., Allen, M. B., Whatley, F. R., Capindale, J. B., and Rosenberg, L. L. (1956) *Proc. Intern. Cong. Biochem. (3rd Congress)* Brussels, 1955, p. 277.
29. Trebst, A. V., Tsujimoto, H. Y., and Arnon, D. I. (1958) *Nature, 182*:351.
30. Park, R. B., and Pon, N. G. (1961) *J. Molec. Biol., 3*:10.
31. Nir, J., and Pease, D. C. (1973) *J. Ultrastruct., Res., 42*:534.
32. Park, R. B., and Pon, N. G. (1963) *J. Molec. Biol., 6*:105.
33. Sauer, K., and Calvin, M. (1962) *J. Molec. Biol., 4*:451.
34. Park, R. B., and Beggins, J. (1964) *Science, 144*:1009, 201a.
35. Branton, D., and Park, R. B. (1967) *J. Ultrastruct. Res., 19*:283.
36. Howell, S. H., and Moundrianakis, E. M. (1967) *J. Molec. Biol., 27*:323.
37. Howell, S. H., and Moundrianakis, E. M. (1967) *Proc. Natl. Acad. Sci. USA, 58*:1261.
38. Park, R. B., and Pheifhofer, A. O. A. (1968) *Proc. Natl. Acad. Sci. USA, 60*:337.
39. Green, P. (1964) *Amer. J. Botany, 51*:334.
40. Gibor, A., and Granick, S. (1962) *J. Cell Biol., 15*:599.
41. Gunning, B. E. S. (1965) *J. Cell Biol., 24*:79.
42. Kislev, N., Swift, H., and Bogorad, H. (1965) *J. Cell Biol., 25*:327.
43. Bisalputra, T., and Bisalputra, A. A. (1967) *J. Ultrastruct. Res., 17*:14.
44. Woodcock, C. L. F., and Fernández-Morán, H. (1968) *J. Molec. Biol., 31*:627.
45. Stutz, E., and Noll, H. (1967) *Proc. Natl. Acad. Sci. USA, 57*:774.
46. Bager, R., and Hamilton, M. G. (1967) *Science, 157*:709.
47. Bruskov, V. I., and Odintsova, M. S. (1968) *J. Molec. Biol., 32*:471.
48. Brawerman, G., and Eisenstadt, J. M. (1968) In: *Organizational Biosynthesis*, p. 419. (Vogel, H. J., et al., eds.) Academic Press, Inc., New York.
49. Pogo, B. G. T., and Pogo, O. (1965) *J. Protozool., 12*:96.
50. Hoober, J. K., Siekevitz, P., and Palade, G. E. (1969) *J. Biol. Chem. 244*:2621.
51. Kirk, J. T. O. (1966) In: *Biochemistry of Chloroplasts*, p. 319. (Goodwin, T. W., ed.) Academic Press, Inc., New York.

ADDITIONAL READING

Amelunxen, F., and Gronau, G. (1969) Electronen mikroskopische Untersuchungen an den Ölzellen von *Aconis calamus L. Pflanzen Physiol., 60*:156.

Arnon, D. I. (1967) Photosynthetic activity of isolated chloroplasts. *Physiol. Rev., 47*:317.

Buvat, R. (1959) Recherches sur les infrastructures du cytoplasme dans les cellules du méristème apical des ébauches foliaires et des feuilles dévelopées de *l'Elodea canadensis. Ann. Soc. Nat. Bot.,* 11e série, 121.

Erickson, R. O. (1959) Patterns of cell growth and differentiation in plants. In: *The Cell.,* Vol. 1, p. 497. (Brachet, J., and Mirsky, A. E., eds.) Academic Press, Inc., New York.

Granick, S. (1961) The chloroplasts: inheritance, structure and function. In: *The Cell,* Vol. 2, p. 489. (Brachet, J., and Mirsky, A. E., eds.) Academic Press, New York.

Hoober, J. K., Siekevitz, P., and Palade, G. E. (1969) Formation of chloroplast membranes in *Chlamydomonas reinhardi* y-1. *J. Biol. Chem., 244*:2621.

Kreger, D. R. (1969) Cell walls. In: *Handbook of Molecular Cytology.* p. 1444 (Lima-de-Faria, A., ed.) North-Holland Publishing Company, Amsterdam.

Lehninger, A. L. (1970) *Biochemistry: The Molecular Basis of Cell Structure and Function.* Worth Publishers. Inc., New York.

Mühlethaler, K. (1961) Plant cell walls. In: *The Cell,* Vol. 2, p. 85. (Brachet, J., and Mirsky, A. E., eds.) Academic Press, Inc., New York.

Newcomb, E. H. (1969) Plant Microtubules. *Ann. Rev. Plant Physiol., 20*:253.

Rabinowitch, E. (1945) *Photosynthesis.* Inter-Science Publishers, New York.

Stubbe, W. (1971) Origin and continuity of plastids. In: *Origin and Continuity of Cell Organelles.* (Reinert, J., and Ursprung, H., eds.) Springer Verlag, Berlin.

Part Six

CELLULAR BASIS OF CYTOGENETICS

In the following five chapters the nucleus and the chromosomes are presented as entities involved in genetic activity at the cellular level. This general topic is also called the *chromosomal basis of genetics*. The study, begun at the end of the last century, developed so rapidly that for many years it was the best known field of cytology. The development of cariology (Gr., *carion*, nucleus) was somewhat detrimental to the study of the cell as a whole and of its molecular and biochemical aspects, which are now included within the realm of cell biology.

Chapter 12 gives a general account of the progress made in recent years on knowledge of the ultrastructure of the interphase nucleus, the chromosomes, and the nucleolus. The special permeability properties of the nuclear envelope are also discussed here. Knowledge of the localization of the DNA, RNA, and nuclear proteins, as well as of the macromolecular organization of the chromosomes, is of paramount importance in the interpretation of their duplication. (Within the chromosome there are elementary microfibrils and more complex fibrils which produce the folded fiber structure during mitosis and meiosis.) It is astonishing to learn that within a single human chromosome several centimeters of DNA are tightly packed, and that within the nucleolus lies the entire mechanism of ribosomal RNA formation.

Chapter 13 is dedicated to *mitosis*, the basic process by which cellular material is equally divided between daugher cells. The process implies a series of complex changes involving the nucleus and the cytoplasm. It is directly related to the problem of continuity of chromosomes as entities capable of autoduplication and of maintaining their morphologic characteristics and function through successive cell divisions. The importance of isolating the mitotic apparatus, as a method of learning more about its composition and functioning, is emphasized.

Cytogenetics can be understood only with a clear comprehension of meiosis as the division that brings about the reduction in the number of chromosomes and the recombination and interchange of blocks of genes by way of crossing over. Crossing over is cytologically expressed by the chiasma. Meiosis is presented in Chapter 14 along with a discussion of the giant polytenic and the lampbrush chromosomes.

Chapter 15 presents in a very general way the chromosomal basis of Mendel's principles of heredity and the linkages between different genes, which depend on their position in the chromosome and the presence or absence of crossing over. The bases on which genetic maps of the chromosomes are built are mentioned in relation to these concepts. An important part of this chapter is devoted to the different chromosomal aberrations that can be produced either spontaneously or by radiation and chemical agents. The chromosomal aspects of evolution are considered briefly.

In the past decade the study of the normal and abnormal human karyotype has developed considerably and has acquired great importance in cytogenetics. This material has been incorporated in Chapter 16 along with a discussion of chromosomal sex determination. These studies have considerable theoretical and applied value because they include investigations of congenital and hereditary diseases and the varied sexual alterations that can be produced in the human. This material is of great importance to students of medicine; with it they will be better able to interpret the pathogenic mechanism of numerous hereditary diseases and congenital malformations.

twelve

THE INTERPHASE NUCLEUS
AND THE CHROMOSOMES

An introduction to the study of the interphase nucleus and the chromosomes, based on light microscopy, was given in Chapter 2. It was noted that the interphase nucleus contains the *nucleolus,* composed principally of ribonucleoproteins, and *chromatin* which contains deoxyribonucleoproteins in a condensed *(heterochromatin)* or uncondensed *(euchromatin)* form. It was also stated that the nucleus is surrounded by the *nuclear envelope,* which has an important role in the molecular exchanges between the nucleus and cytoplasm.

Some of the morphological details used in the identification of the chromosomes at metaphase were described in Chapter 2. For example, the presence of centromeres or primary constrictions, secondary constrictions, satellites, and nucleolar organizers were considered. In the present chapter the structure of the various components of the interphase nucleus and of the chromosome, as revealed principally by studies of fine structure with the electron microscope, will be described.

Since the gene is a portion of the DNA molecule (see Chapters 17 and 19), the study of the macromolecular arrangement of this molecule in the nucleus and chromosomes is of particular importance in the interpretation of its genetic function. The same considerations apply to the RNA-containing structures within the interphase nucleus.

THE NUCLEAR ENVELOPE AND NUCLEAR PERMEABILITY

As mentioned in Chapter 2, the light microscope gave little information about the *nuclear envelope.* Experiments have shown that isolated nuclei possess osmotic properties. Microsurgically it was observed that a puncture of the nucleus results in rapid lysis. This is at variance with a similar situation in the plasma membrane, which is capable of self-repair in the presence of $Ca.^{2+}$

One of the most interesting discoveries made with the electron microscope is that the nuclear envelope is a dependency of the cytoplasmic vacuolar system. This has been verified not only by observing the continuities of the two elements at many points, but also by studying these structures in different stages of cellular activity and by following them phylogenetically. For example, most bacteria lack internal membranes and the nuclear region has no envelope; mycobacteria have a few cytoplasmic membranes, but have none around the nucleus.[1] Only in fungi and in cells of higher organisms does a definite nuclear envelope appear.[2]

Probably one of the clearest demonstrations that the nuclear envelope is a derivative of the endoplasmic reticulum is observed during mitosis. At telophase, cisternae of the endoplasmic reticulum collect around the chromosomes to re-form the nuclear envelope.[3]

Pore Complexes

The nuclear envelope consists of two concentric membranes separated by a perinuclear space of 10 to 15 nm in width. These membranes have a basic unit structure similar to that of the plasma membrane. The nuclear envelope is interpreted as flattened cisternae of the endoplasmic reticulum having ribosomes only on the outer surface.[4-6] As shown

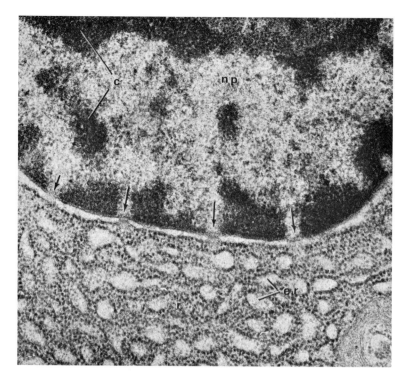

Figure 12-1. Electron micrograph of a portion of the nucleus and cytoplasm from a pancreatic cell of the mouse. The pores in the nuclear envelope are indicated by arrows within the interchromatin channels; *c*, chromatin; *er*, endoplasmic reticulum; *np*, nucleoplasm; *r*, ribosomes. ×48,000. (Courtesy of J. Andrè.)

in Figure 12-1, at certain points the nuclear envelope is interrupted by structures called *pores*. Around the margins of these nuclear pores both membranes are in continuity. At their nuclear side the pores are generally aligned with channels of nucleoplasm situated between more condensed lumps of chromatin (Fig. 12-1). The number of pores varies from 40 to 145 per square micrometer in nuclei of various plants and animals. At their highest density the pores are packed in hexagonal array with center-to-center pore distances of about 150 nm.

In nuclei of *Mammalia* it has been calculated that nuclear pores account for 10 per cent of the surface area; membranes constitute the remaining surface area. The pores are enclosed by circular structures called *annuli*. The pores and annuli, together, are designated the *pore complex*. Negative staining techniques (Chapter 6) have demonstrated that the nuclear pores are octagonal in shape and are about 60 nm in diameter.[7] The *annular material* is a kind of amorphous cylinder that extends through the pore into either side of the pore margin. It has an outer diameter of about 120 nm, which exceeds that of the pore (Fig. 12-2). The central hole of the annulus varies from 30 to 50 nm. This material is digested by trypsin and remains untouched when exposed to ribonuclease and deoxyribonuclease. These findings suggest that the annular material is protein in nature. The pore complex is apparently a rather rigid structure present in a fixed number according to cell type. In certain physiological stages, however, they may change in number. For

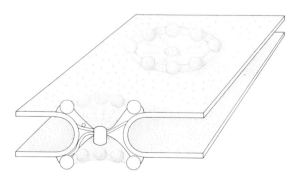

Figure 12-2. Diagram of the pore complex. According to Franke (1970) the octagonal structure of the pore is a result of the presence of eight granules of about 15 nm in diameter evenly spaced around the annulus on both the cytoplasmic and nuclear surfaces. Observe that the size of the pore (60 nm) is considerably reduced by the annulus. (From Franke, W. W., *Z. Zellforsch Mikrosk Anat., 105*:405–429, 1970.)

example, they are reduced in number in maturing erythroblasts and in spermatids.

In *amebae* there is a layer of electron-dense material with a honeycomb structure at the inner surface of the nuclear envelope. This component is made of prisms that are attached at one end to the nuclear envelope at the site of each pore. In other cells, a layer of fibrous material known as *fibrous lamina*, has also been observed on the nuclear side of the envelope.[8]

Permeability of the Nuclear Envelope

Several experiments suggest that the pore complexes may be temporary or permanent openings in the nuclear envelope. By injecting colloidal gold particles, varying in size from 2.5 to 17 nm, into the cytoplasm of amebae, it was found that those with diameters up to 8.5 nm rapidly entered the nucleus. Particles with diameters of 8.9 to 10.6 nm penetrated more slowly, and the larger ones did not enter at all. These results indicate that the openings are smaller than the pore size would indicate.[9, 10]

Evidence has been obtained with these techniques, suggesting that the pores are pathways for the exchange of macromolecules. The annuli may regulate the exchange in relation to the size and possibly to the chemical nature of the penetrating substance. It is important to consider that the permeability of the nuclear envelope is not fixed, but varies in different cell types and within a given cell—at least during the division cycle. Such differences are attributable to changes in the nature of annular material (Feldher, 1971).

The presence of pores in the nuclear envelope should be correlated with some of the electrochemical properties of this structure, which can be investigated with fine microelectrodes (Fig. 12–3).[11] Two types of nuclear envelopes have been recognized with this technique. When giant cells from the salivary gland of *Drosophila* are penetrated with a microelectrode, there is an abrupt change in potential at the plasma membrane (−12 mV); then, as the microelectrode enters

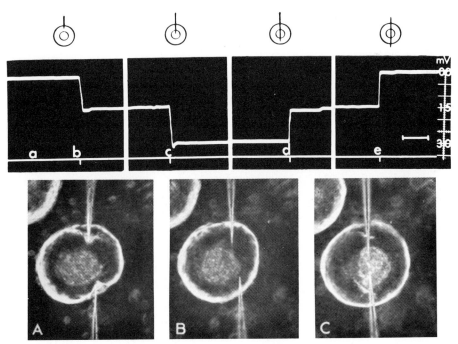

Figure 12–3. Experiment in microsurgery to study the potential of the nuclear envelope in giant nuclei of the salivary gland of *Drosophila*. **Above,** a diagram of the penetration of the microelectrode into the cell is shown along with the membrane or steady potentials registered at each position. **Below,** photomicrographs of cells penetrated by two microelectrodes. **A,** penetration of the membrane; **B,** into the cytoplasm; **C,** into the nucleus. (Courtesy of W. R. Loewenstein and Y. Kanno.)

the nucleus, there is another drop in negative potential at the nuclear membrane (− 13 mV). These results suggest that the nuclear envelope may be a diffusion barrier for ions as small as K⁺, Na⁺, or Cl⁻. In the nuclear envelope present in oöcytes, however, there is no detectable potential, thus indicating a free interchange of ions between the nucleus and the cytoplasm.[12]

There are several morphological observations that suggest there is passage of ribonucleoproteins and other macromolecules across the nuclear pores. The relationship of the pores with channels in the nucleoplasm was mentioned earlier (Fig. 12–1). Dense material extending through the nuclear pores has been observed in oöcytes of amphibia (Fig. 18–12).[13] Some of the material may correspond to ribosomal subunits, other to messenger RNA. The mechanism by which the passage of these substances is achieved is unknown. Cytochemical studies, however, have revealed the presence of an ATPase in the pores,[14] which may provide the necessary energy for the transfer of macromolecules.

Annulated Lamellae

In developing oöcytes and spermatocytes, as well as in certain embryonic and tumor cells, the so-called *annulated lamellae*

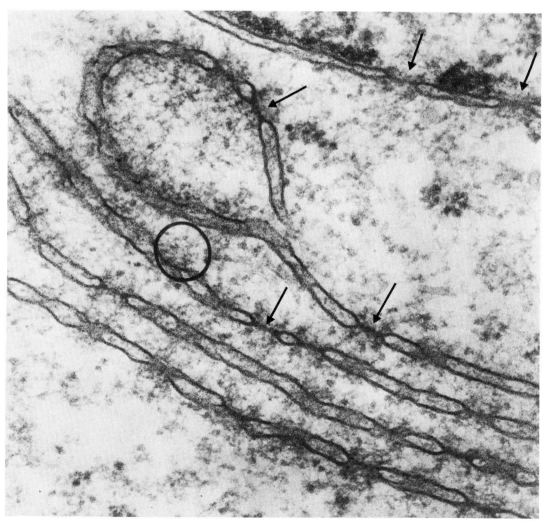

Figure 12–4. Electron micrograph of annulated lamellae observed in a human melanoma cell cultured in vitro. **Upper right,** the nucleus, with the nuclear envelope, showing two pore complexes (arrows). Similar complexes are observed in the lamellae present in the cytoplasm. ×80,000. (Courtesy of G. G. Maul.)

may be found in the cytoplasm. These membranous structures appear as stacks of cisternae or flattened sacs with an intrasacular space 10 to 20 nm wide. At regular intervals these sacs are traversed by pore complexes that are similar in morphology and structure to those found in the nuclear envelope. The pores of the cisternae have an octagonal shape and contain an annular material with the same characteristics as the pore complexes.[15] These are closely spaced and may occupy up to 50 per cent of the surface of the membrane (Fig. 12–4).

The similarity between the nuclear envelope and the annulate lamellae has led to the interpretation that the two structures are related in their morphogenesis. The lamellae have been interpreted as budding off the nuclear envelope. More exact work has demonstrated that first there is the budding of the outer membrane of the nuclear envelope into flattened sacs; later the pore complexes appear.[16] Stacks of annulated lamellae are sometimes found inside the nucleus, but their function is unknown. They are usually associated with rapidly proliferating cells. Sometimes they are associated with ribosomes and RNA in high concentration, and there is also an ATPase activity in the pore complexes of these lamellae.[13] The presence of RNA may be related not only to the ribosomes, but also to the pore complexes. In this case the lamellae could serve as a storage site of RNA in the cytoplasm. Finally, it is known that late in oögenesis these membranes may break down and disappear.

SUMMARY:
The Nuclear Envelope

The nuclear envelope, a differentiation of the vacuolar system of the cytoplasm, is composed of two membranes and a perinuclear space. Nuclear pores represent openings in this envelope at sites where the two membranes are in contact. The pores are octagonal orifices of about 60 nm in diameter. They are not, however, freely communicating openings, but are plugged by a cylinder of protein material—the so-called *annulus*. The pore and the annulus constitute a *pore complex*.

The nuclear envelope regulates the passage of ions and small molecules. By way of the pore complex, the envelope may have an important role in the transfer of macromolecules between the nucleus and the cytoplasm—and vice versa. There is an apparent upper limit of about 15 nm for the size of the macromolecules passing through the pores. The annulus may function as a kind of diaphragm covering part of the nuclear pore and may serve as a special device for selective permeability.

The nuclear envelope may determine the existence of membrane potential. The passage of macromolecules (ribosomal subunits, messenger RNA, etc.) may be seen by electron microscopy.

Some cells (oöcytes, embryonic cells and cancer cells) show *annulated lamellae* that appear to be formed by the budding off of the nuclear envelope. It has pore complexes similar to those of the nuclear envelope.

ULTRASTRUCTURE OF THE INTERPHASE NUCLEUS

Nucleoprotein components may be present either in a condensed state (*heterochromatin*) or in a more dispersed form (*euchromatin*). Both types of chromatin give a positive Feulgen reaction for DNA. In Figure 6–10, euchromatin occupies the light areas of the nucleoplasm, and the heterochromatin constitutes the heavily stained chromocenters. (Heterochromatin and the concept of a satellite DNA will be introduced in a later section.)

In addition to DNA-containing structures, the nucleus contains other components composed mainly of ribonucleoproteins. The most conspicuous one is the nucleolus, whose structure will be presented in a special section. Intermingled between the areas of chromatin there are several ribonucleoprotein structures that have been recognized with the electron microscope. By using uranyl ions and certain extraction methods, it has been possible to differentiate the DNA- from the

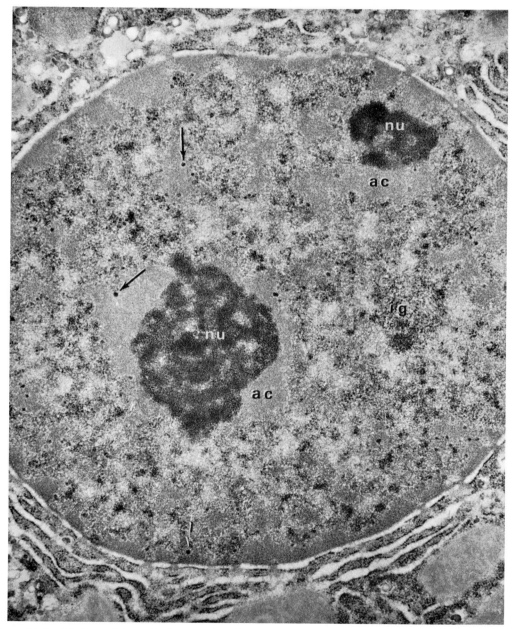

Figure 12–5. Electron micrograph of a normal rat liver nucleus stained with uranyl acetate and treated with EDTA to visualize the ribonucleoprotein components; *nu*, nucleolus; *ac*, associated chromatin; *ig*, interchromatin granules; perichromatin granules (arrows). (See the description in the text.) ×25,000. (Courtesy of W. Bernhard.)

RNA-containing regions of the interphase nucleus. (This technique will be described in greater detail later.)

Figure 2–9 shows the typical image of a nucleus stained with uranyl acetate. Both types of nucleic acid are heavily stained, but DNA is most conspicuous.[17] The placement of the chromatin in clumps near the nuclear envelope and the interchromatin channels leading to the nuclear pores are clearly visible. The mass of the nucleolus is of a lesser density and is surrounded by an associated chromatin of a greater density; the nucleoplasm is light and reveals fine granules.

In Figure 12–5 the nucleus has also been stained with uranyl acetate, but it was then treated with EDTA (ethylenediaminetetraacetic acid) which preferentially removes the uranyl ions from the DNA. EDTA is a chelating agent which binds to the phosphate groups of DNA, thus displacing the UO_2^{2+} ions. In this way, the RNA-containing structures of the nucleus and the ribosomes in the cytoplasm become more conspicuous. By comparing Figure 12–5 with Figure 2–9, the student will recognize that the former is, in a certain manner, a reverse image of the latter. In Figure 2–9 the DNA structure shows a high electron density, whereas in Figure 12–5 the DNA structures are pale and the RNA-protein components appear more dense. Other techniques involving special extractions or enzymatic treatments have also contributed to our knowledge of the fine structure of the nucleus.

In the following sections the ribonucleoprotein structures, beginning with the nucleolus, will be considered; then the fine structure of the deoxyribonucleoproteins and the chromosomes will be presented.

THE NUCLEOLUS

The presence of nucleoli, which appeared as dense granules within the nucleus, was first described by Fontana in 1781. By the end of the 19th century a relationship between the size of the nucleolus and the synthetic activity of the cell was postulated. It was found that nucleoli were small or absent in cells exhibiting little protein synthesis (sperm cells, blastomeres, muscle cells, and so forth), whereas they were large in oöcytes, neurons, and secretory cells—those in which protein synthesis is a prominent feature. In the living cell, nucleoli are highly refringent bodies (Fig. 2–4). This is the result of a large concentration of solid material, which can be measured by interference microscopy (see Chapter 5), and which may constitute 40 to 85 per cent of the dry mass.[18] The light microscope generally reveals the nucleolus as structurally homogeneous, although small corpuscles or vacuoles are sometimes noted.

The nucleolus is frequently attached to the nuclear membrane, and some of these vacuoles and material from the dense part of the nucleolus seem to pass into the cytoplasm.[19] In some living cells, particularly after fixation and silver staining, a filamentous structure called a *nucleoloneme* has been described.[20]

Cytochemistry

Nucleoli have been isolated from oöcytes of marine animals and from liver cells (Fig. 17–2); they contained 3 to 5 per cent RNA. This is less than the amount indicated by cytochemical observations, but some loss may occur during extraction.[21, 22] In nucleoli from pea embryos, the RNA content is 10 per cent, or 20 per cent of the total nuclear RNA.[23] The base composition of nucleolar RNA is very similar to that of ribosomal RNA. The protein content of the nucleolus is high, and, according to some investigators, the main protein components are phosphoproteins.[24] No histones have been found in isolated nucleoli, and the fast green staining test is negative. There is cytochemical evidence for the existence of a high concentration of orthophosphate, which may serve as a precursor of the RNA phosphorus.[25]

Little is known about the enzyme content of the nucleolus. Isolation techniques have verified the presence of acid phosphatase, nucleoside phosphorylase, and NAD^+ synthesizing enzymes. The last two enzymes are important because they are involved in nucleotide and coenzyme synthesis. RNA methylase, an enzyme that transfers methyl groups to the RNA bases, has been localized in the nucleolus of certain cells.[26]

After fixation, the nucleolus is Feulgen-negative; this indicates an absence of deoxyribonucleic acid. The nucleolus stains with pyronine and other stains and absorbs ultraviolet light at 260 nm. Treatment with ribonuclease shows that the capacity to absorb this basophilic stain and ultraviolet radiation depends on the presence of RNA.

The nucleolus may be surrounded by a ring of Feulgen-positive chromatin (Fig. 6–10), which represents heterochromatic regions of the chromosomes associated with the nucleolus. In large nucleoli some Feulgen-positive granules can be seen in portions of the chromosomes that penetrate the nucleolus. Also, after plant roots are treated with ribonuclease, the nucleoli become acidophilic — an indication that this enzyme may have penetrated into the living cell.

Fine Structure

Electron microscopy has confirmed the existence of a definite submicroscopic organization within the nucleolus. In some cells nucleoli have a compact structure; in others the structure may be more or less open and include clear regions that communicate with the nucleoplasm (Fig. 12–6).

Four characteristic components may be recognized: granular zone, fibrillar zone, matrix or amorphous zone, and nucleolar-associated chromatin.

The *granular zone* consists of electron-dense granules of 15 to 20 nm, i.e., smaller in diameter than the ribosomes. Frequently the granular zone occupies the more peripheral region of the nucleolus, which is surrounded by the associated chromatin (Fig. 12–6). Both the granular and the fibrillar zones are clearly distinguished in Figure 18–12. The *fibrillar zone* consists of fine fibers of 5 to 10 nm in diameter. It generally occupies the central region of the nucleolus. Both the granular and fibrillar zones are digested by ribonuclease. The *matrix* is an amorphous background in which the granular and fibrillar components may be suspended. It is thought that the matrix is protein in nature, since it is digested by pepsin.[27]

The nucleolar-associated chromatin is composed of fibers 10 nm in thickness situated around the nucleolus and extending into it. These fibers may occupy special areas within the nucleolus or may be spread rather diffusely. They contain DNA and are removed after treatment with deoxyribonuclease.

Results of studies with radioautography and ultracentrifugation, described in Chapter 18, are consistent with the view that the fibrillar portion is a precursor of the granular portion, and that both contain ribonucleoproteins that are precursors of the cytoplasmic ribosomes.

Experiments performed on cultured cells exposed to a supramaximal temperature (43° C for one hour) have demonstrated that the nucleolus is thermosensitive. The most noticeable change is a loss of the granular portion and of the intranucleolar chromatin. The heat sensitivity probably involves the DNA-dependent RNA synthesis.[28] A similar inhibition of the DNA → RNA transcription is produced with small doses of actinomycin D,[29] but in this case a separation of the *granular* and *fibrillar* ribonucleoproteins is induced. Several antimetabolites have been found to produce a similar separation of the nucleolar components.

It is now well established that the nucleolus is related to the biogenesis of the cytoplasmic ribosomes (see Chapter 18). In nucleoli having an open structure, the less dense parts should be considered nucleoplasm pervading the nucleolar mass.

Evidence of the passage of nucleolar material into the cytoplasm has been observed in fixed cells, as well as in living cells. This migration is particularly evident in amphibian oöcytes, in which nearly 1000 nucleoli gather at the periphery of the nucleus after the pachynemic stage. A material, probably of nucleolar origin, may be seen passing through the pores of the nuclear envelope into the cytoplasm (Fig. 18–12).[30]

Nucleolar Cycle. During mitosis nucleoli undergo cyclic changes. Early studies revealed that (1) nucleoli seem to disappear at the beginning of cell division (prophase) and (2) reappear at the end of division (telophase). The relationship between the nucleolar and chromosomal cycles has been clarified, in part, by the demonstration in plant cells that nucleoli are intimately related to certain chromosomes. Each nucleolus is in contact with a pair of chromosomes; the point of union is a special region called the *nucleolar organizer* (see Chapter 2). At telophase the nucleolar substance may originate from the fusion of small "prenucleolar bodies," which are collected in relation to the nucleolar organizer.[31]

In meristematic cells of *Allium cepa* it has been observed that the nucleolus contains

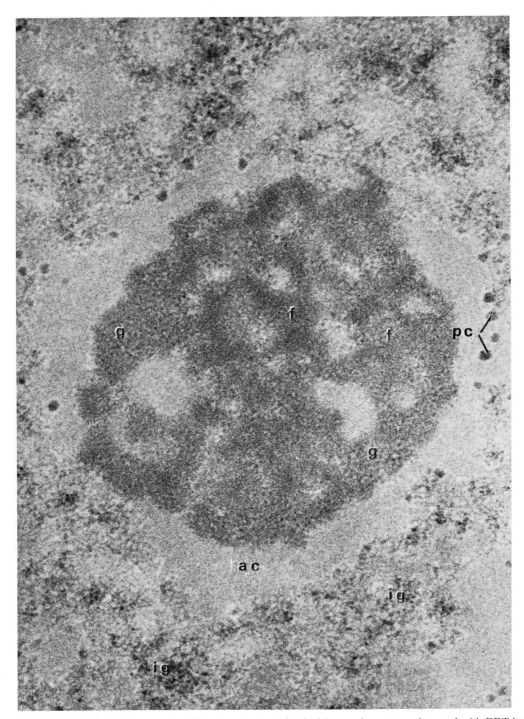

Figure 12–6. Nucleolus from a normal rat hepatocyte stained with uranyl acetate and treated with EDTA which bleaches the DNA. In the nucleolus the fibrillar (*f*) and granular (*g*) portions and the associated chromatin (*ac*) are seen; *ig*, interchromatin granules; *pc*, perichromatin granules. ×60,000. (Courtesy of W. Bernhard.)

a loop of chromatin (DNA) which extends from the nucleolar chromosome. During late prophase this chromatin loop gradually retracts and coils inside the nucleolar zone of the corresponding chromosome. Simultaneously, there is a gradual dispersion of the fibrillar and granular components of the nucleolus into the nucleoplasm. During telophase, nucleolar reconstitution follows two steps: at early and middle telophase, the emerging nucleolus consists of a convoluted chromatin loop which uncoils from the nucleolar organizer and becomes surrounded by fibrillar and granular material; at late telophase, it exhibits all the characteristics of a mature interphase nucleolus. These findings suggest that the only permanent component of the nucleolus is the chromatin loop which contains the genetic information for the synthesis of the nucleolar material (Chouinard, 1971).

Ribonucleoprotein Structures

With the techniques mentioned above (see Fig. 12–5) other nucleoprotein-containing structures—besides the nucleolus—have been recognized. These include: the perichromatin granules, the interchromatin granules, the perichromatin fibers, and the so-called coiled bodies. The *perichromatin granules* are electron-dense granules of about 40 to 45 nm, surrounded by a clear halo and usually present along the edge of masses of condensed chromatin (Fig. 12–6). The number of these granules varies with the physiological state of the cell. For example, they increase in number with the administration of small doses of actinomycin D, which interferes with the synthesis of ribosomal RNA. The *interchromatin granules* are much smaller (i.e., 20 to 25 nm) and more widely distributed, forming clusters in the less-condensed regions of chromatin (Figs. 12–5 and 12–6).[17] The *perichromatin fibers* are found at the edges of masses of chromatin. They may represent a fibrous RNA product of the chromosome that is readily attacked by ribonuclease. These RNA-containing fibers may be precursors of the perichromatin and interchromatin granules, which are more resistant to ribonuclease—probably because of the presence of a higher content of protein. The *coiled bodies* are composed of rather coarse masses of fiber and are sparsely distributed in certain nuclei.

Fine Structure of the Deoxyribonucleoproteins

The electron microscopic study of deoxyribonucleoproteins in the interphase nucleus has made little progress because of technical difficulties. For example, it is difficult to interpret the three-dimensional, arrangement of the nucleoproteins when they are observed in a thin two-dimensional section. In a higher cell the DNA molecule may be very long—sometimes as long as several centimeters in a chromosome. It is understandable that this must be tightly folded and packed to occupy the space of a few microns. Furthermore, in contrast to the prokaryons, the DNA is always associated with histones and with non-histone proteins (see Chapter 17).

In the literature it is possible to read reports in which *microfibrils,* ranging from 3.0 to 17 nm, have been found in the areas occupied by chromatin. The finest microfibrils have been postulated to represent single nucleoprotein molecules.[32, 33]

In general, the dimensions of the microfibrils observed in sections of interphase nuclei are much greater than those expected for a single nucleohistone complex, which should be about 3.0 nm thick. An explanation for this discrepancy may be the presence of a superstructure in the way of a "supercoil" of the original molecule. An alternative explanation may be that each microfibril is surrounded by a matrix of non-histone protein (Wolfe, 1972). In studies of the spermatids of locust, fine microfibrils of about 5.0 nm were observed. These became oriented away from the centriolar region along the axis of the future spermatozoon head. This orientation was accompanied by thickening of the microfibrils up to 15 nm.[32]

A Folded Fiber Model of Chromosome Structure

The problems involved in the sectioning of nuclei and chromosomes were partially solved with the introduction of a spreading technique somewhat similar to that used for viral and bacterial DNA (see Fig. 5–9). Interphase nuclei and metaphase chromosomes from different sources were spread on in an air-water interphase.[34–37] The forces of surface tension tend to separate the ground substance from the chromosome microfibrils. To avoid disruption by drying of the speci-

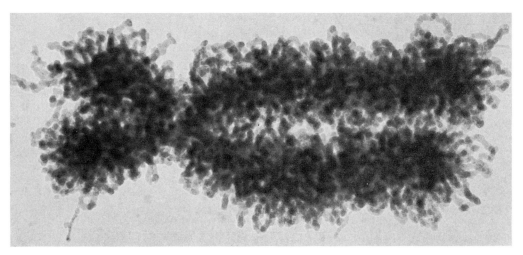

Figure 12–7. Electron micrograph of a whole mounted human chromosome 12 showing the two chromatids composed of fibrils 30 nm thick. (Courtesy of E. J. DuPraw.)

men, Gall[34] used the *critical point method,* in which the water content was replaced with liquid CO_2 in a pressure chamber; then, by heating above the *critical temperature* (i.e.,

31° C), the liquid CO_2 was converted into gas without a change in volume.

As shown in Figure 12–7, the excellent preservation obtained permits the recog-

Figure 12–8. Diagram of the folded fiber model of chromatin in the interphase chromosomes (**A** and **B**) and in the metaphase chromosome (**C**). (See the description in the text.) (Courtesy of E. J. DuPraw.)

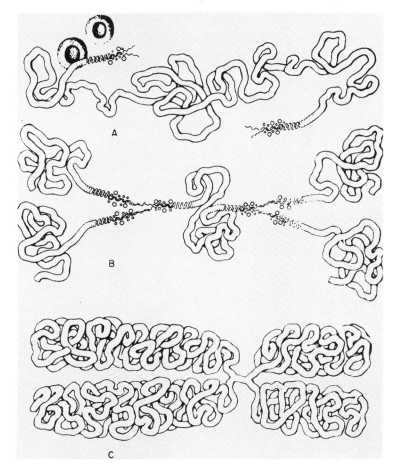

nition of a particular chromosome, with its two chromatids and the primary constriction, in the human karyotype. The bulk of the chromosome appears to be composed of a tightly folded fiber which has a rather homogeneous diameter of 20 to 30 nm.[38] This folded fiber is supposed to contain the DNA histone helix in a supercoiled condition (Fig. 12–8). A "folded fiber model" of chromosome structure has been proposed in which the body of the *chromatid* of classic cytology is represented by a single DNA-protein fiber, first coiled to form the 25 to 30 nm fiber, and then folded back longitudinally and transversely (Fig. 12–7). The folded fiber model applies to both the interphase and the metaphase chromosome (Fig. 12–8). In the latter case, however, it is thought that the two sister chromatids are held at the centromere until anaphase by an unreplicated fiber segment of DNA.

Some of the evidence for the folded fiber model with a single DNA molecule per chromosome stems from the discovery of extremely long segments of DNA in the interphase nucleus. For example, in sea urchin sperm, molecules 50 to 90 μm were observed under the electron microscope,[39] and continuous DNA molecules 1 to 22 mm long have been detected in various mammalian nuclei by radioautography.[40] It has been calculated that the length of the DNA contained in a single chromosome of the first pair from the human karyotype is 7.3 cm. This could be coiled into a 23 nm thick fiber of 1306 μm per chromatid. This figure, compared with the length of the metaphase chromosome, indicates that at this stage a packing ratio of 122:1 must be achieved by the folding of the interphase fiber. Interestingly, in the head of a bacteriophage the DNA is packed at a ratio of 520:1.[38]

Heterochromatin and Satellite DNA

In 1928 Heitz defined *heterochromatin* as those regions of the chromosome that remain condensed during interphase and early prophase and form the so-called *chromocenters* or false nucleoli. The rest of the chromosome, which remains in a non-condensed state, was called *euchromatin* (Gr., *eu*, true). Frequently there is a mass of *heterochromatin* in close contact with the nucleolus, cor-

responding to the *nucleolar organizer*. During mitosis the heterochromatic regions appear localized at the centromeres, at the telomeres, or are intercalated in other parts of the chromosomes. Such regions may stain more strongly or weakly than the euchromatic regions, showing what is called a *positive or a negative heteropyknosis* (Gr., *hetero–* + *pyknosis*, different staining).

Two types of heterochromatin are generally recognized: *facultative* and *constitutive*.[41] In the case of *facultative heterochromatin* one chromosome of the pair becomes either totally or partially heterochromatic. The best known case is that of the X chromosomes in the mammalian female, one of which is active and remains euchromatic, whereas the other is inactive and forms the sex-chromatin, or Barr body, at interphase. This type of heterochromatin is acquired during embryogenesis; in the human, the sex chromatin appears in the embryo between the sixteenth and eighteenth days. Before that time, both X chromosomes are euchromatic (see Chapter 16).

Constitutive heterochromatin is the most common type of heterochromatin. It may be found in the centromeric region, in the telomeres, or as bands in other parts of the chromosome (Fig. 12–9). A general characteristic of heterochromatin is the late replication of its DNA, as can be demonstrated by autoradiography with ^{3}H-thymidine (see Chapter 17). For many years these regions of the chromosome were considered to be devoid of genetic activity. In recent times, however, important genes are recognized to be present in heterochromatin. For example, the polygenes that codify for the ribosomal RNA in the nucleolar organizer and those making the 5S and the transfer RNA are localized in heterochromatic regions (see Chapters 18 and 19). Heterochromatin is also related to the *repetitious or redundant* DNA, also called satellite DNA.

From erythrocytes or liver cells of salamanders it is possible to extract the DNA and then, by ultracentrifugation on a gradient of cesium chloride, to separate it into two components. Most of the DNA appears in a band at a buoyant density of 1.705, whereas a small peak—the satellite—containing only about 2 per cent of the total DNA, has a density of 1.728.[42] The higher density of this satellite DNA is due to the higher content

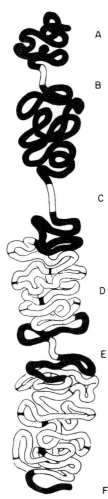

Figure 12–9. Model of constitutive heterochromatin in a mammalian metaphase chromosome. **A,** constitutive heterochromatin; **B,** secondary constriction I or nucleolar organizer; **C,** primary constriction or centromere; **D,** euchromatin; **E,** secondary constriction II, possible site of 5S rRNA cistrons; **F,** telomere. (From Yunis, J. J., and Yasmineh, W. G., *in* DuPraw, E. J. (Ed.), *Advances in Cell and Molecular Biology,* Vol. 2, p. 1, New York, Academic Press, Inc., 1972.)

regions of all the chromosomes during the period of meiosis and can be detected by autoradiography (Fig. 12–11). This finding demonstrates that the satellite DNA is present in the centromeric heterochromatin.

In recent years other techniques to detect constitutive heterochromatin have been developed. These techniques represent a powerful tool to characterize the various chromosomes within the karyotype (see Chapter 16). For example, after heating the section or after treating it with NaOH to denature the DNA, the Giemsa stain, so much used in hematology, demonstrates centromeric and other types of constitutive heterochromatin (Fig. 16–11).[44] Acridine derivatives, particularly the quinacrine mustards, are also useful to detect the heterochromatin regions under the fluorescence microscope and to observe the banding pattern, which is similar to that which develops with the Giemsa stain (Fig. 16–9).[45] The antimalarial compound quinacrine dihydrochloride (Atebrine) stains rather specifically the heterochromatin of the Y chromosome of the human and of large apes.[46] With this stain the Y chromosome can be detected as a small, highly fluorescent body during interphase in lymphocytes and even in human spermatozoa (Fig. 16–10).[47, 48] It has been postulated that

(about 68 per cent) of guanine + cytosine in the molecule (Fig. 12–10). (Note that this satellite DNA does not correspond to the ribosomal DNA present in the nucleolar organizer [Fig. 18–7].) By using the satellite DNA as template for an in vitro synthesis, it is possible to obtain a ³H-RNA complementary to the satellite DNA. This ³H-RNA can be used to hybridize the satellite DNA peak not only in the test tube, but also on a slide directly on the cells that contain this type of DNA.[43]

This hybridization, also called *annealing,* is specifically localized at the centromeric

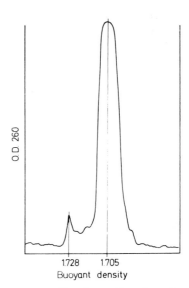

Figure 12–10. Density gradient in cesium chloride of the DNA extracted from cells of the salamander. Observe the main peak of DNA with a buoyant density of 1.705 and the small satellite peak of higher density (1.728). (From MacGregor, H. C., and Kezer, J., *Chromosoma, 33*:167–182, 1971.)

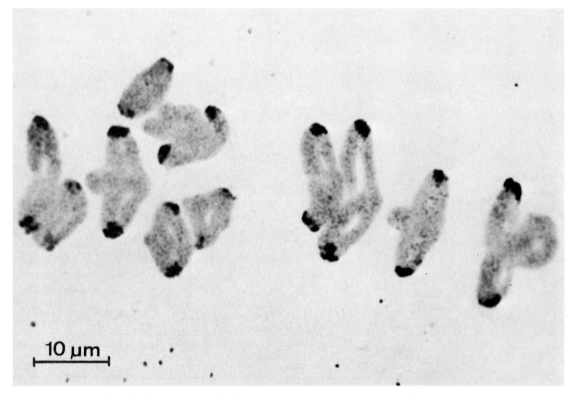

Figure 12–11. Meiotic chromosomes of salamander processed for in situ hybridization. Incubation with tritiated RNA complementary to the heavy satellite DNA shown in Figure 12–10. Observe that the labeling has occurred in the centromeric region of the chromosomes. (Courtesy of H. C. MacGregor.)

quinacrine stains the adenine-thymine rich portions of the DNA more intensely.[49]

In plant cells there are constitutive heterochromatin and large amounts of repetitive DNA; however, there is no facultative heterochromatin.[50] The recent progress made on the study of heterochromatin and chromosome banding is having a tremendous impact on human cytogenetics (see Chapter 16). Furthermore, it is providing renewed interest in the study of problems of taxonomy, phylogenesis, and the evolution of species.

Function of Heterochromatin

Several possible roles have been suggested for constitutive heterochromatin and satellite DNA. Some of these have been mentioned earlier (i.e., nucleolar organizer, polygenes for tRNA and 5S, etc). Centromeric heterochromatin may be involved in the separation of chromosomes during cell division. It has been postulated that oncogenic viruses (i.e., viruses capable of inducing cancer) may be inserted in the heterochromatic regions. Because of the repetitive nature, these regions could also be less susceptible to mutagens (substances that induce mutation). Note that in prokaryons there is no need for satellite DNA and constitutive heterochromatin. Apparently, this type of DNA appears with evolution in the higher cells—probably to compartmentalize the genome into several functional parts, as for example, the nucleolus, the centromere, and the multiple initiation points of DNA replication. Some of the repeated sequences are kept as spacers between genetically active regions. Within the euchromatin these sequences may serve as transcriptional stops, initiation points for RNA polymerase, or sites of attachment to the nuclear membrane. The sequences may also be used in pairing of homologous chromosomes and crossing over sites in meiosis. (For more details see Yunis and Yasmineh, 1971.)

The interphase nucleus contains DNA- and RNA-containing structures in which the nucleic acids form complexes with proteins. These two types of structures may be recognized under the electron microscope by staining with uranyl acetate. Treatment with EDTA removes the UO_2^{2+} from the DNA, thereby leaving the ribonucleoproteins more heavily stained (Fig. 12–5).

The principal RNA-containing structure is the *nucleolus*. This is a highly refringent structure with a large concentration of solid material, in particular, proteins (phosphoproteins). The RNA content ranges from 3 to 10 per cent. The nucleolus is Feulgen-negative, but is surrounded by a ring of heterochromatin which may penetrate into its structure. The fine structure of the nucleolus reveals the presence of four components: the *fibrillar and granular zones*, the *matrix*, and the *nucleolar-associated chromatin*. The fibrillar and granular zones are composed of ribonucleoproteins and, as will be shown in Chapter 18, are related to the biogenesis of the nucleolus. The matrix is composed of protein, and the associated chromatin is related to the nucleolar organizer — that portion of the chromosome that contains the genetic information to synthesize the other components of the nucleolus. Physiological changes of these various components may be observed, and are particularly striking during mitosis.

Other *ribonucleoprotein structures* are the *perichromatin* granules (40 to 45 nm), the *interchromatin* granules (20 to 25 nm), the *perichromatin fibers,* and the *coiled bodies.*

The fine structure of the DNA-containing structures is difficult to assess in thin sections. Better results have been obtained by the use of a spreading technique followed by the *critical point drying method* (Fig. 12–7).

The most general conclusion derived from these studies is that interphase nuclei and metaphase chromosomes contain fibrils of about similar size. The average diameter of the metaphase fibrils has been found to exceed 30 nm (Fig. 12–7), whereas those of the interphase chromatin were found to be in the range of 23 to 25 nm. Thinner microfibrils have been seen in chromosomes of spermatocytes and in oöcytes that were actively engaged in RNA synthesis. The metaphase chromosome is composed of a rather thick fiber that is tightly folded and packed. The thickness of this fiber is much greater than the DNA-histone molecule (2 to 3 nm). Although the large fiber is probably the result of the supercoiling of the DNA-histone molecule, some contributions from a matrix of non-histone material cannot be disregarded.

Heterochromatin represents condensed regions of the chromosome that are late in replicating. *Facultative* heterochromatin corresponds to the X chromosome that becomes genetically inert during embryogenesis. *Constitutive* heterochromatin is found in the centromeric regions and the

SUMMARY:
Organization of the Interphase Nucleus

telomeres and is sometimes intercalated as bands in the chromosomes. Part of this chromatin corresponds to repeated sequences of DNA, which, when isolated, represent the so-called *heavy satellite DNA*. Centromeric heterochromatin may be stained with Giemsa stain or with fluorescent dyes, and may be labeled with ^{3}H-RNA complementary to satellite DNA. Although most of the heterochromatin may be genetically inert, it contains certain polygenes, such as ribosomal DNA, transfer DNA, and 5S DNA; it probably plays an important role in replication and transcription of the DNA.

REFERENCES

1. Fauré-Fremiet, E., and Roullier, C. (1958) *Exp. Cell Res., 14*:29.
2. Mirsky, A. E., and Osawa, S. (1961) The interphase nucleus. In: *The Cell,* Vol. 2, p. 677. (Brachet, J., and Mirsky, A. E., eds.) Academic Press, New York.
3. Barer, R., Joseph, S., and Merck, G. A. (1959) *Exp. Cell Res., 18*:179.
4. Gall, J. G. (1956) *Brookhaven Symposia in Biology, 8*:17.
5. Wischnitzer, S. (1958) *J. Ultrastruct. Res., 1*:201.
6. Watson, M. L. (1959) *J. Biophys. Biochem. Cytol., 6*:147.
7. Gall, J. G. (1967) *J. Cell Biol., 32*:391.
8. Fawcett, D. W. (1966) *Amer. J. Anat., 119*:129.
9. Feldherr, C. M. (1965) *Exp. Cell Res., 38*:670.
10. Mauld, G. G. (1971) *J. Cell Biol., 51*:558.
11. Loewenstein, W. R., and Kanno, Y. (1962) *Nature* (London), *195*:462.
12. Loewenstein, W. R., Kanno, Y., and Ito, S. (1966) *Ann. N.Y. Acad. Sci., 137*:708.
13. Stevens, B. J., and Swift, H. (1966) *J. Cell Biol., 31*:55.
14. Scheer, U., and Franke, W. W. (1969) *J. Cell Biol., 42*:519.
15. Maul, G. G. (1970) *J. Cell Biol., 46*:604.
16. Kessel, R. G. (1968) *J. Ultrastruct, Res.,* Suppl 10:1–82.
17. Bernhard, W. (1968) *Excerpta Medica International Congress Series, 166*:20.
18. Vincent, W. S. (1955) *Int. Rev. Cytol., 4*:269.
19. González Ramirez, J. (1963) Considerations on nucleolar physiology. In: *Cinemicrography in Cell Biology,* p. 429. (Rose, G. G., ed.) Academic Press, Inc., New York.
20. Estable, C., and Sotelo, J. R. (1950) *Inst. C. Biol.,* Montevideo, *1*:105.
21. Birstiel, M. L., and Chipchase, M. I. H. (1963) *Fed. Proc., 22*:473.
22. Maggio, R., Siekevitz, P., and Palade, G. E. (1963) *J. Cell Biol., 18*:267, 293.
23. Stern, H., Johnston, F., and Seeterfield, G. (1959) *J. Biophys. Biochem. Cytol., 6*:57.
24. Vincent, W. S. (1952) *Proc. Natl. Acad. Sci. U.S.A., 38*:139.
25. Tandler, C. J., and Sirlin, J. L. (1962) *Biochim. Biophys. Acta, 55*:228.
26. Sirlin, J. L., Jacob, J., and Tandler, C. J. (1963) *Biochem. J., 89*:447.
27. Marinozzi, V., and Bernhard, W. (1963) *Exp. Cell Res., 32*:595.
28. Simard, R., and Bernhard, W. (1967) *J. Cell Biol., 34*:61.
29. Perry, R. P. (1964) *Natl. Cancer Inst. Monogr., 14*:73.
30. Miller, O. L. (1962) In: *Fifth International Congress for Electron Microscopy,* Vol. 2, p. NN-8. (Breese, S. S. Jr., ed.) Academic Press, Inc., New York.
31. Lafontaine, J. G., and Chouinard, L. A. (1963) *J. Cell Biol., 17*:167.
32. De Robertis, E. (1964) *Natl. Cancer Inst. Monogr., 14*:33.
33. Sotelo, J. R., and Wettstein, R. (1965) *Natl. Cancer Inst. Monogr., 18*:133.
34. Gall, J. G. (1963) *Science, 139*:120.
35. DuPraw, E. J. (1965) *Proc. Natl. Acad. Sci. U.S.A., 53*:161.
36. Gall, J. G. (1966) *Chromosoma, 20*:221.
37. Wolfe, S. L. (1965) *J. Ultrastruct. Res., 12*:104.
38. DuPraw, E. J. (1968) *Cell and Molecular Biology.* Academic Press, Inc., New York.

39. Solari, A. J. (1965) *Proc. Natl. Acad. Sci. U.S.A., 53*:503.
40. Sasaki, M. S., and Norman, A. (1966) *Exp. Cell Res., 44*:642.
41. Brown, S. W. (1966) *Science, 151*:417.
42. MacGregor, H. C., and Kezer, J. (1971) *Chromosoma, 33*:167.
43. Pardue, M. L., and Gall, J. G. (1970) *Science, 168*:1356.
44. Schnell, W. (1972) *Chromosoma, 38*:319.
45. Caspersson, T., Zech, L., Johanson, C., Lindsten, J., and Hultén, M. (1970) *Exp. Cell Res., 61*:473.
46. Pearson, P. L., Bobrow, M., Vosa, C. G., and Barlow, P. W. (1971) *Nature* (London), *231*:326.
47. George, K. P. (1970) *Nature* (London), *226*:80.
48. Barlow, P., and Vosa, C. G. (1970) *Nature* (London) *226*:961.
49. Weisblum, B., and de Haseth, P. L. (1972) *Proc. Natl. Acad. Sci. USA, 69*:629.
50. Britten, R. J., and Kohne, D. E. (1968) *Science, 161*:529.

Birnstiel, M. (1967) The nucleolus in cell metabolism. *Ann. Rev. Plant Physiol., 18*:25.
DuPraw, E. J. (1968) *Cell and Molecular Biology.* Academic Press, Inc., New York.
Engelhardt, P., and Pusa, K. (1972) Nuclear pore complexes. *Nature [New Biol]., 240*:163.
Feldher, C. M. (1971) Structure and function of the nuclear envelope: nucleocytoplasmic exchanges. In: *Advances in Cytopharmacology,* Vol. 1, p. 89, Raven Press, New York.
Feldher, C. M. (1972) Structure and function of the nuclear envelope. In: *Advances in Cell and Molecular Biology, 2*:273.
Lima-de-Faría, A., ed. (1969) *Handbook of Molecular Cytology.* North-Holland Publishing Co., Amsterdam.
Mirsky, A. E., and Osawa, S. (1961) The interphase nucleus. In: *The Cell,* Vol. 2. (Brachet, J., and Mirsky, A. E., eds.) Academic Press, Inc., New York.
Monneron, A., and Bernhard, W. (1969) Fine structural organization of the interphase nucleus in some mammalian cells. *J. Ultrastruct. Res., 27*:266.
Vincent, W. S., and Miller, O. L., eds. (1966) The nucleolus: its structure and function. *Natl. Cancer Inst. Monogr., 23.*
Wagner, R. P., ed. (1969) Nuclear physiology and differentiation. *Genetics, 61*:1.
Wischnitzer, S. (1973) The submicroscopic morphology of the interphase nucleus. *Int. Rev. of Cytology, 34*:1.
Wolfe, S. L. (1972) *Biology of the Cell.* Wadsworth Publishing Co., Belmont, California.
Yunis, J. J., and Yasmineh, W. G. (1971) Heterochromatin, satellite DNA, and cell function. *Science, 174*:1200.
Yunis, J. J., and Yasmineh, W. G. (1972) Model of mammalian constitutive heterochromatin. In *Advances in Cell and Molecular Biology,* Vol. 2. (DuPraw, E. J., ed.) Academic Press, Inc., New York.

ADDITIONAL READING

thirteen

MITOSIS

Cell division is the complex phenomenon by which cellular material is divided equally between daughter cells. This process is the final, and microscopically visible, phase of an underlying change that has occurred at molecular and biochemical levels. Before the cell divides by mitosis, its fundamental components have duplicated—particularly those involved in hereditary transmission. In this respect, cell division, or mitosis, can be considered as the final separation of the already duplicated macromolecular units.

The essential features of cell division by mitosis were considered in Chapter 2. Here a detailed cytologic analysis of the process will be made, followed by a consideration of the fine structure of the chromosomes and the mitotic apparatus during the mitotic cycle, as well as the function of these structures in cell division.

The molecular mechanisms leading to the duplication DNA, the transcription of RNA molecules, and the synthesis of proteins during the cell cycle will be considered in Chapter 17.

MITOSIS

Figure 13–1 is a general diagram of the different stages of mitosis. These are considered as phases of a cycle that begins at the end of the intermitotic period *(interphase)* and ends at the beginning of a new interphase. The main divisions of this cycle are: *prophase, metaphase, anaphase,* and *telophase. Cytokinesis,* a process of separation of the two cytoplasmic territories, is simultaneous with anaphase, telophase, or a later stage. The time of occurrence of this process depends on

the cell type; this process is quite different in animal and plant cells. The various phases of mitosis will be described in a way that should give an idea of the sequence of events that occur in the nucleus and the cytoplasm. Some of the underlying processes will be considered in terms of their fine structure, biochemistry, and physiological significance in the second part of the chapter.

Prophase

The beginning of *prophase* (Gr., *pro,* before) is indicated by the appearance of the chromosomes as thin threads inside the nucleus. In fact, the word "mitosis" (Gr., *mitos,* thread) is an expression of this phenomenon, which becomes more evident as the chromosomes start to condense. The condensation occurs by a process of coiling or folding and is simultaneous with a change in the blocks of heterochromatin (i.e., chromocenters) which causes them to be less apparent (Fig. 13–1, *A*). At the same time the cell becomes spheroid, more refractile, and viscous.

Each prophase chromosome is composed of two coiled filaments, the *chromatids,* which are a result of the replication of the chromosome during the S period. As the prophase progresses, the chromatids become shorter and thicker, and the primary constrictions, which contain the centromeres or kinetochores, become clearly visible (Fig. 13–1, *B*). During early prophase, the chromosomes are evenly distributed in the nuclear cavity; as prophase progresses, the chromosomes approach the nuclear envelope, causing the central space of the nucleus to become empty. The centrifugal movement of the chromosomes indicates that the disintegration of the

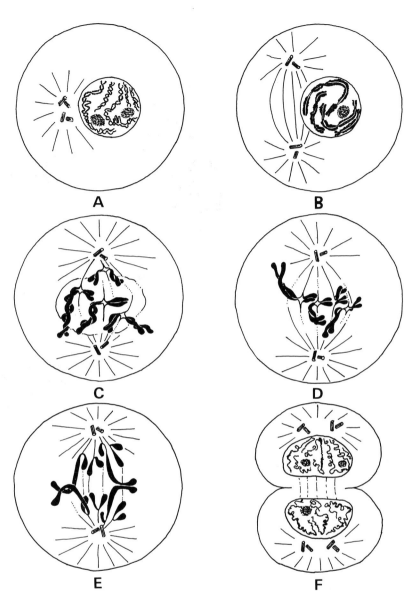

Figure 13-1. General diagram of mitosis. **A,** *prophase,* the nucleoli and chromosomes, shown as thin threads; in the cytoplasm the aster with the pairs of centrioles are shown. **B,** *prophase,* a more advanced stage of this phase in which the chromosomes have shortened. The primary constriction with the centromere is shown; in the cytoplasm the spindle is formed between the asters. **C,** *late Prophase or Prometaphase,* the nuclear envelope disintegrates and the chromosomes become attached to the spindle fibers. **D,** *metaphase,* the chromosomes are arranged along the equatorial plane. **E,** *anaphase,* the daughter chromosomes, preceded by the centromeres, are moving toward the poles. **F,** *telophase,* the daughter nuclei are in the process of reconstitution; cell cleavage has started. Observe that the centrioles have duplicated.

nuclear envelope is approaching, and with it, the end of prophase. At this time each chromosome appears to be composed of two cylindrical, parallel elements that are in close proximity. Not only the primary constrictions, but also the secondary constrictions along some chromosomes may be observed (Fig. 13-1, *B*). Other changes taking place within the nucleus are the reduction in size of the nucleoli, and their disintegration within the nucleoplasm. At the end of prophase, with the rapid fragmentation and disappearance of the nuclear envelope, the nucleolar material is released into the cytoplasm.

While these processes are taking place in the nucleus, in the cytoplasm the most conspicuous change is the formation of the spindle which, essentially, is made of microtubules (see section on fine structure of the mitotic apparatus). In the diagram of Figure 13-1 the appearance of the spindle is related to the existence of centrioles. In early prophase there are two pairs of centrioles, each one surrounded by the so-called *aster,* composed of short microtubules that radiate in all directions. (The name "aster" refers to the starlike aspect of this structure.) The two pairs of centrioles migrate along with the asters, describing a circular path toward the poles, while the spindle lengthens between. The migration of the asters continues until they become situated in antipodal positions (Fig. 13–1, C).

Centrioles replicate during interphase, generally during the S period. At the beginning of prophase there is a single aster surrounding the two pairs of centrioles. One of the pairs remains in position, with half the original aster, while the other pair, along with the other half aster, migrates about 180° around the periphery of the nucleus to reach the opposite pole.

Metaphase

Sometimes the transition between prophase and metaphase is called *prometaphase* (Gr., *meta*, between). This is a very short period in which the nuclear envelope disintegrates and the chromosomes are in apparent disorder (Fig. 13–1, C). After that the spindle fibers invade the central area and their microtubules extend between the poles. The chromosomes become attached by the kinetochores to some of the spindle fibers, and they undergo oscillatory movements until they become radially oriented in the equatorial plane and form the *equatorial plate* (Fig. 13–1, D). (In plant cells the chromosomes occupy the entire surface of the equatorial plane of the spindle.) If small chromosomes are in the group, they are commonly situated toward the interior; the larger ones are customarily found at the periphery. Those fibers of the spindle that connect to the chromosomes are called the *chromosomal fibers;* those that extend without interruption from one pole to the other are the *continuous fibers.* The connection between the centromere and the chromosomal fiber is faintly resolved with the optical microscope; however, each fiber is formed by 4 to 10 microtubules attached to the kinetochore.

Mitosis in which the spindle has centrioles and asters is called *astral* or *amphiastral* and is found in animal cells and some lower plants. Mitosis in which centrioles and asters are absent is called *anastral* and is found in higher plants, including all angiosperms and most gymnosperms. Centrioles and asters are not indispensable to the formation of the spindle. In a certain way, in *astral mitosis* the formation of the spindle is a mechanism which

leads to the distribution of the centrioles between the two daughter cells. (The fine structure of the centriole will be presented in Chapter 22.)

Anaphase

At anaphase (Gr., *ana*, back) the equilibrium of forces that characterizes metaphase is broken by the separation of the centromeres —a process which is carried out simultaneously in all the chromosomes. The centromeres move apart, and the chromatids separate and begin their migration toward the poles (Fig. 13–1, E). The centromere always leads the rest of the chromatid, or *daughter chromosome,* as if this is being pulled by the chromosomal fibers of the spindle. The chromosome may assume the shape of a **V** with equal arms, if it is metacentric, or with unequal arms, if it is submetacentric (Fig. 2–11). During anaphase the microtubules of the chromosomal fibers of the spindle shorten one third to one fifth of the original length. Simultaneously, the microtubules of the continuous fibers increase in length. Some of these stretched spindle fibers now constitute the so-called *interzonal fibers.*

Telophase

The end of the polar migration of the daughter chromosomes marks the beginning of telophase (Gr., *telo*, end). The chromosomes start to uncoil, and become less and less condensed by a process that, in some ways, recapitulates prophase, but in reverse direction. At the same time, the chromosomes gather into masses of chromatin which become surrounded by discontinuous segments of nuclear envelope. Such segments fuse to make the two complete nuclear envelopes of the daughter nuclei. During the final stages the nucleoli reappear at the sites of the nucleolar organizers.

Simultaneous with the uncoiling of the chromosomes, and the formation of the nuclear envelope, *cytokinesis* occurs. This is the process of segmentation and separation of the cytoplasm. In animal cells the cytoplasm constricts in the equatorial region, and this constriction is accentuated and deepened until the cell divides (Fig. 13–1, F). (This process can be followed in living cells with the phase microscope.) In cells of higher animals the period of cytokinesis is marked by active movements at the cell surface that are best described as "bubbling." This typical movement can be induced in non-dividing cultured cells by adding substances that bind divalent cations (e.g., Ca^{2+}). Some investigators suggest that bubbling may reflect the activity of a rapidly expanding membrane.[1] In ameboid cells at telophase both daughter cells have active movements, which appear

to pull them apart. This is best observed in films of dividing cells. The high viscosity of the cytoplasm, characteristic of metaphase and anaphase, decreases during telophase. In most cases asters become reduced and tend to disappear. During cytokinesis the cytoplasmic components are distributed, including the mitochondria and the Golgi complex. Cytokinesis in plant cells will be considered later on in this chapter.

Mitosis is a mechanism by which the cell distributes, in equivalent amounts, the different components that have been duplicated during the interphase. Prophase, metaphase, anaphase, and telophase are characterized by morphological changes that take place in the nucleus and the cytoplasm (Fig. 13–1).

In *prophase* chromosomes appear as thin threads that condense by coiling and folding. Each chromosome contains two *chromatids* which will be the future *daughter chromosomes*. With condensation each chromatid shows the *centromere* or *kinetochore*. The nucleolus tends to disintegrate and disappears at the end of prophase. In the cytoplasm the spindle is formed between the asters (and centrioles) that move toward the poles. Centrioles replicate at interphase during the S period (see Chapter 22).

At the beginning of *metaphase* (prometaphase) the nuclear envelope disintegrates and there is mixing of the nucleoplasm with the cytoplasm. Chromosomes become attached to the microtubules of the spindle and are oriented at the *equatorial* plate. The spindle has continuous microtubules and chromosomal ones. Animal cells have the spindle shown in Figure 13–1 (*astral mitosis*). In plant cells centrioles and asters are absent (*anastral mitosis*).

In *anaphase* the daughter chromosomes, led by the centromere, move toward the poles. The spindle fibers shorten one third to one fifth the original length.

In *telophase* chromosomes again uncoil; the nuclear envelope is reformed from the endoplasmic reticulum; and the nucleolus reappears.

Cytokinesis is the process of separation of the cytoplasm. In animal cells there is a constriction at the equator that finally results in the separation of the daughter cells.

CONDENSATION CYCLE OF CHROMOSOMES

Under the optical microscope during metaphase and anaphase chromosomes generally appear as compact structures (Fig. 13–1). During prophase and late telophase chromosomes become less condensed, and a coiled filament may be visible. First observed by Baranetzky in 1880 in the pollen mother cells of *Tradescantia*, this filament was called *chromonema* (Gr., *nema*, thread) by Vejdosky in 1912. During both mitosis and meiosis, a coiling cycle of the chromonema in which the length of the chromatid becomes progressively reduced between prophase and early anaphase, has been described. At the same time, the number of coils that can be detected with the optical microscope becomes progressively reduced (Table 13-1).

In recent years the classic concept of chromosomal coiling has undergone important changes as a result of the advances made in knowledge of the ultrastructure of the chromosome (see Chapter 12). The old concept of a series of regular coils, starting with the major and minor coils and proceeding to the macromolecular level, must be changed. The basic structure of the chromosome, represented by the folded unit fiber of nucleoprotein of about 10 to 30 nm in diameter, does not show a regular pattern of coiling during prophase and metaphase (Fig. 13-2).

The condensation mechanism is very complex and cannot be described as resulting from the regular folding of the unit fiber. This does not mean that there are no orderly principles guiding the mechanism. In fact, it is well known that each chromosome in the condensed state has a characteristic morphology, which is the final result of a particular folding pattern. While the principles that govern the condensation cycle of chromosomes are unknown, it seems likely that the "information" required must be built into the

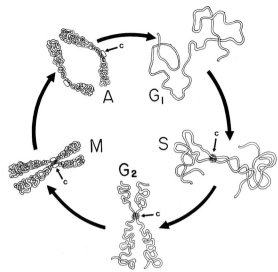

Figure 13-2. The condensation-decondensation cycle of the chromosomes. G_1, chromosomes are completely dispersed; **S**, duplication occurs; and G_2, condensation starts and the chromomeres (c) become apparent. At metaphase, **M**, and anaphase, **A**, the condensation is maximal and the two centromeres are clearly visible.

unit fiber. The existence of special "folders" along this fiber which are progressively activated as condensation progresses has been postulated. These folders could be initially attractive regions spaced along the unit fiber (see Wolfe, 1972).

Number of Strands in the Chromosome

The number of threads or strands constituting the chromosome has been a matter of debate for several decades. Classic morphological observations favored the idea that in a chromosome each chromatid contained two, four, or more strands. In anaphase chromosomes of plant cells, a longitudinal split of the chromatid (half chromatids) has been observed in the living condition,[2] and some electron micrographs of the same material have also suggested that the chromatid may be double. Electron microscopic observations of whole mounted metaphase chromosomes sometimes show the presence of an apparent multi-strandedness with several parallel fibers that may even pass across the primary constriction.

These morphological observations may be more the result of a longitudinal folding of the unit fiber than a true multi-strandedness. Most recent arguments favor the idea that

TABLE 13-1. Length in Micrometers and Number of Coils of the Chromatids of *Trillium grandiflorum* During Mitosis of Pollen Grains (Microspores)*

Condition	Length of the Chromatid	Number of Turns
Prophase	346	554
Middle prophase	202	242
Middle prophase	205	276
Final prophase	173	151
Final prophase	154	170
Final prophase	142	187
Metaphase	77	130
Anaphase (15 cells, average)	95.0 ± 2.9	130 ± 3.3

*Data from Sparrow, A. H., Huskins, C. L., and Wilson, G., Studies on the chromosome spiralization cycle in *Trillium. Can. J. Res. C., 19*:323, 1941. Reproduced by permission of the National Research Council of Canada from the Canadian Journal of Research, Vol. 19, Sec. C., 1941.)

each chromatid is made of a single strand. It is assumed that during the G_1 period of interphase each chromosome contains a single DNA duplex, whereas in the G_2 period, after replication, it consists of two DNA duplexes, each one constituting one chromatid or daughter chromosome (Fig. 13–2). The concept of a single-stranded structure is the best interpretation of the results obtained by Taylor on the semiconservative duplication of DNA in eukaryotic cells (Fig. 17–10). This concept also explains why a cell, irradiated with x-rays during the G_1 period, frequently bears chromosomal aberrations in which the two chromatids are broken. If irradiation occurs at the G_2 period, on the other hand, chromatid aberrations involving the break of a single chromatid are observed. These important concepts can easily be deduced by the close observation of Figure 13–2 in which the structure of the chromosomes at G_1 and G_2 is interpreted. The results obtained by the action of the enzyme deoxyribonuclease also favor the single strand model. The kinetics of the breaks produced by this treatment lead to the interpretation that the most probable structure of the chromosome is the single DNA duplex.[3]

Structure of the Primary Constriction and Centromeres

The fact that the shape of each chromosome is determined by the position of the primary constriction has been mentioned before (Fig. 2–11). The primary constriction is a region in which there is a narrowing of the chromosome, causing a division of the chromosomes into two arms. The primary constriction shows a faintly positive Feulgen reaction, indicating that there is DNA extending through it.[4] The presence of DNA of the repetitive type has been also demonstrated by hybridization experiments[5] constituting the so-called centromeric heterochromatin (Fig. 12–12). Electron microscopy also demonstrates the passage of nucleoprotein fibers from one arm of the chromatid to the other. Five to seven fibers have been observed to pass through it (Fig. 13–3).

In the primary constriction there is a special differentiation of the chromosome, called the *centromere* or *kinetochore*. The centromere is the site of implantation of the microtubules that constitute the chromosomal spindle fibers; for this reason it has been considered as functionally related to the chromosomal movements during mitosis. Usually two kinetochores, one in each chromatid, are observed, but in some cases a large number may be found (*diffuse centromeres*). In these instances, the microtubules are implanted at many points along the chromosomal arms. Diffuse kinetochores have been found in groups of plants (e.g., *Luzula*), in insects (e.g., Hemiptera), and in some algae. The centromere may be detected during prophase, before there is a connection with the microtubules (Fig. 13–1, *B*). Under the electron microscope the centromere appears as a plate or cup-like disc, 0.20 to 0.25 μm in diameter, plastered upon the primary constriction. In cross section it usually consists of an electron-dense layer 30 to 40 nm thick, and having a convex outer surface (Fig. 13–3, *B*). Between this layer and the underlying chromatin fibers

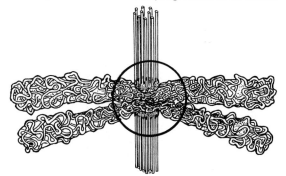

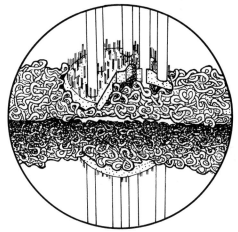

Figure 13–3. Diagram of a metaphase chromosome showing the folded fiber structure and the centromere with implanted microtubules. **Below,** an inset at higher magnification, showing the convex electron-dense layer and the fibrillar material forming the "corona" of the centromere. Several microtubules of the spindle are shown penetrating the various layers reaching the chromosome fibers.

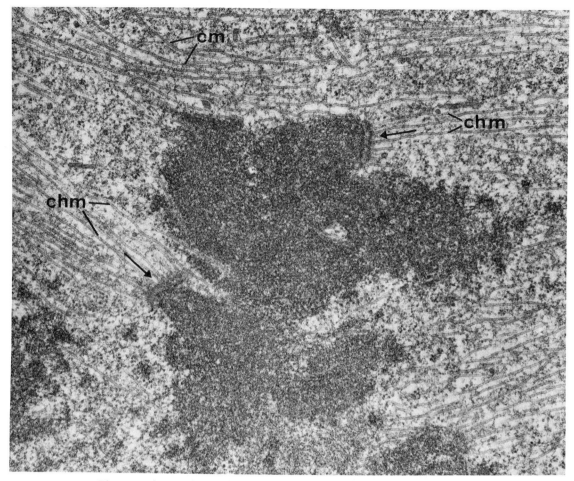

Figure 13–4. Electron micrograph of a Chinese hamster fibroblast in tissue culture at metaphase. The arrows point to kinetochores where the chromosomal microtubules (*chm*) are implanted. **Above,** interzonal or continuous microtubules (*cm*) are observed. The chromosomes show a tightly packed structure. ×40,000. (From Brinkley, B. R., and Cartwright, J., Jr., *J. Cell Biol., 50*:416–431, 1971.)

there is a less dense zone 15 to 30 nm thick. On the convex side there is fibrillar material which forms a kind of electron-dense "corona."[6, 7, 8] Several microtubules are attached to each kinetochore (Fig. 13–4). In general, they penetrate the various layers, sometimes reaching the chromatin fibers (Fig. 13–3, *B*).

The exact chemical nature of the centromere is unknown, although it is thought that it represents a non-chromatin material added to the surface of the primary constriction. It has been suggested that the kinetochore is a

gene product and that its activity is genetically controlled.[9] At present there is no evidence that the kinetochore itself contains genetic material.

The main function of kinetochores seems to be related to the attachment of the chromosomal microtubules. A related function may be to serve as nucleation centers for the polymerization of tubulin, the microtubular protein. In this way the structure might direct the formation of the chromosomal spindle fibers during prometaphase and metaphase.[10]

SUMMARY:
Structure of the
Chromosome

During prophase and metaphase the chromosomes undergo a process of condensation which is reversed during telophase. This process is related to the folding and unfolding of the unit fiber that constitutes the basic structure of the chromosome (see Chapter 12). The condensation cycle is specific for each chromosome and probably depends on special regions (folders) spaced along the unit fiber. It is generally believed that each chromosome is made of a single strand of DNA during the G_1 period. The single strand duplicates during the S period, and later on (G_2 and prophase) each chromosome is double (two chromatids) (Fig. 13–2). Experiments involving x-irradiation and the action of DNAase tend to confirm the single strand model of the chromatid.

The *centromere* or *kinetochore* is a special differentiation of the chromosome, situated at the primary constriction. It is the site of implantation of the microtubules of the spindle, it has a cup-like shape (0.20 to 0.25 μm); and it is made of non-chromatin material. The kinetochore may be a nucleation center for the polymerization of tubulin, the protein of microtubules.

FINE STRUCTURE AND FUNCTIONAL ROLE OF THE MITOTIC APPARATUS

The term *mitotic apparatus* has been applied to the asters that surround the centrioles and the mitotic spindle.[11] In fixed preparations the aster appears as a group of radiating fibers that converge toward the centrioles.

Around the centrioles there is often a clear zone called the *microcentrum* or *centrosome* into which the fibers of the aster do not penetrate (Fig. 13–5).

The spindle has the so-called *chromosomal fibers,* joining the chromosomes to the poles; the *continuous fibers,* extending pole to pole; and the *interzonal fibers,* observed

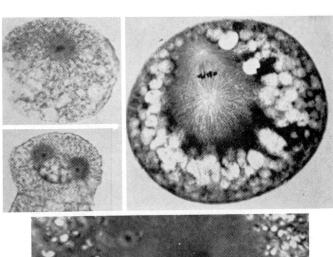

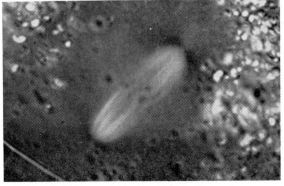

Figure 13–5. **Upper left,** photomicrograph of the cell center in the egg of *Ascaris megalocephala.* **Upper right,** section of an egg of *Nereis limbata,* showing the spindle and the asters. ×610. (Courtesy of D. P. Costello.) **Below,** mitotic apparatus (asters and spindle) of the first division of the oöcyte of *Chaetopterus pergamentaceus* observed with the polarizing microscope. The spindle fibers show a positive birefringence. The aster rays appear dark because they are perpendicular to the spindle. ×1500. (Courtesy of S. Inoué.)

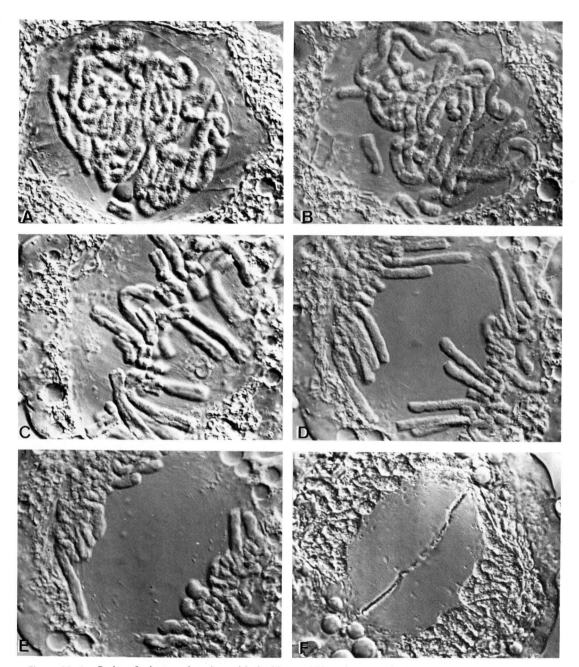

Figure 13-6. Series of micrographs taken with the Nomarski interference microscope of the same living cell of the endosperm of *Haemanthus*. The micrographs were taken at the following times: **A,** 5:19; **B,** 5:39; **C,** 7:03; **D,** 7:45; **E,** 7:53; **F,** 8:09.

   A, *late prophase,* showing the condensed chromosomes; the nuclear envelope is breaking at certain points. **B,** *late prophase,* showing the disapperance of the nuclear envelope. **C,** *metaphase,* with the chromosomes in the equatorial plane; some spindle fibers are observed. Note that this is an anastral mitosis, lacking asters at the poles. **D,** *anaphase,* with the chromosomes moving toward the poles. **E,** *late anaphase,* showing some interzonal spindle fibers. **F,** *telophase,* showing the phragmoplast at the equatorial plane. ×2000. (Courtesy of A. S. Bajer.)

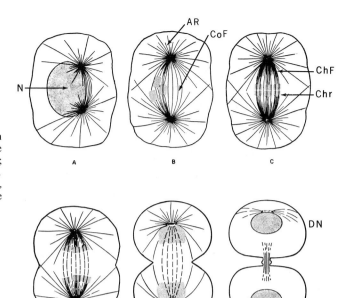

Figure 13-7. Diagram of mitosis in an animal cell showing the changes in birefringence of the various regions of the spindle. *AR,* aster; *Cen,* centriole; *ChF,* chromosomal fibers; *Chr,* chromosomes; *CoF,* continuous fibers; *DN,* daughter nucleus. (See the description in the text.) (Courtesy of S. Inoué.)

between the daughter chromosomes and nuclei in anaphase and telophase. All these fibers, including those of the aster, are composed of microtubules.

The fine structure of the mitotic apparatus is studied principally with the electron microscope (Fig. 13-4). Interesting observations may also be made using polarization microscopy[12] and the interference microscope provided with the Nomarski contrast optics (Fig. 13-6).[13] Figures 13-7 and 13-8 summarize the cyclic changes in birefringence

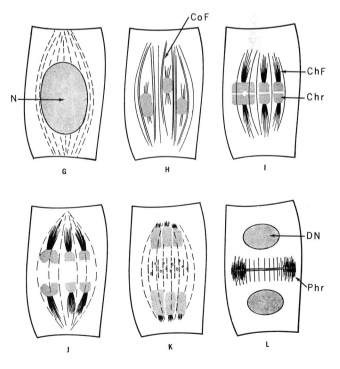

Figure 13-8. Diagram of mitosis in a plant cell showing the changes in birefringence of the spindle fibers. Note the absence of centrioles and asters. Abbreviations as in Figure 13-7. *N,* nucleus; *Phr,* phragmoplast. (See the description in the text.) (Courtesy of S. Inoué.)

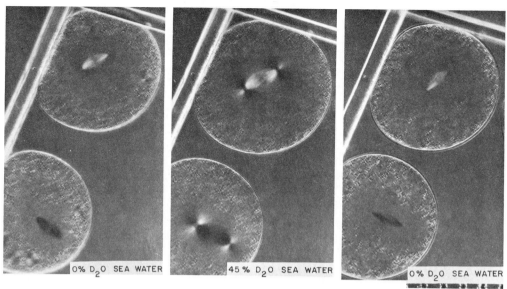

Figure 13–9. Action of heavy water (D_2O), added to the sea water, on the birefringence of the mitotic apparatus of sea urchin eggs. (See the description in the text.) (Courtesy of S. Inoué and H. Sato.)

observed in animal and plant cells during mitosis. Because of their positive birefringence, spindle and aster fibers are readily observed with the polarization microscope which has the advantage of allowing for the studies to be carried out on living cells. In both types of cells the chromosome fibers and the continuous fibers may be distinguished. In plant cells, which are devoid of centrioles and asters, the first spindle fibers appear in a clear zone surrounding the nucleus at prophase (Fig. 13–8, A). Birefringence is strongest near the kinetochores, but becomes weaker toward the poles (Fig. 13–8, B–D). During anaphase the chromosomes are led by intensely birefringent chromosomal spindle fibers (Figs. 13–7, C–D and 13–8, C–D). The continuous fibers, whose birefringence is very low in early anaphase, become more conspicuous in late anaphase and telophase. In animal cells such fibers form a kind of bundle that maintains a connection between the two daughter cells for some time (Figs. 13–7, F and 13–8, E–F). Under the electron microscope this bundle of microtubules appears surrounded by a dense material and by some vesicles, constituting the so-called *midbody*.[14]

In plant cells the interzonal region of the spindle is transformed into the *phragmoplast,* which, in some ways, is similar to the midbody of animal cells (Fig. 13–8, K, L). The phragmoplast is responsible for the formation of the

cell plate, which ultimately becomes the new *cell wall* that separates the daughter plant cells (see below).

The cyclic changes in birefringence are interpreted as reflecting the systematic assembly and disassembly of the microtubules that make up the spindle fibers. Studies employing micromanipulation have shown that, within certain limits, the spindle fibers resist extension, maintain mechanical integrity, and are instrumental in the movement of the chromosomes. If for example, the chromosomal fibers are cut and a kinetochore is oriented toward the opposite pole, that chromosome may acquire a new fiber and may move toward that pole.[15]

Studies in living cells also reveal that the spindle fibers represent a very dynamic structure. Their birefringence is abolished in a matter of seconds by low temperature; but after return to normal temperature, the cell recovers in a few minutes, with continuation of the arrested mitosis. Intense hydrostatic pressure, microbeam ultraviolet irradiation, and certain drugs, such as colchicine, Colcemid, and others, also induce disappearance of the birefringence and the microtubules. One interesting change is produced with heavy water (Fig. 13–9). When dividing sea urchin oöcytes are placed in 45 per cent D_2O, the birefringence increases twofold, and the volume of the spindle increases about tenfold in 1 to 2 minutes. After returning to H_2O the birefringence reverts to normal within a few minutes.[12] These experiments imply that in the cytoplasm there is a great excess of building material (see below) for the mitotic apparatus. It is also known that the effect of D_2O is independent of protein synthesis.[16]

Fine Structure of the Spindle and Aster. Microtubules

The spindle and asters are made of microtubules, which are cylindrical structures 19 to 23 nm in diameter and several micrometers in length. There may be as few as 16 microtubules in the spindle of yeast cells[17] and as many as 5000 in the spindle of a higher plant cell. A precise study of the number of microtubules has been carried out in dividing animal cells in the various regions of the spindle.[14] It is believed that many of the

microtubules of the continuous fibers of the spindle may span poles without branching. Spindle microtubules are generally arranged in parallel groups, following a relatively straight course with little bending (Fig. 13–4).

By the use of negative staining the microtubules have been shown to be true hollow tubes. In cross section they may show as many as 13 subunits of about 3.3 nm in diameter forming the wall of the cylinder.[18, 19] These and other observations suggest that microtubules are a result of the assembly of protein monomers. (In Chapter 22 the similarity between these microtubules and those found in cilia and flagella will be shown.)

The mitotic apparatus may be isolated from dividing sea urchin eggs[11] (Fig. 13–10) and studied biochemically.[20] A microtubular protein, called *tubulin*, has been isolated from these eggs. This protein is composed of a structural unit of 110,000 to 120,000 daltons—

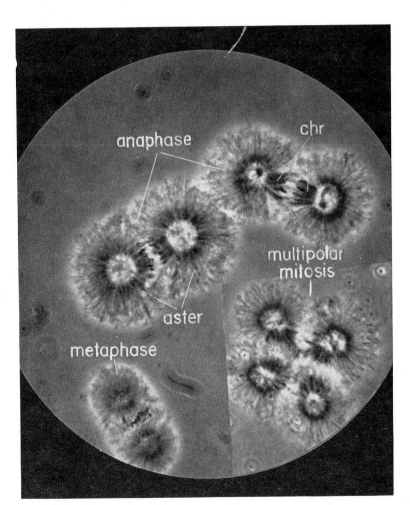

Figure 13–10. Mitotic apparatus isolated from the sea urchin egg in division. *chr*, chromosomes. (Courtesy of D. Mazia.)

a dimer that can be split into two monomers of 55,000 to 60,000.[21, 22] The dimer of tubulin has a dimension of 4×8 nm and may be detected by negative staining on the wall of the microtubule. Chapter 22 will indicate that the biochemistry of microtubules is better known in the case of cilia and sperm tails, in which tubulin A and B have been isolated (see Stephens, 1971).[23]

For the formation of mitotic microtubules, a dual mechanism, involving first the polymerization of subunits and then the completion of the tubules by secondary bonds, has been postulated. This hypothesis is supported by the fact that colchicine, a mitotic inhibitor, produces an amorphous gel without birefringence or evidence of microscopic fibers. A number of drugs, such as podophyllin, vincristine, vincaleukoblastine, and Colcemid—a derivative of colchicine—have a similar action, binding to the subunits of tubulin. In cultured cells treated with Colcemid and observed under the electron microscope, an aberrant mitotic apparatus is formed in relation to the centrioles, which do not migrate toward the poles. This inhibitor may prevent the formation of the continuous spindle microtubules necessary for centriole movement.

Temperature has an effect on the formation of microtubules. For example, in amebae undergoing mitosis it has been found that cooling at 2° C for five minutes produces degradation of the spindle microtubules. These are re-formed when the temperature is raised, but there is a temporary disruption of their parallel alignment.

During the various phases of mitosis the number and size of the microtubules may change considerably. The molecular mechanism of elongation or shortening of microtubules is not well understood. There may be addition or subtraction of new tubulin subunits from the ends or along the length of the microtubule. The rate of growth of the microtubule in the aster and the spindle may vary from 1 to 11 μm per minute. The rate may be the same, or faster, than the actual displacement of the chromosome during anaphase (see Bajer and Molé-Bajer, 1972).

Role of the Mitotic Apparatus

Current hypotheses about the role of the mitotic apparatus are based on the idea that the microtubules can generate some sort of mechanical force either by "pushing" or by "pulling" the other cell components. These two mechanisms involve the elongation or the shortening of the microtubules. In prophase, the migration of centrioles toward the poles is probably caused by the "pushing" which is a result of the elongation of the continuous fibers. In fibroblasts, separation of centrioles occurs at a rate of 0.8 to 2.4 μm per minute. As soon as the nuclear envelope begins to disintegrate, the nuclear region is invaded by microtubules that establish pole-kinetochore attachments before the chromosomes move toward the metaphasic plate. The characteristic shapes assumed by the chromosomes in metaphase and anaphase and the anaphasic bridges that may result from the stretching of dicentric chromosomes suggest that the forces responsible for the pulling of the chromosomes toward the poles is transmitted to the kinetochores. These forces may be great enough to produce the rupture of these dicentric chromosomes. It has been calculated that to move a chromosome, a force of about 10^{-8} dynes is needed, and that the entire displacement—from the equator to the pole of a chromosome—may require the use of about 30 ATP molecules (see Forer, 1969).

In isolated spindles a specific ATPase, which is more concentrated near the poles, has been detected. This enzyme could provide the mechanism for releasing the energy from ATP. It has been postulated that the chromosomal spindle fibers are developed from the centromere or kinetochore of each chromosome and grow toward the poles of the cell. The fact that chromosomal fragments having no centromere do not undergo anaphase movements confirms the importance of these spindle fibers in chromosomal movement.

Essentially, three types of mechanisms have been proposed for the anaphase movement of chromosomes: contraction, sliding, and assembly-disassembly of the microtubules. Contraction as a mechanism can be discarded, since the microtubules do not become thicker with the polar movement of the chromosomes. The sliding hypothesis, which was conceived by analogy with that of muscle contraction (see Chapter 22), also has little support. Thus, the hypothesis of *assembly-disassembly* is most favored at the present time. According to this theory, there is an equilibrium between a large pool of tubulin monomers and the oriented polymerized tubulin that forms the microtubules.[12] Upon polymerization, some structured water is believed to dissociate—a phenomenon which may explain the enhancement of polymerization induced by heavy water (Fig. 13–9). The dynamic equilibrium is also very sensitive to changes in temperature and non-aqueous solvents, indicating the importance of hydrophobic interactions between the nonpolar groups of the protein monomers.

During anaphase there are two kinds of movements that participate in differing proportions according to the cell type,[24, 25] but both contribute to the displacement of the chromosomes toward the poles. One of these movements, which is due to the shortening of the chromosomal microtubules, is demonstrated by the fact that the microtubules become considerably shorter during anaphase. The other movement, which is due to the lengthening of the continuous microtubules, also contributes to displacement.

Cytokinesis in Animal Cells

The separation of the daughter nuclei and *cytokinesis* or *cell cleavage* may be two separate processes. For example, the eggs of most insects undergo division of the nucleus to form a multinucleate *plasmodium* without separation of the cytoplasmic territories. Cleavage of animal cells has been studied mainly in tissue culture and dividing eggs and is produced by the furrowing of the cell. The first visible changes consist of the appearance of a dense material around the microtubules at the equator of the spindle at either mid or late anaphase. Then, although spindle microtubules tend to disorganize and disappear during telophase, they usually persist, and may even increase in number at the equator, frequently being intermingled with a row of vesicles and the dense material; the entire structure is called the *midbody* (Fig. 13–11). Simultaneously, there is a depression on the cell surface—a kind of constriction that deepens gradually until reaching the mid-body. With the completion of the furrowing, the separation of the cell is concluded.[26]

In a normal division there is perfect co-ordination between the movement of the chromosomes and the position of the furrow. In fragments of eggs in which the nucleus was eliminated by centrifugation (see Fig. 2–5), cleavage may take place. The importance of the spindle was inferred by the study of multipolar cell divisions in which a furrow is formed between each pair of asters. Even the removal of the whole mitotic apparatus of the sea urchin egg, however, does not inhibit cell cleavage.

Current hypotheses suggest that cell cleavage by furrowing may be the result of a contractile mechanism that involves the cell cortex.[26] A contractile protein exhibiting ATPase activity and properties similar to myosin was isolated from dividing sea urchin eggs.[27] These experiments should be correlated with other results showing that ATP may produce cytoplasmic contraction in different cell types. A system of fine microfilaments may be detected under the plasma membrane at the region of the furrow (Fig. 13–12). It can be demonstrated that such filaments are composed of an actin-like protein. Another indication that a contractile mechanism is acting on cell cleavage is provided by the action of the drug *cytochalasin*, which is able to inhibit such a process as well as other contractile mechanisms in cells. Cytochalasin produces disruption of the microfilaments (see Allison, 1973 and Chapter 22).

Cytokinesis in Plant Cells

Phragmoplast, Cell Plate, and Cell Wall

The phragmoplast begins to form in mid-anaphase of plant cells. Under the electron microscope it is possible to observe that the interzonal microtubules at the equator plane have scattered patches of vesicles and of dense material applied to their surface. The vesicles are derived from the Golgi complexes which are found in the regions adjacent to the phragmoplast and which migrate into the equatorial region to be clustered around the microtubules. Although the phragmoplast initially is found as a ring in the periphery of the cell, with time it grows centripetally by addition of microtubules and vesicles until it extends across the entire equatorial plane (Fig. 13–6 *E* and *F*). The vesicles increase in size and then fuse until the two daughter cells become separated by fairly continuous plasma membranes. At this time the phragmoplast has been transformed into the *cell plate*, a

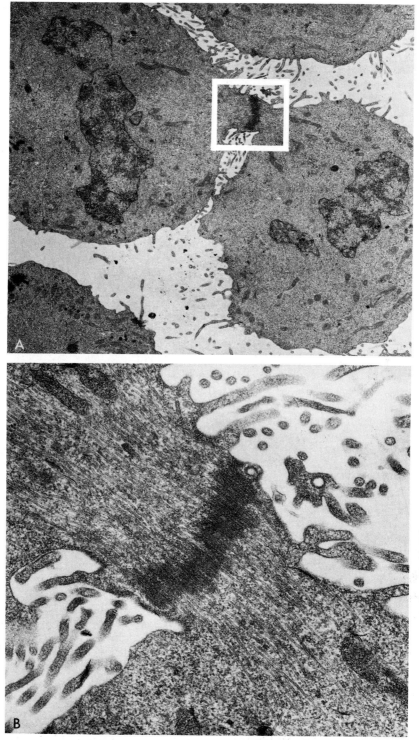

Figure 13-11.   Electron micrograph of a HeLa cell at the completion of cytokinesis. **A,** the two daughter cells are still joined by a small bridge which contains the interzonal microtubules and the electron-dense midbody. **B,** an inset at higher magnification. **A,** ×10,000; **B,** ×30,000. (Courtesy of B. R. Brinkley.)

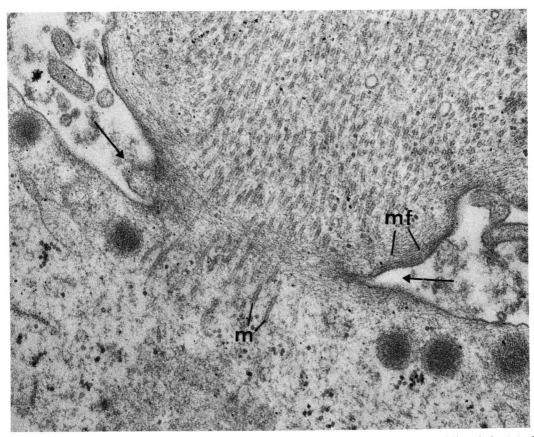

Figure 13–12. Electron micrograph of the advancing furrow (*arrows*) in a cleaving rat egg. Micotubules (*m*) of the interzonal fibers of the spindle are observed in the bridge between the daughter cells. Below the plasma membrane a network of fine microfilaments (*mf*) is observed. (See the description in the text.) ×34,000. (Courtesy of D. Szollosi.)

process that requires between 30 and 120 minutes. Thin cytoplasmic connections — the *plasmodesmata* — (see Chapter 12) traverse the cell plate and remain in place for communication between the adjacent daughter cells.

The formation of the cell plate also leads to the synthesis of the *cell wall*. The *Golgi* vesicles in the phragmoplast are already filled with a secretory material consisting mainly of *pectin*. The fusion of the vesicles results in the combining of the pectin in the extracellular space between the daughter cells, thereby forming the main body of the *primary cell wall*. (As mentioned in Chapter 12 pectin is an amorphous polymer made of galacturonic acid.) Later on, microfibrils of cellulose are laid down in a semi-crystalline lattice on the two surfaces of the facing daughter cells (see Chapter 12).

The *mitotic apparatus* comprises the spindle and the asters which surround the centrioles. The spindle is made of the *chromosomal* fibers, the *continuous* fibers, and the *interzonal* fibers; the latter are observed at anaphase and telophase between the daughter chromosomes. Study of the mitotic apparatus can be performed by electron microscopy, but in the living condition polarization microscopy is most useful in following the development of the various fibers (Fig. 13–7).

SUMMARY:
The Mitotic Apparatus

In plant cells the interzonal region of the spindle is transformed into the *phragmoplast,* which is the precursor of the cell plate. Spindle fibers are dynamic structures that are assembled and disassembled. Microtubules are depolymerized in the cold and become more developed in heavy water.

The protein *tubulin* is composed of dimers of 110,000 to 120,000 daltons. Colchicine and other drugs bind to tubulin, preventing its polymerization and the formation of microtubules. In this way mitosis is stopped at metaphase. The mechanism by which some microtubules shorten during anaphase is not completely known, but is probably by depolymerization at the polar ends.

Elongation and shortening of microtubules seem to be the two major mechanisms by which the chromosomes are moved toward the poles. The microtubules attached to the kinetochores exert considerable pulling force. It is calculated that 30 ATP molecules of energy are required for a chromosome to move from the equator to the pole. The most favored explanation for the mechanical action of the spindle is the so-called *assembly-disassembly* hypothesis. During anaphase the assembly of microtubules of the continuous fibers may cause elongation, with separation of the poles and "pushing" of the chromosomes. The *disassembly* of microtubules of the chromosomal fibers may cause shortening and "pulling" of the chromosomes.

Cytokinesis or cell cleavage differs considerably in animal and plant cells. In the former, separation of daughter cells is produced by an equatorial constriction which involves a contractile mechanism at the cell cortex. This is achieved by a system of actin-like microfilaments. A dense structure called the *midbody* may be formed.

In plant cells cytokinesis starts with the formation of the *phragmoplast,* which comprises the interzonal microtubules and Golgi vesicles. This structure is transformed into the *cell plate,* which separates the territories of the daughter cells. Within the cell plate the primary cell wall is produced by a secretory mechanism consisting mainly of the production of pectin, which is contained in Golgi vesicles (see Chapter 12).

REFERENCES

1. Swann, M. M., and Mitchison, J. M. (1958) *Biol. Rev., 33*:103.
2. Bajer, A. (1965) *Chromosoma, 17*:291.
3. Gall, J. G. (1963) *Science, 139*:120.
4. Kezer, J., and MacGregor, H. C. (1971) *Chromosoma, 33*:146.
5. MacGregor, H. C., and Kezer, C. (1971) *Chromosoma, 33*:167.
6. Bajer, A. (1968) *Proc. 22nd Symp. Soc. Exper. Biol.,* p. 285, Oxford, England.
7. Brinkley, B. R., and Stubblefield, E. (1966) *Chromosoma, 19*:28.
8. Jokelainen, P. T. (1967) *J. Ultrastruct. Res., 19*:19.
9. Luykx, P. (ed.) (1970) *International Review of Cytology,* Supplement 2, Academic Press, Inc., New York.
10. Dietz, R. (1972) *Chromosoma, 38*:11.
11. Mazia, D., and Dan, K. (1952) *Proc. Natl. Acad. Sci., USA, 38*:826.

12. Inoué, S., and Sato, H. (1967) *J. Gen. Physiol., 50*:259.
13. Bajer, A., and Allen, R. D. (1966) *Science, 151*:572.
14. Brinkley, B. R., and Cartwright, J. (1971) *J. Cell Biol., 50*:416.
15. Nicklas, R. B. (1967) *Chromosoma, 21*:1.
16. Wilt, F. H., Sakai, H., and Mazia, D. (1967) *J. Molec. Biol., 27*:1.
17. Moor, H. (1967) *Protoplasma, 64*:89.
18. Kiefer, B., Sakai, H., Solari, A., and Mazia, D. (1966) *J. Molec. Biol., 20*:75.
19. Barnicot, N. A. (1966) *J. Cell Sci., 1*:217.
20. Kane, R. E. (1967) *J. Cell Biol., 32*:243.
21. Borisy, G. G., and Taylor, E. W. (1967) *J. Cell Biol., 34*:525.
22. Stephens, R. E. (1970) *J. Molec. Biol., 47*:353.
23. Everhart, L. P. (1971) *J. Molec. Biol., 61*:745.
24. Ris, H. (1949) *Biol. Bull., 96*:90.
25. Hughes, A. F., and Swann, M. M. (1948) *J. Exp. Biol., 25*:45.
26. Szollosi, D. (1970) *J. Cell Biol., 44*:192.
27. Oknishi, T. (1962) *J. Biochem., 52*:145.

Allison, A. C. (1973) The role of microfilaments and microtubules in cell movement, endocytosis, and exocytosis. *Ciba Symp., 14*:109, Elsevier Publishing Co., Amsterdam.
Bajer, A. S., and Molè-Bajer, J. (1972) Spindle dynamics and chromosome movements. In: *International Review of Cytology,* Supplement 3, Academic Press, Inc., New York.
Brinkley, B. R., and Cartwright, J. (1971) Ultrastructural analysis of mitotic spindle elongation in mammalian cells in vitro. *J. Cell Biol., 50*:416.
Dan, K. (1966) Behavior of sulphydryl groups in synchronous division. In: *Cell Synchrony.* (Cameron, I. L., Inc., and Padilla, G. M., eds.) Academic Press, Inc., New York.
Forer, A. Chromosome movements during cell division. In: *Handbook of Molecular Cytology.* p. 553. (Lima-de-Faría, A., ed.) North-Holland Publishing Co., Amsterdam.
Harris, R. J. C., ed. (1963) *Cell Growth and Cell Division. Symp. Internat. Soc. Cell Biol.,* Vol. 2, Academic Press, Inc., New York.
Hughes, A. (1952) *The Mitotic Cycle.* Academic Press, Inc., New York.
Inoué, S., and Sato, H. (1967) Cell motility by labile association of molecules. *J. Gen. Physiol., 50*:259.
John, B., and Lewis, K. R. (1969) The chromosome cycle *Protoplasmatologia, 6b*:1.
Lafontaine, J. G., and Lord, A. (1970) Organization of nuclear structures in mitotic cells. In: *Frontiers of Biology: Handbook of Molecular Cytology.* p. 381. (Lima-de-Faría, A., ed.) American Elsevier Publishing Co., Inc., New York.
Mazia, D. (1961) Mitosis and the physiology of cell division. In: *The Cell.* Vol. 3, p. 77. (Brachet, J., and Mirsky, A. E., eds.) Academic Press, Inc., New York.
Rappaport, R. (1971) Cytokinesis in animal cells. *Int. Rev. Cytol., 31*:169.
Schrader, F. (1953) *Mitosis.* Columbia University Press, New York.
Stephens, R. E. (1971) Microtubules. In: *Fine Structure of Proteins and Nucleic Acids* (Biological Macromolecular Series, Vol. 4), pp. 355–391. Timasheff, S. N., and Fasman, G. D. eds.) John H. Dekker and Sons, New York.
Wolfe, S. L. (1972) *Biology of the Cell.* Wadsworth Publishing Co., Belmont, California.
Wolff, S. (1969) The strandedness of chromosomes. *Int. Rev. Cytol., 25*:279.
Wolper, L. (1960) The mechanics and mechanism of cleavage. *Internat. Rev. Cytol., 10*:163.

ADDITIONAL READING

fourteen

MEIOSIS

The historic aspects and an introduction to the important process of division by meiosis have been presented in earlier chapters (see Chapters 1 and 2). In essence, in mitosis there is a single duplication of the chromosomes, followed by the separation of the daughter chromosomes; in meiosis (Gr., *meioun*, to diminish) the single duplication is followed by two divisions, with each of the four cells formed having a haploid number of chromosomes. This change is also reflected in a reduction in the DNA content of the cell; in fact, as shown in Table 17–2, if the diploid value of DNA is 2n, by duplication, there is 4n, and after meiosis, only 1n in each haploid nucleus. The other fundamental difference between mitosis and meiosis is that the genetic material remains constant in mitosis, (i.e., with only rare mutations or chromosomal aberrations), whereas genetic variability is one of the main features of meiosis.

Meiosis occurs in the germ cells of sexually reproducing organisms. In both plants and animals, germ cells are localized in the gonads. The time at which meiosis takes places varies among the different organisms, and on this basis the process can be classified into: *terminal, intermediate,* or *initial.*

Terminal meiosis, also called *gametic meiosis,* is found in animals and a few lower plants. A typical example is represented for the human in Figure 14–1, in which spermatogenesis and oögenesis are shown diagrammatically.

(In terminal meiosis, note that the meiotic division occurs immediately before the formation of the gametes.) After several divisions, the zygote produces somatic and germ cells. These, by repeated divisions, give rise to several generations of *gonocytes,* which after a variable period become primary germ cells that are transformed into *spermatogonia* in the male and *oögonia* in the female.

Later, by division of the primary spermatogonia cell, secondary spermatogonia develop. Each secondary spermatogonium gives rise, in a final division, to two daughter cells, which begin to increase in volume and are called primary spermatocytes or *spermatocytes I.* At division (first meiotic division) the *primary spermatocyte* gives rise to two daughter cells or *secondary spermatocytes,* which divide again (second meiotic division), resulting in four cells called the *spermatids.* These cells, by differentiation (spermiogenesis) are transformed into spermatozoa. In the female, the successive stages are oögonia, primary oöcytes, secondary oöcytes, polar bodies, and ova. In place of four functional gametes, as in the male, there is only one, the mature *ovum,* since the other three become infertile *polocytes,* or *polar bodies.*

Intermediary or *sporic meiosis* is characteristic of flowering plants. As shown in Fig. 14–2, meiosis takes place at some intermediate time between fertilization and the formation of the gametes.

In higher plants the reproductive organs —anthers in male and ovary or pistil in female —produce microspores and megaspores, respectively. The cells that undergo meiosis to produce megaspores are called megasporocytes. Microspores are produced by microsporocytes (pollen mother cells). Each microsporocyte gives rise, by meiosis, to four functional microspores. Each megasporocyte produces four megaspores by meiosis, of which three degenerate. The remaining megaspore develops into the female gametophyte, which gives rise to the egg cell.

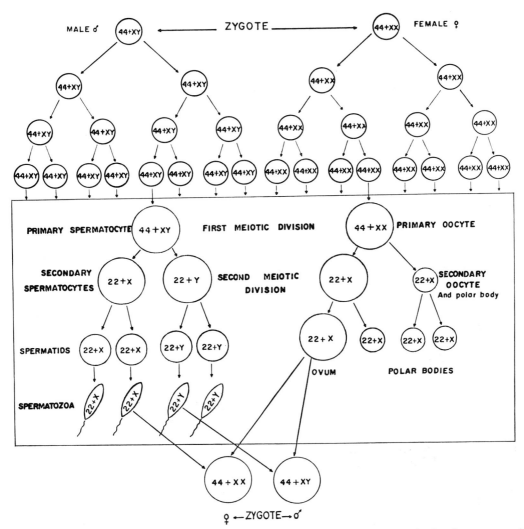

Figure 14–1. Diagram of spermatogenesis and oogenesis in the human. **Above,** mitosis of spermatogonia and oögonia. **Middle** (within the box), the meiotic divisions. **Below,** fertilization and zygote. Notice the 44 autosomes and the XY sex chromosomes.

In plants, microspores and megaspores are not the final gametes. Before fertilization, they undergo two mitotic divisions in the anther or three in the ovary to produce the male and female gametophytes, respectively.

Initial or *zygotic meiosis* occurs in some algae, fungi, and diatoms. Meiotic division occurs immediately after fertilization; in this case, only the egg is diploid. These organisms are very valuable for cytogenetic studies because in them the processes of DNA duplication and recombination may be clearly separated (see Fig. 15–5).

ANALYSIS OF MEIOSIS

Meiosis starts after an interphase that is not very different from that of an intermitotic interphase. During the premeiotic interphase DNA duplication has occurred at the S period. In the G_2 period of interphase apparently there is a decisive change that directs the cell toward meiosis, instead of toward mitosis. Some experiments in cultured cells of lily anthers tend to show that this change takes place at the beginning of G_2, but its intimate nature is still unknown.[1]

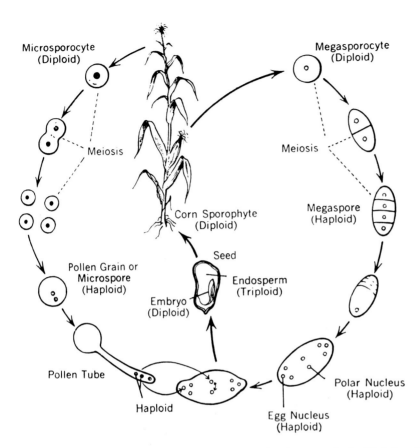

Figure 14–2. Life cycle of a plant. (Modified from Sinnott, E. W., Dunn, L. C., and Dobzhansky, T., *Principles of Genetics,* 5th ed., New York, McGraw-Hill Book Co., 1958.)

The first meiotic division is characterized by a long prophase during which homologous chromosomes pair closely and interchange hereditary material.

The classic stages of mitosis do not suffice to describe the complex movements of the chromosomes in meiosis. The successive meiotic stages are the following:

Meiotic Division I

Preleptonema corresponds to the early prophase of meiosis. Chromosomes are extremely thin and difficult to observe. Only the sex chromosomes may stand out as compact heteropyknotic bodies.

During *leptonema,* (Gr., *leptos,* thin;

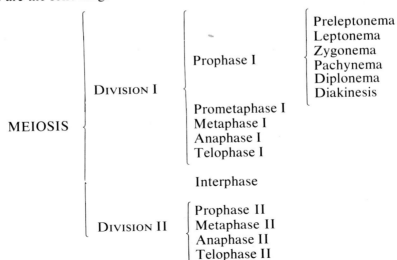

nema, thread) (Fig. 14–3), the chromosomes become more apparent as long filaments showing bead-like thickenings—the so-called *chromomeres* (Fig. 14–4, *A*). The morphologic interpretation of the chromomeres has varied widely. While some investigators believed they represent condensations of nucleoprotein material, others favored the view that they are regions of superimposed coils. Some support for this last concept comes from elec-

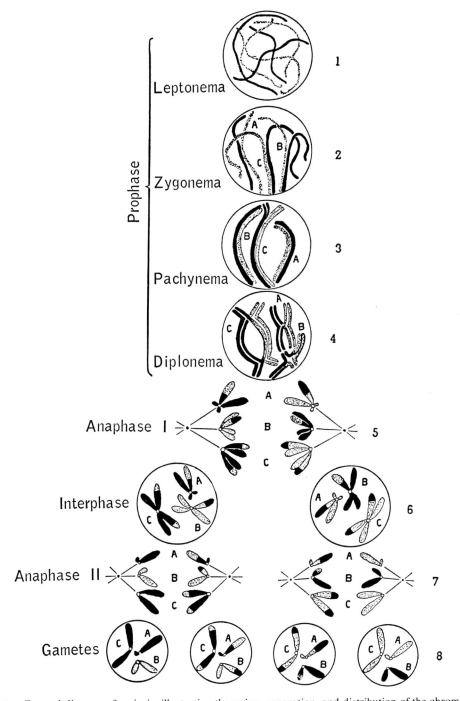

Figure 14–3. General diagram of meiosis, illustrating the union, separation, and distribution of the chromosomes.

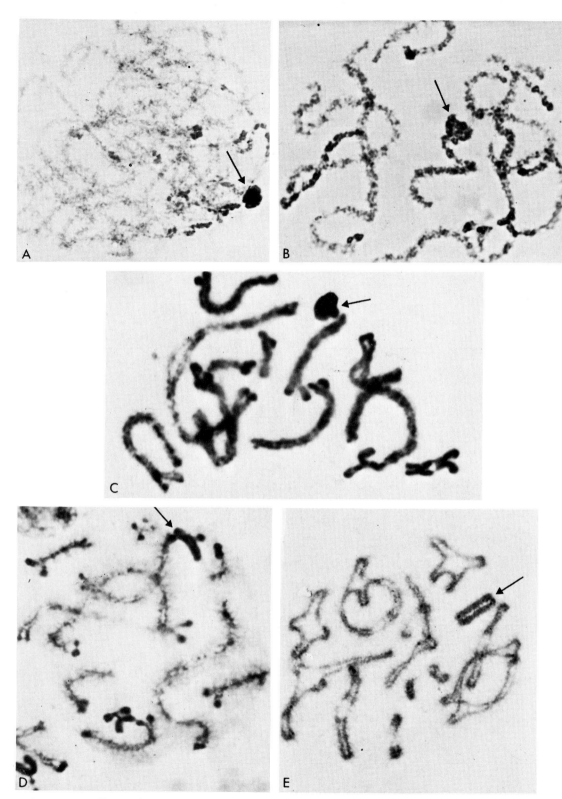

Figure 14–4. Stages of meiosis in the South American grasshopper, *Laplatacris dispar* (2n = 22 + X). **A,** *leptonema,* showing the long thin filaments with the chromomeres. (The X chromosome is indicated by an arrow in this and in subsequent micrographs.) **B,** *pachynema,* showing thick filaments in which the homologous chromosomes have paired. **C,** *early diplonema,* showing the way in which the homologous chromosomes have shortened considerably and have begun to separate. **D,** *mid-diplonema,* showing the configuration of the bivalents with the chiasmata. **E,** *late diplonema,* showing the chiasmata in distinct form. (From F. A. Saez.)

tron microscopic observations of leptonemic chromosomes, which show the strands of the chromosomes folded back and forth in the chromomeres. Chromomeres were once thought to represent the genes. However, they are too few in number (1500 to 2500 in the lily) to include all the genes. At leptonema each chromosome consists of two chromatids, but this is not visible under the light microscope. Frequently, the leptonemic chromosomes have a definite polarization and form loops whose ends are attached to the nuclear envelope at points near the centrioles, contained within an aster. This peculiar arrangement is often called the "bouquet."

During *zygonema* (Gr., *zygon,* adjoining) the homologous chromosomes begin to pair (Fig. 14–3). Sometimes the chromosomes unite at their polarized ends and continue pairing at the antipodal extremity; in other cases, fusion occurs simultaneously at various places along the length of the thread. Polarization seems to favor regularity in pairing. Pairing is remarkably exact and specific; it takes place point for point, and chromomere for chromomere, in each homologue. The two homologues do not fuse during pairing, but remain separated by a space of about 0.15 to 0.2 μm which, under the electron microscope, appears occupied by the so-called *synaptonemal complex* (see Fig. 14–10). This structure is in some way related to the close pairing and recombination of the allelic genes during the meiotic prophase.

During *pachynema* (Gr., *pachus,* thick), the pairing of the chromosomes has reached completion (Figs. 14–3 and 14–4, *B*). The chromosomes contract longitudinally, resulting in shorter and thicker threads. At this moment, the double constitution of the filament can be observed (Fig. 14–4, *B*). By middle pachynema, the nucleus appears to contain half the number of chromosomes. Each unit is a *bivalent* or *tetrad* composed of two homologous chromosomes in close longitudinal union which contain four chromatids.

Each homologous chromosome has its independent centromere; thus, each bivalent has four centromeres. At about late pachynema a longitudinal line of separation becomes apparent in each homologue in a plane perpendicular to that of the pairing. This means that at this stage each pachynemic element consists of four chromatids (Fig. 14–5, *1, 2*). The chromatids of each homologue are called sister chromatids.

The space between the homologues occupied by the synaptonemal complex is

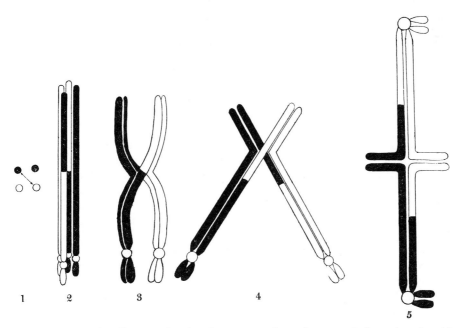

Figure 14–5. **Above, 1** and **2,** diagrams showing the process of crossing over; **3,** formation of a chiasma; **4,** terminalization; **5,** rotation of the chromatids of one bivalent.

maintained during this period. Experimental evidence suggests that during pachynema two of the chromatids of the homologues exchange segments (i.e., recombine). It is thought that transverse breaks occur at the same level of the chromatids, and that this event is followed by interchange and final fusion of the chromatid segments (Fig. 14–5, 2). All the changes involved in recombination, however, occur at a molecular level and are not detectable by either the light microscope or the electron microscope.

Pachynema is usually a long-lasting stage of the prophase, whereas leptonema and zygonema may last only a few hours. Pachynema may last for days, weeks, or even years.

At *diplonema* the intimately paired chromosomes repel each other and begin to separate (Fig. 14–4, *C, E*). However, this separation is not complete, since the homologous chromosomes remain united by their points of interchange, or *chiasmata* (Gr., *chiasma,* cross piece). Chiasmata are generally regarded as the expression of the phenomenon called *crossing over,* or *recombination,* by which chromosomal segments with blocks of genes are exchanged between homologous members of the pairs. With few exceptions, chiasmata are found in all plants and animals. At least one chiasma is formed for each bivalent. Their number is variable, since some chromosomes have one chiasma and others have several. During diplonema the four chromatids of the tetrad become visible; under the electron microscope the synaptonemal complex can no longer be observed. (Fig. 14–5).

Diplonema is a long lasting period. In the fifth month of prenatal life, for example, human oöcytes have reached the stage of diplonema and remain in it until many years later, when ovulation occurs. In most species there is uncoiling of the chromosomes during diplonema. In fish, amphibian, reptilian, and avian oöcytes, however, the uncoiling becomes so marked that the greatly enlarged nucleus assumes an interphase appearance. In these cases the bivalent chromosomes may attain a special configuration known as the *lampbrush chromosome,* in which the chromonema uncoils into loops that converge upon a more coiled axis (see Fig. 14–14). It will be shown later that the presence of these lampbrush chromosomes is related to an intensive RNA synthesis and to the enormous growth of the oöcyte.

In *diakinesis* (Gr., *dia,* across) the contraction of the chromosomes again becomes accentuated (Fig. 14–6, *B*). The tetrads are more evenly distributed in the nucleus, and the nucleolus disappears. Meanwhile, there is the movement of the chiasmata from the centromere toward the ends of the chromosome arms. By this process, called *terminalization,* the number of chiasmata diminishes; at the end of this period the homologues are held together only at their ends (Fig. 14–5, *4, 5*).

In *prometaphase I* coiling of the chromosomes reaches its maximum. The nuclear envelope breaks down, and the spindle microtubules become attached to the centromeres. With the electron microscope it has been observed that each homologue has two centromeres; in a tetrad, therefore, there are four centromeres.[3] In spite of this, each homologue is attached to one of the poles, and the two chromatids behave as a functional unit.

In *metaphase I* the chromosomes become arranged at the equator. The centromeres, pulled by the spindle fibers, are ready to separate (Fig. 14–6, *C*). If the bivalent is long, it presents a series of annular apertures between the chiasmata in perpendicularly alternating planes. If the chromosomes are short, they have a single annular aperture.

In *anaphase I* the daughter chromatids of each homologue, united by their centromeres, move toward their respective poles (Fig. 14–3). The short chromosomes, generally connected by a terminal chiasma, separate rapidly. Separation of the long chromosomes, which have interstitial and unterminalized chiasmata, is delayed. In side view, anaphase chromosomes show different shapes, depending on the position of the centromere.

It should be recalled that, by way of the chiasmata, segments were transposed between two of the chromatids of each homologue. Thus, when the homologous paternal and maternal chromosomes separate in anaphase, their composition is different from that of the originals. Two of their chromatids are mixed; the other two maintain their initial nature with reference to a single locus (Figs. 14–3 and 14–5).

Telophase I begins when the anaphase

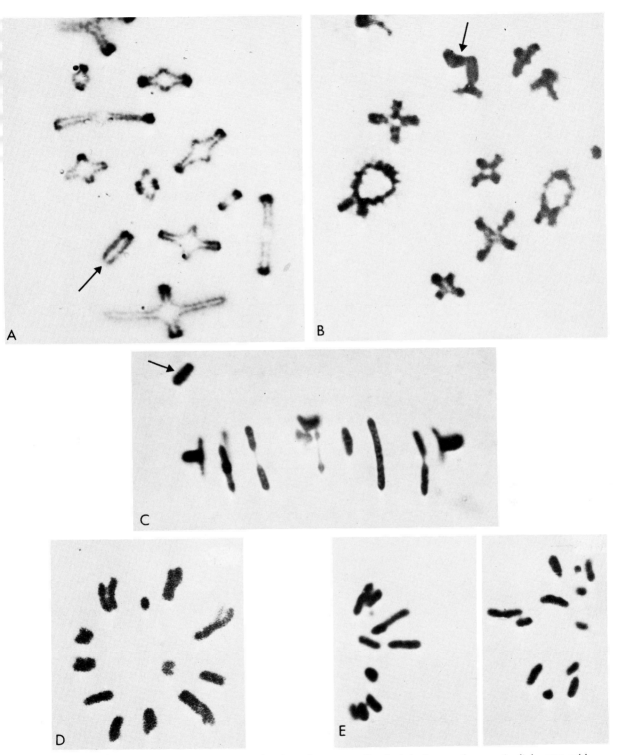

Figure 14–6. Continuation of Figure 14–4, showing other stages of meiosis in the grasshopper. **A,** *diplonema,* with Giemsa stain showing the constitutive heterochromatin localized at the centromere regions (C bands). **B,** *diakinesis,* showing the condensation of the bivalents. **C,** *metaphase I,* side view. The arrow points to the X chromosome that is advancing toward the pole; all the autosomes are still in the equatorial plane. **D,** *metaphase II,* polar view. **E,** *anaphase II,* side view. At this moment each chromosome consists of a single chromatid, and each daughter cell has the haploid number of chromosomes. (From F. A. Saez.)

groups arrive at their respective poles. Chromosomes may persist for some time in a condensed state, showing all their morphologic characteristics. Following telophase is a short *interphase* which has characteristics similar to mitotic interphase. Sometimes interphase may persist for a considerable length of time.

The result of the first meiotic division is the formation of the daughter nuclei, which in animals are called spermatocytes II (in the male) and oöcyte II plus the first polar body (in the female).

At the interphase between the two meiotic divisions there is no replication of the chromosomes. These are now haploid in number, although each one consists of two chromatids.

Meiotic Division II

A short *prophase II* is followed by the formation of the spindle, which marks the beginning of metaphase II.

At *metaphase II* (Fig. 14–6, *D*) chromosomes become arranged on the equatorial plane, the centromeres separate, and the two sister chromatids go toward the opposite poles during *anaphase II* (Fig. 14–3). Since in this division the longitudinal halves of each parental chromosome (chromatids) separate, each of the four nuclei of *telophase II* has one chromatid, which is now called a chromosome. Each nucleus has a haploid number of chromosomes (Figs. 14–3 and 14–6, *E*).

SUMMARY:
Events in Meiosis

Meiosis is a special type of cell division present in germ cells of sexually reproducing organisms. It consists of a single duplication of the chromosomes, followed by two consecutive divisions. The result is four haploid cells. Meiosis may be: (1) *terminal or gametic,* occurring immediately before the formation of the gametes (Fig. 14–1); (2) *intermediary or sporic,* taking place sometime between fertilization and the formation of the gametes (found in higher plants); or (3) *initial or zygotic,* occurring immediately after fertilization (found in fungi; see Figure 15–5).

Meiosis is divided into Division I and Division II. Division I has a long prophase during which the homologous chromosomes pair closely and interchange hereditary material. The stages of *leptonema, zygonema, pachynema, diplonema,* and *diakinesis* are recognized. Chromosomal pairing occurs at zygonema and is completed at pachynema, giving rise to the *bivalent* that is formed of four chromatids *(tetrad).* Between the homologues there is a 0.1 to 0.2 μm space occupied by the synaptonemal complex (see below). At pachynema, transverse breaks are produced and there is interchange (i.e., recombination) between two of the *homologous chromatids,* while the other two remain intact (Fig. 14–5). At *diplonema,* the phenomenon of recombination, expressed morphologically as *chiasmata,* is also referred to as *crossing over.* The number of chiasmata varies in different chromosomes. In *diakinesis,* contraction of the chromosomes is accentuated, and chiasmata move toward the ends of the chromosomes *(terminalization).* The rest of the meiotic process is very similar to mitosis.

The essence of the meiotic process is the formation of four nuclei, each differing from one another, in which each chromosome of the parent is represented once. As a result of the *chiasmata* in crossing over, the chromosomes usually do

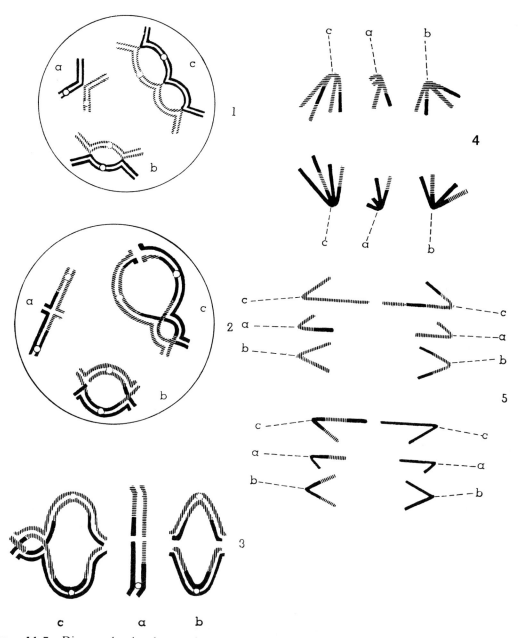

Figure 14–7. Diagram showing the genetic consequences of the meiosis of three pairs of chromosomes with (*a*) one chiasma, (*b*) two chiasmata, and (*c*) three chiasmata. **1,** diplonema; **2,** advanced diplonema showing the process of terminalization; **3,** metaphase I; **4,** anaphase I; **5,** anaphase II, showing the distribution of the chromosomes in the four nuclei formed. **Solid line,** the paternal chromosomes; **dashed line,** the maternal chromosomes. The centromere is represented by a circle.

not consist of either completely maternal or completely paternal material, but of alternating segments of each. For example, in Figure 14–7 all segments of chromosome A, between the centromere and the chiasma, effect a reductional division in anaphase I and an equational division in anaphase II. On the other hand, the segments located between the distal end of the chromosome and the chiasma effect a reduction in anaphase II (Fig. 14–7).

Meiosis is, therefore, a mechanism for distributing the hereditary units (genes), permitting their random independent recombination. Crossing over provides a means by which genes of different chromosomes can be brought together and recombined. If this process did not take place, the evolution of the species would be suspended by unalterable chromosomes, and organisms would not have their characteristic diversity.

The study of meiosis is a prerequisite for the understanding of the chromosomal basis of genetics. Only after the process of meiosis is understood will its significance in hereditary phenomena become apparent (see Chapter 15).

Synaptonemal Complex and Recombination

Two of the most important phenomena to take place during meiotic prophase are the linear pairing of the chromosomes and the interchange of segments between two of the homologous chromatids by *crossing over* or *recombination*. Studies of spermatocytes and oöcytes with the electron microscope have demonstrated the presence of an axial differentiation of the meiotic chromosomes at the time of pairing. This special structure is generally called the *synaptonemal complex (SC)* (other designations are *chromosomal core, synaptinemal complex,* or *axial com-*

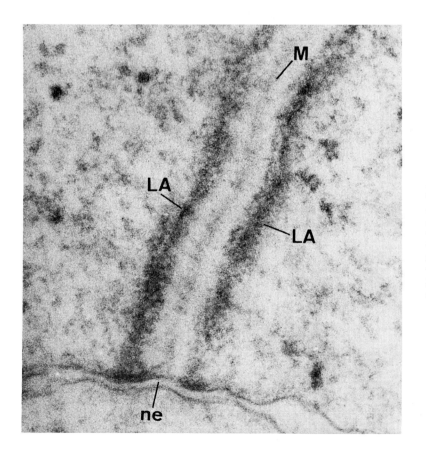

Figure 14–8. Electron micrograph of the synaptonemal complex in a dog spermatocyte. The two lateral arms (*LA*), corresponding to the homologous chromosomes, are shown parallel to each other and extending into the nuclear envelope (*ne*). The medial element (*M*) is simpler than in certain invertebrates (see Fig. 14–9). ×125,000. (Courtesy of J. R. Sotelo.)

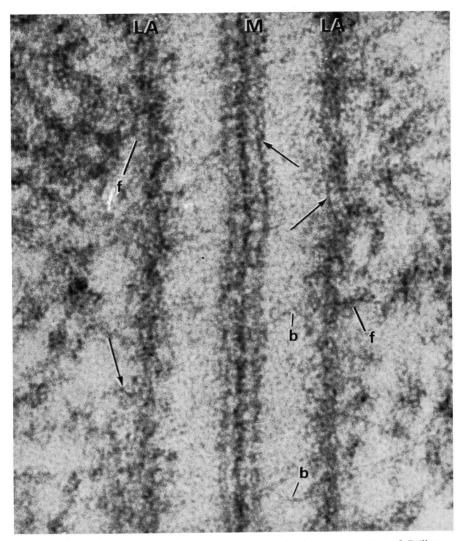

Figure 14–9. Longitudinal section along the synaptonemal complex in a spermatocyte of *Grillus argentinus*. The tripartite structure of the complex consists of a medial component (*M*) and lateral arms (*LA*). *b*, a bridge between these two components; *f*, fibrils. Microfibrils (2 nm), which form the structure of the chromosome, are marked with arrows. ×400,000. (Courtesy of J. R. Sotelo.)

plex).[2] By examining sections with both the light and electron microscope, it has been shown that the SC occupies the thin space between the paired homologues.

The morphology of the SC, which is very similar in plants and animals, consists of three parallel components. There are two dense *lateral lines*, also called *arms*, varying in width from 20 to 80 nm in the various species. They are formed of electron-dense coarse granules or fibers. These arms are joined to the adjacent chromosomes by fine fibrils. In most plants and animals — with the

exception of the insects — the two side arms are separated by an axial space of lower density (Fig. 14–8). As shown in Figure 14–9, in insects the central component may be very complex. In this case, it has the aspect of a ladder with three dense parallel lines and bridges crossing at intervals of 20 to 30 nm. These bridges are formed by fine fibrils that span the central and lateral components and are arranged perpendicularly to them. Study of cross sections give the impression that the SC is a ribbon-like structure interposed between the homologues.

As noted before, at zygonema and pachynema the chromosomes are frequently polarized and attached to the nuclear envelope. This same situation exists with the SC. By using serial sections it has been possible to reconstruct all the bivalents of a spermatocyte.[4, 5] In the human spermatocyte it has been found that the SC is present only in the region of pairing between the lateral components of the X and Y chromosomes.[6] Researchers employing cytochemical studies using deoxyribonuclease to digest DNA and the heavy metal *indium* to stain nucleic acids have suggested that the main component of the SC is protein in nature.[7] This protein is basic and probably similar to histones.[8] The fine fibers that cross between the two lateral arms and connect with the chromosomes probably contain DNA.

Function of the Synaptonemal Complex

The fact that the appearance and disappearance of the SC coincides with the stages of meiosis in which pairing and recombination occur has led to the interpretation that they are functionally related. In leptonema before pairing, in fact, single elements of the SC are observed. Complete synaptonemal complexes are seen at zygonema in the regions of pairing, and these are even more conspicuous at pachynema. The diagram of Figure 14–10 is an interpretation of the dynamic role of the SC during meiotic prophase.[9]

Several pieces of evidence indicate that the SC is more directly related to the process of recombination. Some evidence, for example, is provided by the action of inhibitors of DNA synthesis. It will be mentioned in Chapter 17 that at meiotic prophase there is a small amount of DNA synthesis which, if inhibited, can arrest the formation of the synaptonemal complex.[10–12] During zygonema there is also a distinctive synthesis of protein which, if inhibited, halts pairing and impairs DNA synthesis.[13]

The SC has been interpreted as a protein framework that permits the proper alignment of the homologues at exactly identical chro-

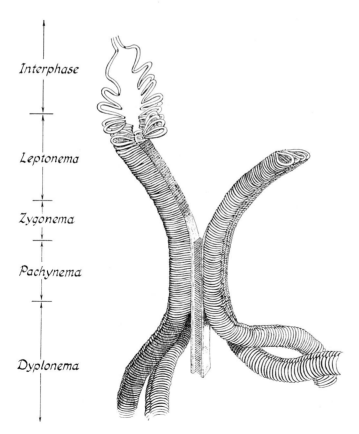

Interphase

Leptonema

Zygonema

Pachynema

Dyplonema

Figure 14–10. Three-dimensional diagram of the synaptonemal complex (*SC*) at different stages of meiosis. In pachynema the homologous chromosomes, now fully paired, are joined by the synaptonemal complex. This complex separates at diplonema; the chromatids show a relational coil. (From Roth, T. F., *Protoplasma, 61*:346, 1966.)

momeres. However, since recombination by crossing over occurs at the molecular level, it is necessary to assume that DNA fibers of the paired chromatids should reach the central component of the SC within distances of at least 1.0 nm in order for the breakage and recombination to take place.

This mechanism could be achieved by the fine DNA fibrils that are arranged like transverse bridges and that connect the homologous chromatids with the lateral and central elements of the SC (Fig. 14–9). It is thought that the homologous sequences of nucleotides search for each other and achieve a close pairing inside the central component of the SC. It is within these special regions that recombination, with the exchange of DNA segments, occurs. (The molecular mechanisms that may be acting at this point will be described in Chapter 17.)

The synaptonemal complex is a special structure, protein in nature, that develops during meiotic prophase in all eukaryotes. It is composed of two lateral and one central component. The lateral components of the SC appear in each of the homologues, and the central component becomes evident during pairing. It seems possible that it is inside this central component that the close pairing needed for the recombination of the DNA molecules of homologous chromatids takes place. This pairing may be achieved via thin DNA fibers that pass across the SC. At diplonema the SC is shed from the bivalents, with the exception of the regions in which the repelling homologues are held together by a chiasma. Thus, a chiasma contains a piece of SC that will ultimately disappear and be replaced by a chromatin bridge (for further details on the SC see Moses, 1968; Sotelo, 1969; Westergaard and Von Wettstein, 1972; Comings and Okada, 1972; and Moens, 1973).

**SUMMARY:
The Synaptonemal
Complex**

Giant Chromosomes

In certain cells, particularly at certain stages of their life cycle, special types of giant chromosomes may be observed. These are characterized by their enormous size and by a corresponding increase in volume of the nucleus and the cell. These special chromosomes include the so-called *polytenes* found in dipteran larvae—particularly in the salivary glands—and the *lampbrush chromosomes* observed in oöcytes of different vertebrates and invertebrates.

Inclusion of these two types of giant chromosomes in a chapter on meiosis is justified, because the lampbrush chromosomes are meiotic chromosomes remaining in a long diplonema, whereas polytenes are chromosomes in a permanent prophase and in somatic pairing (see below).

Polytene Chromosomes

In tissues of dipteran larvae, such as the salivary glands, gut, trachea, fat body cells, and malpighian tubules, some chromosomes are strikingly different from the somatic chromosomes of the same organisms. First observed by Balbiani in 1881, polytene chromosomes received little attention until after 1930 when their cytogenetic importance was demonstrated by Kostoff, Painter, Heitz, and Bauer.

In *Drosophila melanogaster* the volume of polytene chromosomes is about 1000 times greater than that of the somatic chromosomes. The total length of the four-paired set is 2000 μm, compared to 7.5 μm in somatic cells. Figure 14–11, 7 indicates (at the same magnification) the entire somatic set as compared with the smallest pair (IV) of giant chromosomes.

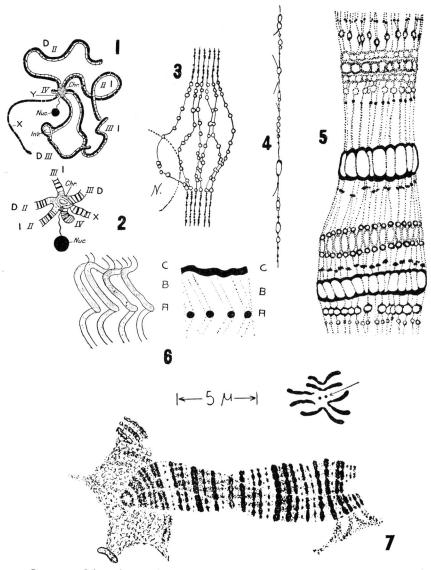

Figure 14–11. Structure of the polytene chromosomes.

1, general schematic aspect of the chromosomes of the salivary gland of a male of *Drosophila melanogaster* after they have been spread out by crushing the nucleus. The paternal chromosome (in white) and the maternal one (in black) are paired. *Chr.,* chromocenter; *D II* and *II I,* right and left arms of chromosome II; *D III* and *III I,* right and left arms of the third chromosome; *IV,* the fourth chromosome; *Inv.,* an inversion in the right arm of the third chromosome; *Nuc.,* nucleus; *X* and *Y* indicate the sex chromosomes respectively.

2, the chromocenter (*Chr.*) formed by the union of the heterochromatic parts of all the chromosomes in a female of *D. melanogaster.* (The other symbols are the same as for 1.)

3, a heterochromatic region of the X chromosome of *D. pseudoobscura,* showing its relations with the nucleolus (*N.*) and the filamentous (chromonemic) constitution of the chromosome.

4, detail of a component chromonema of the polytene chromosome in which the different chromomeres are seen.

5, schematic structure of the chromosome of *Simulium virgatum,* showing the organization of the chromonemata, chromomeres, and vesicles, which together give the appearance of the bands. The segment drawn corresponds to a euchromatic zone.

6, diagram to illustrate the interpretation of the helicoidal chromonema and the false chromomeres produced by the turns of the spiral. A zone (*B*) with four chromonemata is shown between two consecutive bands (at the left). To the right is the aspect of the same region when observed in a different focusing plane. *A* has a granular aspect, which simulates chromomeres. *C* appears as a continuous solid line.

7, the fourth polytene chromosome of *D. melanogaster,* adhering to the chromocenter, which is at the left. Above, at the right, the somatic chromosomes of the same fly as they appear in mitosis. The difference in size between the giant chromosome IV and the somatic chromosome IV is indicated by the arrow and drawn to the same scale. (**1** and **2,** after White, 1942; **3,** after Bauer, 1936; **4,** after Painter and Griffen, 1937; **5,** after Painter, 1946; **6,** after Ris and Crouse, 1945; **7,** after Bridges, 1935.)

Consequently, the volume of the cytoplasm also increases considerably. This phenomenon is shown diagrammatically in Figure 14–12, in which a typical somatic cell is compared with a similar giant cell of the same animal. Notice that the entire nucleus of a somatic cell corresponds in volume to a small portion of a polytene chromosome.

Another characteristic of polytene chromosomes is that the homologous pairs are closely associated, as in meiotic prophase. This phenomenon is called *somatic pairing,* and the chromosomes are considered to be in a permanent prophase (see Figure 14–11, *1*).

Along the length of the chromosome a series of dark *bands* alternates with clear zones called *interbands* (Fig. 14–13). The dark bands stain intensely, are Feulgen-positive and absorb ultraviolet light at 260 nm. These bands may be considered as disks of varying sizes that occupy the whole diameter of the chromosome. The larger bands have a more complicated structure. They often form *doublets.* The interbands are fibrillar, do not stain with basic stains, are Feulgen-negative and absorb little ultraviolet light. The constancy in number, localization, and distribution of the disks or bands in the two homologous (paired) chromosomes is notable. It is easy to construct, from a giant chromosome, topographic maps of the bands and interbands and to verify any disarrangement or alteration in the order of their linear structure. There are over 5000 bands in the four chromosomes of *Drosophila.*

Before the recognition of these giant chromosomes, genetic maps of the chromosomes of *Drosophila* were made by crosses (see Chapter 15). After their discovery, it was possible to compare the rather abstract genetic map with the topographic map of each chromosome. Combined genetic and cytologic studies have permitted the identification of many of the bands with specific genetic loci.

Mainly through the work of Beerman, on developing salivary glands of *Chironomus,* it has been possible to obtain definite evidence that the peculiar constitution and diameter of these chromosomes is due to their formation from a number of fibers (four at the origin for each chromosome) which multiply many times, remaining together like the threads of a rope (Fig. 14–11). Each fiber, delicate and difficult to perceive, may be considered as a chromonema.

In thin sections observed under the electron microscope, it is possible to observe parallel fibers running longitudinally. These fibers are thought to extend through the bands and interbands, being more tightly coiled in the bands than in the interbands.

The process of reduplication of the strands is called *endomitosis* (see Chapter 15). About ten reduplications are probably produced, resulting in about 1000 fibers (i.e., $2^{10} = 1024$), with the corresponding increase in DNA.[14] The length of a polytene chromosome is more or less the same as a mitotic chromosome during prophase. Some investigators have postulated that the chromonemata are drawn out without spiralization. In addition to these observations on the development of the giant chromosome, the polytene constitution can be verified morphologically by the fact that certain regions, called *puffs,* may split into numerous sub-units.[15]

In Chapter 20 the formation of puffs and Balbiani rings in polytene chromosomes will be discussed. These are cyclic phenomena that are the expression of gene activity (see Berendes, 1973).

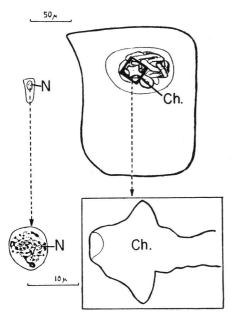

Figure 14–12. Diagram indicating the relationship of cellular and nuclear volume in polyteny. **Left,** diploid somatic cell of normal size. **Right,** similar cell but from the salivary gland of a larva. **Below,** notice that the volume of the nucleus of a diploid cell is similar to that of a small segment of one giant polytene chromosome. *N,* nucleus; *Ch.;* chromosome. (Courtesy of C. Pavan.)

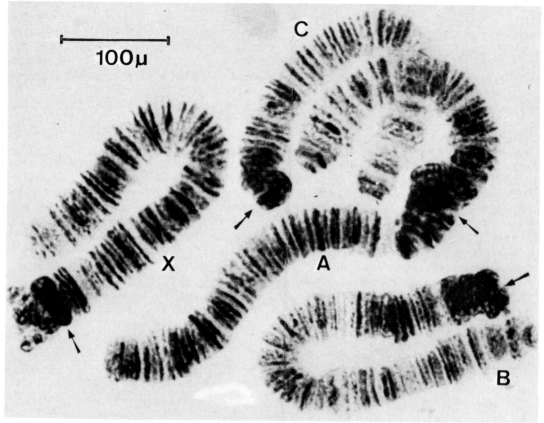

Figure 14–13. Polytenic chromosomes of *Rhincosciara angelae*, showing the three autosomes (*A, B,* and *C*) and the X chromosome. Arrows indicate the sites of puffs. (Courtesy of C. Pavan.)

Lampbrush Chromosomes

The lampbrush chromosomes were discovered by Ruckert in 1892, but only recently have they been interpreted accurately. These are even longer than the polytene chromosomes and are found in oöcytes during the extended diplonemic phase of the first meiotic division. In general, this phase corresponds to a period of maximum synthesis in which the yolk is produced.

Maximum size is reached in some urodele oöcytes, in which the total length of the chromosomal set may be 5900 μm, or three times longer than that of the polytene chromosomes.[16]

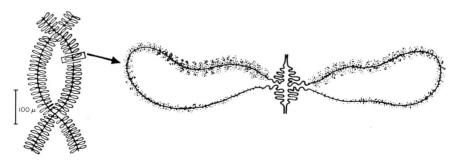

Figure 14–14. Diagram of the lampbrush chromosomes of the oöcyte of *Triturus.* **Left,** low magnification. **Right,** higher magnification, showing the lateral expansions in the form of a handle and the spiralization of the chromonemata. (From Gall, J. G., *Brookhaven Symp. in Biol., 8:*17, 1956.)

The growth of lampbrush chromosomes is the result of an increase in the size of the chromonemata. The chromosomes have many fine lateral projections, giving them the appearance of a test tube brush or lampbrush. Each bivalent chromosome consists of two homologues held together at "contact points" or chiasmata (Fig. 14–14).

At early diplonema the four chromatids of the tetrad are visible, but later on they become difficult to detect. The axis of each homologue consists of a row of granules or chromomeres from which pairs of loop-like lateral projections extend. The chromomeres represent regions in which the chromatids are tightly coiled (Fig. 14–14). There are four loops with similar morphology, corresponding to the four chromatids, at each level of the chromosome. The continuity between the loop and the chromatid can be demonstrated by stretching the chromosomes by micromanipulation.[17] The pairs of loops can be distinguished by the size, thickness, and amount of matrix that coats each of them. Cytochemical studies using ribonuclease and proteases show that the loop is coated with RNA and protein.[18] Furthermore, by using deoxyribonuclease, the loops break, thereby suggesting that they contain an axis of DNA.

Observation of isolated chromosomes under the electron microscope[19] reveals that the loops contain a fine fiber at the axis, corresponding to the DNA molecule, and fine fibrils of ribonucleoproteins inserted per-

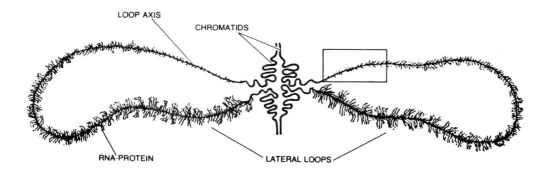

LOOP AXIS
CHROMATIDS
RNA-PROTEIN
LATERAL LOOPS

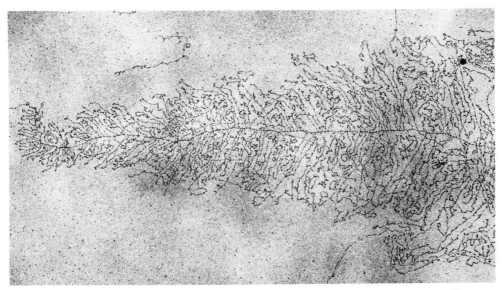

Figure 14–15. **Above,** diagram of the structure of a lampbrush chromosome, from an oöcyte of *Triturus*. **Below,** an electron micrograph that represents a view of the region in the rectangular inset. Note the axis of DNA upon which fine fibrils of ribonucleoprotein are inserted perpendicularly and which increase in length toward the right. (See the description in the text.) ×15,000. (Courtesy of O. L. Miller, Jr.)

pendicularly (Fig. 14–15). The appearance is very similar to that of the electron micrograph shown in Figure 18–15 for the formation of ribosomal RNA from DNA in nucleoli of amphibian oöcytes. At the point of attachment of these fibrils there is, presumably, an enzyme (RNA polymerase) that synthesizes the RNA (Fig. 14–15).

In general, the matrix is asymmetrical, being thicker at one end of the loop than at the other. Experiments with RNA precursors (e.g., ³H-uridine) show that the radioactivity appears at the thin end and progresses in a few days toward the thicker end. The synthesis of this RNA material can be inhibited by actinomycin; it appears to be messenger RNA. By using labeled amino acids researchers have found that the label appears in all regions of the loop at the same time.[16] To explain the findings with ³H-uridine it has been proposed that the loop is continuously spinning out from the chromomere at the thin end and recoiling at the thick end. Gall estimated that the DNA forming the loops represents about one twentieth of the total DNA. In *Triturus* the length of all the loops has been calculated to be about 50 cm; if this be so, the total DNA length per cell would be 10 meters (see Chapter 17).

In *Triturus cristatus* this stage lasts for about 200 days; when the synthetic activity ends, the RNA material is given off and the loop collapses. There is evidence that some of the RNA synthesized on these chromosomes is stored in the cytoplasm during oögenesis, and is used later on in embryogenesis.

SUMMARY:
Giant Chromosomes

Polytenes are giant chromosomes in which there have been about ten reduplications by endomitosis resulting in the production of about 1000 threads (*chromonemata*). These chromosomes are in a permanent prophase and the homologous chromosome pairs are closely associated (somatic pairing).

Lampbrush chromosomes are observed during the long period of diplonema in the meiosis of oöcytes of invertebrates and vertebrates. The special morphology of the bivalents, with their numerous loops, represents the spinning out of the DNA from the axis into these regions. Such a mechanism is coupled with active RNA synthesis and the accumulation of proteins. The RNA produced is transported into the cytoplasm and probably is used during embryogenesis.

REFERENCES

1. Stern, H., and Hotta, Y. (1969) In: *Handbook of Molecular Cytology*, p. 520, (Lima-de-Faría, A., ed.) North-Holland Publishing Co., Amsterdam.
2. Moses, M. J. (1964) In: *Cytology and Cell Physiology*. (Bourne, G. H., ed.) Academic Press, Inc., New York.
3. Luykx, P. (1965) *Exp. Cell Res., 39*:658.
4. von Wettstein, R. and Sotelo, J. R. (1967) *J. de Microscopie, 6*:557.
5. Moens, P. B. (1968) *Chromosoma, 28*:1.
6. Solari, A. J., and Tress, L. L. (1970) *J. Cell Biol., 45*:43.
7. Coleman, J. R., and Moses, M. J. (1964) *J. Cell Biol., 23*:63.
8. Sheridan, W. F., and Barnett, R. J. (1967) *J. Cell Biol., 35*:125a.
9. Roth, T. F. (1966) *Protoplasma, 61*:346.
10. Hotta, Y., Ito, M., and Stern, H. (1966) *Proc. Natl. Acad. Sci., USA, 56*:1184.
11. Roth, T. F. and Ito, M. (1967) *J. Cell Biol., 35*:247.
12. Sotelo, J. R. (1969) In: *Handbook of Molecular Cytology*, p. 412,

(Lima-de-Faría, A., ed.) North-Holland Publishing Co., Amsterdam.

13. Parchman, L. G. and Stern, H. (1969) *Chromosoma, 26*:298.
14. Kurnick, N. B., and Herskovitz, I. (1952) *J. Cell Comp. Physiol., 39*:281.
15. Bauer, H., and Beermann, W. (1952) *Chromosoma, 5*:246.
16. Gall, J. G. and Callan, H. G. (1962) *Proc. Natl. Acad. Sci. USA, 48*:562.
17. Callan, H. G. (1963) *Int. Rev. Cytol. 15*:1.
18. Gall, J. G. (1956) *Brookhaven Symp. in Biology, 8*:17.
19. Miller, O. L., Hamkalo, B. A., and Thomas, C. A. (1970) *Science, 169*:392.

Berendes, H. D. (1973) Synthetic activity of polytene chromosomes. *Int. Rev. Cytol., 35*:61.

Callan, H. G. (1969) Biochemical activities of chromosomes during the prophase of meiosis. In: *Handbook of Molecular Cytology* p. 540. (Lima-de-Faría, A., ed.,) North-Holland Publishing Co., Amsterdam.

Comings, D. E., and Okada, T. A. (1972) Architecture of meiotic cells and mechanisms of chromosome pairing. *Adv. Cell Molec. Biol., 2*:310.

Moens, P. B. (1973) Mechanism of chromosome synapsis at meiotic prophase. *Int. Rev. Cytol., 35*:117.

Moses, M. J. (1968) Synaptinemal complex. *Ann. Rev. Genet., 2*:363.

Sotelo, J. R. (1969) Ultrastructure of chromosomes at meiosis. In: *Handbook of Molecular Cytology* p. 412. (Lima-de-Faría, A., ed.) North-Holland Publishing Co., Amsterdam.

Stubblefield, E. (1973) The structure of mammalian chromosomes. *Int. Rev. Cytol., 35*:1.

Swanson, C. P. (1967) *Cytology and Cytogenetics.* Prentice-Hall, Inc., Englewood Cliffs, New Jersey.

Taylor, J. H. (1969) The structure and duplication of chromosomes. In: *Genetic Organization,* Vol. 1, p. 163. (Caspari, E. W., and Ravin, A. W., eds.) Academic Press, Inc., New York.

Westergaard, M., and von Wettstein, D. (1972) The synaptinemal complex. Ann. Rev. Genet., 6:71.

**ADDITIONAL
READING**

fifteen

CYTOGENETICS. CHROMOSOMAL BASES OF GENETICS

Cytogenetics has emerged from the convergence of cytology and genetics (see Chapter 1). This discipline is concerned with the cytologic and molecular bases of heredity, variation, mutation, phylogeny, morphogenesis, and evolution of organisms. Cytogenetics also deals with important problems applicable to medicine and agriculture.

This chapter is concerned with the cytogenetic aspects of heredity, mutation, and evolution. The interpretation of genetic phenomena at a molecular level will be dealt with in Chapter 19.

LAWS OF HEREDITY

In 1865 Gregor Johann Mendel, while studying crosses between peas (*Pisum sativum*), discovered the laws of hereditary transmission. Mendel selected several varieties of sweet peas that have pairs of differential or *contrasting* characteristics. For example, he used plants that have white and red flowers, smooth and rough seeds, yellow and green seeds, long and short stems, and so forth. After crossing the parental generation (P_1), he observed the resulting *hybrids* of the first filial generation, F_1. Then he crossed the hybrids (F_1) among themselves and studied the result in the second filial generation, F_2.

In a cross between parents with yellow and green seeds, in the first generation he found that all the hybrids had yellow seeds and, thus, the characteristic of only one parent. In the second cross (F_2), the characteristics of both parents reappeared in the proportion of 75 per cent to 25 per cent, or 3:1.

Law of Segregation

Mendel postulated that the color of the seeds was controlled by a "factor" that was transmitted to the offspring by means of the gametes. This hereditary factor, which is now called the *gene*, could be transmitted without mixing with other genes. He postulated that the gene could be segregated in the hybrid into different gametes to be *distributed* in the offspring of the hybrid. For this reason this is called the *law* or *principle of segregation of the genes*. Mendel found that the plants with yellow seeds in F_2, in spite of showing the yellow color, had different genetic constitutions. One-third of this group always gave yellow seeds, but the other two-thirds of the F_2 generation produced plants with yellow and green seeds in the ratio of 3:1. When the 25 per cent of plants in F_2 with green seeds were crossed among themselves, they always produced green seeds. This shows that they were a pure strain for this character. If we represent the genes in the crossing by letters, designating by *A* the gene with yellow character and by *a* the gene with green character, we have the following:

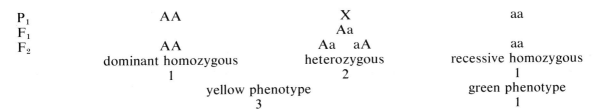

P$_1$	AA	X	aa
F$_1$		Aa	
F$_2$	AA	Aa aA	aa
	dominant homozygous	heterozygous	recessive homozygous
	1	2	1
		yellow phenotype	green phenotype
		3	1

In the first generation (F$_1$) both *A* and *a* genes are present, but only *A* is revealed because it is *dominant*; gene *a* remains hidden and is called *recessive*. In the hybrid F$_1$ both genes are segregated (separated) and enter different gametes. Half of them will have the gene *A*, and the other half, *a*. Since each individual produces two types of gametes in each sex, there are four possible combinations in F$_2$. This gives as a result the proportion

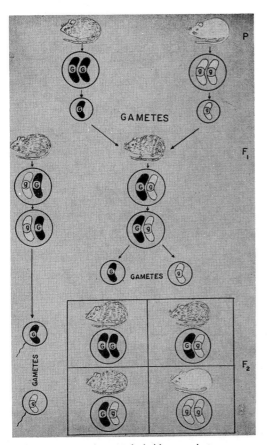

Figure 15–1. A monohybrid cross between a gray mouse (dominant) and a white mouse (recessive). The parallelism between distribution of genes and chromosomes is indicated, as well as the resulting phenotypes in the F$_1$ and F$_2$ generations.

1:2:1, corresponding to 25 per cent of plants with pure yellow seeds (*AA*), 50 per cent with hybrid yellow seeds (*Aa*), and 25 per cent with pure green seeds (*aa*).

Mendel confirmed that F$_1$ hybrids produce two classes of gametes in equal numbers. Backcrossing the F$_1$ hybrid (heterozygous *Aa*) with the homozygous recessive *aa*, he obtained a 1 *Aa*:1 *aa* ratio.

Mendel's results can now be explained in terms of the behavior of chromosomes and genes. The genes present in the chromosomes are found in pairs called *alleles*. In each homologous chromosome the gene for each trait occurs at a particular point called a *locus* (plural *loci*). In the case illustrated in Figure 15–1, the gray mouse will have two *GG* genes, one in each homologue. Since the two homologues pair and then separate at meiosis, the two *GG* genes must also separate to enter the gametes. The mechanism is the same in a dominant as in a recessive white mouse. In the hybrid F$_1$, one chromosome bears gene *G* and the homologous chromosome bears gene *g*.

When hybrids are self-fertilized, the gametes unite in the combinations shown by the checkerboard method illustrated in Figure 15–1.

Genotype and Phenotype

In 1911 Johansen proposed the term *genotype* for the genetic constitution and *phenotype* for the visible characteristics shown by the individual. For example, in the case of the peas with green or yellow seeds there are two phenotypes in F$_2$: yellow seeds and green seeds in the ratio of 3:1, respectively. However, according to the genetic constitution, there are three different genotypes: *AA*, *Aa*, and *aa* in the ratio of 1:2:1. This means that there are two Mendelian proportions, the phenotypic (3:1) and the genotypic (1:2:1). The phenotype includes all the characteristics of the individual that are an expression of gene activity. For example, in the human phenotypic characteristics include the different hemoglobins or the blood groups or a difference in taste toward thiourea.

In crossings of certain plants that have white and red flowers, such as *Mirabilis jalapa,* it is possible to find in F$_2$ three phenotypes (red, pink and white flowers), which correspond to the three genotypes. This is due to incomplete dominance. The rule of dominance and recessiveness is not always accomplished completely; dominance may be complete in most cases, but incomplete in others. In this case there is a mixture of characteristics, called *intermediary heredity.*

Law of Independent Assortment

Whereas the law of segregation applies to the behavior of a single pair of genes, the *law of independent assortment* describes the simultaneous behavior of two or more pairs of genes located in different pairs of chromosomes. Genes that lie in separate chromo-

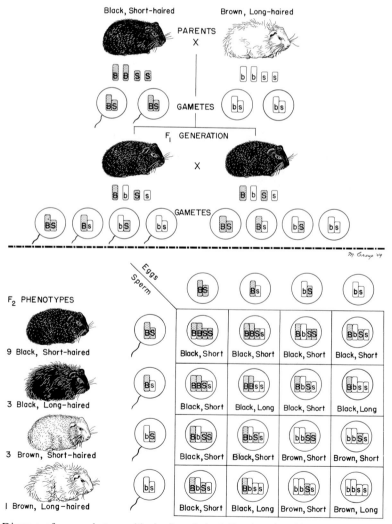

Figure 15–2. Diagram of a cross between black, short-haired (dominant) and brown, long-haired (recessive) guinea pigs. The independent assortment of genes is evident. (See the description in the text.) (From Villee, C. A., *Biology,* 6th ed., Philadelphia, W. B. Saunders Co., 1972.)

somes are independently distributed during meiosis. The resulting offspring is a hybrid (also called a dihybrid) at two loci.

Figure 15–2 diagrams the cross between a black, short-haired guinea pig (*BBSS*) and a brown, long-haired guinea pig (*bbss*). The BBSS individual produces only *BS* gametes; the bbss guinea pig produces only *bs* gametes. At F_1 the offspring are heterozygous for hair color and hair length. Phenotypically they are all black and short-haired. However, when two of the F_1 dihybrids are mated, each produces four types of gametes (*BS, Bs, bS, bs*), which by fertilization result in 16 zygotic combinations. As shown in F_2 there are nine black, short-haired individuals, three black, long-haired, three brown, short-haired, and only one brown, long-haired individual. This phenotypic proportion (9:3:3:1) is characteristic of the second generation of a cross between two allelic pairs of genes.

Linkage and Recombination

Studies of the fly *Drosophila melanogaster* by Morgan and his collaborators between 1910 and 1915 demonstrated that the law of independent assortment was not universally applicable and that in certain crosses of two or more allelic pairs of genes, there was a certain limitation of the free segregation.

In each case there was a marked tendency for parental combinations to remain linked and to produce a lesser proportion of new combinations.

If two genes (A and B, or a and b) are in the same chromosome, only two classes of gametes will be obtained:

$$ABab \times abab = 1\ ABab : 1\ abab$$

Figure 15–3 illustrates the mechanism of meiosis and the formation of the gametes in this hybrid. The coexistence of two or more genes in the same chromosome is called *linkage*.

After studying a considerable number of different crosses in *Drosophila*, Morgan reached the conclusion that all genes of this fly were clustered into four linked groups, corresponding to the four pairs of chromosomes. The first chromosome has many hundreds of genes; the fourth chromosome, and the smallest, has only a few. Further studies showed that the linkage is not absolute and that it may be broken with a certain frequency. For example, if a hybrid female of this insect with the genes "gray" and "long wings" (double dominant) is crossed with a male having the genes "black" and "vestigial wings" (double recessive), four classes of descendants are obtained instead of the expected two (Fig. 15–4). The first two, or

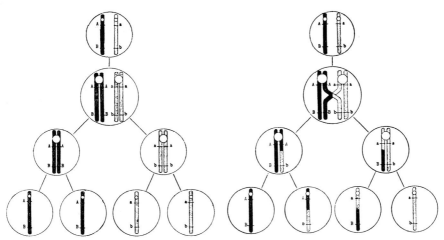

Figure 15–3. **Left,** diagram of the segregation of two pairs of allelic genes localized on the same pair of chromosomes without crossing over. The result is two types of gametes, AB and ab. A case of linkage. **Right,** diagram of the segregation of two pairs of allelic genes on the same chromosome between which crossing over takes place during meiosis. Four types of gametes result: AB, ab, Ab, aB. A case of linkage with crossing over.

parental, combinations are those expected from the linkage and they appear in 83 per cent of cases; the other two are new combinations ("gray, vestigial wings" and "black, long wings") and appear in 17 per cent of cases. (In Figure 15–4 dominant genes are marked ++ and recessive genes *bv*.)

Morgan concluded that the flies composing this 17 per cent are the product of a rupture of the linkage and that the recombination must come about through an interchange of parts between the two homologous chromosomes of the hybrid (Fig. 15–3, *right*).

Thorough studies have been made of all the classes of recombinations of a great number of genes. The results can be represented graphically by maps of each chromosome, showing the topography and respective loca-

tions of the genes. Corn (*Zea*), which is perhaps the most thoroughly studied of plants, has been mapped very completely, with localization of several hundred genes. Chromosome maps also have been constructed for the hen, mouse, sweet pea, and others. In the human species a series of genes has been localized in the sex chromosomes and in the autosomes.

Recombination, Chiasmata, and Genetic Maps

It is generally accepted that there is a correspondence between the presence of chiasmata, occurring at meiotic diplonema, and the genetic crossing over, or recombination, which takes place at the molecular level. The frequency of recombination of two linked genes is a function of the distance which separates them along the chromosome. When two genes are close to one another, the probability of crossing over is less than when they are far apart. If the distance between genes is estimated by linkage analysis, it is possible to construct a map indicating the relative position of each gene along the chromosome.

The distance between genes is expressed in units of recombination. Since in crossing over only two of the four chromatids interchange (see Fig. 15–5), the percentage of recombination will be half the average frequency of chiasmata. If, for example, in 100 meioses there are only 10 in which a chiasma between two genes is formed, of the 400 resulting gametes 360 will have parental combinations, and the other 40 will have 20 parental combinations and 20 recombinations.

The ratio of recombinations will be

$$\frac{20}{360 + 20} = 5\%$$

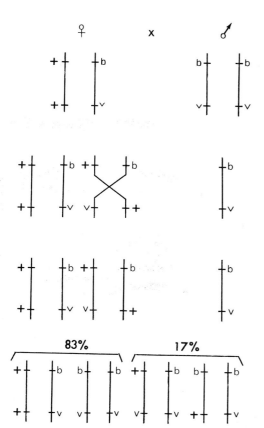

Figure 15–4. Cross involving two linked genes. Of the four types of resulting offspring, two are the expected combinations (83 per cent of the individuals) and two are new recombinations (17 per cent of the individuals). (See the description in the text.)

and the distance between the two genes will be 5 units.

In the case of a cross experiment involving three genes (1, 2, 3), if the distance between 1 and 2 is *x* units and between 2 and 3 is *y* units, the distance between 1 and 3 will be x + y units.

An appraisal of the possible number of new recombinations may be obtained by counting the number of chiasmata during meiosis. The so-called *recombination index* is calculated by adding to the number of bi-

	Diploid number	Bivalent number	Chiasmata per cell	Recombination index
Schistocerca cancellata (locust)	23	11	17	28
Zea mays (corn)	20	10	27	37

valents the number of chiasmata detected in the same cell at diplonema. In a species with a higher index, the possibility of new recombinations is higher, and this implies a greater possibility of variation. Two examples are given at the top of the page.

In general, genetic maps represent the relative order of genes along the chromosome; however, the frequency of crossing over varies for different points of the chromosome and for different organisms. The concept of a linear arrangement of genes in specific loci

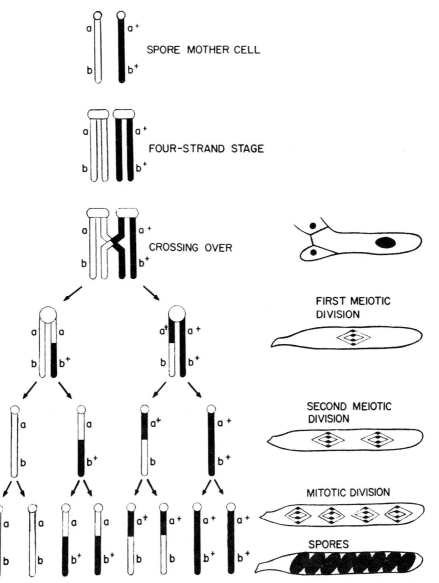

Figure 15–5. Diagram of the formation of ascospores in *Neurospora crassa*. A single crossing over between genes a and b, the behavior of one pair of chromosomes during the first and second meiotic divisions, and the division by mitosis of each of the four products are shown.

is in accordance with present knowledge of the structure of the DNA molecule and its function in genetic phenomena.

There are several other pieces of evidence to support the relationship between chiasmata and crossing over. The number of chiasmata is related to the length of the chromosomes in the bivalent. The presence of one chiasma reduces the possibility of another occurring in the immediate vicinity. This phenomenon has been called *positive interference*. The chiasma frequency is roughly constant in a given species, but can be modified by genetic or environmental action. Chromosome pairing, chiasma formation, and crossing over are under genetic control. Lines of rye with high and low chiasma frequency have been obtained by inbreeding. Among environmental factors that may affect the number of chiasmata are: temperature, radiation, chemicals, and nutrition (see Henderson, 1969). That the chiasma formation involves the exchange of segments of chromatids can be demonstrated in the case of bivalents that are heteromorphic (i.e., the two homologous segments have different size or morphology [see Fig. 16–1]). The exchange has also been demonstrated using pulse labeling with ^{3}H-thymidine in meiosis of a grasshopper.[1] Actual exchange between labeled and unlabeled segments was observed at the metaphase of the second meiotic division. The study of recombination at the molecular level will be dealt with in Chapter 17.

Recombination in Neurospora

Among the different organisms studied in genetics, the mold *Neurospora* occupies a special place. The advantage of this material is twofold: (1) it is possible to identify and to follow the fate of each of the four chromatids present in the bivalent meiotic chromosome and thus to determine whether the crossing over involves two, three, or all four chromatids; and (2) it is possible to make a close correlation between genetic constitution and biochemical expression of genes.

As shown in Figure 15–5, the four cells resulting from the two meiotic divisions undergo a mitotic division, which gives rise to eight haploid ascospores. Each of these ascospores can be isolated by dissection and cultured separately, giving rise to haploid individuals having the genetic constitution carried in each of the four original chromatids of the bivalent chromosome.

Figure 15–5 indicates a single crossing over between genes *a* and *b* and the resulting products. Analysis of the eight ascospores shows that only two of the chromatids interchange segments while the other two remain intact. It is also observed that the segregation of genes may occur during either the first or the second meiotic division, depending on the position of the locus concerned in relation to the point of crossing over and the centromere. Thus in Figure 15–5 genes aa^+ separate in the first meiotic division and genes bb^+ in the second meiotic division.

SUMMARY:
Fundamental Genetics

Mendel (1865) discovered the laws of heredity by studying crosses between peas having pairs of *contrasting* characteristics (i.e., allelic). In a cross between parents having yellow and green seeds he found that in F_1 all the *hybrids* had yellow seeds (dominant gene). In the F_2 generation 75 per cent of the plants had yellow seeds and 25 per cent had green seeds (recessive gene). Mendel postulated that the genes are transmitted without mixing (i.e., via the *law of segregation*). He demonstrated that in F_2 there are 1 dominant homozygous, 2 heterozygous, and 1 recessive homozygous offspring. The *genotypic* segregation is 1:2:1 in spite of the fact that the *phenotypic* proportion is 3:1. The law of segregation can be explained in terms of behavior of chromosomes during meiosis (see Chapter 14). At times there is incomplete dominance (i.e., intermediary heredity).

The behavior of two or more pairs of allelic genes follows

the *law of independent assortment.* Genes that lie in different chromosomes are independently distributed during meiosis. For two pairs of alleles the phenotypic proportion is 9:3:3:1 (Fig. 15–2).

Studies by Morgan and collaborators (1910 to 1915) demonstrated that the law of independent assortment may be limited. In *Drosophila* it was found that all genes were clustered into four linked groups corresponding to the four chromosome pairs. *Linkage* is not absolute and may be broken by *recombination* (Fig. 15–4) during the meiotic prophase (see Chapter 14). There is a correspondence between the number of chiasmata at diplonema and that of recombinations (crossing over) taking place at the molecular level. The distance between genes in a chromosome may be measured in *units of recombination,* and *genetic maps,* that represent the relative order of genes along the chromosomes may be constructed. The number of chiasmata (and recombinations) is related to the length of the chromosome. There is a *positive interference* that reduces the possibility of one chiasma from occurring near another. In a species chromosome pairing, chiasmata and crossing over are rather constant and are under genetic control. The mold *Neurospora* is ideal for the study of recombination and its relationship to the biochemical expression of genes. After the two meiotic divisions, there is a mitotic one and eight ascospores are formed. Each one is haploid and contains a single chromatid in which the genetic recombination may be studied by separating and culturing each of the ascospores.

Chromosomal Changes

The normal functioning of the genetic system of an organism is maintained by the constancy of the hereditary material carried in the chromosomes. Sometimes changes may occur in chromosomes that are brought about spontaneously or by experimental accidents, producing structural disarrangements. Knowledge of such changes has been favored by experimental methods that increase the frequency of changes and provide valuable means for analyzing the genetic and structural organization of the chromosome.

Some of the principal *structural changes* in the chromosome will be mentioned in the following sections.

Deficiency or Deletion

A *deficiency* is a chromosomal change in which a segment—either *interstitial* or *terminal*—is missing (Fig. 15–6). The deleted segment does not survive if it lacks a centromere. Terminal deficiency results from a single break in a chromosome. Interstitial deficiency results from two breaks followed by a union of the broken ends. Terminal deficiencies have been reported in maize, but are rare in *Drosophila* and other organisms. In heterozygous deficiency, one chromosome is normal, but its homologue is deficient (Fig. 15–6).

Animals with a homozygous deficiency usually do not survive to an adult stage because a complete set of genes is lacking. This suggests that most genes are indispensable, at least in a single dose, to the development of a viable organism. Deficiencies are important in cytogenetic investigations of gene location for determination of the presence and position of unmated genes.

Duplication

Duplication occurs when a segment of the chromosome is represented two or more

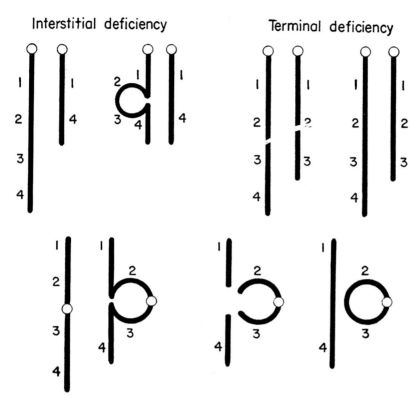

Figure 15-6. Diagram illustrating the origin of various types of deficiencies.

Formation of an acentric rod and deletion ring

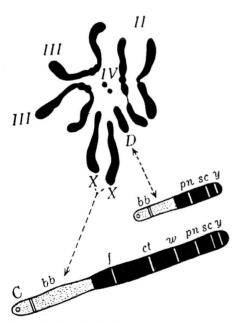

Figure 15-7. Duplication of a segment of the X chromosome in *Drosophila melanogaster*. The duplicated element (*D*) appears as a separate extra chromosome. *In black*, the euchromatic region; *stippled*, the heterochromatic region.

times in the chromosome. This may be a free fragment with a centromere or a chromosomal segment of the normal complement. If the fragment includes the centromere, it may be incorporated as a small chromosome (extra chromosome, Fig. 15-7). If the duplication occurs in an added segment, the disposition may be in tandem. An example of tandem duplication is the well-known *Bar* in *Drosophila*. Duplications make it possible to investigate the effects of an extra complement of genes in corresponding loci. In general, duplications are less deleterious to the individual than deficiencies.

Translocation

A translocation is a chromosomal rearrangement in which (1) segments are exchanged between nonhomologous chromosomes (reciprocal translocation) or (2) a segment of one chromosome is transferred to a different part of the same chromosome or to another chromosome (simple translocation). Reciprocal translocations may be both homozygous and heterozygous (Fig. 15-8).

Figure 15–8. Schematic representation of homozygotic and heterozygotic reciprocal translocations compared with the normal arrangement.

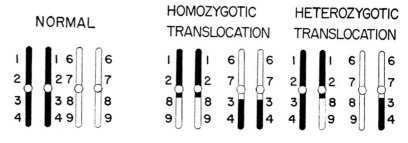

Cytologically, a translocated homozygote cannot be distinguished from a normal pair of chromosomes, but it can be detected by genetic experiments. Heterozygote translocations give rise to special pairing configurations in meiosis.

An interesting result occurs when, during translocation, both chromosomes are broken very close to their centromeres. The fusion creates a metacentric chromosome with two arms in the form of a **V** and a small fragment, which tends to be eliminated. Figure 15–9 illustrates the mechanism of *centric fusion*, which has occurred during the phylogeny of *Drosophila*, grasshoppers, reptiles, birds, mammals, and other groups. It is a process that establishes a new type of chromosome and reduces the somatic chromosome number of the species.

Inversion

An inversion is a chromosomal aberration in which a segment is inverted 180 degrees. Inversions are called *pericentric* when the segment includes the centromere and *paracentric* if the centromere is located outside the segment.

When a chiasma occurs within the inverted segment the result will be different in each case. In pericentric inversion (Fig. 15–10, *A*), at pachynema a loop is formed to allow for the pairing of the inverted segment. At anaphase I the chromosomes show duplications and deficiencies (Fig. 15–10, *B*). In a paracentric inversion at pachynema a loop is also formed (Fig. 15–10, *C*), but at anaphase I dicentric and acentric chromatids are produced (Fig. 15–10, *D*). The dicentric chromatids form a bridge that can easily be recognized. This bridge breaks when the anaphase chromosomes separate toward the poles. The acentric fragment is lost because of the lack of a centromere. Inversions are the cause of evolution in certain species of *Drosophila* and *Orthoptera*.

Isochromosomes. A new type of chromosome may arise from a break (i.e., a misdivision) at the centromere. As shown in Figure 15–11, the two resultant telocentric chromosomes may open up to produce chromosomes with two identical arms (i.e., isochromosomes). this type of chromosome has been produced in irradiated material. At meiosis they may pair with themselves or with a normal homologue.

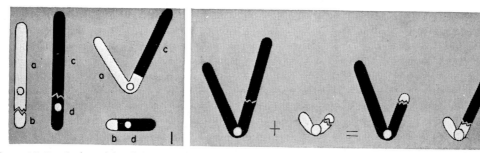

Figure 15–9. **1,** the origin of a new V-shaped (metacentric) chromosome by *centric fusion* of two nonhomologous acrocentric chromosomes. Segment *bd* is lost. **2,** *dissociation.* A metacentric and a small, supernumerary chromosomal fragment undergo a translocation, which results in two chromosomes (acrocentrics or metacentrics).

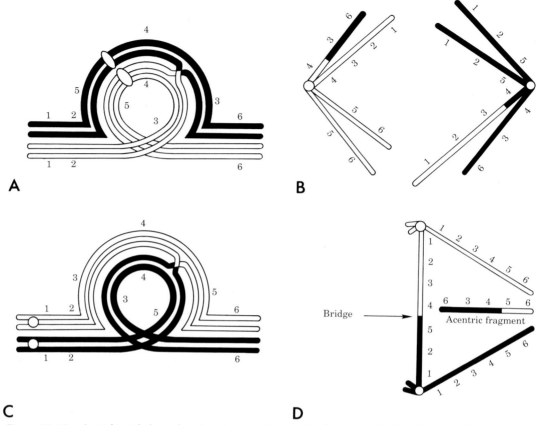

Figure 15-10. **A,** pericentric inversion at pachynema showing the formation of a loop during pairing; **B,** the results at anaphase I consist of duplication and deficiencies in the chromatids; **C,** paracentric inversion at pachynema with formation of a loop; and **D,** the result at anaphase I showing an acentric fragment and a dicentric chromatid with a bridge. (See the description in the text.)

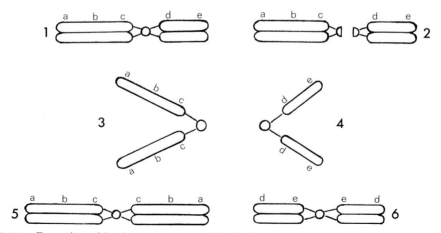

Figure 15-11. Formation of isochromosomes. **1,** original chromosome; **2,** misdivision of the centromere at the beginning of mitotic anaphase; **3** and **4,** the chromatids unfold into two isochromosomes; **5** and **6,** in the next division, two complete isochromosomes are present. Note that in each isochromosome, the arms exhibit genetic constitution.

TABLE 15-1. Chromosome Complements in Euploids and Aneuploids

Type	Formula	Complement*
EUPLOIDS		
Monoploid	n	(ABCD)
Diploid	2n	(ABCD) (ABCD)
Triploid	3n	(ABCD) (ABCD) (ABCD)
Tetraploid	4n	(ABCD) (ABCD) (ABCD) (ABCD)
Autotetraploid	4n	(ABCD) (ABCD) (ABCD) (ABCD)
Allotetraploid	4n	(ABCD) (ABCD) (A'B'C'D') (A'B'C'D')
ANEUPLOIDS		
Monosomic	2n − 1	(ABCD) (ABC)
Trisomic	2n + 1	(ABCD) (ABCD) (B)
Tetrasomic	2n + 2	(ABCD) (ABCD) (B) (B)
Double trisomic	2n + 1 + 1	(ABCD) (ABCD) (AC)
Nullisomic	2n − 2	(ABC) (ABC)

*A, B, C, D are nonhomologous chromosomes.

Changes in Chromosome Number

The number of chromosomes is generally constant for plant and animal species and may serve as an aid in determining their phylogeny and taxonomic relationships. Table 2–1 shows a list of plants and animals with their diploid (2n) number of chromosomes. The lowest chromosome number is found in the nematode *Ascaris megalocephala univalens* with two chromosomes in somatic cells and n = 1 in germ cells. The number of chromosomes may change in some organisms; this change may be of two main kinds (see Table 15–1). In *euploids* the set of chromosomes is kept balanced, whereas in *aneuploids* there is a loss or gain or one or more chromosomes, causing the set to become unbalanced. Chapter 16 will show that in humans aneuploidy may cause severe alterations of the phenotype.

Haploidy

Some exceptional plants and animals have a *monoploid* (or *haploid*) chromosome set. In these organisms meiosis is irregular because of the absence of homologous chromosomes. As a result, gametes with varying numbers of chromosomes may be formed. Examples of haploids are found in *Sorghum, Triticum, Hordeum,* and *Datura*. In animals, one sex may normally be haploid, as is the male of *Hymenoptera*.[2]

Polyploidy

A plant or animal that has more than two haploid sets of chromosomes is called a *polyploid* (Fig. 15–12). This change is common in nature, especially in the flowering plants. A diploid organism has two similar genomes; a triploid has three, an autotetraploid has four; and so on. Polyploids may originate either by reduplication of the chromosome number in somatic tissue with suppression of cytokinesis or by formation of gametes with an unreduced number of chromosomes.

Meiosis in a triploid is more irregular than in a tetraploid. In general, polyploids of uneven number are sterile because the

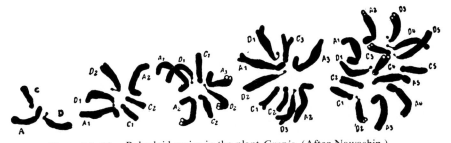

Figure 15–12. Polyploid series in the plant *Crepis*. (After Nawashin.)

gametes have a more unbalanced number of chromosomes. At pachynema only two chromosomes are always paired in a particular segment, whatever may be the number of homologues and the place of contact during pairing.

The scarcity of polyploids among animals is due to the mechanism by which sex is determined. Because one of the two sexes has two different types of gametes, sterility or sexual abnormalities may result. If polyploidy occurs, the genic balance between the sex chromosomes and the autosomes is disturbed, and the race or species may disappear because of sterility.

Several species of amphibians show spontaneous polyploidy; species with 104 chromosomes in mitosis and octovalent chromosomes in meiosis have been described. In such polyploids the DNA content is correspondingly increased.[3] Polyploids are useful in the study of the expression of genes in multiple dosage. Studies of this type have been made on amphibians for serum albumin, hemoglobin, and various enzymes.[4]

Polyploidy has been induced experimentally by temperature shock.[5] An example of polyploidy in mammals is the hamster *Cricetus cricetus*.[6] Cells with reduplicated chromosome complexes are frequently observed in animals and in pathologic tissues.

It is also possible to induce polyploidy

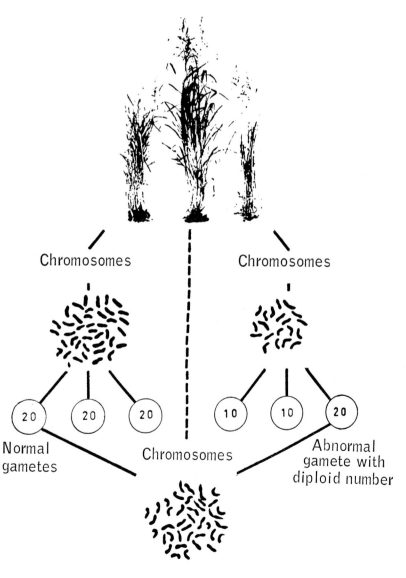

Figure 15–13. The origin of the Argentine *Sorghum almum* by crossing *Sorghum halepense* (2n = 4x = 40) with a diploid *Sorghum* (2n = 2x = 20) in which the fertilization occurred between one gamete not reduced (x = 20) and a normal gamete of *S. halepense*. Somatic chromosomes of the parents and the hybrid allopolyploid are illustrated. (After Saez and Nuñez, 1949.)

with substances such as colchicine, acenaphthene, heteroauxin, and veratrine. The most frequently employed drug is alkaloid colchicine. Seeds are immersed in colchicine at the beginning of germination. This substance may also be injected into young plants.[7]

These substances inhibit the formation of the spindle, and thus cell division is not completed. After a time the cells recover their normal activity, but have double the number of chromosomes. From the standpoint of pure and applied scientific work innumerable possibilities are offered by the experimental production of polyploids.

Allopolyploidy

This is a type of chromosomal variation that is produced in crosses between two species having different sets of chromosomes. The resulting hybrid has a different number of chromosomes than the parents. For example, the Argentine black *Sorghum* (*S. almum*) is an allotetraploid ($2n = 4x = 40$) originated in nature by an interspecific cross between *S. helepense* ($2n = 4x = 40$) with *S. sudanense* ($2n = 2x = 20$) (Fig. 15–13). In this case the fertilization occurred between one abnormal diploid gamete of *S. sudanense* and a normal gamete of the other species. In most cases crosses between distantly related species produce sterile diploid hybrids.

The study of meiosis in allopolyploids is of importance in determining the species which have taken part in the formation of the hybrid; it may also furnish a key to the probable phylogeny.

Aneuploidy

When one or more chromosomes reduplicate, the organism is said to be *polysomic*. This is a special kind of aneuploidy caused by faulty separation of chromosomes during meiosis. One of the chromosomes, along with its homologue, passes to the same pole and is contained in the same gamete. This phenomenon is also called *nondisjunction*. Such a gamete, upon union with any normal gamete, gives rise to a *trisomic* individual ($2x + 1$). Table 15–1 indicates other types of aneuploidy, such as those found in *monosomic* and *nullisomic* organisms.

From the genetic point of view, monosomic organisms are interesting because they have genes without alleles. This allows one to follow the distribution of the recessive gene located in the unpaired element and to determine the values of linkages and of crossing over in the progeny.

Endomitosis, Polyteny, Polysomaty, and Somatic Reduction

Geitler found chromosomal reduplication and subsequent separation, without formation of a spindle and disappearance of the nuclear *envelope*. He called this process *endomitosis*. Reduplication may be so marked that the diploid chromosome number (normally 21 in *G. lateralis*) may reach 1024 or 2048. In the epithelial sheath of the testis of *Orthoptera* there are nuclei with a high degree of ploidy.

In tumor cells of mammals, such as in ascitic and solid tumors, reduplication and polyploidy are common. Reduplication of chromosomes may take place by two mechanisms, *polyteny* and *polysomaty*. In polyteny the sister chromatids do not separate, and a multistranded polytene chromosome is formed (see Chapter 14). In polysomaty, the separation of the sister chromatids results in a somatic polyploid with chromosomes that have normal strands. Between these extremes the differences are only of degree, since polyteny is a special case of the general phenomenon of *endopolyploidy*. There are cases in which polyteny and polysomaty coexist in the same cell, as, for example, in the genus *Lestodisplosis*.

As a result of the study of polysomaty in relation to differentiation, it can be shown that tissues having a high mitotic activity also have a constant chromosome number. If cell division is slower than reduplication, polyteny and polysomaty may occur. On the other hand, if cell division is faster than reduplication, *somatic reduction* may take place, in which the chromosome number of somatic cells is reduced. This type of change has been observed in higher plants and in some insects. Somatic reduction can be increased experimentally by various treatments.

Somatic Variation in Chromosome Number

A variation, called *somatic aneuploidy*, has been observed in various tissues of vertebrates, including man.

An example of chromosomal variation is

the so-called *somatic segregation,* in which two different daughter cells arise from a somatic mitosis. This process may produce individual cells in various tissues or parts of the body that have different chromosome numbers, e.g., in mosaic individuals, variegations, gynandromorphs, and others (see Chapter 16). There are several causes for this curious change, including endomitosis, somatic reduction, somatic crossing over, fragmentation or deletion of chromosomes, and so forth.

Chromosomal Variations in Cancer. In cultures of normal cells, chromosomal variations are common, particularly after several transplants. These changes may lead to malignancy, but this is not the case in all cultures. The cytogenetic analysis of mammalian tumors has led investigators to consider them as altered karyotypes, which are genetically and cytologically unstable. Most ascitic tumors of rats have been found to consist of a mixture of cell types. The chromosomal content of these cell populations is variable, and the *modal number* is different from the diploid number of the species (aneuploidy). Since cells having the same chromosome mode can be perpetuated by transplantation, it was thought that these were the stem lines of the tumor, and the cell populations were considered as variants of this "stem line."[8, 9]

The concept of stem lines has been extended to primary tumors of mice.[10] Serologic methods[11] and cultures of isolated cell clones have been also used on this problem.[12]

SUMMARY:
Variations
in Chromosomes

Changes in the number and structure of chromosomes may occur spontaneously or experimentally by the action of radiation or chemicals. Some of the main *structural changes* are: (1) *Deficiency or deletion,* in which a part (either *interstitial* or *terminal*) of the chromosome is missing. The parts of the chromosome lacking the centromere generally are lost. Deficiency may be *heterozygous* (one chromosome is normal) or *homozygous* (both chromosomes are deficient). The latter generally do not survive (Fig. 16–6). (2) *Duplication,* in which a chromosome segment is represented two or more times (tandem duplication). (3) *Translocation,* in which there is an exchange of segments between nonhomologous chromosomes (*reciprocal* type) or between different parts of the same chromosome (simple type). Sometimes *centric* fusion occurs when the two chromosomes are broken near the centromere and form a metacentric (V-shaped) chromosome (Fig. 15–9). (4) *Inversion,* in which there is breakage of a segment, followed by its fusion in a reverse position. It is *pericentric* if it includes the centromere, and *paracentric* if the centromere is outside. *Isochromosomes* may arise from a break at the centromere, resulting in two chromosomes with identical arms (Fig. 15–11).

The number of chromosomes is generally constant for plant and animal species. *Chromosomal changes in number* are of two main kinds: in *euploids* the set is kept balanced; in *aneuploids* there is a loss or gain of one or more chromosomes. In exceptional cases there are *haploid* organisms. In plants, *polyploids* (triploid, tetraploid, etc.) are rather common. They originate by reduplication without cytokinesis. In animals, polyploids are scarce, because sex is frequently determined by a pair of different chromosomes (XY). Polyploidy can be induced by colchicine; this substance is used in agriculture to improve certain plant species. *Allopolyploidy* consists of the formation of a hybrid with different sets of chromosomes. Sometimes a diploid gamete fertilizes a normal haploid gamete, producing a new species with a triploid number (Fig. 15–13).

Among the aneuploid organisms there are the *trisomic* (i.e., three similar chromosomes) and the *monosomic* (only one of the pair of chromosomes). These two conditions are important in the human (see Chapter 16) and may arise by nondisjunctional division.

Endomitosis consists of the duplication of chromosomes without subsequent cytoplasmic division. Large polyploid nuclei may be formed as a result. *Polyteny* consists of the duplication of the sister chromatids, forming the giant polytenic chromosomes (see Chapter 14).

Somatic reduction in the number of chromosomes may occur in certain cells of higher plants. Sometimes there is somatic aneuploidy. These processes may lead to the formation of *mosaic* individuals having cells with different chromosome numbers. Such chromosome variations are common in cancer cells and may be a cause of malignancy.

CYTOGENETIC EFFECT OF RADIATION

Muller, Stadler, and Altenburg[13–16] independently discovered the mutagenic effect of radiation. Experimenting with x-rays on *Drosophila melanogaster*, barley, and maize, they found a considerable increase in the frequency of mutation. Radiation—x-rays, γ-rays, β-rays, fast neutrons, slow neutrons, and ultra-violet rays—and any other kind of mutagenic agents can induce point mutations at the DNA level or chromosomal aberrations.

In various organisms it was demonstrated that the number of mutations induced by radiation is proportional to the dose, i.e., the intensity of irradiation. Figure 15–14 shows, in the case of *Drosophila*, that the relation is a linear function over the range 25 to 9000 roentgen units (R).*

The effects of radiation are cumulative over long periods of time. For example, 0.1 R per day for 10 years is enough to increase the mutation rate to about 150 per cent of the spontaneous level.

---

*A roentgen (R) is the amount of radiation sufficient to produce two ions per μm^3 and is defined as the radiation required to liberate one electrostatic unit in 0.001293 gm of air.

In mammals the effects of "acute" (i.e., a period of seconds or minutes) and "chronic" (i.e., continuous slow exposure separated by days or weeks) irradiation show different rates of mutation. An acute exposure of 20 R to human cells in culture is sufficient to induce one chromosomal break in each cell. Chronic irradiation of the mouse testis produces fewer mutations in the spermatogonia than an acute dose at high intensity.[17, 18] For the molecular mechanism of ultraviolet irradiation in bacteria, see Chapter 19.

Chromosomal Aberrations

Radiations may induce chromosomal fragmentation and thus alter the structure of chromosomes. In contrast to gene mutations, chromosomal aberrations do not increase in direct proportion to the dose but rather increase exponentially (Fig. 15–15). More chromosomal aberrations are produced by continuous than by intermittent treatment.

A low dose of radiation may not be enough to cause fractures in one chromosome. As the dose increases, the number of breaks increases, and "aberrant" fusions become more and more likely. If the dose is intermittent or of low intensity, there is a greater chance that the broken ends of the chromo-

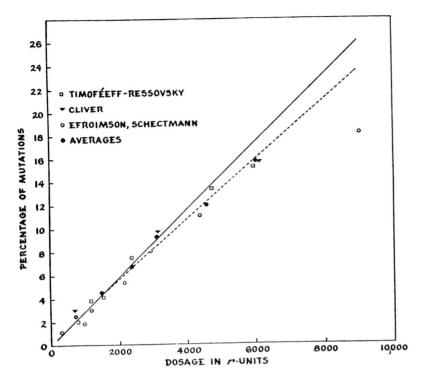

Figure 15–14. Relationship between doses of x-rays and frequency of lethal mutations in *Drosophila melanogaster.* (From Timoféeff-Ressovsky, Zimmer, and Delbrück.)

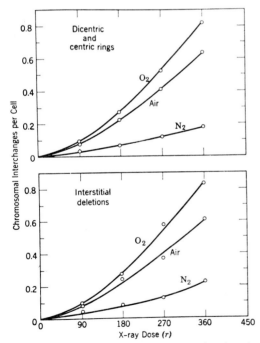

Figure 15–15. X-ray dosage curves, showing chromosomal aberrations included in *Tradescantia* in the presence of oxygen, air, and nitrogen. (From Giles, N. H., and Riley, H. P., *Proc. Natl. Acad. Sci. U.S.A.*, *35*:640, 1949.)

some will rejoin, or "heal," in the original chromosome structure before a second break could cause aberration.

There are three main types of structural alterations that may be induced by ionizing radiation:

(1) a *chromosomal type,* in which the two chromatids are fractured; (2) a *chromatid type,* in which fracture occurs in one chromatid; and (3) a *subchromatid type,* in which half a chromatid is involved. Chromatid aberrations are induced only in those chromosomal regions which were affected by irradiation after DNA synthesis; chromosomal aberrations are produced in regions that have not yet duplicated the DNA content. Irradiation generally produces localized lesions which may stabilize and form the so-called gaps (Fig. 15–16).[19] After some time, such gaps may be repaired. (For the mechanism of DNA repair, see Chapter 17.) Such a repair may result in complete restitution of the original structure *(true repair).* If there is no repair, the lesion becomes stabilized and cytologically visible. The process of restitution is inhibited by cold, cyanide, and dinitrophenol. Oxygen increases the number of

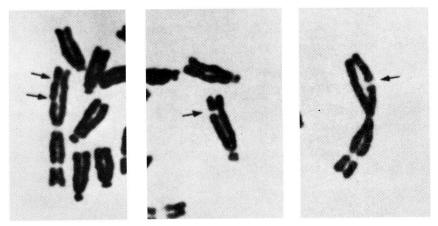

Figure 15–16. Chromatid gaps and breaks (arrows) induced by radiation. (From H. Evans.)

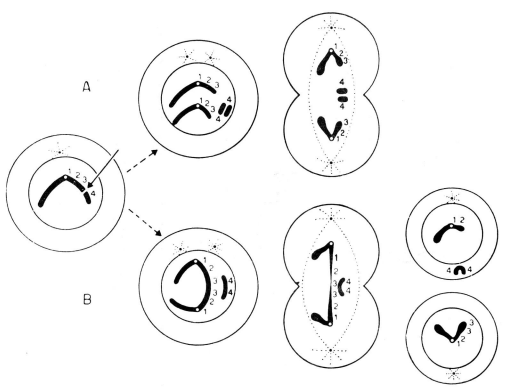

Figure 15–17. A single break in a chromosome between loci 3 and 4. **A,** the two parts of the broken chromosome reduplicate. The fragments with the centromere go to opposite poles. The fragments without the centromere remain in the equatorial region and are eliminated. **B,** the two parts of the broken chromosome reduplicate and the broken ends unite, forming a chromosome with two centromeres and another chromosome without a centromere. During mitosis the two centromeres move to opposite poles and the chromosome section between them breaks. The two daughter cells receive chromosomes with different constitutions. The fragment without a centromere is eliminated in the cytoplasm. (From Stern, C., *Principles of Human Genetics,* 2nd ed., San Francisco, W. H. Freeman & Co., 1960.)

fractures and chromosomal interchanges (Fig. 15–15).

If only one break is induced in one chromosome, a slight deficiency is observed. Aberrations resulting from two breaks (translocations, inversions, and large deletions) depend on the dose. The frequency of two simultaneous breaks is proportional to the square of the dose of radiation.[20]

When a chromosome breaks, the two fragments may either reunite or may remain separated permanently. If, instead of restituting at once, the broken ends reduplicate to form two chromatids, the fragments with the centromere migrate to different poles, and the fragments without a centromere (acentric) are eliminated in the cytoplasm (Fig. 15–17, A). Sometimes the sister fragments reduplicate and unite, forming a dicentric chromosome and another acentric chromosome. During mitosis the centromeres of the dicentric chromosome move to opposite poles and form a "bridge" between the daughter nuclei that finally breaks at some point (Fig. 15–17, B).

In man, double or multiple chromosomal breaks are induced by acute exposures (e.g., heavy medical irradiations, atomic accidents, or atomic warfare), whereas single breaks are produced at low doses. Chromosomal aberrations have been observed in blood cultures of humans who have had radiation treatments or injections of radioactive substances.[21, 22] The genetic effect of radiation has been studied in space flights. After Geminis III and IV, no increase in chromosomal aberrations was found in the blood cells of the astronauts who made the flights.[22]

Somatic Mutations

Somatic mutations are not transmitted from generation to generation but may be cumulative and produce severe changes in the individual, depending on the type of cell affected and the time at which the mutation occurs. Radiation can affect tissues that undergo mitosis as well as tissues in which cell division no longer takes place. If mutation occurs during early embryonic development, a large number of cells are affected. The majority of mutated genes are recessive and thus have no effect as long as the individual is heterozygous. However, if, in a descendant

cell, it becomes homozygous, its phenotype will be immediately manifested. It is probable that some cases of cancer produced in irradiated individuals are caused by somatic mutation.

Germ Cell Mutations

In contrast to somatic mutation, mutations in the germ cells may be transmitted to the offspring. However, in most cases of mutation caused by irradiation, both somatic and germ cells are frequently affected.

Even the lowest doses of radiation are genetically harmful, and the effects are dangerous to all organisms, from the simplest to man. In microorganisms and plants, however, a few useful mutations may be obtained by irradiation. For example, irradiation is one method of producing new antibiotics and plants with high economic value.

CYTOGENETIC ACTION OF CHEMICAL AGENTS

The discovery of the effects of colchicine on cell division in animals (Lits, 1934) and in plants (Dustin, Havas, and Lits, 1937; Gavaudan, P., Gavaudan, N., and Pompriaskinsky-Kabozieff, 1937) increased interest in the effects of chemical agents in general.

The effect of colchicine on the mechanism of mitosis has led to experimentation with many other substances to elucidate the physiology of normal and abnormal mitosis as well as the mechanism of mutagenesis.[23–27] The most frequently found effect of chemical agents is inhibition of mitosis and meiosis; in a few cases chemical agents have a stimulating effect.

The inhibitory action may result in: (1) chromosomal configurations that facilitate the counting and analysis of the karyotype; (2) polyploidy having innumerable practical applications; (3) the discovery of mutagens and anticancer substances; and (4) information about the basic mechanisms of inhibitory actions. Similar studies of meiosis have provided indirect information about the physiology of pairing and crossing over. In general, the specimen is observed immediately after the chemical action takes place. If the specimen survives, it can be maintained in a physiologic medium for a certain time for a study of residual effects or recuperation.

Inhibitors of the mitotic and meiotic processes are generally called "mitotic poisons."[28] These active agents can be endogenous (naturally occurring in tissues) or exogenous (developed or originating outside the organism).

Mitotic poisons can be grouped according to their chemical structure, or, more advantageously, according to their actions and the possible mechanisms involved. The substance may inhibit different phases of mitosis, or all phases of mitosis with different intensity. One or several mitotic mechanisms may be involved, e.g., nuclear membrane cycle, chromosomal condensation, behavior of the centromere, formation of the spindle and chromosomal movement, chromosomal duplication, nuclear cycle and metabolism of nuclear DNA, nuclear and cytoplasmic RNA, proteins, and the energy-producing processes for mitosis and meiosis.

The chemical agents that inhibit mitosis are here grouped as either (1) those that act at prophase and interphase, or (2) those that act at metaphase and the following phases.

Chemical Agents That Act at Prophase and Interphase

Some substances produce a change at the critical stages during which the chromosomes duplicate. Certain chemicals inhibit oxidation (cyanide, azide) or uncouple oxidative phosphorylation processes (2,4-dinitrophenol) that provide the energy for mitosis. Therefore, they prevent the mitotic process, but not chromosomal duplication. The mitotic phase does not take place, but the nuclear volume increases.

Other agents affect carbohydrate metabolism (e.g., the adrenal glucocorticoids) or interfere with chromosomal replication by changing the metabolism of DNA and protein.

Important substances are those that produce a chromosomal fragmentation similar to that caused by ionizing radiation. Well-known examples are the nitrogen mustards. A chromosomal fracture followed by reorganization, as in the situation induced by irradiation, is frequently observed. Research with chemical agents indicates that some areas of a chromosome are more sensitive than others.

Among these sensitive areas the centromere is one that is most frequently broken. (This effect has not been observed in similar research with radiation.) A highly active chemical agent can cause nuclear disintegration. In this process "DNA droplets" are forced out of the nucleus into the cytoplasm, and nuclear vacuolation followed by nuclear lysis finally takes place.[29]

Chemical Agents That Act at Metaphase and the Following Phases

This type of action is called *mitosis C* because it is produced principally by colchicine. This alkaloid, which affects the formation of the spindle, also acts on the chromosomes. The action on the spindle leads to different degrees of blockage of chromosomal division in metaphase and anaphase. Since chromosomal duplication is not affected, polyploidy may result. Treated with colchicine, chromosomes may continue the spiralization cycle; the two chromatids are contracted and repel one another, but remain united by the centromeres (ski configuration). The action of colchicine on the spindle microtubules is presented in Chapter 13. (For the mutagenic action of chemicals, see Chapter 19.)

Agents That Stimulate Mitosis

More is known about agents that may stimulate mitosis in plant than in animal cells. The plant hormones gibberellin, indoleacetic acid, and kinetin induce cell division in plant tissues. *Kinetin* (6-furfurylaminopurine) increases the mitotic rate in meristems of *Allium,* and at low concentrations, generally reduces the duration of interphase and increases the mitotic rate.

In animal cells under certain experimental conditions special hormones may act as stimulating agents (e.g., insulin and adrenocorticotropin). One of the most frequently used mitogenetic agents is *phytohemoagglutinin* (PHA) a glycoprotein extracted from *Phaseolus* spp. When applied to a blood sample, this substance agglutinates erythrocytes. After 48 to 72 hours the mononuclear leukocytes enter into mitosis.[30, 31] In Chapter 16 the use of PHA for the study of the human karyotype will be mentioned.

CYTOGENETICS AND EVOLUTION

The development of comparative cytology and cytogenetics has brought about great progress toward an understanding of evolution. McClung and S. Navashin were the first to emphasize the importance of cytogenetics to taxonomy and to the study of evolution by comparing genomes of related species. Systematics has been greatly advanced by cytogenetic investigation, which now provides many of the best methods for elucidating correlations between different taxonomic categories. In general, families, genera, and species are characterized by different genetic systems.

The study of the karyotype of different species has revealed interesting facts about both the plant and animal kingdoms (see Chapter 2 and Table 2–1). It has been demonstrated that individuals in wild populations are, to some extent, heterogeneous cytologically and genetically. In some cases, even if the genes are identical they may be ordered in a different way, owing to alterations of the chromosomal segments. These changes have an important bearing on the evolution of species.

The majority of plant species originate from an abrupt and rapid change in nature, and aneuploidy or polyploidy are the prime sources of variation. In the animal kingdom polyploidy is not so important. Among vertebrates, different species of fishes have a different number of chromosomes. Amphibians are generally characterized by a special number for each family. Reptiles and birds have large chromosomes (macrochromosomes) and small chromosomes (microchromosomes) that serve to differentiate them cytologically.

Owing to structural alterations, the number of centromeres may increase or decrease. Navashin's hypothesis that the variation in chromosome number is due to the fact that centromeres cannot originate *de novo* has been confirmed by experiments in both kingdoms. Matthey distinguishes between the basic chromosome number and the number of chromosomal arms, also called the fundamental number (FN). According to this concept, the metacentric chromosome has *two* arms and acrocentric and telocentric chromosomes have *one*. This is an important distinction in a group having both acrocentric and metacentric chromosomes, and the number of arms in each of the different species can be compared.

Another method used to study the cytogenetics of evolution is the application of measurements of total chromosomal area and DNA content.

With regard to the absolute size of their chromosomes mammals and birds constitute two independent groups. These two orders have different DNA contents and different sex-determining mechanisms. Speciation depends more on chromosomal rearrangements and mutation of individual genes than on changes in the total amount of genetic content.[32]

Two opposite changes in the number (and configuration) of chromosomes are of particular importance in evolution (Fig. 15–9). In *centric fusion*, a process that leads to a decrease in chromosome number, two acrocentric chromosomes join together to produce a metacentric chromosome (Fig. 15–9, 1). In *dissociation,* or *fission,* a process that leads to an increase in chromosome number, a metacentric (commonly large) and a small, supernumerary metacentric fragment become translocated, so that two acrocentric or submetacentric chromosomes are produced (Fig. 15–9, 2).

Fusion and dissociation are the main mechanisms by which the chromosome number can be decreased and increased during evolution of the majority of animals and in some groups of plants.

Studies of somatic and polytene chromosomes in several hundred species of *Drosophila* have elucidated the formation and evolution of this genus, which has been thoroughly analyzed from genetic, ecologic, and geographic standpoints.

Observation of chromosomal organization and of the different karyotypes in the individual, the species, genera, and the major systematic groups indicates that a chromosomal mechanism is involved in the process of evolution.

The problem of evolution should be considered from the different biochemical, cytologic, genetic, ecologic, and experimental aspects. All these methods and approaches should be used to analyze the intricate relationships between groups of organisms, particularly those that show marked variations.

Ionizing radiation—x-rays, γ-rays, β-rays, fast neutrons, slow neutrons, and ultraviolet light—can produce *point mutations* or *chromosomal aberrations*. The number of mutations increases proportionally with the dose of x-rays. The effect of radiation is cumulative. An exposure of cultured cells to 20 R (roentgen units) is sufficient to produce one chromosome break per cell. In contrast to mutations—which increase proportionally with the dose, chromosomal aberrations increase exponentially with the dose (Fig. 15–15). Breakage can be followed by "healing" of the broken end (see DNA repair in Chapter 17). (The breaks may be at the chromosome, the chromatid, or subchromatid level.) Chromatid breaks are produced in cells irradiated after the S period. If two breaks are produced, translocations, inversions, and large deletions may be induced. Dicentric chromosomes may be produced, which form a bridge at anaphase (Fig. 15–17). In the human, heavy medical irradiation, atomic accidents, or radioactive substances may produce chromosomal aberrations. In the production of new antibiotics and plants, irradiation is being used to economic advantage. *Somatic mutations* are not transmitted from one generation to another, whereas germ cell mutations may be passed to the offspring.

Mitotic poisons are classified according to the effect they have on the different phases of the mitotic cycle. Certain chemicals that inhibit oxidations or uncouple oxidative phosphorylation inhibit mitosis at interphase and prophase. Some chemicals act at metaphase, interfering with the formation of the spindle. (Colchicine is the best known in this group.) Mitosis stops at metaphase, and the chromosomes continue to condense, although they remain joined by the centromeres (ski configuration). Some chemicals stimulate mitosis. In plants, gibberellin, indoleacetic acid, and kinetin increase the rate of mitosis. In animals, certain hormones and *phytohemoagglutinin* (PHA) stimulate mitosis. (See Chapter 16 for an explanation of the way in which PHA is used in the study of the human karyotype.) There are also certain drugs (e.g., nitrogen mustards) which produce chromosomal fragmentation.

Cytogenetic studies have provided excellent methods for establishing taxonomic interrelationships and have, thereby, contributed to studies of *evolution* and *systematics*. One of the most frequent causes of evolution is changes in the order of genes as a result of chromosomal aberrations. In plants, aneuploidy and polyploidy are frequent sources of variation, whereas in mammals and birds, speciation depends more on chromosomal rearrangement and point mutations.

In the study of evolution, the number of chromosomes, the characteristics of the karyotype, the total chromosomal area, and the content of DNA are investigated. The presence of metacentric chromosomes may, in some cases, result from fusion of two acrocentric chromosomes. The contrasting

SUMMARY:
Mutagens, Mutations, and Evolution

phenomenon (i.e., dissociation) may lead to an increase in chromosome number. The problem of *evolution* is very complex and beyond the scope of this book, however, knowledge of cytogenetics is most fundamental to its understanding.

REFERENCES

1. Taylor, J. H. (1965) *J. Cell Biol., 25*:57.
2. White, M. J. D. (1954) *Animal Cytology and Evolution.* Cambridge University Press, London.
3. Beçak, W. (1969) *International Symposium on Nuclear Physiology and Differentiation. Genetics, 61*:183.
4. Beçak, W., Beçak, M. L., and Rebello, M. N. (1967) *Chromosoma, 22*:192.
5. Fankhauser, G. (1945) *Quart. Rev. Biol., 20*:20.
6. Sachs, L. (1952) *Heredity, 6*:357.
7. Eigisti, O. J., and Dustin, P. (1955) *Colchicine.* Iowa State College Press, Ames, Iowa.
8. Levan, A. (1956) *Ann. N.Y. Acad. Sci., 63*:774.
9. Makino, S. (1956) *Ann. N.Y. Acad. Sci., 63*:818.
10. Ford, C. E., Hamerton, J. L., and Mole, R. H. (1958) *J. Cell. Comp. Physiol., 52*:235.
11. Hauschka, T. S. (1958) *J. Cell. Comp. Physiol., 52*:197.
12. Puck, T. T. (1959) Quantitative studies on mammalian cells "in vitro." *Biophysical Science* p. 433. (Oncley, J. L., ed.) John Wiley and Sons, New York.
13. Muller, H. J. (1927) *Science, 66*:84.
14. Muller, H. J. (1928) *Z. Abstam. verebungsl.* suppl. *1*:234.
15. Stadler, L. J. (1928) *Science, 68*:186.
16. Altenburg, E. (1928) *Amer. Nat., 62*:540.
17. Russell, W. L. (1954) In: *Radiation Biology,* Vol. 1, p. 825. (Hollaender, A., ed.) McGraw-Hill Book Co., New York.
18. Russell, W. L., Russell, L. B., and Kelly, E. M. (1958) *Science, 128*:1546.
19. Evans, H. J. (1967) *Radiation Research.* North-Holland Publishing Co., Amsterdam.
20. Giles, N. H. (1955) *Brookhaven Symp. Biol., 8*:103.
21. Bender, M. A., and Gooch, P. C. (1962) *Radiat. Res., 16*:44.
22. Bender, M. A., Gooch, P. C., and Kondo, S. (1968) *Radiat. Res., Res., 34*:228.
23. Muller, H. J. (1959) *Acta Genet. Stat. Med., 6*:157.
24. Neel, J. V. (1958) *Amer. J. Hum. Genet., 10*:398.
25. Turpin, R., Lejeune, J., and Rethore, M. O. (1956) *Acta Genet. Stat. Med., 6*:204.
26. Puck, T. T. (1959) *Rev. Mod. Physics, 31*:433.
27. Stern, C. (1973) *Principles of Human Genetics.* W. H. Freeman Co., San Francisco.
28. Biesele, J. J. (1958) *Mitotic Poisons and the Cancer Problem.* Elsevier Publishing Co., New York.
29. Saez, F. A., and Drets, M. (1958) *Port. Acta Biol. A., 5*:287.
30. Nowell, P. C. (1960) *Cancer Res., 20*:462.
31. Agrell, I. P. S., and Karlsson, B. W. (1967) *Exp. Cell Res., 48*:634.
32. Beçak, W., Beçak, M. L., Mazanth, H. R. S., and Ohno, S. (1964) *Chromosoma, 15*:606.

ADDITIONAL READING

Ashton, B. G. (1967) *Genes, Chromosomes, and Evolution.* Longmans, Green, and Co., London.
Auerbach, C. (1967) The chemical production of mutations. *Science, 158*:1145.
Biesele, J. J. (1958) *Mitotic Poisons and the Cancer Problem.* Elsevier Publishing Co., New York.

Bodmer, W. F., and Darlington, A. J. (1969) Linkage and recombination at the molecular level. In: *Genetic Organization*, Vol. 1, (Caspari, E. W., and Ravin, A. W., eds.) Academic Press, Inc., New York.

Darlington, C. D., and Bradshaw, A. D. (1964) *Teaching Genetics*. Oliver & Boyd, Edinburgh.

Dobzhansky, T. (1958) *Genetics and the Origin of Species*. 3rd Ed. Columbia University Press, New York.

Eigisti, O. J., and Dustin, P. (1955) *Colchicine*. Iowa State College Press, Ames, Iowa.

Evans, H. J. (1967) *Radiation Research*. North-Holland Publishing Co., Amsterdam.

Hayes, W. (1964) *The Genetics of Bacteria and Their Viruses*. Blackwell Scientific Publications, Oxford.

Henderson, S. A. (1969) Chromosome pairing, chiasmata, and crossing-over. In: *Handbook of Molecular Cytology*, p. 327. (Lima-de-Faría, A., ed.) North-Holland Publishing Co., Amsterdam.

Hollaender, A. (1954) *Radiation Biology*. 3 volumes. McGraw-Hill Book Co., New York.

Jacobs, P. A., Price, W. H., and Lou, P. (1970) *Population Cytogenetics*, Edinburgh University Press, Edinburgh.

John, B., and Lewis, K. R. (1965) The meiotic system. *Protoplasmatologia*, VIF. I, 1–335, Berlin.

Lea, D. E. (1955) *Actions of Radiations on Living Cells*. 2nd Ed. Cambridge University Press, London.

Levan, A. (1967) Some current problems of cancer cytogenetics. *Hereditas, 57*:343.

Lewis, K. R., and John, B. (1963) *Chromosome Marker*. J. & A. Churchill, Ltd., London.

Lewis, K. R., and John, B. (1964) *The Matter of Mendelian Heredity*. J. & A. Churchill, Ltd., London.

Neel, J. V. (1963) *Changing Perspectives on the Genetic Effects of Radiation*. Charles C. Thomas, Springfield, Ill.

Stern, C. (1973) *Principles of Human Genetics*. Freeman & Co., San Francisco.

Stern, C. (1967) Genes and people. *Perspect. Biol. Med., 10*:500.

Swanson, C. P. (1957) *Cytology and Cytogenetics*. Prentice-Hall, Inc., Englewood Cliffs, New Jersey.

Taylor, J. H. (1963) *Molecular Genetics*, Part I. Academic Press Inc., New York.

Taylor, J. H. (1967) *Molecular Genetics*. Part II. Academic Press Inc., New York.

Taylor, J. H. (1967) Meiosis. *Encyclopedia of Plant Physiology, 18*:344, Springer-Verlag, Berlin.

Wagner, R. P., and Mitchell, H. K. (1964) *Genetics and Metabolism*. 2nd Ed. John Wiley & Sons, New York.

White, M. J. D. (1961) *The Chromosomes*. 5th Ed. Methuen & Co., London.

Wilson, G. B., and Morrison, J. H. (1961) *Cytology*. Reinhold Publishing Corp., New York.

Wolff, S. (1963) *Radiation-induced Chromosome Aberrations*. Columbia University Press, New York.

sixteen

SEX DETERMINATION AND HUMAN CYTOGENETICS

SEX DETERMINATION

The fact that male and female individuals are found in about equal numbers led to the belief that sex determination is directly related to heredity. Studies of sex determination have demonstrated that the male and female characteristics are transmitted from one generation to the next in the same way as any other hereditary characteristics.

There is physiologic and cytologic proof that sex is determined as soon as the egg is fertilized and that it depends on the gametes. Among the physiologic evidence is the finding that identical twins—which originate from a single zygote—are always of the same sex. Furthermore, in certain species having polyembryonic development (e.g., armadillo), all the embryos that have developed from a single fertilized egg are of the same sex. Cytologic evidence was first obtained by McClung,[1] who demonstrated that the karyotype of a cell is composed of not only common chromosomes (autosomes) but also of one or more special chromosomes that are distinguished from the autosomes by their morphologic characteristics and behavior. These were called accessory chromosomes, allosomes, heterochromosomes, or sex chromosomes.

In certain species the gametes are not identical with respect to the sex chromosomes. One of the sexes is heterozygous, producing two types of gametes. The other is homozygous, producing only one type of gamete. Therefore only two combinations of gametes are possible in fertilization, and the result is 50 per cent males and 50 per cent females (Fig. 14-1).

Sex Chromosomes

The majority of organisms have a pair of sex chromosomes, which, in the course of evolution, have been specialized for sex determination. One of the sexes has a pair of identical sex chromosomes (XX), the other may have a single sex chromosome, which may be unpaired (XO) or paired with a Y chromosome (XY) (Fig. 14-1). The XY pair is also called heteromorphic because of the different morphology of the chromosomes.

In many species spermatogenesis produces two kinds of spermatozoa in similar proportion. Oögenesis, on the other hand, produces only one kind of gamete (ova) (see Figure 14-1). This type of sex determination is found in mammals, including the human, and in certain insects, such as Drosophila. The male is heterogametic, whereas the female is homogametic.

In other vertebrates (birds, some reptiles, and fishes) and invertebrates (e.g., insects of the order Lepidoptera), the female is heterogametic and the male is homogametic. In this instance there are two kinds of ova (X and Y) and only one kind of spermatozoa (X). In Orthoptera, males are XO and females XX. In some cases sex is determined by the Y chromosome, as in the case of the axolotl in which sex depends on the presence or absence of this chromosome.[2]

In the human the Y chromosome determines the male sex. Thus, an XO individual (lacking the Y chromosome) resembles a female but lacks ovaries (Turner's syndrome). On the other hand, an XO mouse is a normal female. The Y chromosome probably determines the male sex in all mammals.

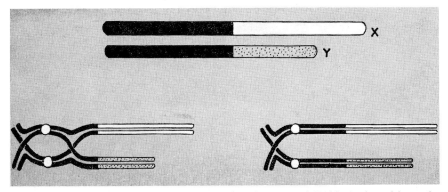

Figure 16-1. Diagram of a pair of XY sex chromosomes of a mammal. In black, the pairing or homologous segments; in white, the differential segments of the X chromosome; stippled, the differential segments of the Y chromosome. The configuration of the bivalent depends on the position of the chiasmata, which is produced only in the homologous segment.

Sex chromosomes are not distinguishable in some animals. In such cases, the sex-determining genes are probably confined to a short region of a pair of chromosomes.

The sex chromosomes can be thought of as composed of a *homologous* and a *differential region*. The homologous region corresponds to the pairing segment, and when recombination takes place, it is limited to this part (Fig. 16-1). The differential region influences sex determination.

Sex Chromatin and Sex Chromosomes

Barr and Bertram[3] opened an important field by their discovery in 1949 of a small chromatin body (a chromocenter) in nerve cells of the female cat—a structure which is absent in the male. These observations were then made in other tissues and animals, including the human. In nuclei of the epidermis of females, this chromatin body, called the sex chromatin, or Barr body, is found in much higher proportion than in males (Fig. 16-2 C, D).[3, 4]

The frequency with which sex chromatin can be detected in the female varies from tissue to tissue. In nervous tissue the frequency may be 85 per cent, whereas in whole mounts of amniotic or chorionic epithelium it may be as high as 96 per cent. In oral smears the frequency may vary between 20 and 50 per cent in normal females (see Hamerton, 1969).

Sex chromatin appears in the interphase nucleus as a small chromocenter heavily stained with basic dyes; it gives a positive Feulgen reaction (Fig. 16-2, C). It can be found attached to the nucleolus, as in nerve cells of certain species (Fig. 16-2, A); attached to the nuclear membrane, as in cells of the epidermis or of the oral mucosa; free in the nucleoplasm, as in neurons after electric stimulation (Fig. 16-2, B); and as a nuclear expansion, the best-known example being that of the neutrophil leukocyte in which the sex chromatin appears as a small rod called the drumstick (Fig. 16-2, D). This characteristic of the leukocyte has been utilized as a test of sex determination, along with the investigation of the basal cells of the epidermis and smears of the oral mucosa (Fig. 16-2, F).

The study of sex chromatin has a wide field of medical applications and offers the possibility of relating the origin of certain congenital diseases to chromosome anomalies. Among these applications is the diagnosis of sex in intersexual states in postnatal and even in fetal life (see the sections dealing with human cytogenetics in this chapter).

The relationship between sex chromatin and sex chromosomes has been elucidated.[5] Sex chromatin is derived from only one of the two X chromosomes; the other X is not heteropyknotic at interphase. The number of corpuscles of sex chromatin at interphase is equal to $nX - 1$. This means that there is one Barr body fewer than the number of X chromosomes. This relationship between sex chromatin and sex chromosomes is particularly evident in some humans who have an abnormal number of sex chromosomes (see Table 16-1).

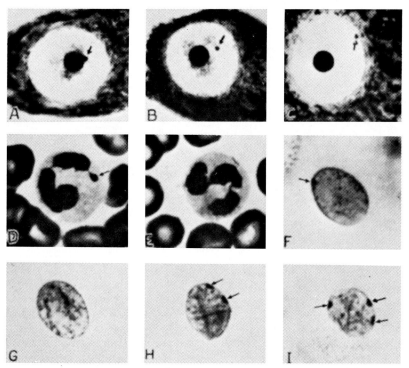

Figure 16–2. Sex chromatin in a nerve cell of a female cat. **A,** near the nucleolus; **B,** in the nucleoplasm; **C,** under the nuclear membrane (from M. L. Barr); **D,** normal leukocyte with a drumstick nuclear appendage from a human female. ×1800; **E,** same as D, in a male. (Remember that 90 per cent of females also lack the drumstick, as in the male.) ×1800. **F,** one sex chromatin corpuscle (*arrow*) in a nucleus from an oral smear. ×2000; **G,** same as *F*, from a male. Notice the lack of sex chromatin. ×1800. **H,** nucleus from the XXX female with two sex chromatin bodies. Vaginal smear. ×2000. **I,** similar, from an XXXX female. The three Barr bodies are indicated by arrows. ×2000. (From Barr, M. L., and Carr, D. H., *in* Hamerton, J. L. (Ed.), *Chromosomes in Medicine,* Medical Advisory Committee of the National Spastics Society in association with Wm. Heinemann, Little Club Clinics in Developmental Medicine, No. 5, 1963.)

The Single X Nature of Sex Chromatin

The differential behavior of the two X chromosomes in the female led Lyon (see Lyon, 1966, 1972) to the so-called *inactive X hypothesis* by which: (1) only one of the X chromosomes is genetically active; (2) the X undergoing heteropyknosis may be either of maternal or paternal origin, and the decision by which X becomes inactive is taken at random (Fig. 16–3); and (3) the inactivation occurs early in embryonic life and remains fixed. It is now admitted that only a part of the X chromosome condenses into a Barr corpuscle; this condensed chromosome is said to contain *facultative heterochromatin,* as opposed to the *constitutive heterochromatin* found in other chromosomes (see Chapter 12).

Inactivation of the X chromosome takes place even in 3X and 4X individuals (Fig. 16–2, *H* and *I*). The inactivation starts in the human in the late blastocyst stage, on about the sixteenth day of embryonic life; the inactivated X chromosome then remains heterochromatic in the somatic cells (Fig. 16–3).[6, 7]

The fact that X inactivation occurs at random has been demonstrated in human diseases linked to the X chromosome. The *Lesch-Nyhan syndrome,* in which a deficiency of one enzyme of the purine metabolism (i.e., hypoxanthine-guanine phosphoribosyl transferase) produces mental retardation and increased uric acid levels, results from a recessive mutation in the X chromosome. When fibroblasts of heterozygous women are cultured in vitro, two types of cell clones are obtained. Half the clones contain the enzyme, whereas the other half lack the enzyme.[8]

TABLE 16-1. Sex Aneuploids in Man

X \ Y	O	Y	YY	Sex Chromatin
x	Monosomic XO Turner's syndrome 2X − 1 2n = 45	Disomic XY Normal 2X 2n = 46	XYY	0
xx	Disomic XX Normal 2X 2n = 46	Trisomic XXY Klinefelter's syndrome 2X + 2 2n = 47	Tetrasomic XXYY Klinefelter's syndrome 2X + 2 2n = 48	1
xxx	Trisomic XXX Metafemale 2X + 1 2n = 47	Tetrasomic XXXY Klinefelter's syndrome 2X + 2 2n = 48		2
xxxx	Tetrasomic XXXX Metafemale 2X + 2 2n = 48	Pentasomic XXXXY Klinefelter's syndrome 2X + 3 2n = 49		3
Phenotype	♀	♂	♂	

Heteropyknosis of the Sex Chromosomes and the Sex Vesicle

During meiosis the XY pair is embedded in the so-called sex vesicle in most mammalian species. This vesicle is apparent during the zygonema and pachynema stages, at which time the XY chromosomes are not heteropyknotic. At the end of prophase, when the vesicle disintegrates, the XY bivalent becomes heteropyknotic again.

In the mouse the sex chromosomes, and

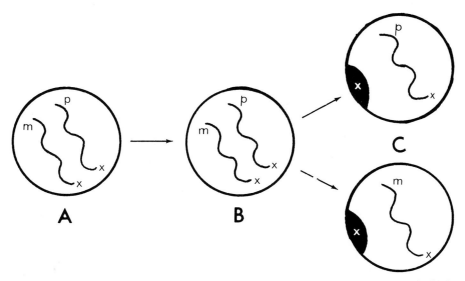

Figure 16-3. Diagram of the evolution of XX chromosomes. **A,** in the zygote both the paternal (*p*) and maternal (*m*) X are euchromatic. **B,** in the early blastocyst the same is true as in **A. C,** in late blastocyst 50 per cent of the cells have a maternal heterochromatic X chromosome, and the other 50 per cent have the paternal heterochromatic X chromosome.

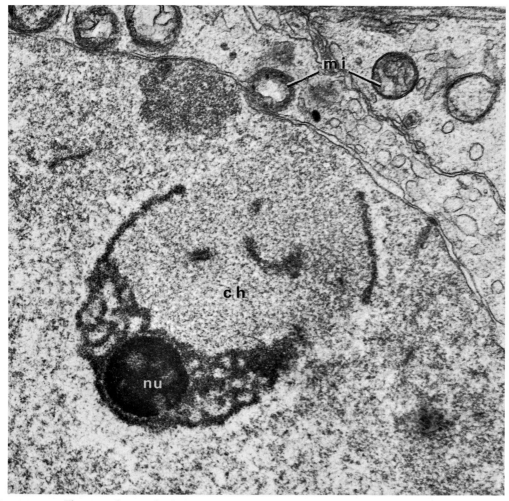

Figure 16–4. Electron micrograph of a sex vesicle of a mouse showing the crescent shaped nucleolar zone containing a dense body; *ch*, chromatin part of the vesicle; *nu*, nucleolus; *mi*, mitochondria. ×20,000. (Courtesy of A. J. Solari and L. Tres.)

hence the sex vesicle, are concerned with the formation of the nucleolus (i.e., they contain the nucleolar organizers). In this case the vesicle has two zones: (1) the *chromatic* zone, oval in shape and attached to the nuclear membrane, and (2) the *RNA-containing- or nucleolar-zone,* which has a crescent shape. The chromatic zone contains thin, convoluted DNA microfibrils and has some filamentous cores that represent the lateral components of the synaptonemal complex (see Chapter 14). The nucleolar zone contains RNA and is Feulgen negative. In the center of this zone there is a dense round body; the rest is a spongelike structure made of granules of about 15 nm (Fig. 16–4).[9] The human

sex vesicle contains no RNA, and the nucleolar zone is absent.[10]

Neo-XY System of Sex Determination

In addition to the common type of sex determination based on XY chromosomes, in several species a special type of sex determination has been observed and designated the *neo-XY system.* As shown in Figure 16–5 this arises from the fracture of the X chromosome, followed by fusion of the main fragment to one autosome. This association constitutes the *neo-X* chromosome. At meiosis the other autosome of the pair forms

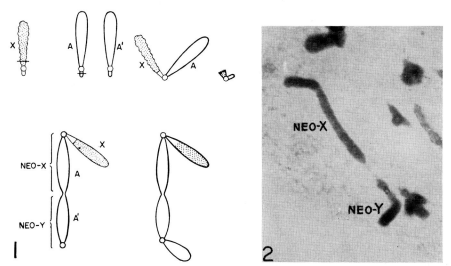

Figure 16–5. **1,** the mechanism of centric fusion between a member (A) of a pair of autosomes (AA′) and the sex chromosome (X). The small fragment with the centromere at the right is lost. **Below, left,** the neo-X – neo-Y chromosome at metaphase 1; **right,** the same element, showing the neo-Y with a submedian centromere. **2,** the neo-X – neo-Y chromosome during the first meiotic metaphase in *Aleuas lineatus*. (After F. A. Saez.)

the so-called *neo-Y* chromosome and remains confined to the male sex.

There is evidence that the neo-Y chromosome gradually becomes heterochromatic. It was demonstrated by radioautography with ^{3}H-thymidine that. as in the case of the X chromosome, the neo-Y chromosome replicates late in the synthetic period.[12]

Gynandromorphs

Gynandromorphism is a genetic mosaic of both male and female sexual characters present in the same individual. A gynandromorph possesses chromosomes of both sexes.

In *Drosophila* a gynandromorph is produced by the elimination of one of the X chromosomes during division of the egg.[11] Figure 16–6 shows an individual in which the right half is male and the left half, female. Gynandromorphs are common among silkworms and bees. The occurrence of gynandromorphism in vertebrates is difficult to assess because it depends on the hazardous distinction between gynandromorphism and intersexuality due to hormonal effect (see below).

Sex Differentiation

Although the primary determination of sex is made at fertilization (Fig. 14–1) the

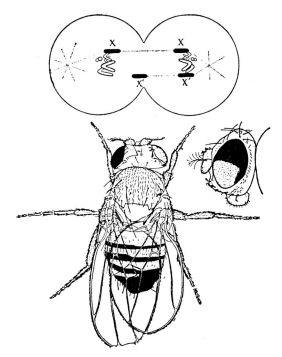

Figure 16–6. Gynandromorph of *Drosophila.* **Above,** first division in the segmentation of the egg, showing the elimination of an X chromosome. **Below,** the resulting gynandromorph individual, the left side of which is female (XX) and the right side male (XO). **Right,** head of a fly. The X chromosome has been eliminated in one of the last somatic mitoses, showing a red color spot in the eye. (After Morgan, Bridges, and Sturtevant; taken from Waddington, 1939.)

embryo acquires its definite sex characteristics by a more complex mechanism. An epigenetic factor (i.e., hormonal) may assume control of the genetic determination during development, thereby changing the phenotypic direction of sex. Among vertebrates a condition of bisexuality may exist (e.g., the coexistence of structures of the functional sex, together with primordia of the heterologous sex). For example, male amphibians have a rudimentary ovary (Bidder's organ) and vestigial oviducts.

In the human embryo until the sixth week the gonads and the primordia of the urogenital tract are identical in males and females. At this time the gonad has already been invaded by the primary XX or XY germ cells. The gonad differentiates into a definite testicle at the seventh week, whereas the female gonad differentiates between the eighth and ninth weeks of development. An important epigenetic factor at the time of differentiation is the production of androgens by somatic cells in the embryonic male gonad; in the female this hormone is lacking. Administration of androgen to the mother at this time may produce a shift in the differentiation of the genitalia into a male type (feminine pseudohermaphroditism). Gonadal differentiation in the human male probably depends on the production of a local hormone correlated with the presence of the Y chromosome. This hormone accelerates the development of the testis, whereas in the female the absence of the hormone permits the slower development characteristic of the ovary (see Jost, 1970).

SEX-LINKED INHERITANCE

Sex-linked genes are those carried by the sex chromosomes and yet not involved in sex determination.

The following is an example of sex-linked inheritance in *Drosophila:* when a homozygous red-eyed female (dominant) is crossed with a white-eyed male (recessive), all individuals in the F_1 are red-eyed (Fig. 16–7, *1*), but when the cross is between a white-eyed female and a red-eyed male, male offspring in the F_1 have white eyes. These experiments demonstrate that the gene for red eye in this case is carried by the X chromosome, but not by the Y.

In organisms with an XY type of sex determination, genes may be present in the differential segments of the X and Y (Fig. 16–1). Such genes are not alleles, since they are in a nonhomologous section of the chromosome. They are completely linked and crossing over cannot occur in them.

There are three types of sex-linked inheritance: (1) *X-linked,* by genes localized in the nonhomologous section of X and which

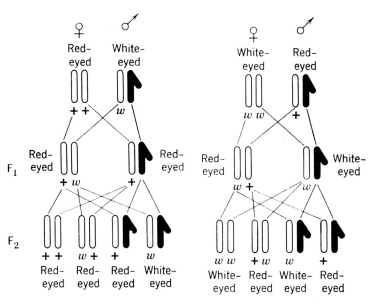

Figure 16–7. Sex-linked inheritance of eye color in *Drosophila.* Reciprocal crosses of: **1,** a wild-type, red-eyed female to a white-eyed male and **2,** a white-eyed female to a wild-type, red-eyed male. (See the description in the text.) (From Morgan, Sturtevant, Muller, and Bridges, 1919.)

have no corresponding alleles in Y; (2) *Y-linked*, by genes localized in the nonhomologous section of Y and which have no alleles in X; and (3) *XY-linked*, by genes localized in one chromosomal segment which is homologous in both X and Y (the so-called *incomplete* linkage).

Genes Linked to the X Chromosome

The classic example of this form of sex-linkage is the red color in *Drosophila*. In man the genes that determine *daltonism* (i.e., red-green color blindness) and *hemophilia* are linked to the X chromosome. Eight per cent of males have daltonism, whereas this is found in only 0.5 per cent of females. In the latter, both X chromosomes are altered at the same locus. Hemophilia (a defect of blood clotting) is inherited as a sex-linked recessive gene. Rarely is a female a hemophiliac. In such a case the father is a hemophiliac and the mother is a carrier of hemophilia.

Other X-linked genes produce the following conditions in the human: *ichthyosis,* *myopia, Gower's muscular atrophy,* and one type of color blindness. All these anomalies are transmitted in the same way as the "white-eyed" trait in *Drosophila;* the same reasoning can be followed to obtain F_1 and F_2.

Genes Linked to the Y Chromosome

Genes in the nonhomologous region of the Y chromosome pass directly from father to son. For example, *ichthyosis hystrix gravis* and other diseases follow the male line.

Genes Localized in the Homologous Segments of Both X and Y Chromosomes

These genes are inherited as the autosomal genes. They are *partially sex-linked.* In the human there are several defects of this type, among which are total color blindness, two skin diseases (*xeroderma pigmentosum* and *epidermolysis bullosa*), *retinitis pigmentosa, spastic paraplegia,* and other diseases.

Sex determination is transmitted from one generation to the next by a hereditary mechanism. In many species, as in the human, the gametes are not identical with respect to the sex chromosomes. One of the sexes is heterozygous; the other is homozygous. Upon fertilization the result is 50 per cent males and 50 per cent females (Fig. 14–1). In most cases a pair of sex chromosomes is involved in sex determination (XX in females, XY or XO in males). Frequently the male is heterogametic, but in certain cases (birds, fishes) the female is heterogametic. In the human and most mammals the male character is determined by the Y chromosome. Sex chromosomes have a homologous region in which recombination may occur and a differential region that is related to sex determination (Fig. 16–1).

Sex chromatin is a small chromatin body observed in the interphase nucleus of females and has a medical application in the diagnosis of intersexual states. Sex can be determined in the fetus by the study of smears of amniotic epithelium. Sex chromatin is derived from only one of the two X chromosomes, which becomes heteropyknotic. The number of chromatin bodies is nX − 1. It is thought that the X chromosome forming the sex chromatin is genetically inactive — an inactivation which occurs early in embryonic life (i.e., *facultative heterochromatin*). This inactivation occurs at

SUMMARY:
Cytogenetics of Sex
Determination

random, as can be demonstrated in cell clones of a human hereditary disease involving an enzyme deficiency in purine metabolism.

In certain species there is a so-called *neo-XY* system of sex determination in which the X chromosome has fractured and is fused to an autosome. A *gynandromorph* is a mosaic individual having male and female sexual characteristics and chromosomes of both sexes (Fig. 16–6). In the human, sex is determined primarily by the sex chromosomes; then an epigenetic factor (hormonal) assumes control of *sex differentiation*. Until the sixth week of embryonic life, both male and female gonads are identical. At this time they are invaded by XX or XY *germ cells*. The testicle differentiates in the seventh week, whereas the ovary begins to develop one or two weeks later. The early differentiation in the male depends on a local production of androgen. In the female the lack of androgen results in the slower development of the ovary.

Sex-linked inheritance is brought about by genes carried in sex chromosomes. In an XY individual the genes present in the differential segment of both sex chromosomes are not allelic and are linked (i.e., do not recombine). There are X-linked and Y-linked genes and genes having an incomplete linkage. A classical type of X-linked inheritance is the red eye in *Drosophila* (Fig. 16–7). In the human, *daltonism* and *hemophilia* are linked to the X chromosome. A few diseases are Y-linked and follow the male line, whereas others are partially linked (*xeroderma pigmentosum, spastic paraplegia,* and others).

HUMAN CYTOGENETICS

In recent years the advances in genetics and cytology have been applied to man, opening new fields with important biological and medical implications. These advances have resulted from the use of more refined techniques for studying chromosomes, by which the human karyotype of a relatively large number of patients and normal individuals from the general population have been studied in detail.

Since 1959, when Lejeune *et al.* detected an extra chromosome in mongoloid patients,[13] human cytogenetics has received a considerable degree of attention. Three years before, Tjio and Levan[14] had demonstrated in fibroblasts from normal human embryos that the diploid number was 46 (i.e., 44 autosomes + XY in the male and 44 + XX in the female) (Fig. 16–8).

Tissue cultures of fibroblasts, bone marrow, skin, and peripheral blood combined with the action of colchicine and hypotonic solutions to block mitosis at metaphase and to separate the chromosomes were used to study the human karyotype. An important technical advance has been the introduction of *phytohemagglutinin*, which induces lymphocytes to transform into "lymphoblast-like" cells that start to divide 48 to 72 hours after exposure. The strong mitogenic properties of this substance allowed the development of microtechniques which employ small amounts of blood. More recently, cultures of amniotic fluid have been used in chromosome diagnosis. This technique has been useful for karyotyping those cases of mothers who have borne children with chromosome abnormalities or those who are carriers of a balanced translocation in whom the risk of conceiving an affected child is high.

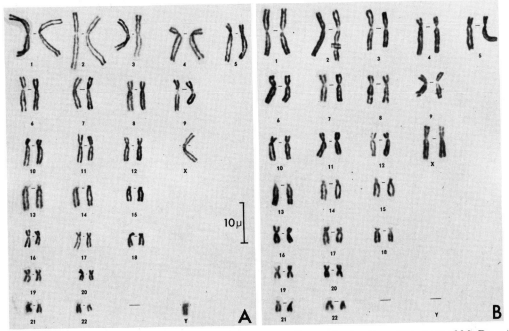

Figure 16-8. Human male (**A**) and female (**B**) karyotypes from a blood culture. (Courtesy of M. Drets.)

The Normal Human Karyotype

In 1960 a study group meeting in Denver devised a standardized system of nomenclature for human chromosomes. The 22 pairs of autosomes were numbered in descending order of length and classified, according to the position of the centromere, as *metacentric, submetacentric,* and *acrocentric* chromosomes (see Chapter 2). Patau[15] suggested that similar groups of chromosomes be designated by a capital letter, and this suggestion was followed in the London report (1963). Table 16-2 shows the nomenclature of chromosome groups proposed by the Denver and London reports and gives a brief description of chromosomes in each group.

Banding Patterns in Human Chromosomes

In recent years special techniques have been developed to identify accurately each chromosome pair in the human karyotype. A

TABLE 16-2. Characteristics of the Chromosomes in the Human Karyotype

Denver Report	*London Report*	*Description*
Group 1-3	Group 1-3 (A)	Large chromosomes with approximately median centromeres; 1, 2, and 3 can usually be identified morphologically
Group 4-5	Group 4-5 (B)	Large submetacentric chromosomes
Group 6-12	Group X, 6-12 (C)	Medium-sized submetacentric chromosomes
Group 13-15	Group 13-15 (D)	Large acrocentric chromosomes
Group 16-18	Group 16-18 (E)	No. 16 is metacentric; No. 17-18 are small submetacentric chromosomes
Group 19-20	Group 19-20 (F)	Small metacentric chromosomes
Group 21-22	Group 21-22 + Y (G)	Short acrocentric chromosomes. (The Y chromosome belongs to this group, but has no satellites; it is of variable size and can usually be identified morphologically)
Sex chromosomes Y X		

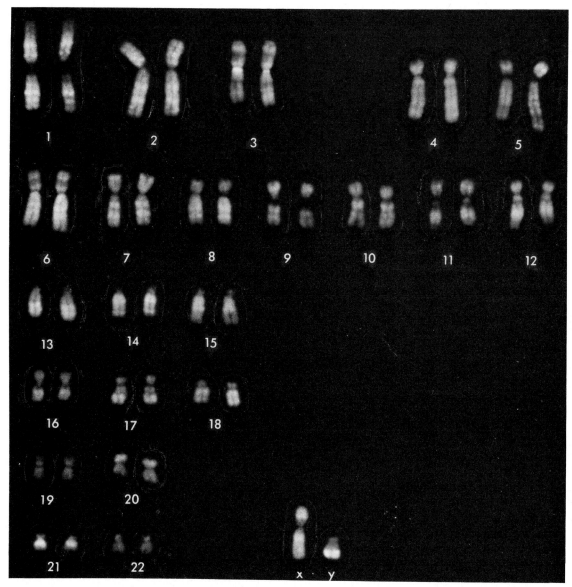

Figure 16–9. Human karyotype showing the fluorescent bands produced by staining with quinacrine mustard. Q bands are shown here. (Courtesy of T. Caspersson.)

banding pattern was observed by using fluorescent staining with quinacrine mustards[16] (Fig. 16–9). It was also found that quinacrine specifically stains the Y chromosome, not only at metaphase, but also during interphase. (This fact has helped in finding numerical variations of the Y chromosome [Fig. 16–10].) Other techniques are based on the treatment of the cells with alkali or acids and on staining with Giemsa (a classic blood stain). Methods have been developed to detect constitutive heterochromatin—mentioned in Chapter 12—in human chromosomes.[17] With Giemsa, there is a bright staining of bands near the centromere (C banding); however, with changes in the technique, a more complex banding pattern along the chromosome arms has been obtained[18] (Fig. 16–11). Other procedures involve enzymatic digestion of chromosomes with trypsin, DNA denaturation and reannealing, and, more recently, the use of antinucleoside antibodies.

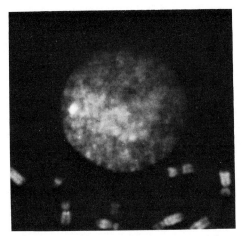

Figure 16-10. Staining, by quinacrine mustard, of the Y chromosome in a human interphase cell. The strongly fluorescent body is clearly distinguished within the nucleus. (Courtesy of T. Caspersson.)

These techniques have led to the observation of several types of banding patterns [e.g., quinacrine (Q bands); centromeric (C bands); Giemsa (G bands); telomeric (T bands); and reverse (R bands)]. These patterns were used in the Paris conference of 1971 to establish a new map of human chromosomes. Since the banding is specific for each chromosome pair, a more refined analysis of normal and abnormal chromosomes can be made with these methods.

Abnormal Human Karyotypes

Deviations from the normal karyotype are found in autosomes, sex chromosomes, or in both. They generally consist of aneuploidy, such as monosomy or trisomy. Structural aberrations, such as translocations, deficiency, duplication, and other more complex alterations have also been observed (see Chapter 15).

The most common structural aberration in the human is *reciprocal translocation.*

Various mechanisms are involved in the production of abnormal karyotypes. A frequent one, which may give rise to different types of aneuploidy, is *nondisjunction.*

Mitotic nondisjunction may occur at the mitotic division that precedes the formation of germ cells or during cell division of the zygote. In the first case the effects are similar to those occurring in meiotic nondisjunction;

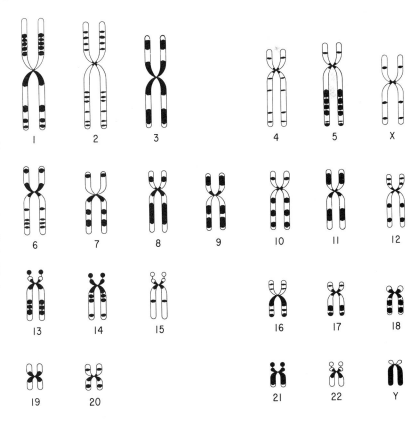

Figure 16-11. Diagram showing the banding pattern of human chromosomes revealed with the Giemsa stain. Since the banding is specific for each chromosome pair, this method, as well as that involving fluorescent stains (Caspersson, T., et al., 1970), is of great help in preparing a more refined analysis of normal and abnormal chromosomes. (Modified from Drets, M. E., and Shaw, M. W., *Proc. Natl. Acad. Sci. USA*, 68:2073, 1971.)

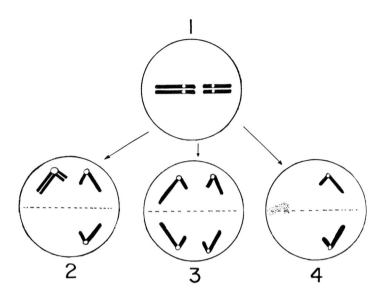

Figure 16–12. Mitotic nondisjunction and chromosome loss. **1**, normal metaphase. **2**, nondisjunction anaphase giving rise to monosomic and trisomic nuclei. **3**, normal anaphase. **4**, a chromosome loss results in two monosomic nuclei.

but in the second case, since the alteration occurs early in embryonic development, a *mosaic* of different cell lines occurs.

The immediate cause of nondisjunction is the lagging of one sister chromatid in anaphase, which, at telophase, remains in one of the cells together with the other sister chromatid (Fig. 16–12). This change gives rise to a cell line that lacks one chromosome or has one chromosome in excess in the pair (monosomy and trisomy).

Meiotic nondisjunction, in which the pair of homologous chromosomes fails to separate during meiosis, may give rise to an aneuploid ovum, which, when fertilized by a normal spermatozoon, results in a zygote with chromosomal abnormality. In some cases fertilization may take place between abnormal gametes from both parents, thus producing more complex types of aberrations.

Although most human genetic diseases are caused by changes at the level of the DNA molecule, in the case of the chromosomal aberrations in man the genetic message contained in each chromosome is maintained intact (Chap. 19). Here the alteration is quantitative; it resides in disequilibrium established by the excess (trisomy) or defect (monosomy) in the amount of genetic material. Such a dosage effect may be dangerous to the organism and may produce severe anatomical and functional anomalies (i.e., malformations). It will be shown later that trisomy produces changes that are characteristic for each chromosome present in excess; however, in

all instances of trisomy the tendency is toward involution of the nervous system, resulting in a more or less severe mental defect. Another important consequence of chromosomal excess is spontaneous abortion. It has been found that a large proportion of aborted fetuses show a trisomy in one of the larger chromosomes in the karyotype. Malformation, mental retardation, sterility, and spontaneous abortion operate as strong selective mechanisms tending to eliminate from the general population those individuals carrying deleterious genetic imbalances.

Numerical Autosomal Aberrations

21-Trisomy (Mongolism). Among the most important autosomal aberrations is *mongolism,* which is characterized by multiple malformations, mental retardation, and markedly defective development of the central nervous system. Since in identical twins both individuals are generally affected, but in fraternal twins only one is affected, mongolism originates from defective gametes.

It was discovered that the mongoloid has an extra chromosome. Pair 21 is trisomic instead of normal. This aberration probably originates from nondisjunction of pair 21 during meiosis.

The extra chromosome of pair 21 in some cases may become attached to another autosome (translocation), usually to pair 22.

The phenotype of a mongoloid is recognizable at birth in most cases. The face of such a patient has a special moon-like aspect, with oblique palpebral fissures, increased separation between the eyes, and a skin fold (epicanthus) at the inner part of the eyes. The nose is flattened; the ears are malformed; the mouth is constantly open; and the tongue protrudes.

Mongolism is the most common congenital disease and is present in more than 0.1 per cent of births. Its frequency increases as the mother's age exceeds 35. The occurrence of this trisomy is sporadic, and in general there is no recurrence in the family. However, in the rarer cases of mongolism by translocation, the disease may affect siblings and may appear in successive generations. Fortunately this "translocation trisomy" represents only 3 to 4 per cent of all cases of mongolism. In this type there is no change in frequency with the age of the mother, and when the aberration is properly determined by karyotype analysis, the parents should be warned of a repetition of this defect.

21-Monosomy. Complete deletion of one of the chromosomes in pair 21 is apparently lethal, but there is a syndrome in which a large part of one is lacking. Children with this condition have a morphologic aspect which is, to some extent, the opposite of mongolism. The nose is prominent, the distance between eyes shorter than normal, the ears are large and the muscles contracted. It seems that in trisomy and monosomy of the 21st, the phenotype shifts to one or the other side of normal.[13]

18-Trisomy. In this case the child is small and weak, the head is laterally flattened, and the helix of the ear scarcely developed. The hands are short and show little development of the second phalanx; the digital imprints are rather simple. These children are very retarded mentally and usually die before one year of age.

18-Monosomy. This is the opposite syndrome in which a partial deletion of one chromosome of the pair occurs.[19] The ears are voluminous, the fingers long, and the digital imprints are complex and convoluted.

13-Trisomy. Multiple and severe body malformations, as well as profound mental deficiency, are characteristic features of Patau's syndrome, in which there is a trisomy of the 13th chromosome. The head is small and the eyes are often small, or absent. Harelip, cleft palate, and malformations of the brain are frequent. The internal organs are also severely malformed, and in most cases death occurs soon after birth. Meiotic nondisjunction is thought to be the cause of this chromosomal anomaly.

Structural Aberrations

Cases carrying all types of structural aberrations (see Chapter 15) have been detected in man. Translocations of several types, deletions, duplications, ring chromosomes, inversions, and isochromosomes have frequently been associated with congenital diseases. These chromosomal changes appear sporadically in the general population, and they usually lead to a phenotypic change that is more or less characteristic.

One of the most remarkable phenotypes associated with a chromosome structural change is the condition known as *cri du chat* syndrome.[20] The affected baby has a strange mewing cry, multiple malformations, and mental retardation associated with partial deletion of the short arm of chromosome 5.

Another type of patient showing facial alteration, skeletal and ophthalmologic abnormalities, along with profound mental retardation is the carrier of a deletion of the long arm of chromosome 18.

Aberrations of Sex Chromosomes

These aberrations differ from autosomal aberrations in that the X and Y chromosomes are different genetically. In addition, the regulatory mechanism which inactivates the extra X chromosomes should be taken into account.

Table 16–1 shows some of the most common abnormalities found in sex chromosomes together with the clinical syndromes and an indication of the presence or absence of sex chromatin. The female gamete is probably more involved than the male gamete in these aberrations, since for each ovum millions of spermatozoa are produced. The most important aberrations of this type follow.

Klinefelter's Syndrome. Most affected individuals are practically normal except for minor phenotypical anomalies. They have small testes, frequently gynecomastia (enlarged breasts), tendency to tallness, obesity, and underdevelopment of secondary sex characteristics. Spermatogenesis does not occur, thereby resulting in complete infertility. The testes contain sclerosed and hyalinized seminiferous tubules and aggregated Leydig cells. These individuals have a positive sex chromatin and 47 chromosomes (44 autosomes + XXY).

Males with Multiple Sex Chromatin Corpuscles. Several males with 48 chromosomes (44 autosomes + XXXY) and two Barr corpuscles have been described. These individuals have features of Klinefelter's syndrome and are mentally retarded. Patients with 49 chromosomes (44 autosomes + XXXXY) have also been reported. They display extensive skeletal anomalies, extreme hypogenitalism, and low-grade mental deficiency. These persons have three sex chromatin masses (Table 16–1).

XYY Syndrome. These males have a tendency to tall stature, mental defects, and strong antisocial behavior; they do not show any other morphologic alterations. Carriers of XYY chromosomes have a greater inclination toward early delinquency and aggression than males of normal karyotype, and several of them have been found in maximum security institutions[21, 22] (see Hook, 1973).

Turner's Syndrome (Gonadal Dysgenesis). Patients with Turner's syndrome usually have a female appearance with short stature, webbed neck (folds of skin extended from the mastoid to the shoulders), and generally infantile internal sexual organs. The ovary does not develop and shows complete absence of germ cells. As a result of this ovarian dysgene-sis, menstruation does not occur, and secondary sexual characteristics do not develop. The karyotype shows 45 chromosomes (44 autosomes + X), and there is no sex chromatin. In this syndrome there is little difference in the gonads up to the third month of gestation; the ovaries contain approximately the normal number of germ cells (about 2 or 3 × 10⁶ cells). Later on there is a rapid atresia of the germ cells, leading to their virtual disappearance after puberty. It is probable that the lack of one X chromosome does not interfere with the formation and migration of the primordial germ cells into the germinal ridge. It does determine, however, the progressive ovarian atresia by failure of primordial follicles—a process which would normally be regulated by the heterochromatic X lacking in this syndrome.[23]

Females with X Polysomy. Triplo-X constitution (47 chromosomes; 44 autosomes + XXX) was detected in phenotypically near-normal females. A number of these females are mentally subnormal or psychotic, and some of them menstruate. Two sex chromatin bodies are found in cells from these women. A few severely retarded patients with three corpuscles of sex chromatin and 48 chromosomes (44 autosomes + XXXX) have been found.

Mixed Chromosomal Aberrations. Klinefelter's syndrome may be found combined with mongolism. Such a person has 48 chromosomes: 45 autosomes (including a trisomic pair 21) + XXY.

Influence of the X and Y Chromosome. In contrast to *Drosophila*, the presence of a Y chromosome in man suffices to produce male organs, even in the presence of several X chromosomes. The absence of Y, as in Turner's syndrome, produces a female, but with gonadal dysgenesis.

In man an XXY is a sterile male (Klinefelter's syndrome) and an XO is a sterile female (Turner's syndrome) (Table 16–1), but in the mouse XO is a fertile female. The X chromosome is strongly feminizing in mammals, including man, and probably has genes that are indispensable for development. So far, an individual without the X chromosome has not been found.

Mosaics. Sometimes chromosomal aberrations are produced during development of the embryo. One interesting example is induced by the loss of one Y chromosome in the

TABLE 16–3. Sex Chromosome Mosaics

	Clinical Syndrome	Sex Chromatin
Females		
XO/XY	Turner	–
XO/XX	Turner	–
XO/XYY	Turner	–
XO/XXX	Variable	–
Males		
XX/XXY	Klinefelter	+
XY/XXY	Klinefelter	+
XXXY/XXXXY	Small gonads and immature sexual characteristics, mental disorder	3+
XO/XY	Hermaphrodite	

first division of the zygote. This may result in twins, of which one has *normal* male characters and the other has Turner's syndrome.

Sometimes chromosomal aberrations due to nondisjunction are produced during development of the embryo. These individuals possess different chromosomal complements from cell to cell within the same tissue or between tissues, depending on the embryonic stage at which nondisjunction occurred. Table 16–3 illustrates the most frequently found mosaics of the sex chromosomes in females and males. Since sex mosaics having the constitutions YO, YO/XY, YO/XXY,

and so forth, have not yet been found, it seems that the combination YO is not viable.

Sex Chromosomes and True Hermaphroditism. A true hermaphrodite is an individual who has both ovarian and testicular tissue. The two types of gonadal tissue may be separated or in close proximity (ovotestis). Frequently, true hermaphrodites have a 46, XX chromosome complement and are chromatin-positive. Others may have a 46, XY chromosome set. However, 46, XX/46, XY mosaicism has also been detected in this condition, as have other mixoploid chromosome complements.

SUMMARY:
Aberrations of Human
Chromosomes

Human cytogenetics has expanded in recent years by the use of techniques for the study of the karyotype. Tissue culture, "blastlike" transformation of blood lymphocytes by phytohemoagglutinin, hypotonic treatment, and use of colchicine have been major technical advances. In the normal human karyotype the 22 pairs of autosomes are numbered in descending order of length and are classified according to the position of the centromere (Table 16–1). In the last few years considerable progress has been made in the identification of each chromosome by use of various techniques that show the banding pattern of chromosomes. Quinacrine fluorescent stains demonstrate the Y chromosome at interphase (Fig. 16–10) and show a special banding pattern for each chromosome pair. Several techniques, based on the use of the Giemsa stain, are very useful (Fig. 16–11). By these, and other techniques, several kinds of banding may be determined: quinacrine (Q bands), centromere (C bands), Giemsa (G bands), telomeric (T bands), and reverse (R bands). These advances are leading to a revolution in the study of human cytogenetics.

Abnormal karyotypes can be produced by *mitotic or meiotic nondisjunction* resulting in aneuploidy. There are also structural aberrations such as translocations (Chapter 15). The presence of one chromosome in excess (trisomy) or one lacking (monosomy) causes a genetic disequilibrium with severe anatomical and functional anomalies. These are particularly evident in the development of the central nervous system. Examples of numerical autosomal aberrations are: *mongolism,* generally caused by a trisomy of the 21st pair (sometimes the extra chromosome is translocated upon another autosome); *monosomy of the 21st pair* produces individuals in whom the appearance of the face is approximately opposite to that of a mongoloid; *trisomy* and *monosomy* of the 18th pair; and *trisomy* of the 13th pair. All these cases have profound alterations of the CNS and lethal malformations.

Aberrations of the sex chromosomes lead to the production of several syndromes. In *Klinefelter's* syndrome there are 47 chromosomes (44 + XXY), and the sex chromatin is positive. The individual has male characteristics, but has a small testis and underdeveloped secondary sex characteristics. Cases with 48 and 49 chromosomes (44 + XXXY and 44 + XXXXY) and having 2 and 3 corpuscles of sex chromatin have been described (Table 16–1).

In Turner's syndrome the individual has a female appearance with dwarfism. The ovaries do not develop; there are 45 chromosomes (44 + X), and sex chromatin is lacking.

REFERENCES

1. McClung, C. E. (1902) *Biol. Bull., 3*:343.
2. White, M. J. D. (1961) *The Chromosomes.* 5th Ed. Methuen & Co., London.
3. Barr, M. L., and Bertram, E. G. (1949) *Nature, 163*:676.
4. Barr, M. L. (1955) *Anat. Rec., 121*:387.
5. Ohno, S., Kaplan, W. D., and Kinosita, R. (1959) *Exp. Cell Res., 18*:415.
6. Russell, L. B. (1961) *Science, 133*:793.
7. Ohno, S. (1967) *Sex Chromosomes and Sex-linked Genes.* Springer-Verlag, Berlin.
8. Migeon, B., Der Kaloustian, V., Nyham, W., Young, W., and Childs, B. (1968) *Science, 160*:425.
9. Solari, A. J., and Tres, L. L. (1967) *Exp. Cell Res., 47*:86.
10. Solari, A. J., and Tres, L. L. (1967) *Chromosoma, 22*:16.
11. Saez, F. A. (1963) *Port. Acta Biol. A, 7*:11.
12. Diaz, M. O., and Saez, F. A. (1968) *Chromosoma, 24*:10.
13. Lejeune, J. (1964) In: *Progress in Medical Genetics*, Vol. 3, p. 144. (Steinberg, A. G., and Bearn, A., eds.) Grune and Stratton, Inc., New York.
14. Painter, T. S. (1923) *J. Exp. Zool., 37*:291.
15. Patau, K. (1960) *Amer. J. Hum. Genet., 12*:250.
16. Caspersson, T., Farber, S., Foley, G. E., Kudynowski, J., Modest, E. J., Simonsson, E., Wagh, U., and Zeck, L. (1968) *Exp. Cell Res., 49*:219.
17. Arrighi, F. E., and Hsu, T. C. (1971) Localization of heterochromatin in human chromosomes. *Cytogenetics* (Basel) *10*:81.
18. Drets, M. E., and Shaw, M. W. (1971) Specific banding patterns of human chromosomes. *Proc. Nat. Acad. Sci. USA, 68*:2073.
19. Grouchy, J., Bonnette, J., and Salmon, C. (1966) *Ann. Génét., 9*:19.
20. Lejeune, J., Lafourcade, J., Berger, R., and Rethoré, M. O. (1965) *Ann. Génét., 8*:11.
21. Jacobs, P. A., Brunton, M., Melville, M., Brittain, R. P., and McClemont, W. F. (1965) *Nature, 208*:1351.
22. Price, W. W., and Whatmore, P. B. (1967) *Nature, 213*:815.
23. Singh, R. P., and Carr, D. H. (1966) *Anat. Rec., 155*:369.

ADDITIONAL READING

Barlow, P. W. (1967) Sex chromosomes. *Nature, 216*:5118, 892.
Bartalos, M., and Baranki, T. A. (1967) *Medical Cytogenetics.* The Williams and Wilkins Co., Baltimore.
Bergsma, D., ed. (1973) *Birth Defects Atlas and Compendium.* National Foundation on Birth Defects, Original Articles Series. 1006 pp., The Williams & Wilkins Co., Baltimore.
Brown, S. W. (1966) Heterochromatin. *Science, 151*:417.

Caspersson, T. and Zech, L. (1973) *Chromosome Identification*. p. 355. Nobel Foundation, Academic Press, Inc., New York.

Drets, M. E., and Shaw, M. W. (1971) Specific banding patterns of human chromosomes. *Proc. Natl. Acad. Sci. USA., 68*:2073.

Ford, E. H. R. V. (1973) *Human Chromosomes*, p. 381. Academic Press, Inc., London.

Gallien, L. (1959) Sex determination. In: *The Cell,* Vol. 1, p. 399. (Brachet, J., and Mirsky, A. E., eds.) Academic Press, Inc., New York.

Gardner, L. I. (1961) *Molecular Genetics and Human Disease*. Charles C Thomas, Springfield, Ill.

Hamerton, J. L. (1963) *Chromosomes in Medicine*. Medical Advisory Committee of the National Spastics Society in association with Wm. Heinemann. Little Club Clinics in Developmental Medicine, No. 5.

Hamerton, J. L. (1969) Sex chromosomes and their abnormalities in man and animals. In: *Handbook of Molecular Cytology*, p. 751. (Lima-de-Faría, A., ed.) North-Holland Publishing Co., Amsterdam.

Hamerton, J. L. (1971) *Human Cytogenetics*. Vol. 1, p. 422; Vol. 2, p. 474, Academic Press, Inc., New York.

Hook, E. B. (1973) Behavioral implications of the human XYY genotype. *Science, 179*:139.

Jost, A. D. (1970) Development of sexual characteristics. *Sci. J., 6*:67.

Levine, H. (1971) *Clinical cytogenetics*. Little, Brown and Co., Boston.

London Conference on the Normal Human Karyotype. (1964) *Ann. Hum. Genet., 27*:295.

Lyon, M. F. (1966) X chromosome inactivation in mammals. In: *Advances in Teratology*, Vol. 1, p. 25. (Woolam, D. H. M., ed.) Logos Press, London.

Lyon, M. F. (1972) X chromosome inactivation and developmental patterns in mammals. *Biol. Rev. 47*:1.

McKusick, V. A. (1971) The mapping of human chromosomes. *Sci. Am, 224*:104.

Mittwoch, U. (1967) *Sex Chromosomes*. Academic Press, Inc., New York.

Mittwoch, U. (1967) Sex differentiation in mammals. *Nature, 214*:554.

Ohno, S. (1967) *Sex Chromosomes and Sex-linked Genes*. Springer-Verlag, Berlin.

Paris Conference on standardization in Human Cytogenetics. (1973) Birth defects Atlas and Compendium. Vol. 8, pp. 1–43, National Foundation on Birth Defects. Original Article Series. The Williams & Wilkins Co., Baltimore.

Renwick, J. H. (1971) The mapping of human chromosomes. *Ann. Rev. Genet., 5*:81.

Taylor, A. I. (1969) Autosomal abnormalities. In: *Handbook of Molecular Cytology*, p. 804. (Lima-de-Faría, A., ed.) North-Holland Publishing Co., Amsterdam.

Part Seven

MOLECULAR BIOLOGY

The following four chapters comprise the main topics
currently included within the general heading of molecular
biology. These are the fields in which progress has been
rapid and remarkable in recent years and where the future
of many aspects of cell biology lies. The studies are
intimately related to molecular genetics, a discipline that
attempts to explain hereditary phenomena as the result of
specific chemical components localized or formed in the
chromosomes. Experimental studies have provided
conclusive evidence that genetic information is initially
dictated by the arrangement of bases in deoxyribonucleic
acid (DNA), and that this information is then transcribed
in the various molecules of ribonucleic acid (RNA) (i.e.,
messenger RNA, transfer RNA, and ribosomal RNA). The
genetic information is finally translated into various specific
proteins and enzymes. Knowledge of the Watson-Crick
model of DNA presented in Chapter 3 and of the exact
disposition of the base pairs in this molecule is
prerequisite to understanding this part of the book.

In Chapter 17 the cytochemical aspects of the nucleus,
DNA, the RNAs, and nuclear proteins will be presented.
An important concept is that the DNA content of the
nucleus is generally constant for a species. The molecular
mechanism of DNA duplication will be explained primarily
as it occurs in the bacterial cell. It will be shown that in
higher cells DNA duplication takes place during a specific
period of interphase called the synthetic or S phase. This
is preceded and followed by two gaps, G_1 and G_2, during
which DNA synthesis does not take place.

Using tritiated thymidine, which specifically labels
synthesized DNA, it is possible to observe by
radioautography the duplication of the chromosome,
which in bacteria is made of a single circular DNA molecule.
The result is extraordinarily simple; and amazingly, during

DNA duplication, the unwinding of the two polynucleotide chains and the corresponding rupture of hydrogen bonds must occur at 10,000 revolutions per minute.

Chapter 18 is focused on the study of ribosomes as macromolecular organelles presiding over the many steps of protein synthesis. Knowledge of the chemical organization of ribosomes has progressed considerably now that these organelles have been separated into their RNA and protein components and the functional units have been reconstituted. Another fundamental point elucidated is the biogenesis of ribosomes from the nucleolar organizer. The function of the nucleolus should become clear when it is recognized that this nuclear element is the site of the origin and processing of ribosomes.

Chapter 19, which covers protein synthesis and molecular genetics, introduces some simple examples of human diseases that show how the genes control biochemical reactions and how they are able to produce molecular changes that result in the absence of certain enzymes or in alterations of other proteins. The central dogma of molecular biology — DNA duplication, transcription into RNA, and translation into proteins — is here explained by a consideration of the relationship between genes and protein synthesis. The existence of long-lived mRNAs in higher cells is noted, and the complex processing of these molecules in eukaryons is described. Studies that have elucidated the genetic code are briefly presented. Knowledge from these studies supplies the necessary background for understanding the role of the ribosomes and polyribosomes in protein synthesis. The existence of a cycle in which the ribosomal subunits are associated and dissociated is one of the most interesting new features. The presence of special initiation, elongation, and termination factors, which play specific roles in the synthesis of proteins, is emphasized. Finally, there is a discussion of the regulation of genic action in lower cells. The existence of regulatory and operator genes and the operon concept are of great importance in this relation.

Chapter 20 treats the fundamental problems of cell differentiation and cellular interaction. This chapter is of special importance in cell biology because it brings together the recent knowledge of how regulation of genic action takes place in higher cells. This is where important research in molecular biology will certainly be concentrated in the future. Experiments on nuclear control of the cytoplasm and cytoplasmic action on the nucleus illustrate the continuous interrelationship that exists between the two main territories of the cell. One of the most interesting concepts is that in higher cells a large part of the genome is redundant and does not function in transcription at a certain moment of the life cycle. The possible role of histones, acidic proteins, and RNA on the regulation of gene function is mentioned, and examples of genic regulation in polytene chromosomes are presented. These concepts are also related to the study of the differentiation, growth, renewal, and aging of cells. The study of the life cycle of the different cell types that constitute a higher organism is of considerable theoretical and practical interest, especially as it relates to embryonic and cancerous cells. These concepts are valuable as an introduction to the study of the different tissues, each of which has a different cycle of division and differentiation.

Finally, the new concepts of cellular interaction and communication, the electrical coupling between cells, and the important concept of contact inhibition are introduced.

seventeen

CYTOCHEMISTRY OF THE NUCLEUS, THE CELL CYCLE, AND DNA DUPLICATION

In Chapter 12 background information was presented on the nucleus, the chromosomes, and their ultrastructure. The chemical and macromolecular organization of these structures can now be discussed. This is an introductory chapter on the field now known as *molecular genetics,* in which control and regulation of cellular functions are investigated. In these new fields the principal investigations concern the role of nucleic acids in genetic functions. Some of the early evidence of this role came from the work on *bacterial transformation,* in which a strain of bacteria was changed genetically by the action of extracts of another strain (Fig. 19–1). In 1944 Avery and his collaborators demonstrated that the substance responsible for this transformation is deoxyribonucleic acid (DNA).

DNA can be considered as the main genetic constituent of cells, carrying information in a coded form from cell to cell and from organism to organism. Ribonucleic acid (RNA) can also carry genetic information, replacing DNA in some viruses or serving as an intermediary in the transcription of genetic information, which is finally expressed in the formation of the specific proteins of the cells. Figure 17–1 introduces these concepts and represents what at present is considered the *central dogma in molecular biology.* According to the diagram, genetic information is transferred in three steps: (1) *duplication* of the DNA molecule and, thus, of its genetic

information by a template mechanism; (2) *transcription* of this information into different RNA molecules; and (3) *translation* of this information into the various protein components (including the enzymes) of a cell.

These fundamental processes are usually studied in bacteria and viruses, in which the mechanisms follow a similar, although simpler, pattern than those of higher plant and animal cells. For this reason, reference is made in this and other chapters to studies of these simple micro-organisms.

In this chapter after some consideration of the cytochemistry of the nucleus and the cell cycle, *DNA duplication* will be discussed. In Chapter 19 the processes of *transcription* and *translation* will be analyzed.

An exception to the central dogma shown in Figure 17–1 has been found: Transcription may be reversed. In other words, *RNA may be transcribed into DNA.* This is discussed at the end of the section on enzymes and DNA synthesis under the heading *Reverse Transcriptase.*

Figure 17–1. Diagram of the flow of information from the genome (DNA). (See the description in the text.) (From S. Spiegelman.)

CYTOCHEMISTRY OF THE NUCLEUS

The study of the chemical organization of the nucleus has followed two main lines. The first, which is essentially biochemical, consists of isolating a large enough number of nuclei to permit analysis by biochemical methods. The second approach, which is essentially cytologic, uses the cytophotometric and radioautographic methods described in Chapter 6. The results of both approaches are complementary and should be integrated within the discussion of the chemical organization and physiology of the nucleus.

This fundamental field was begun in 1869 by Miescher, who analyzed the chemical composition of pus cells, spermatozoa, and hemolyzed nucleated red cells of birds and other organisms, and demonstrated that nucleic acids are one of the main components of the nucleus.

A great deal of new information is now being provided from the use of HeLa cells (a tumor of human origin) and other cultured cells. To isolate the nuclei, the cells are first swollen hypotonically and then gently homogenized. The nuclei are centrifuged and treated with some surface-active agents to free them from adherent cytoplasm. This treatment strips off the outer nuclear envelope together with the few remaining ribosomes. A nucleolar fraction may be obtained by treating the nuclei with highly ionic solutions and digesting the chromatin with DNAse.[2] Figure 17-2 illustrates isolated nuclei and nucleoli from liver cells.

The number of nuclei can be measured easily in a chamber similar to that used for blood counts. The amount of a chemical substance per nucleus (viz., DNA) can then be determined.

The main result of these biochemical studies is the discovery that the nucleus has a complex chemical organization in which the *nucleoproteins* are the most important components. Nucleoproteins result from the combination of nucleic acids and proteins, and in certain cells constitute the major part of the solid material (96 per cent of the trout spermatozoon and almost 100 per cent in certain erythrocyte nuclei.)

The protein part of the nucleus is complex and has several components. Of these, the best known are two strongly basic and simple proteins: the protamines and the histones. In addition to these there are several acidic proteins, the so-called nonhistone proteins, which may constitute the most abundant component of the interphase nucleus.

Early studies indicated that DNA was the only nucleic acid present in the nucleus. Later on, however, RNA-protein complexes were extracted. In liver nuclei it was shown that less than 5 per cent of the dry mass is RNA, whereas 85 to 90 per cent of the total RNA is in the cytoplasm. The nuclear RNA is distributed in the nucleolus, chromatin, and nuclear sap.

Metaphase chromosomes from cultured cells have been isolated with a high degree of purity and shown to contain 16 per cent DNA, 12 per cent RNA, and 72 per cent protein. At least 50 per cent of the protein is of the acidic type.[3]

Thus, the chemical composition of the nucleus includes: (1) nuclear proteins (i.e., nucleoprotamines, nucleohistones, nonhistonic or acidic proteins, and enzymes), (2) DNA, and (3) RNA.

DNA Content of the Nucleus

The use of biochemical and cytophotometric methods supplied the first information about *DNA constancy* of the cell.[4, 5] For example, diploid cells of various tissues of the fowl contain approximately 2.5 picograms (1 pg = 10^{-12} g) of DNA, whereas the haploid sperm has half that amount. Analyses of DNA content in several different mammalian species have shown that the variations in DNA content are small but characteristic for that species. Greater variation was detected in species of birds and fishes than in mammals.[6]

Among invertebrates, the lowest DNA values are found in the most primitive animals, such as sponges and coelenterates. In fishes, the DNA content per cell tends to remain constant within the different species of a family. In amphibians, the DNA varies from 168 pg in *Amphiuma* to 7.33 pg in the toad.

Cytophotometry has produced accurate measurements in a large variety of cell types of a species, showing that the DNA content is practically constant.[7] This is certainly related to the number of chromosome sets (n), which is 2n in most somatic cells. In the

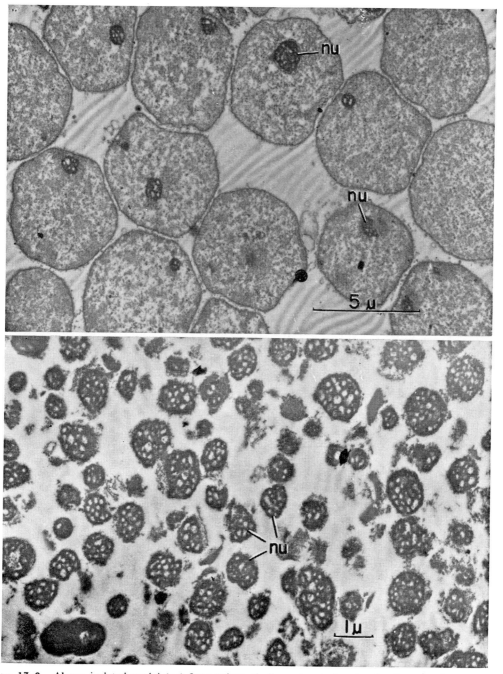

Figure 17–2. **Above,** isolated nuclei (*nu*) from guinea pig liver observed under the electron microscope. ×3000. **Below,** isolated nucleoli (*nu*) from guinea pig liver. ×16,000. (Courtesy of R. Maggio, P. Siekevitz, and G. E. Palade.)

TABLE 17–1. DNA Content and Chromosome Complement*

Cells	Mean DNA-Feulgen Content†	Presumed Chromosome Set
Spermatid	1.68	haploid (n)
Liver	3.16	diploid (2n)
Liver	6.30	tetraploid (4n)
Liver	12.80	octoploid (8n)

*From Pollister, A. W., Swift, H., and Alfert, M., *J. Cell. Comp. Physiol., 38* (suppl. 1): 101, 1951.

liver there are large nuclei which contain two and four times as much DNA as the diploid nuclei (Table 17–1). This duplication or quadruplication of DNA content obviously corresponds to polyploidy (see Chapter 15 and Table 17–1).[8]

Another interesting example is found in the study of spermatogenesis (Table 17–2). Before meiosis occurs (see Chapter 14), there are two classes of cells (spermatogonia) having different DNA contents (2n and 4n). The early primary spermatocyte is tetraploid (4n). After the first maturation division, the secondary spermatocyte contains half the DNA content, corresponding to a diploid (2n) condition. Finally, the second maturation division results in four spermatids; these have the DNA of only one chromosomal set (haploid cell) (Table 17–2). Similar results have been observed in oögenesis during maturation of the oöcytes.

DNA Content and Length of the DNA Molecules

Since DNA is a highly polymerized macromolecule (see Chapter 4), it is interesting to compare the DNA content of a cell

TABLE 17–2. DNA Content at Different Stages in Spermatogenesis*

Cell Type	DNA-Feulgen†
Premeiotic { Class 2n	3.28 ± 0.07
{ Class 4n	5.96 ± 0.07
Primary spermatocyte	6.28 ± 0.07
Secondary spermatocyte	3.35 ± 0.04
Spermatid	1.68 ± 0.02

*From Pollister, A. W., Swift, H., and Alfert, M., *J. Cell. Comp. Physiol., 38* (suppl. 1): 101, 1951.

nucleus to the probable length of the molecule. We have already mentioned that mitochondrial DNA is a circle with a circumference of about five micrometers; in viruses it may be several micrometers long (Fig. 5–9) and bacterial DNA may reach a length of one millimeter (Fig. 17–8). The length of a DNA molecule in eukaryons is more difficult to determine. Isolated DNA fibers appear as long unbranched microfibrils with a diameter of 20 μm. From the Watson-Crick model (Fig. 3–13) it is known that each nucleotide occupies a space of 0.34 nm and that a complete turn of the two strands is composed of ten such nucleotides; thus one complete turn covers a distance of 3.4 nm.

Considering that one picogram of DNA is equivalent to 31 cm of DNA, it is possible to calculate that there are about 174 cm of DNA in the human diploid cells (5.6 pg), 37 meters in *Trillium* (120 pg), and 97 meters in polytenic chromosomes of *Drosophila* (293 pg).[9] The DNA content in the 46 human chromosomes has been estimated by ultraviolet cytophotometry,[10] and from these measurements it appears that the DNA content is proportional to the size of the chromosome. The largest chromosome (1), which is 10 μm long, should accommodate about 7.2 cm of DNA in a tightly packed form.

Nucleoprotamines

Protamines are simple, basic proteins with a very low molecular weight (on the order of 4000 daltons). They are very rich in the basic amino acid arginine and thus have an isoelectric point of pH 10 to 11. They are found in spermatozoa of some fishes and are tightly bound to DNA by salt linkages. In general, they consist of a 28 residue polypeptide, with a total length of 10 nm, and contain 19 arginines and 8 or 9 nonbasic amino acids.[11] In the nucleus of the trout spermatozoon the DNA-protamine complex accounts for 91 per cent of the dry weight. This complex gives x-ray diffraction patterns which seem to indicate that the protamine wraps helically around the DNA molecule.[12] Using cytochemical techniques it has been shown that during the development of the spermatozoon there is a progressive replacement of the histone by protamine.[13] This may be due to the higher affinity of protamine for DNA. Biochemically there are at least three classes of spermatozoa, depending on whether the DNA is in complex with protamine (salmon, herring, trout, rooster, snail, squid), with histones (plants, carp, sea urchin) or with other types of protein (most mammals).[14]

Nucleohistones

Histones are also basic proteins, but they exhibit greater heterogeneity than protamine. Isolation of histones from the DNA complex is difficult because they easily become associated with the DNA-bound acidic proteins (Table 17–3). The molecular weight of histones is greater (10,000 to 18,000 daltons) than that of protamines. In addition to about 13 per cent arginine, histones contain other basic residues, including lysine and histidine. Several histones of different composition have been isolated, and three main types have been characterized: (1) very lysine-rich, (2) arginine-rich and (3) slightly lysine-rich.[15]

Histones have been separated by gel electrophoresis from various animals and tissues into six different bands that in every case are rather similar. For example, the amino acid sequence of a particular histone has been compared in pea seedlings and in calf thymus, and they differ only in two amino acid residues.

Like protamines, histones are bound to DNA by ionic bonds.[16] The ratio of histone to DNA in chromatin is nearly 1:1 over a wide range of plants and animals (Table 17–4). The association between DNA and histone can also be demonstrated cytochemically with the Feulgen technique for DNA and the fast green technique for histones, both applied to the same material.

In Chapter 20 the possible role of histones in gene regulation will be discussed.

It may be advanced here, however, that the small degree of heterogeneity, the lack of tissue specificity, and the high degree of conservation of histones during evolution are evidence contrary to a simple mechanism of gene control involving different molecular species (see Elgin et al., 1971).

Nonhistone or Acidic Proteins

Interphase nuclei, as well as chromatin (Table 17–4), contain abundant nonhistone proteins, some of which may be linked to DNA.[17] Among those not linked to DNA are the so-called *residual proteins* which remain after extraction of the other proteins with salt solutions (Table 17–3). Some acidic proteins that are rather soluble may be lost during the isolation and purification procedure. Employing nonpolar media in this process will prevent the loss of nuclear proteins.

The considerable amount of nonhistone protein present in a metabolically active cell contrasts markedly with the composition of the spermatozoon, which is much less active. In the spermatozoon the nucleus may be made up entirely of nucleoprotamine or nucleohistone, indicating that the other proteins have left the nucleus. The volume of a nucleus is proportional to its protein content; for example, in neurons, the large nucleus contains 20 times more protein than the sperm head.

In contrast to histones, acid proteins have a rapid turnover, are very heterogeneous, and show phylogenetic as well as tissue specificity.[18, 19] Acidic proteins show selective binding to DNA and have stimulatory effects on transcription.[20] Furthermore, certain protein fractions are selectively influenced by hormones such as hydrocortisone and estradiol. Another difference is in the time of synthesis within the cell cycle. Whereas histones are synthesized during the S period,

TABLE 17–3. Nuclear Proteins From Liver *(Data expressed as per cent dry weight)*

Extraction	Protein	DNA	RNA	Fraction
0.14 M NaCl	17.0	1.5	3.4	Soluble "nuclear sap"
0.10 M Tris†	5.3	6.8	10.9	Ribonucleoprotein
2.0 M NaCl-1	54.0	31.0	5.2	Basic histones linked to DNA
2.0 M NaCl-2	10.0	12.2	5.1	Nonhistone proteins linked to DNA
0.05 N NaOH	5.6	0.3	8.9	Acidic protein not linked to DNA
Residual	2.2	0.0	0.6	Residual

*From Steele, W. J., and Busch, H., *Cancer Res., 23*:1153, 1963.
†Tris (hydroxymethyl) aminomethane.

TABLE 17-4. Composition of Chromatin (*Data expressed as per cent dry weight*)

Component	Liver*	Pea Embryo†
DNA	31	31
RNA	5	17.5
Histone	36	33
Acidic protein	28	18

*Data from Steele, W. J., and Busch, H., *Cancer Res., 23*:1153, 1963.

†Data from Huang, R. C., and Bonner, J., *Proc. Natl. Acad. Sci. USA, 48*:1216, 1962.

the synthesis of acidic nuclear proteins goes on throughout the entire cell cycle.[21] In both cases the synthesis is in the cytoplasm, and the proteins are then transported into the nucleus.

One important component of the acidic proteins is the *nuclear phosphoproteins,* which in certain nuclei may account for 5 to 10 per cent of the total protein content. This component undergoes rapid phosphorylation and dephosphorylation and is found mainly in the diffuse or active chromatin (i.e., euchromatin).[22, 23] In lymphocytes stimulated by phytoagglutinins it was found that together with the increase in RNA synthesis there was also phosphorylation of the acidic proteins.[24]

Recent studies indicate that acidic proteins may play a fundamental role in the control of genic expression by reactivating regions of the genome that are masked by histones. This important problem will be considered later on and in Chapter 20.

Nuclear Enzymes

Enzymes are also nonhistone proteins. Of the nuclear enzymes the most important are those involved in the synthesis of nucleic acids. (These will be described in the section on DNA duplication and in Chapter 19.)

Some enzymes related to nucleoside metabolism are found in high concentration in the nucleus (e.g., adenosine diaminase,

nucleoside phosphorylase, and guanase). Bound to DNA are DNA polymerase, nucleoside triphosphatase, and the histone acetylases.

Although the nucleus does not contain cytochrome oxidase or succinic dehydrogenase, it does contain the soluble enzymes of anaerobic glycolysis, including aldolase, enolase, 3-phosphoglyceraldehyde dehydrogenase, and pyruvate kinase.[25] These findings suggest that the cell nucleus uses glycolysis as the main source of energy. However, it has been found that isolated nuclei can synthesize ATP by an aerobic process accompanied by the uptake of oxygen.[26]

Other Nuclear Components

Nuclei contain large amounts of cofactors, precursor molecules, and minerals. NAD, ATP, and acetyl-CoA are present, as are other nucleosides, mono- and triphosphates, and the intermediates of glycolysis (see Chapter 5).

The distribution of *minerals* can be studied by microincineration. The spodogram (ash picture) shows that the ash of the nucleus is more concentrated than that of the cytoplasm. (The ash is composed of phosphorus, potassium, sodium, and, particularly, calcium and magnesium.) Apparently minerals are found in greater proportion in chromatin. Calcium and magnesium have been localized in the nucleus by means of the emission electron microscope.

Calcium may play a significant structural role when it is bound to DNA.

Phosphate is present in considerable amounts in the nucleolus, where the existence of a phosphate pool has been verified by electron microscopy and electron diffraction. The phosphate pool may be involved in the cycle of chromatin condensation during mitosis.[27]

The *lipid* content of the nucleus has been investigated in isolated nuclei.[28] Direct staining of the nuclei with lipid reagents, such as Sudan black, is generally negative.

SUMMARY:
Chemistry of the Nucleus

The central dogma of *molecular biology* states that genetic information is transferred by three steps: (1) duplication of DNA, (2) transcription of DNA into RNA, and (3) translation of genetic information into proteins (*protein synthesis*). Under some conditions, however, RNA may be transcribed into DNA (*reverse transcription*).

Cytochemical study of the nucleus can be carried out after its isolation or within the cell by using cytophotometry and radioautography (Chapter 6). Cultured cells are treated with hypotonic solutions, then gently homogenized, and the nuclei are separated by centrifugation. Nucleoli may also be separated (Fig. 17–2).

The *nucleoproteins* are the main components of the nucleus. These proteins include the nucleohistones, nucleoprotamines, and acidic proteins and enzymes. DNA and RNA are combined with these proteins. The DNA content of a cell is a constant for a species (e.g., in the fowl a diploid cell contains 2.5 pg of DNA per nucleus and half that amount in the haploid cell). Polyploid cells contain two or four times the diploid amount of DNA (Table 17–1). In amphibians, the DNA content may vary from 168 pg to 7.3 pg; this great difference is due to the presence of considerable amounts of repetitive DNA. The length of DNA per nucleus or chromosome may be calculated. One picogram of DNA is equivalent to 31 cm of DNA. The human contains 174 cm of DNA; in a single chromosome there are 7.2 cm of tightly packed DNA. The chromosome of *Escherichia coli* is 1.1 mm long, and DNA of mitochondria is 5 μm long.

The *protamines* are basic proteins of low molecular weight (4000 daltons) present in certain spermatozoa. *Histones* are also basic (10,000 to 18,000 daltons) and are present in most nuclei. Histones are tightly bound to DNA in a 1:1 ratio (Table 17–4). There are about six different histones in most nuclei. In the interphase nucleus the *acidic proteins* are abundant; among these molecules are the *residual protein* and all the enzymes. Acidic proteins are very heterogeneous and of rapid turnover. They are specific for the species and the tissue. Histones are synthesized in the S period and acidic proteins are synthesized throughout the cell cycle. The possible role of histones and acidic proteins in gene regulation will be considered in Chapter 20.

DNA AND THE CELL CYCLE

In Chapter 3 it was mentioned that a growing cell undergoes a cell cycle that comprises essentially two periods: the *interphase* (period of non-apparent division) and the period of *division*. Division may take place by mitosis, meiosis, or other mechanisms of cell replication. For many years the interest of cytologists was concerned mainly with the period of division in which dramatic changes visible under the light microscope could be observed. The cell cycle can be considered as the complex series of phenomena by which cellular material is divided equally between daugher cells. Cell division is only the final and microscopically visible phase of an underlying change that has occurred at the molecular level. Before the cell divides by mitosis, its main molecular components have already been duplicated. In this respect, cell division can be considered as the final separation of the already duplicated molecular units.

The introduction of cytochemical methods, such as the Feulgen stain, followed by a cytophotometric quantitative assay, first suggested that the doubling of DNA takes place during interphase. The studies done by autoradiography with labeled thymidine, however, were the most important in deter-

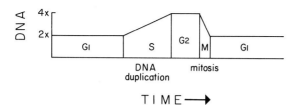

Figure 17-3. Life cycle of a cell showing the changes in DNA content during the various periods as a function of time. *2X* corresponds to the diploid content of DNA; *4X* corresponds to the tetraploid content.

mining the exact period in which DNA replication takes place in a eukaryotic cell. These studies demonstrated that the synthesis occurs only in a restricted portion of the interphase—the so-called S period (i.e., *synthetic period*) and that this is preceded and followed by two periods—G$_1$ and G$_2$ (*G, gap*) in which there is no DNA synthesis[29] (Fig. 17-3).

In general, these studies are carred out on cell cultures in which the cell cycles of each individual cell are synchronized by various methods (see Mitchison, 1971). A typical experiment consists of exposing cultured cells for 10 minutes to ^{3}H-thymidine, a nucleoside that enters only the DNA molecule. The culture is then washed thoroughly, and samples are fixed for radioautographic studies at one- to two-hour intervals for 24 hours. Generally, the first labeled mitotic chromosomes appear four hours later, indicating that the G$_2$ period is about that long. The proportion of labeled divisions in this particular case enables one to calculate an S period of eight and one-half hours and a G$_1$ period of four hours. During the G$_2$ period

the cell contains two times (4n) the amount of DNA present in the original diploid cell (2n). Following mitosis and cytokinesis, the resulting cells that again enter the G$_1$ period have a DNA content equivalent to 2n (Fig. 17-3).

Of the various periods of the cell cycle the G$_1$ period is the most variable; in most eukaryotic cells it lasts a minimum of three to four hours. Depending on the physiological condition of the cells, it may last days, months, or years. The S and G$_2$ periods, on the contrary, are relatively constant. In most cells DNA synthesis takes place in seven to eight hours and the G$_2$ period lasts two to five hours, whereas the period of mitosis requires only about one hour. In Table 17-5 the times of mitosis and interphase in different cell types are indicated. In higher plants the total cycle time is about the same as in mammalian cells in spite of the lower temperature at which they divide.

The cell cycle is exceptionally rapid during the first cleavages of the egg cell; the blastomeres divide without any intervening growth of the cell. In this case the S period is very short, and the G$_1$ period may even be absent.

Initiation of DNA Synthesis

Since in eukaryotes nuclear DNA synthesis is a discontinuous process, one of the important problems in cell biology is to establish the causes of the initiation of the S period (see De Terra, 1969). Experiments of nuclear transplantation into amphibian oöcytes[30] and of fusion of two different cell

TABLE 17-5. Mitotic and Intermitotic Times in Various Cell Types*

Cell	Times in Minutes	
	Intermitotic	*Mitotic*
Vicia faba root meristem (19° C.)	1300	150
	1400	186
Pisum sativum (peas) root meristem (20 ° C.)	1350	177
Chick fibroblasts (38° C.)	660–720	23
Mouse spleen cultures	480–1080	43–90
Rat jejunum (in animal)	2000	28
Jensen's sarcoma (in animal)	720	27
Rat corneal epithelium (in animal)	14,000	70
Chrotophaga (grasshopper) neuroblast	27	181
Drosophila egg	2.9	6.2
Psammechinus (sea urchin) embryo, two to four cell stage (16° C.)	14	28

*From Mazia, D. *In* Brachet, J., and Mirsky, A. E. (eds.), *The Cell,* Vol. 3, New York, Academic Press, Inc., 1961.

types[31] have shown that nuclei that normally do not divide are able to start DNA synthesis. Other experiments suggest that during the G_1 period some kind of triggering substance is produced which induces the continuation of the cell cycle. Although the exact cause of this triggering effect still remains unknown, it seems possible that a protein synthesized by the activation of certain genes may have special significance. In microspores of *Lillium* it has been found that if protein synthesis is inhibited during G_1 the cell fails to enter the S period. If the drug that inhibits protein synthesis is applied at the beginning of the S period it will continue without being affected.[32] The species of protein produced are not known. They might be enzymes involved in DNA synthesis, histones, or nonhistone proteins that change the structure of the chromosomes, making the initiation points of the replication accessible to the replicating enzyme. In fact, structure of the chromosome may be a third factor involved in the initiation of the DNA cycle.

During the entire cell cycle the chromosome participates in at least three different activities: (1) self-duplication, (2) transcription and transfer of genetic information to the rest of the cell, and (3) the coiling and uncoiling cycle associated with the separation of the duplicated chromosomes or daughter chromatids. Self-duplication and transcription occur at the moment when chromosomes are most dispersed (i.e., uncoiled).

Asynchrony of DNA Duplication

Studies have shown that during the S period DNA synthesis may be asynchronous, and that one or more of the homologous chromosomes may duplicate at either an earlier or later time than the others. This phenomenon is called *asynchrony within the chromosome set.* An example present in the human female and in other mammals is that one of the two X chromosomes is heterochromatic and late replicating, whereas the other is euchromatic and replicates earlier.[33] In other instances, portions of a single chromosome duplicate at different times; this is called *intrachromosomal asynchrony.* Thus, a chromosome may duplicate as a single or as a complex unit. Asynchrony was first found in chromosomes of cultured cells of the Chinese hamster, in which five or six chromosomes

have segments that duplicate late.[34] An important finding in the asynchrony of DNA replication is that the heterochromatic regions of the chromosomes (i.e., those regions that remain condensed during interphase) are always late replicating (see Lima-de-Faría, 1969).

The asynchrony in DNA synthesis accounts for the different degrees of vulnerability of individual chromosomes to damage by chemicals that may affect DNA synthesis. For example, 5-fluorodeoxyuridine (FUDR) (after being converted into 5-fluorodeoxyuridine monophosphate) is an irreversible inhibitor of thymidilate synthetase (see O'Donovan and Neuhard, 1970). This drug acts during the S period and affects the chromosomes that are being duplicated.

DNA Duplication in Prokaryons

To gain a better understanding of the mechanism of DNA duplication in higher cells it is convenient to summarize what is known about this process in a prokaryote cell such as *E. coli.* As mentioned in Chapter 2, this cell contains a single circular chromosome which, upon unfolding, has a length of about 1.1 mm. The chromosome is made of a single molecule of DNA without associated histone, and it is attached at a point of the cell membrane (Fig. 2–1). The mechanism of DNA replication may be considered as a direct consequence of the molecular model proposed by Watson and Crick in 1953 (Fig. 3–11). This mechanism involves the unwinding of the two polynucleotide strands, followed by the duplication of two complementary new strands via a template mechanism (Fig. 17–4) in which each DNA strand acts as a template for the newly synthesized molecule. DNA replicates only once in the course of a cell cycle, and the immediate stimulus to DNA synthesis could be the unwinding of the double helix of the DNA molecule. The mechanism could be set into action simply by the separation of the two DNA strands, which permits the nucleotides to fall into phase and to be linked by the action of the enzyme DNA-polymerase. Since the polynucleotide strands are joined by relatively weak hydrogen bonds, no enzymatic mechanism is required for the separation. Furthermore, the question of unwinding (there are

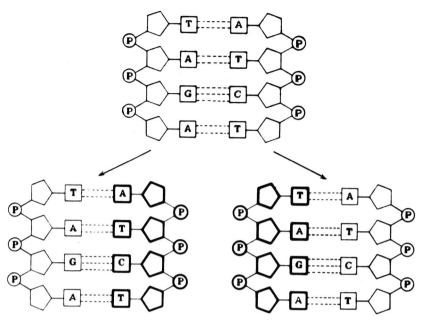

Figure 17–4. Diagram showing the mechanism of DNA duplication. **Above,** the two standard parent molecules which separate by opening of hydrogen bonds. **Below,** the two new strands that have been synthesized and that have a complementary base composition with respect to the parent DNA strands are indicated by bold outlines. (From Kornberg, A., *Ciba Lecture in Microbial Biochemistry,* New York, John Wiley and Sons, 1962.)

about 1000 turns in 3.4 μm of DNA) can be simplified by assuming that the unwinding and the rewinding of the DNA can occur simultaneously. Replication may be visualized as an advancing fork forming a Y with the parent strands (Fig. 17–5). In the chromosome of *E. coli* there are, in general, two replication forks, whereas in a eukaryotic cell there may be several thousand forks.

Semiconservativeness of DNA Duplication

The Watson-Crick model also suggests that replication is semiconservative, which means that half of the DNA is conserved (i.e., only one strand is synthesized; the other half of the original DNA is retained) (Fig. 17–6). This has been verified by several demonstra-

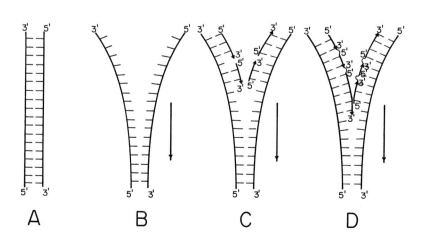

Figure 17–5. Diagram of DNA duplication according to the model of discontinuous synthesis. **A,** the two antiparallel strands of DNA; **B,** separation of the strands; **C,** beginning of replication by the synthesis of short chains in the 5′ → 3′ direction; and **D,** joining of the segments by a DNA ligase.

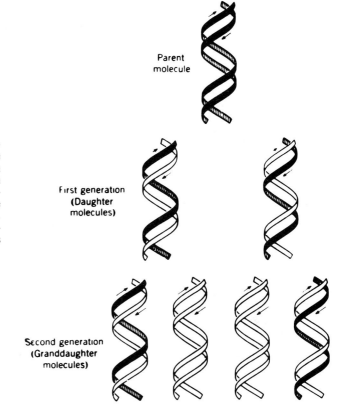

Parent molecule

First generation (Daughter molecules)

Second generation (Granddaughter molecules)

Figure 17–6. Diagram interpreting the experiment of Meselson and Stahl described in the text. **Above,** parent DNA molecule with both strands labeled with ¹⁵N. **Middle,** first generation shows the daughter molecules (in white) synthesized in a medium containing ¹⁴N. (Note that the DNA molecules are hybrids of ¹⁵N and ¹⁴N DNA strands.) **Below,** at the second generation (granddaughter molecules) two molecules are hybrids and two are not.

tions. In their classic experiment Meselson and Stahl made use of the heavy isotope ^{15}N. The DNA containing ^{15}N is more dense than the DNA containing ^{14}N. *E. coli* was grown in a medium containing ^{15}N and was then passed to another medium containing ^{14}N. The DNA was isolated and its density was determined by ultracentrifugation on a cesium chloride gradient (Fig. 6–8). It was found that after the first division cycle there is only one DNA peak corresponding to the hybrid molecule (i.e., one strand is labeled with ^{14}N, the other with ^{15}N). At the second generation two peaks of DNA appear — one in which the two DNA strands contain ^{14}N and the other still corresponding to hybrid molecules (Fig. 17–6).[35]

The semiconservative nature of DNA replication has been demonstrated at both the molecular and the cellular level. After treatment with ^{3}H-thymidine and by use of a special radioautographic technique, it has been possible to demonstrate that duplication starts at a fixed point of the chromosome (Fig. 17–7). At the site of duplication the two parent strands separate, and the daughter

molecules lie alongside.[36] During the entire duplication period the daughter molecules remain attached to each other and to the far end of the parent molecule. Finally, the two circular DNA molecules separate and a new cycle begins (Fig. 17–8). Since the entire process of DNA duplication in *E. coli* may take place in 20 to 30 minutes, the helix in the molecule should unwind at the extraordinary rate of 10,000 revolutions per minute![37]

It has been suggested that the point of attachment of the chromosome to the membrane (Fig. 2–1) represents the enzymatic complex that constitutes the replicating apparatus where duplication originates. The attachment to the membrane may also explain the mechanism of separation of the two daughter chromosomes. After a process of condensation by the folding of the DNA, this could take place by the growth of the membrane between the points of attachment of the two chromosomes. The result is that the nuclear areas, containing the daughter chromosomes, separate prior to the formation of a partition membrane between the two daughter bacteria. Thus, in bacteria, cell division is not by

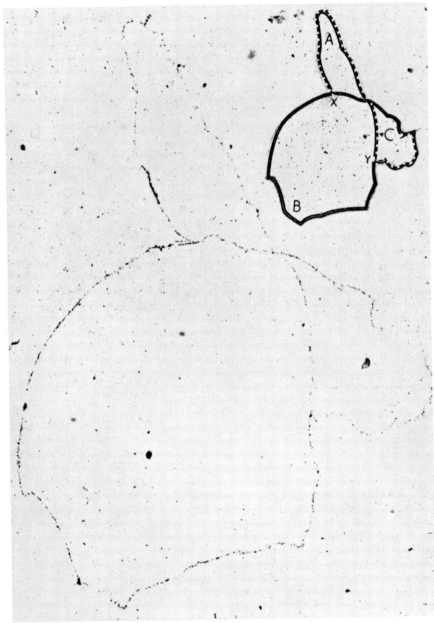

Figure 17–7. Radioautograph, observed with the light microscope, of a chromosome of *E. coli* K 12 *Hfr* labeled with tritiated thymidine for two generations. The bacterium has been gently lysed, and the entire duplicating chromosome is observed. Notice the circular disposition of the single DNA molecule that constitutes the chromosome and that is being duplicated. In segment *B* of the molecule there are about twice the number of grains per μm than in segments *A* and *C*. (See the inset and Figure 17–8.) (Courtesy of J. Cairns.)

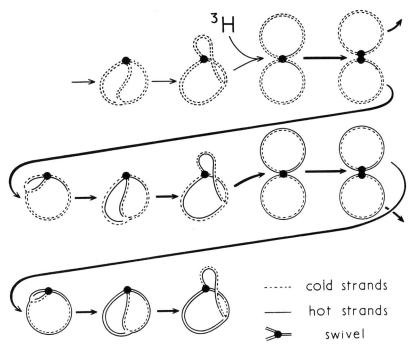

cold strands

hot strands

swivel

Figure 17–8. Diagram that interprets the experiment of Cairns (Fig. 17–7). At point ^3H the labeled thymidine is added, and the DNA synthesis of hot strands starts and proceeds from the swivel onward. This point of the DNA molecule is supposed to divide only when the duplication of the circular DNA is completed. In the second generation, there is a stage which corresponds to the autoradiograph of Figure 17–7. In this, point X is supposed to correspond to the swivel. (Courtesy of J. Cairns.)

mitosis, but by *amitosis*. As in the case of eukaryotic cells, protein synthesis is needed to start a new cycle of DNA duplication; if the duplication is already under way, the inhibitors of protein synthesis have no effect. (For more details see Mitchison, 1971.)

From the autoradiographic work illustrated in Figures 17–7 and 17–8 it was concluded that the duplication of the DNA molecule proceeds unidirectionally. The two forks observed were interpreted by Cairns;[36] one represents the advancing replicating point, and the other represents the origin of replication which remains at a fixed point. Recently, various experiments favor the idea that in *E. coli,* as in eukaryons (see below), DNA duplication proceeds bidirectionally.[38] In *Bacillus subtilis* it has been observed that the number of replicating forks may increase when the culture medium is enriched and the cell divides at a faster rate. Figure 17–9 shows an interpretation of these findings. From the initiation point of the circular chromosome, two replicating forks proceed bidirectionally; at a certain stage — also starting from the same point of origin — two new replication forks are

initiated, thus producing four forks that advance bidirectionally. Bidirectional DNA replication accounts for at least a portion of the circular genome[39] (see Pato, 1972).

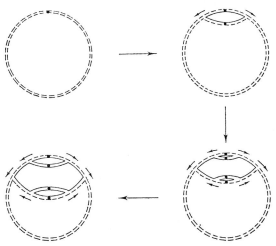

Figure 17–9. Diagram of the formation of a symmetric chromosome from a circular structure containing a single initiation site. The small arrows indicate the directions of movement of replication forks. (From Wake, R. G., *J. Molec. Biol.,* 68:501, 1972.)

DNA Duplication in Eukaryons

The semiconservative replication of DNA has also been demonstrated in higher cells.[40] For example, plant cells were grown in a medium containing radioactive thymidine and then placed in a nonradioactive medium. It was found that the nuclei in the S phase incorporated the tracer, whereas those that had previously completed DNA duplication were not labeled. Then colchicine was used to stop mitosis. This drug prevents cell division, but it does not impair chromosome duplication, thus permitting the accumulation of the chromosomes of the first and second generations within the original cell.

These results are interpreted in Figure 17–10. Prior to replication each chromosome contains only two polynucleotide units. At the time of duplication, two new units (now labeled) are built alongside. Now each chromatid includes an original non-labeled strand and a new labeled one, but under the light microscope the entire chromatid appears labeled. When the second duplication occurs, in the absence of labeled precursor, the labeled and unlabeled strands separate, and the result is a labeled and a nonlabeled sister chromatid (Fig. 17–10), as actually seen in the radioautographs. While the rate of DNA duplication in *E. coli* is estimated at 20 to 30 micrometers per minute in a human cultured cell, in the HeLa cell, for example, the rate is only 0.5 μm per minute.

One of the techniques for the study of DNA replication consists of using synchronized cells that are cultured within a millipore chamber. These are submitted to a pulse of ^{3}H-thymidine and, at different intervals, the cells are treated with the proteolytic enzyme *pronase* and then are dialyzed. In this way only the DNA fibers remain deposited on the millipore and can be detected by radioautography[41] (see Callan, 1972). As shown in Figure 17–11 the DNA duplicates in sections of about 50 to 60 μm that are arranged in tandem, one after the other, each section corresponding to a growing point or fork. DNA duplication in eukaryons is illustrated by the model shown in Figure 17–12 in which the replication starts at the initiation points (0) and proceeds in both directions until it reaches the terminal points (T), where the converging growing forks meet. In cells of the Chinese hamster it has been estimated that there should be about 50,000 initiation points in the entire genome.

Enzymes and DNA Synthesis

It was mentioned above that DNA replication is produced by a template mechanism by which one of the polynucleotide chains

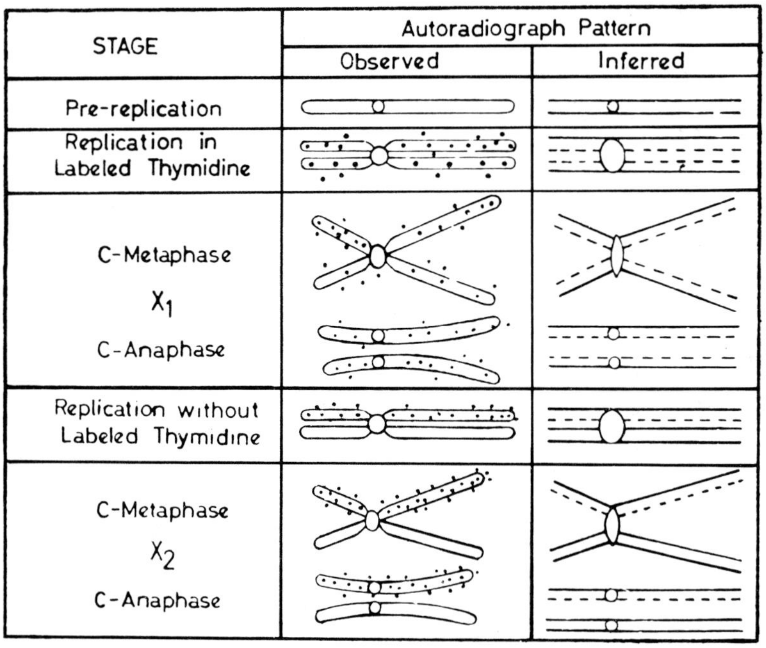

Figure 17–10. Diagram interpreting the experiment of Taylor, Woods, and Hughes, 1957. In the inferred autoradiograph pattern, the broken lines correspond to the tritiated DNA strands. (From Lewis, K. R., and John, B., *Chromosome Marker*, Baltimore, Williams & Wilkins Co., 1963.)

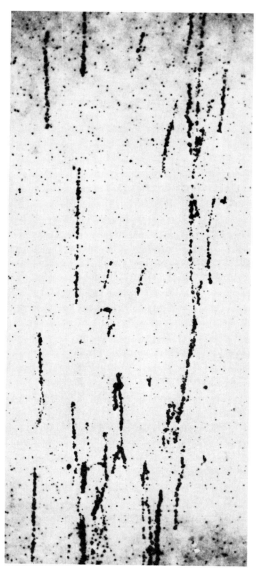

Figure 17-11. Radioautograph of DNA replication in a cell grown in tissue culture. The various segments that are being duplicated are indicated by the parallel lines of silver grains of ³H-thymidine. (See the description in the text.) (Courtesy of H. G. Callan.)

serves as a template for the newly synthesized molecule. This mechanism was initially thought to involve the *DNA polymerase* which was first isolated by Kornberg from *E. coli.* However, Kornberg's enzyme, now called *Pol I,* is no longer considered to be the main synthesizing enzyme, because bacterial mutants lacking this enzyme synthesize DNA.[42] Pol I is now assumed to be concerned mainly with DNA repair. Other DNA polymerases, called *Pol II* and *Pol III,* are probably involved in DNA synthesis.[43] DNA polymerase, with a single polypeptide chain, has a molecular weight of about 100,000. It has been observed under the electron microscope as a roughly spherical body with a diameter of about 6.5 nm attached at regular intervals to the DNA chain.[44] For its action DNA polymerase requires the presence of a *template DNA,* a *primer DNA,* and the four nucleotide triphosphates (ATP, TTP, GTP, and CTP) that are used in the synthesis. It is supposed that the first of the binding sites of the enzyme attaches to the template, the second binds the triphosphate nucleotide, and the third binds to the 3'—OH end of the primer (Fig. 17–13). Under these conditions the enzyme can add the new nucleotides only from the 5' end to the 3' end of the polynucleotide chain. The action of the enzyme produces a linkage between the inner phosphate of the nucleotide and the oxygen attached to the 3' carbon. At the same time the two terminal phosphates are hydrolyzed and released. Under this condition the enzyme is ready to proceed by linking the following nucleotides in a sequential manner. One of the problems, which is difficult to explain, concerns the fact that DNA polymerase acts only in the direction 5' → 3'. (From Chapter 4 the reader should remember that both DNA strands are complementary and antiparallel.) When the two strands unwind, one of them will be facing the DNA polymerase in the right direction (5' → 3'), but in the other, the situation will be unfavorable. Several solutions to this paradox have been advanced. The most favored is that DNA synthesis is discontinuous on the two strands; in other words, after the unwinding of the DNA, the new strands are made in short segments, always in the 5' → 3' direction. These segments are then joined together by the action of another enzyme, *polynucleotide ligase* (Fig. 17–5). The ligase can produce the linkage of the 3' end of one strand with the 5' end of another, provided the two are held in place in a double helix.[45]

The model of discontinuous synthesis of the DNA[46] is supported by experimental evidence. If bacteria are exposed for a few seconds to ³H-thymidine, only short pieces of DNA, having 1000 to 2000 nucleotides, are found. With longer periods of labeling the DNA is already in the form of high molecular weight strands.

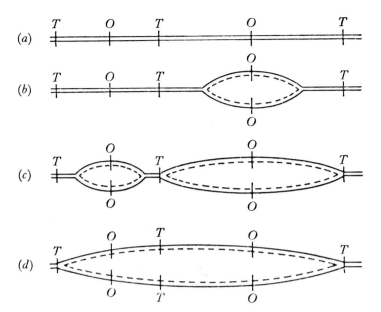

Figure 17–12. Bidirectional model of DNA replication in eukaryons. Each pair of horizontal lines represents a segment of a double helical DNA molecule with two strands. The newly formed chains are indicated by broken lines. *O* and *T* indicate sites of origin and termination of replication. (**a**), two adjacent replication units prior to replication; (**b**), replication started in the unit to the right; (**c**), replication started in the unit to the left; and (**d**), replication completed in both units. (From Huberman, J. A., and Riggs, A. D., *J. Molec. Biol.*, 32:327, 1968.)

Reverse Transcriptase

In recent years an exception to the previously mentioned central dogma of molecular biology has been recognized. In fact, the strict relationship of DNA to RNA, stated in Figure 17–1, may be reversed in certain systems. In other words, *RNA may be transcribed into DNA*. This is done by the so-called *reverse transcriptase,* an *RNA-dependent DNA polymerase* that is able to synthesize DNA on an RNA template.

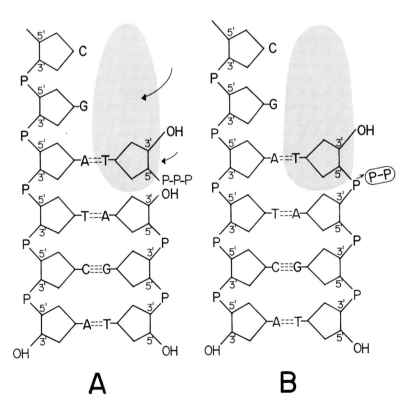

Figure 17–13. Diagram showing the action of DNA polymerase in DNA duplication. **A**, the template chain to the left and a piece of primer DNA to the right. A nucleotide (i.e., thymidine triphosphate) is being put in place along with the enzyme DNA polymerase. **B**, the enzyme has produced the linking of the nucleotide with release of P—P. Observe that the enzyme can add only nucleotides in the direction of $5' \rightarrow 3'$.

Tumor-producing (i.e., oncogenic) viruses containing an RNA genome have been found to act as templates for the synthesis of DNA.[47, 48] This tumor virion contains the reverse transcriptase by which an RNA/DNA hybrid molecule can be produced.[49, 50] In this way RNA viral genes can be integrated in the genome of the host cell. The virion also contains a DNA polymerase which can replicate the double-stranded DNA. These findings are generating considerable interest in the entire field of mammalian — including human — tumors (see Temin, 1972).

Gene Mutation and DNA Repair

Although the gene is generally stable, it may change by a process called *mutation*. The capacity to mutate is a property of the genetic material that is as important as stability. Mutation takes place in all living organisms and is the origin of hereditary variations. A genetic mutation, which may occur spontaneously and without apparent cause, becomes incorporated in the population and is transmitted by sexual reproduction. Asexual unicellular organisms and somatic tissue may also undergo mutation; the rate of spontaneous mutation is generally low. For example, in each generation of *Drosophila melanogaster* mutations occur at the rate of 1:100,000 to 1:1,000,000. Unicellular organisms are more appropriate for the study of mutation frequency because large numbers of organisms can be handled in a short time.

A *mutation* may correspond to the change of a single nucleotide, of a long polynucleotide chain, or even of an entire chromosome arm. When the locus is restricted, the result is called a *gene mutation;* when it is extended, involving the chromosome structure, it is considered a *chromosomal mutation or aberration*. (Examples of the latter were given in Chapter 15.) In gene mutation the change in a single nucleotide may result in the *substitution* of one purine or pyrimidine for another, or even of a purine by a pyrimidine. *Insertions* or *deletions* of one or more nucleotides may also occur, but these are more frequent in meiosis and result in errors in recombination (see Chapter 19).

In Chapter 15 the action of ionizing radiations on the chromosome was presented; here the action of ultraviolet light on bacteria, and the molecular mechanism by which the alteration produced in the DNA molecule can be

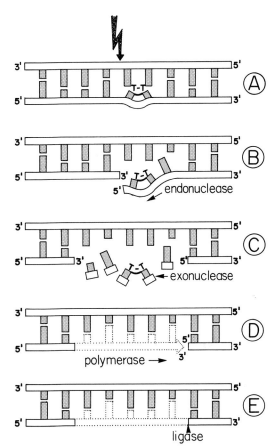

Figure 17–14. Diagram showing the effect of ultraviolet light on the DNA molecule and the mechanism of DNA repair. **A,** under the action of ultraviolet light a dimer of T—T is produced; **B,** the affected DNA strand is recognized and incised by a molecule of endonuclease; **C,** the strand segment is excised by a molecule of exonuclease; **D,** the gap is filled by DNA polymerase; and **E,** the synthesized segment is joined by a DNA ligase.

repaired will be considered. Some of these mechanisms of DNA repair also apply to ionizing radiations.

Mutations can be considerably increased by exposure of bacteria to ultraviolet light. This treatment tends to produce *dimerization* of adjacent pyrimidine bases; for example, —T—T—, C—C, or C—T *dimers* are produced within a single strand of DNA. Dimerization reduces the distance between nucleotides from 0.34 to 0.28 nm and produces other changes that impair the mechanism of DNA duplication. In most cases this type of mutation is corrected by a mechanism of *DNA repair* which involves (Fig. 17–14): (1) the recognition and incision of the affected DNA strand by an endonuclease, (2) excision and broadening of the gap by an exonuclease (e.g., DNA polymerase), (3) filling of the gap by repair replication (this is also done by DNA polymerase), and (4) covalent joining of the polynucleotide by the ligase.[51] In this mechanism, step 1 (i.e., recognition and incision) is very important. *Nucleases* are specific enzymes that can recognize sequences of nucleotides (primary structure) as well as secondary and tertiary structures in the DNA molecule (for example, the existence of loops). For step 1 to occur, the presence of light is required. If after UV irradiation the bacteria are kept in the dark, DNA repair is inhibited. The dose of ultraviolet light is an important factor, because at a low dose there is no inhibition of DNA duplication; when a critical level is reached, however, duplication is impaired. In this case the excision proceeds until duplication can be resumed again. The production of pyrimidine dimers in special regions of the DNA (i.e., the promoter, see Chapter 19) may have important consequences, since it will impair both DNA duplication and transcription.[52]

In humans, a hereditary disease due to an impairment of DNA repair has been described. In fact, cultured fibroblasts from patients with *Xeroderma pigmentosum* cannot replace pyrimidine dimers produced by UV light. From the clinical point of view, the syndrome is characterized by intolerance to sunlight and multiple skin cancer in the sunshine-exposed areas.

At variance with the ultraviolet light, ionizing radiations (x-rays, neutrons, etc.) tend to fracture the DNA strand, producing a so-called *nick* which can be repaired by DNA polymerase and ligase.

DNA Synthesis and Recombination

In Chapters 14 and 15 the morphology of meiosis was presented as well as the processes on which genetic recombination are based—chiasma formation and crossing over. After studying DNA synthesis and DNA repair the reader is in a better position to interpret recombination as a molecular

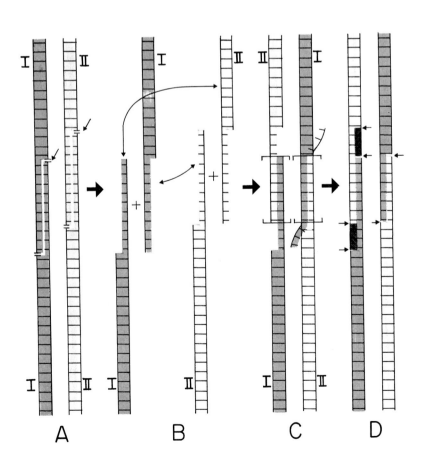

Figure 17–15. Molecular model of recombination during meiotic prophase. **A,** the DNA of the pairing chromatids (I and II) undergoing the effect of endonucleases, producing "nicks" on each of the strands; **B,** unraveling of the strands; **C,** rejoining of the opposite chromatids; and **D,** elimination of excess pieces of DNA and filling of gaps by a process similar to that of DNA repair (see Fig. 17–14). (Modified from a diagram of C. A. Thomas, Jr.)

A   B   C   D

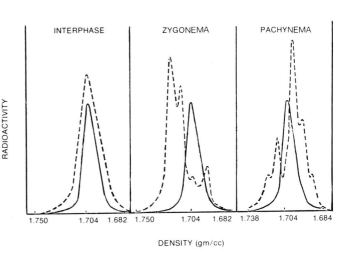

Figure 17–16. Ultracentrifugation patterns of DNA from meiotic cells of *Trillium* after equilibration in a CsCl gradient. Observe that the buoyant density of the DNA synthesized at zygonema is different from that at both interphase and pachynema. (See the description in the text.) (Courtesy of H. Stern and Y. Hotta.)

mechanism that takes place at the level of the DNA molecule. In prokaryons there is no meiosis, but recombination may be achieved between parental DNA molecules; as for example, between two bacteriophages simultaneously infecting a bacterium (see Chapter 19).

The molecular model for recombination that is presently favored can easily be understood if one remembers the mechanism of DNA repair (Fig. 17–14). In this case recombination takes place between the DNA of two homologous chromatids in eukaryons or between two parental DNA molecules in prokaryons. It is necessary to assume that regions of the two chromatids undergoing recombination should be closely paired and in register. This could be achieved with the help of the synaptonemal complex appearing in zygonema and pachynema of meiosis in eukaryons (see Chapter 14). At the points of close alignment of the two DNA molecules, "nicks" are produced by the action of endonucleases on each of the four strands of DNA (Fig. 17–15, *A*). Then the two DNA molecules unwind and unravel in such a way as to expose four portions of single-stranded DNA (Fig. 17–15, *B*). After this, rejoining occurs between the opposite chromatids, again forming two complementary double helixes (Fig. 17–15, *C*). Close observation of Figure 17–15 permits one to recognize that at the region of recombination there is a short region of a hybrid molecule originating from opposite parents. Recombination may not be completely exact, and gaps or overlaps may be produced during the rejoining. The overlaps are excised by exonucleases and the gaps are filled in, as in the mechanism of DNA

repair, by the intervention of DNA polymerase and ligase (see Fig. 17–14).

The above mechanism of recombination requires the presence of a small amount of DNA synthesis during meiotic pachynema. In fact, in microsporocytes of *Trillium* it has been found that during zygonema and pachynema, coincident with chromosomal pairing and recombination, there is a small DNA synthesis that amounts to 0.3 per cent of the total. As shown in Figure 17–16 the DNA synthesized in these two stages of meiosis has a buoyant density different from the large amount of DNA produced during premeiotic interphase.[53] At zygonema, inhibitors of DNA synthesis can arrest pairing and the formation of the synaptonemal complex (see Chapter 16). In addition, during zygonema there is a distinctive synthesis of protein, the inhibition of which, brought about by low doses of cycloheximide, halts pairing and impairs DNA synthesis.[54] During zygonema DNA is associated with protein and phospholipids. This complex may play a role in pairing which appears to be initiated at the nuclear membrane.[55] At this stage there is also an increase in the DNA polymerase bound to the chromatin.[56] The DNA synthesis produced in pachynema is associated with recombination and DNA-repair mechanisms (see Kihlman, 1971).

RNA Synthesis and the Cell Cycle

The study of the nuclear and cytoplasmic RNAs has come to be of primary importance because of their role in genetic transcription (Fig. 17–1) and protein synthesis. Since these

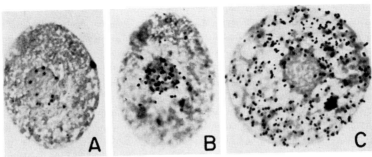

Figure 17–17. **A,** radioautograph of a *Tetrahymena* incubated in ³H-cytidine for 1.5 to 12 minutes. Notice that all labeled RNA is restricted to the nucleus. **B,** the same as in **A,** after 35 minutes. RNA begins to enter the cytoplasm. **C,** the same, incubated for 12 minutes in ³H-cytidine and then for 88 minutes in a nonradioactive medium. Notice that while the nucleus has lost all labeled RNA, the cytoplasm is heavily labeled. (Courtesy of D. M. Prescott.)

studies were started in bacteria, the knowledge of the RNA species is more complete in these than in other organisms.

Three main types of RNA are recognized in bacteria: messenger RNA (mRNA), transfer RNA (tRNA), and ribosomal RNA (rRNA), all of which are coded on the DNA molecule within the chromosome.

When an RNA radioactive precursor is incorporated into a eukaryotic cell, it is consistently found first in the nucleus and then in the cytoplasm. This has been interpreted as a demonstration that RNA synthesis takes place in the nucleus and then this molecule is transferred to cytoplasm. For example, if the protozoa *Tetrahymena* are incubated in ³H-cytidine for 1½ to 12 minutes, the labeled RNA appears in the nucleus (Fig. 17–17, *A*). After 35 minutes both the nucleus and cytoplasm contain about the same amount (Fig. 17–17, *B*). Finally, if the cell is subjected to ³H-cytidine for a few minutes and then incubated with nonradioactive precursor (a chase) for a longer period, the cytoplasm is labeled, but the nucleus is not (Fig. 17–17, *C*).[57]

In a eukaryotic cell about 80 per cent of the total RNA is ribosomal RNA, and most of the remainder is transfer RNA. Both are rather stable molecules that are involved in protein synthesis and will be presented in detail in Chapters 18 and 19. The other classes of RNA, which represent a minor proportion, are more unstable and constitute the so-called *heterogeneous nuclear RNA* and the *messenger RNA*, both of which will be discussed in Chapter 19. An additional complexity regarding RNA types is added by the presence of RNA in cytoplasmic organelles such as mitochondria and chloroplasts.

When mammalian cultured cells are pulse-labeled with uridine and examined in an autoradiograph, only the cells undergoing mitosis do not incorporate the label. All the other cells present in the G_1, S, and G_2 phases do synthesize RNA. In other words, unlike DNA, RNA synthesis is a continuous process that stops only when the chromosomes are highly condensed—i.e., from late prophase to telophase. The inhibition of RNA synthesis during mitosis is probably due to the fact that the DNA cannot be transcribed when it is highly condensed. The minor synthesis taking place during this period may be attributed to the RNA of the cytoplasmic organelles. In fact, there is evidence that this type of RNA synthesis does continue during mitosis.[58]

Protein Synthesis and the Cell Cycle

Proteins constitute the largest macromolecular component of a growing animal cell and can be used as a measure of cell growth. In general terms, protein synthesis is similar to RNA synthesis in that it is a continuous process during the interphase of the cell. During mitosis there is a reduction in amino acid incorporation.[59] This can be more precisely determined by the use of autoradiography that permits the detection of protein synthesis in single cells.[60] By using labeled amino acids one may observe that at metaphase chromosomes contain small amounts of protein. In fact, shortly before, with the breakage of the nuclear envelope, large amounts of nuclear protein are released into the cytoplasm.

An important occurrence at late telophase is the return of some labeled proteins from the cytoplasm into the nucleus. In nuclei of amebae, two types of nuclear proteins have been demonstrated.[61] One of them migrates continuously between the nucleus and the cytoplasm, and the other, apparently, is nonmigratory and remains in the nucleus. During mitosis both these nucleus-specific proteins are released into the cytoplasm and return to the nucleus when division is over. These findings are of great interest in dealing with the general problem of the nucleo-cytoplasmic relationship and the nature of the agents that may carry messages from the nucleus to the cytoplasm and vice versa (see Chapter 20).[62]

In certain human diseases, such as *lupus erythematosus,* auto-antibodies are produced against certain components of a person's own cells. If these antibodies are labeled with a fluorescent dye, they may serve to localize the intracellular antigen (see Chapter 6). With this method an antibody for deoxyribonucleoprotein that "stains" the chromosome is obtained.[63] In addition, there are two other antigens in the nucleus, which during cell division enter the cytoplasm and return to the nucleus at telophase.

The association between DNA and histones is an important factor during the cell cycle; in fact, these two components of the chromosome do not dissociate during mitosis. Also related to the cycle of the nuclear proteins are the observations that during spermiogenesis (in the grasshopper) histone is replaced by a protamine-like histone that is richer in arginine and is more basic. Studies with labeled amino acids (^{3}H-arginine) have shown that this basic protein is synthesized in the cytoplasm, using the ribosomal machinery, and that it then migrates into the nucleus and combines with DNA.[64]

Early cytological work using fast green stain showed that histones are doubled during the S period in which the DNA replicates. This has been observed in both animal and plant cells.[65] More recent work using radioactive arginine and lysine also tends to support this view.[66, 67] In some cases, however, there is some histone synthesis outside the S period. In contrast, the acid nuclear proteins are synthesized throughout the entire interphase (see above).

Other interesting studies refer to synthesis of individual proteins such as various enzymes.[68] It has been shown in studies with yeast that each enzyme appears at a definite moment of the G_2 period after DNA replication and may show a special quantitative profile. Since these enzymes represent the final translation of the genetic code (Fig. 17–1) the transcription into the mRNA must occur ahead of time. The lag period may be investigated by using inhibitors of the DNA-RNA transcription, such as actinomycin D. The lag period might also result from the time involved in the transport of mRNA from the nucleus to the cytoplasm. In mammalian cells some enzymes are synthesized continuously during the cell cycle; however, in some cases the rate of synthesis of an enzyme may double at a characteristic part of the cycle. These doubling points appear to correspond to the functional replication of the corresponding genes. (For more details on the cell cycle, see Mitchison, 1971.)

SUMMARY: The Cell Cycle

The cell cycle comprises a long *interphase* and a period of *division* (mitosis, meiosis). During interphase there is duplication of all the main components of the cell, i.e., DNA, RNA, and proteins. Using radioautography with ^{3}H-thymidine it has been found that DNA duplication occurred in a restricted part of the interphase—the *synthetic,* or S period. This is preceded and followed by the G_1 and G_2 *(gap)* periods. For example, in a cell dividing every 24 hours G_2 is four hours long and S is eight hours. In G_2 the cells contain 4n times the amount of DNA (Fig. 17–3). The cause of initiation of DNA synthesis is not well known. A triggering substance, probably a protein produced during the G_1 period, determines the beginning of the S period. Self-duplication and transcription of DNA take place when the DNA is most dispersed (uncoiled) in the chromosomes. The S period varies for different chromosomes or for parts of a chromosome. For example, the heterochromatic X chromosome is late replicating, and the heterochromatic regions of other chromosomes also replicate late in the S period.

   DNA duplication is better known in bacteria. In *E. coli* there is a naked circular chromosome containing 1.1 mm of DNA. DNA duplication is via a template mechanism by which the two DNA strands unwind and two new strands are formed by free nucleotides that fall into phase and are linked by DNA polymerase. Replication may be visualized as an advancing fork forming a **Y** with the parent strands (Fig. 17–5).

   DNA duplication is *semiconservative*. In bacteria this has been demonstrated experimentally at the biochemical level (Fig. 17–6). By radioautography with ³H-thymidine the duplication may be made visible. Complete DNA duplication in *E. coli* takes place in 20 to 30 minutes and the DNA helix is unwound at 10,000 revolutions per minute. The attachment of the DNA to the membrane (Fig. 2–1) is important in initiating DNA duplication and in the mechanism of separation of the two daughter chromosomes and cells (bacteria divide by amitosis). More recent work suggests that in bacteria there are two – instead of one – replicating forks advancing bidirectionally (Fig. 17–9); there are times when four forks may be advancing concurrently.

   In *eukaryons* DNA duplication is also semiconservative, as can be demonstrated experimentally in plant cells (Fig. 17–10). While the DNA of *E. coli* duplicates at 20 to 30 micrometers per minute, in a cultured human cell duplication proceeds at only 0.5 μm per minute. It may be demonstrated that there are numerous replicating forks advancing bidirectionally from initiation points toward termination points (Fig. 17–11). In the hamster there are some 50,000 initiation points in the entire genome.

   DNA duplication is achieved on each of the parent polynucleotide strands by the intervention of DNA polymerases (Pol II and III) and the precursor nucleotides (i.e., adenosine, guanosine, cytosine, and thymidine triphosphates). The synthesis is in the $5' \rightarrow 3'$ direction and is in short segments on both strands. DNA ligase is the enzyme involved in the covalent joining of the DNA segments.

   In certain systems RNA may be transcribed into DNA. This is done by an RNA-dependent DNA polymerase (i.e., *reverse transcriptase*). Such an enzyme has been found in tumors produced by viruses (i.e., *oncogenic viruses*).

   The mechanism of DNA replication is also related to that

of DNA repair and recombination. DNA repair may occur when an alteration has been produced in the DNA molecule, as, for example, under the action of radiations or certain chemicals (see Chapter 15).

A well known case is the DNA repair after UV irradiation of a bacterium. This treatment produces dimers between adjacent pyrimidine bases (e.g., T—T, C—C, or C—T). In the dimer the distance between nucleotides is reduced from 0.34 nm to 0.28 nm. DNA repair involves: (1) an *endonuclease* that produces incision of the affected DNA strand, (2) an *exonuclease* that produces excision of the gap, (3) a *DNA polymerase* that fills the gap by repair replication, and (4) *a ligase* that causes covalent joining of the polynucleotide (Fig. 17–14). The presence of light is required for reaction (1) to be initiated. In the dark DNA repair is inhibited.

In *recombination,* which takes place at pachynema of meiosis, the homologous chromatids undergo a process similar to that of DNA repair in which endonucleases, exonucleases, DNA polymerase, and ligase are involved. In *Trillium* it has been found that during recombination there is limited DNA synthesis (0.3 per cent of the total). At zygonema there is a special protein synthesis whose inhibition halts pairing and recombination. The molecular model of recombination that is favored at present involves the close pairing of the DNA molecules of the homologous chromatids. This pairing is followed by breaks in the four DNA strands, by unraveling of the strands, and by reassociation of the two recombining molecules (Fig. 17–15). At the sites where there are gaps or overlaps in the strands excision and DNA repair follow. Such mechanisms involve the synthesis of DNA and proteins.

RNA synthesis is continuous during the entire cell cycle, but it stops during mitosis, presumably because the condensation of the chromosomes prevents DNA transcription. The main RNA species—i.e., ribosomal, transfer, messenger, and nuclear heterogenous RNA—probably follow a similar pattern of synthesis. In prokaryotic cells, on the other hand, RNA synthesis occupies the entire cell cycle.

Protein synthesis is a continuous process during interphase, but it is inhibited at metaphase of mitosis. During mitosis many nuclear proteins are released into the cytoplasm, and they return at telophase. The nucleocytoplasmic interrelationships are based mainly on the transfer of proteins between these two cellular compartments (see Chapter 20). The histones remain associated with the DNA during the entire cycle, and both are synthesized during the S period. The acidic proteins are produced during the entire interphase.

In yeasts it has been found that individual proteins (enzymes) appear at a definite moment of the G_2 period. The rate of synthesis of an enzyme may double at that moment of the cycle when there is a functional replication of the corresponding genes.

REFERENCES

1. Avery, O. T., McLeod, C. M., and McCarthy, M. (1944) *J. Exp. Med., 79*:137.
2. Penman, S., Smith, J., Holtzman, E., and Greenberg, H. (1966) *Natl. Cancer Inst. Monogr., 23*:489.
3. Maio, J. J., and Schildkraut, C. L. (1967) *J. Molec. Biol., 24* :29.
4. Boivin, A., Vendrely, R., and Vendrely, C. (1948) *C. R. Acad. Sci. (Paris), 226*:1061.
5. Mirsky, A. E., and Ris, H. (1948) *J. Gen. Physiol., 31*:1.
6. Vendrely, R. (1955) In: *The Nucleic Acids,* Vol. 2, p. 155. (Chargaff, E., and Davidson, J. N., eds.) Academic Press, Inc., New York.
7. Swift, H. (1950) *Anat. Rec., 105*:56.
8. Pollister, A. W., Swift, H., and Alfert, M. (1951) *J. Cell. Comp. Physiol., 38* (suppl. 1):101.
9. Du Praw, E. J. (1968) *Cell and Molecular Biology.* Academic Press, Inc., New York.
10. Rudkin, G. T. (1967) In: *The Chromosome.* p. 12. (Yerganian, G., ed.) Williams and Wilkins Co., Baltimore.
11. Callanan, M. J., Carroll, W., and Mitchell, E. (1957) *J. Biol. Chem., 229*:279.
12. Feughelman, M., Langridge, R., Seeds, W. E., Stokes, A. R., Wilson, H. R., Hooper, C. W., Wilkins, M. H. F., Barclay, R. K., and Hamilton, L. D. (1955) *Nature, 175*:834.
13. Alfert, M. (1956) *J. Biophys. Biochem. Cytol., 2*:109.
14. Vendrely, R., Knobloch-Mazen, A., and Vendrely, C. (1960) *Biochem. Pharmacol., 4*:19.
15. Johns, E. W., Phillips, D. M. P., Simson, P., and Buttler, J. A. V. (1960) *Biochem J., 77*:631.
16. Lindsay, D. T. (1964) *Science, 144*:420.
17. Steele, W. J., and Busch, H. (1963) *Cancer Res., 23*:1153.
18. Busch, H. (1965) *Histones and Other Nuclear Proteins.* Academic Press, Inc., New York.
19. Wang, T. J. (1967) *J. Biol. Chem., 242*:1220.
20. Teng, C. S., Teng, C. T., and Allfrey, V. G. (1971) *J. Biol. Chem., 246*:3597.
21. Borun, T. W., and Stein, G. S. (1972) *J. Cell Biol., 52*:308.
22. Frenster, J. H. (1965) *Nature, 206* :680.
23. Kleinsmith, L. J., and Allfrey, V. G. (1969) *Biochim. Biophys. Acta, 175*:123, 136.
24. Kleinsmith, L. J., Allfrey, V. G., and Mirsky, A. E. (1966) *Science, 154*:780.
25. McEwen, B. S., Allfrey, V. G., and Mirsky, A. E. (1963) *J. Biol. Chem., 238*:2571 and 2579.
26. Allfrey, V. G., and Mirsky, A. E. (1961) In: *Protein Biosynthesis,* p. 49. (Harris, R. J. C., ed.) Academic Press, Inc., New York.
27. Tandler, J. C., and Solari, A. J. (1968) *J. Cell Biol., 39*:134a.
28. Dounce, A. J. (1955) In: *The Nucleic Acids,* Vol. 2, p. 93. (Chargaff, E., and Davidson, J. N., eds.) Academic Press, Inc., New York.
29. Howard, A., and Pelc, S. R. (1953) *Heredity,* (suppl. 6):261.
30. Gurdon, J. B., and Woodland, H. R. (1968) *Biol. Rev., 43*:233.
31. Harris, H. (1970) *Cell Fusion.* Oxford University Press, London.
32. Hotta, Y., Parchman, L. G., and Stern, H. (1968) *Proc. Natl. Acad. Sci. USA, 60*:575.
33. German, J. (1964) *J. Cell Biol., 20*:37.
34. Taylor, J. H. (1960) *J. Biophys. Biochem. Cytol., 7*:455.
35. Meselson, M., and Stahl, F. W. (1958) *Proc. Natl. Acad. Sci. USA, 44*:671.
36. Cairns, J. (1963) *J. Molec. Biol., 6*:208.
37. Cairns, J. (1966) *J. Molec. Biol., 15*:372.
38. Masters, M., and Broda, P. (1971) *Nature* [*New Biology*] *232*:137.
39. Wake, R. G. (1972) *J. Molec. Biol., 68*:501.
40. Taylor, J. H., Woods, P. S., and Hughes, W. L. (1957) *Proc. Natl. Acad. Sci. USA, 43*:122.
41. Huberman, J. A., and Riggs, A. D. (1968) *J. Molec. Biol., 32*:327.
42. Elson, D. (1965) *Ann. Rev. Biochem., 34*:449.
43. Kornberg, T., and Gefter, M. L. (1971) *Proc. Natl. Acad. Sci. USA., 68*:761.
44. Griffith, J., Huberman, J. A., and Kornberg, A. (1971) *J. Molec. Biol., 55*:209.

45. Gillert, M. (1967) *Proc. Natl. Acad. Sci. USA., 57*:148.
46. Okazaki, R., Okazaki, T., Sakabe, K., Sagimoto, K., and Sugino, A. (1968) *Proc. Natl. Acad., Sci. USA., 59*:598.
47. Baltimore, D. (1970) *Nature* (London) *226*:1209.
48. Mizushima, S., and Temin, H. (1970) *Nature, 226*:1211.
49. Spiegelman, S., Burny, A., Das, M. R., et al. (1970) *Nature* (London), *227*:563.
50. Spiegelman, S., Burny, A., Das, M. R., et al. (1970) *Nature* (London), *227*:1029.
51. Howard, F. P. (1968) *Ann. Rev. Biochem., 37*:175.
52. Radman, M., Cordone, L., Krsmanovic-Simic, D., and Errera, M. (1970) *J. Molec. Biol., 49*:203.
53. Stern, H., and Hotta, Y. (1969) *Genetics, 61*:27.
54. Parchman, G. L., and Stern, H. (1969) *Chromosoma, 26*:298.
55. Moses, M. (1968) *Ann. Rev. Genet., 2*:363.
56. Hecht, N. B., and Stern, H. (1971) *Exp. Cell. Res., 69*:1.
57. Prescott, D. M. (1961) In: *Cell Growth and Cell Division*, Vol. 2, *Internat. Soc. Cell Biol.* (Harris, R. J. C., ed.) Academic Press, Inc., New York.
58. Fan, H., and Penman, S. (1970) *Science, 168*:135.
59. Martin, D. W., Tomkins, G. M., and Bresler, M. A. (1969) *Proc. Natl. Acad. Sci. USA., 63*:842.
60. Prescott, D. M., and Bender, M. A. (1962) *Exp. Cell Res., 26*:260.
61. Goldstein, L. (1963) In: *Cell Growth and Cell Division*, Vol. 2, p. 129. *Internat. Soc. Cell Biol.* (Harris, R. J. C., ed.) Academic Press, Inc., New York.
62. Goldstein, L., and Prescott, D. M. (1967) *J. Cell Biol., 33*:637.
63. Beck, J. S. (1962) *Exp. Cell Res., 28*:406.
64. Hancock, R. J. (1969) *J. Molec. Biol., 40*:457.
65. Woodard, J., Rasch, E., and Swift, H. (1961) *J. Biophys. Biochem. Cytol., 9*:445.
66. Das, N. K., and Alfert, M. (1968) *Exp. Cell Res., 49*:51.
67. Bloch, D. P., and Teng, C. (1969) *J. Cell Sci., 5*:321.
68. Mitchison, J. M. (1968) *Excerpta Medica Internat. Cong. Ser., 166*:26.

ADDITIONAL READING

Chargaff, E., and Davidson, J. N. (1955) *Nucleic Acids.* Academic Press, Inc., New York.
Coming, D. E., and Okada, T. A. (1973) DNA replication and the nuclear membrane *J. Molec. Biol., 75*:609.
Crick, F. H. C. (1957) Nucleic acids. *Sci. Am., 197*:188.
Davidson, J. N. (1960) *The Biochemistry of the Nucleic Acids.* 4th Ed. John Wiley & Sons, New York.
De Terra, N. (1969) *Int. Rev. Cytol., 25*:1.
Elgin, S., Froehner, S. C., Smart, J. E., and Bonner, J. (1971) In: *Advances in Cell and Molecular Biology*, Vol. I, p. 2. (Du Praw, ed.) Academic Press, Inc., New York.
Kihlman, B. A. (1971) In: *Advances in Cell and Molecular Biology*, Vol. I, p. 59. (Du Praw, ed.) Academic Press, Inc., New York.
Lima-de-Faría, A. (1969) In: *Handbook of Molecular Cytology*, p. 278. North-Holland Publishing Co., Amsterdam.
McElroy, W. D., and Glass, B., eds. (1957) *The Chemical Basis of Heredity.* The Johns Hopkins Press, Baltimore.
Goulian, M. (1971) *Ann. Rev. Biochem., 40*:855.
Miller, O. L., Beatty, B. R., Hamkalo, B. A., and Thomas, C. A., Jr., (1970) *Cold Spring Harbor Symp. Quant. Biol., 35*:505.
Mirsky, A. E. (1968) The discovery of DNA. *Sci. Am., 218*:78.
Mitchell, J. S., ed. (1960) *The Cell Nucleus* (Symposium), Academic Press, Inc., New York.
Mitchison, J. M. (1971) *The Biology of the Cell Cycle.* Cambridge University Press, New York.
O'Donovan, G. A., and Neuhard, J. (1970) *Bact. Rev., 34*:278.
Pato, M. J. (1972) *Ann. Rev. Microbiol., 26*:347.
Stein, G. S., Spelsberg, T. C., and Kleinsmith, L. J. (1974) Non-histone chromosomal proteins and gene regulation. *Science, 183*:817.
Strauss, B. S. (1960) *An Outline of Chemical Genetics.* W. B. Saunders Co., Philadelphia.
Taylor, J. H. (1958) The duplication of chromosomes. *Sci. Am., 198*:36.
Temin, H. M. (1972) RNA-directed DNA synthesis. *Sci. Am., 226*:24.

eighteen

RIBOSOMES: STRUCTURE AND BIOGENESIS

The concept of the *ribosome* as a definite submicroscopic organelle composed of ribonucleic acid and protein was introduced in Chapter 2. First observed under the electron microscope as *dense particles* or *granules*,[1] ribosomes were then isolated and their RNA content was demonstrated. The rapid advances in the analysis of submicroscopic organelles have led to the concept that ribosomes are universal components of biological organisms.

From a physiological viewpoint, ribosomes are used by the cell for protein synthesis—the process by which amino acids are assembled in a definite sequence to produce the polypeptide chain. Protein synthesis requires the ordered interaction of three types of RNA molecules of nuclear origin in addition to activated amino acids: ribosomal, transfer, and messenger RNA.

The function of the ribosome, along with the process of protein synthesis, will be discussed in Chapter 19. In this chapter the major emphasis will be on morphology, structure, macromolecular organization, and biogenesis of ribosomes. Even though this book is devoted to the eukaryotic cell, some aspects of the study of ribosomes will be in relation to the bacterial cell, since the molecular organization of ribosomes has been studied principally in prokaryons (see Kurland, 1972).

Free and Membrane-bound Ribosomes

Ribosomes are either free in the cytoplasmic matrix or attached to the membranes of the endoplasmic reticulum (see Chapter 9). It will be shown later that these two types are somewhat interchangeable; there is a cycle of both the ribosomes and their subunits that is related to their function in protein synthesis (see Chapter 19).

A bacterium may contain as much as 25 to 30 per cent of its weight in ribosomes—all of which are free in the protoplasm. Most ribosomes are also free in the matrix of yeast cells, reticulocytes, meristematic plant tissues (Fig. 11–5), and embryonic nerve cells (Fig. 9–1). In plant cells it can be observed that ribosomes precede the development of membranes.

In cells that are engaged in protein synthesis, such as enzyme-secreting cells and plasma cells, most ribosomes are attached to the membranes of the endoplasmic reticulum. This relationship suggests that these membranes assist in the removal of the newly synthesized proteins from the ribosomes and in their transport and secretion. The two-dimensional arrangement of ribosomes on the membrane may also facilitate protein synthesis. Most of the ribosomes shown in Figure 18–1 are attached to membranes. Since the section is, for the most part, tangential to the cisternae of the endoplasmic reticulum, it is possible to observe the two-dimensional disposition of the attached ribosomes.[2, 3] Recurrent patterns that form discrete coils which correspond to groups of functionally related ribosomes may also be observed. These structures are called *polyribosomes* or *polysomes*.

The proportion of membrane-bound ribosomes varies widely. In cells engaged in producing proteins for export, such as those of the exocrine pancreas, which synthesize

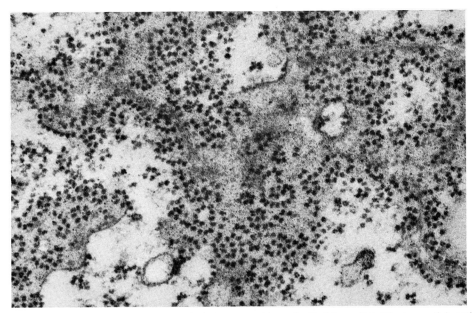

Figure 18–1. Electron micrograph of a root hair from an epidermal cell of the radish. The tangential section through the membrane of the endoplasmic reticulum shows groups of ribosomes (i.e., polyribosomes) disposed in recurrent patterns. ×57,000. (Courtesy of H. T. Bonnett, Jr., and E. H. Newcomb.)

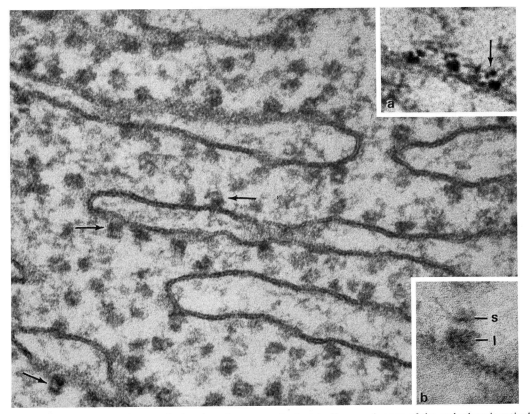

Figure 18–2. Electron micrograph showing ribosomes attached to the membranes of the endoplasmic reticulum. Observe the "unit membrane" structure. Arrows indicate ribosomes where the attachment of the large (60S) subunit is best observed. ×208,000. (Courtesy of G. E. Palade.) Inset **a,** attachment of the large and small subunits forming a "cap," which is subdivided into two portions (arrow). Insert **b,** at higher magnification, the small (s) and large (l) subunits appear to be separated by a clear cleft. **a,** ×200,000; **b,** ×410,000. (Both insets courtesy of N. T. Florendo.)

zymogen granules, or the plasma cells, which produce immunoglobulins, most ribosomes are associated with the membranes. The opposite situation exists in a rapidly growing embryonic cell or in a dedifferentiated tumor cell. The liver is an intermediary type in which about 75 per cent of ribosomes are bound, with the other 25 per cent remaining free in the cytoplasmic matrix. In Figure 18–2 free and bound ribosomes may be observed; the latter are attached to the membrane via the large subunit of the ribosome.

The *HeLa cell* is a dedifferentiated human tumor cell in which only about 15 per cent of the ribosomes are membrane-bound.[4] In these cells, by the use of various treatments that dissociate the ribosomes from the membrane, two classes of bound ribosomes have been recognized. Some ribosomes are more loosely bound and belong to a class having a relatively high buoyant density. The others are more tightly attached by the large subunit and have a lower buoyant density.[4]

In chick embryo cells subjected to hypothermia, free ribosomes become organized into sheets with a crystal-like arrangement with a square unit cell. These crystals are interpreted as groups of inactive ribosomes that are not associated with messenger RNA and which lack nascent protein.[5]

Ribosomal Subunits

The ribosome is approximately spheroid in shape, with a diameter of 23 nm and is composed of a large and a small subunit (Table 18–1). In eukaryotic cells it was shown that the ribosome is attached to the microsomal membrane by the large subunit, and a cleft separating the two subunits and parallel to the membrane has been observed[6, 7] (Fig. 18–2).

Ribosomes require low concentrations of Mg (0.001 M) for structural cohesion. If the Mg concentration is increased tenfold, two ribosomes combine to form a "dimer" with twice the molecular weight of the individual

ribosome. If the Mg concentration is lowered, the single ribosome can be dissociated reversibly into *subunits* (Fig. 18–3). For an 80S ribosome the two subunits produced are 60S and 40S, and in the presence of this divalent cation they can re-form the 80S unit (Fig. 18–3). Likewise, the dimer can be converted back into two ribosomes by lowering the Mg^{2+} concentration. The binding with Mg^{2+} seems to be through the RNA phosphate and not through the protein. Notice in Table 18–1 that the 70S ribosomes have subunits of 50S and 30S.

The fine structure of the ribosome is very complex and is not yet fully elucidated. Since the ribosome is highly porous and hydrated, the RNA and the protein are probably intertwined within the two subunits. In sections stained with uranyl ions, which stain RNA selectively, each ribosome appears as a star-shaped body with four to six arms implanted on a dense axis.

Negative staining of isolated ribosomes and their subunits has led to a better understanding of the fine structure of these organelles. From these and other studies several models for the large and small subunits have been proposed (see Nonomura et al., 1971). In liver ribosomes the 60S subunit appears either round, with a diameter of 23 nm, or triangular, with two convex sides and a third more flattened side that has a narrow depression or notch in the middle. The 40S subunit is elongated and has a curved profile of about 23 nm × 12 nm, with a convex and a concave side. There is a line or partition that divides

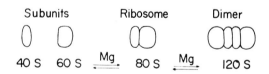

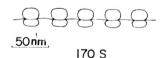

Figure 18–3. Diagram of the subunit structure of the ribosome and the influence of Mg. A polyribosome formed by five ribosomes is indicated. The filament uniting the ribosomes is considered to be messenger RNA. The sedimentation constants (S) of the different particles are indicated.

TABLE 18–1.   Size of Various Ribosomes

Ribosomes	Size	Subunits		RNAs	
Eukaryons	80S	60S	40S	28S + 5S	18S
Bacteria	70S	50S	30S	23S + 5S	16S
Mitochondria (in mammals)	55S	35S	25S	21S	12S

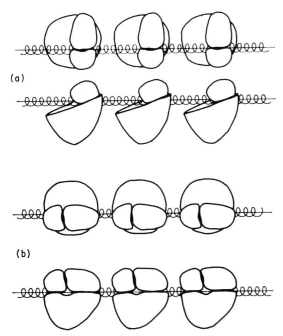

(a)

(b)

Figure 18–4. Diagram showing the possible relationship of ribosomes with messenger RNA (mRNA) in polyribosomes. Each ribosome is formed of a large and a small subunit, the latter having two portions. **a**, the mRNA runs between the subunits with all the ribosomes in the same position; **b**, the ribosomes are in a position perpendicular to that of **a**. (From Nonomura, Y., Blobel, G., and Sabatini, D., *J. Molec. Biol., 60*:303, 1971.)

the small subunit into two unequal portions (see inset, Fig. 18–2).[8] In the complete ribosome the small subunit is attached by the concave side to the more flattened side of the large subunit and the partition in the small subunit is in coincidence with the notch of the 60S subunit, forming a kind of tunnel between the two (Fig. 18–4).

Figure 18–4 shows a schematic view of polyribosomes along with two possible interpretations of the position of the messenger RNA. In both models it may be observed that a portion of the messenger RNA is held between the two subunits, thus being protected from degradation by nucleases (see Chapter 19). Some structural and biochemical studies support the view that the large subunit is traversed by a kind of channel which contains the nascent polypeptide chain. This portion of the chain is protected from the action of proteolytic enzymes (Fig. 19–14).

CHEMICAL COMPOSITION

Ribosomal RNAs

The major constituents of ribosomes are RNA and proteins present in approximately equal proportions; there is little or no lipid material. The positive charges of proteins are not sufficient to compensate for the many negative charges in the ester-phosphate chain of RNA, and for this reason ribosomes are strongly negative and bind cations and basic dyes.

Ribosomal RNA is contained in the two subunits of the ribosome forming a 28S and an 18S particle, with a molecular weight of 1.3×10^6 and 6×10^5 daltons for the 60S and 40S subunits, respectively. (In bacteria the RNAs are contained in 23S and 16S subunits; see Table 18–1.) It is thought that each ribosomal subunit contains a highly folded ribonucleic acid filament to which the various proteins adhere.[9]

The internal organization of the ribosome is not well understood. Because ribosomes easily bind basic dyes, it is thought that the RNA is exposed at the surface of the subunit, and the protein is assumed to be in the interior in relation to nonhelical parts of the RNA. About 60 per cent of the RNA is helical (i.e., double stranded) and contains paired bases. These double-stranded regions are due to hairpin loops between complementary regions of the linear molecule. The base composition of ribosomal RNAs of different origin varies, and it does not follow the base rule characteristic of the Watson-Crick model for DNA (see Chapter 3). The most abundant bases are guanine and adenine. Both the 28S and the 18S ribosomal RNA contain a characteristic number of methyl groups, mostly as 2′-O-methyl ribose. The sequence of the 23S and 16S RNA from *Escherichia coli* has been analyzed. Both are specific, showing no apparent homology.[11]

The large subunit of prokaryotic and eukaryotic ribosomes contains a small RNA consisting of 120 nucleotides[12–14] and having a sedimentation of 5S. This 5S RNA has not been found in mitochondrial ribosomes (Table 18–1). The complete sequence of the 5S RNA of *E. coli* is known.[15] In reconstitution experiments the 5S RNA is needed for the correct assembly of the large subunit (see following page).

Ribosomal Proteins

In recent years our knowledge of ribosomal proteins has made remarkable progress, particularly with respect to bacterial ribosomes (see Nomura, 1973). In *E. coli* some 20 different protein molecules have been isolated from the 30S subunit, and some 30 to 35 proteins, from the 50S. Therefore, a total of 50 or more proteins have been isolated in a single ribosome. The primary structure of several of these proteins and of the ribosomal RNAs has been elucidated. Of particular interest, from the standpoint of cell and molecular biology, is the work in which ribosomal proteins have been partially or totally dissociated. This dissociation has been followed by the reconstitution of their structure and function. These studies have led to knowledge of the role played by some of the proteins in ribosomal function. For example, if ribosomes or ribosomal subunits are centrifuged in a gradient of 5M cesium chloride they lose 30 to 40 per cent of their proteins.[16, 17] In this way, both the 50S and the 30S subunits may be dissociated into two inactive core particles (40S and 23S, respectively), which contain the RNA and some proteins; at the same time, several other proteins—the so-called *split proteins* (SP)—are released from each particle (Fig. 18–5). There are SP50 and SP30 proteins which may reconstitute the functional subunit when added to the corresponding core. This partial reconstitution is rapid and may be achieved in a few minutes at 37° C.[18] Some of the split proteins are apparently specific for each ribosomal subunit. The split proteins have been further fractionated and divided into acidic (A) and basic (B) proteins. Figure 18–5 shows at least six different groups of proteins in the ribosome.[19] In reconstitution experiments in which one protein at a time is omitted, three of these proteins have been recognized to be essential for ribosomal function. As will be shown in Chapter 19, the testing of ribosomal function can be carried out on a cell-free system containing ribosomes, messenger RNA, amino acids, and a supernatant containing transfer RNA and amino acid–activating enzymes.

One interesting conclusion of these

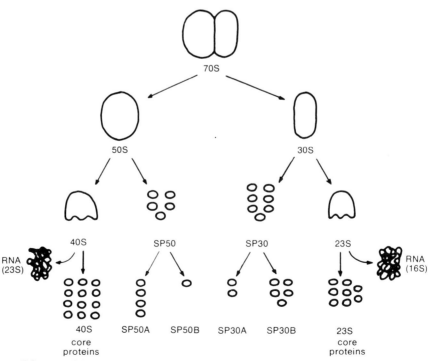

Figure 18–5. Diagram of the protein components of a 70S subunit. (See the description in the text.) (From Nomura, M., and Traub, P., *J. Molec. Biol., 34*:609, 1968.)

studies is that both the core and the split proteins may be present with a different stoichiometry. The so-called *unit proteins* are present in one copy per ribosome; the *fractional proteins,* in 0.5 copy per ribosome; and the *marginal proteins,* in 0.7 to 0.8 copy per ribosome (see Kurland, 1972). Therefore, not all ribosomes are identical. For example, one protein identified as S14 is essential for ribosomal function, and at the same time is represented by only 0.5 copy per ribosome. The conclusion is that normally there are two types of ribosomes: S14 (+), that are active, and S14 (−) that should be inactive. However, this conclusion would be difficult to demonstrate experimentally.

Total Reconstitution

Functional 30S subunits have been reconstituted by the addition of 16S rRNAs to a mixture of proteins from the 30S subunit.[19, 20] Interestingly, the reconstitution of both ribosomal subunits and the complete ribosome takes place spontaneously. The subunits are re-formed by the principle of "self-assembly," as are the subunits of some proteins (see Chapter 7). The complex organization of this organelle may be dissociated and regenerated by simple physicochemical interactions of the component macromolecules; this may be performed in vitro and without pre-existing cell structures.[21]

Dissociation of the 30S subunit may be achieved by treatment with four molar urea and two molar LiCl, which separate the proteins. If the 16S, previously extracted with phenol, is placed in the presence of the 20 proteins obtained from the 30S, the reconstitution takes place in two steps.

The following reconstitution scheme has been proposed: [22]

$$16S\ RNA + \xrightarrow[\text{proteins}]{R_1} RI\ \text{particles} \xrightarrow{\text{heat}}$$

$$RI^*\ \text{particles} + \xrightarrow{S\ \text{proteins}} 30S\ \text{ribosomal subunit}$$

In the first step, performed at a low temperature, the 16S RNA binds some of the 30S ribosomal proteins, forming an RI particle (i.e., a reconstitution intermediate) that is inactive. In the second step, the RI particles are heated at 40° C in the presence of the other proteins that have remained in the supernatant (i.e., S proteins) thereby forming an excited intermediate, RI*. Within 20 minutes fully active 30S ribosomal subunits are formed.

The reconstitution of the 30S subunit is highly specific. It can be achieved with 16S RNA of other bacteria, but not with 16S RNA from yeast or the 23S RNA from *E. coli.*

This type of study, which has also been carried out on the 40S subunit (see Nomura, 1973), is leading to interesting conclusions about the function of some of the specific proteins and their three-dimensional organization in the ribosome. For example, an assembly map of the 30S subunit has been made in which some proteins are shown to be directly bound to the 16S RNA with various binding affinities. Certain proteins require the prior attachment of other proteins in order to be incorporated in a stepwise and cooperative manner. (In other words, the previous attachment of certain proteins facilitates the binding of others.)

This type of analysis has demonstrated that whereas some of the proteins probably play a structural role in the assembly of the ribosome, others are definitely engaged in the specific ribosomal functions (see Chapter 19).

SUMMARY: Structure and Composition of Ribosomes

Ribosomes are organelles composed of ribonucleic acids and proteins and have a fundamental role in protein synthesis. Ribosomes may be free or membrane-bound. Although they are found as single units, in most cases they form groups of functionally related ribosomes that constitute *polyribosomes or polysomes.* In *E. coli* 25 to 30 per cent of the weight is represented by free ribosomes or polysomes. Membrane-bound ribosomes are found principally in the endoplasmic reticulum of protein-secreting cells. The size of ribosomes varies from that of eukaryons (80S), to that of bacteria (70S), to that of most mitochondrial ribosomes (55S). Each ribosome is composed of two subunits of different size. For example, the 80S ribosome contains 60S and 40S subunits. The subunits may be dissociated by lowering the Mg^{2+} concentration. The 60S subunit, which is about 23 nm in diameter, has a spheroidal or triangular profile, with two convex sides and a more flattened one that has a narrow depression. The 40S subunit is elongated (23 nm $\times$ 12 nm), has a convex and a concave side, and is divided into two unequal parts by a partition. A tunnel containing the messenger RNA is formed between the two subunits, and another channel for the polypeptide chain is postulated in the 60S subunit. The binding of the ribosome to the membrane of the endoplasmic reticulum is by way of the large subunit.

Ribosomal RNA represents 45 per cent of the mass in the 80S ribosomes and 65 per cent in the 70S ribosomes; the rest is composed of proteins. In eukaryons the large subunit contains a molecule of 28S RNA and one of 5S RNA. The small subunit contains one 18S RNA. In bacteria the 50S subunit contains a 23S and 5S RNA, and the 30S subunit contains a 16S RNA. Mitochondrial ribosomes lack 5S RNA. Ribosomal RNAs contain a number of methyl groups, for the most part as 2'-O-methylribose.

Bacterial ribosomes contain more than 50 different proteins. About 40 per cent of the proteins can easily be separated from both subunits (split proteins), leaving a core with the RNAs and the more tightly bound proteins. Reconstitution of functionally active subunits can readily be obtained by addition of the split proteins to the core. Total reconstitution of each subunit may be achieved by addition of the corresponding RNA and a pool of the proteins. The reconstitution of the 30S subunit is as follows:

$$16S \text{ RNA} + \xrightarrow[\text{proteins}]{R_1} \text{RI particles} \xrightarrow{\text{heat}}$$

$$\text{RI* particles} + \xrightarrow[\text{proteins}]{S} 30S$$

A reconstitution intermediate (RI) is formed in the cold; the final reconstitution needs activation by heat. By performing reconstitution experiments in which one protein is omitted at a time, it is possible to gain information about the role of each protein in the ribosome. Many proteins are represented in single units within the ribosome (*unit proteins*), but others are *marginal* or *fractional*, being in fewer than one copy per ribosome. The reconstitution of ribosomes is an excellent example of the principle of self-assembly, which was discussed in Chapter 7. The function of the ribosomes in protein synthesis will be discussed in Chapter 19.

BIOGENESIS OF RIBOSOMES

The complex molecular organization of the ribosomes, with their special RNA molecules and their 50 or more specific proteins, implies that more than 50 genes are involved in their formation. Ribosomal biogenesis in bacterial cells differs from that in eukaryotic cells; the emphasis here will be on biogenesis in eukaryotic cells.

In bacteria the RNA genes coding for the 5S, 23S, and 16S ribosomal RNAs are tightly clustered in a region of the chromosome and are present in only a few copies.[23] In other words, the ribosomal genes are in a single operon (see Chapter 19) that is transcribed as a unit, being the RNA molecules immediately released from the DNA. This process has been followed under the electron microscope (see Fig. 19–15).[24]

In eukaryons the biogenesis of ribosomes is much more complex and involves a long-lasting process in which several regions of the cell are involved. The 18S and 28S RNAs are transcribed as a much larger molecule in the nucleolar organizer and are represented by many copies of ribosomal DNA (i.e., there is gene redundancy or amplification). The DNA coding for the 5S RNA is also highly repetitive, but the molecule is synthesized outside the nucleolus. The formation of the various RNA molecules is coupled with that of the numerous proteins that are made in the cytoplasm by the usual mechanism of protein synthesis. Most of the processing of the subunits of the ribosomes occurs within the nucleolus, and from there they are finally delivered to the cytoplasm of the cell. Thus, the complex mechanisms involved in the biogenesis of ribosomes represent a very interesting example of coordination at the cellular and molecular levels.

Historically, this study is related to the early cytochemical work of Caspersson with ultraviolet absorption and to that of Brachet with specific staining for ribonucleic acids. The work of both researchers demonstrated the presence of large nucleoli, rich in RNA, in protein-synthesizing cells. The discovery of ribosomes and their role in protein synthesis promptly suggested a relationship between the nucleolus and these organelles. The nucleolar organizers were interpreted as the chromosomal sites containing the genes that code for rRNA. Finally, the introduction of new methods in molecular biology demonstrated that the nucleolus plays a key role in the biogenesis of ribosomes.

Nucleolar Organizer and Ribosomal DNA

Direct evidence that the nucleolus is responsible for the synthesis of rRNA was obtained in 1964, when it was discovered that an anucleolate mutant of the amphibian *Xenopus laevis* was incapable of rRNA synthesis.[25] Diploid cells of the wild type of *X. laevis* have two nucleoli; the heterozygous mutant contains only one nucleolus per diploid cell and the homozygous mutant is anucleolate. This condition is lethal and the embryo dies at the tail bud stage. Until then, homozygous mutants rely on maternal ribosomes for protein synthesis.

It was possible to demonstrate with the DNA-RNA hybridization technique[26] (see Chapter 3) that the DNA associated with the nucleolus is responsible for coding rRNA.

The maximal amount of rRNA bound to

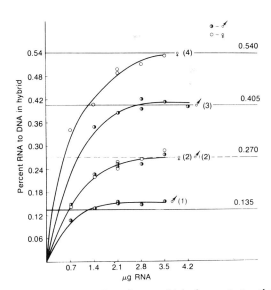

Figure 18–6. Experiment which demonstrates that the DNA, complementary with ribosomal RNA, is associated with the nucleolus organizer. The saturation level for the DNA-rRNA hybrids increases with the number of nucleolar organizers. Strains of *Drosophila* having one, two, three, and four nucleolar organizers were used. (From Ritossa, F. M., and Spiegelman, S., *Proc. Natl. Acad. Sci. USA*, 53:737, 1965.)

a definite amount of DNA (i.e., the saturation level) is a measure of the number of specific duplex DNA-RNA molecules formed. By using this technique researchers have found that nucleolar organizers contain about 1000 rDNA cistrons. In the heterozygous mutant the saturation was reached at half the level of the wild type, and in the homozygous mutant there were no rDNA cistrons at all.[27]

A similar type of experiment was performed with *Drosophila melanogaster,* in which the nucleolar organizer is contained in a heterochromatic region of the X or Y chromosome.[28] Cells with organizers that were either deficient or duplicated were analyzed with the DNA-RNA hybridization technique. As shown in Figure 18–6, the saturation level for the wild type is about 0.27 per cent. Half the saturation level was found in mutants having only one organizer, and this was proportionally higher in those *Drosophila* mutants with three or four organizers. Some 130 rDNA cistrons per organizer were calculated in the haploid nucleus of the wild type. In the so-called "bobbed" mutants of *Drosophila,* the number of rDNA cistrons may be reduced, and it can be calculated that death will occur when fewer than 40 copies of rDNA cistrons are present in the nucleolar organizer.[29] Several experiments with *X. laevis* and *Drosophila* demonstrate that the DNA cistrons for the two rRNAs exist in equal amounts and are strictly alternating.[30–33]

Biogenesis of the 5S RNA

In eukaryons the 5S ribosomal RNA is not related to the nucleolar organizer. In *X. laevis,* by use of the DNA-RNA hybridization technique, it was found that there are many more 5S genes (i.e., about 27,000) than 18S and 28S genes and that they are not spatially related.[34] Because of this it is understandable that the 5S RNA is present in the anucleolate mutant.[30] In HeLa cells there are about 10,000 copies of 5S DNA. In *Drosophila* the number of 5S DNA is about the same as that of 18S and 28S genes.[35] In this case the 5S genes are located in the autosomes while the 28S + 18S genes are in the nucleolar organizer which, as said before, is related to the sex chromosomes (see Pardue, et al, 1973).

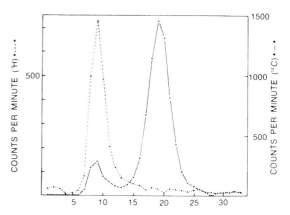

Figure 18–7. Experiment to demonstrate ribosomal DNA in the ovary of *Xenopus.* Solid lines represent the ultracentrifugation pattern of the DNA labeled with ^{14}C in a CsCl gradient. Note the main peak and the small satellite peak. The dotted lines represent hybridization experiments performed with ribosomal RNA labeled with tritium. Hybridization coincides with the minor DNA peak. (From Gall, J. G., *Proc. Natl. Acad. Sci. USA,* 60:553, 1968.)

Isolation of the Satellite Ribosomal DNA

The ribosomal DNA, present in the nucleolar organizer, may be isolated in certain cells by the use of ultracentrifugation techniques on density gradients of cesium chloride, followed by hybridization with ribosomal RNA labeled with radioisotopes. In the case of *X. laevis* it is possible to demonstrate the presence of a small fraction of DNA having a higher buoyant density than the major band (i.e., 1.723, as compared to 1.698). This minor fraction, which represents only 0.15 to 0.20 per cent of the total DNA, is the so-called *satellite ribosomal DNA.* This fraction is completely absent in the case of the anucleolate mutant (see above) of *X. laevis.* The higher buoyant density is related to the higher guanine-cytosine content (70 per cent) in the satellite DNA. By using the hybridization technique with ribosomal RNA it is possible to demonstrate that this hybridization coincides with the satellite DNA[36] (Fig. 18–7). In Chapter 17 another type of satellite DNA, corresponding to the centromeric heterochromatin, was mentioned. In that case the buoyant density of the satellite was lower than that of the major DNA band.

TABLE 18–2. Amplification of rDNA Cistrons in Various Organisms

Organism	% rDNA	rDNA Cistrons Per Haploid Genome
E. coli	0.42–0.65	8–22
B. subtilis	0.38	9–10
HeLa cells	0.005–0.02	160–640
Drosophila (wild type)	0.27	130
Xenopus (wild type)	0.06–0.11	1200–3000

Amplification (Redundancy) of the rDNA

In the previous examples of *Drosophila* and *X. laevis* a considerable degree of *amplification,* or *redundancy,* of the rDNA genes occurs in the DNA. This is particularly evident in many amphibian oöcytes, in which the nucleus may reach a diameter of more than 0.5 mm and contain 1000 or more nucleoli. Such gene amplification may help these oöcytes to accomplish an intense rRNA synthesis that may be 100,000 times greater per genome than in a liver cell. Table 18–2 indicates the percentage of total DNA involved in rRNA synthesis and the degree of amplification per haploid genome in various organisms. It is notable that in bacteria there is only a small degree of rDNA amplification as compared to the other examples mentioned.

Redundancy of rDNA is one way that the cell has to control the production of ribosomes, but apparently it is not the only one. Hybridization experiments with somatic chicken tissues (i.e., kidney, liver, embryo, all with different rates of rRNA synthesis) demonstrated the presence of a similar number of rDNA cistrons of about 100 per haploid genome. Even in sperm cells and erythrocytes in which the production of ribosomes has stopped there is the same level of DNA-rRNA saturation.[37]

In certain plants there is an amplification of ribosomal RNA genes that varies from

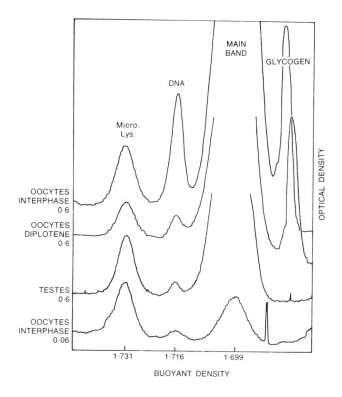

Figure 18–8. DNA components in different tissues of *Acheta* demonstrated by analytical ultracentrifugation. The ribosomal DNA (rDNA), with a buoyant density of 1.716, is most prominent in oöcytes at interphase. The density of rDNA is compared with that of the main DNA band (1.699), DNA from a bacterium, and the glycogen band. *Micro. Lys. Micrococcus lysodeikticus.* (Courtesy of A. Lima-De-Faría.)

1,600 to 13,000 per diploid nucleus, but the number of such genes does not change during the development of a particular plant (see Ingle and Sinclair, 1972).

An interesting example of ribosomal DNA redundancy was found in oöcytes of the insect *Acheta domesticus,* in which a large body containing DNA and histones was observed. By analytical centrifugation (Fig. 18–8) it may be noted that at interphase the satellite rDNA, with a buoyant density of 1.716, is most prominent. This type of DNA, which hybridizes with ribosomal RNA, is almost totally absent in testes.[38]

Processing of the Ribosomal RNA in the Nucleolus

The steps involved in the synthesis and processing of the ribosomal RNAs can be studied best after isolation of the nucleolus from cultured cells.[39] Figure 18–9 shows the variety of ribosomal RNA fractions that are present in such isolated nucleoli. The predominating fractions are 45S, 32S, and 28S, whereas there is very little 18S RNA. This last finding is interpreted as an indication that the 18S RNA is rapidly released into the cytoplasm and is not retained in the nucleolus. To study the time-course of the RNA processing, the cultured cells are studied at various intervals following the application of a pulse with a radio-labeled RNA precursor (e.g., ^{3}H-uridine). From this study the following series of events may be recognized:

(1) The first ribosomal RNA is a large

TABLE 18–3. Approximate Molecular Weights of Different rRNAs and Possible Length of the Extended Molecules.*

rRNA	MW in Daltons	Length in μm
45S	4.5×10^6	~ 4.5
32S	2.2×10^6	~ 2.2
28S	1.6×10^6	~ 1.6
18S	0.6×10^6	~ 0.6

*Compiled from Weinberg, A., et al., *Proc. Natl. Acad. Sci. U.S.A.,* 58:1088, 1967, and Miller, O. L. *in* Lima-de-Faría, A. (Ed.), *Handbook of Molecular Cytology,* Amsterdam, North-Holland Publishing Co., 1970.

molecule of 45S having a molecular weight of 4.5×10^6 daltons and a length of about 4.5 μm (Table 18–3). This 45S RNA is transcribed from the nucleolar organizer (ribosomal DNA) in about 2.2 minutes (in HeLa cells) and is present in the fibrillar region of the nucleolus (see Chapter 12). It is assumed that 20 RNA polymerases are transcribing simultaneously on the ribosomal DNA cistron.

(2) During the process of transcription certain regions of the 45S become methylated in the ribose moiety (producing 2'-o-methyl ribose). (The process of methylation may be studied by pulses of labeled methionine, as donor of methyl groups, followed by a chase in cold methionine.) The regions of the 45S RNA that are methylated correspond to the future 28S and 18S ribosomal RNAs.

(3) The 45S RNA has a lifetime of about 15 minutes and is followed by cleavage and degradation into smaller components (see Perry, 1972). The processing of the 45S RNA is assumed to start by cleavage by an endonuclease into a 41S and a 20S RNA.

(4) The 20S RNA is degraded in its nonmethylated region by an exonuclease, thus producing the 18S. This is rapidly released to the cytoplasm.[40]

(5) The 41S RNA undergoes a series of degradations by exonucleases which attack the nonmethylated regions; by this process the 36S and 32S intermediates are produced.

(6) The 32S remains in the granular part of the nucleolus for about 40 minutes and is then degraded to the 28S RNA (Fig. 18–10). This remains in the nucleolus for another 30 minutes before entering the cytoplasm. From the data shown in Table 18–3 it is evident that about half the 45S molecule is lost by the successive degradations. This occurs in the

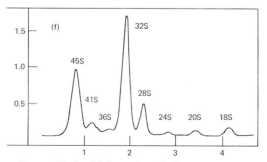

Figure 18–9. Main peaks of ribosomal RNA observed in the nucleolus of *HeLa* cells. (See the description in the text.) Nucleolar RNA was submitted to electrophoresis in polyacrylamide gel. (From Weinberg, A., Loening, V., Willems, M., and Penman, S., *Proc. Natl. Acad. Sci. USA,* 58:1088, 1967.)

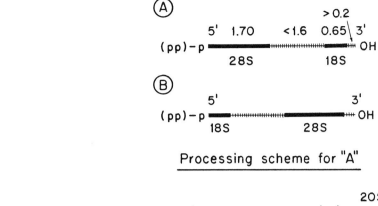

Processing scheme for "A"

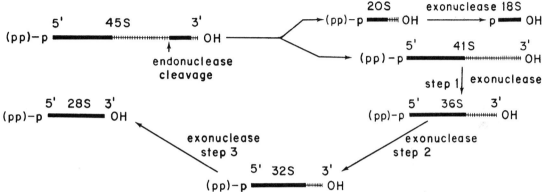

Figure 18–10. Diagram showing possible arrangements of 28S and 18S segments within the 45S ribosomal RNA molecule and possible processing mechanisms. Observe that by the action of endonucleases and exonucleases, the 45S molecule is divided and degraded stepwise to yield the 18S and 28S ribosomal RNA. (See the description in the text.) (From Perry, R. P., and Kelley, D. E., *J. Molec. Biol., 70*:265, 1972.)

regions that are nonmethylated and have a higher content of GC. Thus, the processing of the ribosomal RNA leads to an increase in methyl groups and to a decrease in GC.

Synthesis of Ribosomal Protein and Ribosomal Assembly

The complexity of the ribosome's protein structure, in which about 50 protein species have been demonstrated, increases the difficulty in understanding protein biosynthesis and assembly within the ribosome. The so-called structural core proteins are apparently linked early to the nascent rRNA, and the other proteins are probably bound during a later phase.[41, 42]

The 45S RNA has already been found in association with protein, forming a particle of about 80S.[43] All the transitions between the 45S and the final 28S and 18S occur within ribonucleoprotein particles, and in this way both subunits are exported to the cytoplasm. Several observations indicate that during

maturation of the ribosomal subunits there is a progressive conformational change that starts from extended strands and terminates in relatively compact particles.[44] It has been postulated that, in addition to the progressive loss of RNA, maturation of the ribosome also involves a loss of protein.[45]

Another poorly understood aspect of ribosome assembly is how the extranucleolar synthesis of the 5S RNA is coordinated with that of the 45S, which at the end remains in the large subunit. Those genes coding for 5S ribosomal RNA are not present in the nucleolar organizer. When cultured cells are treated with small doses of actinomycin D, the production of the rRNAs is inhibited, but the synthesis of 5S RNA persists, and this RNA is retained in the nucleoplasm.

The final stages of ribosomal maturation apparently occur in rather rapid succession. Some large ribosomal subunits have been detected in the nucleoplasm before they pass into the cytoplasm to combine with messenger RNA and to become incorporated into poly-

SYNTHESIS AND MATURATION OF RIBOSOMES IN THE NUCLEOLUS

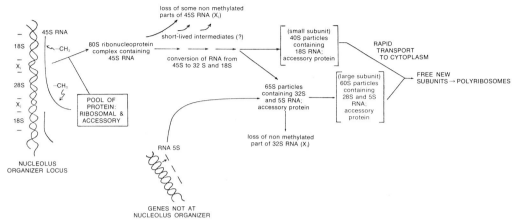

Figure 18-11. Stages in the biogenesis of ribosomes. (From Perry, R. P., *in* Lima-De-Faría, A. (Ed.), *Handbook of Molecular Cytology,* Amsterdam, North-Holland Publishing Co., 1969.)

ribosomes. As previously mentioned, the first to appear in the cytoplasm are the small 40S subunits; these form a larger cytoplasmic pool. The large 60S subunit in HeLa cells may be found within the nucleus, but no true ribosomes are observed.[46] The main features of the synthesis and maturation of ribosomal RNA and the ribosomal assembly of proteins are summarized in Figure 18-11.

Regulation of Ribosomal Assembly

There are several indications that there is a mechanism coordinating the synthesis of the ribosomal RNA and ribosomal proteins. If a cell is treated with *actinomycin D,* the formation of ribosomal RNA in the nucleolus is inhibited, but the synthesis of ribosomal proteins continues. If protein synthesis is inhibited by *puromycin,* the RNA synthesis in the nucleolus drops and fewer ribosomes are formed. In this case, however, the amount of 45S does not diminish because the degradation of this RNA is slowed down. By this treatment, therefore, both the synthesis and the processing of the ribosomal RNAs are reduced. The mitotic cycle is an excellent example of physiological regulation. In metaphase, RNA synthesis stops (see Chapter 17); however, at this stage in the cell there is as much 45S and 32S RNA as in the interphase. In other words, at metaphase both the synthe-

sis and processing of the ribosomal RNA have stopped simultaneously.[47]

Cytochemical and Ultrastructural Analysis of Ribosomal Biogenesis

The biogenesis and processing of ribosomal RNA just discussed can also be demonstrated at the cytological and ultrastructural levels by using cytochemical methods and the electron microscope. This is an excellent example of the convergence of the approaches of molecular biology and cell biology in the study of fundamental cell phenomena.

For an ultrastructural approach to ribosomal biogenesis it is necessary to return to the structure of the nucleolus described in Chapter 12. In the small nucleoli of amphibian oöcytes the *granular region,* containing 15 nm particles, and the *fibrillar region,* forming a central core, may be clearly observed[48] (Fig. 18-12).

Cytochemical studies have suggested that the following dynamic relationship exists between the different portions of the nucleolus:[49]

nucleolar DNA → fibrillar area → granular area.

When cells are labeled with ³H-uridine for five minutes and followed by a chase (in cold uridine), the fibrillar part of the nucleolus

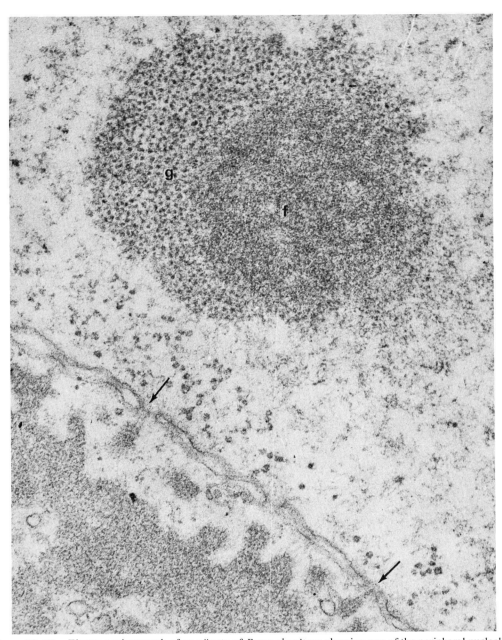

Figure 18–12. Electron micrograph of an oöcyte of *Rana clamitans* showing one of the peripheral nucleoli with a fibrillar (*f*) central portion and a granular (*g*) peripheral portion. Arrows indicate material entering the cytoplasm through the nuclear pores. ×70,000. (Courtesy of O. L. Miller.)

first incorporates the precursor which is later seen in the granular part.[50] These findings suggest that the fibrils contain the 45S ribosomal precursor of RNA.[51]

Cytochemical Demonstration of rDNA in Oöcytes

In *Xenopus* oöcytes at the pachynema stage, excess DNA begins to accumulate around the nucleolus in the form of Feulgen positive granules (Fig. 18–13). This phenomenon of DNA duplication is accompanied by intense ^{3}H-thymidine incorporation. Later on, the nuclear DNA content triples and the redundant DNA appears as a compact mass which is subsequently broken into hundreds of small granules that are then incorporated into the nucleoli.[52, 53]

By using a hybridization technique with ^{3}H-ribosomal RNA on oöcytes that are squashed and treated with alkali to denature the DNA, it is possible by radioautography (Chapter 6) to detect the regions containing the repeating ribosomal DNA.[53] The deposition of the silver grains was found precisely in the region where the DNA had accumulated during oögenesis (Fig. 18–13, *C*). It has been calculated that this DNA mass contains 25 to 30 pg rDNA, equivalent to about 3000 nucleolar organizers.[53]

Electron Microscopic Observations of rDNA–RNA Transcription

When amphibian nucleoli are isolated, under certain conditions the granular component becomes dispersed and the fibrillar portion unwinds and expands into circles resembling beaded necklaces.[48] These structures consist of a single DNA molecule irregularly coated with a ribonucleoprotein matrix. When this DNA axis is maximally stretched it is possible to observe under the electron microscope periodic regions, about 7 to 8 μm long, covered with matrix for 4.3 to 5 μm and having matrix-free spaces in between (Fig. 18–14). This finding suggests that each rDNA cistron coding for a 45S molecule is separated by segments of non-transcribed DNA and that the cistrons are probably read as single units.[54] The matrix covering the DNA molecule is composed of many tiny ribonucleoprotein molecules, probably nascent 45S RNA, that increase in size toward one of the ends of the transcribing region (Fig. 18–14). Based on this observation it is believed that on each rDNA cistron there are at least 100 RNA polymerases acting at the same time, each one transcribing a single 45S RNA (see Fig. 18–15).

Passage of rRNA into the Cytoplasm

With the introduction of thin sectioning in electron microscopy, observation of dense particles on both sides of the nuclear envelope of different cells suggested the passage of ribosomes from the nucleus to the cytoplasm.[55] The transport of ribonucleoprotein particles into the cytoplasm is easily observed in dipterans and in amphibian oöcytes, in which material of nucleolar origin may be seen passing through the pores of the nuclear en-

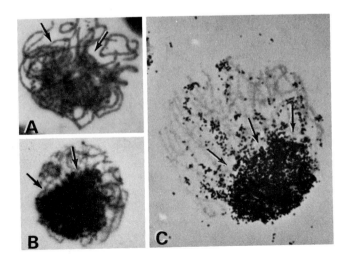

Figure 18–13. *Xenopus* oöcytes during the period of nucleolar DNA synthesis. **A,** at late pachynema, the excess DNA begins to accumulate as granules around the nucleolus (arrows). **B,** later on, the excess DNA appears as a dense mass (arrows). A and B, Feulgen reaction, ×1700. **C,** large pachynema oöcyte prepared with the ^{3}H-rRNA hybridization technique. Observe that the silver grains are mainly deposited on the mass of excess DNA. (See the description in the text.) ×1200. (Courtesy of J. G. Gall.)

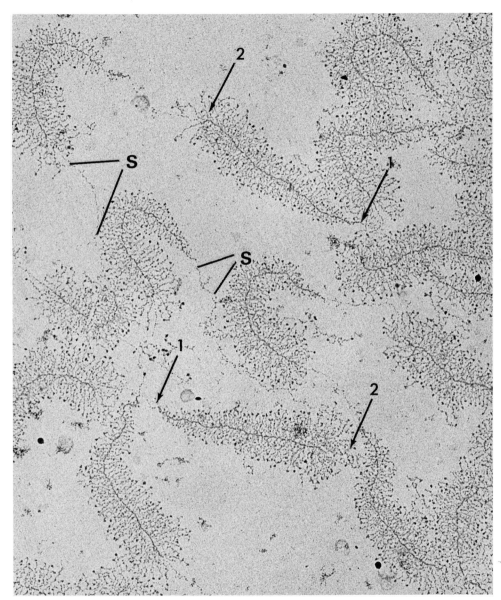

Figure 18–14. Electron micrograph showing nucleolar genes in the process of transcribing ribosomal RNA in oöcytes of *Xenopus laevis*. The fibrillar portion of nucleoli (see Figure 18–12) was isolated and dispersed. The DNA molecule shows free segments (*S*) separating two regions coated with a filamentous material. These are ribosomal RNA molecules which grow from sites 1 to 2. (See the description in the text.) ×25,000. (Courtesy of O. L. Miller and B. R. Beatty.)

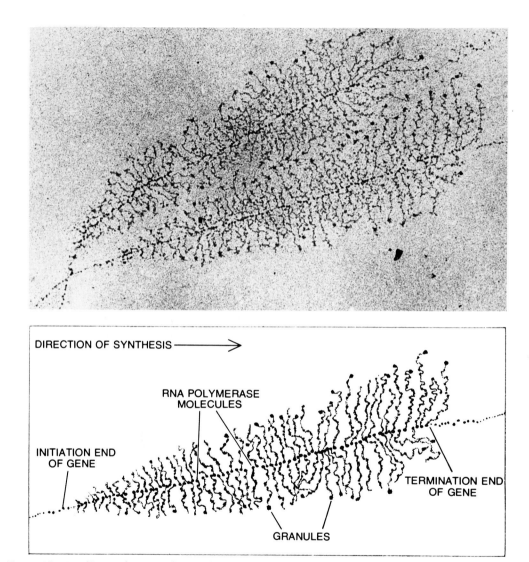

Figure 18-15. **Above,** electron micrograph showing two nucleolar genes in the process of transcription of the ribosomal RNA; **below,** labels are self-explanatory. ×35,000. (Courtesy of O. L. Miller and B. R. Beatty.)

velope into the cytoplasm (Fig. 18–12). These particles apparently go by way of the annuli in an elongate configuration in the salivary glands of the dipteran *Chironomus*,[56] similar observations have been made in lymphocytes. The RNP particles in *Chironomus* probably represent messenger RNA. In the nucleus of *Amoeba proteus* helical structures may be observed traversing the nuclear pores to reach the cytoplasm. These helixes contain RNA that is probably of nucleolar origin and that is believed to be nascent or incomplete ribosome[57] (see Chapter 17).

**SUMMARY:
Biogenesis of Ribosomes**

The biogenesis of ribosomes in bacteria differs from that in eukaryons. In bacteria the genes for the 5S, 23S, and 16S RNAs are clustered, forming an operon that is transcribed as a unit. There are only a few copies of this operon, and the RNAs are immediately released. In eukaryons there are multiple copies of the ribosomal genes (i.e., there is redundancy or amplification). The 18S and 28S RNAs are formed from the DNA of the nucleolar organizer. The 5S RNA, which may be even more redundant, is formed elsewhere in the genome.

The first indication that the nucleolus was involved in ribosomal biogenesis came from the discovery of an anucleolate mutant in *X. laevis*. By use of a ribosomal RNA-DNA hybridization technique, it was shown that the homozygous mutant contains about 1000 copies of ribosomal DNA cistrons; the heterozygous mutant contains about 500 copies; and the anucleolate mutant, none. Using ultracentrifugation techniques researchers have separated a small DNA fraction (0.15 to 0.20 per cent) having a higher buoyant density than the major DNA band. This minor fraction is called the *nucleolar satellite DNA,* since it hybridizes with ribosomal RNA (Fig. 18–7).

In nucleoli isolated from cultured cells it is possible to follow the synthesis and processing of the ribosomal RNAs. The first ribosomal RNA is a large molecule of 45S which is transcribed in about 2.2 minutes. During transcription, portions of this molecule, which will become the 28S and 18S RNA, undergo methylation. The 45S remains in the fibrillar region of the nucleolus for about 15 minutes; it is then cleaved by an endonuclease into a 41S and a 20S RNA. The 20S is degraded to the 18S, which is rapidly released into the cytoplasm. By a series of degradations by exonucleases (i.e., 36S, 32S) the 41S finally forms the 28S RNA. For about 30 minutes this rRNA remains in the granular region of the nucleolus before entering the cytoplasm. There is an early assembly of the 45S RNA accompanied by proteins coming from the cytoplasm. Ribonucleoprotein particles of decreasing size are formed during the degradation of the RNA until the mature ribosomal subunits are ready (Fig. 18–11). There is a regulation of the synthesis of ribosomal RNA and the assembly of the ribosomes. In metaphase, RNA synthesis stops, but the amount of 45S and

32S RNA remains constant, thereby indicating that RNA degradation has also stopped. The biogenesis and processing of the ribosomal RNA can be followed cytochemically and ultrastructurally. For example, the ribosomal DNA amplification occurring during oögenesis may be demonstrated by a hybridization technique at the cytological level. With the electron microscope the nucleolar genes transcribing the ribosomal RNA may be observed in isolated nucleoli of *X. laevis* (Fig. 18–15). The transport of ribonucleoproteins to the cytoplasm may also be detected in various cells. Biogenesis of the ribosomes is the principal function of the nucleolus.

REFERENCES

1. Palade, G. E. (1955) *J. Biophys. Biochem. Cytol., 1*:59.
2. Siekevitz, P., and Palade, G. E. (1960) *J. Biophys. Biochem. Cytol., 7*:619.
3. Bonnett, H. T., and Newcomb, E. H. (1965) *J. Cell Biol., 27*:423.
4. Rosbash, M., and Penman, S. (1971) *J. Molec. Biol., 59*:227.
5. Byers, B. (1967) *J. Molec. Biol., 26*:155.
6. Sabatini, D. D., Tashiro, Y., and Palade, G. E. (1966) *J. Molec. Biol., 22*:23.
7. Shelton, E., and Kuff, E. L. (1966) *J. Molec. Biol., 22*:23.
8. Florendo, N. T. (1969) *J. Cell Biol., 41*:335.
9. Hart, R. G. (1965) *Proc. Natl. Acad. Sci. USA, 53*:1415.
10. Cotter, R., McPhie, P., and Gratzer, W. B. (1967) *Nature* (London), *216*:864.
11. Fellner, P., and Sanger, F. (1968) *Nature* (London), *219*:236.
12. Knight, E., and Darnell, J. (1967) *J. Molec. Biol., 28*:491.
13. Comb, D. G., and Zehavi-Willner, T. (1967) *J. Molec. Biol., 23*:441.
14. Forget, B. G., and Weissman, S. M. (1968) *Science, 158*:1645.
15. Canter, C. R. (1968) *Proc. Natl. Acad. Sci. USA, 59*:478.
16. Staehelin, T., and Meselson, M. (1966) *J. Molec. Biol., 19*:207.
17. Hosokawa, K., Fujimura, R. K., and Nomura, M. (1966) *Proc. Natl. Acad. Sci. USA., 55*:190.
18. Nomura, M., and Traub, P. (1968) *J. Molec. Biol., 34*:609.
19. Nomura, M., Traub, P., and Bechman, H. (1968) *Nature, 219*:793.
20. Traub, P., and Nomura, M. (1968) *Proc. Natl. Acad. Sci. USA, 59*:777.
21. Staechlin, T. H., Raskas, H., and Meselson, M. (1968) In: *Organizational Biosynthesis*, p. 443 (Vogel, H. J., et al., eds.) Academic Press, Inc., New York.
22. Traub, P., and Nomura, M. (1969) *J. Molec. Biol., 40*:391.
23. Smith, I., Dubnau, D., Morell, P., and Marmur, J. (1968) *J. Molec. Biol., 33*:123.
24. Hamkalo, B. A., and Miller, O. L., Jr. (1973) In: *Gene Expression and Its Regulation*, p. 63 (Kenney et al., eds.) Plenum Publishing Corp., New York.
25. Brown, D. D., and Gurdon, J. B. (1964) *Proc. Natl. Acad. Sci. USA, 51*:139.
26. Hall, B. D., and Spiegelman, S. (1961) *Proc. Natl. Acad. Sci. USA, 47*:137.
27. Wallace, H., and Birnstiel, M. L. (1966) *Biochim. Biophys. Acta, 114*:296.
28. Ritossa, F. M., and Spiegelman, S. (1965) *Proc. Natl. Acad. Sci. USA, 53*:737.
29. Ritossa, F. M., (1968) *Excerpta Medica Int. Cong. Ser., 166*:21.
30. Brown, D. D. (1966) *Natl. Cancer Inst. Monogr., 23*:297.
31. Ritossa, R. M., Atwood, K. C., Linsley, L., and Spielgelman, S. (1966) *Natl. Cancer Inst. Monogr., 23*:449.
32. Quagliarotti, G., and Ritossa, F. M. (1968) *J. Molec. Biol., 36*:57.
33. Birnstiel, M. (1967) *Ann. Rev. Plant. Physiol., 18*:25.

34. Brown, D. D., and Weber, C. S. (1968) *J. Molec. Biol., 34*:661.
35. Perry, R. P., Chang, Ty., Freed, J. J., Greenberg, J. R., Kelley, D. E., and Tartoff, K. D. (1970) *Proc. Natl. Acad. Sci. USA., 65*:609.
36. Gall, J. G. (1968) *Proc. Natl. Acad. Sci. USA, 60*:553.
37. Ritossa, F. M., Atwood, K. C., and Spiegelman, S. (1966) *Genetics, 54*:819.
38. Lima-de-Faría, A., Birnstiel, M., and Jaworska, H. (1969) *Genetics, 61*:145.
39. Penman, S., Smith, I., Hottzman, E., and Greenberg, H. (1966) *Natl. Cancer Inst. Monogr., 23*:489.
40. Perry, R. P., and Kelley, D. E. (1972) *J. Molec. Biol., 70*:265.
41. Birnstiel, M. L., Wallace, H., Sirlin, J. L., and Fischberg, M. (1966) *Natl. Cancer Inst. Monogr., 23*:431.
42. Flamm, W. G., and Birnstiel, M. I. (1964) *Exp. Cell Res., 33*:616.
43. Warner, J. R., and Soeiro, R. (1967) *Proc. Natl. Acad. Sci. USA, 58*:1984.
44. Perry, R. P. (1969) In: *Handbook of Molecular Cytology* (Lima-de-Faría, A., ed.) North-Holland Publishing Co., Amsterdam.
45. Liau, M. C., and Perry, R. P. (1969) *J. Cell Biol., 42*:1969.
46. Vaughan, M. H., Warner, J. R., and Darnell, J. C. (1967) *J. Molec. Biol., 25*:285.
47. Penman, S. (1971) *Acta Cientifica Venezolana, 22*: suppl. 2, 57.
48. Miller, O. (1966) *Natl. Cancer Inst. Monogr., 23*:53.
49. Marinozzi, V. (1964) *J. Ultrastruct. Res., 10*:433.
50. Genskens, M., and Bernhard, W. (1966) *Exp. Cell Res., 44*:579.
51. Bernhard, W., and Granboulan, N. (1968) In: *The Nucleus*, p. 81. Academic Press, Inc., New York.
52. McGregor, H. C. (1968) *J. Cell Sci., 3*:437.
53. Gall, J. G. and Pardue, M. L. (1969) *Proc. Natl. Acad. Sci. USA, 63*:378.
54. Miller, O. L. (1969) In: *Handbook of Molecular Cytology* (Lima-de-Faría, A., ed.) North-Holland Publishing Co., Amsterdam.
55. De Robertis, E. (1954) *J. Histochem. Cytochem., 2*:341.
56. Stevens, B. J., and Swift, H. (1966) *J. Cell Biol., 31*:55.
57. Stevens, A. R., (1967) *Machinery for Exchange Across the Nuclear Membrane*, p. 189. Prentice-Hall, Inc., Englewood Cliffs, N.J.

ADDITIONAL READING

Adelman, M. R., Sabatini, D. D., and Blobel, G. (1973) Ribosome-membrane interaction. *J. Cell Biol., 56*:206.
Bonner, J. (1961) Structure and origin of the ribosomes. In: *Protein Biosynthesis.* (Harris, R. J. C., ed.) Academic Press, Inc., New York.
Borst, P., and Grivell, L. A. (1971) Mitochondrial ribosome. *Febs Letters. 13*:73.
Ingle, J., and Sinclair, J. (1972) Ribosomal RNA, genes, and plant development. *Nature* (London) *235*:30.
Küntzel, H., and Noll, H. (1967) Mitochondrial and cytoplasmic polysomes from *Neurospora crassa. Nature* (London), *215*:1340.
Kurland, C. G. (1972) Structure and function of the bacterial ribosomes. *Ann. Rev. Biochem., 41*:377.
Miller, O. L., and Beatty, B. R. (1969) Visualization of nucleolar genes. *Science, 164*:955.
Nanninga, N. (1973) Structural aspects of ribosomes. *Internat. Rev. Cytol., 35*:135.
Nomura, M. (1972) Assembly of bacterial ribosomes. *Fed. Proc., 31*:18.
Nomura, M. (1969) Ribosomes. *Sci. Am., 221* (No. 4):28.
Nononmura, Y., Blobel, G., and Sabatini, D. (1971) Structure of liver ribosomes studied by negative staining. *J. Molec. Biol., 60*:303.
Pardue, M. L., Brown, D. D., and Birnstiel, M. L. (1973) Location of the genes for 5S ribosomal RNA in *Xenopus laevis. Chromosoma, 42*:191.
Perry, R. P. (1972) Regulation of ribosome synthesis. *Biochem. J., 129*:35.
Ts'o, P. O. P. (1962) The ribosomes. *Ann. Rev. Plant Physiol., 13*:45.

nineteen

PROTEIN SYNTHESIS AND MOLECULAR GENETICS

The concept of *genotype* and *phenotype* was first introduced in the discussion of cytogenetics (Chapter 14); however, little has been said of the mechanisms by which the genotype is ultimately expressed in the phenotype. In this chapter an analysis is given of the molecular pathway by which a gene produces phenotypic changes, such as those observed by Mendel in his experiments with peas (Chapter 15).

In recent years the main emphasis in genetics has been on the mechanisms of heredity at the molecular level. Interest has shifted from the chromosome to deoxyribonucleic acid (DNA), ribonucleic acid (RNA), and proteins, all of which constitute the fundamental molecular "machinery" of the cell. This change in emphasis is, in part, the result of genetic experiments with microorganisms, which have revealed the mechanisms of genetic interchange. Large populations of microorganisms can be obtained by culture under precise and reproducible experimental conditions. Furthermore, changes in phenotype are easily related to changes in enzymes and metabolism. Bacteria and viruses are the organisms most often used in these studies.

Bacteria are usually haploid, multiply by simple division, and transmit the hereditary characteristics in this way. In some cases, however, recombination of genetic material may occur in these organisms. The diagram in Figure 19–1 indicates three ways by which DNA can be transferred from one bacterium to another. In *transformation,* the DNA extracted from a strain of bacteria penetrates another strain (Fig. 19–1, *A*). The donor DNA introduces new genetic information providing

for the possibility of new phenotypic characteristics. This DNA is incorporated into the original genome, which becomes diploid for the genes introduced. In this region of partial diploidy, genetic recombination may occur between the two DNA molecules. A transformation experiment of this type provided the first direct evidence that DNA was the mole-

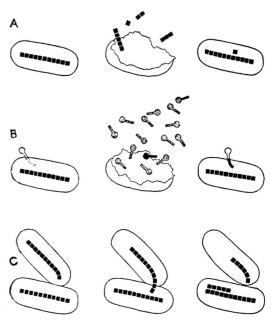

Figure 19–1. Diagram showing the three methods by which DNA can be transferred from one bacterium to another. **A,** in *transformation* the bacteria are destroyed and the DNA that is liberated penetrates another bacterium. **B,** in *transduction* the bacteriophage carries DNA from one bacterium to another. **C,** in *conjugation* the DNA is transferred directly by pairing and sexual recombination.

cule that contained the genetic information. In 1944 Avery and collaborators demonstrated that the DNA from a pathogenic strain of pneumococcus could produce the transformation of a non-pathogenic strain in the culture.

In transduction the DNA is carried from one bacterium to another by means of a bacteriophage (bacterial virus) (Fig. 19–1, *B*). In *conjugation* (Fig. 19–1, *C*) bacteria of different sexes pair and recombine sexually. These and other mechanisms of recombination analysis are widely used in molecular genetics, and have permitted the establishment of very detailed genetic maps of bacterial chromosomes.

Recombination in Bacteriophages

Bacteriophages (Gr., *phagein*, to eat) are the organisms best suited for analysis of the fine structure of genes. Figure 19–2 illustrates the complex molecular structure of a T₄ phage and indicates the probable function of its different parts. The bacteriophage injects its DNA molecule into a bacterium and multiplies, producing hundreds of copies in a few

minutes; this crowding finally causes lysis of the bacterium. The effect of phages on bacteria may easily be observed on agar cultures. Each phage particle produces a region of lysis in the bacterial *lawn*, called a *plaque*, which may show special characteristics for different mutants. Certain mutants can also be recognized by the kinds of bacteria they attack.

A phage is a haploid organism having a single chromosome (Fig. 5–9). The DNA of the T₄ phage has about 200,000 base pairs and contains enough information to code for a number of proteins. Since a bacterium may be simultaneously infected with two different mutants of a phage, recombination between both DNA molecules can take place. Such recombination can be recognized, even if it is present in very small proportions (i.e., one in 10^8 to 10^9 phages). With this type of fine analysis it has been possible to make genetic maps that approach the molecular level and to recognize several different genetic units.[1] For example, a *recon* is the unit of recombination which, minimally expressed, may correspond to the distance between two nucleotides; a *muton* is the unit of mutation

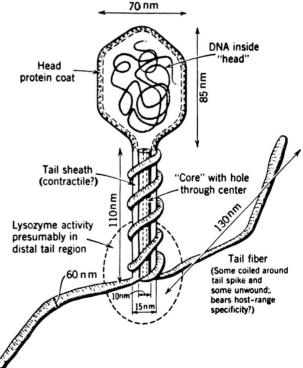

Figure 19–2. Diagram showing the macromolecular organization of a bacteriophage, indicating the probable function of each part.

which may be as small as a nucleotide pair. This is understandable since a change in a base of the *codon* may give rise to a different amino acid.

The knowledge gathered in the study of bacteria, bacteriophages, and other viruses, and in the molds *Neurospora* and *Aspergillus* can be applied to higher cells, since at the molecular level all living organisms are assumed to have similar genetic mechanisms. To illustrate this point, this chapter will begin with a discussion of some human diseases whose causes may be explained in terms of molecular changes in the genes and protein.

Phenylketonuria and Other Human Diseases

By 1930 it was discovered that certain patients who had a severe mental disorder excreted an abnormal compound in the urine (*phenylpyruvic acid*). The disease was called *phenylketonuria* and was found to be associated with a recessive gene. In fact, this disease manifests itself only in homozygous individuals; the probability of this condition appearing is increased by consanguineous unions (e.g., cousin-cousin or uncle-niece). In this disease, phenylalanine, a normal amino acid of the diet, cannot be oxidized to tyrosine and is transformed into phenylpyruvic acid. The primary action of the mutated gene has been to produce an *absence of the enzyme* needed for normal metabolism. The mental disorder, which can lead to idiocy or imbecility, is the result of the accumulation of phenylalanine and phenylpyruvic acid, both of which reach a toxic level in the central nervous system. If the disorder is discovered early enough, the mental disease can be prevented by a special diet that is low in phenylalanine. In humans there are several other "inborn errors of metabolism" due to genic action, such as *tyrosinosis, alkaptonuria, goitrous cretinism,* and *albinism,* in which blocks in the metabolism of phenylalanine and tyrosine are involved. In albinos no melanin is synthesized (see Fig. 19–3).

In *galactosemia,* another hereditary disease, the utilization of galactose is prevented by the lack of a special enzyme; the unmetabolized sugar-phosphates accumulate in the blood and are toxic to the brain. Elimination of galactose sources from the diet can improve the condition. Babies born with galactosemia cannot be fed milk, but are given milk substitutes with casein hydrolysates in its place.

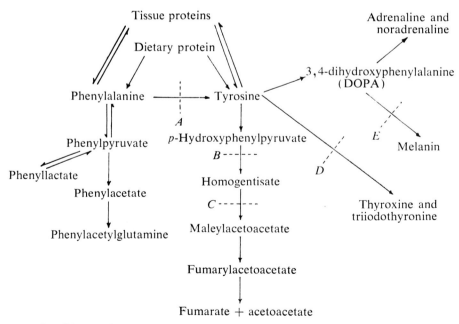

Figure 19–3. Possible genetic blocks in the normal metabolism of the amino acids phenylalanine and tyrosine in humans. These blocks lead to the production of the following genetic syndromes: *A*, phenylketonuria; *B*, tyrosinosis; *C*, alkaptonuria; *D*, goitrous cretinism; *E*, albinism. (From Harris.)

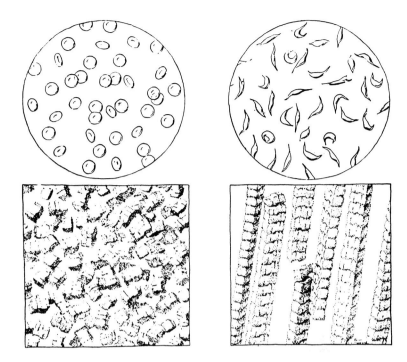

Figure 19-4. Above, left, normal erythrocytes; right, erythrocytes from venous blood of a patient with sickle cell anemia. Below, left, the hemoglobin molecules are randomly distributed in a normal individual; right, disposition of molecules of the venous blood in sickle cell anemia. In this case the hemoglobin molecules are in a crystalline array, which produces birefringence and deformation of the erythrocytes. (From L. Pauling.)

Sickle Cell Anemia

Another excellent example of an inherited disease in humans, in which the genes determine the structure of proteins, is *sickle cell anemia*. An inherited disorder found principally in Negroes, sickle cell anemia is characterized by a change in shape of the red blood cells in the venous blood. Owing to a decrease in oxygen tension, the erythrocytes become sickle-shaped (Fig. 19-4), a condition which may cause rupture of the cell, followed by severe hemolytic anemia. The study of the distribution of this disease within a family shows that it is caused by a recessive gene. A homozygous recessive individual has sickle-shaped cells and suffers from anemia, whereas a heterozygous individual has sickling, but no other symptoms of the disease.

The molecular bases of this genetic disease were discovered through studies of the hemoglobin (Hb) molecule.[2] Abnormal hemoglobin was found to have a different electrophoretic behavior. Normal hemoglobin (HbA) and sickle cell hemoglobin (HbS) differ in their net surface charge and thus move differently in an electric field. Note in Figure 19-5 that a heterozygous individual has both HbA and HbS in about equal quantity.

Hemoglobin is a tetrameric protein with a molecular weight of 64,500 and contains 600 amino acids arranged in four polypeptide chains: two identical α-chains and two identical β-chains. The amino acid sequence of the hemoglobin is controlled by two structural genes, α and β. It has been demonstrated that the chemical abnormality of HbS resides in a change of a single amino acid. As shown

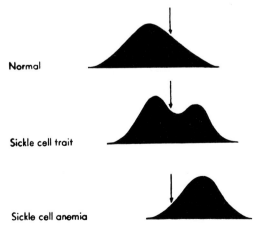

Figure 19-5. Electrophoretic behavior of various human hemoglobins. **Upper diagram,** normal homozygous dominant. **Center diagram,** sickle cell trait heterozygous. **Lower diagram,** sickle cell anemia. The arrows indicate the reference point of origin of the electrophoretic pattern. (After L. Pauling.)

TABLE 19–1. Chemical Differences in the Sequence of Amino Acids in Human Hemoglobins*

	Amino Acid Sequence†	Codon
HbA	$\overset{+}{\text{NH}_3}$-$\overset{+}{\text{Val}}$-His-Leu-Thr-Pro-$\overset{-}{Glu}$-$\overset{-}{\text{Glu}}$-$\overset{+}{\text{Lys}}$. . . .	GAA, GAG
HbS	$\overset{+}{\text{NH}_3}$-$\overset{+}{\text{Val}}$-His-Leu-Thr-Pro-$\overset{-}{Val}$-Glu-$\overset{+}{\text{Lys}}$. . . .	GUA, GUG
HbC	$\overset{+}{\text{NH}_3}$-$\overset{+}{\text{Val}}$-His-Leu-Thr-Pro-$\overset{+}{Lys}$-$\overset{-}{\text{Glu}}$-$\overset{+}{\text{Lys}}$. . . .	AAA, AAG

*From Ingram, V. M., *The Biosynthesis of Macromolecules.* Menlo Park, Calif., W. A. Benjamin, Inc., 1965, p. 160. Copyright © 1965, W. A. Benjamin, Inc., New York and Amsterdam.

†Glu, glutamic acid; His, histidine; Leu, leucine; Lys, lysine; Pro, proline; Thr, threonine; Val, valine.

in Table 19–1, in the sixth amino acid from the N-terminus of the β-peptide chain *glutamic acid* is replaced by *valine*.[3] In another abnormal hemoglobin (HbC) glutamic acid is replaced by *lysine* in the same position. In this case the abnormality leads only to mild anemia. From this analysis it is evident that, at the genetic level, the mutation has occurred only in the β structural gene for hemoglobin. As shown in the same table, the DNA codons

coding for the amino acid have changed only in a single letter (see below). In the heterozygous individual, one β structural gene is abnormal, and the other is normal. This type of mutation—and others that can be found in hemoglobin in which only one amino acid is replaced—correspond to the so-called *point mutations* of bacterial genetics. Table 19–2 shows a list of human hereditary diseases produced by a similar mechanism of mutation.

TABLE 19–2. Some Hereditary Disorders in Man in Which the Specific Lacking or Modified Enzyme or Protein Has Been Identified*

Disorder	Affected Enzyme or Protein
Acatalasemia	Catalase
Afibrinogenemia	Fibrinogen
Agammaglobulinemia	Y-Globulin
Albinism	Tyrosinase
Alkaptonuria	Homogentisic acid oxidase
Analbuminemia	Serum albumin
Galactosemia	Galactose-1-phosphate uridyl transferase
Glycogen storage diseases:	
Type I (von Gierke's)	Glucose-6-phosphatase
Type III	Amylo-1, 6-glucosidase
Type IV	Amylo-(1, 4 → 1, 6)-transglycosylase
Type V (McArdle's)	Muscle phosphorylase
Type VI (Hers')	Liver phosphorylase
Goiter (familial)	Iodotyrosine dehalogenase
Hemoglobinopathies	Hemoglobins
Hemophilia A	Antihemophilic factor A
Hemophilia B	Antihemophilic factor B
Histidinemia	Histidase
Hyperbilirubinemia (Gilbert's disease)	Uridine diphosphate glucuronate transferase
Hypophosphatasia	Alkaline phosphatase
Lesch-Nyhan Syndrome	Xypoxanthine-guanine phosphoribosyl transferase
Maple syrup urine disease	Amino acid decarboxylase
Methemoglobinemia	Methemoglobin reductase
Phenylketonuria	Phenylalanine hydroxylase
Wilson's disease	Ceruloplasmin
Xanthinuria	Xanthine oxidase
Xeroderma pigmentosum	DNA repair enzymes

*Modified from White, A., Handler, P., and Smith E., *Principles of Biochemistry,* New York, McGraw-Hill Book Co., 1964.

The Genetic Code

The two preceding chapters and the examples of human genetic diseases have provided the proper perspective for understanding how the genetic information contained in the DNA molecule can control the synthesis of specific proteins. At the molecular level it has been found that the *codons,* i.e., the hereditary units that contain the information to code for a single amino acid, are made of three nucleotides (a triplet). This information is first *transcribed* into the messenger RNA (mRNA), which has a sequence of bases complementary with DNA, from which it is copied. In fact, mRNA, like DNA has only four bases, whereas proteins may contain up to 20 amino acids. Permutation of the 4 bases yields 4^3 or 64 triplets—more than enough to code for 20 amino acids. If the genetic code consisted of doublets, the number of codons would be insufficient (i.e., $4^2 = 16$). The mRNA in turn serves as an intermediary that contains the same genetic information and *translates* this information into the amino acid sequence of the protein.

The length of a gene is determined by the length of the message to be translated, i.e., the number of amino acids in the protein. For example, 1500 nucleotide pairs contain 500 codons that may code for a protein having 500 amino acids.

In the process of expression of genetic information into proteins, in addition to the information coded in the DNA molecule and the messenger RNA, the following molecules and macromolecular components are involved: (1) *transfer RNAs* (tRNAs) which specifically combine with each amino acid (i.e., AA-tRNA) and recognize the corresponding codon on the messenger RNA; and (2) *the ribosomes,* which constitute the macromolecular machinery involved in many of the coordinated steps of protein synthesis (in Chapter 18 the structure and biogenesis of ribosomes were discussed). Now, the various elements involved in genetic information and expression, starting with the genetic code, will be considered.

It is important to remember some of the fundamental experiments that facilitated the discovery of the genetic code. In 1961 Nirenberg and Matthaei made the basic observation that synthetic polyribonucleotides could act as artificial mRNAs and could stimulate the incorporation of amino acids into polypeptides.[4] The first one used was *polyuridylic acid* (poly U) and the result was

TABLE 19–3. The Genetic Code

1st Base					2nd Base				3rd Base
		U		C		A		G	
U	UUU	Phe	UCU	Ser	UAU	Tyr	UGU	Cys	U
	UUC	Phe	UCC	Ser	UAC	Tyr	UGC	Cys	C
	UUA	Leu	UCA	Ser	UAA	Nonsense	UGA	Nonsense	A
	UUG	Leu	UCG	Ser	UAG	Nonsense	UGG	Trp	G
C	CUU	Leu	CCU	Pro	CAU	His	CGU	Arg	U
	CUC	Leu	CCC	Pro	CAC	His	CGC	Arg	C
	CUA	Leu	CCA	Pro	CAA	Gln	CGA	Arg	A
	CUG	Leu	CCG	Pro	CAG	Gln	CGG	Arg	G
A	AUU	Ile	ACU	Thr	AAU	Asn	AGU	Ser	U
	AUC	Ile	ACC	Thr	AAC	Asn	AGC	Ser	C
	AUA	Ile	ACA	Thr	AAA	Lys	AGA	Arg	A
	AUG	Met	ACG	Thr	AAG	Lys	AGG	Arg	G
	AUG	F-Met							
G	GUU	Val	GCU	Ala	GAU	Asp	GGU	Gly	U
	GUC	Val	GCC	Ala	GAC	Asp	GGC	Gly	C
	GUA	Val	GCA	Ala	GAA	Glu	GGA	Gly	A
	GUG	Val	GCG	Ala	GAG	Glu	GGG	Gly	G

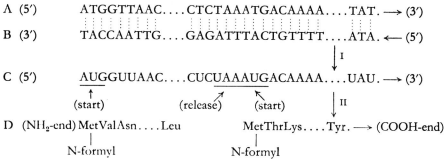

Figure 19–6. Diagram illustrating the transcription and translation steps in the expression of genetic information. **I,** transcription; **II,** translation. **A,** DNA strand 5'→3'; **B,** DNA strand 3'←5'; **C,** polycistronic messenger RNA 5'→3' copied from **B**; **D,** polypeptide chains. The starting and termination codons are underlined. (Courtesy of S. Ochoa.)

the coding of *polyphenylalanine* (a peptide chain made of phenylalanine). Thus, it was deduced that the codon for phenylalanine was UUU. Other homopolymers, such as *poly A,* stimulated the uptake of lysine and *poly C* of proline.

The use of synthetic RNAs of known composition was made possible by a previous discovery by Ochoa that the enzyme *poly-nucleotide phosphorylase* can link the specific nucleotides added to the medium. By 1963, the experiments with synthetic RNAs done in the laboratories of Nirenberg and Ochoa had established most of the codon sequences.[5] The recognition of codons was later made possible by the use of trinucleotide templates of known base composition.[6] When ribosomes are incubated with ^{14}C-AA-tRNA and such trinucleotides, complexes are formed that can easily be detected by filtration. In the laboratory of Khorana, polyribonucleotides with alternating doublets or triplets of known sequences were synthesized and used in cell-free systems.[7]

As shown in Table 19–3, several RNA codons may code for a single amino acid—a fact that is also called *degeneracy of the genetic code.* Leucine, for example, may be coded by CUU, CUC, and CUA. In most cases the synonymous codons differ only in the base occupying the third position of the triplet. The first two bases of the codon are apparently more important in coding. Since the same amino acid is coded by synonymous codons, it is logical to assume that mutations due to replacement of the third base may go unnoticed.

Figure 19–6, *C,* shows that the reading of the message is done in the 5'→3' direction of the mRNA. The polypeptide chain is always assembled sequentially from the end bearing a free —NH₂ group to the end bearing a —COOH group (Fig. 19–6, *D*). It is also known that there are special signals for starting and ending the reading of the individual cistrons.

The *initiation signal* for the synthesis of a protein is the AUG codon. When the AUG codon is at the beginning of the message *(starting codon),* in bacteria, it will code for N-formylmethionine. If the AUG codon is in another position, it will code for methionine. The *termination signal* is provided by the so-called *nonsense* codons UAG, UAA, and UGA (Table 19–3). In Figure 19–6, *D,* UAA is the release or *termination codon* that is situated immediately before the AUG, which begins the reading of the new cistron.

Although most of our knowledge about the genetic code comes from experiments with *E. coli,* essentially similar results have been obtained with other systems such as amphibian, mammalian liver, and plant tissue. It may be said that the *genetic code* is largely universal, i.e., there is a single code for all living organisms. As Nirenberg has pointed out, the genetic code may have developed at the same time as the first bacteria, some three billion years ago, and since then it has changed relatively little throughout evolution of living organisms.[6]

Mutagens and the Genetic Code

There is much genetic evidence that a deletion or insertion of a single base in the

DNA molecule may produce mutations such as those described for the human hemoglobins (Table 19–1). *Mutagens* are chemical substances that may alter the genetic code by a *point* or *single base mutation*. These mutations may be classified into four groups: (1) *transitions*, in which a pyrimidine base is replaced by another pyrimidine base (i.e., $C \rightleftarrows T$) or a purine is replaced by another purine (i.e., $A \rightleftarrows G$); (2) *transversions*, in which a purine is substituted by a pyrimidine or vice versa (i.e., $A \rightleftarrows C$ or $G \rightleftarrows T$); (3) *single base deletions;* and (4) *single base insertions*.

In the first two groups a protein with a change in a single amino acid will be formed (see Table 19–1). In the other two groups the message will be read "out of frame," and the amino acids linked from the point mutation onward will generally be abnormal.

Nitrous acid produces the oxidative deamination of bases in DNA. Thus, adenine is deaminated to hypoxanthine, which is then complementary to cytosine (instead of thymine). Cytosine is deaminated into uracyl, which then pairs with adenine, instead of guanine. Nitrous acid may act on free phage particles, viruses, and on isolated DNA directly. *Hydroxylamine* is another powerful mutagen which may act on isolated DNA. This substance reacts with cytosine, producing a change which causes it to pair with adenine, rather than guanine. In the hamster it has been observed that hydroxylamine acts in the region of the centromere, probably because this is rich in guanine-cytosine pairs.

Other mutagens are *base analogues* of normal nucleic acids that can be incorporated into the DNA without affecting replication. The DNA containing the base analogue has a greater tendency to make a wrong pairing (i.e., mispair) than the normal base. For example, *5-bromouracil* may pair with adenine, and *2-aminopurine* may pair with thymidine.

The thymidine analogue, *5-bromodeoxyuridine*, damages certain specific regions of the chromosomes which are assumed to be rich in adenine-thymine (A-T) pairs (see Chapter 17). This mutagen acts only during the period of DNA replication (i.e., S period). Similarly, *5-fluorodeoxyuridine*, an inhibitor of the enzyme thymidylate synthetase, acts only during the S period.

Alkylating agents, such as *ethyl-methane sulfonate* and *ethyl-ethane sulfonate*, react with the purine bases guanine and adenine, producing ethylation in the purine ring. These bases are then lost by hydrolysis, and gaps are produced on the DNA strands. *Nitrogen mustards* and ethylene oxides also remove bases from DNA.

Some fluorescent dyes such as acridine orange and *proflavin* may cause *intercalation* or *deletion* of a base in the nucleotide sequence and in this way may disrupt the reading of the genetic message. Some of these fluorescent dyes may sensitize DNA to the action of light. Some of the mutagens experimentally used to change the genetic code may be employed in cancer chemotherapy (see Auerbach, 1967).

SUMMARY:
Molecular Genetics

This chapter is dedicated to the study of gene expression at the molecular level—the mechanisms by which the *genotype* is expressed in terms of the *phenotype*. The use of microorganisms such as bacteria, viruses, and certain molds has permitted genetic experiments in large populations. Although bacteria are haploid, recombination may be produced by transformation, transduction, or conjugation. The study of bacteriophages (bacterial viruses) has permitted a finer analysis of gene structure. Bacteriophages inject their DNA molecule into the bacterial host. This molecule multiplies, forming hundreds of copies that code for their own proteins and cause bacterial lysis. Since a bacterium can simultaneously be injected with two different mutants, recombinations are produced which may be followed by plaque analysis. In this way, genetic maps that approach the molecular level can be made. The genetic mechanisms demonstrated in bacteria and viruses are found throughout the plant and animal kingdom. Excellent examples are phenylketonuria, cretinism, albinism, galactosemia, and many other human hereditary diseases in which there are "inborn errors of metabolism" caused by the absence of a single enzyme. *Sickle cell anemia* is another example in which the mutation produces a change of a single amino acid in the β chain of the hemoglobin molecule.

These genetic diseases are produced by a *point mutation,* i.e., a change in a single unit of genetic information that codes for one amino acid *(codon)*. The genetic code is made of triplets of bases in the DNA and messenger RNA, each triplet corresponding to a single codon. For example, the information to code for a protein of 500 amino acids is contained in 500 codons, corresponding to 1500 nucleotides. The genetic code was discovered by the use of cell-free systems in which the synthesis of proteins was directed by an artificial mRNA. For example, with poly U, polyphenylalanine was obtained (i.e., the codon is UUU). Several RNA codons (Table 19–3) may code for a single amino acid. This is known as the degeneracy of the code (leucine is coded by CUU, CUC, and CUA). The difference is only in the third letter of the triplet. AUG is the *starting codon* and UAA, UGA, and UAG are nonsense, or *termination codons.* The genetic code is largely universal; there is a single code for all living organisms.

The genetic code may be changed by the action of chemicals called *mutagens.* Some mutagens change one base into another (e.g., nitrous acid, hydroxylamine). Others are base analogues (e.g., 5-bromouracil, 2-aminopurine, 5-bromodeoxyuridine, and 5-fluorodeoxyuridine) that act during the replication period of DNA.

Alkylating agents, such as ethylene oxides and nitrogen mustards, cause removal of bases from DNA. Some fluorescent dyes (acridine, proflavine) cause intercalation or deletion of a base. Some mutagens are used in cancer chemotherapy.

TRANSCRIPTION OF GENETIC INFORMATION

Messenger RNA

The demonstration of a template RNA that carries genetic information from DNA emerged from work with bacteria; now this research is being carried on with higher organisms. The term "messenger" RNA (mRNA), proposed by Jacob and Monod in 1961, refers to the fact that this is a template molecule copied from DNA and has a rapid turnover. As shown in the experiment of Figure 19–7, *I*, the rapidly synthesized mRNA could be detected after five seconds of incubation with ^{14}C-uridine.[8] After longer time periods (15 minutes), the mRNA disappeared and the ribosomal RNAs became labeled (Fig. 19–7, *II*).

The average life span of some mRNAs in *E. coli* is about two minutes, after which the molecules are broken down by ribonucleases. In fact, in bacteria mRNA may be read on one end while the other end is still being transcribed. It may also disintegrate at the starting

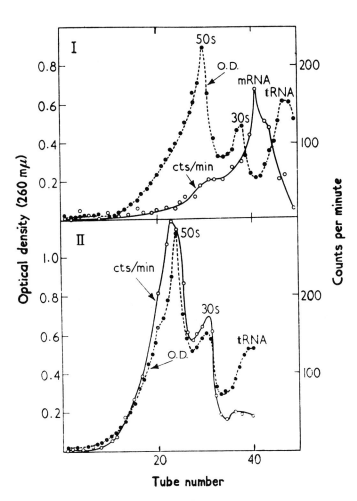

Figure 19–7. Diagram showing the rapid turnover of messenger RNA (*mRNA*). *E. coli* were incubated for 5 seconds with ¹⁴C-uridine and then washed in nonradioactive uracil. In **I** the cells were rapidly frozen and in **II** they were incubated for 15 minutes at 37° C. prior to freezing. Ribonucleic acids were extracted in both experiments and ultracentrifuged. The optical density at 260 nm indicated the concentration of different RNA molecules, and the counts per minute indicated the incorporation of ¹⁴C-uridine. Note in **I** that the only labeled RNA is mRNA, whereas in **II** radioactivity is found in two peaks of ribosomal RNA (*rRNA*) (50S and 30S subunits) and in transfer RNA (*tRNA*). (From Gros., F., Hiat, H., Gilbert, W., Kurland, C. G., Risebrough, R. W., and Watson, J. D., *Nature* (London), *190*:581, 1961.)

end, while the reading is terminating in the other. Some electron microscopic observations confirm this finding (Fig. 19–16).

Another characteristic of mRNA is that it is *heterogeneous*. The size of the molecule varies considerably, since it is adapted to the dimension of the polypeptide chain for which it will code. In *E. coli* the average size of a mRNA cistron (i.e., the length of DNA coding for one polypeptide chain) is 900 to 1500 nucleotides, corresponding to peptide chains containing 300 to 500 amino acids. However, when several adjacent cistrons are copied at the same time, the mRNAs are much longer. These are the so-called *polygenic* or *polycistronic* messengers. For example, the ten specific enzymes involved in the metabolism of histidine are coded in a single mRNA molecule. In higher cells, although mRNAs of very high molecular weight are encountered, some mRNAs may be monocistronic.

It was mentioned before that hemoglobin is coded by two (α and β) cistrons.

Messenger RNA is *complementary* to chromosomal DNA; it forms RNA-DNA hybrids after separation of the two DNA strands (see Chapter 4). Synthesis of mRNA is accomplished *with only one of the two strands* of DNA, which is used as a template. The enzyme RNA *polymerase* joins the ribonucleotides, thus catalyzing the formation of the 3′-5′-phosphodiester bonds that form the RNA backbone. In this synthesis the AU/GC ratio of RNA is similar to the AT/GC ratio of DNA. It is now well established that the mRNA synthesis is initiated at the 5′ end, and that the direction of growth is from the 5′ end to the 3′ end. The RNA polymerase attaches to an initiator site of the structural gene, in the "promoter," and it catalyzes mRNA synthesis until a *termination site* is reached. Figure 19–6, *A* and *B* shows the two strands of DNA that are in an antiparallel, or polarized, disposition and the mRNA (Fig. 19–6, *C*), which is copied in the $5' \rightarrow 3'$ direction. Another criterion by which mRNA can be recognized is that it becomes rapidly attached to ribosomes and forms part of the polyribosomes.

The following properties may, therefore, be used to identify mRNA in prokaryons: rapid turnover, heterogeneity, complementary base composition with DNA, and attachment to ribosomes. The most important of these properties is that mRNA functions as a template in protein synthesis.

RNA Polymerase and Factor Rho

The enzyme that synthesizes mRNA in prokaryons, i.e., RNA polymerase, has a large molecular weight and is composed of several subunits. A complete enzyme comprises the following polypeptides: 2α of 40,000 daltons, 1β of about 155,000, $1\beta'$ of about 165,000, and 1σ of 95,000, thereby having a total molecular weight of 495,000.[9] The sigma factor may possibly determine the specificity of the transcription by attaching to the *promoter*. This factor directs the rest of the enzyme (i.e., core enzyme) that is not specific to transcribe a certain gene.[10] As soon as the mRNA chain is started, the sigma factor is released from the core enzyme and can be used in the transcription of other specific mRNA molecules (see Fig. 19–24).

A termination factor, Rho, found in *E. coli,* causes the release of the transcribed RNA molecules. This protein, which has a molecular weight of 200,000 daltons and four polypeptide chains of 50,000, is capable of recognizing some DNA sequences that function as termination signals.[11]

Messenger RNA in Eukaryons

The diagram of Figure 19–8 is a simplified view of the transcription in the nucleus and transport into the cytoplasm of the differ-

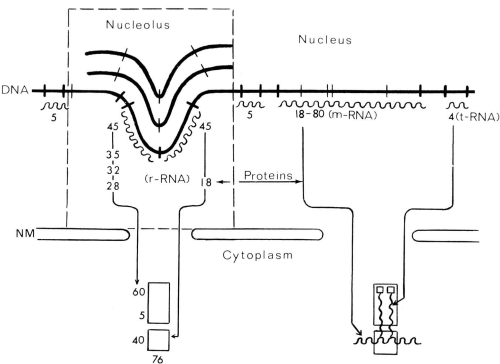

Figure 19–8. Diagram of the transcription and transport of nuclear RNAs in a eukaryotic cell and of mRNA participation in protein synthesis. As indicated in Chapter 18, the 45S nucleolar RNA produces, in a series of steps, the 28S and 18S, which enter the 60S and 40S ribosomal subunits, respectively. Each 45S RNA yields one 28S and one 18S RNA. The 5S RNA is transcribed outside the nucleolus and finally enters the 60S ribosomal subunit. Sites on the genome for the transcription of 4S tRNA and 18–80S mRNA are also indicated. *NM,* nuclear membrane. (Courtesy of W. Bernhard.)

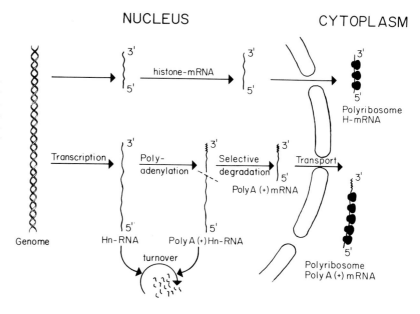

NUCLEUS   CYTOPLASM

Figure 19-9. Diagram showing the synthesis and processing of heterogeneous nuclear-RNA (*Hn-RNA*), Poly A (+) messenger RNA, (*Poly A (+) mRNA*) and histone-messenger RNA (*histone-mRNA*). The steps of transcription, polyadenylation, selective degradation and transport of mRNA into the cytoplasm are indicated. Note the differences between histone-mRNA and the Poly A (+) mRNA. (See the description in the text.)

ent RNA molecules involved in protein synthesis in an eukaryotic cell. All the RNA species are transcribed from the corresponding locus in the DNA molecule. Those regions of the genome transcribing the messenger RNAs are called the *structural genes.* The other regions, coding for the different ribosomal RNAs (see Chapter 18) and transfer RNAs, are generally called the *determinants for RNA.*

The origin and fate of messenger RNA in eukaryotic cells is much more complex than in bacteria. In recent years it has been found that the formation of a functionally active mRNA is the consequence of a complex series of steps that comprise: (1) the actual transcription of DNA into mRNA precursors, (2) the intranuclear processing of these precursors, and (3) the transport of the mRNAs into the cytoplasm and their association with ribosomes to initiate the process of translation or protein synthesis (Fig. 19-9).

Heterogeneous Nuclear RNA

It has been suggested that mRNA is synthesized in the nucleus as part of a heterogeneous population of large RNA molecules which constitute the so-called *heterogeneous nuclear RNAs* (Hn-RNAs). If cultured cells are treated with radioactive RNA precursors (e.g., uridine), rapidly labeled RNAs appear in the nucleus in large amounts. These Hn-

RNAs range in size from 5×10^5 to 10^7 daltons and are degraded, for the most part, within the nucleus at a relatively rapid rate.[12] Only about 20 per cent of the Hn-RNA, in terms of total nucleotide, is not degraded and has been postulated to be converted into mRNAs.

The heterogeneous nuclear RNA is found in isolated nuclei, largely in the nucleoplasm outside the nucleolus. It has a DNA-like base composition and readily hybridizes with DNA. Morphologically, it is possible that the Hn-RNAs correspond to the fibrillar and granular ribonucleoproteins that are found in the nucleoplasm of cells with the electron microscope (see Chapter 12).[13] By using small doses of actinomycin D researchers have found it is possible to suppress selectively the synthesis of ribosomal RNA and to recognize the formation of Hn-RNA and mRNA in eukaryons.[14]

Polyadenylation and Transport of mRNA

A new aspect in the problem of conversion of Hn-RNA into mRNA has been the discovery that both types of RNA contain a sequence of polyadenylic acid (poly A), approximately 200 nucleotides long.[15-18] This poly A sequence is added to the Hn-RNA after the transcription is completed[19] and is attached to the 3′ end of the RNA molecule (see Perry et al., 1973). It is assumed that once the Hn-RNA molecule is transcribed from DNA, the poly A sequence is added, in a stepwise fashion, by the action of a poly A synthetase (Fig. 19–9). Simultaneous with, or after polyadenylation, and starting from the 5′ end, a selective degradation of the Hn-RNA molecules takes place by the action of nucleases. The final product is poly A (+) mRNA, which will finally reach the cytoplasm (Fig. 19–9). The discovery that most messenger RNAs of eukaryons contain a 3′ terminal poly A chain has been a recent major advance in molecular biology. Although the real significance of poly A is still unknown, several possible functions have been hypothesized. The post-transcriptional attachment of poly A may be a special step in the processing of mRNA and may be related to its transport to the cytoplasm. One of the methodological advances is that the presence of poly A confers to the mRNA special physicochemical properties by which it can be readily retained by nitrocellulose filters or by columns having polyuridine or polydeoxythymidine sequences. This has made possible the rapid separation of mRNAs from the nucleus or the cytoplasm, free of other contaminant RNAs and without submitting the cells to the action of actinomycin D. Metabolic studies on the heterogeneous nuclear and messenger RNAs have been made using this technique.

Another interesting finding has been that the *mRNAs for the histones* are directly transcribed from repetitive sequences in the DNA without the need for a giant precursor molecule (Hn-RNA) or polyadenylation (Fig. 19–9). Having no poly A, the histone-mRNA passes through the nitrocellulose filter along with the other RNA species (i.e., 28S and 18S ribosomal RNAs, 4S and 5S) present after phenol extraction of the ribonucleic acids. It has been demonstrated that the histone-mRNAs enter the cytoplasm without delay, whereas the poly A (+) mRNAs have a 15 minute delay before appearing attached to the polyribosomes. This lag period may be attributed to the just mentioned processing steps that these mRNAs probably undergo before passing into the cytoplasm.

Turnover of Eukaryotic mRNAs

The life span of mRNAs in bacteria may be extremely short, but it is not necessarily short in eukaryotic cells. In fact, *metabolically stable* mRNAs are found in eukaryotes. A typical example is that of the mammalian reticulocyte, an immature red cell that has lost the nucleus, but still retains the ribosomes and other components of the biosynthetic machinery that produce hemoglobin. In this cell there is mRNA which continues to code for this protein for a number of hours, and even for days. Also, the mRNAs coding for plasma proteins in liver cells are very stable.

To study the turnover of mRNAs in cultured cells the RNA is labeled with specific precursors for different periods of time. Then

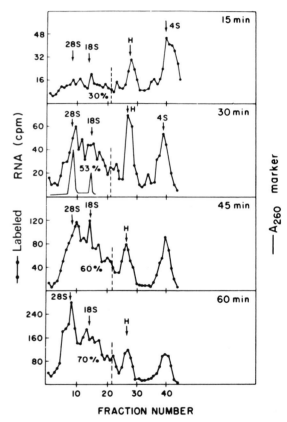

Figure 19–10. Turnover of messenger RNAs in cultured L cells. The culture was treated with a small dose of actinomycin D and was labeled with ³H-uridine. At various intervals RNA was extracted from the polyribosomes and analyzed by polyacrylamide gel electrophoresis. The position of the *28S* and *18S* ribosomal RNA and histone-mRNA (*H*) is indicated. This kinetic study demonstrates that the histone-mRNA is most rapidly metabolized. (See the description in the text.) (From Perry, R. P., Greenberg, J. R., Kelley, D. E., La Torre, J., and Schochetman, G., "Messenger RNA: its origin and fate in mammalian cells," *in* Kenney *et al.* (Eds.), *Gene Expression and Its Regulation*, Vol. I, New York. Plenum Publishing Corporation, 1973, p. 149.)

the polyribosomes are separated by gradient centrifugation, and the RNAs are separated on polyacrylamide gels (Fig. 19–10). The radioactive profile shows that there are three distinct zones containing: poly A (+) mRNAs, histone mRNAs, and low molecular weight RNAs. At 60 minutes the zone of high molecular weight represents about 70 per cent of the radioactivity and contains several RNA components that are retained by nitrocellulose filters (i.e., they are poly A (+) mRNAs). A second zone that contains 10 per cent of the radioactivity represents principally histone mRNA (Fig. 19–10, *H*) and traverses the filter, together with the low molecular weight RNA corresponding to 4S and 5S.

Kinetic studies (Fig. 19–10) demonstrate that the histone mRNA is more rapidly labeled and metabolized, whereas the higher molecular poly A (+) mRNAs increase in their specific radioactivity for periods as long as 60 minutes. In long-term kinetic studies, carried out in exponentially growing cultured cells, it has been found that the half-life of the poly A (+) mRNA is about one generation time (i.e., 10 hours for L cells that are transformed fibroblasts). On the contrary, the histone-mRNAs have a shorter half-life. Under the same conditions the ribosomal RNAs have an infinite half-life (see Perry et al., 1973).

Free Ribonucleoproteins in the Cytoplasm

There is evidence that in eukaryotic cells the mRNAs do not enter the cytoplasm as naked strands of RNA, but in association with proteins, thereby forming ribonucleoprotein complexes.[20–22] Some of these complexes may remain free in the cytoplasm without being associated with ribosomes—a characteristic specifically associated with mRNA. Spirin has coined the term *informosomes* to refer to these free nucleoprotein complexes to convey the idea that they may represent mRNAs kept in the cytoplasm for long periods of time before being genetically expressed. Some of the studies were carried out on eggs of fish or sea urchins at a stage of rapid cleavage when no ribosomal RNA was being synthesized. This RNA is uridine-labeled, and the complex is preserved by stabilization with formaldehyde. In reticulocytes a portion of the messenger RNA that codes for the α-chain of globin is

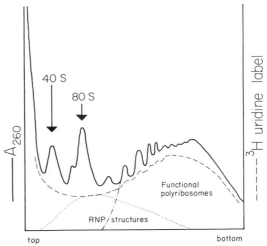

Figure 19–11. Distribution of ribosomes, functional ribosomes, and ribonucleoprotein (RNP) structures in a sucrose gradient. The distribution is demonstrated by absorption at 260 nm (A_{260}) and by radioactivity with ^{3}H-uridine. (See the description in the text.) (From Perry, R. P., Greenberg, J. R., Kelley, D. E., La Torre, J., and Schochetman, G.: "Messenger RNA: its origin and fate in mammalian cells," in Kenney, *et al.* (Eds.), *Gene Expression and Its Regulation*, Vol. I, New York, Plenum Publishing Corporation, 1973, p. 149.)

present as free RNP structures in the post-ribosomal fraction. In a sucrose gradient these ribonucleoprotein structures settle in the region of free ribosomes, whereas some of them settle in that of polyribosomes (Fig. 19–11). The difference between ribosomes and the RNP structures is in the higher protein:RNA ratio (i.e., 4:1) and the buoyant density (1.55 for ribosomes, and about 1.40 to 1.47 for the RNP structures). In the RNP structures the content of poly A is considerably lower than in the functional polyribosomes. Studies suggest that some of the free RNP structures normally found in the cell cytoplasm probably do not constitute true mRNA[23] (see Perry et al., 1973).

Transfer RNA

Transfer or soluble RNAs are a group of small ribonucleic acids that in some ways act as interpreters of the genetic code. In fact, tRNAs are able to read the message expressed as codons in mRNA, and at the same time are able to recognize the amino acids that the codons specify. The need for this transfer RNA molecule (or adaptor) is easily understood, since there is no specific affinity between the side groups of many amino acids and the bases in the mRNA. This fundamental step in translation is performed by specific aminoacyl-activating enzymes that attach the specific amino acids to one end of the respective tRNA molecule, and by specific triplets of bases *(anticodons)* that are able to bind to the corresponding codons of mRNA with hydrogen bonds. For each of the 20 naturally occurring amino acids there must be at least one or more specific tRNAs. Transfer RNA accounts for 10 to 15 per cent of the total RNA in *E. coli*. Each tRNA has a sedimentation constant of 4S and contains 75 to 85 nucleotides. The sequence of alanine tRNA was elucidated by Holley and collaborators in 1965.[23]

The primary structure of some 30 tRNAs, from *E. coli* to mammalian liver cells

to higher plants cells, is already known (see Smith, 1973). For example, the tyrosine tRNA has 85 nucleotides, but it is formed from a larger precursor having 126 nucleotides, which is cleaved and methylated by specific enzymes. The structure is very similar in all the species studied. One characteristic in common is the presence of several unusual bases, most of which are derived from methylation: pseudouridine (abbreviated ψ), inosinic acid, methylguanine, methyl aminopurine, methylcytosine, ribothymine and others (inosinic acid has the same base-paring properties as guanine). Some of the common features of tRNAs are the 3' end of the chain which always terminates in CCA and attaches to the specific amino acid; and the 5' end which terminates with guanine (Fig. 19–12). In most tRNAs there is a sequence G-T-ψ-C-G shown in a special region, i.e., the *ribosomal recognition site* (Fig. 19–12, *4*). Other residues—the dehydrouridines—are in an-

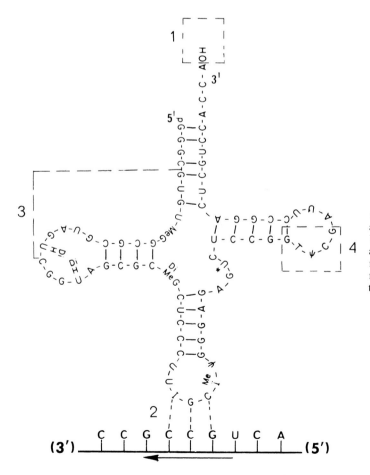

Figure 19–12. Diagram of the alanine transfer RNA from *E. coli* showing: **1,** the aminoacyl binding site; **2,** the anticodon site bound by hydrogen bonds to the codon for alanine; **3,** the region of the tRNA recognized by the alanine activating enzyme; **4,** the ribosomal recognition site common to all tRNAs.

other region of the molecule (Fig. 19–12, *3*).

The secondary structure of tRNA is determined by the fact that the complementary bases are in groups of four to seven, thus causing the double-stranded regions to be short and interspersed with loops in which the RNA is single-stranded. Several possible models have been proposed; the so-called *clover leaf* (Holley) is most widely accepted (Fig. 19–12).[23] In this model the *anticodon* contains three unpaired bases (a triplet) capable of recognizing and forming hydrogen bonds with the codon in the mRNA. The anticodon is the most specific end of the molecule, that is, the one that reads the message. Its position is on the arm directly opposite the amino acid acceptor end. Other functionally important portions of the molecule are the ribosome recognition site, which is common to all tRNA molecules, and the site related to the recognition of the specific amino acid-activating enzyme (Fig. 19–12).

The clover leaf secondary structure raises the possibility of a tertiary structure in tRNA. X-ray diffraction data are consistent with a structure in which the tRNA has a rather compact configuration, such as one in which some arms are folded together, rather than one with an open configuration such as the one shown in Figure 19–12. In a recent model coming from the x-ray diffraction analysis the tRNA had the shape of a letter L. The two major portions of the L consist of base-paired double helixes that are joined by the unpaired segments of the molecule, whereas the anticodon is in one extremity of the L and the amino acid acceptor is in the other. This acceptor end sticks out from the rest of the molecule and may be capable of some degree of movement.[24]

In a given tRNA molecule, therefore, four special sites may be recognized: (1) the amino acid attachment site, (2) the codon recognition site (anticodon), (3) the specific site for recognizing the amino acid-activating enzyme, and (4) the ribosome recognition site.

Formation of the Aminoacyl-tRNA Complex. For each amino acid there is a specific activating enzyme (i.e., tRNA-aminoacyl synthetase) which activates the carboxyl group of that amino acid for covalent bonding with the adenylic acid residue of the 3′ end of tRNA. The first step of the reaction involves the use of ATP:

$$AA + ATP \xrightarrow[\text{synthetase}]{\text{aminoacyl}} \tag{1}$$

$$AA \sim AMP + P \sim P$$

The $AA \sim AMP$ intermediate formed remains bound to the enzyme until the proper tRNA arrives, and at that moment the formation of the $AA \sim tRNA$ complex occurs:

$$AA \sim AMP + tRNA \xrightarrow[\text{synthetase}]{\text{aminoacyl}} \tag{2}$$

$$AA \sim tRNA + AMP$$

In order to perform these two functions, the activating enzyme must have two sites: one to recognize the amino acid and another to recognize the specific tRNA.

Codon-anticodon recognition. One of the features of the genetic code is that the third base of the codon is less important in the coding, thereby giving rise to the degeneracy of the genetic code. Since there are several codons for each amino acid (Table 19–3) the question arises as to whether each of them has a different tRNA or whether a single tRNA can recognize several codons. Crick[25] proposed that the third base of the anticodon (Fig. 19–12) might have a certain amount of "wobble" by which it could pair to more than one base (for example, I in the anticodon could pair with U, C, or A in the codon). The presence of this wobble would permit the use of a lesser number of tRNAs than the 61 code words shown in Table 19–3. However, both from yeast and *E. coli*, a large number of tRNAs have been isolated (66, in the case of *E. coli*.) In fact, there is *redundancy* in the sense that several tRNAs may respond to the same codon. The number of tRNAs for each amino acid varies from two to five, and each has more than one tRNA. This arrangement may have evolutionary advantages in that it may cause resistance of damage by mutation (see Lewin, 1971).

The Synthesis of Transfer RNA Genes

One of the most important recent achievements of molecular biology has been the complete synthesis of the genes coding for yeast *alanine transfer RNA* and bacterial *tyrosine transfer RNA* (tyr-tRNA) by Khorana and associates. This feat has been possible because of the previous knowledge of the complete nucleotide sequence of these transfer RNAs.

It was mentioned earlier that bacterial tRNAs are formed from larger precursors, which in the case of the tyr-tRNA, has 126 nucleotides. This chain is then cleaved by a nuclease, and 41 nucleotides are removed from the initiation end of this tRNA, which then contains 85 nucleotides.

As shown in Figure 19–13, not only has the complete sequence of the DNA duplex coding for the precursor of the tyr-tRNA of *E. coli* been synthesized, but also adjacent parts that correspond to the *promoter* and *terminator*. These parts do not belong to the structural gene proper, but to regions involved in its regulation, i.e., in the turning on and off of this particular gene.

The painstaking work of assembling the 126 nucleotides in exact order involved first the *chemical synthesis* of short deoxynucleotide sequences, which then became joined by hydrogen bonds to form complementary strands. Second, *enzymatic synthesis* was used to join the double-stranded pieces. For this purpose the DNA repair enzyme *(polynucleotide ligase)* mentioned in Chapter 17 was employed.

Figure 19–13 shows that the structural gene was made by the joining of 26 pieces of single-stranded oligonucleotides which first formed four double-stranded segments. These segments were finally sealed to form the entire gene with the 126 pairs of complementary nucleotides. The figure also shows a sequence of 29 nucleotides of the promoter region, as well as a sequence of 23 nucleotides of the terminator region that have been synthesized and joined to the structural gene. So far it is still not known if these sequences represent part or all of the segments involved in the regulation of this particular gene (Khorana, 1975).

Synthetic Tyrosine tRNA Gene

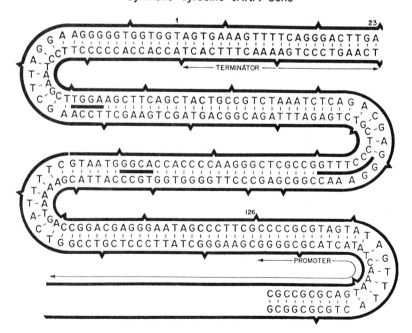

Figure 19–13. Diagram showing the entire sequence (1 to 126) of nucleotides corresponding to the synthesized structural gene for tyrosine transfer RNA precursor. A sequence of 29 nucleotides of the *promoter* and 23 nucleotides of the *terminator* have also been synthesized. Along the two chains of the duplex the short oligonucleotides that were joined by the ligase are indicated. The three lines within the duplex mark the four segments of DNA which were finally joined to form the entire tyrosine transfer RNA gene. (Courtesy of H. G. Khorana.)

The transcription of the genetic information contained in DNA is carried out by messenger RNA (mRNA). In bacteria mRNA is a template molecule copied from a gene (a cistron) or a group of related genes (polygenic or polycistronic messengers) that has a rapid turnover; the average life span of the template is two minutes. The synthesis of mRNA is accomplished by using only one of the DNA strands as template and is carried out from the 5' end toward the 3' end. At the DNA cistron RNA polymerase attaches to the initiator site or *promoter* and catalyzes the synthesis until the termination site is reached. Messenger RNAs are heterogeneous because they carry messages of different length according to the protein or proteins they code. Bacterial RNA polymerase is a complex enzyme containing five subunits and has a total molecular weight of 495,000. One of the subunits, the *sigma factor*, determines the specificity of the core enzyme by attaching to the promoter region of the structural gene. In *E. coli* a termination protein factor, Rho, has been demonstrated.

In *eukaryons* the transcription of genetic information is much more complex; several of the concepts derived from studies with bacteria do not apply. Messenger RNA is synthesized in the nucleus as part of *heterogeneous nuclear RNA* (Hn-RNA). About 80 per cent of Hn-RNA is degraded in the nucleus; the rest is left as mRNAs. The Hn-RNA is found largely in the nucleoplasm outside the nucleolus, as ribonucleoprotein particles (see Chapter 12). After the transcription of the Hn-RNA a sequence of about 200 nucleotides of polyadenylic acid (poly A) is added to the 3' end. Simultaneous with this *polyadenylation*, the *degradation* of the Hn-RNA starts from the 5' end. The final product is the poly A (+) mRNA, which penetrates the cytoplasm (Fig. 19–9). The presence of poly A in most mRNAs of eukaryons has facilitated the separation of mRNAs. The mRNA coding for histones lacks poly A and enters the cytoplasm without delay. An important difference between the mRNA of prokaryons and that of eukaryons is the long life of most of the eukaryotic mRNAs. Here the mRNA is metabolically stable and can be used for a number of hours and even for days. In eukaryons mRNA is associated with proteins forming ribonucleoprotein complexes. Some of these complexes may remain free in the cytoplasm without being attached to the polyribosomes. Of these free particles, some may correspond to the so-called *informosomes*.

Transfer RNAs (tRNAs) are a group of small RNAs (4S) that recognize the codons on the mRNA and carry the specific amino acid for the formation of the polypeptide at the acceptor end. From *E. coli* more than 60 different tRNAs have been isolated that are able to transfer the 20 amino acids. Transfer RNAs contain many methylated bases: the 3' end terminates in CCA and is attached to the specific amino acid; the 5' end terminates in guanine. The *anticodon* has three unpaired bases (triplet) that recognizes the various codons. Its position is opposite to the *acceptor* end. In tRNA there is also a specific *recognition site* for the *amino acid–activating* enzyme and a *ribosome recognition site* that has a sequence of bases similar in all tRNAs.

A tertiary structure for tRNA has been postulated. The formation of the aminoacyl-tRNA is by specific enzymes that activate the carboxyl group of the amino acid, binding it to the adenylic acid residue in the 3′ end of tRNA. The activating enzyme has one site to recognize the amino acid, and another to recognize the specific tRNA.

RIBOSOMES AND PROTEIN SYNTHESIS

The ribosome can be visualized as a macromolecular machine, with many precisely matched parts, that selects and presides over all the components involved in the synthesis of protein. Ribosomes must be able to recognize and guide the multiple interactions of a large array of molecules (including aminoacyl-tRNA, and peptidyl-tRNA), the various other factors involved in the initiation, elongation, and termination of the polypeptide chain, and to hold and "move" the mRNA containing the genetic message. The concepts dealing with the structure and biogenesis of ribosomes, discussed in Chapter 18, are indispensable to understanding the role of ribosomes in protein synthesis.

Polyribosomes and Protein Synthesis

The early electron microscopic observations of cell sections revealed that ribosomes were frequently associated in groups, occasionally forming recurrent patterns (Fig. 18–1). It was not until 1962 that the function of these *polyribosomes,* or *polysomes,* in protein synthesis was discovered.[26-27]

After treating reticulocytes with ^{14}C-labeled amino acids and using gentle methods for disruption (e.g., lysis in hypotonic solutions) researchers found that in addition to the typical sedimentation band of single ribosomes (see Figure 19–14) larger units were present. These particles ranged from 108S to 170S, or more. At the same time, the maximum radioactivity, indicating the synthesis

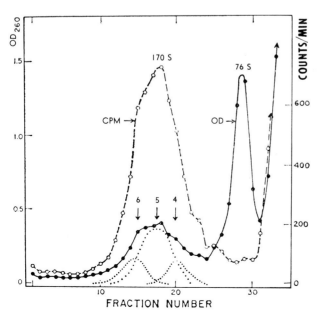

Figure 19–14. Experiment done to demonstrate that the synthesis of hemoglobin occurs in polyribosomes. Reticulocytes were incubated for 45 seconds in the presence of a pool of amino acids labeled with [14]C. The cells were lysed osmotically and the soluble part (hemoglobin) was centrifuged for two hours on a continuous density gradient of sucrose. After this the tube was punctured and thirty fractions were collected and observed under the electron microscope. Similar fractions were analyzed for optical density at 260 nm for RNA and for the number of counts. While the optical density (OD) shows a peak at 76S (sedimentation constant corresponding to single ribosomes), the peak of radioactivity corresponds to the ribosomal tetramers, pentamers, and hexamers indicated with arrows. (From Warner, J. R., Rich, A., and Hall, C. E., *Science, 138*:1399–1403, 1962.)

of hemoglobin, was detected in the 170S fraction; this corresponded to a polyribosome of five units (a pentamer) (Fig. 19–14). It was confirmed by electron microscopy that about 75 per cent of the ribosomes in the 170S peak were present as pentamers.

Figure 19–11 shows the complex pattern of ribosomes obtained from cultured eukaryotic cells (i.e., L fibroblasts) upon ultracentrifugation on a continuous gradient of sucrose. The absorbance at 260 nm shows the high peak of the single ribosomes (i.e., 80S) preceded by the 60S and 40S subunits, and followed by a series of peaks of larger size, corresponding to the functional polyribosomes. In the diagram the region corresponding to free ribonucleoprotein structures is indicated as well as the pattern of labeling with a short pulse of ^{3}H-uridine, which corresponds to the messenger RNAs that are associated with the polyribosomes.

Ultrastructure of Polyribosomes and the Transcription Process

By observing isolated polyribosomes under the electron microscope one may detect a thin filament, corresponding to the mRNA, between the ribosomes. In polyribosomes of hemoglobin the distance between the centers of the individual ribosomes is about 34 nm and the total length of the mRNA is about 150 nm. The number of ribosomes in a polyribosome may vary considerably and seems to be related to the length of the mRNA that should be "read" in the translation process. The longer the message to be read the greater is the number of ribosomes used simultaneously in its translation. Note that each ribosome of the polyribosome carries a nascent polypeptide chain. In the case of hemoglobin, therefore, five chains are formed at the same time along the mRNA. (A detailed electron microscopic study of the ribosomes and polyribosomes was mentioned in the previous chapter.) As shown in the diagram of Figure 19–15, it is assumed that the mRNA is held between the two subunits of each ribosome. This may explain why segments of 25 nucleotides of the mRNA are protected from exogenous ribonucleases by the ribosome.

Experiments performed on bacteria treated with lysozyme to render them osmotically sensitive have permitted the study of the bacterial chromosome and the identification of the ribosomal genes, as well as structural genes, during the process of transcription.[28–30] In the DNA fibers there are stretches in which polyribosomes appear attached to nascent mRNA chains positioned at right angles. As shown in Figure 19–16, in such regions there is a gradient of short to long polyribosomes, suggesting that the transcription of the mRNA is coupled with translation. In other words, as soon as the mRNA is being transcribed by RNA polymerase, the ribosomes become attached to the mRNA to initiate protein synthesis. Some electron micrographs suggest the presence of the RNA polymerase (Fig. 19–17) molecules as

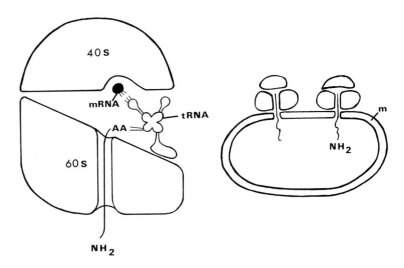

Figure 19–15. **Left,** diagram of a ribosome showing the two subunits and the probable position of the messenger RNA and the transfer RNA. The nascent polypeptide chain passes through a kind of tunnel within the large subunit. **Right,** diagram of the relationship between the ribosomes and the membrane of the endoplasmic reticulum and the entrance of the polypeptide chain into the cavity. *m,* Membrane of endoplasmic reticulum. (Courtesy of D. D. Sabatini and G. Blobel.)

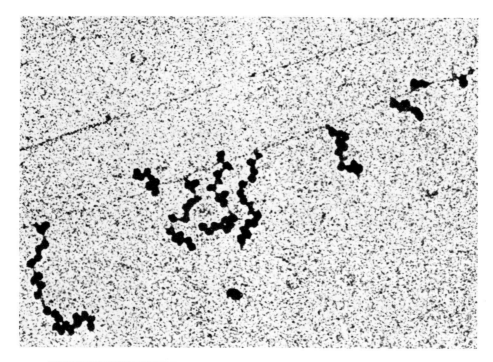

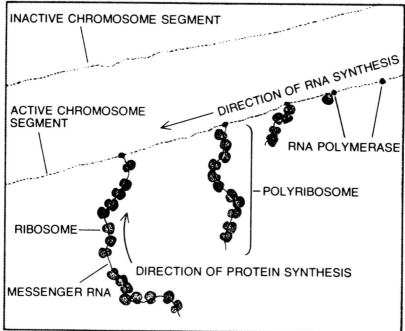

Figure 19–16. Electron micrograph of the transcription and translation processes in bacteria. Two stretches of DNA, one naked, the other with nascent messenger RNA arranged at right angles, are observed. The bottom diagram facilitates the interpretation of the electron micrograph. (See the description in the text.) (Courtesy of O. L. Miller, Jr.)

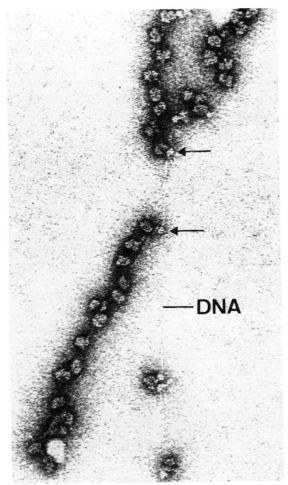

Figure 19–17. Electron micrograph of bacterial DNA with messenger RNA and polyribosomes. Arrows indicate the RNA polymerases that synthesize the messenger RNA. In both Figures 19–15 and 19–16 the macromolecules are observed with negative staining. (Courtesy of O. L. Miller, Jr.)

Ribosome-Polysome Cycle. Dissociation Factor

There is evidence that the two subunits of the ribosome may exist freely and that they may associate during the process of protein synthesis and dissociate at the end of the process.[31] The essential features of this ribosomal cycle are indicated in Figure 19–18, in which a pool of various ribosomal subunits and a polyribosome with initiator and terminator ends are shown.[32] Two different subunits are seen entering the ribosome at *i*. (The black subunits are from heavy ribosomes tagged with ^{13}C, ^{15}N, and ^{2}H, whereas the white subunits contain the normal ^{12}C, ^{14}N, and H atoms). At the end of the cycle two subunits separate and become incorporated into the pool of subunits.[33]

Exchange of ribosomal units also occurs in eukaryotic cells. In reticulocytes, for example, the combination of the 40S and 60S subunits takes place at the beginning of the synthesis and dissociation at its completion.[34] The same exchange of subunits has been observed in yeast cells.[35]

In *E. coli* it has been found that a dissociation factor (protein in nature) is required for the separation of the subunits at the end of the cycle.[36, 37] This dissociation factor can be extracted from the 30S subunit and apparently is identical to the initiation factor IF$_3$, which is involved in the binding of mRNA to the 30S subunit (see below and Table 19–4). A model of the ribosome-polysome cycle in which the dissociation factor splits the 70S by forming a complex with the 30S subunit has been postulated. This factor should be released at a later stage at which

a granule smaller than the ribosome at the DNA template. Possible degradation of the mRNA is suggested by the finding of shorter polyribosomes distal to the longest ones in the gradient. There are long stretches of DNA that have no polyribosomes but show the granules corresponding to RNA polymerase; therefore, it has been suggested that these may be associated with inactive parts of the genome awaiting a proper initiation signal. Furthermore, these findings suggest that only a small portion of the bacterial genome is active in transcription at any one moment.[30]

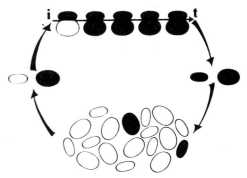

Figure 19–18. Ribosomal cycle in protein synthesis. (See the description in the text.) (From Kaempfer, R., *Proc. Natl. Acad. Sci. USA, 61*:106, 1968.)

TABLE 19–4. Protein Factors in Protein Synthesis

Phase	Factor	Source	Function
1° Initiation	F_3	High salt 30S	Dissociation of ribosomal subunits.
	IF_1 IF_2 IF_3 } GTP	ribosomal wash	Binding of mRNA and initiator tRNA to 30S subunit.
2° Elongation	Tu Ts } T + GTP	Supernatant fraction	Binding of aminoacyl-tRNA Tu-GTP complex to ribosome.
	Peptidyl transferase	50S ribosomal subunit	Peptidyl transfer from peptidyl-tRNA to aminoacyl-tRNA.
	G + GTP	Supernatant fraction	Translocation of peptidyl-tRNA; release of free tRNA.
3° Termination	R_1 R_2	Supernatant fraction	Release of protein at UAA, UAG, or UGA codons.

*From Lipmann, F., "What Do We Know About Protein Synthesis?" *in* Kenney, F. T., Hamkalo, B., Favelukes, G., and August, J. T. (eds.), *Gene Expression and Its Regulation,* New York, Plenum Publishing Corp., 1973. (See Fig. 19–19.)

time the 30S and 50S subunits reassociate to begin a new ribosome-polysome cycle (Fig. 19–19).

Role of the Small Ribosomal Subunit

The findings just described suggest that the two subunits of the ribosome may have different functions. For the reading of the genetic message, the first step appears to be the binding of the small 30S subunit to the first codon of the mRNA to form the *initiation complex.*[38] In this model the successive steps would be the binding of the first aminoacyl tRNA, followed by the coupling of the 50S subunits (Fig. 19–19). It has been demonstrated in bacteria that the first aminoacyl tRNA to initiate the synthesis is a formyl (−CHO) derivative of methionine, coded by the AUG codon (see Table 19–3). This codon

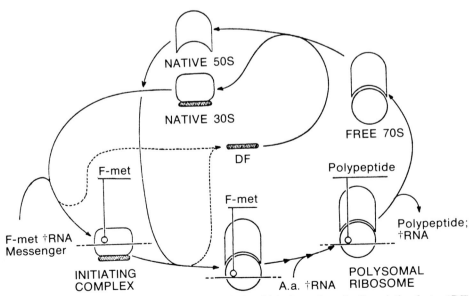

Figure 19–19. A model of the ribosome-polysome cycle in which the action of a dissociation factor (*DF*) is postulated. (See the description in the text.) (From Subramanian, A. R., Ron, E. Z., and Davis, B. D., *Natl. Acad. Sci. USA, 61*:761, 1968.)

should be present at the beginning of every mRNA molecule.[39-42]

Since in protein synthesis the peptide chain always grows sequentially from the free terminal —NH$_2$ group toward the —COOH end, the function of formylmethionine tRNA (F-met-tRNA) is to ensure that proteins are synthesized in this direction. In F-met-tRNA, the —NH$_2$ group is blocked by the formyl group, leaving only the —COOH available to react with the —NH$_2$ of the second amino acid (Fig. 19–6); in this way, the synthesis follows in the correct sequence. Later, the first amino acid is separated from the protein by a hydrolytic enzyme.[40] F-met-tRNA has been found in mitochondria, but it is not involved in the general cytoplasmic protein synthesis of eukaryons.[43]

In eukaryotic cells the initiation of the synthesis is by a special met-tRNA which is not formylated. There is another met-tRNA which is specific for the placement of methionine inside the polypeptide. In both cases,

however, the coding triplet is AUG. In the case of eukaryons the initial methionine is generally split off from the finished polypeptide (see Lucas-Lenard and Lipmann, 1971).

Initiation Factors

In Ochoa's laboratory three initiation factors (IF$_1$, IF$_2$, and IF$_3$) have been separated by washing ribosomes with NH$_4$Cl and submitting them to column chromatography.[41, 44, 45] These initiation factors are proteins loosely associated with the 30S subunit (Table 19–4). The IF$_2$ factor has a molecular weight of 80,000 daltons and contains essential —SH groups upon which its binding with GTP depends (Fig. 19–20). In eukaryotic cells a factor similar to IF$_2$ but which does not require GTP has been found.[46] The IF$_1$ factor is a basic protein with a molecular weight of 9200 daltons that is involved in the binding of F-met-tRNA. The IF$_3$ factor has a molecular weight of 30,000 daltons and is involved

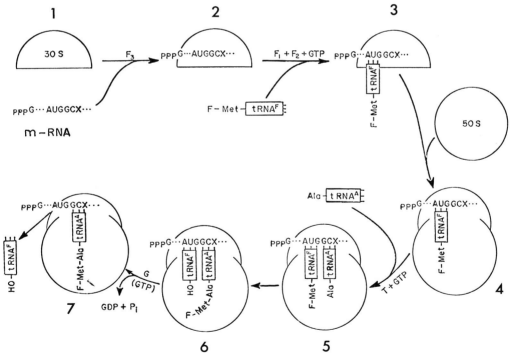

Figure 19–20. General diagram of the initiation steps in protein synthesis involving 30S and 50S ribosomal subunits, messenger RNA, the initiation factors F$_3$ and F$_1$ + F$_2$, GTP; the elongation factors T and G, formylmethionine-tRNA (F-met-tRNA) and alanine-tRNA. **1,** isolated 30S; **2,** binding of m-RNA and F$_3$ to the 30S; **3,** binding of the F$_1$ + F$_2$ + GTP and F-met-tRNA to make the initiation complex; **4,** binding of the 50S subunit to make the complete 70S ribosome; **5,** binding of the second aminoacyl-tRNA; **6,** synthesis of the first peptide bond; **7,** liberation of the free tRNA after translocation.

in the binding of the mRNA to the 30S subunit (Fig. 19–20). It is a basic protein rich in arginine and lysine. All three factors are essential for initiation of protein synthesis when natural mRNAs are used, but they are not required for artificial mRNAs, such as poly U.[38] IF_1 + IF_2 are required to bind F-met-tRNA to the 30S subunit in response to codon AUG.[47] It has been noted that IF_3 also may function as a dissociation factor for the 70S ribosome; this factor stimulates the binding of the 30S subunit to the specific initiation sequence of mRNA.[48]

The upper part of Figure 19–20 shows the steps in the formation of the *initiation complex*. The specificity of the 30S subunit suggests that this particle has an *initiation site* which recognizes the unique structural feature of F-met-tRNA.[45]

Recent studies suggest that the translation of the genetic message may be regulated at its initiation by a change in specificity of factor IF_3. In *E. coli* so-called *interference factors (i factors)* have been isolated which bind to factor IF_3 and change its specificity. In this way the translation of certain mRNAs may be enhanced or inhibited.[49, 50] Since there are several different *i factors*, their relative amounts may regulate the selective affinity of the ribosome for certain initiation signals.

Besides the 30S subunit and the IF factors, the secondary structure of the mRNA molecule may also be important for the correct initiation of protein synthesis. It has been shown that at certain regions the single-stranded mRNA molecule has *loops* or *hair pins* of double-stranded RNA in which both chains are held together by hydrogen bonds. If this secondary structure is destroyed (e.g., by the action of formaldehyde) the specificity of the initiation is lost and abnormal proteins are synthesized in vitro.

Elongation of the Polypeptide Chain

In Lipmann's laboratory soluble protein fractions present in the post-ribosomal supernatant were found to be active in the elongation of the polypeptide chain[51–54] (Table 19–4). At first, only the EFT and EFG factors were isolated, but later the *EFT factor* was found to comprise two proteins: Tu, which is temperature-unstable, and Ts, which is temperature-stable. The function of the EFTu + EFTs

factors is in the binding of the aminoacyl-tRNA to the ribosome (Fig. 19–20, 5). Initially, EFTu + GTP form a complex with the aminoacyl-tRNA prior to its binding to the aminoacyl or acceptor site; the formation of this complex is catalyzed by factor Ts.

The *EFG factor*, also called *translocase*, is involved in the translocation of the mRNA (Fig. 19–20, 6–7). The isolated translocase from *E. coli* has a molecular weight of 72,000 and appears to consist of a single polypeptide chain.[55] This factor is required to split the GTP in contact with the ribosome, thereby yielding GDP and inorganic phosphate. The energy released is needed for the removal of the deacylated tRNA from the ribosome and for the translocation process (Fig. 19–20, 7). Chain elongation can, therefore, be considered as a kind of a cyclic reaction in which T + GTP stimulates the binding of aminoacyl-tRNA and G + GTP promotes the translocation of the newly elongated peptidyl-tRNA (see Laszlo and Lucas-Lenard, 1973).

Role of the Large Ribosomal Subunit

The large subunit contains the enzyme *peptidyltransferase*, or peptide synthetase, which is involved in the formation of the peptide bond. Another function of the 50S subunit is to provide two binding sites for the two tRNA molecules, i.e., the *aminoacyl* or *acceptor site* and the *peptidyl* or *donor site* (Fig. 19–21). These two sites should be next to each other to permit the formation of the peptide bond.[56]

The stepwise growth of the polypeptide chain involves: (1) the entrance of an aminoacyl-tRNA into the aminoacyl site; (2) the formation of a peptide bond and consequent ejection of the tRNA that was in the peptidyl site; and (3) the movement of the tRNA (now carrying the peptide chain) from the aminoacyl to the peptidyl site. This process should be coupled with the simultaneous movement of the mRNA to place the following codon in position (Fig. 19–21).[57] This translocation, in which the ribosome moves along the mRNA in the 5′–3′ direction, requires the G factor and GTP.

It has been found that an acidic protein present in the 50S subunit of bacteria is essential to bring about the effect of the G factor; such a protein is considered to be contrac-

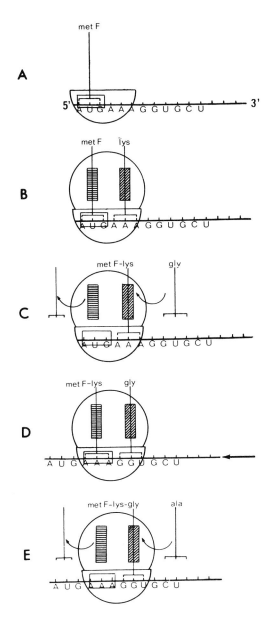

Figure 19–21. Diagram representing the early stages of translation of messenger RNA (5'—3'). The *initiation site* in the 30S subunit is indicated by a white rectangle. The *aminoacyl*, or acceptor, and the *peptidyl*, or donor sites on the 50S subunit, are indicated by horizontal and oblique stripes, respectively. **A,** initiation complex in which formylmethionine-tRNA (F-met-tRNA) binds to the first codon in mRNA (AUG). **B,** the 70S ribosome has been formed, and the second aminoacyl-tRNA (lys-tRNA) binds the second codon (AAA). **C,** the tRNA is eliminated from the peptidyl site, and the first peptide bond is formed. **D,** translocation of the mRNA and of the peptidyl-tRNA has occurred, and a new aminoacyl-tRNA (gly-tRNA) binds to the third codon (GGU). **E,** the molecular events of **C** are now repeated. (Adapted from Ochoa, S., *Naturwissenschaften, 11*:505, 1968.)

tile.[58] Furthermore, there are physicochemical indications that there is a conformational change of the ribosome in parallel with the translocation.[59, 60]

The velocity with which these coordinated processes are produced may be illustrated by the fact that it takes only about one minute to construct a single hemoglobin chain carrying 150 amino acids. In *E. coli* only 10 to 20 seconds are required to form a 300- to 500-amino acid protein. After the polypeptide has grown completely, the terminal tRNA must be split off, thereby creating a free terminal —COOH group. Until this happens, the chain remains attached to the 50S subunit.

Chain Termination

The termination of the polypeptide chain occurs when the 70S ribosome carrying the peptidyl-tRNA reaches a termination codon, presumably located at the end of each cistron. Chain termination leads to the release of the free polypeptide and tRNA, as well as to the dissociation of the 70S ribosome into 30S and 50S subunits (Fig. 19–19).

The UAA, UGA, and UAG codons are involved in chain termination (Fig. 19–6). Two releasing factors, R_1 and R_2, of protein nature are needed.[61, 62] R_1 is specific for UAG and UAA and R_2 for codons UAA and UGA (Table 19–4). The complex formed by the termination factor with the codon in some manner induces the peptidyl transferase to catalyze the transfer to water, rather than to an amino group, thereby releasing the polypeptide chain from the terminally added tRNA (see Lipmann, 1973).

Role of the 60S Subunit in Membrane-Bound Ribosomes

The ribosomes bound to the endoplasmic reticulum were discussed in Chapter 18. It was mentioned that these ribosomes are related to the synthesis of protein for export, whereas free ribosomes form the proteins remaining in the cytoplasmic matrix.[63] Other points mentioned were that the binding to the membrane is by way of the large subunit and that a central channel has been proposed to account for the vectorial flow of the newly formed polypeptide into the cisternae of the endoplasmic reticulum[64] (Fig. 19–15).

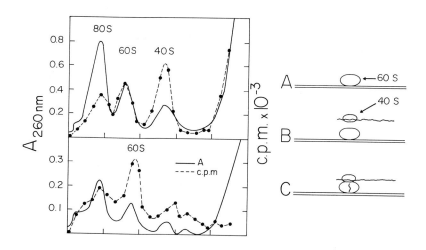

Figure 19-22. Experiments on the assembly of membrane-bound ribosomes. **Upper left,** free ribosomes show the 80S monomers and the 60S and 40S subunits by absorption at 260 nm (A_{260}). In the cells labeled with ³H-uridine there are three peaks of radioactivity corresponding to the 80S, 60S, and 40S. **Lower left,** membrane-bound ribosomes show that only the 60S subunit is labeled with high specific activity.

To the right, **A,** the 60S binds directly to the membrane; **B,** the 40S is complexed with the messenger RNA; **C,** the 40S joins the 60S subunit. (See the description in the text.) (From Baglioni, C., Bleiberg, I., and Zauderer, M., *Nature* [*New Biol.*], *232*:8–12, 1971.)

This model could also be related to the fact that there is a portion of the nascent polypeptide chain that is not digested when polysomes are treated with proteolytic enzymes. This proteolysis-resistant polypeptide fragment consists of about 30 to 35 amino acids and has a total length of 14 nm. Studies have been carried out on the dynamics of the assembly of membrane-bound ribosomes. Free and membrane-bound ribosomes were placed on sucrose gradients and then compared. The pattern of polyribosomes was similar in both cases. In the membrane-bound polyribosomes, however, only 80S ribosomes and few 60S subunits were present, whereas in the free polyribosomes 60S and 40S subunits were evident. In cells labeled with ³H-uridine for 90 minutes it was found that the 60S subunit is the one that labels with higher specific activity in the membrane-bound polyribosomes.[65]

These results favor the model of assembly indicated in Figure 19–22 in which the 60S subunit binds directly to the membrane (*A*); then the 40S subunit, forming a possible complex with the mRNA (*B*), joins the 60S (*C*) to start the synthesis of protein. In this model the attachment of the 60S to the membrane does not depend on the protein and, in fact, is produced even if protein synthesis is inhibited.

It has been demonstrated that the small subunit of the membrane-bound ribosome is capable of being exchanged with small subunits of free ribosomes in vitro. However, under the same conditions, the large subunits do not exchange with the addition of free large subunits. This may reflect the presence in the cell of a stable large-subunit-membrane complex, which is independent of protein synthesis.[66]

Action of Antibiotics in Protein Synthesis

Although this topic is beyond the primary interest of this book, it is important to have an idea of how antibiotics may act at different steps of protein synthesis and to understand that they are important tools for studying this process at the molecular level. *Tetracycline,* for example, preferentially inhibits the binding of AA-tRNA at the "aminoacyl" site whereas *puromycin* attacks the bound AA-tRNA at the "peptidyl" site. The latter acts as a kind of terminator by substituting for tRNA and forming a polypeptide-puromycin chain which is released. Puromycin may interrupt protein synthesis even at 0° C, and tends to dissociate polysomes, thus leading to an accumulation of ribosomal subunits. An interesting example is the antibiotic *fusidic acid,* which blocks the translocation induced by the G factor. *Chloramphenicol* inhibits protein synthesis of bacteria, chloroplasts, and mitochondria, but not the general cytoplasmic ribosomal system. *Cycloheximide,* on the other hand, affects eukaryotic cells, and its application results in a complete re-formation of the polysomes in spite of the inhibition of cytoplasmic protein synthesis normally induced by this antibiotic. *Streptomycin* may cause a misreading of the genetic message at the level of the interaction of tRNA with ribosomes (see Kaji, 1971).

Other drugs that are useful in the study of protein synthesis are: *actinomycin,* an antibiotic that alters the structure and function of DNA; *rifampicin,* which specifically inhibits bacterial RNA-polymerases; and α-amanitin, a toxic peptide used to differentiate between the RNA polymerase of the nucleolus and that of the nucleoplasm in eukaryons (the latter is more sensitive to the drug) (see Kersten, 1971).

SUMMARY:
Protein Synthesis

Ribosomes are macromolecular structures that guide the multiple interactions involved in the synthesis of proteins. To direct these interactions ribosomes function in groups called *polyribosomes* or *polysomes.* The number of ribosomes in a polysome is related to the length of the protein to be synthesized. For example, for the polypeptides of hemoglobin a pentamer (i.e., five ribosomes) is used. Each ribosome carries a nascent polypeptide chain. In the case of hemoglobin, therefore, five chains are formed at the same time along the mRNA.

The polysomes and their mRNA may be observed under the electron microscope. In bacteria the process of transcription of mRNA is simultaneous with the translation, i.e., ribosomes become attached to mRNA while this is being transcribed. Under the electron microscope the points of attachment of RNA polymerase at the DNA template may be observed.

In free polysomes of bacteria and eukaryons a *ribosome-polysome cycle* may be demonstrated. The two ribosomal subunits exchange at the site of initiation and dissociate at the terminator end. Thus, subunits are exchanged and there is a pool of free subunits. A dissociation factor, protein in nature, is involved in the separation of the two subunits at the end of the translation. This factor is similar to initiator factor IF_3.

The two subunits of the ribosome perform different functions in protein synthesis. The small subunit attaches to the first codon of the mRNA, forming the *initiation complex.* The first codon is always AUG and codes for formylmethionine in bacteria and mitochondria, and for methionine in eukaryons. These initial amino acids are later split off by enzymes.

Three protein factors, IF_1, IF_2, and IF_3, loosely associated with the small ribosomal subunit, are involved in the *initiation* of protein synthesis. $IF_1 + IF_2$ and GTP are required to bind F-met-tRNA to the small subunit in response to codon AUG. IF_3 factor stimulates the binding of mRNA to the initiation site of the small subunit.

Protein factors used in the *elongation* of the polypeptide chain are found in the post-ribosomal supernatant. Factor EFT comprises proteins Tu and Ts. EFTu + GTP forms a complex with the aminoacyl-tRNA, and this complex is catalyzed by factor EFTs. The G factor, *translocase,* which is in contact with the ribosome, splits GTP into GDP and P_1. The energy released is used for the removal of the empty tRNA and the translocation process.

The large ribosomal subunit contains the enzyme *peptidyltransferase,* involved in the formation of the peptide bond. This subunit also carries two binding sites, i.e., the *aminoacyl* or *acceptor site,* and the *peptidyl* or *donor site.* After the binding of F-met-tRNA to the AUG codon on the small subunit, the large subunit of the first ribosome becomes attached. The growth of the polypeptide chain involves: (1) the entrance of an aminoacyl-tRNA into the aminoacyl site, (2) the formation of the peptide bond and the release of the free tRNA that was present at the peptidyl site, (3) the movement of the peptidyl tRNA from the aminoacyl to the peptidyl site, and (4) the entrance of a new aminoacyl tRNA into the aminoacyl site (Fig. 19–21). Simultaneously, the mRNA moves to the following codon in the $5' \rightarrow 3'$ direction. All these coordinated steps are accomplished at a fast rate (one globin chain of 150 amino acids per minute). Termination of the polypeptide chain occurs when a termination codon (UAG, UAA, or UGA) is reached. Two termination protein factors, R_1 and R_2, are involved in this process. The 60S subunit attached to the endoplasmic reticulum plays a special role in the vectorial flow of the newly formed peptide chain into the cavity of the reticulum. Experiments on the dynamics of the assembly suggest that the 60S subunit binds directly to the membrane; then the 40S subunit, forming a complex with the mRNA, joins the 60S. In vitro experiments showed that the small subunits exchange with one another, but the large subunit probably forms a stable complex with the membrane and does not exchange.

GENETIC REGULATION

This chapter has shown, to this point, that the structural gene is a DNA molecule whose coded information can be translated into the specific structure of a protein molecule. The impression given is that of a unidirectional, stable phenomenon centrally controlled by DNA and started by the synthesis of an RNA template. Consideration will now be given to the way in which the expression of these structural genes can be modified by other parts of the genome or by external (environmental) agents that may act by way of the cytoplasm. This field is of great importance, since it concerns the different mechanisms by which genetic action is regulated.

Regulation, a fundamental property of living matter, is related to adaptation to the changing environment and to the phenomena of growth and differentiation. Here some of the mechanisms by which genic action is regulated in lower organisms will be briefly considered.

Enzyme Induction and Repression

Escherichia coli is especially appropriate for this type of analysis since its circular chromosome contains the information to code for a total of only 3000 proteins. The DNA is transcribed in discrete segments that cover one to twenty genes at the same time and whose initiation and termination points are fixed. The initiation point, corresponding to the site of attachment of RNA polymerase, is called the *promoter.*

Normally, only some genes of the total genome are active. In fact, if all of them were operative at the same time, the cell would become filled with unnecessary proteins. By changing the nutrients in the culture medium it is possible to *activate* or *repress* certain genes and, in this way, to allow the cell to adapt to the change in environment. For example, *E. coli* growing on glycerol increases the production of enzymes breaking down this substrate, whereas other enzymes are kept at a minimum. On the other hand, if the bacteria are exposed to lactose, the enzyme β-galactosidase, which hydrolyzes this sugar into galactose and glucose, may increase a thousand-fold. These changes, which involve a true enzyme synthesis, are referred to as *enzyme induction.* They have been found in many catabolic systems that act on sugars, amino acids, lipids, and also in many anabolic systems that synthesize amino acids or nucleic acid precursors.

One of the best known induction systems is the just mentioned β-galactosidase, which will be discussed in detail later on.

In examples just mentioned the nutrients glycerol and lactose act simultaneously as *substrate* for the corresponding enzyme and as *inducer* of the enzyme synthesis. An important concept to remember, however, is that not all inducers are substrates and vice versa. In the case of β-galactosidase, isopropylthiogalactoside (IPTG) is an inducer, but it does not serve as a substrate, and phenyl β-D-galactoside is a substrate, but it is not an inducer. In other words, both substances should be present to have normal growth of the *E. coli.*

In opposition to enzyme induction is the process of *enzyme repression* in which the synthesis of a certain enzyme is selectively inhibited by the end product of its metabolic chain. In *E. coli* an example is tryptophan synthetase, which is inhibited by tryptophan and certain analogues. Both enzyme repression and induction, in general, involve more than a single enzyme in a metabolic chain.

Regulatory Genes and the Lactose or Lac Operon

In 1961 Jacob and Monod[67] postulated the existence of a new genetic unit, the *operon.* This is a group of closely related genes in the chromosome that can be controlled (i.e., activated or inactivated) in a coordinated way. The best known operon is the *lac operon,* involved in the utilization of lactose. As shown in Figure 19–23 the lac operon comprises three structural genes (*z, y,* and *a*) and produces a polycistronic mRNA (lac mRNA).

This messenger RNA codes for β-galactosidase, lac permease, and transacetylase. The lac permease is a single-unit protein that acts as a carrier for the entrance of lactose into the cell. The transacetylase is an enzyme of two subunits that catalyzes the transfer of one acetyl group from acetyl-CoA to galactose.

The group of structural genes is controlled by three DNA segments that act as regulatory elements. In Figure 19–23 these are shown as the *i gene* or *regulatory gene* which codes for the *lac repressor,* the *promoter,* and the *operator.*

The i Gene and the Repressor. Studies on bacterial genetics, particularly in relation to the lac operon, have shown that there are genes that have a regulatory function. A mutation in the regulatory gene does not produce an alteration in the amino acid sequence of an enzyme, but causes a marked change in its activity. For example, the β-galactosidase content of *E. coli* can be changed by the regulatory gene which acts by way of a diffusible gene product called the *repressor.* In the case of the lac operon, the *i* gene codes for a protein repressor having four subunits of 40,000 daltons each (total: 160,000 daltons) (Fig. 19–23). This repressor is a soluble protein that binds strongly and specifically to a segment of the DNA known as the *lac operator,* which is composed of 12 to 15 nucleotides.[68, 69] Each of the subunits of the repressor has one binding site for the inducer.

The natural inducer of the lac operon is lactose, which upon entering the cell is modified in some unknown way and binds to the repressor. Experimentally, it is much better to use an artificial inducer such as isopropylthiogalactoside (IPTG), which without further change binds to the repressor and acts as an inducer. Upon binding of IPTG, the repressor undergoes a conformational change by which it becomes unable to bind to the operator (Fig. 19–24), thus inducing the transcription of lac operon. Although there are only ten repressor molecules for each cell of *E. coli,* as was mentioned before, the lac operon may be activated 1000 times after exposure of the cells to lactose,

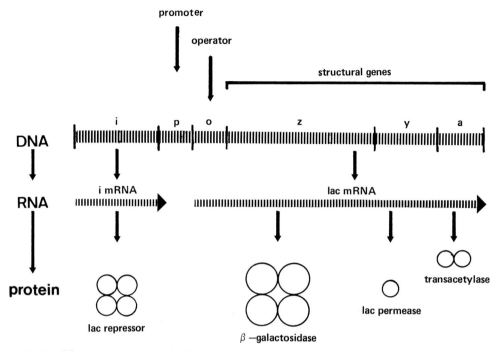

Figure 19–23. Diagram representing the *lac operon*. *i*, the regulatory gene that produces the *i mRNA* that codes for the *lac repressor*, a protein with four subunits. The *promoter* (*p*) is the region of attachment of RNA polymerase; the *operator* (*o*) is the region where the repressor binds (see Fig. 19–24). The *i gene*, the promoter, and the operator are regulatory elements. The structural genes, *z, y,* and *a,* produce a polycistronic lac mRNA and the three proteins indicated below. (See the description in the text.)

and the content of β-galactosidase may attain 3 per cent of the total protein.

The Promoter. The promoter (*p*) is the DNA segment to which the RNA polymerase first becomes attached to initiate the transcription of the structural genes.[70] In Figure 19–23 the promoter is shown immediately to the left of the operator. The promoter is a key regulatory element, since it controls the rate of mRNA synthesis of a given operon. Both the structural genes and the i gene have their own promoter. Mutants of the i gene promoter which allow a fiftyfold over-production of the lac repressor have been isolated.

The Operator. The operator (*o*) is the segment of DNA upon which the repressor binds. The operator requires a close linkage with the structural genes under its control, whereas the regulator gene (i gene) may be situated at a distance from them. In the lac operon the operator is situated at the left of the structural genes *z, y,* and *a* (Fig. 19–23). In this position the RNA polymerase that has attached to the promoter must first pass the operator before transcribing the structural

genes. If the operator is bound to the repressor, the lac genes will not be expressed (Fig. 19–24). Thus, the lac operon is regulated by the repressor (from the i gene), the promoter, and the operator.

Transcriptional Control of Gene Expression

The Jacob and Monod[67] model of the operon is based on a mechanism of *negative control* that operates at the level of transcription and has been found, in many experiments, to be correct. For example, if a culture of *E. coli* is growing in a medium without lactose, the repressor molecules, being in an active conformation, will be able to bind to the operator and will prevent the RNA polymerase from transcribing the structural genes (Fig. 19–24). The result is that β-galactosidase and the other proteins will not be synthesized. On the other hand, if a culture of *E. coli* is exposed to lactose or IPTG, the *inducer* binds to the lac repressor, which by a

NO INDUCER

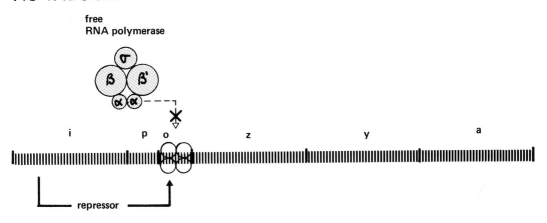

INDUCTION

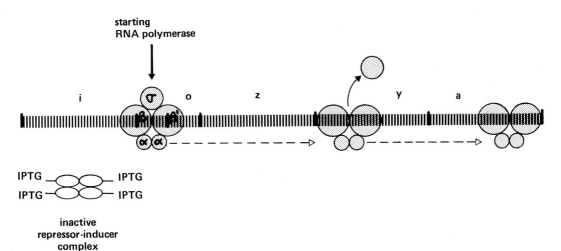

Figure 19–24. Diagram representing the regulation of the lactose operon in the absence and in the presence of the inducer. In the absence of the inducer, the *lac repressor* binds tightly to the operator, interfering with the transcription of the structural genes. The repressor is a tetramer with subunits of 40,000 daltons. When the inducer (IPTG) is present, it binds to the repressor, eliciting a conformational change that prevents its binding to the operator; as a consequence, RNA polymerase is free to transcribe the structural genes. Observe that *E. Coli* RNA polymerase has five subunits of different size: σ (95,000 daltons); β/β' (about 160,000); and two α (40,000). Note that after transcription has started the sigma (σ) subunit is released and only the core enzyme remains bound to the DNA.

conformational change loses its affinity for binding the operator. In this condition the operator is free, and RNA polymerase can progress to transcribe the entire lac genes thereby producing a polycistronic lac mRNA that will code for the three enzymes of the system (Fig. 19–23). Under these conditions large amounts of β-galactosidase, permease, and transacetylase are produced, and the bacteria are able to metabolize lactose.

The case of *enzyme repression* just described, in which the tryptophan operon is selectively inhibited by tryptophan, can also be explained on the basis of a similar model. In this case, the regulatory gene produces a *repressor* protein which is normally in an *inactive* form. This repressor, upon binding with the metabolite (a *co-repressor*), undergoes a conformational change by which it can bind the operator and thereby inhibit the transcription of the entire operon involved in the synthetic pathway of tryptophan. This type of control also applies to the synthetic chain for histidine. In cases such as these it may be postulated that the affinity of the repressor for binding the operator is normally low and that this increases under the action of the co-repressor (see Stent, 1967).

Role of Cyclic AMP in Bacterial Transcription

Cyclic AMP (cAMP) has a regulatory role in bacteria activating and deactivating the inducible operons at the level of the promoter. In *E. coli* a cyclic AMP *receptor protein (CRP)* that is able to bind cyclic AMP with high affinity has been found. This is a soluble protein with two subunits (i.e., a dimer) of 45,000 daltons. After the CRP-cAMP complex is formed, it binds to the promoter and favors the recognition of this site by RNA polymerase, thus enchancing the transcription of the inducible operon. In the case of the lac operon, in addition to the *negative control* by the repressor, there is *positive control by the CRP-cAMP complex* (Fig. 19–25). This last mechanism is also involved in the utilization of other metabolites such as galactose, maltose, and arabinose. In spite of this general role of cAMP in regulation, it is not essential for the life of the bacterium. Thus, mutants of *E. coli* that cannot synthesize cAMP (i.e., that lack adenylcyclase) can grow on a glucose substrate.[71]

It has been shown that a normal culture of *E. coli* grown in glucose has a lower cAMP

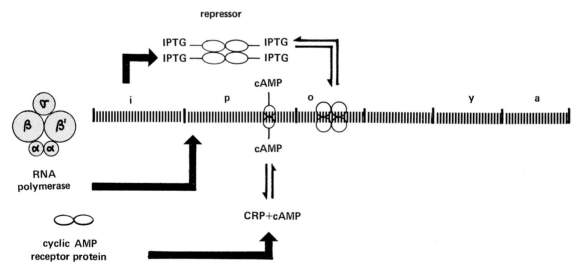

Figure 19–25. Diagram representing the mechanism by which cyclic AMP regulates the lac operon (see Fig. 19–23). The cyclic receptor protein (CRP) with its two subunits of 45,000 daltons is indicated. This protein binds cAMP, forming the CRP-cAMP complex which favors the action of RNA polymerase by binding to the promoter. In addition to the *negative control* by the repressor (see Fig. 19–24), there is this *positive control* by the CRP–cAMP complex. (Modified from Pastan, I., "Current directions in research on cyclic AMP," in Anfinsen, C. B., Goldberger, R. F., and Schechter, A. N. (Eds.), *Current Topics in Biochemistry,* New York, Academic Press, Inc., 1972.)

content than when it is exposed to a poorer energy source, such as lactose or succinate.[72] When the intracellular cAMP is low, as in glucose-grown cultures, the cyclic AMP receptor protein will not bind to the promoter of the lac operon and this will not be turned on. Therefore, if *E. coli* is grown in the presence of glucose and lactose, it will use only the richest carbon source. It is understandable that by this mechanism of regulation the *E. coli* can adapt more efficiently to a changing environment, such as that present in the human intestine.

Molecular biology, therefore, has provided an explanation for a long known phenomenon: when bacteria are grown in glucose, the rate of synthesis of inducible enzymes is very low. This phenomenon is known in the literature as *catabolite repression*. In fact, this repression is reversed, upon addition of exogenous cAMP.[73]

Genetic regulation is a fundamental property of living matter by which the organism adapts to a changing environment. In *E. coli* the circular DNA codes for some 3000 proteins in discrete segments corresponding to from one to twenty genes. In each segment the initiation point is the *promoter,* which is the point of attachment of RNA polymerase. Normally, not all genes are fully active, and some of them become activated when there is a change in the medium. For example, if the *E. coli* is grown on lactose there is an induction of β-galactosidase. *Enzymatic induction* implies an actual synthesis of the corresponding enzyme; *enzymatic repression* is an inhibition of the synthesis of the enzyme.

Regulation of gene activity is by way of *regulator* genes. These produce the corresponding mRNA and a *repressor protein* which binds to a special site of the genome called the *operator.* This may control a single gene or a group of them called an *operon.* In the case of the *lac operon* there are three structural genes (i.e., β-galactosidase, permease, and transacetylase) that are regulated by the same operator, and all of them are involved in the utilization of lactose. It is thought that the operator is at the initiation of the transcription near the promoter (Fig. 19–23).

Regulation of the lac operon is by way of the repressor which binds to the operator, thereby interfering with the transcription by RNA polymerase. In this condition the lac operon is inactive and no enzymes are produced. In the presence of an inducer (suitable metabolite), the repressor is inactivated and the operator becomes free. In this condition the entire operon may be transcribed by the RNA polymerase. The lac repressor is a protein of 160,000 daltons with four subunits of 40,000 daltons, each of which has a binding site for the inducer.

The lac operon also has a positive control mechanism provided by cAMP, which interacts with a *receptor protein* (CRP). The CRP-cAMP complex binds to the promoter, thus allowing the RNA polymerase to recognize the promoter. When cAMP levels are low (i.e., when cells grown are in glucose), β-galactosidase levels will decrease, by a phenomenon known as *catabolite repression*.

**SUMMARY:
Genetic Regulation**

REFERENCES

1. Benzer, S. (1962) The fine structure of the gene. *Sci. Am.*, *206*:70.
2. Pauling, L. (1952) *Proc. Amer. Phil. Soc.*, *96*:556.
3. Ingram, V. M. (1966) *The Biosynthesis of Macromolecules.* W. A. Benjamin, Inc., New York.
4. Nirenberg, M., and Matthaei, H. (1961) *Proc. Natl. Acad. Sci. USA*, *47*:1588.
5. Nirenberg, M., and Ochoa, S. (1963) *Cold Spring Harbor Symp. Quant. Biol.*, *28*:549, 559.
6. Nirenberg, M. (1967) In: *The Neurosciences*, p. 143. (Quarton, G. C., Melnechuk, T., and Schmitt, F. O., eds.) The Rockefeller University Press, New York.
7. Khorana, H. G. (1965) *Fed. Proc.*, *24*:1473.
8. Gros, F., Hiatt, H., Gilbert, W., Kurland, C. G., Risebrough, R. W., and Watson, J. D. (1961) *Nature* (London), *190*:581.
9. Travers, A. A., and Burgess, R. R. (1969) *Nature* (London), *222*:537.
10. Bautz, E. K., Bautz, F. A., and Dunn, J. J. (1969) *Nature* (London), *223*:1022.
11. Roberts, J. W. (1969) *Nature* (London), *223*:480.
12. Darnell, J. E. (1968) *Bacteriol. Rev.*, *32*:262.
13. Bernhard, W., and Granboulan, N. (1968) *The Nucleus*, p. 81. Academic Press, Inc., New York.
14. Penman, S., Vesco, C., and Penman, M. (1968) *J. Molec. Biol.*, *34*:49.
15. Kates, J. (1970) *Cold Spring Harbor Symp. Quant. Biol.*, *35*:743.
16. Lim, L., and Canellakis, E. S. (1970) *Nature* (London), *227*:110.
17. Edmons, M., Vaughan, M. H., and Nakazato, H. (1971) *Proc. Natl. Acad. Sci. USA*, *68*:1336.
18. Darnell, J. E., Wall, R., and Tushinski, R. J. (1971) *Proc. Natl. Acad. Sci. USA*, *68*:1321.
19. Darnell, J. E., Philipson, L., Wall, R., and Adesnik, M. (1971) *Science*, *174*:507.
20. Spirin, A. S., Beltisina, N. V., and Lerman, M. I. (1965) *J. Molec. Biol.*, *14*:611.
21. Perry, R. P., and Kelley, D. E. (1968) *J. Molec. Biol.*, *35*:37.
22. Henshaw, E. C. (1968) *J. Molec. Biol.*, *36*:401.
23. Holley, R. W., Apgar, J., Everett, G., Madison, J., Marquisee, M., Merrill, S., Penswick, J., and Zamir, A. (1965) *Science*, *147*:1462.
24. Kim, S. H., Quigley, G., Suddath, F. L., McPherson, A., Sneden, D., Kim, J. J., Weinzierl, J., Blattmann, H., and Rich, A. (1972) *Proc. Natl. Acad. Sci. USA*, *69*:3746.
25. Crick, F. H. C. (1966) *J. Molec. Biol.*, *19*:548.
26. Warner, J. R., Rich, A., and Hall, C. E. (1962) *Science*, *138*:1399.
27. Wettstein, F. O., Staehelin, T., and Noll, H. (1963) *Nature* (London), *197*:430.
28. Miller, O. L., Jr., and Beatty, B. R. (1969) *Science*, *164*:995.
29. Miller, O. L., Jr., Hamkalo, B. A., and Thomas, C. A. (1970) *Cold Spring Harbor Symp. Quant. Biol.*, *35*:505.
30. Hamkalo, B. A., and Miller, O. L., Jr. (1973) In: *Gene Expression and Its Regulation*. (Kenney et al., eds.) Plenum Publishing Corp., New York.
31. Mangiarotti, G., and Schlessinger, D. (1966) *J. Molec. Biol.*, *20*:123.
32. Kaempfer, R. (1968) *Proc. Natl. Acad. Sci. USA*, *61*:106.
33. Kaempfer, R., Meselson, M., and Raskas, H. J. (1968) *J. Molec. Biol.*, *31*:277.
34. Colombo, B., Vesco, C., and Baglioni, C. (1968) *Proc. Natl. Acad. Sci. USA*, *61*:651.
35. Kaempfer, R. (1969) *Nature* (London), *222*:950.
36. Subramanian, A. R., Ron, E. Z., and Davis, B. D. (1969) *Proc. Natl. Acad. Sci. USA*, *61*:761.
37. Algranati, I. D., Gonzalez, N. S., and Bade, E. G. (1969) *Proc. Natl. Acad. Sci. USA*, *62*:574.
38. Nomura, M., and Lowry, C. V. (1967) *Proc. Natl. Acad. Sci. USA*, *58*:946.
39. Marcker, K., and Sanger, F. (1964) *J. Molec. Biol.*, *8*:835.
40. Clark, B. F. C., and Marcker, K. (1968) *Sci. Am.*, *218* (No. 1):36.
41. Stanley, W. M., Salas, M., Wahba, A. J., and Ochoa, S. (1966) *Proc. Natl. Acad. Sci. USA*, *56*:290.
42. Thach, R. E., Dewey, K. F., Brown, J. C., and Doty, P. (1966) *Science*, *153*:416.
43. Smith, A. E., and Marcker, K. A. (1968) *J. Molec. Biol.*, *38*:241.

44. Iwasaki, K., Sabol, S., Wahba, A. J., and Ochoa, S. (1968) *Arch. Biochem. Biophys., 125*:542.
45. Ochoa, S. (1968) *Naturwissenschaften, 11*:505.
46. Zasloff, M., and Ochoa, S. (1972) *Proc. Natl. Acad. Sci. USA, 69*:1769.
47. Anderson, J. M., Bretcher, M. S., Clark, B. F. C., and Marcker, K. A. (1967) *Nature* (London), *215*:490.
48. Brown, J. C., and Doty, P. (1968) *Biochem. Biophys. Res. Commun., 30*:284.
49. Lee-Huang, S., and Ochoa, S. (1972) *Biochem. Biophys. Res. Commun., 49*:371.
50. Revel, M., Pollack, Y., Groner, Y., Scheps, R., Inouye, H., Berissi, H., and Zeller, H. (1973) *Biochimie, 55*:41.
51. Nishizuka, Y., and Lipmann, F. (1966) *Proc. Natl. Acad. Sci. USA, 55*:212.
52. Allende, J. E., Seeds, N. W., Conway, T. W., and Weissbach, H. (1967) *Proc. Natl. Acad. Sci. USA, 58*:1566.
53. Ertel, R., Brot, N., Redfield, B., Allende, J. E., and Weissbach, H. (1968) *Proc. Natl. Acad. Sci. USA, 59*:861.
54. Lucas-Lenard, J., and Haenni, A. L. (1968) *Proc. Natl. Acad. Sci. USA, 59*:554.
55. Leder, P., Skogerson, L. E., and Nau, M. W. (1969) *Proc. Natl. Acad. Sci. USA, 62*:454.
56. Monro, R. E. (1967) *J. Molec. Biol., 26*:147.
57. Watson, J. D. (1970) *Molecular Biology of the Gene.* W. A. Benjamin, Inc., New York.
58. Kischa, K., Möller, W., and Stöffler, G. (1971) *Nature [New Biol.], 233*:62.
59. Schreier, M. H., and Noll, H. (1971) *Proc. Natl. Acad. Sci. USA, 68*:805.
60. Chuang, D. M., and Simpson, M. V. (1971) *Proc. Natl. Acad. Sci. USA, 68*:1474.
61. Capecchi, M. R. (1966) *Proc. Natl. Acad. Sci. USA, 55*:1517.
62. Capecchi, M. R. (1967) *Proc. Natl. Acad. Sci. USA, 58*:1144.
63. Siekevitz, P., and Palade, G. (1960) *J. Biophys. Biochem. Cytol., 7*:619.
64. Redman, C. M., and Sabatini, D. D. (1966) *Proc. Natl. Acad. Sci. USA, 56*:608.
65. Baglioni, C., Bleiberg, I., and Zauderer, M. (1971) *Nature [New Biol.], 232*:8.
66. Borghese, D., Blobel, G., and Sabatini, D. D. (1973) *J. Molec. Biol., 74*:415.
67. Jacob, F., and Monod, J. (1961) *J. Molec. Biol., 3*:318.
68. Gilbert, W., and Müller-Hill, B. (1966) *Proc. Natl. Acad. Sci. USA, 56*:189.
69. Gilbert, W., and Müller-Hill, B. (1967) *Proc. Natl. Acad. Sci. USA, 58*:2415.
70. Ippen, K., Miller, J., Scaipe, J., and Beckwith, J. (1968) *Nature* (London), *217*:825.
71. Brickman, E., Soll, L., and Beckwith, J. (1973) *J. Bacteriol., 116*:582.
72. Makman, R., and Sutherland, E. W. (1965) *J. Biol. Chem., 240*:1309.
73. Pastan, I., and Perlman, R. (1970) Cyclic AMP in bacteria. *Science, 169*:339.

Anfinsen, C. B. (1961) *The Molecular Basis of Evolution.* John Wiley & Sons, New York.
Auerbach, C. (1967) The chemical production of mutations. *Science, 158*:1145.
Basilio, C. (1967) RNA code words in several species. *Natl. Cancer Inst. Monogr., 27*:181.
Beckwith, J. (1967) Regulation of the lac operon. *Science, 156*:597.
Boulter, D., Ellis, R. J., and Yarwood, A. (1972) Biochemistry of protein synthesis in plants. *Biol. Rev., 47*:113.
Bretscher, M. S. (1968) How repressor molecules function. *Nature* (London), *217*:509.
Clark, B. F., and Marcker, K. A. (1968) How proteins start. *Sci. Am., 218*:36.
Crick, F. H. C. (1966) The genetic code: yesterday, today and tomorrow. *Cold Spring Harbor Symp. Quant. Biol., 31*:3.

ADDITIONAL READING

Crick, F. H. C. (1967) Origin of the genetic code. *Nature* (London), *213*:119, 5072.

DuPraw, E. J. (1968) *Cell and Molecular Biology.* Academic Press, Inc., New York.

Epstein, W., and Beckwith, J. (1968) Regulation of gene expression. *Ann. Rev. Biochem., 37*:411.

Goulian, M. (1969) Synthesis of viral DNA. *Sci. Am., 220* (No. 3):35.

Haselkorn, R., and Rothman-Denes, L. (1973) Protein synthesis. *Ann. Rev. Biochem., 42*:397.

Kaji, A., Igaraschi, K., and Shull, R. H. (1971) Mode of action of antibiotics on various steps of protein synthesis. In: *Advances in Cytopharmacology,* Vol. 1, p. 99. Raven Press, New York.

Kersten, H. (1971) Use of inhibitors to study the structure and function of nucleic acids and ribosomes. *FEBS Letters, 15*:261.

Khorana, H. G. (1975) A bacterial gene for tyrosine transfer RNA: promoter and terminator sequences and progress in total synthesis. Internat. Symp. on Macromolecules IUPAC. Elsevier, Amsterdam, p. 371.

Laszlo, B., and J. Lucas-Lenard, J. (1973) Studies on the binding of bacterial elongation factors EFTu and EFG to ribosomes. *Arch. Biochem. Biophys., 154*:555.

Lipmann, F. (1969) Polypeptide chain elongation in protein synthesis. *Science, 164*:1024.

Lipmann, F. (1973) What do we know about protein synthesis? In: *Gene Expression and Its Regulation,* pp. 1–12. (Kenney, F. T., Hamkalo, B., Favelukes, G., and August, J. T., eds.) Plenum Publishing Corp., New York.

Lewin, B. M. (1970) Second golden age of molecular biology. *Nature* (London), *237*:1009.

Lewin, B. M. (1970) *The Molecular Basis of Gene Expression.* John Wiley & Sons Ltd., New York.

Lucas-Lenard, J., and Lipmann, F. (1971) Protein synthesis. *Ann. Rev. Biochem., 40*:409.

Müller-Hill, B. (1971) Lac repressor. *Angewandte Chemie, 10*:160.

Nirenberg, M. (1967) The genetic code. In: *The Neurosciences: A Study Program.* (Quarton, G. C., et al., eds.) The Rockefeller University Press, New York.

Ochoa, S. (1963) Synthetic polynucleotides and the genetic code. *Fed. Proc., 22*:62.

Ochoa, S. (1968) Translation of the genetic message. *Naturwissenschaften, 55*:505.

Pastan, I. (1972) Current directions in research on cyclic AMP. In: *Current Topics in Biochemistry,* p. 65. (Anfinsen, C. B., Goldberger, R. F., and Schechter, A. N., eds.) Academic Press, Inc., New York.

Paul, J. (1967) *Cell Biology.* Heinemann Educational Books, Ltd., London.

Perry, R. P., Greenberg, J. R., Kelley, D. E., Latorre, J., Schochetman, G. (1973) Messenger RNA: its origin and fate in mammalian cells. In: *Gene Expression and Its Regulation,* Vol. 1, p. 149. (Kenney et al., eds.) Plenum Publishing Corp., New York.

Philipps, B. R. (1969) Primary structure of transfer RNA. *Nature* (London), *223*:374.

Proteins at Cold Spring Harbor (1969) *Nature* (London), *223*:133.

Sager, R., and Ryan, F. J. (1963) *Cell Heredity.* John Wiley & Sons, New York.

Smith, J. D. (1973) Nucleotide sequence and function of transfer RNA and precursor transfer RNA. In: *Gene Expression and Its Regulation.* (Kenney et al., eds.) Plenum Publishing Corp., New York.

Spiegelman, S. (1963) Genetic mechanisms. Information transfer from the genome. *Fed. Proc., 22*:36.

Staehelin, T. H., Raskas, H., and Meselson, M. (1968) In: *Organizational Biosynthesis.* (Vogel, H. J., ed.) Academic Press, Inc., New York.

Stent, G. S. (1967) Induction and repression of enzyme synthesis. In: *The Neurosciences: A Study Program.* (Quarton, G. C., et al., eds.) The Rockefeller University Press, New York.

Watson, J. D. (1970) *Molecular Biology of the Gene.* W. A. Benjamin, Inc., New York.

Yanofsky, C. (1967) Gene structure and protein structure. *Sci. Am., 216*:80.

twenty

CELL DIFFERENTIATION AND CELLULAR INTERACTION

In multicellular organisms most cells are adapted to specialized functions, and the morphology of the cell is modified accordingly. For example, nerve cells assume a shape and structure adapted to the functions of *irritability* and *conductivity*, and these modifications enable them to react to stimuli and to transmit signals from one part of the organism to another. The progressive specialization in structure and function constitutes, in a restricted sense, *cell differentiation*.

Cell differentiation may be defined as the process by which stable differences arise between cells of an individual (see Gurdon, 1973). One of the general principles of cell differentiation in animal tissues is that once a cell has specialized it becomes a *stable type*, which cannot revert back to the undifferentiated state. For example, a cultured cell from the skin epithelium remains epithelium and does not change into a muscle or another cell type. Cell differentiation is a persistent change very different from the changes in enzyme induction or repression in bacteria, which were discussed in the previous chapter. The phenomena involving enzymes depended on the nutrient added to the medium and were entirely reversible.

Cell differentiation occurs throughout the life of the organism. However, during the embryonic period it reaches a maximum and becomes one of the most important processes. Most organisms develop from a single cell — the *fertilized ovum* — that gives rise to all tissues and organs. This cell divides actively to form the embryonic structure known as the *blastula*, in which tissues are not yet defined. In many species, up to this time, this process is mainly quantitative and involves only an increase in the number of cells; but after the blastula is formed, the process becomes qualitative, as well. The cells of the blastula begin to rearrange themselves by a process called *gastrulation*, at which time the three germ layers are formed and the future organs are determined. Cell differentiation begins during gastrulation and continues through the process of tissue formation (*histogenesis*), which is followed by the formation of organs (*organogenesis*).

In many cases, before it is possible to recognize that a cell has differentiated, there is a period during which the cell is already committed to the change. In this state, called *determination*, the cell has already stabilized its capacity to follow a certain pattern of differentiation.

One of the classic examples of determination is provided by the *imaginal discs* of *Drosophila*. In the larva, such discs contain cells that are apparently undifferentiated, but which, with time, give rise to legs, wings, antennae, and other body parts. Experiments in which a disc is transplanted into another larva or is serially grafted demonstrate that this capacity is already determined. These cells do not lose their determination and are still able to produce the corresponding part in the adult.

Only in recent years has the field of *developmental biology* (which includes cell differentiation as one of its main topics) abandoned its purely morphologic approach and developed into the cellular and molecular levels.

In molecular terms cell differentiation

implies the preferential synthesis of some specific proteins, e.g., hemoglobin in erythrocytes, gamma globulin in plasmocytes, actin and myosin in muscle. To induce such synthesis, certain genes must be activated at a definite time in particular cells. Cell differentiation can be explained in molecular terms only when the complex mechanisms of gene regulation in the higher cells are understood.

In this chapter the problem of cell differentiation will be considered after a discussion of some selected examples that are based mainly on experimental work. These examples will be interpreted in molecular terms, based on the genetic mechanisms described in the previous chapter.

NUCLEOCYTOPLASMIC INTERRELATIONS IN PROTOZOA

Throughout previous chapters numerous examples of nucleocytoplasmic interrelations have been cited. The nucleus and the cytoplasm are interdependent; one cannot survive without the other. The cytoplasm provides most of the energy for the cell through oxidative phosphorylation (in mitochondria) and anaerobic glycolysis, and the cytoplasmic ribosomes contain most of the "machinery" for protein synthesis. On the other hand, the nucleus provides templates for specific synthesis (mRNA) and also supplies the other important RNA molecules (rRNA and tRNA). Any discussion of nucleocytoplasmic interrelations must consider (1) the mechanisms by which the genes contained in the chromosomes exert their control on the metabolic processes of the cytoplasm and (2) the mechanisms by which the cytoplasm influences gene activity.

Nuclear control of the cytoplasm was first considered during the last century by Balbiani in the so-called *merotomy* experiments (Gr., *meros,* part), in which protozoa were enucleated and studied. Such enucleated fragments are able to sustain most cellular activities; e.g., they can form a cellulose membrane and carry on photosynthesis (plant cells), react to stimuli and ingest food (amebae), activate cilia (ciliated cells), undergo cytoplasmic streaming, and so forth. However, these cells generally survive only a short time and are incapable of growth and reproduction. Within 5 to 10 minutes after being enucleated by micromanipulation, an ameba loses its surface tension, becomes spheroid and develops numerous blunt pseudopodia. Its movements are slowed down and it can no longer digest foods. In this state it may survive for about 20 days. If a nucleus is successfully implanted within three days, the ameba becomes extended, starts to move normally, and digests foods. Finally, it may even divide and eventually produce a culture mass.[1] This "reactivation" produced by the implanted nucleus may take place in a few seconds or several minutes. When a nucleus is transferred from one ameba to another of the same species (homotransfer), activation of the cytoplasm, division, and mass culture may occur readily. When a nucleus of another species is implanted heterotransfer), the cytoplasm is activated, but cell division occurs less readily.[2] Detailed analysis of cultured enucleate human cell fragments have also been made.[3]

Nucleocytoplasmic Relationships in Acetabularia

The unicellular marine alga *Acetabularia* is a unique system for the study of nucleocytoplasmic relationships. The only nucleus in this giant cell, which is about six centimeters long, is located in the basal or rhizoid end. The basal portion can be easily amputated, and in this enucleate condition, survival of the organism may be prolonged for months. During part of this time the enucleate cell can carry on synthetic and morphogenetic activities, indicating that either the cytoplasm is autonomous or that it stores a large reserve of nuclear products. Protein synthesis in enucleate cells is essentially normal only during the first week. In contrast, photosynthesis and respiration are not affected for several months. The synthesis of individual enzymes (e.g., phosphorylase and phosphatase) diminishes at various times; this suggests that the nuclear products may be specific for each protein. One can assume, therefore, that in the cytoplasm there would be smaller or larger pools of different mRNAs which have come from the nucleus, depending on the amount of protein desired.[4, 5]

Acetabularia is also an excellent material in which to study aspects of cell differentiation. The apical end of this marine alga forms a cap that is characteristic of the

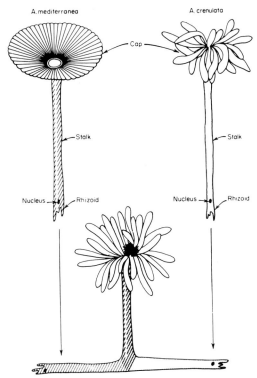

Figure 20–1. Experiment in nuclear grafting between two species of the unicellular alga *Acetabularia*. (See the description in the text.) The type of cap formed appears to depend on substances synthesized by the implanted nucleus. (From Hämmerling, J., *Ann. Rev. Plant Physiol.*, 1963.)

species. The morphology of the cap is determined by the nucleus as can be demonstrated by nuclear transplants. For example, *A. mediterranea* and *A. crenulata* have very different caps (Fig. 20–1). If a nucleus from *A. mediterranea* is implanted in an enucleate *A. crenulata*, during the second cycle after the nuclear transplant an intermediary type of cap appears, and in the third cycle the cap will be of the *A. mediterranea* type. This experiment and its reciprocal, the nuclear transplant from *A. crenulata* into *A. mediterranea*, demonstrate that a diffusible substance has been produced by the nucleus and that after some time, it has exerted a morphogenetic effect at the apical end. If two nucleate cells of different species are grafted together, a hybrid having a cap with intermediary shape is formed (Fig. 20–1). Although the nature of the substance inducing cap differentiation is unknown, it is assumed that it may be a kind of stable messenger RNA that controls the production of specific proteins.[6] In the presence

of ribonuclease, the enucleate portion of the alga becomes unable to regenerate the cap.

Molecular Events in Amphibian Eggs

One of the systems in which advances have been made in the study of cell differentiation is that of amphibian eggs, especially in *Xenopus laevis*. Since this material has been widely used in various experiments, it is important to note some of the molecular events that occur before and after fertilization[7] (Fig. 20–2). During oögenesis the synthesis of the various RNA species and proteins is considerable, but when the egg has matured, synthesis ceases almost completely. Soon after fertilization an enormous DNA synthesis begins; this is particularly impressive because up to this point, both the sperm and the egg nucleus were synthetically inert for weeks (sperm) or months or even years (oöcytes). Later on during cleavage, the production of transfer RNA and of nuclear heterogeneous RNA (this includes mRNA) is initiated. So far, the maternal ribosomes have been used for protein synthesis, but during gastrulation the new ribosomes become available (Fig. 20–2).

The induction of DNA synthesis is attributed to the entrance of a cytoplasmic protein, probably DNA polymerase, into the egg pronuclei. This enzyme is absent or inactive in the cytoplasm of the oöcyte before the rupture of the germinal vesicle, but it appears soon after the rupture occurs. The egg is now ready for fertilization.

Quantity of Genetic Information. Nuclear Transplantation

One of the fundamental problems in the study of cell differentiation is to determine if the specialized cells contain the same quantity of genetic information as the original egg cell. The fact that all diploid cells of an organism have the same DNA content (Chapter 17) suggests that cell differentiation does not depend on a gain or loss of genetic information, but on other mechanisms, probably acting at the transcription or translation levels. This problem may be tested by means of *nuclear transplantation*.

Unfertilized eggs of *Xenopus laevis* are irradiated with ultraviolet light to destroy the

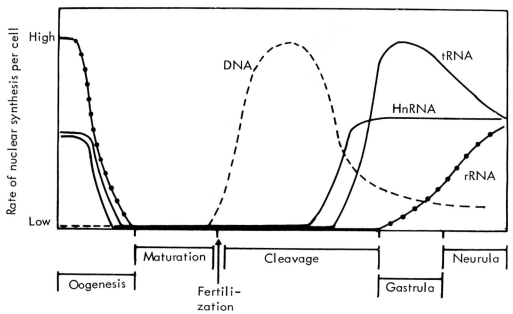

Figure 20–2. Molecular events occurring in amphibian eggs before and after fertilization. *HnRNA* corresponds to the so-called nuclear heterogeneous RNA (see Chapter 17), which presumably includes also messenger RNA; *tRNA*, transfer RNA; and *rRNA*, ribosomal RNA. (See the description in the text.) (Courtesy of J. B. Gurdon.)

nucleus. Cells from different tissues of a tadpole, e.g., cells from the brain, skin or the intestine, are drawn into a micropipette which permits the entrance of only the nucleus and a small amount of surrounding cytoplasm (Fig. 20–3).

When the nuclear transfer is carried out successfully, a normal adult frog may develop from the combination of the egg cytoplasm and the transplanted nucleus. To have better control of the experiment it is possible to use as donor nucleus one from a mutant having only one nucleolus. This nucleus may then be transferred to a recipient cell of the wild type (i.e., having two nucleoli). The fact that a nucleus from the skin or the intestine can produce a normal tadpole is a clear demonstration that *during embryonic differentiation there is no permanent loss or inactivation of genes.*[7]

Cytoplasmic Control of Gene Expression

The technique of nuclear transplantation may be used to examine the mechanisms by which genes are activated and expressed during cell differentiation. Nuclei from blas-

tula, gastrula, neurula, or even intestinal epithelium or brain undergo considerable changes when implanted into the egg's cytoplasm. The results obtained depend on the time at which the implantation was made (see Fig. 20–2). For example, if brain nuclei are injected into the growing oöcyte (i.e., during oögenesis) they enlarge, and RNA synthesis is activated within them. If the transfer is made in oöcytes with condensed chromosomes (i.e., during maturation) there is no change in the implanted nucleus. If the transplant is made at a later period, when DNA duplication has already started, then the nucleus enlarges and starts to synthesize DNA. These experiments demonstrate that the implanted nucleus responds according to the state of the cytoplasm. It is thought that the enzyme DNA polymerase penetrates the transplanted nucleus and induces the DNA synthesis. At the same time, there is a *repressor of RNA synthesis* which may suppress it in the implanted nucleus. These experiments demonstrate that in the *egg cytoplasm there are molecules that can regulate gene expression.*

Another interesting observation is the production of heterokaryons by introduction

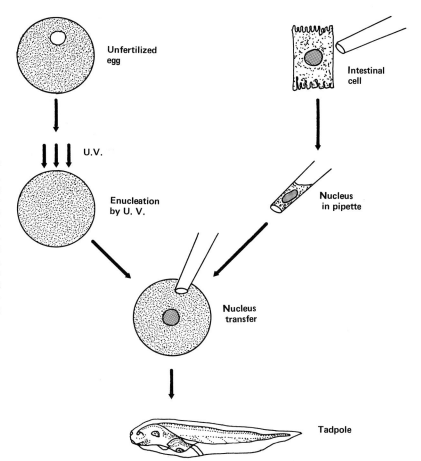

Figure 20–3. Schematic diagram of the experiment on nuclear transplantation in an unfertilized egg of *Xenopus laevis*. The egg is treated with ultraviolet irradiation to destroy the nucleus. The nucleus of an intestinal cell is drawn into a micropipette and then implanted in the egg. The result is a normal tadpole. (From Gurdon, J. B., "Gene expression during cell differentiation," in *Readings in Genetics and Evolution,* Oxford, The Clarendon Press, Chap. 4, 1973.)

Unfertilized egg

Intestinal cell

U.V.

Enucleation by U. V.

Nucleus in pipette

Nucleus transfer

Tadpole

of mature nuclei of chicken erythrocytes into HeLa cells.[8] The primary event is the enlargement of the transplanted nucleus accompanied by loosening of the condensed chromatin. This nucleus, which had lost the capacity for synthesizing RNA and DNA, resumes both functions in the presence of the HeLa cell's cytoplasm.

The first RNA produced is polydispersed and remains in the nucleus. Two to three days after the transplant, when nucleoli have appeared, the RNA is transferred into the cytoplasm. At this time the information carried by the chicken erythrocyte nucleus may be expressed and produces specific surface antigens in the HeLa cell.

Cytoplasmic Action During Mitosis. Similar mechanisms may function during a normal mitotic cycle. At metaphase, along with the condensation of the chromosome, there is loss of DNA polymerase[9] and nonhistone proteins.[10]

At telophase, DNA polymerase and other molecules present in the cytoplasm may become associated with expanding chromatin threads. The mixing of chromosomes and cytoplasm during mitosis may thus effectively reprogram the future genetic activity of the cell.[7]

Mechanisms of Gene Expression

There is much evidence that the diversity of cell phenotypes in metazoa is derived from the fact that each cell expresses only a limited amount of its full genetic potential and that different cell types express different portions of their genome.

As was mentioned previously, specialized cells contain the same quantity of genetic information as those that are still undifferentiated. This implies that cell *differentiation does not result in the loss or permanent in-*

activation of genes. Having settled this question, one must assume that cell differentiation depends on: (1) a differential *transcription* of the genes, (2) a differential *translation* of particular genes, or (3) both mechanisms of control of gene expression.

There is much direct evidence to demonstrate that there is a differential transcription of genes. One of the most obvious is provided by the polytene chromosomes of insects in which certain bands may show RNA synthesis and other signs of differential gene activity. (The importance of these phenomena war-

rants their being included in a special section of this chapter.)

Mention has already been made of the molecular events that occur in amphibian eggs after fertilization which indicate that the genes for transfer RNA, ribosomal RNA, and messenger RNA may be transcribed at different times and rates (Fig. 20–2). Cells specialized in the production of certain proteins are the best suited to isolate a particular mRNA. This isolation has been facilitated by techniques based on the polyadenylation of mRNAs in eukaryons. Muscle cells, antibody-

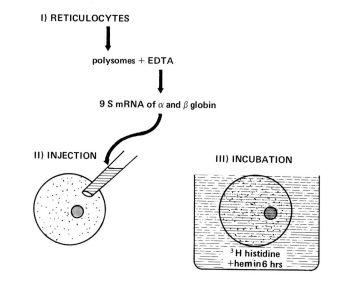

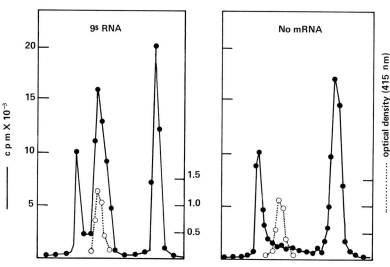

IV) HOMOGENIZATION AND PROTEIN ANALYSIS ON SEPHADEX G 200

Figure 20–4. Diagram of the experiment of the injection of hemoglobin messenger RNA into an amphibian egg and demonstration of its translation into hemoglobin molecules. (See the description in the text.) (From Gurdon, J. B., "Gene expression during cell differentiation," in *Readings in Genetics and Evolution,* Oxford, The Clarendon Press, Chap. 4, 1973.)

producing cells, silk glands, hen oviducts and others may be used, but the reticulocytes (i.e., young erythrocytes of mammals) have been preferred. Reticulocytes may be produced in large quantities in animals subjected to induced hemorrhage or to drugs that destroy red cells (e.g., phenylhydrazine). Reticulocytes are characterized by the presence of a basophilic substance which contains the polysomes and the mRNA involved in hemoglobin synthesis (Fig. 19–7). After hemolysis, the polysomes are separated and may be dissociated by a chelating agent (e.g., EDTA), releasing a 9S RNA which is the mRNA coding for the α and β globin chains of hemoglobin.[11]

This mRNA can be used in a cell-free system to produce hemoglobin. However, this can also be achieved by a very ingenious technique consisting of the injection of the 9S RNA into frog oöcytes. These cells normally do not synthesize hemoglobin, but after injection of this mRNA, they can translate it into specific hemoglobin molecules.[12]

As shown in Figure 20–4, after the mRNA injection the oöcytes are incubated in the presence of ³H-histidine and hemin (to provide for the heme group). After six hours, they are homogenized, and the proteins are separated by column chromatography. It may be observed that the oöcytes injected with 9S mRNA contain a special protein peak which coincides with the absorption of true hemoglobin used as a marker (Fig. 20–4). The newly formed hemoglobin contains both α and β globin chains and is specific for the message inoculated. In other words, the 9S

mRNA from a rabbit, mouse, or duck produces the corresponding hemoglobin molecule.[13] These experiments demonstrate that a living cell is capable of translating a message provided by another cell without loss of specificity (see Chantrenne and Marbaix, 1972). Furthermore, they show that the injected mRNA is very stable (having a half-life of several days), and that the egg has a spare capacity to translate a foreign mRNA. This capacity can be brought to the point of saturation by increasing the amount of injected 9S RNA without changing the ability of the egg to react to the injection of a second type of mRNA. This technique is certainly one of the most useful to study the translation of specific messenger RNAs, since the living cell has all the molecular machinery to achieve the specific synthesis of proteins. The translation step seems to be rather non-specific, since Xenopus eggs can translate besides hemoglobin almost any mRNA that is injected into them.

It must be concluded that the *synthesis of different kinds of proteins* in a differentiating cell *does not depend* on DNA replication (since there is no loss of genes) or on translation, but on *a differential transcription of certain genes.*

Later in this chapter mention will be made of several cases in which there may be post-transcriptional control of gene expression. These mechanisms, however, are involved more in reversible metabolic changes than in the permanent ones occurring in differentiation.

**SUMMARY:
Cell Differentiation**

Cell differentiation may be defined as the process by which stable differences arise among cells of an individual. In Metazoa, cell differentiation is a persistent change that implies the preferential synthesis of some specific proteins, e.g., hemoglobin, myosin, and gamma globulin, by the specific cells (e.g., red cells, muscle cells, and plasmocytes). Prior to the overt differentiation there may be a period of *determination* in which the capacity to differentiate in a particular direction has already been established. In a eukaryotic cell there is a continuous nucleocytoplasmic interrelationship. This can be demonstrated in *protozoa* that are enucleated in the so-called *merotomy* experiments. The alga *Acetabularia,* often used in the study of nucleocytoplasmic relationships, has a single nucleus and a large cytoplasm differentiated into a rhizoid, a stalk, and a cap, which is characteristic of a species. Appropriate experimentation may demonstrate the existence of a nuclear substance, diffusing into the cytoplasm and controlling the morphogenesis of the cap.

Amphibian eggs are most useful for the study of the mechanism of cell differentiation. During oögenesis, there is a high rate of RNA synthesis (i.e., tRNA, rRNA, and heterogeneous nuclear RNA, which includes mRNA). RNA synthesis ceases during maturation and is reinitiated after cleavage and gastrulation. DNA synthesis begins slightly before fertilization and increases enormously during cleavage (Fig. 20–2). Nuclear transplantation into amphibian eggs may be used to demonstrate whether the amount of genetic information remains unchanged or changes during cell differentiation. Since normal tadpoles may be produced with nuclei from differentiated cells (e.g., from brain, skin, intestine, etc.), it may be concluded that during differentiation there is no permanent loss or inactivation of genes.

Experiments of nuclear transplantation made in oöcytes, mature eggs, or during a period in which DNA synthesis starts, show different effects on the transplanted nucleus. This finding suggests that there is a cytoplasmic control of gene expression. Similar findings have been observed in heterokaryons, such as in hybrids between two cells. For example, the nucleus of a chicken erythrocyte may be activated in the cytoplasm of a mammalian cell. Cytoplasmic control is normally exerted in cells and varies according to the period of the cell cycle.

One of the best characterized mRNAs is the 9S RNA from reticulocytes that codes for the α and β globin chains. Injection of this mRNA into frog oöcytes leads to the production of specific hemoglobin molecules. Experiments of this kind demonstrate that a living cell is capable of translating a message provided by another cell by using its own machinery of protein synthesis.

It is concluded that the synthesis of specific proteins in cell differentiation does not depend on DNA replication or on the translation of mRNA, but on a differential transcription of certain genes.

GENE REGULATION IN EUKARYONS

One of the greatest challenges of modern biology is interpreting higher cells and organisms in the light of molecular mechanisms known to act in viruses and bacteria. The simpler structural organization of these lower organisms is conducive to study by genetic and biochemical methods under precise and reproducible conditions. As seen in the previous chapter, these investigations have disclosed the fundamental processes by which genes are duplicated and the genetic expression is regulated. Such mechanisms also function, in a general sense, in higher cells; and the fact that a single genetic code exists for all living organisms has already been mentioned.

The much greater complexity exhibited by a plant or animal cell, relative to a bacterium, implies that they should contain a much larger store of genetic information. It is already known that the DNA of *E. coli* is over one millimeter long and contains approximately 4.5×10^6 nucleotides. On the other hand, the DNA of a bull sperm (haploid) contains 3.2×10^9 nucleotides, which corresponds to about a seven hundredfold increase in coding information. This quantitative difference should be reflected, at the transcription and translation levels, in an increased number of messenger RNAs and structural proteins and enzymes. It will be shown later on that the increase in size of the genome is actually not as great as indicated by the DNA content of certain cells. For example, in *Amphiuma*, the creature having the largest known genome (168 pg of DNA and about 8×10^{10} nucleotides), the DNA content is 80 per cent *redundant* or *repetitious* DNA.

Regulation of gene function must necessarily be much more complex in higher cells than in bacteria. Only the proper balance of genes can produce the normal development of a complex organism. Chapter 16 showed that the increase of a single chromosome, resulting in trisomy, may result in an abnormal development leading to severe alterations of the nervous system (mongolism). At the chromosome level, gene regulation may be related to the existence of proteins tightly bound to the DNA, histones, and some acidic proteins, which are lacking in bacterial DNA. These proteins may shuttle between the cytoplasm and the nucleus and influence genetic expression.

DNA Redundancy and Gene Amplification

The concept of DNA redundancy and gene amplification was introduced in Chapter 18, and in Table 18–2 the number of DNA cistrons involved in the synthesis of ribosomal RNA in different organisms was recorded. It was mentioned that in some amphibian oöcytes there are thousands of similar genes that can amplify considerably the coding of ribosomal RNA.

Redundant DNA involved in the production of transfer RNA has been described by using hybridization experiments. These repeated determinants of tRNA are scattered among the various chromosomes. This

is at variance with ribosomal DNA which is localized in the nucleolar organizer (see Chapter 18). The DNA molecules present in mitochondria and chloroplasts may also be considered repetitious, since they are very similar in each type of organelle.

DNA redundancy is a constant characteristic of all eukaryotic cells, from protozoa to the higher plants and animals; this represents another important difference between eukaryons and prokaryons, in which there are few repeated DNA sequences in the genome.

The demonstration of repeated sequences of DNA in higher cells was a consequence of the remarkable properties of separated complementary strands of DNA to recognize each other and to reassociate. This process can be measured by the techniques of denaturation and renaturation (Chapter 3).[14] Essentially, these methods are based on the fact that the greater the number of repeated sequences in the molecule, the faster is the reassociation of DNA strands. In the DNA extracted from mouse cells, 10 per cent of the total was found in short nucleotide sequences that were repeated in about 1 million copies.[15] Another 20 per cent of the genome consisted of repeated sequences of 1000 to 100,000 copies; the remaining 70 per cent was nonrepetitious DNA, i.e., it was represented in only one copy (Fig. 20–5).

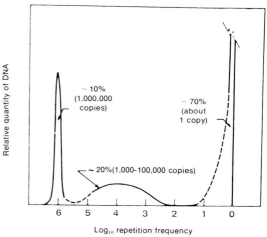

Figure 20–5. Frequency of repeated DNA sequences present in DNA from a mouse. (See the description in the text.) (From R. J. Britten and D. E. Kohne.)

The presence of such repeated sequences may have two different genetic effects. It may amplify certain specific genes, as in the case of the ribosomal cistrons. On one hand, it may reduce the amount of total genetic information. However, the effect may be more complex and the information content of the repetitious DNA may be greater than previously estimated because the nucleotide sequences of a family of these molecules may be similar, but not identical. Families of some of these repeated sequences are common among related species and become less common in more divergent species. This type of analysis may have considerable importance from the viewpoint of taxonomy and evolution.[16] It is postulated that during the evolution of the organisms these repeated families of DNA might have arisen by a manifold replication, and that these copies then must have been integrated within the chromosome and disseminated through the species by natural selection.[14] The relationship of DNA redundancy to heterochromatin and satellite DNA and some of the possible roles of DNA redundancy, in addition to gene amplification, were mentioned in Chapter 12.

Gene Transcription in Eukaryons

One of the most interesting conclusions drawn by modern researchers on the molecular biology of chromosomes is that in eukaryotic cells most of the DNA does not function in transcription; i.e., it is not active in RNA synthesis.

Chapter 12 showed that a portion of the chromatin in the interphase may be present in a dense, supercoiled, heterochromatic state; the rest exists as more loosely extended euchromatin. Using radioautography with ^{3}H-uridine it was demonstrated with the electron microscope that most of the RNA is synthesized in euchromatin, whereas the condensed heterochromatin is virtually inactive in RNA synthesis.[17] During spermiogenesis RNA synthesis ceases almost completely. In fact, the spermatozoon can be thought of as a device that injects tightly packed and inactive DNA into the egg, very much like a phage that injects DNA into a bacterium.

Another example of DNA inactivation and chromatin condensation is the metaphase chromosome in which RNA synthesis is greatly reduced. At metaphase there is also a drastic reduction in protein synthesis (see Fan and Penman, 1970). Note that one of the two X chromosomes in the female becomes heterochromatic and genetically inactive.

The restricted RNA synthesis can be demonstrated by using isolated chromatin and DNA in experiments of RNA synthesis. Using bacterial RNA polymerase as the synthesizing enzyme, it was found that the synthesis obtained with isolated chromatin was only a fraction of that obtained with DNA. Furthermore, the RNA formed by chromatin of different tissues varied, indicating that some of the DNA masking is organ specific.[18]

Role of Histones

Several investigators have thought that *histones,* because of their close association with DNA at certain stages of the cell cycle, could regulate genetic activity. Histones are able to completely inhibit the in vitro production of RNA by DNA and also hinder the hybridization of DNA by RNA. If histones were the main components masking DNA function, the regulation of genetic activity could result from changes introduced into these proteins by acetylation, methylation, or phosphorylation.[19] All of these chemical changes would tend to reduce the degree of binding between DNA and histones.

A very interesting system is provided by the small lymphocytes from human blood which normally have a dense, rather inactive (i.e., repressed) chromatin. After a few minutes' exposure to *phytohemagglutinin,* these cells show signs of activation and become large "blastoid" cells, which may undergo mitosis. (Phytohemagglutinin is used in blood cultures to study the karyotype; see Chapter 16.) A series of events follows, of which a first step seems to involve an action on the cell membrane, causing it to become more permeable. The RNA synthesis in the nucleus greatly increases, and this phenomenon appears to be preceded by acetylation of the chromosomal histones.[20] It has also been postulated that the activation of the nucleus may be due to the entrance of a neutral lysosomal protease (histonase) which digests the histone coat of the DNA.[21]

Role of Nonhistone Proteins and RNA

It has been suggested that histones probably mask DNA in a rather general and non-specific manner.[22] More important are some acidic proteins that could activate the regions of DNA repressed by their association with histones. As mentioned in the preceding section on cytoplasmic control of DNA synthesis, proteins coming from the cytoplasm (i.e., DNA and RNA polymerase) are probably involved in the regulation of the genome; later on the importance of acidic proteins in polytenic chromosomes will be discussed. More recently it has been suggested that an RNA fraction bound to proteins and present in chromatin is required to produce a specific control at the level of the genome.[23, 24]

Gene Regulation in Eukaryons as Compared with Bacteria

Although bacteria and higher cells have the same basic mechanisms of genetic information, they differ considerably in the way this information is regulated. The previous chapter has shown that in bacteria the control of transcription is exerted by repressor proteins that combine with specific regulatory sites (i.e., operator), thus blocking transcription (Fig. 19–23). Furthermore, it was mentioned that an inducer acts by combining with the repressor, thus preventing its binding to the operator and allowing gene transcription.

In eukaryons *derepression* of inactive (i.e., repressed) regions of the genome seems to be the most frequently found method of regulation at the transcription level. In Chapter 17 it was shown that although there is much RNA transcribed in the nucleus (i.e., heterogeneous nuclear RNA), there is only a small fraction that reaches the cytoplasm and may be used as messenger RNA. It has been proposed that part of this excess RNA could be used as an intranuclear *activator* to produce the activation of several genes situated in non-contiguous positions in the genome and even in different chromosomes. According to this hypothesis, in higher cells, rather than a large increase in the number of structural genes, there is a considerable increase in the number and complexity of the regulatory genes (see Britten and Davison, 1969).

An interesting example of regulation is provided by the hepatoma cells (i.e., liver tumor cells). If cell cultures are treated with corticosteroids (e.g., dexamethasone), the synthesis of the enzyme tyrosine-amino-transferase increases up to 15 times with no effect on the rest of the cell proteins. It may be demonstrated that the production of the enzyme is induced only in part of the G_1 and in the S period, but not in the rest of the cell cycle. In this case, the hormone does not induce the synthesis of a protein that was not formed before, but rather increases the rate of translation of the message. It is thought that the steroid antagonizes a post-transcriptional repressor that normally inhibits mRNA translation and promotes degradation of the messenger (see Tomkins et al., 1970).

In the complex post-transcriptional mechanisms that take place in eukaryons there are several potential sites of regulation. For example, this may be exerted: (1) on nascent mRNA by protecting it or by enhancing degradation; (2) on the transport of mRNA from the nucleus to the cytoplasm; (3) in the association of mRNA to the polyribosomes; (4) in the activation or inactivation of ribosomes; or (5) in the processes of initiation, elongation, and termination of the polypeptide chain (see Tomkins et al., 1970).

Gene Activity in Polytene Chromosomes

In Chapter 13 the morphology of polytene chromosomes was discussed. In the past few years these chromosomes have become a most valuable material with which to study the expression of genetic activity. Their usefulness was first recognized when, at specific times during the growth of insect larvae, local variations in size and the degree of condensation of the bands on the polytene chromosomes were detected. These bands appear as localized swellings and are called "puffs." The so-called Balbiani rings are puffs of a similar nature, only much larger. The formation of puffs, called "puffing," may occur on single bands, or it may include adjacent bands. In the Balbiani rings the chromonemata running through a certain band may be spun laterally to form a series of loops that gives the rings the appearance of a lampbrush chromosome (see Chapter 13).[25, 26] Puffing is a cyclic and reversible phenomenon; at definite times, and in different tissues of the larvae, puffs may appear, grow, and disappear.

In 1952, Beerman interpreted the puffing of polytene chromosomes as an expression of gene activity. The uncoiling of the DNA within the puff was thought to be a prerequisite for transcription. In 1955, Pavan and Breuer, using ^{3}H-thymidine, found that in certain puffs in polytene chromosomes of *Rhynchosciara angelae* there was an extra replication of DNA.[27] This phenomenon can now be interpreted as DNA redundancy.

In most cases, RNA and proteins are synthesized or accumulated at the puff. When short pulses of ^{3}H-cytidine are applied, the puff and the nucleoli are almost exclusively labeled.[28, 29]

The study of puffing may be performed experimentally with factors that induce their formation or change their activity. The steroid hormone *ecdysone,* which is secreted by the prothoracic gland and induces molting in insects, may induce puff formation when injected into larvae[30] and it has been observed that ^{3}H-ecdysone is bound to the sites of puffing in the chromosomes. Temperature shock and several other means may also induce puffing.

An important finding was that by staining the polytene chromosome with light green at low pH, the first phenomenon to appear in puff induction was the accumulation of an acidic protein. This may be detected at the presumed site of the puff only three minutes after temperature shock (Fig. 20–6). The protein is not synthesized locally, as may be demonstrated with labeled amino acids, but it is not known if it comes from the cytoplasm or the nucleoplasm.[31] Interestingly, there is no incorporation of ^{3}H-uridine during this short period; this appears at a later time and increases considerably, indicating RNA synthesis. These findings bring to the forefront the possible role of acidic proteins as one of the main factors in the control of genetic expression at the chromosome level.

In the formation of a complete puff the following processes are successively involved: accumulation of acidic protein, despiralization of DNA, synthesis of RNA, and storage of the newly synthesized RNA. It is assumed that cells contain a pool of these

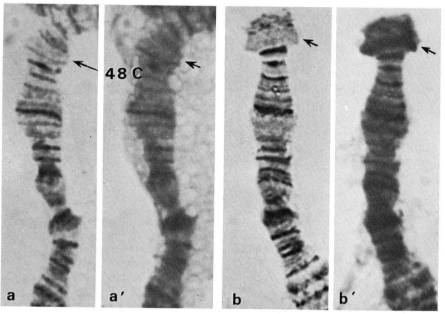

Figure 20–6. Polytenic chromosomes of *Drosophila hydei* showing the accumulation of an acidic protein in the presumed region of puff 48 C; **a** and **a'** correspond to the controls; **b** and **b'** photographed 3 minutes after a temperature shock induced by transferring the larvae from 25 to 37° C.; **a** and **b** photographed through a green filter to demonstrate the DNA bands; **a'** and **b'** photographed through a red filter to show the acidic proteins. Note that these acidic proteins have accumulated in 3 minutes. (Courtesy of H. D. Berendes.)

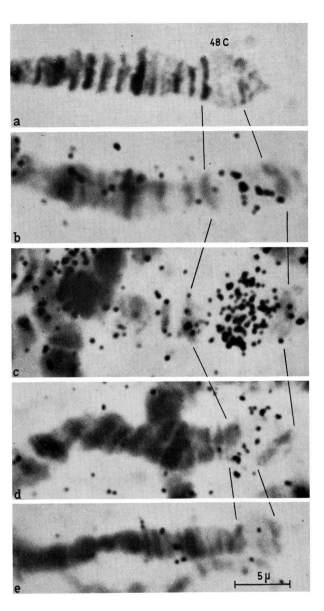

Figure 20–7. Experiment to demonstrate how ³H-uridine is incorporated into puff 48 C of *Drosophila hydei*. **a,** control; **b,** after 3 minutes of temperature shock (a few silver grains have been deposited); **c,** 15 minutes after temperature shock. Observe the greatly increased incorporation; **d,** 15 minutes after injection of actinomycin D in larvae having puffs as in **c; e,** 30 minutes after injection of the antibiotic in animals with puffs as shown in **c.** These results indicate that in 15 to 30 minutes most of the synthesized RNA has been eliminated from the puff. (Courtesy of H. D. Berendes.)

proteins which mobilize upon stimulation and which are able to recognize specific regions of the genome.[31]

By injecting actinomycin D at a time when the puff was fully developed and had accumulated considerable amounts of ³H-cytidine, it was possible to determine the time in which the synthesized RNA is released. As shown in Figure 20–7, *d* and *e,* most of this RNA was eliminated from the puff in 15 to 30 minutes.[31]

Puffs and Balbiani rings can be interpreted as sites in which genes may be active

in duplication (redundant DNA) or, more frequently, in transcription of specific RNA molecules. This genetic activity may be correlated with the production of polypeptides in the salivary glands and in other larval tissues in which polytenic chromosomes are present.

Molecular Model of Chromosomes

Some of the concepts that are analyzed in this section, along with the data presented in Chapter 12, have led to the postulation of

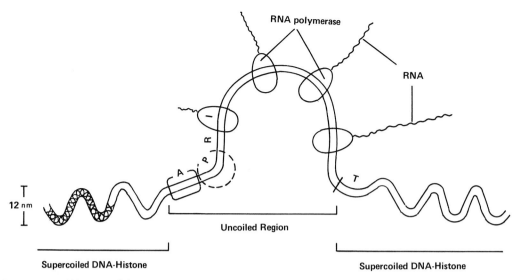

Figure 20–8. Diagram of the structure of a region of a eukaryotic chromosome undergoing transcription. A region of supercoiled DNA-histone becomes uncoiled and RNA is transcribed by RNA polymerase. (*A*) Address site closely related to the promoter (*P*) where the RNA polymerase first attaches to (*R*) regulatory and (*I*) initiation sites. (From Paul, J., *Nature* (London), *238*:944, 1972.)

molecular models for eukaryotic chromosomes (see Crick, 1971; Paul, 1972). In these models it is recognized that DNA is a continuous molecule throughout the chromatid and has regions with different degrees of condensation, as in the case of the polytene chromosomes. Crick (1971) proposes that the compact regions are involved in control and regulation, whereas the loose regions (i.e., fibrous DNA) are involved in coding.

According to Paul (1972) the nucleohistone forms a supercoiled structure of about 12 nm, which may actually be observed under the electron microscope (see Fig. 12–8). In this model, at the region to be transcribed, there are controlling elements similar to those found in bacteria, i.e., a *promoter, regulator sites,* and an *initiator site* (operator). In addition, an *address locus* has been postulated in which a polyanionic molecule — probably a non-histone protein — binds and produces uncoiling of the DNA. This uncoiling permits the approach of the RNA polymerase to the promoter and the start of the transcription. Once the RNA polymerase reaches the initiation site, it could cause further unwinding of the DNA by accumulation of the nascent RNA (also a polyanion) and of RNA binding protein.

A model such as that shown in Figure 20–8 accounts for the appearance of puffs in giant chromosomes and the loops in lampbrush chromosomes (see Paul, 1972).

Cell Division and Cell Differentiation

It is generally accepted that there is a certain antagonism between cell division and cell differentiation, but this is not always the case. Differentiation generally takes place during mitotic interphase, or after cell divisions have definitely ceased. A nerve cell loses not only its capacity to be transformed into other types of cells, but also its capacity to divide.

Interesting examples of the antagonism between cell differentiation and cell division are found in the so-called *stem cells.* It has been observed in various tissues that after one division, one of the two sister cells starts to differentiate and no longer divides, while the other remains undifferentiated and continues to divide. In the intestinal epithelium of the mouse the fate of the cell will apparently depend on the position of the cell in relation to the others in the same villus.[32]

The specific synthesis of a protein, such as myosin, apparently excludes the synthesis of DNA. At a definite moment there is a kind of "switch" in the cell,[33] by which one or the other synthetic pathway is selected.

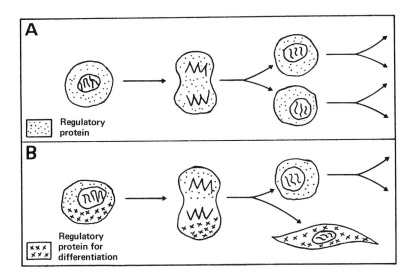

Figure 20–9. Diagram showing a possible mechanism for cell differentiation by regulatory proteins. **A**, the regulatory protein is divided equally between the two daughter cells; the cells continue to be undifferentiated; and **B**, the regulatory protein for differentiation accumulates on one side of the cell; one of the daughter cells differentiates. (See the description in the text.) (From Gurdon, J. B., "Gene expression during cell differentiation," in *Readings in Genetics and Evolution*, Oxford, The Clarendon Press, Chap. 4, 1973.)

Although the intimate mechanism of this selection is still unknown, it implies a choice of the genome either for replication or for transcription.

One of the explanations of the regulation of cell differentiation is illustrated in Figure 20-9. It is assumed that genes are regulated by their association with regulatory proteins produced in the cytoplasm and associated with chromosomes at the end of mitosis, thereby causing the reprogramming of the cell for the same activity. In the example shown in Figure 20–9, *A*, the sister cells will continue to be in the undifferentiated state.

In Figure 20–9, *B*, regulatory proteins accumulate on one side of the cell and are distributed unequally during mitosis because of some environmental factor that affects the two sides of the cell differently (e.g., the action of an embryonic inducer). The effect of this environmental factor is the production of an unspecialized stem cell and a differentiated cell that no longer divides. This type of mechanism, in which regulatory proteins are distributed unequally, is especially applicable to the embryonic development of eggs having a mosaic distribution of cytoplasmic material (see Gurdon, 1973).

Eukaryotic cells have the same basic molecular mechanisms of genetic information, transduction, and translocation found in bacteria and viruses. The main difference, however, is in the much more complex way in which gene regulation takes place. The genome of a eukaryon is much larger than that found in a bacterium and it is divided into several chromosomes, which should be in proper balance to produce normal development. A large proportion of the DNA of eukaryons is *redundant* (i.e., repetitious), and is formed by repeated nucleotides sequences. *DNA redundancy* is demonstrated by techniques involving denaturation and reassociation of the DNA strands. In the mouse 10 per cent of the DNA is formed of about 1,000,000 copies of short nucleotide sequences; 20 per cent, of longer repeated sequences; and 70 per cent, of single copies. DNA redundancy may produce gene amplification, as in the case of genes for the 28S and 18S ribosomal RNA located in the nucleolar organizer. The 5S

SUMMARY:
Regulation of Genetic
Information

ribosomal RNA and the 4S transfer RNA are also repetitious and are scattered throughout the genome. The cytoplasmic DNA of mitochondria and chloroplasts is also redundant. In most cases, the function of repetitious DNA is unknown. Note that the relationship between DNA redundancy, heterochromatin, and satellite DNA was discussed in Chapter 12.

In eukaryons, at any given time, most of the DNA is not genetically functioning. The condensed chromatin found in heterochromatin, metaphase chromosomes, and the sperm is virtually inactive in transcription. Histones, because of the close association with DNA, may inhibit transduction, but the effect seems to be rather nonspecific. A more specific role is ascribed to nuclear acidic proteins and to RNA. In eukaryons, *derepression* of inactive (i.e., repressed) regions of the genome seems to be acting at the transcription level. According to some investigators the excess nuclear heterogeneous RNA may play a role in regulation. It was pointed out in Chapter 19 that most of this RNA is degraded inside the nucleus.

Regulation of gene expression frequently occurs post-transcriptionally at different steps of the translation. One interesting example is provided by the action of a steroid hormone on tyrosine-aminotransferase of hepatoma cells. In polytene chromosomes of insect larvae, the expression of gene activity is observed in the so-called *puffs* and *Balbiani rings*. At the puff there is uncoiling of the chromosome bands, followed by synthesis of RNA and accumulation of protein. Puff formation is stimulated by the steroid hormone *ecdysone,* which becomes bound to the sites of puffing. Experimenters who have used temperature shock to induce certain puffs have produced rapid accumulation of an acidic protein, followed by RNA synthesis. In some puffs there may be DNA duplication, but more frequently they are engaged in the synthesis of specific RNA molecules. The presence of a certain polypeptide in the secretion of the salivary glands has been correlated with the activity of a single puff. Molecular models of chromosomes based on some of these concepts have been postulated.

A certain antagonism exists between cell division and cell differentiation. In the case of "stem" cells, after mitosis one cell differentiates and no longer divides, while the other remains undifferentiated. An explanation based on the production of regulatory proteins and their cyclic association with chromosomes is presented. If the regulatory protein is divided equally, the two sister cells continue to be undifferentiated and divide. If the regulatory protein for differentiation is divided unequally, one sister cell may remain as a stem cell and the other may undergo differentiation. During mitosis, therefore, there is a cyclic reprogramming of gene activity.

Renewal of Cell Populations

Estimation of the number of cells in a tissue may be made by determining the tissue's DNA content (see Chapter 17). By analyzing the DNA content of an entire rat from the embryo to the adult, one can calculate that at 10 days before birth, i.e., at 12 to 13 days of embryonic life, the organism contains 50 million cells, at birth 3 billion, and at 90 days after birth, 67 billion.[34] With this method one can also calculate the *increase in the number of nuclei per day* and the amount of material associated with one nucleus. This index, called *weight per nucleus* (organ weight divided by the number of nuclei), increases with the enlargement of the cell or with accumulation of intercellular material (Fig. 20–10).

The *mitotic rate* of a tissue may be calculated after injection of *colchicine,* which arrests cell division at metaphase. However, the best method for studying dividing cells is radioautography with ³H-thymidine.[35] In a 16 day old rat embryo such a method reveals that most cells are dividing, and that only some nerve cells have stopped dividing at that time. Since the rat embryo contains 50 million cells 10 days prior to birth, assuming that the cell divisions are regular and without cell loss, one can calculate that there have been 25 to 26 cell generations with a half day division time. In other words, in the early embryo the number of cells should double every half day. In the last 10 days of embryonic life the doubling time increases to 1.7 days.

Classification of Cell Populations

Studies of growth and renewal of cell populations during postnatal life has led to the classification of the different cell types into three groups:[35]

Static Cell Populations. These are *homogeneous groups of cells in which no mitotic activity can be detected* and in which *the total DNA content remains constant.* Only the nerve cells were found not to divide after the age of seven days. In the nervous system not all the cells belong to the static population. In fact, some neuroglial cells and

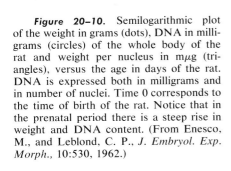

Figure 20–10. Semilogarithmic plot of the weight in grams (dots), DNA in milligrams (circles) of the whole body of the rat and weight per nucleus in mµg (triangles), versus the age in days of the rat. DNA is expressed both in milligrams and in number of nuclei. Time 0 corresponds to the time of birth of the rat. Notice that in the prenatal period there is a steep rise in weight and DNA content. (From Enesco, M., and Leblond, C. P., *J. Embryol. Exp. Morph.,* 10:530, 1962.)

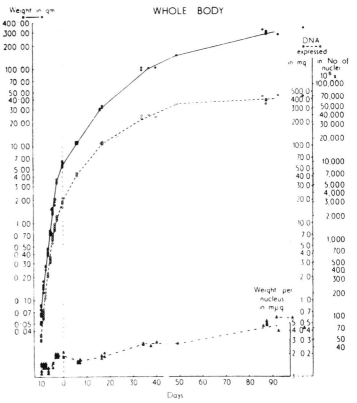

subependymal cells of young animals may be labeled with thymidine. One of the most important characteristics of static cells is the constant increase in cell volume that runs parallel to the growth of the organism and is proportional to the total volume of the body.

Expanding Cell Populations. These are *homogeneous groups of cells showing scattered mitoses in numbers that account for the increase in the total DNA content.* This implies that the life of each cell continues for as long as that of the individual and that new cells are produced by division only to cover the growth of the tissue. By labeling the cells with ³H-thymidine, it can be observed that a few scattered cells are undergoing mitosis. Contrary to earlier theories, the dividing cells are fully differentiated. Examples of expanding cell populations are found in: the pancreas, the thyroid, the kidney, the adrenal, the salivary glands, and muscle.

One impressive characteristic of expanding cells is that the mitotic index increases dramatically with adequate stimuli. For example, after partial extirpation of the liver, or unilateral nephrectomy, there is a rapid regeneration. It may be assumed that the liver normally produces a substance that inhibits its own growth. Partial extirpation would reduce the number of circulating inhibitors, thus allowing for more cell divisions.

Renewing Cell Populations. These are *homogeneous groups of cells in which mitosis is abundant and exceeds that required for the total increase in DNA content.* In renewing populations, the high production of cells is balanced by a corresponding cell loss. Red blood cells, which have an average life of 120 days, are good examples. The rapid renewal of cells in the intestine can be demonstrated with ³H-thymidine (see Figure 6–22). In the epidermis cells are renewed rapidly to replace the dead ones that are shed at the surface. The renewing process is even more marked during wound healing.

Other renewing populations are the testis (which continuously sheds spermatozoa), the lung (which produces "dust cells"), the thymus, the lymph ganglia, the bone marrow, and all hematopoietic organs in general.

CELLULAR INTERACTION

Intimately related to the molecular biology of cell differentiation is cellular inter-

action. In a multicellular organism the cells of the different tissues cooperate in a harmonious way, and among them there is a high degree of functional interdependence. Cells may interact over long distances by way of diffusion of some chemical substances, but in many cases their interaction depends on a short range action, as by *cell contact.*

During embryonic development there is considerable cell motility and mass movement of cells. One of the most interesting examples of cell migration is the case of the germinal cells. In avian embryos these cells differentiate in the extraembryonic endoderm while the rest of the gonad develops in the embryonic mesodermal germinal ridge. The germinal cells enter the blood stream and are carried through all parts of the embryo, but they selectively adhere and accumulate at the germinal ridges. It may be postulated that there is a *specific recognition* between these cells. This sort of recognition is of paramount importance in the development of the nervous system, in which millions of neurons must find their specific partners to establish synaptic junctions.

Cell Adhesion

In order to aggregate, cells must first come in contact with one another. There are three main types of contact between cells which lead to their aggregation in a tissue:[36]

1. *The aggregation may form by the inclusion of the cell in a common matrix.* In most cases this extracellular matrix consists of the *extraneous coat* of the cell. These substances (cellulose, hyaluronic acid, and so forth) accumulate after the cells have made contact with each other.

2. *The aggregation of cells may have little or no demonstrable intercellular material.* Electron microscopy shows the gap between cells to be frequently on the order of 15 nm (Fig. 8–3). The possibility that such close contacts are due to the physical properties of the membrane (i.e., attractive and repulsive forces) has been considered.[37] However, electron microscopy favors the view that even in these small intercellular spaces there is some substance, possibly a mucoprotein, that accounts for the specificity of cell association. Furthermore, the enzymes that are more effective in separating cells are proteases and mucases.

3. *The aggregation may imply the presence of intercellular channels.* Cells in a tissue use these channels to interchange more or less freely. In plant cells, cytoplasmic bridges or plasmodesmata (Fig. 2–6) have been recognized for a long time. These provide narrow connections between the cells and across the cellulose walls and permit the free passage of ions and probably macromolecules.

Intercellular Communications

By introducing intracellular microelectrodes into adjacent cells of a tissue investigators have been able to demonstrate that in many animal cells there are intercellular communications.[38] In this case the cells are electrically coupled and have regions of low resistance in the membrane through which there is a rather free flow of electrical current carried by ions.[39] The other parts of the cell membrane, which are not coupled, show a much higher resistance. This type of coupling, called *junctional communication*, is found extensively in embryonic cells.[40] In adult tissue it is usually found in epithelia,[41] cardiac cells, and liver cells.[42] Skeletal muscle and most neurons do not show electrical coupling.

The giant cells of the salivary glands of *Chironomus* are excellent for demonstrating electrical coupling.[43] Between these cells there are *septate junctions* that establish a direct communication between two adjacent cell membranes. In most other junctional communications, the so-called *gap junctions,* mentioned in Chapter 8, are found.

The essential components of the coupling between two cells are represented in the diagram of Figure 20–11. Each passageway unit consists of a pair of membrane regions of high permeability (*junctional membranes*) circumscribed by a kind of seal (*junctional seal*) that insulates the junctional region from the extracellular space.

It was mentioned above that ions freely traverse the junctional communications. Fluorescent dyes are frequently used to trace them. After microinjection of cultured cells it is possible to observe the passage of the dye into adjacent cells. In *Chironomus* the cell to cell passageways have been traced by the injection of *peroxidase.* This enzyme traverses the septate junctions and gives a

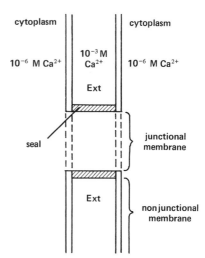

Figure 20–11. Diagram showing the essential components of the coupling between cells. (See the description in the text.) (From Loewenstein, W. R., *Fed. Proc., 32*:60, 1973.)

reaction in the adjacent cells. Molecules as large as 10,000 daltons go through this type of communication (see Loewenstein, 1973).

Junctional communications can be formed in a matter of seconds when two embryonic cells are pressed together. This formation is dependent on the presence of Ca^{2+} ions in the extracellular space. Normally, the gradient of Ca^{2+} is $10^{-3}M$ on the outside, to $10^{-6}M$ on the inside (Fig. 20–11). Uncoupling results from injecting Ca^{2+} into the cell or from cell injury.

It has been found that the junctional communications depend on the energy provided by oxidative phosphorylation (i.e., ATP). Treatments that inhibit cell metabolism, such as cooling to 8° C., dinitrophenol, cyanide, and olygomicin, produce uncoupling between the cells; this may be reversed by injection of ATP. It is thought that the uncoupling may be due to the influx of Ca^{2+}, either from the medium or from mitochondria, during the action of the inhibitors.[44]

Functions of Intercellular Communications

Several functions may be ascribed to junctional communications. In cardiac muscle, and in electrical synapses between certain neurons they are related to communication of electrical signals between these cells. Junctional interconnections are particularly

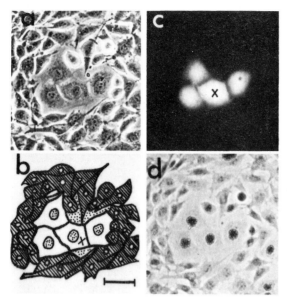

Figure 20–12. Experiment that demonstrates the lack of coupling between normal cells and cancerous cells. **a,** phase contrast micrograph of a culture having four normal liver cells surrounded by cancerous cells; **b,** tracing of micrograph in **a; c,** the cell marked with X was injected with fluorescein; it is observed that the stain has diffused into the three other normal cells, but not into the cancerous ones; **d,** since the fluorescein is labeled with tritium after the radioautograph, only the four normal cells are ^{3}H-labeled. (From Azarnia, R., and Loewenstein, W. R., *J. Membr. Biol.,* 6:368, 1971.)

widespread during embryonic development at stages in which cell differentiation takes place. Because diffusion of molecules can readily take place through them, these junctions are well suited for the dissemination of signals controlling cellular growth and differentiation at close range (see Loewenstein, 1973).

An interesting finding has been that certain cancer cells show no intercellular coupling and that they fail to communicate with normal cells[45] (Fig. 20–12). It is assumed that these cancer cells have a genetic defect which has interrupted the passage of growth-controlling molecules between them. Recent experiments in which heterokaryons have been produced between a cancer cell and a normal cell (i.e., cell fusion) have shown that it is possible to obtain a non-cancerous hybrid.[46] In this hybrid the intercellular communications, which were absent in the original cancerous cell, were established between the hybrid cells.[47] These and other experiments

point to the possible regulatory role of molecules of small molecular weight — probably nucleotides — between cells controlling growth and differentiation.[48] Such regulatory agents could not be transferred to the cancerous cells having no intercellular communication.

Cell Dissociation and Reassociation

In 1908 Wilson described how living sponges forced through a fine silk mesh disaggregate into isolated motile cells, and then, upon standing, reaggregate to form fresh sponges. Much later it was found that embryonic tissues treated with trypsin dissociate into individual cells and then reaggregate to form the specific patterns of the original tissue.[49]

The process of reaggregation depends on the motility of cells. If all the cells of a chick embryo are dissociated with trypsin and allowed to stand, they will reaggregate. When similar cells attach to each other, they form aggregates that are characteristic for a given cell population, e.g., retinal cell, kidney or bone.[50]

The process of reaggregation is not species specific. If cells of chick and mouse embryos are mixed, they reaggregate according to the cell population rather than the species. Reaggregation does not take place in the absence of calcium ions or if the treatment used to disaggregate the tissue selectively blocks certain carbohydrates and glycoproteins. Reaggregation of retinal cells is inhibited by puromycin.[51]

The mechanism by which a cell can "recognize" and aggregate with another of similar kind takes place at the cell surface, but its intimate nature is unknown. It has been suggested that the surface of the cell is highly ordered in the tangential direction and that this is reflected in the spatial organization of ionized acid groups that bind Ca^{2+} and Mg^{2+}.[52] The surface material of the adjacent cells may have a molecular fit of reactive groupings.

In Chapter 8 it was noted that glycoproteins and glycolipids present in the cell coat are involved in the "molecular" recognition of cells of similar type and that immunochemical interactions have been postulated.[53] This property is gene-dependent and is expressed at the cell surface. Molecular recognition is highly developed between neurons of the central nervous system where specific neuronal circuits of immense complexity are formed (see Singer and Rothfield, 1973).

Contact Inhibition and Cancer

Certain cancerous cells lack intercellular communications. This property, however, is not found in all malignant cells. Another interesting phenomenon is the absence of the so-called *contact inhibition*. This refers to *the inhibition of cell motility and* also of *mitotic activity* in normal cells grown in tissue culture at the time they come into contact.[54] This is frequently observed in cultures growing on a solid support such as a glass surface or a millipore. As long as cells float freely in the nutrient medium, they generally divide every 24 hours. However, when they come in close contact in a monolayer, the rate of mitosis slows down and there is inhibition of cell division. This process can be easily observed when ^{3}H-thymidine incorporation is used as a measure of the rate of DNA synthesis. The inhibition depends on some unknown signal between cells in contact, and not on a diffusible substance acting at a distance.

In cancer cells, the mitotic rate is not inhibited and in cultures the cells tend to pile up, forming irregular masses several layers deep. These cells show less adhesion to the solid support or among themselves and motility is more pronounced. These properties may explain why neoplasms invade other tissues and follow an uncontrolled growth. The loss of contact inhibition is easily studied in normal cultured cells that are "transformed" into cancerous cells by oncogenic viruses (i.e., viruses capable of inducing cancer) such as the polyoma or the SV40 (i.e., simian virus 40). Changes in the surface properties of these cells have been observed; for example, using specific staining procedures for mucopolysaccharides (i.e., Hale's reaction or ruthenium red) a considerable increase of this substance was discovered in cells infected with polyoma virus.[55, 56] These cells also showed an increased electrophoretic mobility.[57] Both the staining properties and the mobility are reduced by the action of neuraminidase, an enzyme that splits sialic acid (neuraminic acid) from glucoproteins. It was observed under the electron microscope that the gap junctions tend to disappear in cells transformed by oncogenic viruses. This may explain the lack of electrical coupling and contact inhibition between cancer cells mentioned earlier.

The changes in the surface of the cell may also be studied by using concavallin coupled with ferritin. Under the electron microscope it is possible to observe, in an isolated cell membrane, that the cancer-transformed cell has a much larger number of receptors for concavallin than the normal cell.[58]

Virus and Cancer

In recent years considerable progress has been made on the mechanisms by which a normal cell may be transformed into a cancerous cell that shows all the characteristics just mentioned. More than 50 years ago Peyton Rous demonstrated that a chicken sarcoma could be reproduced by inoculation of a free cell extract. Viruses that produce leukemia and other tumors in mice and birds belong to the so called *myxoviruses* which have a single-strand chain of RNA in their genome.

The small viruses of polyoma and SV40 are the most often used in cell transformation. Polyoma virus has a circular double-stranded DNA genome of only 1.6 μm in length (Fig. 20–13) and a capsid with 72 capsomeres. When added to a culture of normal cells, the virus multiplies and kills only a small proportion of the cells. Most of the cells become transformed into cancerous cells and the virus apparently disappears. There are many indications that the DNA of the virus becomes integrated into the genome of the cell in regions in which there is some genetic homology. This can be demonstrated by studies of hybridization between the DNAs of the two genomes.

The viral genome contains only 5000 nucleotides – enough information to codify for only a few proteins. For example, the genes to code for DNA replication, for the protein of the capsid, for an internal protein and for two antigens are present in the polyoma DNA.

With cell transformation, the host cell shows some of the changes that were just mentioned, i.e., lack of control of cell division and of cell motility, and changes in cell contact and cell surface. Other changes that may

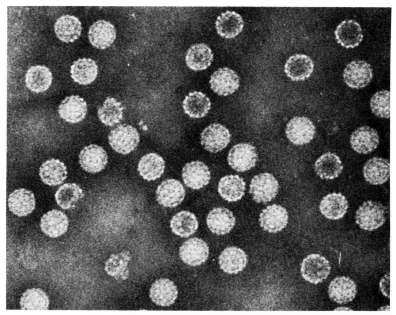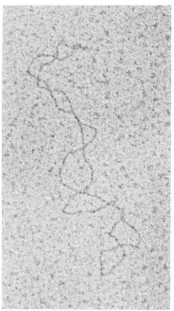

Figure 20–13. **Left,** electron micrograph of polyoma virus. The virus has a diameter of 45 nm with 72 subunits in the capsid; **Right,** polyoma DNA, which appears as a twisted filament. This DNA has a length of 1.6 μm, a sedimentation constant of 20S, and a molecular weight of 3×10^6 daltons. In both micrographs the magnification is approximately ×150,000. (Courtesy of C. Vásquez.)

be demonstrated by fluorescent antibodies are the appearance of the two antigens that are specific of the virus—one in the nucleus, the other in the cell membrane.

The use of oncogenic viruses in cancer research is of fundamental importance in reaching a better understanding of the molecular biology of cancer—an understanding that is prerequisite to definite progress in its treatment.

SUMMARY:
Intercellular Interactions

Cellular interaction is intimately related to the molecular biology of cell differentiation. Specific cell contacts imply short range chemical interactions between cells. Such interactions are of great importance in embryogenesis for *specific recognition* of cells in a tissue (e.g., case of germinal cells).

Cell adhesion may result in the inclusion of similar cells in a common extracellular matrix or in the production of intercellular channels, as in the plasmodesmata of plant cells. In many tissues there is a space of 15 nm between adjacent cell membranes that contains mucoproteins.

Special *intracellular communications* are found in many animal cells, especially in the embryo. Adjacent cells are *electrically coupled*, i.e., there is current flow in regions of low resistance. Morphologically, *septate* and *gap* junctions are present in these junctional communications. Each passageway unit consists of a pair of membrane regions of low resistance and a junctional seal that insulates the junctional region from the extracellular space. Ions, fluorescent dyes, peroxidase, and other molecules up to 10,000 daltons may traverse these junctional regions. These communications are produced only in the presence of a Ca^{2+} gradient, and they require chemical energy for their maintenance. Junctional communication may convey electrical signals between certain neurons (i.e., electrical synapses) and in cardiac cells; however, most neurons and skeletal muscle lack electrical coupling. Junctional communications are probably used for the transfer of substances that control growth and differentiation in cells.

Some cancerous cells show no intercellular coupling, but they may regain it by fusion with normal cells (producing hybrids). Experiments on cell dissociation and reaggregation show that cells of a given tissue "recognize" each other. Glycolipids and glycoproteins (including surface antigens) are probably involved in cell to cell recognition.

Contact inhibition refers to the inhibition of cell motility and mitotic activity that is found in normal cultured cells. Cancerous cells do not show contact inhibition. In them, there are changes in the cell coat that may be observed with cytochemical methods and the electron microscope. Normal cultured cells, transformed by oncogenic viruses (i.e., polyoma or SV40), lose the contact inhibition. The DNA of polyoma is a double-stranded circular DNA of 1.6 μm containing genetic information to code for a few proteins. The viral genome becomes integrated in certain regions of the lost genome that have some genetic homology. The viral genome expresses itself in the many changes that are characteristic of a cancerous cell. Among these is the appearance of two antigens (one at the nucleus, the other at the cell membrane) that are characteristic of the virus.

REFERENCES

1. Fonbrune, P. (1949) *La Technique de Micromanipulation.* Masson & Cie., Paris.
2. Lorch, I. J., and Danielli, J. F. (1960) *Nature, 166:*329.
3. Goldstein, L., Cailleau, R., and Crocket, T. T. (1960) *Exp. Cell Res., 19:*332.
4. Hämmerling, J. (1963) *Ann. Rev. Plant Physiol.,* p. 14.
5. Keck, K. (1963) The nuclear control of synthetic activities in *Acetabularia. Proc. XIV Internat. Cong. Zool. (Wash.), 3:*203.
6. Hämmerling, J. (1953) *Int. Rev. Cytol., 2:*475.
7. Gurdon, J. B., and Woodland, H. R. (1968) *Biol. Rev. 43:*233.
8. Harris, H. (1967) *J. Cell Sci., 2:*23.
9. Mazia, D. (1966) Biochemical aspects of mitosis. In: *The Cell Nucleus: Metabolism and Radiosensibility,* p. 15. Taylor and Francis, Ltd., London.
10. Himes, M. (1967) *J. Cell Biol., 35:*175.
11. Lebleu, B., Marbaix, G., Huez, G., Temmerman, J., Burny, A., and Chantrenne, H. (1971) *Europ. J. Biochem., 19:*264.
12. Gurdon, J. B., Lane, C. D., Woodland, H. R., and Marbaix, G. (1971) *Nature* (London), *233:*177.
13. Lane, C. D., Marbaix, G., and Gurdon, J. B. (1971) *J. Molec. Biol., 61:*73.
14. Britten, R. J., and Kohne, D. E. (1968) *Science, 161:*529.

15. Waring, M., and Britten, R. J. (1966) *Science, 154*:791.
16. Martin, M. A., and Hoyer, B. H. (1966) *Biochemistry, 5*:2706.
17. Frenster, J. H., Allfrey, V. G., and Mirsky, A. E. (1963) *Proc. Natl. Acad. Sci. USA, 50*:1026.
18. Paul, J., and Gilmour, R. S. (1968) *J. Molec. Biol., 34*:305.
19. Allfrey, V. G. (1968) *Excerpta Medica Internat. Cong. Ser., 166*:28.
20. Pogo, B. G., Allfrey, V. G., and Mirsky, A. E. (1966) *Proc. Natl. Acad. Sci. USA, 55*:805.
21. Weissmann, G., and Hirschhorn, R. (1971) In: *Advances in Cytopharmacology:* Proceedings, Vol. 1, p. 191. (Clementi, F., and Ceccarelli, B., eds.) Raven Press, New York.
22. Comings, D. E. (1967) *J. Cell Biol., 35*:669.
23. Bekhor, I., Kung, C. M., and Bonner, J. (1969) *J. Molec. Biol., 39*:351.
24. Huang, R. C. C., and Huang, P. C. (1969) *J. Molec. Biol., 39*:365.
25. Beerman, W. (1961) *Chromosoma, 12*:1.
26. Beerman, W. (1962) *Protoplasmatologia* (04). Springer-Verlag, Vienna.
27. Pavan, C., and Breur, M. (1955) *Symp. Cell Secretion. Bello Horizonte,* p. 90.
28. Pelling, G. (1959) *Nature* (London), *184*:655.
29. Sirlin, J. L. (1960) *Exp. Cell Res., 19*:177.
30. Clever, U., and Karlson, P. (1960) *Exp. Cell Res., 20*:623.
31. Berendes, H. D. (1968) *Chromosoma, 24*:418.
32. Quastler, H., and Sherman, F. (1959) *Exp. Cell Res., 17*:420.
33. Abbott, J., and Holtzer, H. (1966) *J. Cell Biol., 28*:473.
34. Enesco, M., and Leblond, C. P. (1962) *J. Embryol. Exp. Morph., 10*:530.
35. Leblond, C. P. (1964) *J. Natl. Cancer Inst., 14*:119.
36. Harris, R. J. C. (1961) In: *Cell Movement and Cell Contact,* Internat. Soc. Cell Biol. (Harris, R. J. C., ed.) Academic Press, Inc., New York.
37. Curtis, A. S. G. (1962) *Biol. Rev., 37*:82.
38. Loewenstein, W. R., and Kanno, Y. (1964) *J. Cell Biol., 22*:565.
39. Loewenstein, W. R. (1967) *Develop. Biol., 15*:503.
40. Potter, D. D., Furshpan, E. J., and Lennox, E. S. (1966) *Proc. Natl. Acad. Sci. USA, 55*:328.
41. Loewenstein, W. R. (1966) *Proc. N. Y. Acad. Sci., 137*:441.
42. Penn, R. D. (1966) *J. Cell Biol., 29*:171.
43. Oliveira-Castro, G. M., and Loewenstein, W. R. (1971) *J. Membr. Biol., 5*:51.
44. Politoff, A. L., Socolar, S. J., and Loewenstein, W. R. (1969) *J. Gen. Physiol., 53*:498.
45. Azarnia, R., and Loewenstein, W. R. (1971) *J. Membr. Biol., 6*:368.
46. Harris, H. (1971) *Proc. Roy. Soc. Lond., Ser. B., 179*:1.
47. Azarnia, R., and Loewenstein, W. R. (1973) *Nature* (London), *241*:455.
48. Subak-Sharpe, H., Buck, P., and Pitts, T. D. (1969) *J. Cell Sci., 4*:353.
49. Moscona, A. (1957) *Proc. Natl. Acad. Sci. USA, 43*:184.
50. Moscona, A. (1962) *J. Cell Comp. Physiol., 60*:65.
51. Moscona, M., and Moscona, A. (1963) *Science, 142*:1070.
52. Steinberg, M. S. (1962) *Exp. Cell Res., 28*:1.
53. Burnet, F. M. (1961) *Science, 133*:307.
54. Abercrombie, M. (1966) *Conference on Tissue and Organ Culture,* p. 249. Bedford, Pennsylvania.
55. Defendi, V., and Gasic, G. (1963) *J. Cell. Comp. Physiol., 62*:23.
56. Martinez-Palomo, A., and Brailowsky, C. (1968) *Virology, 34*:379.
57. Forrester, J. A., Ambrose, E. J., and Stoker, M. (1964) *Nature, 201*:945.
58. Nicolson, G. L. (1971) *Nature* (New Biol.), *233*:244.

ADDITIONAL READING

Abercrombie, M. (1967) Contact inhibition. The phenomena and its biological implications. *Natl. Cancer Inst. Monogr., 26*:249.
Brachet, J. (1967) Exchange of macromolecules between nucleus and cytoplasm. *Protoplasma, 63* (1–3):86.

Brachet, J. (1973) Embryologie moléculaire et differentiation cellulaire. *La Recherche, 3*:43.

Britten, R. J., and Davison, E. H. (1969) Gene regulation for higher cells: a theory. *Science, 165*:349.

Chantrenne, H., and Marbaix, G. (1972) Traduction des "ARN 9S" messagers de l'hémoglobine, dans des systèmes hétérologues. *Biochimie, 54*:1.

Crick, F. (1971) General model for chromosomes of higher organisms. *Nature* (London), *234*:25.

Fan, H., and Penman, S. (1970) Regulation of protein synthesis in mammalian cells. *J. Molec. Biol., 50*:655.

Flamm, W. G. (1972) Highly repetitive sequence of DNA in chromosomes. *Int. Rev. Cytol., 32*:2.

Grobstein, C. (1967) Mechanisms of organo-genetic tissue interaction. *Natl. Cancer Inst. Monogr., 26*:279.

Gurdon, J. B. (1968) Nucleic acid synthesis in embryos and its bearing on cell differentiation. *Essay in Biochemistry, 4*:26.

Gurdon, J. B., and Woodland, H. R. (1968) The cytoplasmic control of nuclear activity in animal development. *Biol. Rev., 43*:233.

Gurdon, J. (1973) Gene expression during cell differentiation. In: *Readings in Genetics and Evolution.* Chap. 4, Oxford Biology Readers, Oxford University Press, New York.

Harris, H., Sidebottom, E., Grace, D. M., and Bramwell, M. E. (1969) The expression of genetic information. *J. Cell Sci. 4*:449.

Loewenstein, W. R. (1973) Membrane functions in growth and differentiation. *Fed. Proc., 32*:60.

McCormick, W., and Penman, S. (1969) Regulation of protein synthesis in HeLa cells. *J. Molec. Biol., 39*:315.

Paul, J. (1972) General theory of chromosome structure and gene activation in eukaryotes. *Nature* (London) *238*:444.

Tomkins, G., Gelehrter, T. D., Granner, D., Martin, D., Samuels, H. H., and Thompson, E. B. (1970) Control of specific gene expression in higher organisms. *Science, 166*:1474.

Wagner, R. P., ed. (1969) Nuclear physiology and differentiation. *Genetics, 61*:No. 1.

Part Eight

CELL PHYSIOLOGY

The following five chapters bring together some of the most important functions of the cell that have not been discussed in previous chapters. The intimate relationship of these physiologic processes with the structure and chemical organization of the cell are discussed.

In Chapter 21 under the heading Cell Permeability are grouped all the processes by which the cell regulates the entrance and exit of different ions and molecules. The important concepts of passive diffusion and active transport and their relationships to the osmotic and ionic concentration of the cell and to the membrane potentials are introduced. The existence of uncharged and charged pores in the membrane is postulated, as is the concept of permease systems that may serve as the specific chemical transport through the membrane. The importance of the ATPases in ionic transport is emphasized. In addition to the entrance of molecular material, there are less specific mechanisms of bulk ingestion of solids and fluids generically designated endocytosis, which includes phagocytosis and pinocytosis. These processes are intimately related to digestion, to the lysosome, a particle that contains hydrolytic enzymes, and to the peroxisomes, which contain some oxidases. The reader should recognize the importance of lysosomes in the functioning of the cell and in many pathologic conditions.

Chapter 22 is dedicated to the more primitive types of cell motion, including movement of cilia and flagella, cytoplasmic streaming or cyclosis, and ameboid movement. The elaborate ultrastructure of cilia and of the basal bodies, which are similar to centrioles, is described. The biochemistry of cilia comprises an important section, and the concepts explored here serve as an introduction to the study of ciliary movements, ciliary derivatives, and new knowledge concerning the origin and multiplication of cilia.

The microtubules and microfilaments are considered in this chapter as fibrillar structures of the cytoplasm of most cells because of their possible implication in cell contraction and motion, circulation and transport of products, and cell shape and differentiation. Because of these correlations, all the above processes and structures will be studied in an integrated fashion.

In Chapter 23 the study of muscle presents an extraordinary example of macromolecular machinery adapted to the work of contraction, a phenomenon that can be explained as resulting from the interaction of fibrous protein molecules—actin, myosin, and tropomyosin—which are recognizable with the electron microscope in the intimate structure of the muscle fiber. Also, on the structural base of the sarcoplasmic reticulum it is possible to explain conduction of action potentials to the interior of the muscle fiber and the functional synchronization of myofibrils. Muscle is an admirable example of physiologic and structural integration comparable only to that of mitochondria and chloroplasts.

Chapter 24 introduces the cellular bases of nerve conduction and synaptic transmission. Both muscle and nerve tissues conduct impulses by way of action potentials, but nerve tissue is specially adapted to receive stimuli, to transmit impulses acrosss the synapse, and to induce a response at the effectors. Special emphasis is placed on nerve conduction and synaptic transmission. In unmyelinated fibers conduction implies a change in membrane potential which produces the nerve impulse; this is conducted without decrement in an *all-or-none* fashion. In myelinated fibers the action potential apparently jumps from one node to the next (saltatory conduction) and is conducted electrotonically along the internode. The ultrastructure of the axon, the biosynthetic properties of the perikaryon, and the nature of axon flow are emphasized.

Synaptic transmission can be electrical, but more frequently it is chemical; this implies a neurochemical mechanism and the production of transmitter substances at the nerve ending. Both types of transmission have a structural foundation. Chemical synapses have a complex structure, both at the membranes and at the endings. The main presynaptic component is represented by the synaptic vesicles, the true quantal units of the transmitter. Nerve endings and synaptic vesicles have been isolated by cell fractionation techniques, and their content in transmitters and related enzymes has been studied. In general terms, in chemical synapses a localized process of neurosecretion takes place that is similar to the production of other neurohumors. Work on the separation of nerve-ending membranes, junctional complexes, and receptor proteins is briefly presented.

Chapter 25 is a discussion of cell secretion—a process that is highly developed in numerous kinds of cells. The study of the secretory cycle is interesting because it implies the coordinated intervention of all cellular components both in time and in space. This chapter is a review of the numerous processes and structures that have been studied in previous chapters.

It will be shown that cell fractionation techniques, the use of labeled substances, and the electron microscope are now permitting an integrated study of the synthesis of the secretion products and of their passage through the vacuolar system to ultimate expulsion from the cell. The adrenal, the pancreatic and the parotid cells are used as central examples simply because the several stages of the secretory cycle are most characteristic and best studied in them.

twenty-one

CELL PERMEABILITY, ENDOCYTOSIS, LYSOSOMES, AND PEROXISOMES

Permeability is fundamental to the functioning of the living cell and to the maintenance of satisfactory intracellular physiologic conditions. This function determines which substances can enter the cell, many of which may be necessary to maintain its vital processes and the synthesis of living substances. It also regulates the outflow of excretory material and water from the cell.

The study of permeability should be based on an analysis of the chemical and molecular organization of the cell membrane (see Chapter 8).

The presence of a membrane establishes a net difference between the *intracellular* fluid and the *extracellular* fluid in which the cell is bathed. This may be fresh or salt water in unicellular organisms grown in ponds or the sea, but in multicellular organisms the internal fluid, i.e., the blood, the lymph, and especially the *interstitial* fluid, is in contact with the outer surface of the cell membrane.

Osmotic Pressure and Physiologic Solutions

One of the functions of the cell membrane is to maintain a balance between the osmotic pressure of the intracellular fluid and that of the interstitial fluid. When plant cells are placed in a solution that has an osmotic pressure similar to that of the intracellular fluid, the cytoplasm remains adherent to the cellulose wall. When the solution of the medium is more concentrated, the cell loses water, and the cytoplasm retracts from the rigid cell wall. On the other hand, when the solution of the medium is less concentrated than the intracellular fluid, the cell swells and eventually bursts.

At the end of the last century, Hamburger demonstrated that maintenance of osmotic pressure plays an important role in the life of the cell. He found that the cell membrane behaves like an osmotic membrane and that a solution of 0.9 per cent sodium chloride maintains mammalian erythrocytes intact, whereas in less concentrated solutions they are hemolyzed. In a medium of higher concentration, the erythrocytes retract, owing to loss of water.

From the biological viewpoint, solutions can be grouped into three classes: (1) *Isotonic* solutions, which have the same osmotic pressure as that of the cells. For example, 0.3 M solutions of nonelectrolytes are isotonic in relation to mammalian cells. (2) *Hypotonic* solutions, which have a lower osmotic pressure than that of the cells. For example, a 0.66 per cent solution of sodium chloride, which is isotonic for amphibian erythrocytes, is hypotonic for mammalian cells. (3) *Hypertonic* solutions, which have a higher osmotic pressure than that of the cells.

These findings led to the adoption of *physiologic solutions,* which have a total osmotic pressure the same as the blood of animals, and a balanced concentration of different ions. Examples of physiologic solutions are Ringer's and Tyrode's solutions.

In higher organisms, the osmotic pressure of the body as a whole is regulated principally by the kidneys, and the osmotic pressure of the interstitial fluid is about the same as that of the intracellular fluid.

In plants, the intracellular fluid has a higher osmotic pressure than the extracellular fluid. The cell is protected from bursting by a rigid cellulose wall.

Animal cells generally lack the turgidity that characterizes plant cells. The unfertilized eggs of some marine animals, such as the sea urchin, behave like genuine osmometers. Since they are spheroid, one can, by measuring the diameter, determine the volume and the changes that the egg undergoes with changes in the osmotic pressure of the medium. Thus, their internal osmotic pressure can be determined by finding the concentration at which their volume does not change.

In many unicellular organisms the osmotic equilibrium is maintained by means of a contractile vacuole. This "organelle" extracts water from the protoplasm, releasing its contents into the external medium.

Ionic Concentration and Electrical Potentials Across Membranes

In all cells there is a difference in ionic concentration with the extracellular medium and an electrical potential across the membrane. These two properties are intimately related, since the electrical potential depends on an unequal distribution of the ions on both sides of the membrane.

TABLE 21–1. Ionic Concentration[†] and Steady Potential in Muscle[*]

Interstitial Fluid			Intracellular Fluid
Cations	Na$^+$	145	12
	K$^+$	4	155
Anions	Cl$^-$	120	3.8
	HCO$_3^-$	27	8
	A$^-$ and others	7	155
Potential		0	$^-$90 mv

*Modified from Woodbury, J. W., "The Cell Membrane: Ionic and Potential Gradients and Active Transport," in Ruch, T. C., Patton, H. D., Woodbury, J. W., and Towe, A. L. (Eds.), *Neurophysiology*, Philadelphia, W. B. Saunders Co., 1961.

†Ionic concentration in mEq.

As shown in Table 21–1, the interstitial fluid has a high concentration of Na$^+$ and Cl$^-$ and the intracellular fluid a high concentration of K$^+$ and of larger organic anions (A$^-$).

Using fine microelectrodes with a tip of 1 μm or less investigators are able to penetrate through the membrane into a cell and also into the cell nucleus (Fig. 12–3) and to detect an *electrical potential* (also called the *resting*, or *steady*, potential), which is always negative inside. The values of the membrane potential vary in different tissues between −20 and −100 millivolts (mv).

DIFFUSION OR PASSIVE PERMEABILITY

In the absence of an intervening membrane, when two solutions of different concentration are mixed a process of intermixing called *diffusion* occurs. For example, if a concentrated solution of sugar is placed in contact with water, there will be a net movement (flux:M) of the solute from the region of higher concentration to that of a lower concentration. In this case the higher the difference in concentration between the two solutions (i.e., the *concentration gradient*), the more rapid the rate of diffusion.

The presence of a lipoprotein membrane, such as the plasma membrane, greatly modifies this diffusion or passive permeability.

At the end of the last century, Overton demonstrated that substances that dissolve in lipids pass more easily into the cell, and Collander and Bärlund,[2] in their classic experiments with the cells of the plant *Chara*, demonstrated that the rate at which substances penetrate depends on their solubility in lipids and the size of the molecule. The more soluble they are, the more rapidly they penetrate, and with equal solubility in lipids the smaller molecules penetrate at a faster rate (Fig. 21–1).

The permeability (P) of molecules across the membrane is:

$$P = \frac{KD}{t} \tag{1}$$

with K, the partition coefficient; D, the diffusion coefficient, which depends on the molecular weight; and t, the thickness of the membrane. The partition coefficient in most

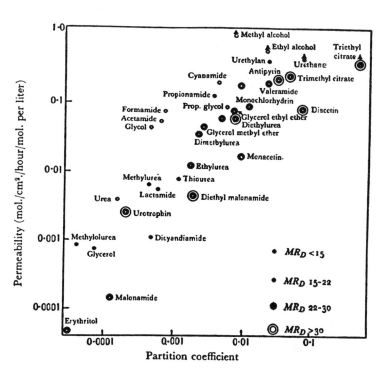

Figure 21–1. Rate of penetration (permeability) in cells of *Chara ceratophylla* in relation to molecular volume (measured by molecular refraction, MR_D), and to the partition coefficient of the different molecules between oil and water. (After Collander, R., and Bärlund, H., *Acta Bot. Fenn., 11*:1, 1933.)

cell membranes is similar to that of olive oil and water. The diffusion of molecules through the membrane is similar to the passage through a polymer having small and large holes and requiring the twisting or breaking of some bonds. Thus, the cell membrane represents a formidable barrier to most types of molecules.

Diffusion of Ions

The diffusion of ions across membranes is even more difficult because it depends not only on the *concentration gradient,* but also on the *electrical gradient* present in the system. It has been shown that within the cell there is a large concentration of non-diffusible anions. In 1911, Donnan predicted that if a theoretical cell, having a non-diffusible negative charge inside, is put in a solution of KCl, K^+ will be driven into the cell by both the concentration and the electrical gradients; Cl^-, on the other hand, will be driven inside by the concentration gradient, but will be repelled by the electrical gradient. As shown by Donnan, the equilibrium concentrations will be exactly reciprocal:

$$\frac{[K^+_{in}]}{[K^+_{out}]} = \frac{[Cl^-_{out}]}{[Cl^-_{in}]} \qquad (2)$$

A Donnan equilibrium involving only physical forces (i.e., without expenditure of energy by the membrane) was confirmed by the demonstration that the membrane potential was negative on the inside and was accompanied by a high K^+ and a low Cl^- concentration (Table 21–1). The relationship between the concentration gradient and the resting membrane potential is given by the Nernst equation:

$$E = RT \log \frac{C_1}{C_2} \qquad (3)$$

where E is given in millivolts, R is the universal gas constant, and T is the absolute temperature.

From (1) and (2) the Donnan equilibrium for KCl can now be expressed as follows:

$$E = RT \log \frac{[K^+_{in}]}{[K^+_{out}]} = RT \log \frac{[Cl^-_{out}]}{[Cl^-_{in}]} \qquad (4)$$

According to (4) any increase in the membrane potential will cause an increase in the ion asymmetry across the membrane, and vice versa. While the first measurements of membrane potentials and ion concentration seemed to confirm this type of *passive* or *diffusion equilibrium,* more precise determinations in different cell types demonstrated that this was not the case. As mentioned in the next section, this discrepancy may be explained by the involvement of the active transport of ions.

CELL PERMEABILITY AND ACTIVE TRANSPORT

In addition to the diffusion or passive movement of neutral molecules and ions across membranes, cell permeability includes a series of mechanisms that require energy. These mechanisms are generally described as *active transport*. Adenosine triphosphate (ATP), which is produced mainly by oxidative phosphorylation in mitochondria, is generally used as the source of energy (see Chapter 10). For this reason, active transport is generally related to, or coupled with cell respiration.

The active transport against a concentration gradient is explained in Figure 21–2 by analogy with a hydrostatic example in which water has to be moved upstream (i.e., against gravity). The osmotic work to be done is expressed by the Nernst equation. A charged molecule crossing through an electrochemical gradient also may imply expenditure of energy. For example, to maintain a low intracellular concentration of Na^+, the cell must extrude sodium against a gradient (i.e., higher Na^+ concentration outside). In addition, it must do this against an electrochemical bar-

rier since the membrane is negative inside and positive outside (Fig. 21–3, *A* and *B*).

Properties of Active Transport

To understand better the criteria that determine whether a substance moves across the cell membrane by active transport, an example from the kidney may be considered. If isolated kidney tubules are immersed in a solution of phenol red, after a certain time the dye passes through the cells and becomes concentrated in the lumen (Fig. 21–4). Thus, the cells are extruding or secreting the dye against a concentration gradient. Other experiments demonstrate that active transport depends on the energy produced by the cell: by cooling the tissue the dye is not concentrated. Certain metabolic poisons that inhibit cell respiration (e.g., cyanide and azide) have the same effect.

The work done by the cell against the concentration gradient can be calculated from the equation in Figure 21–2, in which C_1 is the concentration of the dye inside and C_2, the concentration outside.

Active Transport of Ions and Membrane Potentials

When an ion is transported against an electrochemical gradient, an extra consump-

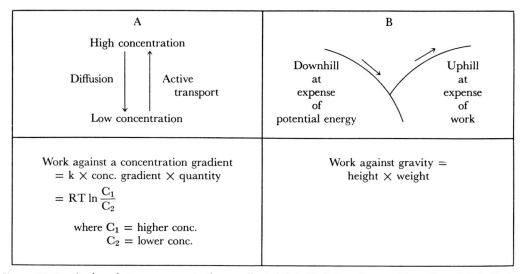

Figure 21–2. Analogy between concentration gradient (left half of **A**) and potential gradient (left half of **B**) and between movement against a concentration gradient (right half of **A**) and work done in moving uphill (right half of **B**). Equations for work against a concentration gradient (osmotic work) and for work in lifting a weight up a height are given at the bottom of the figure. (From Giese, A. C., *Cell Physiology*, 3rd ed., Philadelphia, W. B. Saunders Co., 1968.)

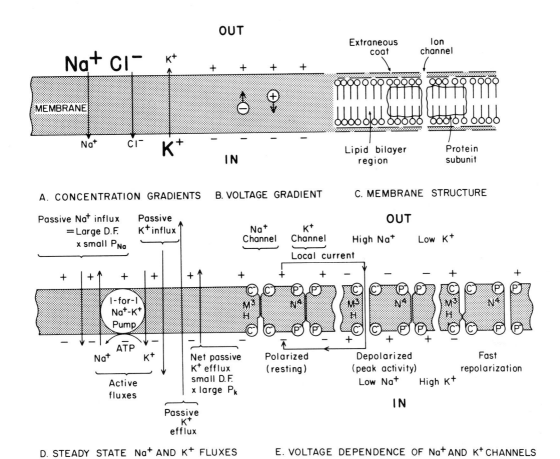

Figure 21–3. Diagram of the molecular structure of the plasma membrane in relation to the transport of ions. (See the description in Chapters 21 and 24.) (Courtesy of J. W. Woodbury.)

tion of oxygen is required. It is calculated that 10 per cent of the resting metabolism of a frog muscle is used for transport of sodium ions. This consumption may increase to 50 per cent in some experimental conditions in which the muscle is stimulated.

That the resting membrane potential is due to active transport may be demonstrated in plant and animal cells that have been metabolically blocked by anoxia or specific poisons. In this case leakage of K^+ occurs and the

potential may decrease to zero. This is clearly observed when anoxia is combined with the poisoning of glycolysis. As suggested by Krogh in 1946, the membrane potential is not really at equilibrium but in a "steady state" involving the constant expenditure of energy.

An interesting example of active transport has been provided by experiments with isolated frog skin. The epithelium is specialized to transport Na^+ from the pond water to the interstitial fluids, and by this mechanism

Figure 21–4. Accumulation of phenol red in sections of proximal kidney tubules of the chick embryo (diagrammatic). Phenol red moves inward until it becomes more concentrated in the inside of the vesicle than on the outside. (From Giese, A. C., *Cell Physiology*, 3rd ed., Philadelphia, W. B. Saunders Co., 1968.)

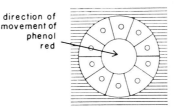

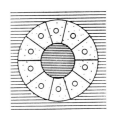

Initial   After lapse of time

the frog can trap this essential ion for use in different tissues. The isolated skin can be kept alive for many hours and used as a wall between two chambers in which the ionic concentration and other factors are changed experimentally.[3] By means of this preparation, a difference in potential across the skin has been demonstrated: the inside surface is positive with respect to the outside. The sodium ions are transported from the outer toward the inner surface, and the current produced is due to the flux of sodium. It was also observed that the antidiuretic hormone of the neurohypophysis stimulates the transport of sodium and water. Similar findings have been observed in the isolated toad bladder[4] and the kidney tubules of *Necturus*.[5]

Ionic transport is intense in the salivary and sweat-producing glands, and even more so in the glands of the stomach that produce H^+ and Cl^- that must be replaced by the blood. This ionic transfer is very marked in the salt-secreting glands of certain marine birds (e.g., albatross).

The active transport of ions is fundamental to maintenance of the osmotic equilibrium of the cell, the required concentration of anions and cations and the special ions needed for the functioning of the cell (see Table 21–1). Together with the extrusion of Na^+, which is continuously pumped out by the cell, there is an exit of water molecules. In this way the cell keeps its osmotic pressure constant.

Potassium ions, which are concentrated inside the cell (Table 21–1), must pass against a concentration gradient. This can be achieved by a "pumping" mechanism at the expense of energy. As explained before, Na^+ also may be transported by an active process, which is sometimes called the "sodium pump."

The diagram in Figure 21–5 summarizes the relationship existing between the transfer of K^+ and Na^+ (i.e., ionic fluxes) by passive and active mechanisms and the resulting steady state potential. The passive (downhill) fluxes are distinguished from the active (uphill) fluxes. Notice that the active pumping out of Na^+ is the main mechanism for maintaining a negative potential inside the membrane of -50 mv.[1,6] This diagram demonstrates that the distribution of ions across the membrane depends on the summation of two distinct processes: (1) simple electrochemical diffusion forces which tend to establish a Donnan equilibrium (i.e., passive transport), and (2) energy-dependent ion transport processes (i.e., active transport).

In Figure 21–3, *D*, it may be observed that the $Na^+ - K^+$ pump drives ions in a 1-to-1 ratio, extruding Na^+ and taking in K^+ (active fluxes). At the same time the passive Na^+ influx depends on a large driving force (D.F.), resulting from the concentration and voltage

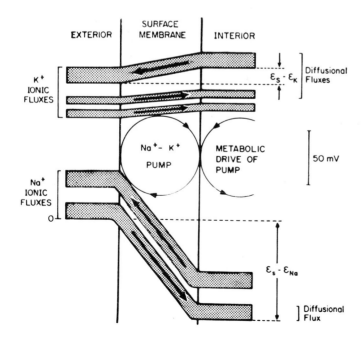

Figure 21–5. Active and passive Na^+ and K^+ fluxes through the membrane in the steady state. The ordinate is the electrochemical potential of the ion ($\epsilon_s - \epsilon_k$ for K^+, $\epsilon_s - \epsilon_{Na}$ for Na^+). The abscissa is the distance in the vicinity of the membrane. The width of the band indicates the size of that particular one-way flux. Passive efflux of Na^+ is negligible and is not shown. (After Eccles, J. C., *Physiology of Nerve Cells*, Baltimore, Johns Hopkins Press, 1957.)

gradients, and only a slight permeability of the membrane to Na^+ (P_{Na}). Similarly, the net passive K^+ efflux results not from a small driving force, but from a greater permeability to K^+ (P_K).

Mechanism of Ionic Transport. Pores in the Cell Membrane

It can now be ascertained that the molecular machinery involved in ionic transport is located within the cell membrane. This has been demonstrated in two key materials. For example, if red blood cells are hemolyzed so that only the cell membrane remains (i.e., a red cell ghost), they can be filled again with appropriate solutions containing ions and ATP, and Na^+ is transported and K^+ is taken up as in a normal cell.

The giant axon of the squid, which has a diameter of about 0.5 mm. can be emptied of the axoplasm and then refilled with solutions of different electrolytes. The transport of ions against a concentration gradient, steady potentials, and even action potentials with the conduction of impulses can be obtained in this preparation in which most of the axoplasm is lacking and the excitable membrane left alone.[7]

The use of radioisotopes demonstrated that ions can enter into the cell rapidly without obvious osmotic effects.[7] It was then suggested that an ionic interchange across the membrane could take place through electrically charged pores.

Knowledge about the diameter of the different ions in the hydrated state is particularly pertinent. In this respect it is interesting to remember that the sodium ion, although smaller than K^+ and Cl^- in weight, is large in the hydrated condition and enters with more difficulty into the cell (see Figure 21-6).

A possible molecular interpretation of the membrane pores is shown in Figure 21-3, *C* and *E*. The presence of embedded protein subunits is postulated within the lipoprotein structure. The pore could be envisioned as the interstice between four adjacent protein subunits which could form a hydrophilic channel across the membrane; two such subunits are shown in Figure 21-3, *C*.[8]

In Figure 21-3, *E*, the ion channels for Na^+ and K^+ are represented in the polarized or resting condition, or in the state of depolar-

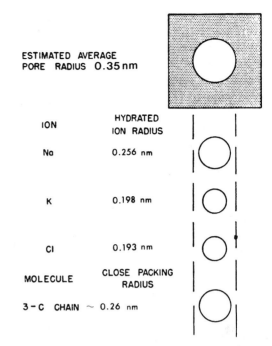

ESTIMATED AVERAGE
PORE RADIUS 0.35 nm

ION	HYDRATED ION RADIUS	
Na	0.256 nm	
K	0.198 nm	
Cl	0.193 nm	
MOLECULE	CLOSE PACKING RADIUS	
3 - C CHAIN	~ 0.26 nm	

Figure 21-6. Schematic representation of the red cell pore. Notice that the hydrated ion radius is larger for Na^+ than for K^+. (From Solomon, A. K., *J. Gen. Physiol.,* *43*:5, part 2, suppl. 1, p. 1, 1960.)

ization that occurs during the action potential. It is postulated that the Na^+ and K^+ channels are closed in the polarized resting membrane. The selectivity of the Na^+ pore will be due to carboxyl groups (C^-) guarding the entrances to the channel (fixed charges). In the case of the K^+ pores, the fixed negative charges would be phosphate groups (P^-). The rest of the diagram shows the opening of the Na^+ pores at the peak of depolarization of the action potential and the opening of the K^+ pores during repolarization (see Chapter 24).

The theory of the existence of pores in the membrane has been strengthened by the study of the penetration of noncharged molecules, which are insoluble in the lipid phase (e.g., urea, formamide, and glycerol). The rate of passage of these substances is related to the size of the molecule and to the area occupied by the pores on the membrane. The equivalent pore radius in different biological membranes has been estimated to range between 0.8 and 0.35 nm (Fig. 21-6).

The total area of the pores in the red blood cell has been estimated to be on the order of 0.06 per cent of the surface area. This means that a 0.7 nm pore would be surrounded by a nonporous square 20 × 20 nm. These findings indicate that the cell uses only a minute fraction of its surface area for ionic interchange.[9]

Permease Systems

The existence of such small pores in the cell membrane would prevent the penetration of some essential molecules. To incorporate them the membrane must develop a specific chemical transport system. In Chapter 19 a description was given of the so-called *permease system* for *β-galactosidase,* which is found in *Escherichia coli* and which determines the penetration of lactose into the bacterium. This genetically determined permease system involves a specific membrane protein specialized for transport and located at the cell membrane. Such a protein has already been labeled and purified.[10] The passage of amino acids and nucleotides would also be controlled by special permeases. *E. coli* is thought to have 30 to 60 such systems specialized in the transport of various molecules.[11]

The inner mitochondrial membrane also contains several permease systems (see Chapter 10). Permeases, like enzymes, only increase the rate at which a reaction reaches an equilibrium. In other words, they do not accumulate or extrude molecules against a concentration gradient, but favor a more rapid equilibrium across the membrane than ordinary passive diffusion.

Active Transport and Na+ K+ ATPase

An important step in the identification of the active transport was carried out in red cell membranes. It was demonstrated that the extrusion of Na+ and the penetration of K+ were related to a special Mg^{2+} activated ATPase. Figure 21–7 shows an idealized diagram of the Na+K+ ATPase within the red cell membrane. It may be observed that the hydrolysis of one ATP provides the energy for the linked transport of 2 K+ ions toward the inside, and 3 Na+ ions toward the outside of the cell.[12] This diagram shows the vectorial characteristics of the enzyme which is sensitive to ATP on the inside of the membrane, but not on the outside. This ATPase is stimulated by a mixture of both Na+ and K+ and is inhibited specifically by the cardiotonic

glycoside *ouabain.* Figure 21–7 also shows that the Na+ and K+ sites are independent and are competitively inhibited by K+ and Na+, respectively. The Na+ K+ ATPase has been found to be concentrated in membranes of nerves, brain, and kidney and is particularly rich in the electric organ of the eel and the salt gland of certain marine birds (e.g., albatross) in which the transport of ions is extremely active.

The vectorial properties of the enzyme can be demonstrated cytochemically; thus, by electron microscopy the liberation of phosphate has been shown to occur exclusively on the inner side of the red cell ghost.[13] The vectorial function of Na+ K+ ATPase can be best explained by the so called "carrier" hypothesis. In its most simple form a carrier consists of: (1) a membrane protein having a specific site to combine with the ligand to be translocated, (2) a mechanism of translocation which moves the ligand from one side of the membrane to the other, and (3) a mechanism for the release of the ligand. Most carrier models involve some kind of conformational change of the protein by which the affinity of the binding site for the ligand changes during the translocation. According to the diagram of Figure 21–7 in the case of Na+ K+ ATPase it is possible to envision that the enzyme has binding sites for Na+ and K+ and a complex carrier mechanism to deliver and release these ions at opposite sides of the membrane. The coupling of the carrier mechanism to the hydrolysis of ATP generates the energy

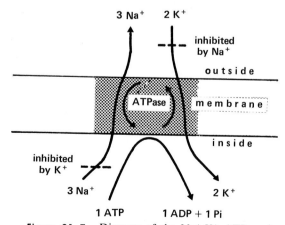

Figure 21–7. Diagram of the Na+-K+ ATPase in the cell membrane. Observe that for each molecule of ATP hydrolyzed at the inner part of the membrane 3 Na+ ions are transported outside and 2 K+ ions are transported inside. (See description in the text.)

needed to move the ions against the concentration gradients.

Studies were carried out on membrane fragments containing the Na^+ K^+ ATPase, using ATP labeled with ^{32}P in the terminal phosphate. Only in the presence of Na^+ did the membrane become labeled with ^{32}P. If ^{32}P-labeled membrane fragments were treated with K^+, the ^{32}P was released as inorganic phosphate. Such findings were interpreted as an indication that the transport reaction takes place in two major steps. Step I consists of the formation of a *covalent phosphoenzyme intermediate* $(E \sim P)$ on the inner side of the membrane and in the presence of Na^+. This first reaction is inhibited by Ca^{2+}, but not by ouabain. In Step II the $[Na^+ \cdot E \sim P]$ complex is hydrolyzed to form free enzyme and Pi. This reaction requires K^+, added to the outside of the membrane, and is inhibited by ouabain, which competes with K^+. Steps I and II can be represented as follows:

Step I $\quad Na^+_{in} + ATP + E \longrightarrow$
$$[Na^+ \cdot E \sim P] + ADP \qquad (5)$$

Step II $\quad [Na^+ \cdot E \sim P] + K^+_{out} \longrightarrow$
$$Na^+_{out} + K^+_{in} + Pi + E \qquad (6)$$

It may be observed that Na^+ and K^+ are ultimately translocated in the opposite directions from which they were bound.

The Na^+ K^+ ATPase is tightly bound to the cell membrane from which it may be released by the use of detergents. The molecular weight has been estimated to be about 670,000 and probably contains several polypeptides. Certain lipids are needed for the activity of this enzyme. In a red cell there are about 5000 enzyme molecules, each of which may extrude 20 Na^+ ions per second. (For a review of this field, see Schwartz et al., 1972.)

Penetration of Larger Molecules

What has been said demonstrates that the general term "cell permeability" comprises a variety of different mechanisms. In addition to certain foreign substances that can penetrate the cell because they are lipid-soluble (e.g., anesthetics), generally ions penetrate through charged pores and other molecules penetrate by permease systems.

There is no doubt that under certain conditions large molecules, such as certain proteins, penetrate the cell. This is the case of ribonuclease, an enzyme that penetrates living plant cells readily and also eggs, flagellates, ascitic tumors, and so forth.

Basic proteins of the protamine and histone types have been reported to enter into living cells. In Chapter 19 it was mentioned that DNA penetrates certain bacteria and produces a genetic change known as *transformation*.

Later on in this chapter *phagocytosis* and *pinocytosis*, by which solid or fluid material in bulk can be ingested by the cell will be discussed.

The study of permeability requires some knowledge of the chemical and molecular organization of the cell membrane (Chapter 8). This membrane regulates the inflow of substances to the cell and the outflow of water ions and other materials. The cell membrane maintains a balance between the osmotic pressure of the intracellular and the interstitial fluid. Solutions may be isotonic, hypotonic, or hypertonic with respect to the intracellular fluid.

The *ionic concentration* in the intracellular fluid differs from that in the interstitial fluid. In the latter, Na^+ and Cl^- are high, and K^+ is low. Inside the cell, K^+ is high, Na^+ and Cl^- are low, and there is a large pool of organic anions (A^-) that do not cross the cell membrane. A *membrane potential* of -20 to -100 millivolts is detected in all cells. Cell membranes are polarized, being negatively charged inside and positively charged outside.

SUMMARY:
Cell Permeability and
Active Transport

Movement of substance across the cell membrane may be by *passive diffusion* or by several mechanisms of *active transport*. *Diffusion* occurs when there is a concentration gradient and takes place from a high to a low concentration. *Diffusion of molecules* across membranes depends on molecular volume and the lipid solubility of the substance. *Diffusion of ions* depends on both the concentration and the electric gradients across the membrane.

The distribution of K^+ and Cl^- generally follows a Donnan equilibrium that depends on the presence of non-diffusible anions inside the cell. The relationship between concentration gradient $\left(\dfrac{C_1}{C_2}\right)$ and membrane potential (E) is given by the Nernst equation:

$$E = RT \log \frac{C_1}{C_2}$$

The *active transport* of neutral molecules and ions requires energy (ATP) and is generally coupled to the energy-yielding mechanisms of the cell. Active transport is blocked by cooling or by certain metabolic poisons of respiration and glycolysis. With complete energy block the membrane potential may decrease to zero and there is leakage of K^+. Active transport may be demonstrated by simple experiments using kidney tubules, frog skin, and toad bladder. Ionic transport is particularly intense in electric tissues of the eel and in salt-secreting glands. By active transport Na^+ and water are pumped out of the cell and K^+ penetrates the cell. Both electrochemical diffusion (i.e., passive transport) and energy-dependent ion transport determine the distribution of ions and the membrane potential across the membrane (Fig. 21–5). The passage of ions is thought to be across charged pores in the membrane. The size of the hydrated ion is important in the transport through the pores.

Cell membranes contain *permease systems* for the penetration of some essential molecules. One of the best known is the permease for *β-galactosidase. Na⁺ K⁺ ATPase* is related to the active transport of Na^+ and K^+ in red blood cells, nerves, brain membranes, electric, and other tissues. The hydrolysis of 1 ATP $\rightarrow$ ADP + Pi in the inner surface of the membrane is coupled with the transport of 3 Na^+ from the inside toward the outside of the membrane and with the transport of 2 K^+ in the opposite direction.

All these properties indicate that the enzyme has a vectorial orientation across the membrane. The Na^+ K^+ ATPase is specifically inactivated with *ouabain*. The function of this enzyme is explained by a complex carrier mechanism which comprises: (1) binding sites for Na^+ and K^+, (2) a mechanism for translocation, and (3) a mechanism for the release of the ligand.

The production of a covalent phosphoenzyme intermediate has led to the concept that the transport is carried out in two steps as follows:

Step I $Na^+_{in} + ATP + E \longrightarrow [Na^+ \cdot E \sim P] + ADP$

Step II $Na^+ \cdot E \sim P + K^+_{out} \longrightarrow Na^+_{out} + K_{in} + Pi + E$

in which $[Na^+ \cdot E \sim P]$ is the phosphorylated complex of the enzyme.

ENDOCYTOSIS

Intimately related to the activity of the plasma membrane are *phagocytosis* and *pinocytosis,* the processes by which solid or fluid material is ingested in bulk by the cell. With the introduction of electron microscopy, it has become apparent that these active processes of penetration are much more developed than was previously thought. The similarities observed between these processes has led to the coining of the term *endocytosis* to include both phagocytosis and pinocytosis. (Exocytosis is the reverse process, by which membrane-lined products are released at the plasma membrane.)

This type of transport is less general than the permeability processes described above, since it is observed only in certain cell types and in some of them only during certain periods of cellular life. As will be mentioned later, it does not replace, but supplements, the other types of permeability that are of more general use to the cell.

The problem of the entrance of solids and fluids in bulk is also related to the formation of digestive vacuoles and granules within the cell and to the more general phenomena of *defense and disposal of ingested material* by the cell. These processes are also associated with a special organelle rich in hydrolytic enzymes called the *lysosome*. Because of this relationship the discussion of lysosomes and of *peroxisomes* is incorporated in this chapter and not in the chapters on the cell organelles, as is generally done.

Phagocytosis (Gr., *phagein*, to eat), is found in a large number of protozoa and among certain cells of the metazoa. In metazoa, rather than serving for cell nutrition, phagocytosis is, in general, a means of defense, by which particles that are foreign to the organism, such as bacteria, dust, and various colloids, are disposed of. Phagocytosis is highly developed in granular leukocytes (first described by Metschnikoff at the end of the last century), and also in the cells of mesoblastic origin ordinarily grouped under the common term *macrophagic* or *reticuloendothelial system*. The cells belonging to this group include the histiocytes of connective tissue, the reticular cells of the hematopoietic organs, and those endothelial cells lining the capillary sinusoids of the liver, adrenal gland, and hypophysis. All these cells can ingest not only bacteria, protozoa, and cell debris, but also smaller colloidal particles. In this instance phagocytosis is called *ultra-phagocytosis*. An example of this process is the capacity of mesoblastic cells to ingest and store vital colloidal dyes.

In protozoa, phagocytosis is intimately linked to ameboid motion. An ameba ingests large particles, including microorganisms, by surrounding them with pseudopodia to form a food vacuole within which the digestion of food takes place. In leukocytes and other phagocytic cells, phagocytosis may be carried out in immobile cells.

Analyzing the process of phagocytosis, one may distinguish two distinct phenomena. First the particle *adheres* (is *adsorbed*) to the mass of the protoplasm, and then the particle actually penetrates the cell. In some cases it has been possible to dissociate these two phases of phagocytosis. For example, at low temperature, bacteria may adhere to the cytoplasm of a leukocyte without being ingested. This phase of adsorption seems to obey physicochemical forces, such as electrostatic surface charges. Macrophages put out hyaline, thin (about 0.25 μm), lamellar pseudopodia that adhere to and extend over the surface of the particle until it is completely surrounded. This phenomenon is comparable to that occurring when a liquid "wets" and extends over a solid surface. Macrophages accumulate the negatively charged vital dyes, such as pyrrole blue, trypan blue, and lithium carmine, and colloidal substances, such as silver, iron saccharate, and India ink.

To be ultraphagocytized, vital dyes must be previously attached to a protein, which acts as "vector." Vital acid dyes or negatively charged colloids, when injected into an animal, accumulate progressively in all the cells of the macrophage system. These substances are deposited, at first, as small granules that increase in size until they constitute true intracellular precipitates. The fundamental characteristic of this system is that it accumulates and concentrates these dyes and colloids even when administered in dilute solutions. After massive injections, other cells that do not belong to this system may also ingest such substances.

Pinocytosis

In addition to the ingestion of solid particles, the uptake of fluid vesicles by the living cell has been observed. This process, first observed by Edwards in amebae and by Lewis[14] in cultured cells, has been called *pinocytosis* (Gr., *pinein*, to drink). As can be readily seen in Lewis's motion pictures, the uptake of fluids is accompanied by vigorous cytoplasmic motion at the edge of the cell, as if vesicles of fluid are being surrounded and engulfed by clasping folds of cytoplasm. Vacuoles taken up at the edge of the cell are then transported to other portions of the cell.

The possibility that pinocytosis is involved in the penetration of proteins into amebae was first suggested by Mast and Doyle in 1934. This was actually demonstrated by using a protein labeled with fluorescein and observing pinocytosis by means of the fluorescence microscope.[15] The presence of the protein seems to act as a stimulus to pinocytosis and the uptake of protein is surprisingly high. During the "feeding period," the ameba ingests approximately one-third its volume of the protein solution. This material is then eliminated in five to six days. The ameba practically "drinks" the protein solution, and with it the organism may absorb other substances that normally do not penetrate. For example, ^{14}C-glucose, if dissolved in the protein solution, can enter the ameba in considerable quantities.[16]

Pinocytosis is induced by certain substances. If an ameba is placed in water, pinocytosis does not take place; if some carbohydrate is added, nothing happens; but if certain amino acids, proteins, and ions are added, pinocytosis begins. The pinocytotic activity, once started, is kept going for about 30 minutes, during which time some 100 fluid channels are formed. Then the process comes to a stop, and the ameba has to wait for two to three hours before starting another pinocytotic cycle. This has been interpreted as an indication that the surface membrane available for invagination is exhausted in the 30-minute period.

That phagocytosis and pinocytosis are essentially similar phenomena can be demonstrated by allowing the ameba to phagocytize some ciliated cells first and then inducing pinocytosis. The number of channels formed is much less under these conditions. In the reverse experiment it has been found that an ameba can ingest much fewer ciliates for food.

Extraneous Coats and Pinocytosis

In Chapter 8 reference was made to the extraneous coats that cover the cell membrane. With a fluorescent protein, it was found that immediately after immersion the cell surface of an ameba becomes covered by a thick layer of protein, the concentration of which may be 50 or more times that of the solution. Then the cell invaginates the membrane heavily encrusted with the protein.[17]

If similar experiments are made with electron-opaque substances, such as ferritin molecules or a suspension of thorium oxide, in the first stage of pinocytosis the concentration of large molecules occurs in the extraneous coat that covers the plasma membrane of the ameba.[18]

The binding to the coat explains why the concentration of the incorporated protein may reach such enormous figures.[19] A kinetic analysis of the uptake by macrophages has been made using radioactive colloidal gold. The interaction involves (1) the reversible adsorption phase, which can be related to the concentration in the extracellular fluid, and (2) the irreversible passage of the surface-bound gold into the cell. The rate of ingestion is proportional to the amount of gold attached to the cell surface. It has been calculated that

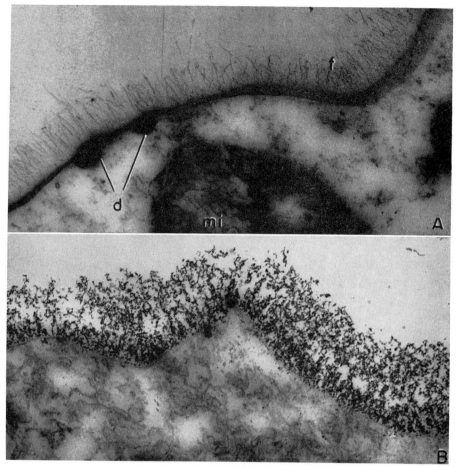

Figure 21-8. **A,** electron micrograph of the cell membrane of the ameba *Chaos chaos,* showing the extraneous coat formed by fine filaments of 5.0 to 8.0 nm in diameter and 100 to 200 nm long. *d,* dense bodies; *f,* filaments; *mi,* mitochondrion. × 55,000. **B,** same as **A,** but after the addition of thorium dioxide particles. These particles are attached to the filaments prior to the formation of channels and penetration of this material into the ameba. × 38,000. (Courtesy of P. W. Brandt and G. D. Pappas.)

in these cells between 2 and 20 per cent of the cell surface may be engulfed in a minute![20]

Figure 21-8, *A,* shows the plasmalemma of an ameba with the membrane proper and the extraneous coat formed by filaments about 6 nm in diameter and 100 to 200 nm in length. After the electron-opaque material is added, it becomes heavily concentrated upon the filaments (Fig. 21-8, *B*). Since acid mucopolysaccharides carry a strong positive charge, they may bind the inducer by electrostatic attraction. The inducer of pinocytosis has been found to be a negatively charged substance, i.e., charged ions, acid dyes, or proteins with an isoelectric point in the acid range. In the ameba the inducer may cause a 50-fold increase in electrical resistance of the

membrane prior to the formation of the typical channels.[21]

Micropinocytosis

The use of the electron microscope demonstrated that the plasma membrane of numerous cells could invaginate, forming small vesicles of about 65 nm, and that this process could be related to pinocytosis but only in smaller vesicles. These vesicles were first found in endothelial cells lining capillaries in which they concentrate in the region adjacent to the inner and outer membranes. The presence of vesicles opening on both surfaces and of others traversing the cytoplasm suggested a possible transfer of fluid across the

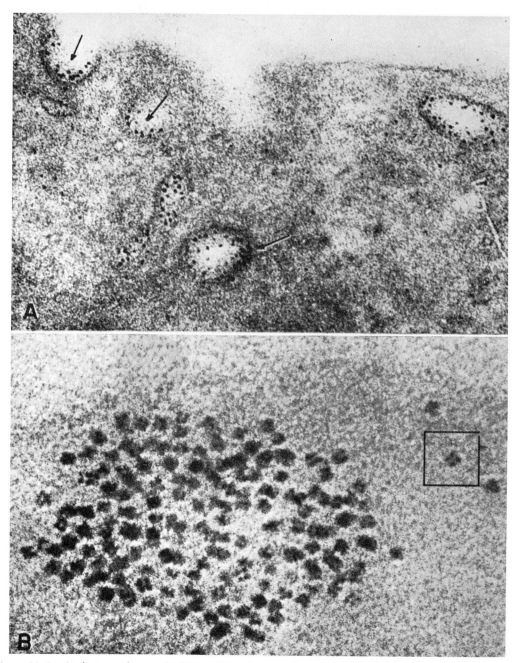

Figure 21–9. **A,** electron micrograph of the peripheral region of an erythroblast, showing the penetration of ferritin molecules by micropinocytosis. The arrows indicate several phases of the process starting at the surface. × 180,000. **B,** molecules of ferritin inside the cytoplasm of a reticular cell. Inset: one molecule with four dense points of about 1.5 nm, each of which contains about 300 atoms of iron. × 850,000. (Courtesy of M. Bessis.)

cell. A similar component was then observed in Schwann and satellite cells of nerve ganglions and in numerous other cell types, particularly in macrophages, muscle cells, reticular cells, etc.

The transport of fluid across the capillary endothelium is assumed to occur in quantal amounts;[22] this process implies the invagination of the plasma membrane, the formation of a closed vesicle by membrane fission, the movement of the vesicle across the endothelial cytoplasm, the fusion of the vesicle with the opposite plasma membrane, and the discharge of the vesicular content. Such a sequence has been corroborated by the discovery of intermediary stages of membrane contact and fusion with progressive elimination of the layers of the unit membrane structure (see Chapter 8). This transendothelial passage of fluid has also been studied by the injection of peroxidase.[23, 24]

An interesting example of the physiologic importance of pinocytosis is found in the bone marrow. In the so-called erythroblastic islands, it is possible to observe reticular cells filled with iron-containing macromolecules of ferritin. These molecules leave the reticular cells and enter the erythroblast, to be used in the manufacture of hemoglobin, by the process of pinocytosis[25] (Fig. 21-9). These findings show how cells can utilize again and again the iron resulting from the destruction of old red blood cells to form new erythrocytes.

Pinocytosis and Active Transport

Phagocytosis and pinocytosis are active mechanisms in the sense that the cell requires energy for their operation. During phagocytosis by leukocytes, oxygen consumption, glucose uptake, and glycogen breakdown all increase significantly. Induction of phagocytosis also produces an increased synthesis of phosphatidic acid and phosphatidyl inositol.[26] In cultured cells, addition of ATP increases the rate of endocytosis,[27] and this is inhibited by respiratory and other metabolic poisons.

Endocytosis is rather unspecific, and sometimes even noxious substances are ingested. It will be shown later that the content of the phagocytized material may be digested by enzymes present in the membrane or added by lysosomes to the ingested vacuole.

Pinocytosis is not an alternative process of active transport, but rather a supporting one. By means of pinocytosis the cell is provided with a much larger interior interface where passive and active transports are carried out more efficiently than at the surface membrane.

THE LYSOSOME

The concept of the lysosome originated from the development of cell fractionation techniques, by which different subcellular components are isolated (see Chapter 6). By 1949, a class of particles having centrifugal properties somewhat intermediate between those of mitochondria and microsomes was isolated by De Duve and found to have a high content of acid phosphatase and other hydrolytic enzymes. By centrifugation it was calculated that the size of these particles ranged from 0.2 to 0.8 μm and because of their enzymatic properties they were named lysosomes (Gr., *lysis*, dissolution; *soma*, body).[28]

Stability and Enzymatic Content of the Lysosome

One important property of the lysosome is its stability in the living cell. The enzymes are enclosed by a membrane and are not readily available to the substrate. After isolation by mild methods of homogenation, the amount of enzyme that can be measured by adding the molecules of phosphate ester is small. This increases considerably if the particles are treated with hypotonic solutions or surface active agents (e.g., *triton*) (see Figure 21-10).

Numerous hydrolytic enzymes are recognized as present in lysosomes. All of them share with acid phosphatase the property of splitting biological compounds in a mild acid medium. In the living cell these enzymes, e.g., phosphatase, glucuronidase, sulfatase, catepsin, etc., are confined within the particle and can act only on material taken along within it, e.g., on phagocytized material. If the lysosome is injured (e.g., by a toxic agent), these enzymes can be released to digest the entire cell. In an isolated lysosome, rupture of the membrane will make the enzymatic content of the lysosome available to the different substrates (Fig. 21-10).

Although the original concept of the

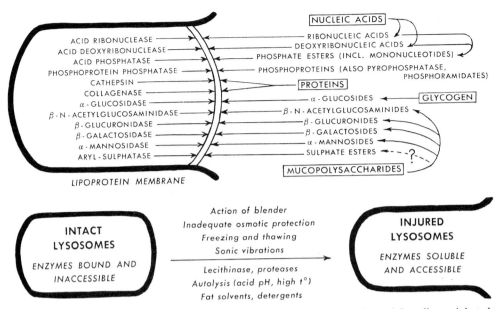

Figure 21–10. Diagram showing the biochemical concept of the lysosome. This model applies mainly to lysosomes of rat liver. **Above,** various hydrolytic enzymes and the substrates on which they act. **Below,** indication of the intact lysosomes and the effect of various agents that disrupt the membrane of the lysosomes. (From De Duve, C., "General Properties of Lysosomes," *in Lysosomes,* Ciba Foundation Symposium, London, J. & A. Churchill, Ltd., 1963.)

lysosome suggests a multienzymatic particle, it is possible that different hydrolytic enzymes may be carried in particles having slightly different sedimentation properties and also different stabilities. Using zonal gradient centrifugation, it has been demonstrated that lysosomes of the rat liver are heterogeneous in terms of their enzyme contents.[29]

Polymorphism of the Lysosome

Observation of the lysosomal fraction of the liver under the electron microscope led to the recognition, among typical mitochondria and contaminating microsomes, of dense bodies about 0.4 μm in diameter, having a single outer membrane and small granules of high electron opacity similar to the ferritin molecules. Rather pure fractions of liver lysosomes have been obtained (Fig. 21–11), and large scale separation of these particles, as well as of peroxisomes, also from liver, has been achieved.[30] Bodies with similar morphologic characteristics were observed in intact liver cells and named "pericanalicular dense bodies" because of their preferential location along the fine bile canaliculi (Fig. 21–12).

Identification of these particles was made

easier when the histochemical techniques for acid phosphatase were carried out at the electron microscopic level (Fig. 21–13). (See Novikoff, 1961.) However, the considerable polymorphism shown by these particles in different cell types and even within a single cell remained as an obstacle to their identification.

According to the current interpretation, the polymorphism is the result of the association of primary lysosomes with the different materials that are phagocytized by the cell. A summary of these concepts is presented in Figure 21–14.

At present four types of lysosomes are recognized, of which only the first is the *primary lysosome;* the other three may be grouped together as *secondary lysosomes.*

(1) The *primary lysosome* (i.e., *storage granule*) is a small body whose enzymatic content is synthesized by the ribosomes and accumulated in the endoplasmic reticulum. From there it penetrates into the Golgi region where the first acid phosphatase reaction takes place.[31] The primary lysosome may be charged preferentially with one type of enzyme or another; it is only in the secondary lysosome that the full complement of acid hydrolases is present. The formation of pri-

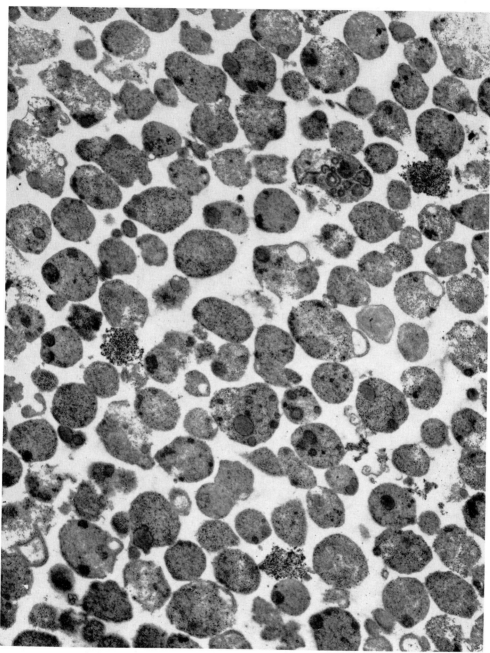

Figure 21–11.   Lysosomes isolated by differential centrifugation from rat liver, showing the very dense particles and the variety of other dense material contained within the single membrane of the lysosome. × 60,000. (Courtesy of C. De Duve.)

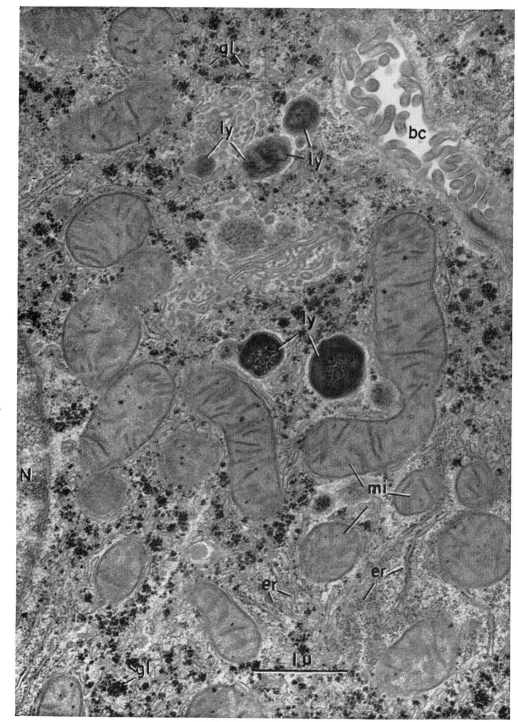

Figure 21–12. Peripheral region of a liver cell, showing a biliary capillary (*bc*) and several bodies interpreted as lysosomes (*ly*). *er*, endoplasmic reticulum; *gl*, glycogen; *mi*, mitochondria; *N*, nucleus. × 31,000. (Courtesy of K. R. Porter.)

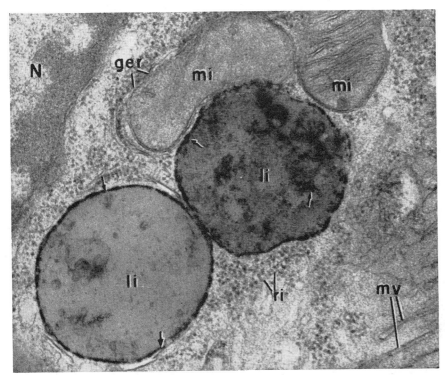

Figure 21–13. Electron micrograph of a proximal convoluted tubule cell of mouse kidney, two hours after injection of crystalline ox hemoglobin. Two absorption droplets (phagosomes, lysosomes) have formed at the apical region, and the acid phosphatase reaction becomes positive at the surface (arrows) and penetrates inside the lysosome. *ger,* granular endoplasmic reticulum; *li,* lysosomes; *mi,* mitochondria; *mv,* microvilli; *N,* nucleus; *ri,* ribosomes. × 60,000. (Courtesy of F. Miller.)

mary lysosomes may be followed in cultures of monocytes, which in the presence of serum proteins become transformed into macrophages. In a short time there is considerable synthesis of hydrolytic enzymes, which may be blocked by puromycin. In these activated cells using ³H-leucine and radioautography at the electron microscope level, the transfer of protein was observed in the following sequence: endoplasmic reticulum → Golgi complex → lysosomes.[32]

(2) The *secondary lysosome* (also called the *heterophagosome* or *digestive vacuole*) results from the phagocytosis or pinocytosis of foreign material by the cell. This body, which contains the engulfed material within a membrane, shows a positive phosphatase reaction, which may be due to the association with a primary lysosome. An interesting method for studying the heterophagosome consists of injecting peroxidase, which is engulfed by the cells and may be detected by a cytochemical reaction (see Chapter 8).[33]

In the macrophages mentioned earlier it may be observed that the *phagosome,* or engulfed vesicle, is surrounded by small Golgi vesicles (primary lysosomes) that fuse with it and form the secondary lysosome. The engulfed material is progressively digested by the hydrolytic enzymes which have been incorporated into the lysosome. The rate and extent of this digestion depends on the amount and chemical nature of the material and the activity and specificity of the lysosomal enzymes. Under ideal conditions digestion leads to products of low molecular weight which pass through the lysosomal membrane and are incorporated into the cell.

(3) *Residual bodies* are formed if the digestion is incomplete. In some cells, such as ameba and other protozoa, these residual bodies are eliminated by *defecation.* In other cells they may remain for a long time and may be important in the aging process. For example, the pigment inclusions found in nerve cells of old animals may be a result of this type of process.

Residual bodies may have important

INTRACELLULAR DIGESTIVE TRACT

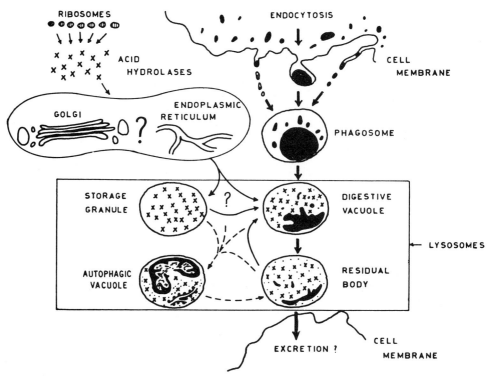

Figure 21–14. Diagram representing four functional forms of lysosomes and their possible interrelationships. (Courtesy of C. De Duve.)

pathologic implications. In some metabolic diseases the absence of some of the lysosomal enzymes may lead to enormous accumulations of products in the cell. This is observed in the various *lipidoses* in which accumulation of phospho- or sphingolipids in membranous formations takes place. In other instances the accumulation of glycogen in lysosomes may lead to severe pathologic disturbances. Residual bodies may be experimentally produced by injection of substances, such as triton WR-1339, dextran sucrose, which are accumulated but not digested; numerous swollen lysosomes result. In addition, about 12 congenital diseases involving lysosome malfunction have been described (Table 19–2).

(4) The *autophagic vacuole, cytolysosome,* or *autophagosome,* is a special case in which the lysosome contains part of the cell in a process of digestion (e.g., a mitochondrion or portions of the endoplasmic reticulum). A large number of these vacuoles are formed in certain physiologic and patho-

logic processes. For example, during starvation the liver cell shows numerous autophagic vacuoles in some of which mitochondrial remnants can be found. This is a mechanism by which the cell can achieve the degradation of its own constituents without irreparable damage. (See De Duve, 1967; Allison, 1967.) In the liver, autophagy may be induced by injection of the pancreatic hormone *glucagon.* This treatment produces a considerable increase in cytolysosomes while the small primary lysosomes diminish in number. This indicates that the preexisting lysosomes are probably the source of hydrolytic enzymes in the autophagic vacuoles.[34]

Lysosomes and Endocytosis

According to the current concepts, primary lysosomes are considered as a secretion product of the cell which, like other secretions (see Chapter 25), is synthesized by ribosomes, enters the endoplasmic reticulum, and reaches the Golgi region for final packaging. Since by this mechanism a cell may produce different types of lysosomes, and also *peroxisomes* and many other secretion products, it is likely that there is a kind of *topological specificity* in this endoplasmic reticulum-Golgi system. In other words, the different secretions may be dispersed through different channels of the intracellular membrane system (see Chapter 9). It is interesting to note that the Golgi complex may add some special products to the secretion such as antigenic proteins and glycoproteins. Using a special chromic acid-phosphotungstic acid mixture and the electron microscope, it was found that glycoproteins are stained in the Golgi region, around dense bodies (presumably lysosomes) and in the cell coat.[35] It has been suggested that glycoproteins are made in the Golgi region and migrate by way of small vesicles toward the lysosomes or the cell surface to constitute the extracellular coat (see Chapter 9).

This system of intracellular secretion for lysosomes is in some way coupled with another system of extracellular origin, formed by the process of endocytosis. It was shown earlier that this process is related to the activity of the plasma membrane. The products trapped by endocytosis probably never enter into the endoplasmic reticulum, and it may be postulated that there is a kind of *unidirectional lock* that allows the products to flow in only one direction. The fusion between the membranes limiting the endocytotic vacuole and the endoplasmic spaces may be prevented because between them there are differences in thickness and fine structure, as well as in chemical composition (see Chapter 9). In Figure 21–14 the thick arrows indicate the flow by way of endocytosis (phagocytosis and pinocytosis), the digestive vacuoles and the excretion of the residues; thinner arrows indicate the unidirectional flow of products.

Numerous examples of the relationship of lysosomes to phagocytosis and pinocytosis can be cited. An interesting case is observed in the kidney tubules when the animal is injected with some foreign material. By injecting the enzyme peroxidase into the animal, phagosomes are produced, and their fate within the cell can be followed by the peroxidase reaction. Injected hemoglobin is engulfed by the kidney tubular cells and the phagosome shows a positive phosphatase reaction, which starts at the periphery and with time "penetrates" to the interior (Fig. 21–13). Autophagic vacuoles surrounding mitochondrial remnants and residual bodies with a layered structure containing undigested material (probably of lipid nature) can also be observed after hemoglobin is injected.

Another interesting example is provided by the leukocytes that contain specific granules, particularly the neutrophils. Electron microscopic and experimental studies on these cells have shown that these granules are really packages of digestive enzymes corresponding to the lysosomes. If these leukocytes are put in contact with bacteria, the bacteria are engulfed and the granules become incorporated and dissolved in the digestive vacuoles. Eventually most granules may be lost by the cell in this process and in some virulent infections the cell may die.[36]

Lysosomes and Cell Autophagy. There is considerable evidence that lysosomes play a role in the removal of parts of cells, whole cells, and even of extracellular material. The case of starvation in the liver cell was noted earlier. During metamorphosis of amphibians there is considerable remodeling of tissues with destruction of numerous cells, and this is accomplished by lysosomal enzymes. For example, the degeneration of the tadpole tail is produced by the action of catepsins (i.e., proteolytic enzymes) contained in the lysosomes. It has been found that with the regression of the tail the concentration of catepsin increases progressively while the total amount of enzyme remains constant.[37] The possible role of lysosomes in regulating the amount of secretion products will be mentioned in Chapter 25.

Evidence indicates that lysosomal enzymes may be discharged outside the cell to produce lytic effects. This may be the mechanism by which *osteoclasts* remove bone. Then the broken parts of the bone may be engulfed and digested by these cells.

In cultured bone tissue receiving an excess of vitamin A, bone removal increases, which is apparently due to the "activation" of the lysosomes. This may be the cause of spontaneous fractures in animals that have vitamin A intoxication.

The opposite effect is observed with cortisone and hydrocortisone. These steroids, which have a well known anti-inflammatory action, also have a stabilizing effect on the membrane of the lysosome.

Peroxisomes and Glyoxysomes (Microbodies)

With improved cell fractionation methods a second group of particles, in addition to the lysosomes, has been isolated from liver cells and other sources. These particles, which were rich in the enzymes peroxidase, catalase, D-amino acid oxidase, and, to a lesser extent, urate oxidase, received the name *peroxisomes*.[38]

In addition to being present in the liver and kidney, *peroxisomes* have been found in protozoa, yeast, and many cell types of higher plants. Although these structures in plant cells show some morphologic similarities to the peroxisomes in animal cells, they have a different enzymatic content, including the enzymes of the glyoxylate cycle, hence their name, *glyoxysomes*.

Morphology

Electron microscopic studies have suggested that these particles correspond, morphologically to the so-called microbodies found in kidney and liver cells.[39, 40] These microbodies, or peroxisomes, are ovoid granules limited by a single membrane; they contain a fine, granular substance that may condense in the center, forming an opaque and homogeneous core (Fig. 9–12). In a quantitative study on rat liver cells the average diameter of peroxisomes was shown to be 0.6 to 0.7 μm. The number of peroxisomes per cell varied between 70 and 100, whereas 15 to 20 lysosomes were found per liver cell.[41] In many tissues peroxisomes show a crystal-like body made of tubular subunits. The number of organelles having these bodies is sometimes correlated with the content of urate oxidase.

Morphogenesis

Peroxisomes are intimately related to the endoplasmic reticulum (Chapter 9). Both in animal and plant cells they appear to be formed as dilatations of this part of the vacuolar system. Regions of endoplasmic reticulum become swollen and filled with an electron-dense substance. Frequently, there are continuities between the endoplasmic reticulum and the membrane of peroxisomes.

The formation of these organelles has been followed in embryonic hepatocytes by using a histochemical reaction that uses the oxidation of 3'-3'-diaminobenzidine (DAB).[42] The enzymes of peroxisomes are synthesized in the ribosomes attached to the endoplasmic reticulum. In fact, catalase was found to be formed in the microsomal fraction of liver.[43] From studies of this type it was assumed that peroxisomes grow slowly and are destroyed after a life span of four to five days.[44] The half-life of catalase was found to be about 36 hours.[45] Studying the catalase content of peroxisomes of different size investigators found that all of them had the same content of enzyme. It has been postulated that peroxisomes are formed very rapidly and that their transport from the endoplasmic reticulum is completed in about one hour. Then, after four to five days, these organelles are probably destroyed by autophagy.

Enzyme Content and Function

Isolation of liver peroxisomes has demonstrated that these organelles contain four enzymes related to the metabolism of H_2O_2. In fact, three of them—urate oxidase, D-amino oxidase, and α-hydroxylic acid oxidase—produce peroxide (H_2O_2), and catalase destroys it. *Catalase* appears to be in the matrix of liver peroxisomes and represents up to 40 per cent of the total protein. Since H_2O_2 is toxic to the cell, catalase probably plays a protective role. The enzyme *urate oxidase* and two other enzymes present in amphibian and avian peroxisomes are related to the catabolism of purines.

Glyoxysomes

Microbodies are a constant feature of plant cells; however, although their morphology is quite similar, they have a more heterogeneous enzyme content than animal cells. For example, during germination of the seeds of *Ricinus*, the fat stored in the endosperm is transformed into carbohydrates by way of the *glyoxylate cycle*, which is a modification of the Krebs cycle (Chapter 4). The overall equation of this cycle is: 2 acetyl CoA $\rightarrow$ succinate + 2H. The difference between the glyoxylate cycle and the Krebs cycle is that the former uses two auxiliary enzymes: *isocitratase* and *malate synthetase* and requires two molecules of acetyl CoA, instead of one. The enzymes of the entire glyoxylate cycle are present in the glyoxysomes.[46, 47] In green leaves there are peroxisomes that carry out a process called *photorespiration.* In this process, *glycolic acid,* a two-carbon product of photosynthesis that is released from chloroplasts, is oxidized by *glycolic acid oxidase,* an enzyme present in peroxisomes. This oxidation, carried out by oxygen, produces hydrogen peroxide, which is then decomposed by catalase inside the peroxisome. Photorespiration is so-called because light induces the synthesis of glycolic acid in chloroplasts. The entire process involves the intervention of two basic organelles: chloroplasts and peroxisomes.

SUMMARY: Endocytosis, Lysosomes, and Peroxisomes

Bulk ingestion of solid or fluid material by the cell is called, respectively, *phagocytosis* and *pinocytosis.* Both processes show similarities and are included within the more general concept of *endocytosis.*

Phagocytosis is found in many protozoa and in certain cells of metazoa where, principally, it plays a defense role against bacteria and colloidal particles. In the ameba, phagocytosis is related to ameboid motion. The particles are surrounded by pseudopodia and are digested inside food vacuoles. In phagocytosis two distinct phases may be distinguished: adsorption of the particle and actual penetration. The macrophagic system in Metazoa accumulates negatively charged vital dyes and colloids, concentrating them into vacuoles. Pinocytosis is easily observed in the ameba as the formation of fluid channels that break up inside the cell into vacuoles. This process is induced by addition of negatively charged proteins, amino acids, and ions. In a fluorescent protein solution the cell coat of an ameba concentrates the protein 50 times or more; then the protein is taken up by invagination of the cell membrane. The relation between pinocytosis and the cell coat is best observed using colloidal gold and the electron microscope. With this instrument it is observed that pinocytosis is widely present in endothelial and other cells as small vesicles of 65 nm (i.e., *micropinocytosis*). These vesicles are engaged in transcellular fluid transport. An important case is found in the erythroblasts in which the ferritin molecules used to manufacture hemoglobin are transferred into the cell by micropinocytosis. Endocytosis, in general, is an active process that requires energy.

Lysosomes are organelles that contain numerous hydrolytic enzymes and can be separated as an intermediate fraction between mitochondria and microsomes. The enzymes, enclosed within a membrane, are maintained in a latent form. They become active if the membrane is broken and the substrates, in acid pH, become accessible. Acid phosphatase, glucuronidase, sulfatase, catepsin, ribonuclease, and α-glucuronidase are some of the lysosomal enzymes. Lysosomes show considerable polymorphism. The *primary lysosomes* (i.e., storage granules) are dense particles of about 0.4 μm surrounded by a single membrane. Their enzymatic content is synthesized by ribosomes in the endoplasmic reticulum and appear in the Golgi region. The formation of primary lysosomes can be blocked by puromycin. *Secondary lysosomes* (i.e., digestive vacuoles) result from the association of primary lysosomes with vacuoles containing phagocytized material. The so-called *phagosome* fuses with lysosomes (i.e., *heterophagosome*) and is digested by hydrolytic enzymes. Sometimes *residual bodies* containing undigested material are formed. These structures may be eliminated, but in most cases they remain in the cell as pigment inclusions and may be related to the aging process.

Several congenital diseases in which some lysosomal enzymes are absent produce the enormous accumulation of undigested products in the cell (e.g., lipidosis, glycogen accumulation). The *cytolysosome* or *autophagic vacuole* consists of the formation of intracellular vacuoles around mitochondria, the endoplasmic reticulum, and other parts of the cell. These parts are then degraded by lysosomes. Cell autophagy plays an important role in metamorphosis of amphibians and in the removal of bone by osteoclasts. Lysosomes are intimately related to the processes of endocytosis.

Peroxisomes are organelles containing peroxidase, catalase, and other oxidases. In liver and kidney they are found as 0.6 to 0.7 μm granules having a single membrane and a dense matrix. Frequently, a crystal-like condensation is observed. The enzymes of peroxisomes are involved in the metabolism of hydrogen peroxide (H_2O_2). Catalase decomposes peroxide ($2H_2O_2 \rightarrow 2H_2O + O_2$) and has a protective effect on the cell. Urate oxidase and other enzymes are related to the purine metabolism.

Some microbodies of plant cells are called *glyoxysomes* because they contain the enzymes of the *glyoxylate cycle*. *Photorespiration,* a process present in green leaves, involves the cooperation of chloroplasts and peroxisomes.

REFERENCES

1. Woodbury, J. W. (1961) The cell membrane: ionic and potential gradients and active transport. In: *Neurophysiology*, p. 2. (Ruch, T. C., Patton, H. D., Woodbury, J. W., and Towe, A. L., eds.) W. B. Saunders Co., Philadelphia.
2. Collander, R., and Bärlund, H. (1933) *Acta Bot. Fenn.*, *11*:1.
3. Ussing, H. H. (1960) Physiology of the cell membrane. *J. Gen. Physiol.*, *43*:5, part. 2, suppl. 1, p. 135.
4. Leaf, A. (1960) Physiology of the cell membrane. *J. Gen. Physiol.* *43*:5, part 2, suppl. 1, p. 175.
5. Wittembury, G. (1960) Physiology of the cell membrane. *J. Gen. Physiol.*, *43*:5, part 2, suppl. 1, p. 43.
6. Eccles, J. C. (1957) *Physiology of Nerve Cells.* Johns Hopkins Press, Baltimore.
7. Backer, P. F., Hodgkin, A. L., and Shaw, T. I. (1962) *J. Physiol.* (London) *164*:330 and 335.
8. Woodbury, J. W. (1969) In: *Basic Mechanisms of Epilepsies*, p. 41. (Jasper, H. H., Ward, A. A., and Pope, A., eds.) Little, Brown and Co., Boston.

9. Solomon, A. K. (1960) Physiology of the cell membrane. *J. Gen. Physiol., 43*:5, part. 2, suppl. 1, p. 1.
10. Kennedy, E. P., Fred Fox, C., and Carter, J. R. (1966) *J. Gen. Physiol., 49*:347.
11. Cohen, G., and Monod, J. (1957) *Bact. Rev., 21*:169.
12. Sen, A. K., and Post, R. L. (1964) *J. Biol. Chem., 239*:345.
13. Marchesi, V. T., and Palade, G. E. (1967) *J. Cell Biol., 35*:385.
14. Lewis, W. H. (1931) *Bull. Johns Hopkins Hosp., 49*:17.
15. Holter, H., and Marshall, J. M., Jr. (1954) *C. R. Lab. Carlsberg,* série chim., *29*:27.
16. Chapman-Andersen, C., and Holter, H. (1955) *Exp. Cell Res.,* suppl. 3, 52.
17. Brandt, P. W. (1962) *Symposium on the Plasma Membrane.* (New York) Heart Association, Inc. *Circulation, 26*:1075.
18. Brandt, P. W., and Pappas, G. D. (1960) *J. Biophys. Biochem. Cytol., 8*:675.
19. Chapman-Andersen, C., and Holter, H. (1964) *C. R. Lab. Carlsberg, 34*:211.
20. Gosselin, R. E. (1967) *Fed. Proc., 26*:987.
21. Brandt, P. W., and Freeman, A. R. (1967) *Science, 155*:582.
22. Palade, G. E. (1968) *Anat. Rec., 130*:467.
23. Palade, G. E., and Bruni, R. R. (1968) *J. Cell Biol., 37*:633.
24. Karnowsky, M. J. (1967) *J. Cell Biol., 35*:213.
25. Bessis, M., and Breton-Gorius, J. (1959) *J. Rev. Hémat., 14*:165.
26. Sastry, P. S., and Hokin, L. E. (1966) *J. Biol. Chem., 241*:3354.
27. Gropp, A. (1963) In: *Cinematography in Cell Biology*, p. 279. (Rose, G. G., ed.) Academic Press, Inc., New York.
28. De Duve, C. (1963) General properties of lysosomes. In: *Lysosomes,* p. 1. Ciba Foundation Symposium. J. and A. Churchill, London.
29. Rahman, V. E., Lowes, J. F., Nance, S. D., and Thomson, J. F. (1967) *Biochim. Biophys. Acta, 146*:484.
30. Leighton, F., Poole, B., Beaufay, H., Baudhuin, P., Coffey, J. W., Fowler, S., and De Duve, C. (1968) *J. Cell Biol., 37*:207.
31. Essner, E., and Novikoff, A. B. (1962) *J. Cell Biol., 15*:289.
32. Cohn, Z. A. (1968) *Excerpta Medica Internat. Congr.,* Ser *166* :6.
33. Straus, W. (1967) *J. Histochem. and Cytochem., 15*:375 and 381.
34. Deter, R. L., Baudhuin, P., and De Duve, C. (1967) *J. Cell Biol., 35*:C11.
35. Rambourg, G. (1966) *Anat. Rec., 154*:41.
36. Hirsch, J. G., and Cohn, Z. A. (1960) *J. Exp. Med., 112*:1005.
37. Weber, R., and Niehus, B. (1961) *Helv. Physiol. Pharmacol. Acta, 19* :103.
38. Beaufay, H., and Berther, J. (1963) In: *Methods of Separation of Subcellular Structural Components,* p. 66. *Biochem. Soc. Symp.,* No. 23 (Grant, J. K., ed.) Cambridge University Press, London.
39. Rodhin, J. (1954) Thesis, Karolinska Institutet, Stockholm.
40. Rouiller, C., and Bernhard, W. (1956) *J. Biophys. Biochem. Cytol., 2*:355.
41. Loud, A. V. (1968) *J. Cell Biol., 37*:27.
42. Essner, E. (1968) *J. Cell Biol., 39*:42a.
43. Higashi, T., and Peters, T. (1963) *J. Biol. Chem., 238*:3952.
44. De Duve, C., and Baudhuin, P. (1966) *Physiol. Rev., 46* :323.
45. Poole, B., Leighton, F., and De Duve, C. (1970) *J. Cell Biol., 41*:536.
46. Breidenbach, R. W., Kahn, A., and Beevers, H. (1968) *Plant Physiol., 43*:705.
47. Vigil, E. L. (1970) *J. Cell Biol., 46*:435.

Allison, A. (1957) Lysosomes and disease. *Sci. Am., 217*:62.
Anderson, B., and Ussing, H. H. (1960) Active transport. *Comp. Biochem. Physiol., 2*:371.
Ciba Foundation Symposium (1963) *Lysosomes.* J. & A. Churchill Ltd., London.
Chapman, D. (1968) *Biological Membranes.* Academic Press, Inc., New York.

ADDITIONAL READING

De Duve, C. (1967) Lysosomes and phagosomes. *Protoplasma, 63*:95.

Eisenman, G., Sandblom, J. P., and Walker, J. L., Jr. (1967) Membrane structure and ion permeation. *Science, 155*:3765.

Epstein, E. (1973) Mechanisms of ion transport through plant cell membranes. *Internat. Rev. Cytol., 34*:123.

Holter, H. (1960) Pinocytosis. *Internat. Rev. Cytol., 8*:481.

Mullins, L. J. (1968) From molecules to membranes. *Fed. Proc., 27*:898.

Ponder, E. (1961) The cell membrane and its properties. In: *The Cell.* Vol. 2, p. 1. (Brachet, J., and Mirsky, A. E., eds.) Academic Press, Inc., New York.

Robertson, R. N. (1960) Ion transport and respiration. *Biol. Rev., 35*:231.

Rothfield, L. I. (1971) *Structure and Function of Biological Membranes.* Academic Press, Inc., New York.

Rustad, R. C. (1961) Pinocytosis. *Scient. Amer., 204,* No. 4:121.

Schwartz, A., Lindenmayer, G. E., and Allen, J. C. (1972) The Na^+, K^+ ATPase membrane transport system: importance in cellular function. In: *Current Topics in Membranes and Transport*, Vol. 3, p. 2. (Bronner, F., and Kleinzeller, A., eds.) Academic Press, Inc., New York.

twenty-two

PRIMITIVE CELL MOVEMENTS. CILIA, CENTRIOLES, MICROTUBULES, AND MICROFILAMENTS

The energy produced by the cell, in addition to being used in chemical transformations, such as protein synthesis, can be consumed in the *mechanical activity* of the cell. Several forms of energy can be included in this type, but the most important becomes apparent as *cell motion*. In certain cases cell movement occurs within the protoplasm and produces no exterior deformation of the cell. This type of motion is called *cytoplasmic streaming,* or *cyclosis.* In other cases, the movement is manifested as the emission of pseudopodia, which leads to displacement of the cell *(ameboid movement).* Furthermore, movements may occur in specially differentiated appendices—*ciliary* and *flagellar motion*—or in specific cytoplasmic fibrils—*muscular motion.*

Ameboid, ciliary, and flagellar types of motion are the main means of locomotion in unicellular organisms. Plant cells are displaced mainly by growth and changes in water content (turgor), but some plant gametes (antherozoids) have flagella. In animals, embryonic cells may move considerably during organogenesis and histogenesis. Cells in tissue culture, in healing wounds, and in cancerous tissue move freely. In mature animals, only gametes, ciliated epithelia, wandering ameboid cells, and the cells of different types of muscle tissue perform visible movement. In this chapter the more primitive types of motion will be considered.

CILIA AND CILIARY MOTION

Ciliary motion is adapted to liquid media and is executed by minute, specially differentiated appendices that vary in size and number. They are called *flagella* if they are few and long, and *cilia,* if short and numerous. In Protozoa, especially the Infusoria, each cell has hundreds or thousands of minute cilia, and their movement permits a rapid progression of the organism in the liquid medium. In some special regions of Infusoria, several cilia fuse and form larger conical appendices, *the cirri,* or membranes known as *undulating membranes.*

One entire class of Protozoa, the Flagellata, is characterized by the presence of flagella. The spermatozoa of metazoans move as isolated cells by means of flagella. On the other hand, epithelial cells that possess vibratile cilia and constitute true ciliated sheets are relatively common. These may cover large areas of the external surface of the body and determine the motion of the animal. Such is the case with some Platyhelminthes and Nemertea and also with larvae of Echinodermata, Mollusca, and Annelida. More often, the ciliated epithelial sheets line cavities or internal tubes, such as the air passages of the respiratory system or various parts of the genital tract. In these organs all the cilia move simultaneously in the same direction, and

497

fluid currents are thus produced. In some cases, the currents serve to eliminate solid particles in suspension (e.g., in the respiratory system). The eggs of amphibians and mammals are driven along the oviduct with the aid of vibratile cilia.

Structure of the Cilia and Flagella

The essential components of the ciliary apparatus are (1) the *cilium,* which is the slender cylindroid process that projects from the free surface of the cell, (2) the *basal body,* or granule, the intracellular organelle similar to the centriole from which it originates, and (3) in some cells fine fibrils—called *ciliary rootlets*—that arise from the basal granule and converge into a conical bundle, the pointed extremity of which ends at one side of the nucleus.

The basal bodies are embedded in the

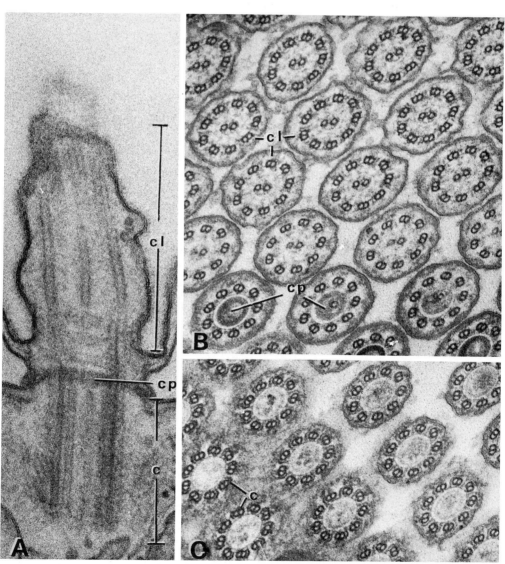

Figure 22–1. Electron micrographs of cilia in longitudinal and cross sections. **A,** cilium of *Paramecium aurelia* showing the centriole or basal body (*c*), the ciliary plate (*cp*), and the cilium (*cl*) proper. **B** and **C,** cross sections through cilia of *Euplotes eurystomes*. **B,** the section passes through the cilium proper showing the typical structure and the ciliary plate (*cp*). **C,** the section passes through the centriole or basal body (*c*). Notice the absence of central tubules and triple number of peripheral tubules. A, × 110,000; B and C, × 72,000. (Courtesy of J. André and E. Fauret-Fremiet.)

ectoplasmic layer beneath the cell surface. In general, they are spaced uniformly and in parallel rows.

Cilia and flagella are extremely delicate filaments whose thickness is often at the limit of the resolving power of the light microscope.

Various epithelia have appendices similar in shape to cilia, but immobile; these are called *stereocilia*. Examples are the processes of the epithelial cells of the epididymis, which seem to intervene in the elimination of cellular secretion. In the macula and crista of the inner ear, there are stereocilia in addition to motile cilia, or *kynocilia*. Stereocilia do not contain microtubules.

Some evidence that cilia and flagella were composed of finer fibrillar elements was obtained by early cytologists at the end of the last century (see Fawcett, 1961). The spermatozoan tail was occasionally seen to fray into minute fibrils; direct evidence of the fibrillar elements was finally obtained with the electron microscope. Both in spermatozoa[1] and antherozoids (plant male gamete) it was found that the flagellum contained 11 fibrils of which two were smaller than the rest.[2] The 9 + 2 fundamental structure was confirmed when their sectioning became available.[3] Furthermore, it was found that the nine thicker fibrils consisted of *doublets or pairs of microtubules* and were arranged along the periphery, whereas the two smaller fibrils were *single microtubules*, centrally located (Fig. 22–1). Thus, cilia and flagella should be considered as organelles composed mainly of microtubules.

Fine Structure of the Axoneme

Axoneme is the term applied to the axial basic microtubular structure of cilia and flagella and is thought to be the essential motile element. In various flagella *accessory structures* made of fibers, microtubules, crystalline bodies, or mitochondrial derivatives may be found, but these are not considered to be an integral part of the motile mechanism. The axoneme of a cilium or flagellum may range from a few microns to 1 to 2 millimeters in length, but the outside diameter is about 0.21 μm. It is generally surrounded by an *outer ciliary membrane,* which is continuous with the plasma membrane. All the components of the axoneme are embedded within the *ciliary matrix* (Fig. 22–1).

Figure 22–2 shows a general diagram of the axoneme, with the typical 9 + 2 microtubular pattern and the other essential structural components of the axoneme. This is the most generally found structure; the main variations so far described refer to the central microtubules (see Warner, 1972). A plane perpendicular to the line joining the two central tubules divides the axoneme into a right and a left symmetrical half. It is now well demonstrated that the plane of the ciliary beat is perpendicular to this plane of symmetry.

The pairs of peripheral microtubules have an ellipsoidal profile, whereas the central ones are circular. The diameter of the peripheral tubules varies from 18 to 25 nm with the central tubules being slightly larger. The two microtubules of each peripheral pair can be clearly distinguished by several morphologic features. The doublets are skewed at about 10 degrees, so that one tubule, designated subfiber A lies closer to the axis than the other (i.e., subfiber B). The microtubule of subfiber A is smaller, but complete, whereas that of subfiber B is larger and incomplete,

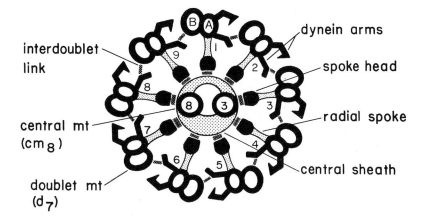

Figure 22–2. Diagram of a cross section of a cilium showing the axoneme with the "9 + 2" microtubular structure. The axoneme is viewed from base-to-tip, with the arms directed clockwise. (See the description in the text.) Modified from Warner, F. D., Macromolecular organization of eukaryotic cilia and flagella, *in* DuPraw, E. J. (Ed.) *Advances in Cell and Molecular Biology,* Vol. 2, p. 193, New York, Academic Press, Inc., 1972.)

interdoublet link

central mt (cm_8)

doublet mt (d_7)

dynein arms

spoke head

radial spoke

central sheath

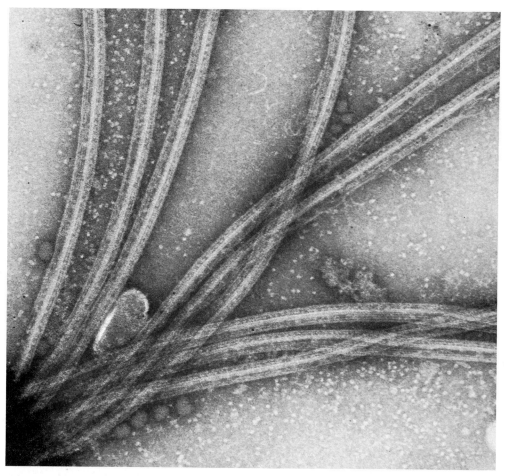

Figure 22-3. Axial filament complex from one flagellum of a spermatozoon of *Childia groenlandica*. The nine pairs of peripheral microtubules are negatively stained with phosphotungstic acid. Observe the fine structure (i.e., subunits) on the wall of the microtubule. × 108,000. (Courtesy of D. P. Costello and C. Henley.)

since it lacks the wall adjacent to A. Furthermore, subfiber A has processes—the so-called *arms*—that are oriented in the same direction in all microtubules. This orientation is clockwise when the axoneme is viewed from base to tip. In Figure 22-2 the nine pairs are numbered, starting from the one lying in the plane of symmetry, and continuing in the clockwise direction.

Microtubular Structure. As in the case of spindle microtubules (see Chapter 13) in cross section, 12 to 13 filamentous subunits have been observed in the wall of ciliary microtubules. Negative staining with phosphotungstic acid or uranyl acetate penetrates the lumen of the tubule and better defines the filaments and bead-like subunits composing the wall.[4, 6, 7] These 4 to 5 nm subunits are

spaced in a lattice of 5 to 6 nm, center to center[8] (Fig. 22-3). Such subunits may correspond to the monomeric form of *tubulin*, the protein forming the wall of the microtubule (see below).

Dynein Arms. In each subfiber A the two *arms* may be observed (Fig. 22-2). The *outer arm* terminates in a hook-like bend, whereas the *inner arm* has no terminal hook. A fine linkage, termed the *peripheral link*, appears to connect the end of the inner arm to the adjacent subfiber B.[9, 10] In longitudinal sections, the arms appear as regularly spaced projections along subfiber A, with a period ranging from 16 to 22 nm. It will be shown later that the arms contain a protein having ATPase activity—the so-called *dynein*.[11]

Radial Linkage. In Figure 22-2 cross-bridges or radial links are seen to connect the peripheral doublet to the axial component which contains the two central microtubules and is surrounded by a *central sheath*.[12, 13] These links are about 36 nm long; they attach to subfiber A on one side,[4, 5] and end in a dense knob or head which may have a fork-like structure. In longitudinal sections the radial links are spaced alternately at 32, 56 or 64 nm. In the diagram of Figure 22–2 the heads of the radial links are seen to join the central sheath by way of a *transitional junction*. In certain axonemes, the central sheath contains a helically wound fiber with a space of 16 nm.[14]

Basal Bodies (Kinetosomes) and Centrioles

Since the classic works of Henneguy and Lenhossek in 1897, it has been suggested that basal bodies (or kinetosomes) of cilia and flagella are homologous with the centrioles found in mitotic spindles (Fig. 13–1). In some cells it was discovered that a centriole engaged in mitosis could carry a cilium at the same time. This homology was fully con-

firmed with the introduction of electron microscopy.

Centrioles are cylinders that measure, on the average, 0.2 μm × 0.5 μm; at times they may be as long as 2 μm. This cylinder is open on both ends, unless it carries a cilium. In the latter case, it is separated from the cilium by a *ciliary plate* (Fig. 22–1, *A*).

The wall of the centriole has nine groups of microtubules arranged in a circle. Each group is a triplet formed of three tubules (rather than two, as in cilia) that are skewed toward the center (Fig. 22–4). Since the angle made by the set of tubules with the tangent to the cylinder varies from 70 degrees in the proximal end to 20 to 30 degrees in the distal end, the tubules twist from one end to the other[15] or describe a helical course.[16] As shown in Figure 22–4, the tubules are designated *subfiber A, B,* and *C,* from the center toward the periphery. Both subfibers A and B cross the ciliary plate and are continuous with the corresponding subfibers in the axoneme; subfiber C terminates near the *ciliary or basal plate*. There are no central microtubules in the centrioles and no special arms; however, they are linked by connectives. The proximal portion of the centriole has a cartwheel appear-

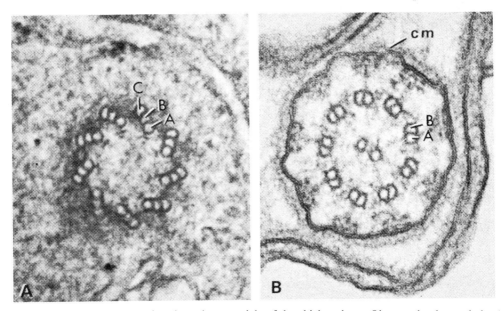

Figure 22–4.   **A,** transverse section through a centriole of the chick embryo. Observe the three tubules (*A, B, C*) present in each of nine groups at the periphery of the centriole. Also note the absence of central tubules and the density of the centriolar wall. **B,** transverse section through a cilium (compare with the centriole structure); *cm,* ciliary membrane. × 150,000. (See Figure 22–1 for further details.) (Courtesy of J. André.)

ance with spokes that radiate from a central hub and connect to subfiber A of each triplet. In longitudinal sections the spokes show a repeating period of 10 to 25 nm. The presence of the cartwheel structure at the proximal end provides the centriole with a structural and functional polarity. The growth of the centriole is from the distal end, and in the case of kinetosomes, it is from this end that the cilium is formed. Furthermore, the *procentrioles,* which are formed at right angle to the centriole, are located near the proximal end (see Wolfe, 1972). It can be demonstrated experimentally that each centriole may be related to the spindle; however, in normal mitosis the spindle microtubules terminate near only one of the centriole of the pair (see Chapter 13). Both centrioles may give rise to a cilium, but in most cases only one does, and the sister centriole remains at a right angle to the kinetosome. In most multiciliated cells there are single basal bodies. It will be shown later that in sensory cells having nonmotile ciliary derivatives and a 9 + 0 configuration, there are two basal bodies (see Fig. 22–5).

The two centrioles of a pair are always located at a distance from each other of at least 0.8 μm and are perpendicular to each other. In cancerous tissue, however, this orientation may be quite disturbed.

Ciliary Rootlets. In some cells, ciliary rootlets originate from the basal body. Most rootlets are striated, having a regular cross-banding with a repeating period of 55 to 70 nm. Five intraperiodic sub-bands have been observed in the cross-striated rootlets.

The striated fibers are composed of parallel microfilaments, 3 to 7 nm in diameter, which in turn are formed of globular subunits. These fibers and filaments may serve a structural role such as anchoring the kinetosomes. By analogy with other microfilaments (see below), a contractile role has been postulated for the ciliary rootlets. Furthermore, ATPase has been found associated with the cross-bands of certain rootlets. *Transitional fibers,* radiating from the triplet and joining the cell membrane, have also been described.

Basal Feet and Satellites. Basal feet are

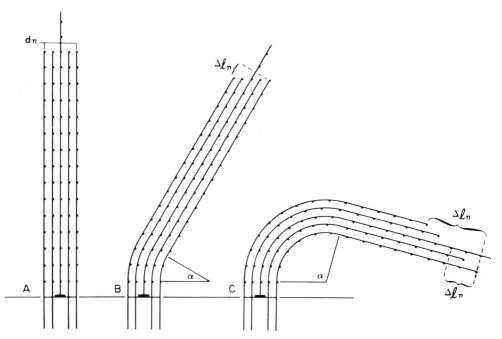

Figure 22–5. Diagram of the sliding model of the axoneme. The amount of displacement (Δ_{ln}) is a function of the distance separating the microtubules (d_n) and the bend angle (α). The central microtubules are represented by the longer lines. (From Sleigh, M. A., The physiology and biochemistry of cilia and flagella, *in* Lima-De-Faría, A. (Ed.), *Frontiers of Biology: Handbook of Molecular Cytology.* Vol. 15, p. 1243. Amsterdam, North-Holland Publishing Co., 1970.)

dense processes that are arranged perpendicularly to the basal body in a particular direction and originate from two or three of the triplets.[17] These processes impose a structural asymmetry on the basal body that has been related to the direction of the ciliary beat. The basal feet, which are composed of microfilaments that terminate in a dense bar, may be a focal point for the convergence of microtubules. *Satellites or pericentriolar bodies* are electron-dense structures lying near the centriole that probably are nucleating sites for microtubules.[18]

Biochemistry of Cilia and Flagella

Cilia and flagella may be isolated by various procedures;[4, 19] the ciliary membrane may then be removed in a calcium free medium or by detergents.[20] Relatively clear axonemes, lacking only the soluble proteins of the matrix, are obtained; these axonemes are still reactivated by ATP.[21] Axonemes may be treated further with detergents to solubilize selectively the structural components, or they may be treated with trypsin, which disrupts the linkages between the microtubules.[22]

Tubulins. Two distinct proteins have been isolated from the ciliary peripheral microtubules (i.e., doublets) by gel electrophoresis.[23] These proteins have been termed *tubulin* A and B and have a molecular weight of 55,000 to 60,000.[24] Molecules of this size correspond to those obtained from other microtubules.[23, 25] It was thought that tubulin A corresponded to the subfiber A, and tubulin B, to the subfiber B;[24] however, further studies in *Chlamydomonas* indicate that both tubulins are present in the various microtubules of the axoneme.[20]

In whole flagella or in the soluble fraction, there are about 20 to 25 secondary proteins in addition to the tubulins; at least ten of these proteins may be structural in nature.

Dynein. An axonemic protein having ATPase activity has been isolated from cilia and flagella.[26] Two fractions sedimenting at 30S and 14S were separated. The latter, which has a molecular weight of about 600,000, is probably a monomer of the 30S (molecular weight of 5×10^6 daltons). Dynein appears to comprise part of the arms of subfiber A. In fact, in reconstitution experiments made with the soluble fraction added to the insoluble component, in the presence of Mg^{2+}, the arms or subfiber A reappears. The

possibility that ATPase may also be related to the radial linkages and their attachment to the core surface has also been suggested (see Warner, 1972).

Physiology of Ciliary Movement

Ciliary movement can be analyzed easily by scraping the pharyngeal epithelium of a frog or toad with a spatula and placing the scrapings in a drop of physiologic salt solution between a slide and a coverglass. On the free surface of the epithelial cell, the rapid motion of the vibratile cilia can be seen. If a row of cilia is observed, the contraction is *metachronic* in the plane of the direction of motion; that is, it starts before or after the contraction of the next cilium. In this way true waves of contraction are formed.

On the other hand, in a plane perpendicular to the direction of motion, the contraction is *isochronic;* all the cilia are observed in the same phase of contraction at a given time. This coordination of the ciliary movement implies the existence of a regulatory mechanism, the nature of which is unknown. The rhythmic contraction of cilia has been interpreted in different ways. A two-step process that involves *intraciliary* excitation followed by *interciliary* conduction has been proposed to explain the metachronic rhythm of the ciliary beat.[27] This mechanism evidently does not depend on the nervous system, since it persists after the epithelium has been separated from the rest of the organism. However, cytoplasmic continuity is indispensable to its maintenance, for if a cut is made in the row of cilia the waves of contraction of the two isolated pieces become uncoordinated.

The direction of the effective ciliary beat appears to be a fixed characteristic that also depends on the underlying cytoplasm. If a piece of epithelium is removed from the pharynx of a frog and implanted with a reversed orientation, the movement is maintained but in a direction opposite to that on the remaining intact epithelium.

Ciliary contractions are generally rapid (10 to 17 per second in the pharynx of the frog). Analysis of the motion has been facilitated greatly by stroboscopic and ultrarapid microcinematography.[28]

Ciliary movement may be pendulous, unciform (hooklike), infundibuliform, or undulant. The first two are carried out in a single plane. In the pendulous movement, typical of the ciliated Protozoa, the cilium is rigid and the motion is carried out by a flexion at its base. On the other hand, in the unciform movement, the most common type in the Metazoa, the cilium upon contraction is doubled and takes the shape of a hook. In the infundibuliform movement, the cilium or flagellum rotates, passing through three mutually perpendicular planes in space, describing a conical or funnel-shaped figure. In the undulant motion, characteristic of the flagella and membranes, contraction waves proceed from the site of implantation and pass to the free border.

Initiation and propagation of the contraction, however, is apparently a property of the entire flagellum. In flagella severed by laser microbeam irradiation it has been found that the isolated pieces were able to initiate and propagate the contraction cycle.[29] It was shown above that the plane of the beat of a cilium is perpendicular to that passing through the central microtubules. In Figure 22–2, this plane passes through Doublet 1 and between Doublets 5 and 6. However, there are many indications that there is a three-dimensional movement during the beat. Although the effective stroke is in a single plane, the cilium moves away from the original plane during the recovery phase. This movement, which can be beautifully demonstrated by scanning electron microscopy,[30] implies that either the whole cilium rotates at its base or that the central microtubules rotate.

Macromolecular Mechanisms. Since the last century, ciliary and flagellar motion have been compared to that of muscle. However, the complex structure of the axoneme suggests that this motion may be more complex than the planar sliding of myofilaments in muscle (see Chapter 23). Two main theories that imply a localized contraction or a sliding mechanism have been postulated. The latter is most favored because the sliding of doublets during the binding of the cilium or flagellum has actually been observed and measured. At present, the molecular mechanisms that generate the sliding and bending are unknown.

The displacement of the doublets during bending has been determined because the central microtubules extend to the very tip, whereas subfibers B terminate at a distance from the tip of 1 to 2 μm, and subfibers A terminate slightly beyond subfibers B. This arrangement has also made it possible to demonstrate that the two subfibers do not move relative to one another within the doublet.[31] In the diagram of Figure 22–5 the amount of sliding displacement is shown to depend on the distance separating the tubules and on the bend angle (see Sleigh, 1971).

Most of the experimental work on ciliary motion has been aimed at demonstrating the involvement of ATP; it was mentioned above that the ATPase activity is associated with the protein *dynein* in the arms of subfibers A. Dynein constitutes about 8 per cent of the total ciliary protein. In glycerin-extracted cilia, flagella, and spermatozoan tails, the addition of ATP produces rhythmic activity that may persist for a few minutes or even hours. In axonemes treated with trypsin, which disrupts the linkages, the addition of ATP still produces the sliding of the doublet microtubules. However, there is no active bend formation and propagation.[22]

These experiments demonstrate that the sliding mechanism of ciliary motion is localized in the fibrils of the axoneme and that the energy source is ATP. Linkages between the axoneme microtubules appear to be needed for complete function.

Ciliary Derivatives

Studies on the structure of retinal rods and cones have shown that the short, fibrous connection found between the outer and inner segments is of a ciliary nature. Cross sections of this so-called connecting cilium have revealed nine pairs of filaments similar to those found in cilia (Fig. 22–6).[32] The filaments of the connecting cilium originate from one of the centrioles, whereas the other centriole is arranged perpendicularly. The outer segment is composed of numerous double membrane rod sacs arranged like a stack of coins. The first stage in the development of a rod is a primitive cilium projecting from the bulge of cytoplasm that constitutes the primordium of the inner segment (Fig. 22–7). This cilium contains the nine pairs of filaments and the two basal centrioles.[32] The apical end is filled with a vesicular material. In the second stage the apical region of the primitive cilium enlarges greatly, owing to the rapid building up of the vesicles and cisternae that constitute the primitive rod sacs. The proximal part of the primitive cilium remains undifferentiated and constitutes the connecting cilium of the adult (Fig. 22–7).

Other structures have also been recog-

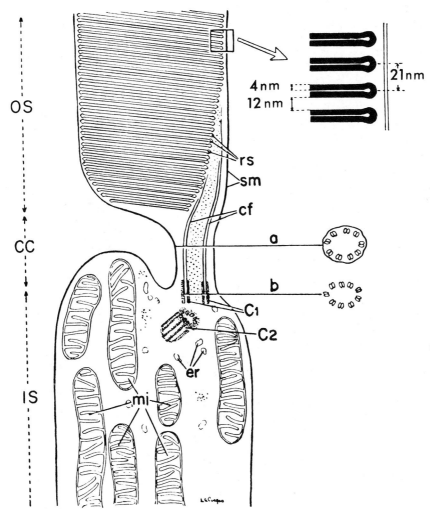

Figure 22–6.   Diagram of a retinal rod cell in the rabbit: the outer segment (*OS*) with the rod sacs (*rs*); the connecting cilium (*CC*) and the inner segment (*IS*) are shown. *a* and *b* correspond to a cross section through the connecting cilium and the centriole (C_1). C_1 and C_2, centrioles; *cf*, ciliary filaments; *er*, endoplasmic reticulum; *mi*, mitochondria; *sm*, surface membrane.

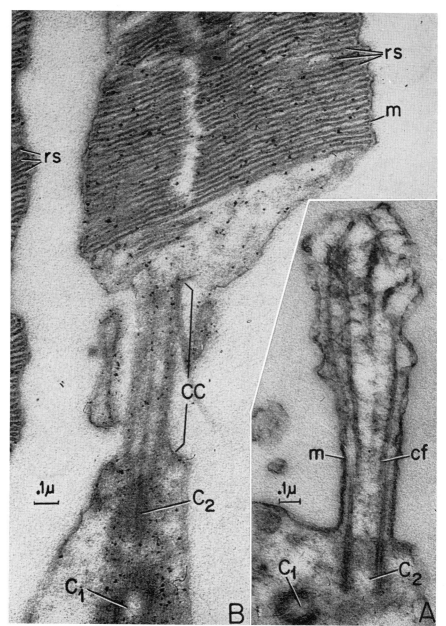

Figure 22–7.   **A,** electron micrograph of a primitive cilium in the retina of an eight day old mouse. C_1 and C_2, centrioles; *cf,* ciliary fibrils; *m,* ciliary membrane. × 72,000. **B,** electron micrograph of an adult rod, showing the outer segment with the rod sacs (*rs*) and the outer membrane (*m*); the connecting cilium (*CC*) and the basal centrioles (C_1 and C_2). × 62,000. (From E. De Robertis and A. Lasansky.)

nized as ciliary derivatives. For example, the so-called crown cells of the saccus vasculosus found in the third ventricle of fishes are modified cilia with swollen ends that are filled with vesicles.[33] Also, the primitive sensory cells of the pineal eye found in certain lizards have a ciliary structure.[34, 35]

Origin of Cilia, Basal Bodies, and Centrioles

The origin of centrioles and basal bodies, or kinetosomes, is viewed from the standpoint that these are possibly semiautonomous cell organelles and in this way similar to mitochondria and chloroplasts, as discussed in Chapters 10 and 11. Isolation of kinetosomes from *Tetrahymena* has verified the presence of RNA and DNA,[36] and this suggests that they are capable of some protein synthesis. More direct evidence of DNA in centrioles was obtained by use of the fluorescent dye acridine orange and by ^{3}H-thymidine incorporation followed by treatment with DNAse.[37, 38] However, most of these studies refer to the pellicle, and the exact localization of DNA has not been determined. Further-

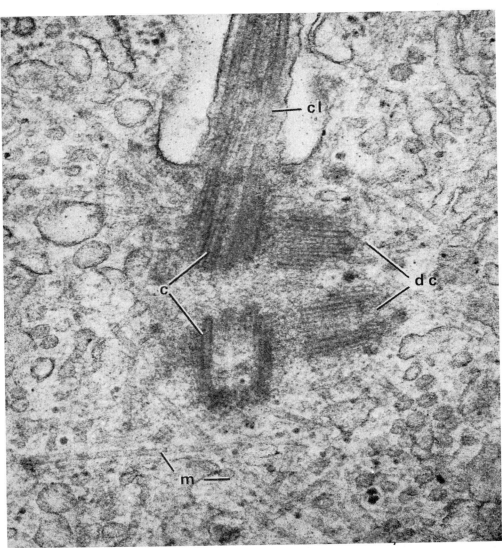

Figure 22–8. Electron micrograph of the pancreas of a chick embryo showing the replication of centrioles; *c*, the two centrioles; *dc*, daughter centrioles; *cl*, cilium; *m*, microtubules. × 50,000. (Courtesy of J. Andrè.)

more, these findings do not resolve the question of whether the synthesis of ciliary proteins depends on the DNA-RNA components of the centriole (see Rattner and Phillips, 1973).

Early cytologists realized that the development of cilia and flagella was directly related to the presence of centrioles. Flagella may be experimentally removed and their regeneration followed; thus, in *Chlamydomonas* it was found that the growth rate of a flagellum is about 0.2 μm per minute.

Studies on the origin of centrioles and kinetosomes were hampered by the small size of these structures. Light microscopists thought that centrioles originate by division from preexisting centrioles and that they possessed a genetic continuity comparable to that of chromosomes. The electron microscope has failed to confirm that such a division process takes place.[39, 40]

There are now several indications that centrioles may be generated by two different mechanisms.[41-43] Centrioles destined to form mitotic spindles or single cilia arise directly from the wall of the preexisting centriole. The *daughter centrioles* appear first as annular structures *(procentrioles)* (Fig. 22–8) which lengthen into cylinders. The groups of three tubules originate from single and double groups that first appear at the base of the procentriole. When they are half grown the daughter centrioles are released into the cytoplasm to complete their maturation. Usually, although not always, daughter centrioles are formed one at a time.[43]

The other mechanism is found in centrioles destined to become kinetosomes, as in a ciliated epithelium. The centrioles are assembled progressively from a precursor fibrogranular material located in the apical cytoplasm.[43, 44] The newly formed centrioles become aligned in rows beneath the apical plasma membrane, and each centriole may then produce satellites from the side, a root from its base and a cilium from its apex.

Development of the cilium begins with the appearance of a vesicle that becomes attached to the distal end of the centriole. The growing ciliary shaft invaginates the vesicular wall which forms a temporary ciliary sheath until the permanent one is formed. In many epithelial cells short cilia enclosed within vesicles may be observed.[45] In Chapter 13 the coordination between the replication of the centriole and the mitotic cycle was mentioned.

Electron microscopic studies of centrioles have revealed that they have a specific tubular structure, that they do not divide, and that they are found in association with other tubular structures (see Microtubules, following). This relationship suggests that the function of the centriole may be to regulate the synthesis and organization of microtubules within the cytoplasm (see Wolfe, 1972).

SUMMARY:
Structure, Motion, and
Origin of Cilia and Flagella

Cilia and *flagella* are motile processes found in protozoa and in many animal cells; in plants, only the antherozoids have flagella. *Kynocilia* (i.e., motile cilia) should be differentiated from *stereocilia*, which are nonmotile and lack a microtubular structure. The ciliary apparatus comprises: (1) the *cilium*, (2) the *basal body*, and (3) the *ciliary rootlets*.

The *axoneme*, which is thought to be the motile element, is the basic microtubular structure of cilia and flagella. It is surrounded by the *ciliary membrane* and is embedded in the *ciliary matrix*. The fundamental structure is 9 + 2 i.e., 9 pairs of peripheral microtubules and 2 single and central microtubules. The peripheral doublets can be numbered from 1 to 9, starting from the doublet cut by the plane perpendicular to the line joining the two central tubules (Fig. 22–2). This plane coincides with the plane of the ciliary beat. In the doublet *subfiber A* and *subfiber B* are distinguished by several morphologic features. Subfiber A has

two *arms* that are oriented in the same direction. The outer
arm is hook-like and the inner arm is linked to the adjacent
subfiber B (i.e., *peripheral link*). With negative staining a
filamentous structure and a lattice of globular components may
be observed in the wall of microtubules. *Radial links* connect
each subfiber A with the central sheath that contains the
central microtubules. These radial links, in longitudinal
sections, are seen as periodically spaced processes. *Basal
bodies or kinetosomes* have the same structure as centrioles. A
centriole is a cylinder (0.2 μm $\times$ 0.5 μm) opened on both
ends; the basal body at the distal end has a *ciliary plate* that
separates it from the cilium. The centriolar wall contains
triplets of microtubules. From the center to the periphery
these tubules are designated A, B, and C. Tubules A and B
traverse the ciliary plate and are continuous with the
corresponding tubules in the cilium, whereas tubule C
terminates near the plate. Centrioles lack the central
microtubules. Centrioles are polarized structures; at the
proximal end they have a cartwheel structure which connects
with the triplets. Procentrioles are formed from the proximal
end, whereas column or flagellum may be formed from the
distal end. Centrioles are generally in pairs and at right
angles.

Basal bodies may be related to cross-banded fibers that
constitute the *ciliary rootlets* and which serve a structural role.
There may be *transitional fibers* joining the triplets to the cell
membrane. Other accessory structures are the *basal feet,* and
satellites or *pericentriolar bodies.* Cilia and flagella may be
isolated, and the axonemes may be separated.

Two distinct proteins, called tubulin A and B, have been
separated from the microtubules; however, whole flagella
contain some 20 to 25 proteins, of which 10 may be
structural. A protein called *dynein,* having ATPase activity,
has been separated. This protein is localized in the arms of
subfiber A.

Ciliary movement may be analyzed from a scraping of frog
pharyngeal epithelium. The contraction of cilia is *metachronic*
in the plane of motion and *isochronic* in a plane perpendicular
to the direction of motion. The macromolecular mechanisms
of ciliary motion are probably more complex than those
taking place in muscle myofilaments. To account for the
movement in cilia, a sliding mechanism is favored. Since the
central microtubules end at the very tip of the cilium, and the
peripheral ones end at a distance, it has been possible to
measure the actual sliding of the doublets and to demonstrate
that it increases with the angle of bend. The ATPase present
in the arms may be important, since ATP is required for
ciliary motion.

The outer segments of the retinal rods and cones are
connected to the inner segment by a *connecting cilium* that
has a 9 + 0 pattern. The photoreceptor develops from a
primitive cilium. Other ciliary derivatives are in the crown

cells of the *saccus vasculosus* and in the pineal eye of certain lizards.

The *centrioles do not divide*. Those which later form mitotic spindles or single cilia arise directly from the wall of pre-existing centrioles. *Procentrioles* and *daughter centrioles* are formed at right angles. In the case of multiple centrioles, there is a fibrogranular material located in the apical cytoplasm from which the rows of basal bodies are formed. Cilia are formed from the distal end of the centriole and may grow at a fast rate.

MICROTUBULES

Microtubules are structures universally present in the cytoplasm of eukaryotic cells and are characterized by their tubular appearance and their uniform properties in the different cell types. Most microtubules are found in the cytoplasmic matrix (see Chapter 9); however, it has just been shown that they may be an integral part of motile organelles such as cilia and flagella and are components of centrioles. Most microtubules are rather labile and do not resist the effects of fixatives such as osmium tetroxide; because of this, intensive studies began only after 1963, when glutaraldehyde fixation was introduced in electron microscopy.[46]

The first observation of these tubular structures, in the axoplasm extruded from myelinated nerve fibers, was made by De Robertis and Franchi[47] (see Figure 24–2). Here the so-called *neurotubules* appeared as elongated, unbranched, cylindrical elements 20 to 30 nm in diameter and of indefinite length. Microtubules were observed in a variety of animal cells studied in sections[48–49] (Fig. 22–8).

Cytoplasmic microtubules are uniform in size and are remarkably straight. They are about 25 nm in outer diameter and several micrometers in length. In cross section they show an annular configuration with a dense wall about 6 nm thick and a light center. Each microtubule is surrounded by a zone of low electron density from which ribosomes or other particles are absent. In Chapter 11 the special orientation of microtubules in plant cells and their possible relationship to cell wall deposition were mentioned.

The wall of the microtubule consists of individual linear or spiraling filamentous structures about 5 nm in diameter, which, in turn, are composed of subunits.[50] In cross section there are about 13 subunits with a center-to-center spacing of 4.5 nm (see Fig. 24–6). Application of negative staining techniques has shown that microtubules have a lumen and a subunit structure in the wall.[51] Occasionally, dense dots or rods have been detected in the center portion of some microtubules (Fig. 24–6).[52]

Solubility of Microtubules

Although all the microtubules studied show approximately the same morphologic characteristics, it is evident that they differ in other properties. For example, microtubules of cilia and flagella are much more resistant to various treatments. The microtubules forming the spindle fibers and the others present in the cytoplasm are, in general, labile and transitory structures. Cytoplasmic microtubules usually disappear if stored at 0° C. or after treatment with colchicine (Chapter 13).

Neither colchicine nor low temperatures can alter the much more resistant tubules of cilia and flagella, but, as was mentioned previously, some of these tubules may disintegrate when treated with chelating agents, salt solutions, or proteolytic enzymes. Based on several of these criteria, four classes of microtubules have been recognized: (1) the cytoplasmic microtubules, (2) the accessory tubules of spermatids and the central pair of the 9 + 2 complex, (3) the B tubules, and (4) the A tubules (of the 9 + 2 complex of cilia and flagella). Stability of the microtubules increases from classes 1 through 4.[53]

Chemical Composition

Microtubules are composed of protein subunits that are rather similar, even though they are found in a variety of cell types. The term *tubulin*, used for the principal protein of cilia and flagella, is also used for the protein of cytoplasmic microtubules. Tubulin is a dimer of 110,000 to 120,000 daltons. The monomers of similar size are believed to be composed of 4 nm × 6 nm subunits. It has been shown that two different monomers— tubulin A and B—have been identified in flagella. In most cases, tubulin is a hetero-dimer having two monomers of different kinds although they are quite similar in molecular weight.

One dimer of tubulin binds to a molecule of ^{3}H-colchicine,[54-56] and this specific property is used for assay of this protein. Tubulin also binds to the Vinca alkaloid *vinblastine,* but at a site other than for colchicine. Vinblastine tends to produce crystal-like structures of tubulin in the cytoplasm, and in homogenates it produces precipitation of this protein, thus allowing rapid purification. The amino acid composition of tubulin from different sources, shows little variation, although some differences are found between the monomers. Some enzymatic activities have been reported for tubulin (e.g., protein kinase activity and ATPase). Some studies indicate that tubulin, in addition to being present as a pool of free dimers and integrating the microtubules, may form an integral part of some membranes.

Assembly of Microtubules

The assembly of microtubules from the tubulin dimers is a non-random-oriented and programmed process. In the cell there are sites of orientation, i.e., centrioles, basal bodies of cilia, and centromeres, from which the polymerization is directed in some way. It has been found that in a concentrated solution of tubulin containing guanosine triphosphate (GTP) and Mg^{2+}, depolymerization can take place only if the level of Ca^{2+} is kept low. This finding suggests that calcium may be a regulating factor in the in vivo polymerization of this protein.[57] Investigators studying the assembly of microtubules have found that the quantity of polymerized tubulin is high at interphase (cytoplasmic microtubules) and metaphase (spindle microtubules), but low at prophase and anaphase. There is also a marked variation in the activity of a calcium-activated ATPase during the cell cycle. This variation may be related to the role of calcium in the assembly of microtubules (see Allison, 1973).

Functions of Microtubules

A rather large list of functions, some of which may still be rather speculative, may be attributed to microtubules:

Mechanical Function. The shape of some cell processes or protuberances has been correlated to the orientation and distribution of microtubules. They are considered as a framework, or *cytoskeleton,* which processes the shaping of the cell and redistributes its content. The integrity of microtubules is necessary for the individual shape of many cells and the rigidity of elongated processes. This is particularly evident in axons (Fig. 24-6) and dendrites, but also in the case of blood cells whose disk shape is dependent on microtubules.

Morphogenesis. Related to their mechanical function is the role that microtubules play in the shaping of the cell during *cell differentiation.* For example, the elongation of the cells during the induction of the lens placode in the eye is accompanied by the appearance of numerous microtubules.[58] The morphogenetic changes that occur during spermiogenesis of the fowl provide another interesting example. The enormous elongation that takes place in the nucleus of the spermatid is accompanied by the production of an orderly array of microtubules that are wrapped around the nucleus in a direction perpendicular to the nuclear axis and that have a double helical arrangement.[59]

Cellular Polarity and Motility. The determination of the intrinsic polarity of certain cells is also related to the mechanical function. It has been found that treatment of various culture cells with *Colcemid,* a substance similar to colchicine, which produces depolymerization of microtubules, results in a change in motion. The following forms of movement have been observed to persist: (1) membrane ruffling, (2) endocytosis (see Chapter 21), (3) attachment to the surface and (4) extension of microvilli. Only a saltatory movement of particles was found to be inhibited after the destruction of the microtubules.[60]

However, the directional gliding of the cell is replaced by a random movement, thus indicating a loss of polarity (see Allison, 1973). Culture cells treated in vitro with dibutyryl cyclic AMP become elongated and the number of microtubules increases. All these observations suggest that microtubules are responsible for maintenance of the shape of the cell, the polarity of movement, and the distribution of certain organelles.

Circulation and Transport. Microtubules may also function as a "microcirculatory system" for the *transport* of macromolecules in the cell's interior; to this end, they probably form limiting channels in the cytoplasm. A very interesting example is the protozoan *Actinosphaerium* (Heliozoia) which sends out long, thin pseudopodia within which cytoplasmic particles migrate back and forth. These pseudopodia contain as many as 500 microtubules disposed in a helical configuration. When these protozoans are exposed to cold,[61] or high pressure, the pseudopodia are withdrawn and the microtubules depolymerize.

Another example of an association between microtubules and transport of particulate material is the melanocyte, in which the melanin granules move centrifugally and centripetally with different stimuli. The granules have been observed moving between channels created by the microtubules in the cytoplasmic matrix. Here, as in other instances, the microtubules may be determining the direction of the movement.[62]

In the erythrophores found in fish scales the pigment granules may move at a speed of 25 to 30 μm per second between the microtubules. The possible role of microtubules in axoplasmic transport will be discussed in Chapter 24.

Contraction. Chapter 13 included a discussion of the role played by microtubules in the formation of the spindle and the movement of chromosomes in mitosis. Microtubules also have a role in the movement of cilia and flagella (see earlier section in this chapter).

Sensory Transduction. Regularly arranged bundles of microtubules are common in sensory receptors, and a possible function in transduction of different incident energy has been speculated.[63]

SUMMARY:
Properties of Microtubules

Microtubules are found in all eukaryotic cells—either free in the cytoplasm or forming part of centrioles, cilia, and flagella. They are tubules 25 nm in diameter, several micrometers long, and with a wall 6 nm thick with 13 subunits. The stability of different microtubules varies. Cytoplasmic and spindle microtubules are rather labile, whereas those of cilia and flagella are more resistant to various treatments. The main component is a protein called *tubulin*. This hetero-dimer of 110,000 to 120,000 daltons, is formed of two different monomers (tubulin A and B) of the same molecular weight (i.e., 55,000). The monomers are 4 nm × 6 nm and probably correspond to the subunit lattice structure seen in the tubular wall. Tubulin binds colchicine and vinblastin at different binding sites. These properties are used in the assay and purification of the protein. The assembly of tubulin in the formation of microtubules is a non-random-oriented and programmed process. Centrioles, basal bodies, and centromeres are sites of orientation for this assembly. Calcium may be a regulating factor in the in vivo polymerization of tubulin. The level of polymerized tubulin is high at interphase and metaphase (i.e., spindle microtubules) and lower at prophase and anaphase.

Several functions, some of which are related to the primitive forms of cell motility described in this chapter, have been attributed to microtubules. These structures play a *mechanical* function in that they form a kind of *cytoskeleton*. The *shape* of the cell and cell processes is dependent on microtubules. This function is particularly apparent during *cell differentiation* of certain placodes, in the nerve cells, and in spermiogenesis. The *polarity* and directional gliding of cultured cells depend on microtubules. These structures are associated with *transport* of molecules, granules, and vesicles within the cell. They play a role in the contraction of the spindle and movement of chromosomes and centrioles, as well as in ciliary and flagellar *motion*. A possible role in *sensory transduction* has been postulated.

CYTOPLASMIC STREAMING

Cytoplasmic streaming, or *cyclosis,* is easily observed in plant cells, in which the cytoplasm is generally reduced to a layer next to the cellulose wall and to fine trabeculae crossing the large central vacuole. Continuous currents can be seen that displace chloroplasts and other cytoplasmic granules. In *Paramecium,* similar but slower movements are seen that displace the digestive vacuoles from the site of ingestion to the site of excretion. In many cells of higher animals, particularly in tissue cultures, intracellular movements can be seen. Mitotic division, with the complex displacement of centrioles, chromosomes and other cell organelles, also belongs to this type of intracellular movement.

The classic experimental work on cyclosis has utilized the cylindroid cells of *Nitella,* which have a thin protoplasmic layer of about 15 μm surrounding a central vacuole of 0.5 mm by 10 cm.

In some plant cells the protoplasmic current can be initiated by chemicals *(chemodynesis)* or by light *(photodynesis).* Cyclosis is modified by temperature, by the action of ions, or by changes in pH. Cyclosis is stopped by mechanical injuries, electric shock, or some anesthetics. Some auxins (plant growth hormones) increase the rate of cyclosis. In general, all the factors that decrease cell viscosity increase the speed of protoplasmic current and vice versa. Cyclosis decreases progressively in cells submitted to increased hydrostatic pressure at the same time that the

protoplasm becomes more liquid. The possible involvement of microtubules in cytoplasmic streaming has been postulated. It has been suggested that microtubules may provide an actual framework upon which the motive force responsible for streaming is generated.[64] In *Nitella,* microtubules are found beneath the plasma membrane at a distance from the cytoplasmic stream.

The actual motive force is, however, provided by microfilaments situated in the region between the stationary ectoplasm and the moving endoplasm.[65] The importance of microfilaments in cyclosis and ameboid motion will be mentioned later.

AMEBOID MOTION

In ameboid motion the cell changes shape actively, sending forth cytoplasmic projections called *pseudopodia,* into which the protoplasm flows. Although this special form of locomotion can be observed easily in amebae, it also occurs in numerous other types of cells. One need only to place a drop of blood between a slide and coverglass to see that the leukocytes, at first spheroidal, change their shape, emit pseudopodia, and move about. In tissue cultures, cells move out actively, forming the zone of migration. In epithelia, the desmosomes connecting the cells disappear. These changes also occur in vivo. For example, in epithelial repair, the cells free themselves and slide along actively toward the depth of the wound. In an inflam-

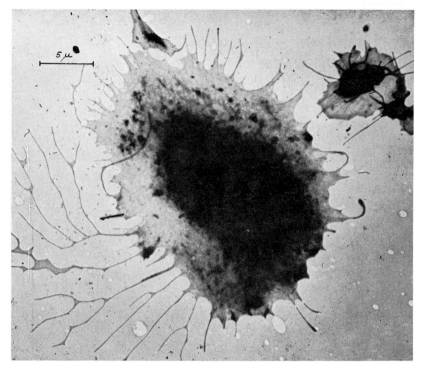

Figure 22–9. Electron micrograph of a polymorphonuclear leukocyte. (See the description in the text.) × 3000. (From E. De Robertis.)

matory process, leukocytes wander out of the blood vessels *(diapedesis)* by active ameboid motion and progress toward the focus of infection.

Some amebae are predominantly *monopodial* (one pseudopodium), but others may be temporarily or permanently *polypodial.* The shape of pseudopodia varies between a stout, almost cylindrical *lobopodium* and a fine filamentous or branching *filopodium* (see Fig. 22–9). Sometimes these fine processes may be anastomosing *(reticulopodia)* as in Foraminifera.

As shown by the classic studies of Mast, the protoplasm of the ameba has a clear ectoplasm, which expands considerably toward the end of the pseudopodium. As shown in Figure 22–10, the axial endoplasm is surrounded by a "shear zone" where particles move more freely. At the advancing end is the hyaline cap and just posterior to it, the "fountain zone," where the axial endoplasm appears to contract actively and flows below the ectoplasmic tube. At the opposite end is the tail process, also called the *uroid,* and near it, the *recruitment zone,* where the endoplasm is recruited from the walls of the ectoplasm in the posterior third of the cell.[66]

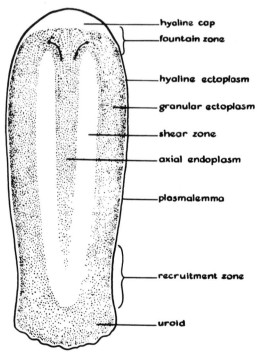

Figure 22–10. A new schema for ameboid structure and movement based on studies of cytoplasmic flow. New terminology is proposed for various regions of the cytoplasm of the ameba. (From Allen, R. D., "Ameboid Movement," *in* Brachet, J., and Mirsky, A. E. (Eds.), *The Cell,* Vol. 2, p. 35, New York, Academic Press, Inc., 1961.)

Properties of Ameboid Motion

In slime molds at times several pseudo-podia move in different directions (Fig. 22–11). There is a pulsating movement, the motive force of which can be measured and recorded as shown in Figure 22–12. By applying sufficient pressure with this apparatus, the movement can be prevented.[67]

When one half the plasmodium is bathed in a solution containing ATP, the motive force developed is much greater. The pulsating motion is quickly abolished by inhibitors of glycolysis, but inhibition of oxidative phosphorylation has little effect.

The rate of progression varies among different amebae between 0.5 and 4.6 μm per second. In the neutrophilic leukocytes it is approximately 0.58 μm per second. This rate is modified by temperature and other environmental factors. Insufficient oxygen supply does not stop the movement, but slows it. Calcium is required for this type of locomotion. In the presence of a substance that extracts calcium (such as an oxalate), motion is stopped. Severe mechanical injury, electric shock or ultraviolet radiation causes retraction of the pseudopodia, and the cell becomes spheroidal.

An important factor in ameboid motion is *adhesion to a solid support.* An ameba that floats freely in the liquid medium can emit pseudopodia, but does not progress; only when it adheres to a solid surface does it commence this type of locomotion. In tissue cultures the fibers of the coagulum serve as support; in connective tissue, the collagenous or reticular fibers may serve this purpose.

There are substances that influence the motion by attracting or repelling the cells. This property, which is called *chemotaxis,* has great importance in defense mechanisms, especially during inflammation.

Injection of ATP into an ameba causes the cortical gel to contract and liquefy. If the microinjection is in the tail, the ameba increases its speed of streaming; if it is in the advancing pseudopodium, it reverses the direction of streaming. With interference microscopy (see Chapter 5), it has been found that in a motile ameba the protein concentration of the cytoplasm is significantly higher at the tail.[68]

Most data favor an active contraction as the motive force for ameboid motion and for protoplasmic streaming as well (see below). The views are divided regarding the most likely site of contraction. Whereas some investigators regard the posterior region of the ectoplasmic tube as more active, others give more importance to the "fountain zone" at the advancing end of the ameba.[66]

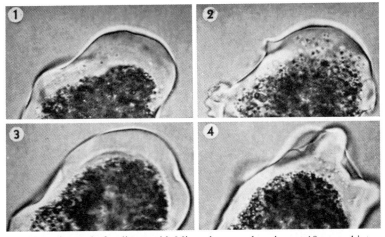

Figure 22–11.   **1–4,** advancing end of a slime mold. Microphotographs taken at 10-second intervals without moving the camera. The base line remained constant so that the advance and retraction could be measured. **1,** hyaline cap at the tip of an advancing pseudopodium, showing the even, bulging contour of the granular gel layer. **2,** same tip, 40 seconds later. The thin gel layer has disintegrated and granules are filling the cap. **3,** same tip, 20 seconds later. Gelation of the periphery of the granular protoplasm and the formation of a new hyaline cap. **4,** same tip, 30 seconds later. Retraction of the pseudopodium and irregular hyaline cap. (Courtesy of Lewis, W. L., and the Iowa State College Press, *in Structure of Protoplasm,* 1942.)

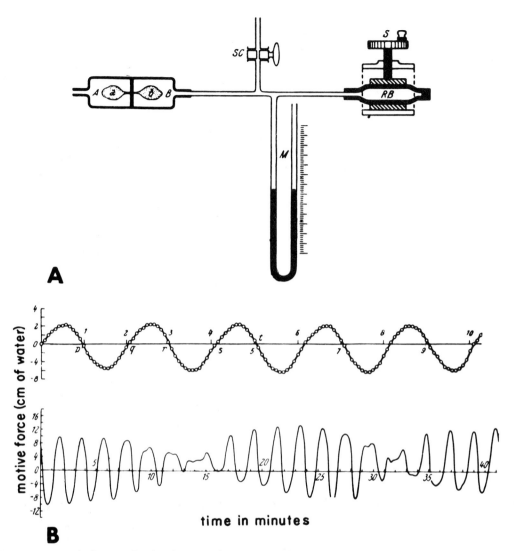

Figure 22–12. **A,** diagram showing the general arrangement for measuring the motive force of protoplasmic streaming in a myxomycete plasmodium. Note the double chamber (*A, B*) with myxomycete (*a, b*) connected through a small hole between the chambers. *M,* water-filled manometer; *S,* screw to control the pressure applied to the bulb (*RB*); *SC,* stopcock. **B,** recording of the motive force expressed in centimeters of water. Note the beatlike wave pattern under normal conditions. The upper line is a recording showing all the points recorded several times a minute. The lower line shows a record in which the points are left out and the abscissa is compressed in order to show the data for a longer period of time. Using this equipment, it is possible to study the effect of temperature, anesthetics, and various injurious agents on the wave pattern. (From Kamiya, N., *Protoplasmatologia,* 8:1, 1959.)

Microfilaments and Cell Motion

Electron microscopy has revealed the presence of *microfilaments* 4 to 7 nm thick in the cytoplasmic matrix of a variety of animal and plant cells (see Chapter 9). Various studies have suggested that such filaments represent contractile systems which may be involved in cytoplasmic streaming, ameboid motion, and other types of cell activities in which a motive force is involved. These studies have recently led to very important conclusions with the demonstration that the contractile proteins of muscle (i.e., *actin* and *myosin*) are also widely found in non-muscular systems and that they may be chemical components of some of the microfilaments (see Chapter 23).

In the study of microtubules the action of colchicine has been of paramount importance. In the case of microfilaments, the action of the drug *cytochalasin B* has had a decisive influence (see Wessells et al., 1971; Allison, 1973). Cytochalasin B has been found to impair numerous cell activities in which some types of microfilaments are involved. For example, it inhibits smooth muscle contraction, beat of heart cells, migration of cells, cytokinesis, endocytosis, exocytosis, and other processes.

When applied to the alga Nitella, cytochalasin B produces rapid cessation of cyclosis, which is reinitiated after removal of the drug (Fig. 22–13). This action is also observed in the presence of cycloheximide, an inhibitor of protein synthesis. There are reports that cytochalasin B disrupts the regular arrange-

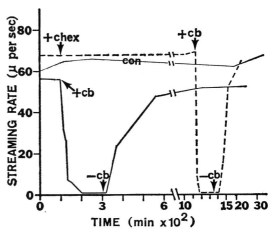

Figure 22–13. Representation of the rate of cytoplasmic streaming as a function of time in the alga, *Nitella. con,* control; *cb,* effect of cytochalasin *B;* and *chex,* lack of effect of cycloheximide. (From Wessells, N. K., et al., *Science, 171*:135, 1971, Copyright 1971 by the American Association for the Advancement of Science.)

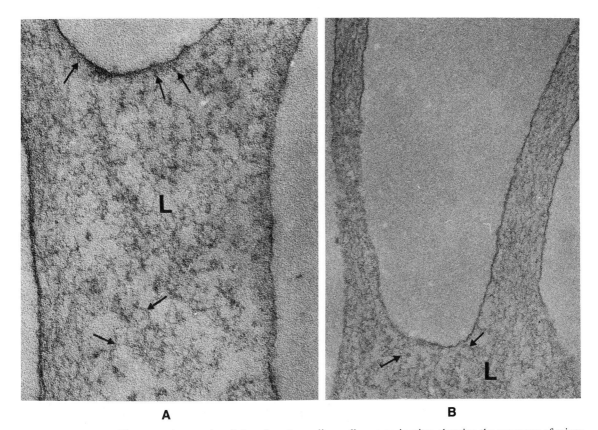

A   B

Figure 22–14. Electron micrographs of dorsal root ganglion cells grown in vitro showing the presence of microfilaments with a lattice arrangement (*L*). **A,** broad process, with some microfilaments inserted in the plasma membrane. × 129,000. **B,** two microspikes with lattice filaments. × 55,000. (Courtesy of N. K. Wessells.)

ment of microfilaments associated with some of these functions. For example, the contractile ring of microfilaments observed during cell cleavage (Fig. 13–12) is altered by this drug.[69]

Networks of microfilaments have been observed in the growth cones of neuroblasts and in the undulating membranes of cultured cells (Fig. 22–14). A thin sheath of microfilaments forming a network is observed below the plasma membrane of fibroblast and glial cells. Under the action of cytochalasin B, these cells stop moving and the network of microfilaments is altered. Not all microfilaments, however, are sensitive to the drug, since in the same cells other filaments that form a sheath are insensitive (see Wessells et al., 1971). These and other studies have led to the general conclusion that the cytochalasin-B–sensitive microfilaments are the contractile machinery of non-muscle cells.

Actin and Myosin in Non-muscular Systems

The presence of actin in a wide variety of non-muscle cells has been demonstrated by biochemical and ultrastructural methods. Actin-like protein has been isolated from the slime mold, *Acanthamoeba,* sea urchin eggs, blood platelets, brain, fibroblasts, and others (see Pollard and Korn, 1972; Allison, 1973). Identification of this protein can be made by its stimulating action on the ATPase of muscle by heavy meromyosin, by co-purification with radioactive actin from muscle, and by amino acid and peptide analysis. With the electron microscope, identification can be made by binding with heavy meromyosin. As in the case of muscle actin (Fig. 23–11) the actin-like microfilaments become "decorated," showing arrowlike complexes.[70] These complexes have been seen in sections of fibroblasts, chondrocytes, macrophages, amebae, and amphibian eggs, as well as in isolated protein. Studies in isolated plasma membranes of amebae show the presence of actin-like filaments attached to the inner surface of the membrane.[71]

Actin is thought to comprise a large proportion of the cytoplasmic proteins of many cells; in developing nerve cells, there

may be up to 20 per cent (see Pollard and Korn, 1971). Actin is present principally in its globular form (G-actin) with a molecular weight of 45,000, and it may quickly polymerize to form the microfilaments of fibrous actin (F-actin), which has a double helical structure. This G- to F-actin transition seems to represent the basis of the classical sol-gel transition in the cytoplasm of moving cells. Although myosin-like proteins have been found in much smaller concentrations than actin, they have been isolated from amebae, blood platelets, and slime molds.[72] These proteins have several properties in common with muscle myosin.[73] For example, they show binding to actin (to produce actomyosin), and they contain a Ca^{2+}-activated ATPase that is inhibited by Mg^{2+}. The molecular weight of 450,000 is similar in muscle, platelets, and slime mold, but it is much smaller (i.e., 140,000) in *Acanthamoeba*.

Identification of myosin filaments with the electron microscope, at present, is less satisfactory. However, filaments that are 13 to 22 nm thick and 0.7 μm long, which are of probable myosin nature, have been observed.[74]

Mechanism of Cell Contraction

The basis of contraction in non-muscular systems is still hypothetical, but the scheme that is evolving is in some ways similar to that of muscle (see Chapter 23). In non-muscular systems it is based on the interaction of actin and myosin filaments, with the consequent production of a shearing force. The differences between the mechanism here and in muscle are in the more random distribution of these proteins and the much smaller concentration of myosin. These factors may help to account for the much slower concentration in these primitive contractile systems. The attachment of active filaments to the plasma membrane, however, is favorable to the production of shearing forces acting at the level of the membrane. Such a relationship makes it possible that these mechanisms may be active not only in cell locomotion, but also in other cell functions, such as endocytosis and exocytosis. (For more details and references, see Komnick et al., 1973; Allison, 1973).

Cytoplasmic streaming or *cyclosis* is found in all kind of cells and produces the displacement of the various organelles. The alga *Nitella* is the organism most often used for this study. Cyclosis may be started by chemicals or by light. Various injuries and *cytochalasin B* stop cyclosis. Microtubules may provide the framework for cyclosis, but the actual motive force is provided by microfilaments.

Ameboid motion is observed in amebae, leukocytes, cultured cells, and in healing wounds. The number and shape of pseudopodia vary (e.g., lobopodial, filopodial, reticulopodial). In amebae there is an axial endoplasm separated by a shear zone from the peripheral endoplasm. The ectoplasm forms a hyaline cap at the advancing end. At the tail there is the *uroid* and the *recruitment zone.* In slime molds there is a pulsating movement that is inhibited by high pressure and is increased by ATP. Other important factors in ameboid motion are the levels of Ca^{2+} and the adhesion to a solid support.

Microfilaments, 4 to 7 nm thick, present in the cytoplasmic matrix, appear to play a major role in cyclosis and ameboid motion. Some of these microfilaments are sensitive to cytochalasin B, an alkaloid that also impairs many cell activities such as the beat of heart cells, cell migration, cytokinesis, endocytosis, and exocytosis, among others. Applied to the alga *Nitella*, cytochalasin B produces a rapid halt to cyclosis; this effect is reversed after removal of the drug. A network of microfilaments is often seen attached to the plasma membrane of moving cultured cells; this is altered by cytochalasin B. It is generally assumed that the cytochalasin-B–sensitive microfilaments are the contractile machinery of non-muscle cells.

By using biochemical and ultrastructural methods investigators have shown the protein *actin* to be found in most non-muscle cells. Under the electron microscope the microfilaments containing fibrous actin *(F-actin)* may be observed because they become "decorated" by their interaction with heavy meromyosin (see Chapter 23). Actin microfilaments are noted to attach to isolated plasma membranes. Actin is one of the most abundant proteins in many cells. The globular *(G-actin)* fibrillar transition is at the base of the classical sol-gel transition in the cytoplasm of moving cells.

Myosin is found in amebae, blood platelets, and slime molds, but in a much lower concentration than actin. It contains a Ca^{2+}-activated ATPase as is found in muscle myosin. Identification of myosin filaments is more difficult.

The *mechanism of contraction* of non-muscle tissue, which includes cyclosis and ameboid motion, is thought to involve the interaction of actin and myosin filaments (as in muscle) and the production of a shearing force. The random distribution of these filaments and the lower content of myosin may explain the slowness of this contraction which requires ATP and the Ca^{2+}-activated ATPase.

SUMMARY:
Cyclosis, Ameboid Motion,
and Microfilaments

REFERENCES

1. Grigg, G. W., and Hodge, A. J. (1949) *Australian J. Sci. Res.*, ser. B., 2:271.
2. Manton, I., Clarke, B., Greenwood, A. E., and Flint, E. A. (1952) *J. Exp. Bot.*, 3:204.
3. Fawcett, D. W., and Porter, K. R. (1954) *J. Morph.*, 94:221.
4. Gibbons, I. R. (1967) In: *Molecular Organization and Biological Function*, p. 211. (Allen, J. M., ed.) Harper and Row, New York.
5. Allen, R. D. (1968) *J. Cell Biol.*, 37:825.
6. Grimstone, A. V., and Klug, A. (1966) *J. Cell Sci.*, 1:351.
7. Henley, C. (1970) *Biol. Bull.*, 139:265.
8. Barton, R. (1969) *J. Cell Biol.*, 41:637.
9. Allen, R. D. (1967) *J. Cell Biol.*, 37:825.
10. Kiefer, B. I. (1970) *J. Cell Sci.* 6:177.
11. Gibbons, I. R. (1965) *Arch. Biol.* (Liege), 76:317.
12. Afzelius, B. (1959) *J. Biophys. Biochem. Cytol.*, 5:269.
13. Gibbons, I. R., and Grimstone, A. V. (1960) *J. Biophys. Biochem. Cytol.*, 7:679.
14. Warner, F. D. (1970) *J. Cell Biol.*, 47:159 and 220.
15. Perkins, F. O. (1970) *J. Cell Sci.*, 6:629.
16. Phillips, D. M. (1970) *J. Cell Biol.*, 44:243.
17. Gibbons, I. R. (1961) *J. Biophys. Biochem. Cytol.*, 7:697.
18. Tilney, L. G., and Goddard, J. (1970) *J. Cell Biol.*, 46:564.
19. Rosenbaum, J. L., and Child, F. M. (1967) *J. Cell Biol.*, 34:345.
20. Witman, G. B., Carlson, K., and Rosenbaum, J. L. (1972) *J. Cell Biol.*, 54:540.
21. Gibbons, B. H., and Gibbons, I. R. (1969) *J. Cell Biol.*, 43:43a.
22. Summers, K. E., and Gibbons, I. R. (1971) *Proc. Natl. Acad. Sci. USA*, 68:3092.
23. Renaud, F. L., Rowe, A. J., and Gibbons, I. R. (1968) *J. Cell Biol.*, 36:79.
24. Stephens, R. E. (1970) *J. Molec. Biol.*, 47:353.
25. Shelanski, M. L., and Taylor, E. W. (1968) *J. Cell Biol.*, 38:304.
26. Gibbons, I. R., and Rowe, A. J. (1965) *Science*, 149:424.
27. Sleigh, M. A. (1957) *J. Exp. Biol.*, 34:106.
28. Gray, J. (1958) *J. Exp. Biol.*, 35:96.
29. Goldstein, S. R., Holwill, M. E. J., and Silvester, N. R. (1970) *J. Exp. Biol.*, 53:401.
30. Tamm, S. L., and Horridge, G. A. (1970) *Proc. Roy. Soc. Lond., Ser. B*, 175:219.
31. Slater, P. (1968) *J. Cell Biol.*, 39:77.
32. De Robertis, E. (1956) *J. Biophys. Biochem. Cytol.*, 2:319.
33. Porter, K. R. (1957) *Harvey Lect.* Ser. 51, p. 175.
34. Steyn, W. (1969) *Nature* (London), 183:764.
35. Eakin, R. M. (1961) *Proc. Natl. Acad. Sci. USA*, 47:1084.
36. Seaman, G. R. (1960) *Exp. Cell Res.*, 21:292.
37. Randall, J., and Disbrey, C. (1965) *Proc. Roy. Soc.*, Ser. B, 162:473.
38. Smith-Sonneborn, J., and Plaut, W. (1967) *J. Cell Sci.*, 2:225.
39. Bessis, M., Breton-Gorius, J., and Thiéry, J. P. (1958) *Rev. Hémat.*, 13:363.
40. Bernhard, W., and de Harven, E. (1958) *Proc. Fourth Internat. Conf. Electron Microscopy*, 2:217.
41. Gall, J. G. (1961) *J. Biophys. Biochem. Cytol.*, 10:163.
42. Mizukanni, I., and Gall, J. G. (1966) *J. Cell Biol.*, 29:97.
43. Sorokin, S. P. (1968) *J. Cell Sci.*, 3:207.
44. Dirksen, E. R., and Crocker, A. (1966) *J. Micros.*, 5:629.
45. Martínez, P., and Drems, F. H. (1968) *Z. Zellforsch.*, 87:46.
46. Sabatini, D. D., Bensch, K., and Barrnett, R. J. (1963) *J. Cell Biol.*, 17:19.
47. De Robertis, E., and Franchi, C. M. (1953) *J. Exp. Med.*, 98:269.
48. Slautterback, M. C. (1963) *J. Cell Biol.*, 18:367.
49. Slautterback, M. C., and Porter, K. R. (1963) *J. Cell Biol.*, 19:239.
50. Ledbetter, M. C., and Porter, K. R. (1964) *Science*, 144:872.
51. Gall, J. G. (1966) *J. Cell Biol.*, 31:639.
52. Echandía, E. L. R., Piezzi, R. S., and Rodríguez, E. M. (1968) *Ann. J. Anat.*, 122:157.
53. Behnke, D., and Forer, A. (1967) *J. Cell Sci.*, 2:169.
54. Borisy, G. G., and Taylor, E. W. (1967) *J. Cell Biol.*, 34:535.

55. Weisenberg, R. C., Borisy, G. G., and Taylor, E. W. (1968) *Biochemistry, 7*:4466.
56. Weisenberg, R. C., and Timasheff, S. N. (1970) *Biochemistry, 9*:4110.
57. Weisenberg, R. C. (1972) *Science, 177*:1104.
58. Byers, B., and Porter, K. R. (1964) *Proc. Natl. Acad. Sci. USA, 52*:1091.
59. McIntosh, J. R., and Porter, K. R. (1967) *J. Cell Biol., 35*:153.
60. Freed, J. J., Bhisly, A. N., and Libowitz, M. M. (1968) *J. Cell Biol., 39*:46a.
61. Tilney, L. G., and Porter, K. R. (1967) *J. Cell Biol., 34*:327.
62. Bikle, D., Tilney, L. G., and Porter, K. R. (1966) *Protoplasma, 61*:322.
63. Moran, D. T., and Varela, F. G. (1971) *Proc. Natl. Acad. Sci. USA, 68*:757.
64. Sabnis, A., and Jacobs, W. P. (1967) *J. Cell Sci., 2*:465.
65. Nagai, R., and Rebhun, L. (1966) *J. Ultrastruct. Res., 14*:571.
66. Allen, R. D. (1961) Ameboid movement. In: *The Cell*, Vol. 2, p. 135. (Brachet, J., and Mirsky, A. E., eds.) Academic Press, Inc., New York.
67. Kamiya, N. (1959) *Protoplasmatologia, 8*:3a, 1.
68. Allen, R. D., and Rolansky, J. D. (1958) *J. Biophys. Biochem. Cytol., 4*:517.
69. Schroeder, T. E. (1972) *J. Cell Biol., 53*:419.
70. Nachmias, V. T., Huxley, H. E., and Kessler, D. (1970) *J. Molec. Biol., 50*:279.
71. Pollard, T. D., and Korn, E. D. (1971) *J. Cell Biol., 48*:216.
72. Hatano, S., and Tazawa, M. (1968) *Biochim. Biophys. Acta, 154*:507.
73. Pollard, T. D. and Korn, E. D. (1973) *J. Biol. Chem., 248*:4682 and 4691.
74. Allira, A., and Wohlfarth-Bottermann, K. E. (1972) *Cytobiologie, 6*:261.

Allen, R. D. (1961) Ameboid movement. In: *The Cell*, Vol. 2, p. 135. (Brachet, J., and Mirsky, A. E., eds.) Academic Press, Inc., New York.
Allen, R. D. (1962) Ameboid movement. *Sci. Am., 206* (No. 2):112.
Allison, A. C. (1973) The role of microfilaments and microtubules in cell movement, endocytosis, and exocytosis. In: *Locomotion of Tissue Cells. Ciba Symposium 14* p. 109, Elsevier Publishing Company, Amsterdam.
Fawcett, D. W. (1961) Cilia and flagella. In: *The Cell*, Vol. 2, p. 217. (Brachet, J., and Mirsky, A. E., eds.) Academic Press, Inc., New York.
Kamiya, N. (1960) Physics and chemistry of protoplasmic streaming. *Ann. Rev. Plant. Physiol., 11*:323.
Komnick, H., Stockem, W., and Wohlfarth-Bottermann, K. E. (1973) Cell motility mechanisms in protoplasmic streaming and ameboid movement. *Int. Rev. Cytol., 34*:169.
Margulis, L. (1973) Colchicine-sensitive microtubules. *Int. Rev. Cytol., 34*:333.
Pollard, T. D., and Korn, E. D. (1972) The contractile proteins of *Acanthamoeba castellani. Cold Spring Harbor Symp. Quant. Biol., 37*:573.
Rattner, J. B., and Phillips, S. G. (1973) Independence of centriole formation and DNA synthesis. *J. Cell Biol., 57*:359.
Sleigh, M. A. (1971) Cilia. *Endeavour, 30*:11.
Sleigh, M. A. (1970) The physiology and biochemistry of cilia and flagella. In: *Frontiers of Biology: Handbook of Molecular Cytology*, Vol. 15, p. 1243. (Lima-de-Faría, A., ed.) North-Holland Publishing Co., Amsterdam.
Stephens, R. E. (1971) Microtubules. In: *Biological Macromolecules*, Vol. 5, (Timasheff, S. N., and Fasman, G. D., eds.) M. Dekker, New York.

ADDITIONAL READING

Warner, F. D. (1972) Macromolecular organization of eukaryotic cilia and
    flagella. In: *Advances in Cell and Molecular Biology,* Vol. 2, p. 193,
    Academic Press, Inc., New York.
Wessells, N. K., Spooner, B. S., Ash, J. F., Bradley, M. O., Ludueña, M. A.,
    Taylor, E. L., Wrenn, J. T., and Yamada, K. M. (1971) Microfilaments
    in cellular and developmental processes. *Science, 171:*135.
Wolfe, J. (1972) Basal body fine structure and chemistry. In: *Advances
    in Cell and Molecular Biology,* Vol. 2, p. 151, Academic Press, Inc.,
    New York.

twenty-three

MOLECULAR BIOLOGY
OF MUSCLE

Cell contractility reaches its highest development in the various types of muscular tissues. The structural organization of muscle is adapted to unidirectional shortening during contraction. Because of this, most muscle cells are elongate and spindle-shaped. The cytoplasmic matrix is considerably differentiated, and the major part of the cytoplasm is occupied by contractile fibrils. In smooth muscle these *myofibrils* are homogeneous and birefringent. In contrast, in cardiac and skeletal muscle the myofibrils are striated and have dark, birefringent (anisotropic) zones alternating with clear isotropic zones (Fig. 23–1). In muscle cells only a small part of the cytoplasm—the *sarcoplasm*—retains its embryonic characteristics. It lies between the myofibrils, particularly around the nucleus.

Some muscle cells are so highly differentiated that they are adapted to produce mechanical work equivalent to 1000 times their own weight and to contract 100 or more times per second.

The different types of muscle cells are included in histology textbooks and the special types of contraction are in physiology textbooks. Here, the emphasis is on the macromolecular organization of the striated skeletal muscle and its relation to the work of contraction.

The study of the molecular biology of muscle is one of the most rewarding examples of the intimate association between structure and function and of the way in which chemical energy is transduced into mechanical work.

Striated skeletal muscles are composed of multinucleate cylindrical fibers, 10 to 100 μm in diameter and several millimeters or

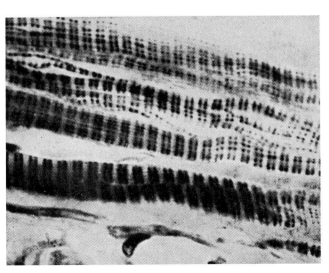

Figure 23–1. Myofibrils of a striated muscle fiber. Iron hematoxylin stain.

centimeters long. These enormous structures arise in the embryo by the fusion of several primordial cells, the so-called *myoblasts*. The entire fiber is surrounded by an electrically polarized membrane with an electrical potential of about -0.1 volt; the inner surface is negative with respect to the outer surface. This membrane, called the *sarcolemma*, becomes depolarized physiologically each time a nerve impulse that reaches the motor innervation of the muscle (*end plate*) activates the membrane. The final result is a coordinated contraction of the entire muscle fiber. Three cytoplasmic components are highly differentiated in the muscle fiber. One is represented by the contractile machinery, which is essentially made of protein myofilaments and is formed embryonically within the cytoplasmic matrix.

The arrangement of these myofilaments determines the different classes of muscle that are now recognized. For example, in striated skeletal and heart muscle of vertebrates the filaments are longitudinally oriented, and there is also a transverse repeating organization. In other types of nonstriated muscles the filaments are oriented in longitudinal or oblique arrays or have a more or less random distribution. The second component of striated muscle is a special differentiation of the vacuolar system, the so-called *sarcoplasmic reticulum*, which is involved with conduction inside the fiber and with coordination of the contractions of different myofibrils, in addition to being related to the relaxation of the muscle after a contraction.

The third component is represented by numerous mitochondria, the so-called *sarcosomes*, which in some cases may attain large dimensions. The abundance of mitochondria may be related to the constancy with which the muscle contracts; for example, there is a greater number in steadily active muscles such as the heart.

Macromolecular Organization of Myofibrils

Myofibrils are long, cylindrical structures about 1 μm in diameter. These striations consist of the repetition of a fundamental unit, the *sarcomere*, which is limited by a dense line called the *telophragma*, or *Z-line*. This line is located in the center of the less dense zone known as the *I-band*, which corresponds to

the relatively isotropic disk (Fig. 23–2). The *A-band*, which is anisotropic with polarized light, has a greater density than the I-band. Under certain conditions, a less dense zone may be observed in the center of the A-band, subdividing it into two dark semidisks (Fig. 23–2). This zone constitutes the *H-disk* (Hensen's disk). In the middle of the H-disk an M-line can be observed.

In a relaxed mammalian muscle, the A-band is about 1.5 μm long, and the I-band, 0.8 μm. The striations of the myofibrils result from periodic variations in density, i.e., in concentration of material along the axis. These striations are in register in the different myofibrils, thus giving rise to the striation of the entire fiber.

Myofilaments are essentially of two kinds: one 10 nm thick and about 1.5 μm long and the other 5 nm thick and about 2 μm long. As shown in Figure 23–3, these two types of filaments are disposed in register and overlap to an extent that depends on the degree of contraction of the sarcomere. In a relaxed condition, the I-band contains only thin filaments; the H-band contains only thick filaments; and within the A-band the thick and thin filaments overlap. In a cross section through the A-band, the regular disposition of the two types of filaments can be observed best (Figs. 23–3 and 23–4). In vertebrate muscle each thick filament is seen to be surrounded by six thin filaments, and each thin filament lies symmetrically among three thick ones (Fig. 23–4).[1] As a consequence of this geometry there are twice as many thin filaments as thick ones.

A cross section through the H-band shows only thick myofilaments and through the I band, only thin myofilaments (Fig. 23–3). This hexagonal paracrystalline organization is different in insect muscle. Here, each thin filament is equidistant from two thick filaments, and each thick filament appears surrounded by 12 thin ones.[2] Figures 23–2 and 23–5 show that the thin filament of one sarcomere apparently passes across the Z-line to the next sarcomere; however, this may not really be true.

Another interesting detail revealed by the electron microscope is that the two sets of filaments are linked together by a system of cross-bridges[3] (Fig. 23–5). These arise from the thick filaments at intervals of about 7 nm. Each bridge is situated along the axis with an

(*Text continued on page 528.*)

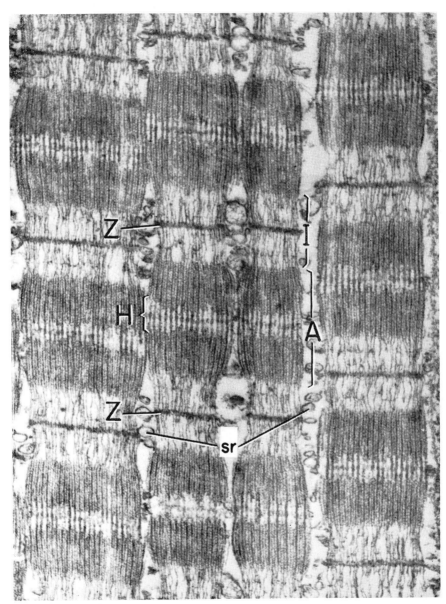

Figure 23–2. Electron micrograph of four myofibrils, showing the alternating sarcomeres with the Z lines and the H, A, and I bands; *sr*, sarcoplasmic reticulum situated between the myofibrils. The finer structure of the myofibril represented by the thin and thick myofilaments is also observed. × 60,000. (Courtesy of H. Huxley.)

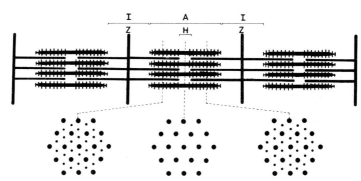

Figure 23–3. Diagram representing the structure of three sarcomeres in striated muscle with the I, A, and H bands. In longitudinal, as well as in cross section, the array of actin and myosin filaments, and their distribution in the various bands are shown. The cross bands in myosin filaments are also indicated. (From Huxley, H. E., *Proc. Roy. Inst. Gr. Br., 44*:274, 1970.)

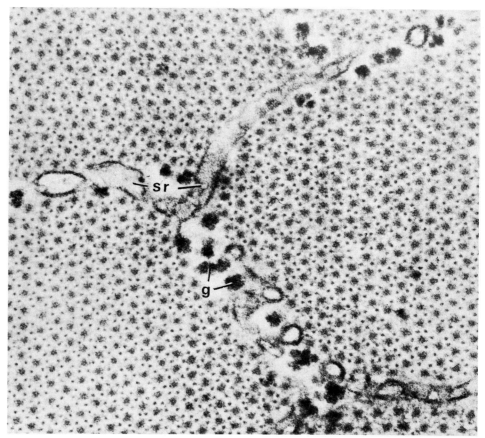

Figure 23–4. Cross section through skeletal muscle of a rabbit showing the arrangement of the myosin (thick) and actin (thin) filaments. Observe the one thin myofilament surrounded by three thick ones. Compare with the diagram in Figure 23–3; *sr*, sarcoplasmic reticulum; *g*, glycogen. × 180,000. (Courtesy of H. E. Huxley.)

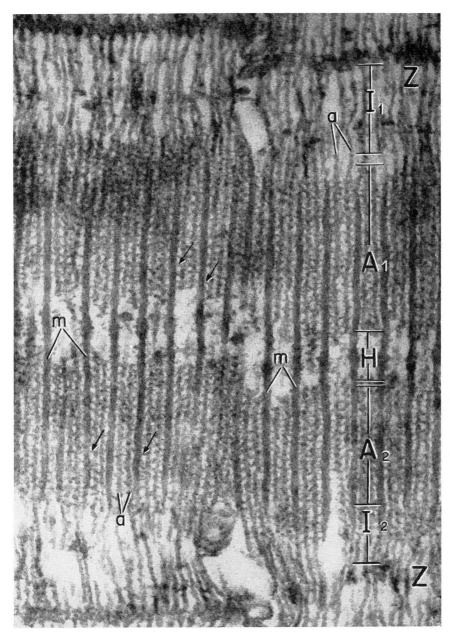

Figure 23–5. Electron micrograph of two sarcomeres in the adjacent myofibrils: Z, Z line; A_1 and A_2, anisotropic half bands; H, Hensen's band; I_1 and I_2, isotropic half bands; m, thick, and a, thin, filaments. The cross band between both types of filaments can be clearly seen. Some of them are indicated by arrows. $\times$ 175,000. (Courtesy of H. Huxley.)

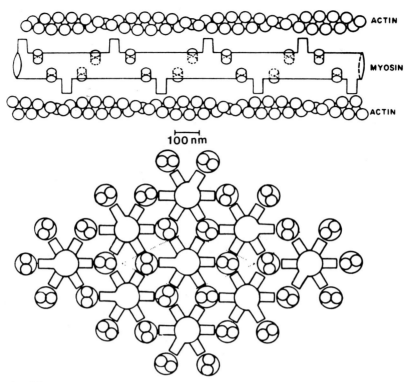

Figure 23-6. Spatial arrangement of myosin and actin filaments in striated muscle. (See the description in the text.) (From Davies, R. E., *Nature* (London), *199*:1068, 1963.)

angular difference of 60 degrees. This means that the bridges describe a helix about every 43 nm. As a result of this arrangement one thick filament joins the six adjacent thin ones every 43 nm (Fig. 23-6).

Changes during Contraction

The observation made in classic histology of a reversal of banding during contraction has been confirmed with modern techniques. These changes can now be studied in the living fiber by means of the phase contrast and interference microscopes. One striking observation is that the A-band remains constant in a wide range of muscle lengths, whereas the I-band shortens in accordance with the contraction.

The shortening of the I-band results from the fact that the thin myofilaments slide farther and farther into the arrays of thick filaments (Fig. 23-7). With the progress of contraction, the thin filaments penetrate into the H-band, and they may even overlap, thereby producing a more dense band in the center of the sarcomere (inversion of the banding). Finally, the thick filaments reach contact and are crumpled against the Z-lines.[4] These findings have been interpreted in the so-called *sliding filament theory* of contraction, which will be discussed in greater detail later. The degree of contraction thus achieved can be measured by determining the length of the sarcomere at rest and when it has shortened. Note that insect muscle, in general, shortens only slightly (about 12 per cent), whereas the shortening in vertebrate muscle may be much greater (about 43 per cent).

Macromolecular Organization in Other Muscles

The electron microscope has revealed that "smooth" muscles may have a varied macromolecular organization. In many cases they contain thin and thick myofilaments, as do striated muscles, but the difference lies in the absence of the Z-line and the lack of periodicity. In mollusks and annelids there are muscles with a helical arrangement that have

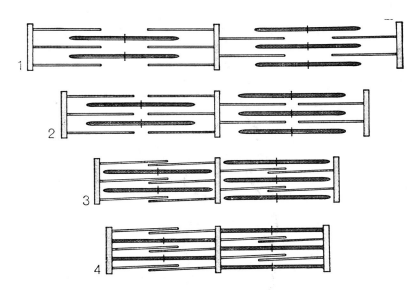

Figure 23–7. Diagram showing the sliding model of contraction. **1,** relaxed condition; the H bands are wide and contain only thick filaments; **2,** beginning of contraction; the thin filaments slide toward the center of the sarcomere, and the H bands become thinner; **3,** further contraction; the thin filaments penetrate the entire H bands (inversion of the banding); and **4,** maximal contraction; thick filaments are crumpled against the Z line. (From Huxley, H. E., *Proc. Roy. Inst. Gr. Br., 44*:274, 1970.)

thin and thick myofilaments linked by cross bridges. In the adductor muscle of the oyster, the so-called paramyosin muscle, each thick filament is surrounded by 12 thin filaments.[5]

The smooth muscle of vertebrates apparently lacks these two types of myofilaments and even myofibrils are difficult to recognize. However, improvements in the preparative techniques have demonstrated coarse myofilaments, resembling the myosin type, in smooth muscle of vertebrates.[6] Furthermore, thick and thin myofilaments were isolated from smooth muscle in chickens. In smooth muscles the contraction is very slow, but extreme degrees of shortening may be achieved.

BIOCHEMISTRY OF MUSCLE

Structural Proteins

The similarities in macromolecular organization of muscle are reflected at the biochemical level by the presence of special structural proteins in the contractile machinery. By the middle of last century Kuhne isolated a muscle protein that he named *myosin*. In 1942, Straub isolated *actin,* and in 1948, Bailey characterized *tropomyosin.* These three proteins are known to be present in the different types of muscles.

Structural proteins account for about 60 to 70 per cent of the total protein in muscle. The remaining 30 to 40 per cent, largely present in the sarcoplasm, includes various soluble proteins, enzymes, and the oxygen carrier *myoglobin.* The three structural or fibrous proteins constitute almost 90 per cent of the myofibril, and in a typical vertebrate muscle they are present in the following proportion:[7] myosin, 54 per cent; actin, 20 to 25 per cent; and tropomyosin, 11 per cent.

Myosin. If a muscle is extracted with a 0.3 M solution of KCl, myosin solubilizes first and can be separated and purified by precipitation at a lower ionic strength. Myosin comprises about half the total protein and has a molecular weight of about 450,000.

Individual myosin molecules are tadpole-shaped with a head and a tail and a total length of about 150 nm (Table 8–1). The head is 15 to 25 nm long and 4 nm thick, and the tail is 1.5 to 2.0 nm thick and occupies the rest of the molecule's length.[8] It has been clearly demonstrated that the thick myofilaments are made of myosin. When isolated, these myofilaments appear as long, spindle-shaped objects 11 nm thick and having the typical side projections (i.e., cross-bridges) described above (Fig. 23–6). Such cross-bridges correspond mainly to the heads of the myosin molecules; most of the tail is in the shaft of the myofilament (see Fig. 23–10, *B*).

Myosin has been fragmented by the action of proteolytic enzymes into the so-called *meromyosins.* With trypsin, each molecule of myosin divides into a *light meromyosin* and a *heavy meromyosin.* The cross-bridge of the thick filament contains the heavy fragment, whereas the light fragment is in the backbone.[9] By using *papain*, investigators can further fractionate heavy meromyosin into an S_1 globular unit and an S_2 helical rod that joins S_1 to the light meromyosin (Fig. 23–8). The most important part of the myosin molecule is the S_1 portion, since it contains the ATPase, as well as the sites for binding ATP and actin. S_1 heavy meromyosin constitutes the side projections of myosin filaments.[10]

Actin. The other major structural protein, *actin*, is less soluble in KCl solutions and dissolves in a 0.6 M concentration. In the absence of salts, actin becomes globular (G-actin) with a molecular weight of about 70,000; however, in the presence of KCl and ATP it polymerizes, forming long fibers (F-actin). This change, which is reversible, is called a globular-fibrillar transformation.

G-actin is approximately spherical, with a diameter of 5.3 nm; it exhibits no ATPase activity. Each molecule contains one ATP and one Ca^{2+}. Polymerization into F-actin is accompanied by dephosphorylation of the bound ATP and release of phosphate. *F-actin* is made of two helical chains twisted around each other; there are about 13 G-actin mole-

cules per turn in each chain (Fig. 23–6).[11, 12] The diameter of the double helix is 7 to 8 nm. As will be mentioned later, F-actin corresponds morphologically to the thin myofilaments.

Tropomyosin. In vertebrates *tropomyosin* is usually present in smaller amounts than the other two structural proteins. It may represent 5 to 10 per cent of the total protein in both striated and smooth muscle. This protein may be extracted in solutions of 1 M KCl or weak acids. It is a molecule about 40 nm long with a molecular weight of about 64,000. Like the light meromyosin, tropomyosin is made of two parallel polypeptide chains twisted in an α-helical configuration. Tropomyosin is intimately associated with actin and probably lies within the grooves of the actin helix (Fig. 23–8, *B*).

Troponin. This is a small globular protein of about 80,000 daltons, having special affinity for Ca^{2+}. According to the model of the thin filament shown in Figure 23–8, *B*, there is a troponin molecule every 40 nm. This molecule is intimately associated with tropomyosin and is intercalated every seven

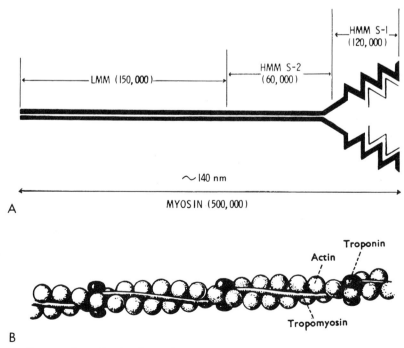

Figure 23–8. **A,** diagram of a molecule of myosin. **B,** diagram of the thin myofilament with the three components actin, tropomyosin, and troponin. (**A,** modified from Lowey, S., Slayter, S., Weeds, A. G., and Baker, H., *J. Molec. Biol., 42*:1, 1969; **B,** from Ebashi, S. E., and Endo. M., *Prog. Biophys. Molec. Biol., 18*:123, 1968.)

or eight G-actin molecules. When the isolated tropomyosin and troponin molecules are put together, they form a crystalline complex which may be observed under the electron microscope.

It is thought that both tropomyosin and troponin play a regulatory role in contraction. As will be mentioned later, when the concentration of Ca^{2+} is raised in the cytoplasmic matrix (myoplasm) to about 10^{-6} M, the contraction is initiated. The binding of 3 to 4 calcium ions to a troponin molecule may induce a conformational change which is transmitted by way of tropomyosin to the molecules of G-actin that are under its control (Ebashi and Endo, 1968; Lowey, 1972). Such changes could affect the tropomyosin-actin interaction within the thin filament (Fig. 23–8, B) and could activate the actin molecules, thereby preparing them to interact with the myosin to achieve contraction (see the following section).

Localization and Development of Structural Proteins

Several methods have permitted the localization of the main structural proteins within the macromolecular architecture of the myofibril. By using muscle fibers treated with glycerol to remove soluble components, and at the same time preserving the ultrastructure, it is possible to make a differential extraction of myosin or actin. After extraction of myosin, the A-bands and the thick filaments disappear.[13] By extracting actin, it is possible to remove a large part of the I-band and the thin filaments. Experimenters using the fluorescent antibody technique with antibodies against myosin (Fig. 6–21) or actin have localized these two proteins in the A- and I-bands, respectively.[14] Workers using a similar immunologic test for tropomyosin have found a positive staining in the I-band, as well as in the Z-line.[15, 16] It was mentioned before that tropomyosin with troponin probably forms a complex with F-actin in the thin filaments. Special extraction methods have been used to remove the Z-line and the M-line and to reconstitute them in glycerinated muscle fibers.[17] Recently, an M-line protein has been identified with the enzyme creatine kinase, and the antiserum against the purified enzyme binds to the center of the sarcomere where the M-line is located.[18]

An interesting approach to the consideration of protein localization is provided by developmental studies.[19] It was observed in chick myoblasts that both thin and thick filaments appear at the same time in the cytoplasmic matrix, but that the thin (actin) filaments are more numerous than the thick (myosin) filaments. This implies that actin predominates at the beginning of the synthesis. The myofilaments become oriented along the long axis before being incorporated into myofibrils. Several models have been proposed to explain how the myofilaments are integrated within the organization of the myofibril.[20]

The Sliding Theory of Muscle Contraction

The shortening of the sarcomere during contraction is due to the sliding of the thin filaments into the A- and H-bands (Fig. 23–7). The sliding theory of muscle contraction postulates that the thin actin filaments are displaced with respect to the thick myosin filaments during each cycle of contraction and relaxation.[11] In the case of a frog sarcomere, 2.5 μm long, there is a shortening of 30 per cent at each contraction. This implies a sliding of 0.37 μm for each half sarcomere. The force generated seems to be proportional to the degree of overlap between thick and thin filaments. This implies that the force is of short range and is produced directly by the cross bridges between the filaments. The sliding movement is thought to result from the repetitive interaction of the cross bridges with the actin filaments. In the diagram of Figure 23–9 it is assumed that each cross bridge represents the active end of the myosin molecule (i.e., the S_1 globular unit) (see Fig. 23–8). The following mechanism takes place: (1) A cross bridge binds to a specific site of the actin filament. (2) The cross bridge undergoes a conformational change which displaces the point of attachment toward the center of the A-band, thereby pulling the acting filament; at the same time a second cross bridge becomes attached. (3) At the end of the cycle the first bridge returns to the starting configuration in preparation for a new cycle. According to this theory, the actin filaments from each half sarcomere are pulled as ropes

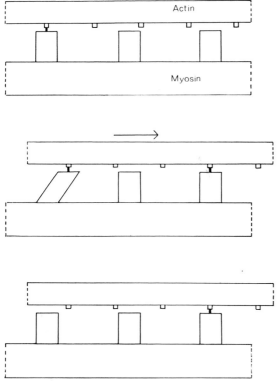

Figure 23–9. Diagram showing, very schematically, the possible mode of action of cross-bridges. A cross-bridge attaches to a specific site on the actin filament, then undergoes some conformational change which causes the point of attachment to move closer to the center of the A band, thus pulling the actin filament along in the required manner. At the end of its working stroke, the bridge detaches and returns to its starting configuration, in preparation for another cycle. During each cycle, it is probable that one molecule of ATP is dephosphorylated. (From Huxley, H. E., *Proc. Roy. Inst. Gr. Br.*, 44:274, 1970.)

toward the center of the sarcomere by the myosin arms that move to and fro.

The energy for this interaction is provided by the splitting of ATP because of the ATPase present in the cross bridge. It is estimated that the splitting of one ATP accounts for a displacement of 5 to 10 nm for each myosin cross bridge.

Polarization of the Myofilaments

To account for the directionality of the movement it is suggested that both the myosin and actin filaments have a definity polarity. In other words, in each half sarcomere the interacting molecules should have a different polarization; furthermore, the forces developed must have opposite directions (Fig. 23–10).[21]

From electron microscopic studies of aggregates of myosin molecules, it is possible to obtain reconstituted myofilaments having the structure shown in the diagram of Figure 23–10, *B*. The molecules of each half of the filament are in two antiparallel sets with a reversal of structural polarity in the center. In the diagram of a sarcomere in Figure 23–10, *A*, the polarity of the cross-bridges is indicated by the arrows. This diagram also emphasizes that all the actin molecules should be polarized in opposite directions in each half sarcomere, for the sliding process to take place. Observe that in the center of the thick myofilaments there is a region devoid of cross bridges.

Since the classic experiments of Szent-Györgyi, it has been known that the mixing of myosin and actin in the test tube results in the formation of the complex *actomyosin*, which contracts in the presence of ATP.[22] Investigators following this interaction under the electron microscope have observed that the myosin molecules bind to the F-actin with a directional orientation. As shown in Figure 23–11, the complex can also be produced by using F-actin and heavy meromyosin. The actin filaments become "decorated" with the myosin, and the complex shows an arrow-like polarity. These findings demonstrate that in each filament of F-actin the G-actin molecules are polarized with the same orientation. Researchers using methods of three dimensional reconstruction[23] have been able to represent the interaction of F-actin with the S_1 heavy meromyosin.[24] Figure 23–12 shows a model illustrating the double helical filament of actin decorated with the S_1 myosin fragments. The arrow-like configuration shown in the electron micrograph (Fig. 23–11) results in the tilting and bending of the S_1 myosin along the helix of F_1 actin (see Huxley, 1971).

The detailed ultrastructural and biochemical information obtained permits an interpretation of the macromolecular mechanisms involved in muscle contraction. It may be postulated that this is a cyclic event involving the repetitive formation and breakdown of actin-myosin linkages at the bridges between thick and thin filaments. At each bridge the following sequence of events is probably produced: (1) formation of a linkage

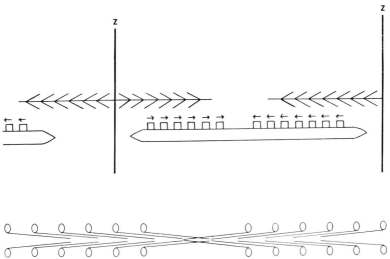

Figure 23–10. **A,** diagram illustrating the polarity of cross-bridges in the myosin filaments and in the actin myofilaments. The sliding forces tend to move the actin filaments toward the center of the sarcomere. (From Huxley, H. E.) **B,** arrangement of myosin molecules within the thick filaments. Each molecule has a tadpole shape with a globular head and a tail. The axis of the filament is formed by the assembly of the tails. Observe that in each half of the myofilament the molecules are polarized. (From Huxley, H. E., *Proc. Roy Inst. Gr. Br., 44*:274, 1970.)

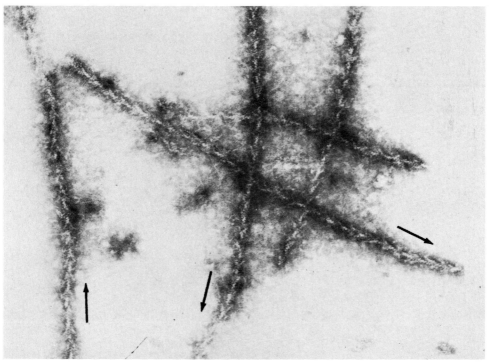

Figure 23–11. Filaments of actomyosin resulting from the interaction of actin and myosin (H-meromyosin). Observe the arrowlike polarity of the actomyosin complex (arrows). Negatively stained; × 155,000. (Courtesy of H. E. Huxley.)

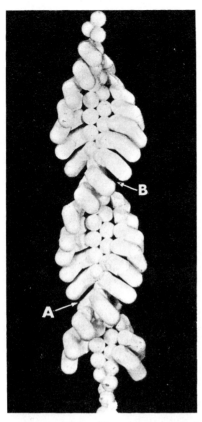

Figure 23-12. Three-dimensional model of an acto-myosin complex. In the axis is the double helical filament of *F*-actin (see Fig. 23-6) made of *G* subunits. The actin has been treated with fragments of heavy meromyosin (fragment S_1, see Fig. 23-8). Points marked *A* and *B* are cross-over points in the actin helix. The arrowhead appearance is produced by the change in position of the point of attachment of the S_1 subunit to the actin molecules. (From Huxley, H. E., *Proc. Roy. Inst. Gr. Br.*, *44*:274, 1970.)

Energetics of Contraction

Whereas the myofibrils constitute the mechanical machinery of the muscle, the fuel needed is produced mainly in the sarcoplasm. In all types of muscle numerous mitochondria, called *sarcosomes,* provide the essential oxidative phosphorylation processes and the Krebs cycle system. These mitochondria are particularly prominent in size and number in heart muscle and in the flight muscles of birds and insects.

The sarcoplasmic matrix contains the glycolytic enzymes as well as other globular proteins, such as myoglobin, salts and high phosphate compounds. Glycogen is present in the matrix as small granules observed under the electron microscope. There are about 1 per cent glycogen and 0.5 per cent creatine phosphate as sources of energy in muscle. Glycogen disappears with contraction through glycolysis, and lactic acid is formed, which can be transformed into pyruvic acid to enter the Krebs cycle (see Chapter 4).

The initial energy source for contraction is ATP. The ADP produced after the initial contraction is again recharged to ATP by glycolysis or from creatine phosphate. Oxidative phosphorylation is the last and most important source of ATP.[25]

Muscle extracted with glycerol leaches out ATP and is converted into a *model* for contraction which can be induced by ATP. Several other types of contractile cells can be induced to contract by ATP, e.g., cilia, flagella, spermatozoan tails, and also dividing cells (see Chapter 22). It is postulated that there is a similar molecular mechanism of contractility in all these cases.

between a heavy meromyosin head and one G-actin (globular) unit; (2) rupture of this linkage by one ATP molecule; (3) hydrolysis of the ATP by the Ca^{2+}-activated ATPase of myosin; (4) formation of a new linkage between the same heavy meromyosin (bridge) and the next G-actin unit (see Fig. 23-9). The relative movement of the thin filament taking place in each sequence would be equivalent to the length of one G-actin unit (5.3 nm). In this mechanism it is thought that *tropomyosin,* forming a complex with the actin helix, may play a regulatory role. *Troponin,* also part of this complex, may regulate the process because of its high affinity for binding Ca^{2+}.

SARCOPLASMIC RETICULUM

The sarcoplasmic reticulum found in skeletal and cardiac muscle fibers is one of the most interesting specializations of the vacuolar system. It was discovered by Veratti in 1902 as a reticulum present in the sarcoplasm of the muscle fiber and extending in between the myofibrils. It was completely neglected until 1953 when the first electron micrographs of this structure were published.[26-27]

The sarcoplasmic reticulum can be considered as an especially differentiated vacuolar system for this cell type. It is a continuous, membrane-limited reticular system

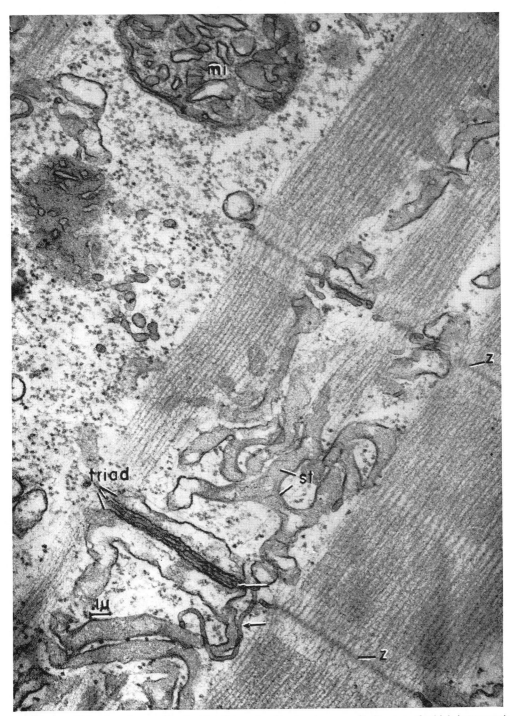

Figure 23-13. Electron micrograph of striated muscle showing two myofibrils, one of which is tangentially cut and shows the arrangement of the sarcoplasmic reticulum. The two components of this system can be seen clearly. The transverse component is represented by the *triad* and especially by the central cisternae of the triad, which continues in special tubules (arrows). Notice the relationship of the triad to the Z line. The longitudinal component of the sarcoplasmic reticulum forms anastomosing tubules (*st*) on the surface of the sarcomere; *mi*, mitochondrion. (Courtesy of K. R. Porter.)

whose organization is superimposed on that of the myofibril. As shown in Figure 23–13, the organization of the vesicles and tubules of the sarcoplasmic reticulum is regular. Special *terminal cisternae* are found at the level of the I-band; between these is a central flattened vesicle forming the *triad*. Between the terminal cisternae the tubules are disposed longitudinally on the surface of the A-band of the sarcomere. This structure is repeated between all myofibrils and also is continuous across the muscle fiber, making connections with the surface membrane at the level of the Z-lines. Earlier, light microscopic findings indicated that the Z-lines, or telophragms, were continuous septa across the fiber reaching the sarcolemma; however, the electron microscope has now clearly demonstrated that their continuity is established by way of the sarcoplasmic reticulum.

The sarcoplasmic reticulum can be divided into two parts. One is longitudinally oriented along the myofibril and would be the equivalent of the endoplasmic reticulum of other cells. The other part is a *transverse component* between the terminal cisternae, which together with the cisternae, constitutes the so-called *triad*. This transverse component (i.e., the T-system) is apparently continuous at certain points with the plasma membrane of the sarcolemma and would be the structure best fitted to conduct impulses from the fiber surface into the deepest portions of the muscle fiber.[28]

In frog muscle immersed for short periods in a solution containing ferritin, it was found that the central vesicle of the triad had filled with these electron opaque molecules. Ferritin was also found in certain tubules that were continuous with the central element of the triad. These findings suggest that at the plasma membrane there are a small number of openings directly communicating through fine tubules with the transverse system of the sarcoplasmic reticulum. Interestingly, the lateral components of the triad never contained ferritin molecules;[29] these, then, are not connected to the plasma membrane.

The disposition of the sarcoplasmic reticulum varies in the muscles of certain invertebrates. In crayfish muscle, invaginations of the plasma membrane penetrate and branch into tubules that form *diads,* instead of triads, at the A-I junction.[30] In muscles of the cockroach, the sarcoplasmic reticulum is abundant and forms a kind of fenestrated envelope around each myofibril. There is a single T-system which forms diads with the sarcoplasmic reticulum in alternating sarcomeres.[31]

Role of the Sarcoplasmic Reticulum

The possible role of the sarcoplasmic reticulum has been suggested by an experiment with microelectrodes in which the stimulation of the sarcolemma at the level of the Z-line produces a localized contraction of the adjacent sarcomeres.[32]

It has been hypothesized that the sarcoplasmic reticulum serves to transmit the excitatory impulse intracellularly.[33] It is thought that the membrane of the sarcoplasmic reticulum is electrically polarized in the same way as the surface membrane of muscle. It has been further assumed that this membrane is capable of conducting impulses inside the muscle fiber by way of the T-system in order to activate the contractile components.

In crayfish, a channeled-current model involving a direct role of Cl^- and a possible interaction between Ca^{2+} and Cl^- (instead of an electrotonic spread) has been postulated.[34] Whatever the intimate mechanism, it is evident that the presence of this intracellular conducting system may explain the physiologic paradox that a fiber 50 to 100 μm in diameter may contract quickly once the activating action potential has passed over the surface.

Muscle Relaxation

Another interesting approach to the study of the sarcoplasmic reticulum has been provided by its isolation. This fraction has been found to contain the relaxing factor, which produces relaxation after contraction.

The relaxation of muscle fibers can be brought about by decreasing the concentration of Ca^{2+} in the intracellular fluid. The sarcoplasmic reticulum in the presence of ATP may actively incorporate Ca^{2+} acting as a pump.[35] This membranous system contains a Ca^{2+}-activated ATPase that is coupled to the Ca^{2+} uptake. The Ca^{2+}-activated ATPase has been purified to homogeneity[36] and has been shown to be the sole protein responsible for the Ca^{2+} transport.[37] Furthermore, a low molecular weight protein called *calsequestrin* is responsible for the binding of Ca^{2+} within the sarcoplasmic reticulum.[38] The Ca^{2+} uptake was found to be inhibited by electrical stimulation and certain drugs that function as β-adrenergic blocking agents.[39] The Ca^{2+} pump of the sarcoplasmic reticulum, like other active transport systems, implies a functional asymmetry of the membrane. Since both the ATPase and the Ca^{2+} uptake are inhibited by agents that block —SH groups, an electron microscopic study was conducted using a cytochemical reagent in which ferritin molecules were attached to an —SH blocking molecule. It was demonstrated that the active groups of the membrane-bound ATPase were localized in the outer surface of the isolated sarcoplasmic reticulum.[40] All these findings suggest that the sarcoplasmic reticulum assumes the important role of relaxing the fiber after contraction and that this is accomplished by the binding of Ca^{2+} at the outer surface and its transport within the sarcoplasmic vesicles.

Within the macromolecular organization of the myofibril the regulatory role is played by the Ca^{2+}-binding protein, troponin, which regulates the activity of the Ca^{2+}-activated ATPase of myosin (see Ebashi and Endo, 1968).

The series of events produced after the arrival of the electrical signal may be the following: the signal is received at the individual Z-band or A-I junction by way of the intermediary vesicles or transverse system. This sets in motion a series of events which may include the release of Ca^{2+} in the vicinity of the triad, the activation of the myofibril ATPase, and the uptake of Ca^{2+} by the elements of the reticulum, which inhibits the ATPase action of the myofibrils and induces relaxation.[27]

All these data, as well as those related to the sliding mechanism of contraction, can be put together in a molecular theory of muscular contraction. This is one of the best examples, so far studied, of a tight coupling between the processes furnishing energy and the actual machinery involved in contraction. In this case, structure and function are so intimately related in the realm of molecular organization that they are an inseparable unit.

SUMMARY:
Molecular Biology of Muscle

Muscle cells are adapted to mechanical work by unidirectional contraction. The functional unit is the *myofibril*, which may be either striated or smooth. In skeletal muscle, myofibrils fill most of the large muscle fiber, leaving small amounts of sarcoplasm, which contains the nuclei, the *sarcoplasmic reticulum*, and large mitochondria, or *sarcosomes*. A *sarcolemma*, with —0.1 volt polarization, surrounds the fiber. This is activated through the *end plate*.

Myofibrils result from the repetition of *sarcomeres*. These are limited by the Z-lines and contain the I-bands, the A-bands and the H-band. The M-line may be observed in the middle of the sarcomere. The myofibril is composed of thick (10 nm) and thin (5 nm) myofilaments. In the relaxed condition, the I-band contains only thin myofilaments, the H-band only thick, and the A-band, both thick and thin. Myofilaments are organized in a paracrystalline hexagonal array. From the thick filaments, at 7 nm intervals, there are cross-bridges extending toward the thin filaments. One thick filament joins to six adjacent thin ones every 43 nm.

During contraction the I-bands shorten as a result of the sliding of the thin filaments into the A- and H-bands. A further reversal of the band may be produced at the center of the sarcomere. In vertebrate muscle, the sarcomere may shorten up to 43 per cent.

The structural proteins *myosin, actin,* and *tropomyosin* constitute 90 per cent of the myofibril and represent the contractile machinery. Myosin molecules are about 150 nm long and appear to be tadpole-shaped with a head and a long tail (Fig. 23–7). Isolated or reconstituted myofilaments show that the heads correspond to the cross bridges and the tails correspond to the shaft of the myofilament. By the action of trypsin, myosin is fragmented into a *heavy* and a *light meromyosin;* within papain, the heavy meromyosin is further divided into S_1 and S_2. S_1 meromyosin contains the Ca^{2+}-activated ATPase and the sites for binding ATP and actin; this particle constitutes the cross bridge.

Actin is composed of globular units (G-actin) 5.3 nm in diameter that polymerize to form a double helical F-actin. F-actin corresponds to the thin filaments. *Tropomyosin* is a thin, double-helical molecule, 40 nm in length, and intimately associated with the helix of F-actin. Associated with actin are small globular molecules of *troponin,* a Ca^{2+}-binding protein. F-actin, tropomyosin, and troponin constitute a complex that intereacts with myosin during contraction. Localization of these proteins in relation to the sarcomere has been achieved by differential extraction and by the use of fluorescent antibodies.

The *sliding theory of muscle contraction* postulates that the thin actin filaments are displaced with respect to the thick filaments at each contraction cycle. Short range forces are generated at the cross-bridges between the filaments. An S_1 portion of the myosin molecules attaches to a binding site of F-actin, and by a conformational change, pulls it toward the center of the sarcomere. The splitting of ATP by the Ca^{2+}-activated ATPase provides the energy. One ATP may account for a 5 to 10 nm displacement of a cross bridge. According to this theory, both the myosin and actin myofilaments should be polarized in each of the sarcomeres (Fig. 23–10). This polarization is actually demonstrated by the electron microscopic observation of reconstituted myosin filaments and of actin filaments treated with heavy meromyosin. The arrow-like configuration of these "decorated" filaments is the result of the F_1-actin helix and the tilting and binding of the S_1 heavy meromyosin (Fig. 23–12). The initial source of energy for contraction is provided by ATP. This energy is later furnished by glycolysis and from creatine phosphate, as well as from oxidative phosphorylation by mitochondria.

The *sarcoplasmic reticulum* is found in skeletal and cardiac muscle fibers. It is a special endoplasmic reticulum whose morphology is adapted to the structure of the contractile apparatus. It has a *transverse* component (T-system), formed by the central vesicle of the *triad* that is continuous by way of fine tubules with the extracellular space. This vesicle is at the level of the Z-line, between the myofibrils. The other two vesicles of the triad, which are connected with the *longitudinal* component of the reticulum, are arranged along the surface of the A-band. Stimulation at the Z-line produces the contraction of the adjacent sarcomeres. It is thought that the sarcoplasmic reticulum, by way of the T-system, conducts the impulses from the sarcolemma into the muscle fiber. In part, muscle *relaxation* is also a function of the sarcoplasmic reticulum. Relaxation is brought about by a decrease in the intracellular concentration of Ca^{2+}. Ca^{2+} is actively incorporated into the reticulum by a Ca^{2+}-activated ATPase. A regulatory role is played by troponin, a Ca^{2+}-binding protein associated with thin myofilaments. The activation of the myofibril at the triad is due to the release of Ca^{2+} from the sarcoplasmic reticulum. This release activates the myosin ATPase and induces the contraction. Relaxation results principally from the re-uptake of Ca^{2+} by the sarcoplasmic reticulum.

Muscle provides one of the best examples of the tight coupling between the processes furnishing energy and the actual mechanical machinery that produces the contraction.

REFERENCES

1. Huxley, H. E. (1966) *Harvey Lect., 60*:85.
2. Hagopian, M. (1966) *J. Cell Biol., 28*:545.
3. Huxley, H. E. (1958) *Sci. Am., 199* (No. 5):67.
4. Huxley, H. E. (1970) *Proc. Roy. Inst. Gr. Br., 44*:274.
5. Hanson, J., and Lowy, J. (1961) *Proc. Roy. Soc.,* Ser. B, *154*:173.
6. Pease, D. C. (1968) *J. Ultrastruct. Res., 23*:280.
7. Mommaerts, W. (1950) *Muscular Contraction.* Interscience Publishers, New York.
8. Zobel, C. R., and Carlson, F. D. (1963) *J. Molec. Biol., 7*:739.
9. Huxley, H. E. (1963) *J. Molec. Biol., 7*:281.
10. Lowey, H., Slayter, S., Weeds, A. G., and Baker, H. (1969) *J. Molec. Biol., 42*:1.
11. Davies, R. E. (1963) *Nature* (London) *199*:1068.
12. Hanson, J., and Lowy, J. (1963) *J. Molec. Biol., 6*:46.
13. Hanson, J., and Huxley, H. E. (1957) *Biochim. Biophys. Acta, 23*:250, 260.
14. Marshall, J. M., Jr., Holtzer, H., Finck, H., and Pepe, F. (1959) *Exp. Cell Res.,* suppl. 7:219.
15. Pepe, F. A. (1966) *J. Cell Biol., 28*:505.
16. Ebashi, S., and Kodama, A. (1966) *J. Biochem., 60*:733.
17. Stromer, M. H., Hartshorne, D. J., Ric, R. V., (1967) *J. Cell Biol., 35*:23.
18. Turner, D. C., Wallimann, T., and Eppenberger, H. M. (1973) *Proc. Natl. Acad. Sci. USA, 70*:702.
19. Yaffec, D., and Feldman, M. (1964) *Dev. Biol., 9*:347.
20. Fischman, D. A. (1967) *J. Cell Biol., 32*:557.
21. Reedy, M. K. (1968) *J. Molec. Biol., 31*:155.
22. Szent-Györgyi, A. (1947) *Chemistry of Muscular Contraction.* Academic Press, Inc., New York.
23. Klug, A., and De Rosier, D. J. (1966) *Nature* (London), *212*:29.
24. Moore, P. B., Huxley, H. E., and De Rosier, D. J. (1970) *J. Molec. Biol., 50*:279.
25. Ruch, T. C., and Fulton, J. F. (1960) *Medical Physiology and Biophysics.* 18th Ed., Chap. 4. W. B. Saunders Company, Philadelphia.
26. Bennett, H. S., and Porter, K. R. (1953) *Amer. J. Anat., 93*:1.
27. Porter, K. R. (1961) *J. Biophys. Biochem. Cytol., 10*, suppl. 219.
28. Andersson-Cedergren, E. (1959) *J. Ultrastruct. Res.,* suppl. *1.*

29. Huxley, H. E. (1964) *Nature* (London), *202*:1067.
30. Brandt, P. W., Reuben, J. P., Girardier, L., and Grunfest, H. (1965) *J. Cell Biol., 25*:233.
31. Hagopian, M., and Spiro, D. (1967) *J. Cell Biol., 32*:535.
32. Peachey, L. D., and Porter, K. R. (1959) *Science, 129*:721.
33. Muscatello, V., Andersson-Cedergren, E., Azzone, G. F., and Der Decken, A. von (1961) *J. Biophys. Biochem. Cytol., 10*, suppl. 201.
34. Reuben, J. P., Brandt, P. W., García, H., Grundfest, H. (1967) *Amer. Zool., 7*:623.
35. Ebashi, S., and Lipman, F. (1962) *J. Cell Biol., 14*:389.
36. MacLennan, D. H. (1970) *J. Biol. Chem., 245*:4508.
37. Racker, E. (1972) *J. Biol. Chem., 247*:8198.
38. MacLennan, D. H., and Wong, P. T. S. (1971) *Proc. Natl. Acad. Sci. USA, 68*:1231.
39. Scales, B., and McIntosh, D. A. D. (1968) *J. Pharmacol. and Exp. Therap., 160*, 249, 261.
40. Hasselbach, W., and Elfvin, L. G. (1967) *J. Ultrastruct. Res., 17*:598.

ADDITIONAL READING

Bourne, C., ed. (1960) *Structure and Function of Muscles.* 2 volumes. Academic Press, Inc., New York.

Ebashi, S. E., and Endo, M. (1968) Calcium ion and muscle contraction. *Progr. Biophys. Molec. Biol., 18*:123.

Huxley, H. E. (1960) Muscle cells. In: *The Cell*, Vol. 4, p. 365. (Brachet, J., and Mirsky, A. E., eds.) Academic Press, Inc., New York.

Huxley, H. E. (1969) The mechanism of muscular contraction. *Science, 164*:1356.

Huxley, H. E. (1971) Muscular contraction. *Proc. Roy. Inst. Gr. Br., 44*:274.

Kelly, R. E., and Rice, R. V. (1968) Localization of myosin filaments in smooth muscle. *J. Cell Biol., 37*:105.

Lowey, S. (1972) Protein interaction in the myofibril. In: *Polymerization in Biological Systems*, Ciba Foundation Symp. 7 (new series), p. 217, Elsevier Publishing Company, Amsterdam.

Murray, J. M., and Weber, A. (1974) The cooperative action of muscle proteins. *Sci. Am., 230*(No.2):59.

Peachey, L. D., and Porter, K. R. (1959) Intracellular impulse conduction in muscle cells. *Science, 129*:721.

Pepe, F. A. (1968) Analysis of antibody staining patterns obtained with striated myofibrils in fluorescence microscopy and electron microscopy. *Int. Rev. Cytol., 24*:193.

Rothfield, L. (1972) In: *The Dynamic Structure of Cell Membranes*, p. 165. (Wallach, D. F. H., and Fischer, H., eds.) Springer-Verlag, Heidelberg.

Rowe, R. W. D. (1973) The ultrastructure of z-disks from white intermediate and red fibers of mammalian striated muscles. *J. Cell Biol., 57*:261.

twenty-four

CELLULAR AND
MOLECULAR NEUROBIOLOGY

One of the most important functions of living organisms is reacting to an environmental change. Such a change, called a *stimulus,* generally elicits a *response.* In its most basic sense this general property is called *irritability.* For example, a unicellular protozoan may react to different stimuli, such as changes in heat or light or the presence of a food particle, by a mechanical response, such as ciliary motion, ameboid movement, etc. *Plants* react by slow responses, which produce differential growth, also called a *tropism.* For example, the responses to the gravitational field, temperature, light, touch, and chemicals are referred to, respectively, as *geotropism, thermotropism, phototropism, thigmotropism,* and *chemotropism.* Irritability reaches its maximal development in animals, and special cells forming the nerve tissue are differentiated to respond rapidly and specifically to the different stimuli.

In these organisms special physiologic *receptors* adapted to "receive" the various types of stimuli are differentiated. Receptors are made of special cells or of the distal endings of neurons, specialized to receive a particular stimulus. For example, the receptors of light, touch, taste, pressure, heat, cold, and others, are characterized by their great sensitivity to the specific stimulus.

At the receptor the *threshold of excitation* is much lower than in any part of the nerve cell.

In an animal, the response to the stimulus may be of a varied nature. Most frequently the animal reacts with a rapid movement by contraction of muscle tissue. However, other types of reactions may be elicited. For example, a hungry dog in the presence of food reacts by secreting saliva; an electric fish, upon being touched, may produce an electrical discharge; and a firefly may give off light quanta. These different types of responses are produced in special tissue (e.g., muscles, glands, electric plates, luminous organs), called *effectors,* that are controlled by efferent neurons.

The Reflex Arc. Action Potentials

In an animal the simplest mechanism of nerve action is represented by the so-called monosynaptic reflex. This consists of a neuronal circuit formed by two *neurons.* One neuron is *sensorial* (afferent) and has a receptor at one end to receive the stimulus. At the other end the sensory neuron makes a special contact, also called a *synapsis,* with a *motor* (efferent) neuron, which in turn acts on the effector (i.e., muscle).

Figure 24–1 is a simplified diagram of the way in which the information *received* at the receptor is *conducted* along the sensory neuron and then *transmitted* at the synapse. Notice that a new wave of information starts in the second neuron, which finally reaches the effector, where the final response is elicited.

As will be shown below, in nerves and muscle information is propagated by changes in the *resting* or *steady potential* at the surface membrane (see Chapter 21). This change originates a *wave of excitation,* which moves along the surface of the cell from one end to the other. In the nerve cell this *propagated* or *action potential* is also known as a *nerve impulse.*

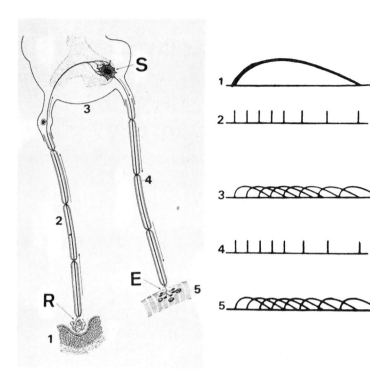

Figure 24–1. Diagram showing the monosynaptic reflex arc. **Left,** one sensory and one motor neuron with the synaptic junction. Notice the receptor and the effector. **Right,** different types of potentials produced at the portions of the reflex arc (1–5), indicated in the figure. *R*, receptor; *S*, synapsis; *E*, effector.

Nerve impulses are *conducted* along the elongated parts of neurons (i.e., the nerve fibers) by way of action potentials. At the receptors, the synapse and the effectors, other electrical potentials having different characteristics are produced. The emphasis of this chapter will be on the *cellular bases* of *nerve conduction* and *synaptic transmission.*

General Organization of a Neuron

The neuron is differentiated for the conduction and transmission of nerve impulses. After embryonic life, neurons do not divide, but remain in a permanent interphase throughout the entire life of the organism. During this time a neuron undergoes changes in volume and in the number and complexity of its processes and functional contacts, but the number of neurons is not increased by cell division. This fact may be of paramount significance, since, in addition to conducting and transmitting impulses, nerve cells store instinctive and learned *information* (e.g., conditioned reflexes, memory)—a property that would be best served by a more permanent system of structures.

Neurons are adapted to their specialized

functions by means of different types of outgrowths. The cell body (perikaryon) may emit one or more short outgrowths, or *dendrites,* which carry nerve impulses centripetally, and a longer one, the *axon,* which carries the impulse centrifugally to the next neuron or the effector. An axon is also called a *nerve fiber* when it is wrapped in the different sheaths after emerging from the neuron. The axon terminates, ramifying in the *telodendrons,* or endings. Some neurons have only one dendrite, and the axon (i.e., *bipolars*) and others have only the axon (i.e., *monopolars*), in addition to the most common *multipolars.* In invertebrates most neurons are monopolar.

The different types of neurons and neuronal interconnections are discussed in histology and neuroanatomy textbooks. Only a few general considerations will be made here.

Structure of the Axon Neurofibrils and Neurotubules

In fixed and stained preparations observed under the light microscope fine filaments, called *neurofibrils,* can be demonstrated in the cytoplasm of the neuron. These

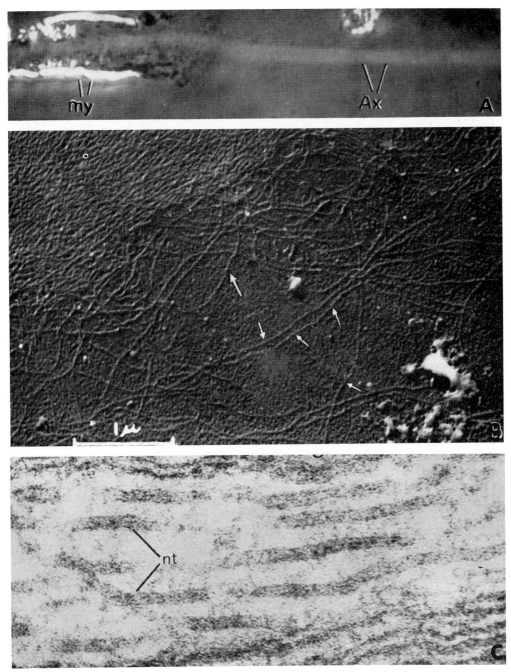

Figure 24–2. **A,** extruded axon (*Ax*) observed under the polarization microscope. Note its weak positive bire-fringence. After appropriate treatment, the axon exhibits both intrinsic and form birefringence; *my,* myelin with a strong birefringence. In the normal fiber, the axon birefringence is obscured by that of the myelin sheath. × 26,000. (From W. Thornburg and E. De Robertis, 1956.) **B,** fibrillar material (neurotubules and neurofilaments) observed under the electron microscope in an axon extruded from the myelin fibers and compressed. The arrows indicate some neurotubules. × 26,000. (From E. De Robertis and C. M. Franchi, 1953.) **C,** section of longitudinally oriented neurotubules (*nt*). × 120,000. (Courtesy of E. L. Rodríguez Echandía, R. S. Piezzi, and E. M. Rodríguez.)

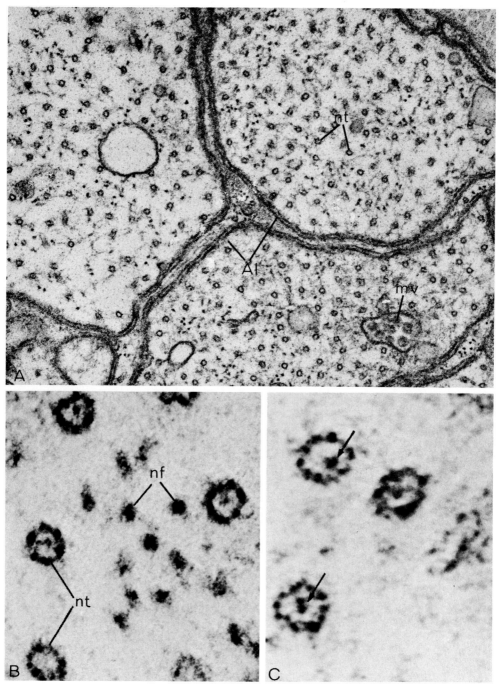

Figure 24–3. **A,** cross section of an unmyelinated nerve showing the axolemma (*Al*), neurotubules (*nt*), and a multivesicular body (*mv*). × 60,000. **B,** same as **A** at higher magnification; *nf,* neurofilaments; *nt,* neurotubules. × 400,000. **C,** neurotubules containing a dense granule (arrows). × 600,000. (Courtesy of E. L. Rodríguez Echandía, R. S. Piezzi, and E. M. Rodríguez.)

neurofibrils run in all directions and continue into the dendrites, axon, and nerve fiber. Better information about the structure of the axon can be obtained by polarization and electron microscopy.

The axoplasm of myelinated nerve fibers may be extruded and separated from the myelin sheath, permitting its study with the electron microscope and with polarization microscopy[1,2] (Fig. 24-2, A). In these studies the partial volume occupied by the axially oriented material has been found to be less than 1 per cent. In the extruded axoplasm a fibrillar material was detected (Fig. 24-2, B).[1] In addition, mitochondria, strands of canaliculi, and vesicles of the endoplasmic reticulum have been observed.[2] The fibrillar material present in the axon, dendrites, and perikaryon is formed principally of long, tubular elements (the neurotubules), 20 to 30 nm in diameter, which were previously seen in nerve homogenates[1,3] (Fig. 24-2, C).

Unmyelinated nerve fibers show numerous neurotubules in cross section (Fig. 24-3, A). Among the neurotubules, there are also cross sections of neurofilaments. These two structural components of the axon are shown at higher magnification in Figure 24-3, B and C. The walls of the neurotubules show a subunit structure similar to that generally found in microtubules (see Chapter 22). Neurotubules containing dense cores (Fig. 24-6, C) have been described.[4] The electron-dense material of the cores has been interpreted as a result of endoluminal migration of some products. In the neurofilaments some lateral outgrowths or side arms have been observed.[5] Neurotubules have been isolated from mammalian brain cells. Most of the biochemical and physiologic concepts described in Chapter 22 for the microtubules also apply to the neurotubules. Neurotubules and neurofilaments, when clumped together under the action of the fixatives and with the addition of colloid silver, form the neurofibrils of classic histology.

Although neurofibrils were described more than a century ago, their significance remained practically unknown. The hypothesis that they are involved in nerve conduction has been disproved. In Chapter 21 the experiments of axoplasm extrusion and replacement by a saline solution with normal conduction of nerve impulses were mentioned. There is now no doubt that nerve conduction takes place at the surface membrane of the axon (see the following section).

Biosynthetic Function of the Perikaryon

The perikaryon is characterized by the presence of considerable amounts of basophilic material—the Nissl substance—which, as in other cells, is composed of ribosomes and endoplasmic reticulum. A well developed Golgi complex is also characteristic of the neuron. The great abundance of ribosomes is related to the biosynthetic functions of the perikaryon, which has to maintain a volume of cytoplasm in its outgrowths that may be considerably greater than its own. (In mammals axons may be as long as 1 meter or longer). If a nerve fiber is cut, the distal part degenerates (wallerian degeneration), and the proximal stump may regenerate later on by a growing process that is dependent on the perikaryon. There is also experimental evidence that the axon is continuously growing and being used at the endings.[6]

Axonal Transport of Macromolecules

The experiments just mentioned—on degeneration—should be correlated with the fact that the axon and the nerve endings are generally devoid of polysomes, and are unable to undertake a significant local protein synthesis. In fact, local synthesis by nerve endings may account for only 2 per cent of the rapidly renewed proteins.[7,8] For this reason most axonal and synaptic proteins should be manufactured in the perikaryon and transported to the nerve endings.

The study of the axonal transport (*axon flow*) of macromolecules has become one of the most intense fields of research in neurobiology (see Grafstein, 1969; Ochs, 1972; Droz et al., 1973). Macromolecules such as proteins, glycoproteins, and enzymes) which are soluble in the axoplasm or integrate the various axonal and synaptic structures (i.e., axonal membrane, neurotubules, neurofilaments, mitochondria, synaptic vesicles, nerve-ending membranes, etc.), are first synthesized in the cell body and then find their way into the dendritic arborizations. They may also pass into the axon and nerve endings, at a distance that may vary from millimeters to meters. The proteins present in nerve endings have half-lives which range from 12 hours to 50 to 100 days;[9] to be replaced, therefore, new proteins should reach the nerve endings to compensate for their loss. Experimental work that has used colchicine and vinblastine has demonstrated that axoplasmic transport is related to the neurotubules.[10] In fact, the integrity of these axonal organelles is critical in maintaining a normal axoplasmic flow. One of the best approaches to the study of axonal transport is the use of radiolabeled precursors for proteins or glycoproteins (e.g., ^{3}H-lysine, ^{3}H-fucose) and the monitoring of the final product by radioautography at the electron microscopic level. In recent experiments the ciliary ganglion of chicken was used because it has definite advantages. In fact, the axons originate from the Edinger-Westphal nuclei, situated near the cerebral aqueduct, where the precursor can be injected. The axons are 10 mm long and terminate in a single giant nerve ending (the presynaptic calyx), which makes contact

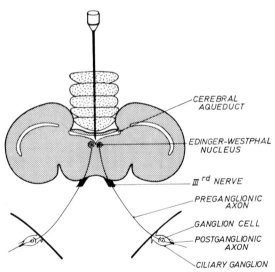

Figure 24–4. Diagrammatic representation of the administration of labeled precursors to the preganglionic neurons of the ciliary ganglion in chickens. The radioactive tracer injected into the cerebral aqueduct reaches the nerve cell bodies of the preganglionic neurons located in the Edinger-Westphal nucleus near the cerebral aqueduct. The preganglionic axons follow the pathway of the oculomotor nerve (III) and terminate in the ciliary ganglion by forming giant presynaptic calices. (From Droz, B., Koenig, H. L., and Di Giamberardino, L., *Brain. Res., 60:93, 1973.)*

with the postsynaptic cell body of the ciliary ganglion (Fig. 24–4).[11] Figure 24–5 shows a radioautograph obtained after the injection of ^{3}H-lysine. In the lower portion of the figure the postsynaptic ganglion cell is shown to be completely devoid of silver grains, indicating that there is no transynaptic migration of proteins. In the upper part, the presynaptic ending is covered with silver grains corresponding to proteins that have migrated from the perikaryon. The radioactive proteins are distributed over synaptic vesicles, mitochondria, and axoplasm.

Fast and Slow Transport

The studies just described, as well as others, have revealed that the proteins may be transported at different rates, and that there are generally two definite groups of proteins — those of fast or those of slow transport. The diagram of Figure 24–6 gives a general idea of these two systems.

Fast Axonal Flow. Axonal flow of about 280 mm per day is found in proteins that are synthesized in polysomes bound to the endoplasmic reticulum (ergastoplasm), that are transferred to the Golgi complex, and then penetrate the axon and dendrites, mainly by way of tubules and vesicles of the smooth endoplasmic reticulum. On their way through the Golgi apparatus, some proteins are converted into glycoproteins by the addition of carbohydrates (see Chapter 9). Figure 24–6 indicates that the proteins of fast transport contribute to the renewal of the axon membrane, the turnover of macromolecules associated with synaptic vesicles, and the presynaptic membrane. Another fraction is related to axonal and nerve ending mitochondria. The fast moving proteins are, thus, assigned mainly to the renewal of various membrane components; by cell fractionation they are recovered in the particulate fractions.[12] Practically all glycoproteins are also transferred by this fast phase of axonal flow.[13] Since these fast moving proteins and glycoproteins can cover 280 mm per day, it is understandable that they begin to appear in the ciliary ganglion within one hour of the injection of the precursors into the aqueduct.

Slow axonal flow. In proteins present in the soluble fraction of the nervous tissue, and to a lesser degree in mitochondria, axonal

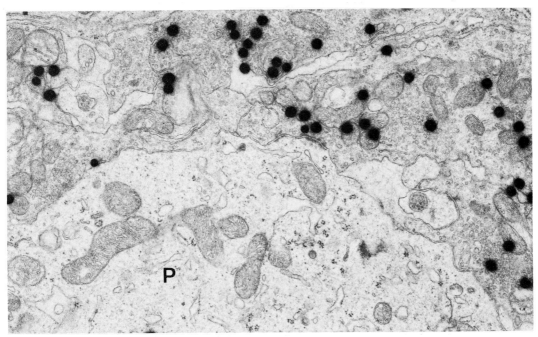

Figure 24–5. Radioautograph of the transport of a protein labeled with ³H-lysine that was injected into the cerebral aqueduct observed under the electron microscope (see Fig. 24–4). The section corresponds to the ciliary ganglion of the chick. In the upper part of the figure the silver grains of the labeled protein are concentrated over the nerve endings, whereas there are none in the postsynaptic perikaryon (*P*). (Courtesy of B. Droz et al., 1973.)

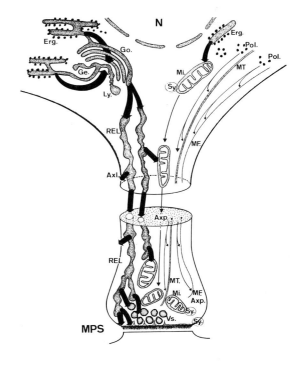

Figure 24–6. Diagram representing the axonal transport of macromolecules. **On the left,** the *fast transport;* **on the right,** the *slow transport.* In fast transport the proteins synthesized in the ergastoplasm (*Erg.*) are transferred to the Golgi complex (*Go.* and *Ge.*) where carbohydrates may be added to make glycoproteins. Along the axon the proteins and glycoproteins are transported by way of the smooth endoplasmic reticulum (*REL*). At the rapid speed of 280 mm per day such proteins participate in the turnover of macromolecules associated with the axolemma (*Axl.*), the synaptic vesicles (*Vs.*), the presynaptic membrane (*MPS*), and mitochondria (*Mi.*). In slow transport, indicated by lightly drawn arrows, the proteins synthesized in free ribosomes (*Pol.*) are slowly transported with the axoplasm (*Axp.*) along the axon. The soluble proteins, as well as the microtubules (*MT.*) and microfilaments (*MF.*), are transported at the slow speed of 1.5 mm per day. *N*, nucleus; *Ly*, lysosome, *Sy*, synaptic protein. (Courtesy of B. Droz et al., 1973.)

flow is slow.[14] Figure 24–6 shows that these proteins are synthesized mainly by free polysomes and probably by-pass the Golgi complex. Soluble enzymes such as choline acetyltransferase, protein subunits of tubulin that assemble to form the neurotubules, or the protein of microfilaments, are transported by slow axonal flow. These enzymes enter the axon at a speed of 1 to 1.5 mm a day and exhibit maximal accumulation in the nerve ending at 6 days. In general, these proteins turn over at a slow rate in the axon, and only a minor fraction of them enters the nerve ending.

The fast and slow moving proteins that enter the axon and nerve endings may compensate for the local breakdown taking place by proteolytic enzymes, as well as the release of proteins that may be related to the function of synaptic vesicles (see below). A possible postsynaptic transfer of macromolecules of small magnitude should also be contemplated.

Nerve Fibers: Diameter and Conduction Velocity

Nerve fibers are *nonmyelinated* when wrapped only in Schwann cells. *Myelinated* nerve fibers also have a myelin sheath that consists of a multilayer lipoprotein system (see Figure 8–6). In the autonomic system of vertebrates, most nerve fibers are unmyelinated and are contained within invaginations of the plasma membrane of the Schwann cells.

TABLE 24–1. Properties of Neurons of Different Sizes (Cat and Rabbit Saphenous Nerves)*

Properties	*Group*		
	A	*B*	*C*
Diameter of fiber (μm)	20–1	3	—
Conduction velocity (m/sec)	100–5	14–3	2
Duration of action potential (msec)	0.4–0.5	1.2	2.0
Absolute refractory period (msec)	0.4–1.0	1.2	2.0

*Modified from Grundfest, H., *Ann. Rev. Physiol.*, 2:213–242, 1940.

The myelin sheath is interrupted at the *nodes of Ranvier*. The distance between nodes varies with the diameter of the fiber. The *internode*, i.e., the distance between successive nodes, is the segment of myelin that is produced and contained within a single Schwann cell. The internode is 0.2 mm in a bull frog fiber of 4 μm, about 1.5 mm in a fiber of 12 μm and 2.5 mm in one of 15 μm. Later the internode will be discussed in relation to saltatory conduction of the myelinated nerve fibers.

Within the internode, obliquitous (conic) *incisures* go across the myelin sheath where the myelin leaflets have a looser disposition. At the node the myelin lamellae are loosely arranged, and a small zone of axon is in direct

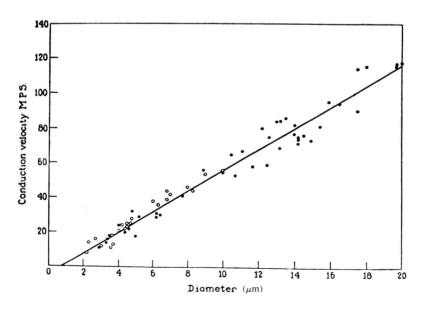

Figure 24–7. The linear relation between diameter and conduction velocity in meters per second (*M.P.S.*) of mammalian myelinated nerve fibers. The dots represent adult nerves; the circles represent immature nerves. (Modified from Hursh, from Gasser, *Ohio J. Sci., 41:*145, 1941.)

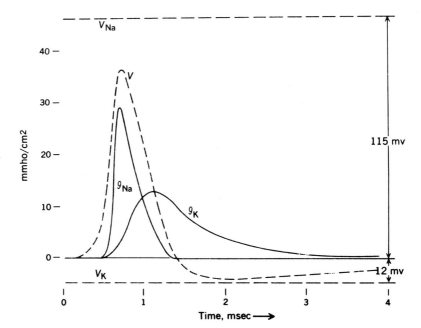

Figure 24–8. Diagram showing a propagated action potential (curve V) and the sodium (^gNa) and potassium (^gK) conductances. Observe that the entrance of sodium coincides with the rising phase of the spike. (From Hodgkin, A. L., and Huxley, A. F., *Cold Spring Harbor Symp. Quant. Biol.*, *17*:43, 1952.)

contact with the extracellular fluid. The myelin sheath acts as an insulator and, as a consequence, myelinated fibers conduct nerve impulses at a much faster rate than unmyelinated fibers. The diameter of the fiber also influences the conduction rate. As shown in Table 24–1, nerve fibers can be classified according to their diameters into groups A, B, and C. C fibers are unmyelinated. The diameter may vary from 20 μm in A fibers to less than 1 μm in C fibers, and the conduction velocity varies from 100 to 2 meters or less per second. As shown in Figure 24–7, the rate of conduction of the nerve impulse follows a linear relationship with the fiber diameter in mammalian myelinated fibers, and it is also related to the internode distance.

Conduction of the Nerve Impulse

For the study of the physicochemical phenomena underlying the conduction of the nerve impulse, consult general physiology textbooks.[15, 16] Here, the subject is discussed briefly and superficially as a continuation of the discussion of *active transport* and *membrane potentials* in Chapter 21.

When a muscle or a nerve fiber is stimulated, a profound change is produced in the electrical properties of the surface membrane and in the steady potential.

As shown in Figure 24–8, experimenters using intracellular recording can demonstrate that with excitation the resting potential is suddenly changed. At the point of stimulation there is not only a depolarization, but also an overshoot, and the potential becomes positive inside. With radioactive tracers it has been found that at the point of stimulation there is a sudden, and several hundredfold increase in permeability to Na^+, which reaches its peak in 100 microseconds.[17] At the end of this period the membrane again becomes essentially impermeable to Na^+, but the K^+ permeability increases, and this ion leaks out of the cell, repolarizing the nerve fiber. In other words, during the rising phase of the spike Na^+ enters, and in the descending phase K^+ is extruded. Complete restoration of the ionic balance takes a longer time after the electrical event (Fig. 24–8).

The action potential that develops in the nerve fiber has several other characteristics: (1) The stimulus produces a slight local depolarization in the fiber, which, after reaching a certain *threshold of activation,* produces spikes of the same amplitude. If the intensity of the stimulus is increased, the height of the spike always remains the same. This is called an *all-or-none response.* (2) The nerve impulse is *nondecremental;* i.e., the amplitude of the spike does not decrease and is the same all along the course of the nerve fiber. This type of action potential is thus well adapted to conduction over long distances without losses (see Figure 24–1). (3) Once a nerve impulse has passed over any point of the fiber, there is a *refractory period* during which it cannot react to another stimulus.

The *propagation* of the nerve impulse is generally explained by the so-called *local circuit theory* (Fig. 24–9). At the point of stimulation the area becomes depolarized (negative outside) and acts as a sink toward which the current flows from the adjacent areas (Fig. 24–9, *B* and *C*). This wave of depolarization advances along the nerve fiber at the rate of conduction that is characteristic for each fiber (Table 24–1 and Fig. 24–7). While this wave of depolarization advances,

repolarization is so rapid that only a fraction of the nerve fiber (a few millimeters or centimeters, depending on the conduction rate) is depolarized at a time. In the recovery period, sodium leaves the cell by the action of the sodium pump (Chapter 21) and potassium re-enters to restore the steady state. This recovery is probably produced at the expense of high energy phosphate bonds. However, impulses continue to discharge for some time in the absence of oxygen, and even when glycolysis is inhibited, which indicates that high energy bonds are stored at the membrane.

Saltatory Conduction

The preceding theory of nerve conduction applies to unmyelinated nerve fibers. In myelinated fibers it is thought that the local circuits occur only at the nodes (Fig. 24–9, *D*). According to this so-called *saltatory theory,* at the internode the impulse is conducted electrotonically, and at each node the action potential is boosted to the same height by ionic mechanisms. In this way the amount of Na^+ and K^+ exchanged is greatly reduced and the net work required is much less. Stimulation of myelinated nerve fibers with fine electrodes has shown that at the nodes the threshold of stimulation is much lower than at the internode.[18] It has been found that the nerve impulse can jump across one anesthetized node, but not two of them.

Graded Responses in the Neuron

Physiologic studies have demonstrated that in addition to the all-or-none response, there is another type of electrical activity in nervous tissue. This is, by far, the most frequent in the central nervous system and is referred to as a *graded response*. In the graded response the impulse is not *propagated* and the *amplitude* varies with the intensity of the stimulus. This type of response is characteristic of the physiologic receptors and synapses (see Figure 24–1). Both the *generator potentials* found at the receptors and the *synaptic potentials* are graded responses.

If a peripheral receptor, such as a Pacini corpuscle or a stretch receptor (neuromuscular spindle), is mechanically stimulated, a local, graded, and decremental potential is recorded, the amplitude and duration of which depend on the intensity and duration of the

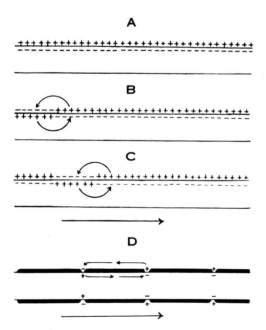

Figure 24–9. Diagram illustrating the local circuit theory of propagation of the action potential (**A, B, C**) in unmyelinated neurons and muscle fibers as compared to saltatory conduction in myelinated neurons (**D**). **A,** the membrane of an unexcited nerve (or muscle) fiber; **B,** the cell membrane excited at one end; **C,** the movement of the action potential, followed by recovery; **D,** node-to-node saltatory conduction. In large nerve fibers less than one hundredth as much ionic exchange occurs during an impulse in saltatory conduction as compared to conduction in an unmyelinated nerve fiber. The arrows in **C** and **D** show the direction of impulse propagation. (After Hodgkin, 1957, *Proc. Roy. Soc. London,* ser. B, *148*:1, 1957.)

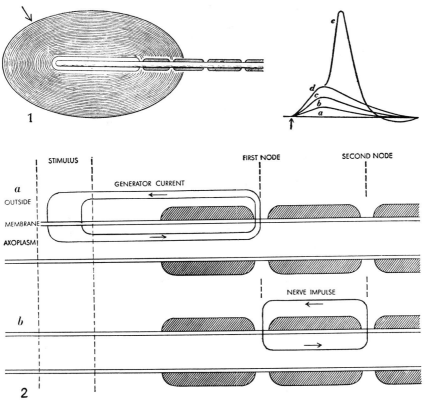

Figure 24–10. **1,** Pacini corpuscle with the nerve ending surrounded by multiple layers. Stimulation at the point marked by the arrow produces a generator potential (**right**) which increases in amplitude (*a-d*) until it fires an all-or-none nerve impulse (*e*). **2,** mechanism of the transducer. The stimulus produces a drop in the resistance of the membrane with ion transfer. Notice the generator current induced by the stimulus (*a*) and the nerve impulse (*b*) originating at the first node. (Courtesy of W. R. Loewenstein.)

stimulus.[19] In the Pacini corpuscle it is possible to remove most of the connective lamellae that surround the nerve ending without impairing the generator potential (Fig. 24–10). It appears that in this case the *biological transducer* capable of transforming the mechanical energy (pressure) into the electrical energy (generator potential) is localized at the sensory part of the ending (Fig. 24–10). Probably the mechanical deformation of the ending produces a change in permeability with entrance of ions and partial depolarization. In the Pacini corpuscle it has been observed that the nerve impulse starts at the first node of Ranvier (Fig. 24–10).[20]

The intensity of the sensory stimulus is reflected in the amplitude of the generator potential and this in turn in the *frequency* of the propagated signal—the stronger the generator potential, the higher the frequency. In this way the information received is coded for conduction along the nerve fiber in the form of a train or volley of impulses (see Figure 24–1).

The nervous tissue is specially differentiated to respond rapidly and specifically to different stimuli. Protozoa and plants contain no nerve differentiation, but have more primitive mechanisms of *irritability*. The simplest mechanism in the nervous tissue is the circuit of the reflex arc, which is

SUMMARY:
General Organization of
the Neuron

integrated by a sensorial *(afferent)* neuron and a motor *(efferent)* neuron. These two neurons are related, respectively, to a physiological *receptor* and an *effector;* they are connected by *synapses.* The nerve impulse is *conducted* along the axon of the nerve fibers and is *transmitted* at the level of the synapse and the effector.

Neurons do not divide during postnatal life, but undergo an increase in volume and in the number and complexity of their processes and functional contacts. Neurofibrils can be observed in the axon and dendrites after fixation and staining of the perikaryon. They result from the clumping together of neurotubules and neurofilaments. Neurotubules are similar to the microtubules of other cells (see Chapter 22). Neurofibrils are not involved in nerve conduction, since this takes place at the axonal membrane.

The perikaryon of neurons is rich in ribosomes and endoplasmic reticulum (Nissl substance). The great abundance of ribosomes is related to its biosynthetic functions. If the nerve fiber is cut, the distal stump degenerates (wallerian degeneration), and the proximal stump may regenerate later on. There is little or no local protein synthesis in the axon and nerve endings, which are devoid of polysomes. Proteins and glycoproteins are synthesized in the perikaryon and are transported to the axon and nerve endings as well as to the dendrites.

The axonal transport of macromolecules depends on the integrity of neurotubules. Colchicine and vinblastine, which destroy these organelles, also stop axonal transport. At the cellular level, this process can best be studied with radioautography. The ciliary ganglion of the chicken has definite advantages in such studies (Fig. 24–4). In the radioautograph the synthesized protein may be observed to enter the nerve ending, but it does not penetrate the post-synaptic ganglion cell.

Axonal transport may be *fast* (280 mm per day) or *slow* (1 to 1.5 mm per day). Fast transport is found in proteins integrating the axonal membrane, synaptic vesicles, presynaptic membrane, and mitochondria, i.e., in membrane-bound proteins. The glycoproteins studied with ^{3}H-fucose are also fast moving. Slow axonal flow is found in soluble proteins (e.g., enzymes, tubulin) that are in the axoplasm, or that integrate the neurotubules, neurofilaments and mitochondria. Few of these proteins reach the nerve ending. The fast and slow moving proteins compensate for the local breakdown and release of proteins taking place within the axon and at the nerve ending.

Nerve fibers are *non-myelinated,* when they are wrapped only in Schwann cells, or they are *myelinated.* Myelin is interrupted at the nodes of Ranvier. The internode distance varies from 0.2 to 2.5 mm and is related to the conduction velocity. The diameter of the nerve fiber is also related to nerve conduction. Nerve conduction is faster with fibers of a larger diameter and which have longer distances between internodes. A, B, and C fibers are distinguished. Nerve conduction is propagated along the axonal membrane by the *action potential*. This consists of a sudden depolarization with increased permeability to Na^+. The membrane potential may depolarize from -90 mV and may overshoot to $+50$ mV. In the ascending phase of the spike there is entrance of Na^+. In the descending phase, K^+ leaks out. The action potential has a threshold of activation, is an all-or-none response, is non-decremental, and has a refractory period. In unmyelinated fibers, *propagation* is accounted for by the *local circuit theory*. In myelinated fibers conduction is considered *saltatory*, from one node to the other. In the internode the impulse is electrotonically conducted.

In receptors and synapses there are *graded* responses that are *not propagated* (generator and synaptic potentials). A good example of a receptor is the Pacini corpuscle, which is a *biological transducer* capable of transforming mechanical energy into electrical energy.

SYNAPTIC TRANSMISSION

The earliest knowledge of *synapses,* or *synaptic junctions,* came from the discoveries at the turn of the last century of the morphologic and physiologic organization of the nervous system. The so-called *neuron theory,* established mainly by Cajal, led to the assumption that the functional interactions between nerve cells was by way of contiguities or *functional contacts.* Different types of nerve terminals on dendrites or perikarya were described by the use of silver staining methods such as the characteristic boutons, the club endings, the so-called baskets or the contacts *en passant.*[21]

In 1897, Sherrington coined the name *synapse* to explain the special properties of the reflex arc, which he considered to be dependent on the functional contact between neurons. He attributed to the synapse a valve-like action, which transmits the impulses in only one direction (see Figure 22–1). In his studies on reflex transmission he discovered some of the fundamental properties of synapses, such as the *synaptic delay* (the delay that the impulse experiences in traversing the junction), the fatigability of the synapse, and the greater sensitivity to reduced oxygen and anesthetics. He also pointed out that the many synapses situated on the surface of a motoneuron could interact, and that some would have an additive excitatory action, whereas others would be inhibitory and antagonize the excitatory ones.[22]

The special zones of contact between two neurons or between a neuron and a nonneuronal element, will be considered as synaptic regions. Also included are the junctions between some receptors and neurons or with an effector cell, i.e., a myoneural junction. Synapses thus embody all the regions "anatomically differentiated and functionally specialized for the transmission of liminal excitations and inhibitions from one element to the following in an irreciprocal direction."[23] These typical polarized synapses comprise the great majority in the nervous system of both vertebrates and invertebrates; however, a more modern definition of the synapse should also include the existence of a complex submicroscopic organization in both the pre- and postsynaptic parts of the junction and of the specific neurochemical mechanism in which transmitter, receptor protein, synthetic and hydrolytic enzymes, and so forth, are involved.

From Figure 24–1, it is clear that the main problem in synaptic transmission consists of finding out by which molecular mechanism the information brought forward by one neuron is transferred to the following. In other words, the problem is how the code of frequency conducted by one neuron originates a new code of frequency in the following neuron.

Chemical and Electrical Transmission

DuBois-Reymond (1877) was the first to suggest that transmission could be either *electrical* or *chemical*. These two types of mechanisms have been observed. However, so far chemical synapses seem to be by far the most frequent in the peripheral and central nervous system.

Electrical transmission was first demonstrated in a giant synapse of the abdominal ganglion of the crayfish cord, and since then, in several other cases.[24] In this type of synapse the membrane contact acts as an efficient rectifier, allowing current to pass relatively easily from the pre- to the post-synaptic element, but not in the reverse direction. In this case the action current of the arriving nerve impulse is passed without delay and can depolarize directly and excite the postsynaptic neuron. Here the one-way transmission is due to the valvelike resistance of the contacting synaptic membranes.

Chemical transmission presupposes that a specific chemical transmitter is synthesized and stored at the nerve terminal and is liberated by the nerve impulse. The transmitter produces a change in ionic permeability at the postsynaptic component with a bioelectrical change. In 1904, Elliot suggested that sympathetic nerves act by liberating adrenalin at the junctions with smooth muscle. Later, von Euler demonstrated that *noradrenalin* was the true adrenergic transmitter. The studies of Dixon (1906) and particularly of Dale (1914) strongly supported chemical transmission in the parasympathetic system. This was finally proved on the heart by Loewi in 1921. Since then, *acetylcholine* has been demonstrated to act in sympathetic ganglia, neuromuscular junctions, and in many central synapses. Modern studies on chemical synaptic transmission have revealed that synapses are the sites of a transducing mechanism in which the electrical signals are converted into chemical signals, and these, in turn, again into electrical signals. It will be shown that active substances of low molecular weight (e.g., acetylcholine, noradrenaline, dopamine, glutamate, γ-aminobutyrate, and others are produced at the nerve endings and packaged in special containers (i.e., synaptic vesicles) in multimolecular quantities. These packages of the so-called *transmitters* are released when the activation is produced by the nerve impulse. The transmitter, in turn, reversibly reacts with special receptor proteins, present at the postsynaptic membrane. This transmitter-receptor interaction produces a change in permeability to certain ions, thereby creating a *synaptic potential* in the postsynaptic cell.

Excitatory and Inhibitory Synapses. Synaptic Potentials

Physiologic studies on synaptic transmission were greatly improved by the use of microelectrodes which could be implanted near the synaptic region or intracellularly in the pre- and postsynaptic neuron.[25] The first synaptic potential to be recorded directly was the *end plate potential* of the myoneural junction.

With intracellular recordings in large nerve cells (e.g., motoneurons, pyramidal cells, invertebrate ganglion cells, etc.),[26–28] it was also observed that the arrival of the presynaptic nerve impulse produces a local synaptic potential. *Synaptic potentials,* as the generator potentials studied above, are graded and decremental and do not propagate. They extend electrotonically only for a short distance with reduction in amplitude.

A typical experiment, shown in Figure 24–11, involves two ganglion cells of *Aplysia* (a marine mollusk), one of which (*P*) acts synaptically upon the other (*F*). Neuron P is impaled with two microelectrodes, one of which is used for stimulation (*St*) and the other for recording (*R*). Neuron F is impaled with one microelectrode (*R*) to register the synaptic potential. Two main types of P cells can be found, one of which produces an excitatory synaptic potential in F (1) and the other an inhibitory postsynaptic potential in F (2).[29]

Excitatory synapses induce a depolarization of the postsynaptic membrane, which upon reaching a certain critical level, causes the neuron to discharge an impulse. The *excitatory postsynaptic potential* (EPSP) is

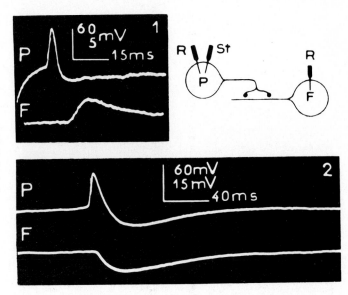

Figure 24–11. Diagram of the experiment in two ganglion cells of *Aplysia* that are related synaptically. (See the description in the text.) 1, excitatory response. Depolarization of the membrane at *F* after arrival of the action potential from *P*. 2, inhibitory response. Hyperpolarization of the membrane at *F* after arrival of the action potential from *P*. (Courtesy of L. Tauc and H. M. Gerschenfeld.)

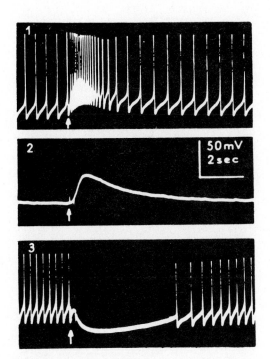

Figure 24–12. 1, intracellular recording in a neuron of *Aplysia* (see Fig. 24–11) that is firing spontaneously. At the point marked by the arrow, acetylcholine is added, producing depolarization and increasing the frequency of discharges (excitatory synapse). 2, same experiment on a neuron, without firing. Only depolarization is produced. 3, neuron in which the action of acetylcholine induces hyperpolarization and inhibition of spontaneous discharges. (Courtesy of L. Tauc and H. M. Gerschenfeld.)

due to the action of the transmitter released by the ending (Fig. 24–11, 1). This causes a change in permeability of the subsynaptic membrane, allowing the free passage of small ions, such as Na^+, K^+ and Cl^- (see also Chapter 21).

Similarly, *inhibitory synapses* affect the subsynaptic membrane. In these instances the transmitter causes a transient increase in membrane potential, the so-called *inhibitory postsynaptic potential* (IPSP) (Fig. 24–11, 2). This hyperpolarizing effect induces a depression of the neuronal excitability and an inhibitory action.

The excitatory or inhibitory action is not dependent exclusively on the type of transmitter substance. For example, acetylcholine is excitatory in the myoneural junction, sympathetic ganglia, and so forth, but inhibitory in the vertebrate heart, in which it reduces the frequency of contraction.

Figure 24–12 shows that in the ganglion cells of *Aplysia* the injection of acetylcholine may also have an excitatory synaptic effect in certain cells producing depolarization and increased frequency of discharges (1) or only a depolarization without firing (2). In other cells the same treatment provokes a hyperpolarization and inhibition of spontaneous discharges (3).

These facts indicate that the nature of a

synapse depends, in particular, on the receptor, i.e., the chemical reactivity of the membrane in the postsynaptic neuron. The use of intracellular recording has greatly contributed to the delineation of some of the basic mechanisms by which the code of signals is transmitted from one cell to the other. All the synaptic potentials from the different excitatory and inhibitory endings impinging upon a neuron are algebraically added. Both types of input will change the electrical properties of the membrane at a critical zone of the cell of low excitatory threshold, which is called the "pacemaker." In this region, which in the motoneurons is located at the initial segment of the axon, new impulses are fired.[22]

Structure of the Synaptic Region

The classic morphologic studies with the light microscope revealed that the size, shape, and distribution of synapses of different regions of the central and peripheral nervous tissue vary considerably. Synapses are classified as *axodendritic, axosomatic,* or *axo-axonic,* according to the relationship of the ending to the postsynaptic component. The endings may have different sizes and shapes, e.g., bud, foot or button ending, club ending, and calix (cup) ending (Fig. 24–13).

In a motoneuron, several thousand nerve endings can be observed to terminate on the surface of the perikaryon and dendrites and a few at the beginning of the axon. As many as 10,000 synapses have been calculated to impinge on a single pyramidal cell of the cortex. This gives one an idea of the extraordinary complexity of the nervous system. This immense number of synapses carry information from numerous other neurons, some of which may have an excitatory, and others an inhibitory effect. Thus, the neuron is a real computation center where all this information is integrated and sent as new nerve impulses along the axon.

With the increased resolution of the electron microscope new structural details became apparent (Fig. 24–14). At the synaptic junction the membranes of the two neurons were seen to be in direct opposition, separated only by a synaptic cleft. Of great physiologic and biochemical interest was the demonstration of a special vesicular component — the *synaptic vesicles* — at the presynaptic endings.[30]

The Synaptic Membranes

Figure 24–14, *A* shows a diagram of one of the most typical synapses found in the cerebral cortex between an axonal terminal

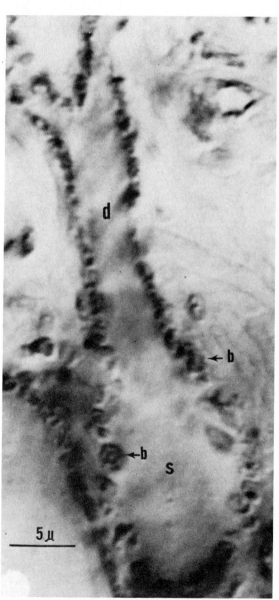

Figure 24–13. Neuron of the *nucleus interpositus* of the cerebellum photographed with Nomarski optics after immunofluorescence staining with an antibody against the enzyme glutamate decarboxylase. The button-like endings (*b*) cover the soma (*s*) and the dendrites (*d*) of the neuron. This method gives a positive reaction with the endings producing γ-aminobutyrate as transmitter. (Courtesy of E. Roberts.)

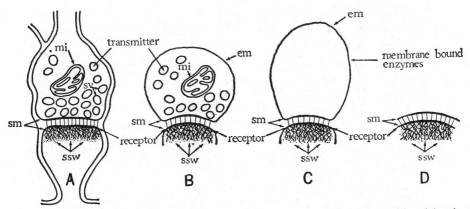

Figure 24–14. Diagram showing the systematic dissection of the synaptic region achieved by the use of cell fractionation methods. **A,** typical synapse of the cerebral cortex showing: *mi,* mitochondria; *sv,* synaptic vesicles; *sm,* synaptic membranes; and *ssw,* subsynaptic web. **B,** isolated nerve ending; *em,* nerve-ending membrane. **C,** after the osmotic shock, only the nerve-ending membrane remains. **D,** after treatment with a mild detergent, only the junctional complex remains. (From De Robertis, E., *Science, 171*:963, 1971. Copyright 1971 by the American Association for the Advancement of Science.)

and the spine of a dendrite (see also Fig. 24–18). At the junction both synaptic membranes appear to be thicker and denser. The synaptic cleft, in between the synaptic membranes, may be about 30 nm, i.e., larger than the spaces between other membranes, and may show a system of fine *intersynaptic* filaments of about 5 nm that join both synaptic membranes.

Another system of filaments or fine canaliculi has been observed to penetrate at a varying distance into the postsynaptic cell. This is the so-called *subsynaptic web.*[31] The demonstration of intersynaptic filaments between the membranes confirms that there is greater adhesion at the junction; this was demonstrated by microdissection experiments. In fact, in an isolated cell the endings break the connection with the axons, but remain attached to the cell.

Presynaptic densities may be observed with special stains[32–33] that suggest the presence of proteins having abundant basic amino acids in the region of the presynaptic contact. The position of the synaptic vesicles and of the pre- and postsynaptic densities has suggested the possible functional polarity of the synapse and the recognition of a series of contacts that were not possible to demonstrate with histological methods. In addition to the *axodendritic, axosomatic,* and *axoaxonic* synapses mentioned above there are evidences of dendrodendritic synapses (i.e.,

between dendrites of different neurons)[34] and also of *reciprocal synapses* in which one neuronal process may be presynaptic at one point and postsynaptic at another.

Furthermore, in the so-called *serial synapses* a nerve terminal may act synaptically on another nerve terminal.[35]

The discovery of these other types of synaptic contacts suggests the possibility of many more functional interactions between neurons than were previously imagined.

Types of Synaptic Vesicles

Most synaptic vesicles have a diameter of 40 to 50 nm and are surrounded by a limiting membrane 4 to 5 nm thick (Fig. 24–15). In general, they are electron-translucent, i.e., not showing dense material inside. They are distributed throughout the nerve ending, but tend to collect near the presynaptic membrane; some of them make close contact with it at certain points called the *active points* of the synapse.[36–38] In the frog, a myoneural junction which has large and extended nerve endings, contains some 1000 synaptic vesicles per μm^3 and a total of about 3×10^{-5}.[39] Approximately 20 per cent of these vesicles are localized near the presynaptic membrane and are readily available for release of the transmitter at the arrival of the nerve impulse (see Hubbard, 1973).

In addition to the small translucent synap-

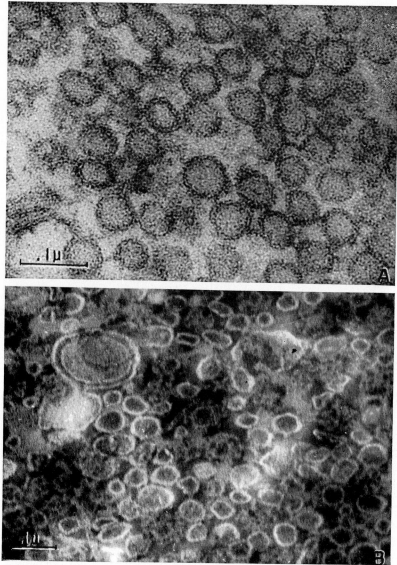

Figure 24–15. **A,** high resolution electron micrograph of synaptic vesicles in the hypothalamus of a rat, showing the fine structure of the vesicular membrane. × 180,000. **B,** isolated synaptic vesicles from rat brain after osmotic shock of the mitochondrial fraction. Negative staining with phosphotungstate. × 120,000. (From De Robertis, E., Rodríguez de Lores Arnaiz, G., Salganicoff, L., Pellegrino de Iraldi, A., and Zieher, L. M., *J. Neurochem., 10*:225, 1963.)

tic vesicles first described by De Robertis and Bennett (1954–55) (Fig. 24–15, *A*), medium-sized vesicles with a *dense core* may be observed in central, as well as peripheral, synapses.[40] In peripheral cholinergic nerve endings only a few per cent of the vesicles have a dense core, whereas most of them are translucent vesicles.

Special mention should be made of the *flattened or elliptical vesicles* found in certain central and peripheral synapses. These structures have been interpreted as corresponding to inhibitory synapses[41] and have been found in nerve endings containing the γ-aminobutyric acid system.[42]

Another type of vesicle is the so-called *complex vesicle* that is similar to the electron-translucent spherical one, but is surrounded

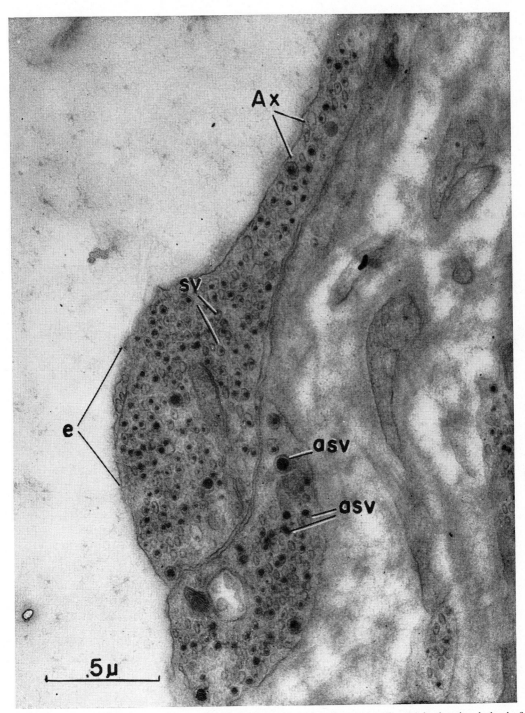

Figure 24–16. Electron micrograph of an adrenergic axon (*Ax*) and nerve ending (*e*) in the pineal gland of a rat. Both are filled with vesicles (*sv*), many of which contain a deposit of reduced osmium (*asv*). These are the adrenergic vesicles. × 60,000. (From Pellegrino de Iraldi, A., and De Robertis, E., *Z. Zellforsch., 87:*330, 1968.)

by a hexagonal array of material that forms a kind of shell.[43] These complex vesicles appear to be similar to other coated vesicles that are produced by the process of endocytosis (see Chapters 9 and 21) and may be related to the process of membrane recycling. In myoneural junctions stimulated to produce a considerable release of transmitter and labeled with peroxidase, it has been found that such complex vesicles increase in number and contain the enzyme,[44, 45] indicating that they have originated by infoldings of the presynaptic membrane.

In sympathetic axons and endings in the pineal gland and in the splenic nerve a special type of synaptic vesicle has been described.[46] These vesicles contain a dense granule formed by a deposit of reduced osmium (Fig. 24–16). They resemble, but are much smaller than, the catechol-containing droplets of the adrenal medulla (see Figure 25–10).

By using pharmacological agents that release catecholamines, one may observe a depletion of the granulated vesicles. These vesicles increase in concentration in the presence of inhibitors of the enzyme monamine oxidase (e.g., iproniazid), or when the animal is given precursors of catecholamine (e.g., dopa, dopamine).[47] All these results indicate that granulated vesicles contain the adrenergic transmitter. Specific cytochemical techniques have been used to demonstrate that the small granulated vesicles of sympathetic nerves are the site of storage of norepinephrine, as well as of 5-hydroxytryptamine, in some cases (see Jaim-Etcheverry and Zieher, 1971).

Role of Synaptic Vesicles in Synaptic Transmission

Several experiments have been carried out to demonstrate the possible role of synaptic vesicles in transmission. In central synapses, cutting the nerve results in early degeneration, with clumping and lysis of the vesicles.[48] Similar observations have been made in the degenerating myoneural junctions.[49]

Electrical stimulation of the nerve endings of the adrenal medulla showed that with certain frequencies of stimuli known to produce maximal output of catecholamines (see Figure 25–3), the number of vesicles increased. With much higher frequencies the

vesicles tended to disappear. All these results indicate that synaptic vesicles play a role in the transmission of the nerve impulse and that a balance exists between the formation of vesicles and their discharge at the synapse.[50]

The relationship between quantal release of transmitter substances and the synaptic vesicles has been confirmed by the observation of nerve terminals that are completely devoid of vesicles when the release has ceased. This has been obtained with β-bungarotoxin,[51] a toxin from certain snakes, or with the toxin of the black widow spider venom,[52] both of which produce a massive release of acetylcholine from the myoneural junction.

Isolation of Nerve Endings and Synaptic Vesicles

A more direct demonstration that synaptic vesicles are the sites of storage of the different transmitter substances came from studies of cell fractionation of the brain. Figure 24–17 shows that by a series of differential and gradient centrifugations it is possible to separate fractions rich in nerve endings from myelin and mitochondria[53] (Fig. 24–18). The nerve endings were submitted to a hyposmotic shock which produced the release of the intact synaptic vesicles (Fig. 24–14, C), thereby leaving the nerve-ending membranes.[54, 55]

Table 24–2 shows that the fraction of synaptic vesicles (M_2) has the highest content of several biogenic amines (i.e., acetylcholine, noradrenaline, dopamine, and histamine), which act as transmitters in the central nervous system.

Such cell fractionation studies have also permitted a systematic dissection of the nerve ending and of its membranes. Figure 24–14 shows that after osmotic shock of the nerve ending (B), the content is lost—including the synaptic vesicles—and only the nerve-ending membranes remain. By further treatment of these membranes with a mild detergent, the junctional complexes composed of the synaptic membranes and related structures are separated (C).[56]

The nerve-ending membranes contain important membrane-bound enzymes such as acetylcholinesterase, Na$^+$-K$^+$-activated ATPase, K + p nitrophenylphosphatase, and adenyl cyclase.[57] This last enzyme synthesizes

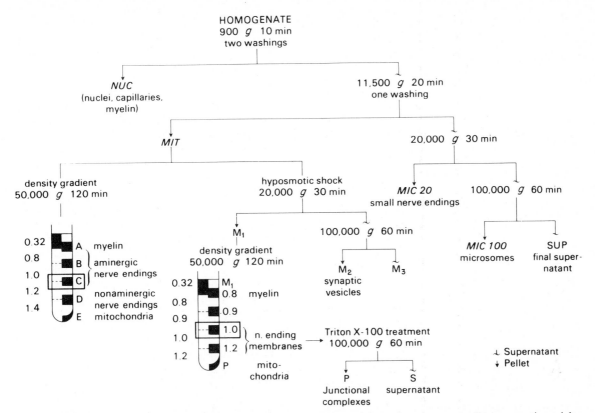

Figure 24–17. Diagram showing the cell fractionation methods used to separate nerve endings, synaptic vesicles, and other components of brain. (From Rodríguez de Lores Arnaiz, G., and De Robertis, E., "Drugs Affecting the Synaptic Components of the CNS." In Dikstein, S. (Ed.), *Fundamentals of Cell Pharmacology,* Springfield, Ill., Charles C Thomas, Publisher, 1973.)

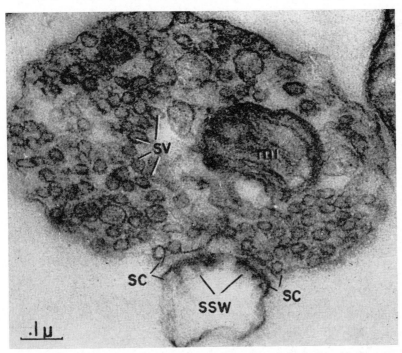

Figure 24–18. Isolated nerve ending in the mitochondrial fraction of the brain with the same components as in Figure 24–14. × 110,000. (From De Robertis, E., Rodríguez de Lores Arnaiz, G., and Pellegrino de Iraldi, A., *Nature* (London), *194*:794, 1962.)

Biogenic Amines	Fraction		
	M_1	M_2	M_3
Acetylcholine	0.55	2.85	1.20
Noradrenaline	0.40	2.56	1.93
Dopamine	0.46	2.46	1.72
Histamine	0.39	2.24	2.27

The crude mitochondrial fraction of the brain was osmotically shocked and then centrifuged. The results are expressed in relative specific concentration. For literature, see De Robertis (1967).

$3'5'$-cyclic AMP, which has been considered an important regulator of various cell activities.[58, 59]

The Acetylcholine System

The various neurons integrating the nervous system can be differentiated not only by their morphology and synaptic relationships, but also by the transmitter they synthesize, store, and release at the nerve endings. Thus, in the central nervous system there are nerve cells that produce various biogenic amines (acetylcholine, noradrenaline or dopamine), or amino acids (glutamate, γ-aminobutyrate (GABA), and glycine). Each of these neuronal types has its own biochemical mechanism which involves several enzymes for the synthesis and catabolism of the particular transmitter.

In addition to the results obtained by cell fractionation, progress has been made on the identification of the various neuronal types. For example, neurons producing catecholamines (i.e., noradrenaline and dopamine) or indolamines (i.e., serotonin) may be identified by the use of fluorescence histochemical techniques.[60] Recently, another powerful technique based on the use of antibodies against some of the specific enzymes is being employed. For example, in neurons producing GABA as transmitter, the essential enzyme is glutamic acid decarboxylase (GAD), which produces GABA by decarboxylation of glutamate. As shown in Figure 24–13, this cytochemical technique permits the identification of the GABA-producing terminals

on soma and dendrites of a neuron of the cerebellum.

For a general description, the acetylcholine system, which is present in neurons of the central nervous system, but which is particularly developed in the preganglionic neurons of the autonomic system and in motoneurons of vertebrates will be considered.

Acetylcholine (ACh) is synthesized via combination of choline and acetylcoenzyme A by the enzyme *cholineacetyltransferase* (Fig. 24–19). This process is most active in the motor nerve terminals of the myoneural junction. The enzyme is produced in the neuronal body and is transported to the nerve terminal by axonal flow. In some species, brain cholineacetyltransferase is associated with the synaptic vesicles,[55] whereas in others, this enzyme is more soluble in the axoplasm.[61] Acetylcholine is stored mainly within the synaptic vesicles, although some ACh that is soluble in the cytoplasm may be present. After its release into the synaptic cleft, acetylcholine interacts with the cholinergic receptor present in the postsynaptic membrane and is degraded into choline and acetate by *acetylcholinesterase,* a hydrolytic enzyme present in nerve ending membranes. As shown in Figure 24–19, choline is again taken up by a transport mechanism into the nerve terminals. By cell fractionation it can be demonstrated that the three main components of the acetylcholine system (i.e., cholineacetyltransferase, acetylcholine, and acetylcholinesterase) are found in a special fraction of nerve endings from brain (see Rodríguez de Lores Arnaiz and De Robertis, 1973).

Acetylcholine Content and Quantal Release

Mention was made before of some of the experimental evidences in favor of the synaptic vesicle as the site of storage of acetylcholine and other transmitters and for its release in quantal units. Direct estimates of ACh content and the number of vesicles in nerve endings of the *Torpedo* electroplax suggest that the content may be very high. In fact, the solution of ACh inside the vesicle is probably isosmotic with plasma (i.e., 0.4 to 0.5M).[62] Although the exact number of molecules per vesicle is still discussed as part

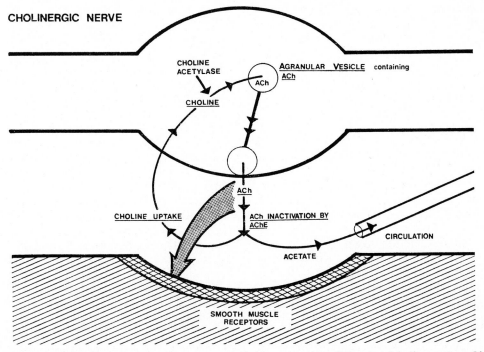

CHOLINERGIC NERVE

CHOLINE
ACETYLASE

CHOLINE

AGRANULAR VESICLE containing
ACh

ACh

CHOLINE UPTAKE

ACh

ACh INACTIVATION BY
AChE

CIRCULATION

ACETATE

SMOOTH MUSCLE
RECEPTORS

Figure 24–19. Diagram of a cholinergic synapse and the components of the acetylcholine system. Observe the storage of the acetylcholine (*ACh*) in the synaptic vesicle and its release. The mechanism of synthesis and inactivation by acetylcholinesterase (*AChE*) is indicated. (From Burnstock, N., *Pharmacol. Rev., 24*:509, 1972. © (1972) The Williams & Wilkins Company, Baltimore.)

of its volume, up to 54,000 molecules may be packed in a saturated solution of ACh. Estimates of the ACh released in each quantum to produce a miniature end-plate potential (see above) has varied in the literature; in the rat diaphragm a fair estimate could be from 12,000 to 21,000 ACh molecules—a number that could easily be packed within a vesicle (see Hubbard, 1973). The presence of ATP and some binding protein within synaptic vesicles has been demonstrated, and it is possible that these substances may play some role in the ACh concentration within the vesicle.[63] Furthermore, it has been found that after nerve stimulation in the presence of hemicholinium, a substance that interferes with the re-uptake of choline, the size of the vesicles and the amplitude of the miniature end-plate potentials are reduced.[64] When a nerve impulse reaches the motor nerve ending, approximately 100 quanta of ACh are simultaneously released, thereby representing the entrance into the synaptic cleft of about 2×10^6 molecules of ACh.

Role of Calcium in ACh Release

The coupling between the depolarization caused by the arrival of the nerve impulse, and the secretion of acetylcholine is mediated by the influx of calcium ions. Extracellular Ca^{2+} is an absolute requirement for the release of ACh from the nerve terminal; in the absence of calcium, there is no release of ACh. When the nerve impulse arrives at the terminal, there is an increase in Ca^{2+} permeability resulting from the depolarization. The entry of calcium into the ending has been demonstrated by using ^{45}Ca.[65] It has been assumed that for the release of each quantum of ACh one to four Ca^{2+} molecules must enter, and that in the release of ACh there is a cooperative action of Ca^{2+}.[66]

Once the Ca^{2+} has penetrated the terminal, the secretion of ACh is probably produced by the opening of the synaptic vesicles that were attached to the presynaptic membrane. Physiological studies have demonstrated that the *synaptic delay* that takes place between the arrival of the nerve impulse and the production of the synaptic potential is due mainly to the interval between depolarization and actual secretion of the transmitter.[67]

Molecular Biology of Receptors

Since the beginning of this century, through the work of Langley, Ehrlich, and others it has been postulated that the transmitter interacts with a specific receptor localized at the chemosensitive sites of the cell membrane. For many years, however, the knowledge of synaptic receptors has principally been indirect — based on the final response obtained from this interaction. For example, it was known that acetylcholine produced contraction of skeletal muscle, or the secretion of a gland, and that these two effects could be blocked with curare and atropine, respectively. Figure 24–20 shows that the primary transmitter receptor interaction is coupled, by way of a conformational change, to the translocation of ions across the cell membrane (i.e., the ionophore), and by several interposed mechanisms can produce a final response. This figure also shows the probable relationship between the primary interaction, the displacement of Ca^{2+}, and certain metabolic processes involving cyclic AMP and other substances.

Isolation of Receptor Proteins

Only in recent years has it been possible to separate synaptic receptor proteins as chemical entities and to analyze, in a more direct way, their interaction with the transmitter.[68] The separation of such substances has been difficult, because they are intrinsic proteins (see Table 8–3) which require strong treatments to be separated from the lipoprotein framework of the membrane. Furthermore, they are localized at the postsynaptic membrane in the sites that show chemosensitivity to the transmitter, and they are present in extremely small concentrations. These properties imply that the membranes should be specially separated by cell fractionation methods. Once they are isolated, such molecules should show a specific and high-affinity binding for the neurotransmitters. Finally, they should be able to undergo conformational changes that are capable of inducing the translocation of ions and producing a bioelectrical response (see De Robertis, 1971, 1973).

Receptor proteins are highly hydrophobic and are intimately related to lipids of the membrane. For the isolation of these proteins, two main procedures are used: extraction with organic solvent[68] or separation by strong detergents.[69, 70] Receptor cholinergic proteins have been isolated not only from the electroplax of *Torpedo* and *Electrophorus* tissues, which have the richest cholinergic innervation, but also from skeletal muscle,[72] smooth muscle,[73] and brain.[68] Adrenergic receptor proteins have been separated from brain,[74] from spleen capsule,[75] and from heart.[76] More recently, receptor proteins to the amino acids glutamate and γ-aminobutyrate have also been isolated from muscle of *Crustacea*

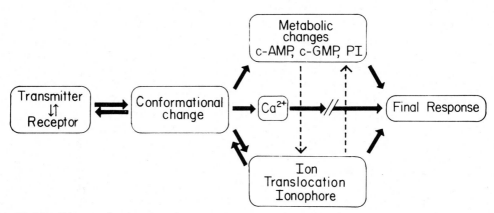

Figure 24–20. Diagram showing the primary interaction between transmitter and receptor and its consequences. (See the description in the text.)

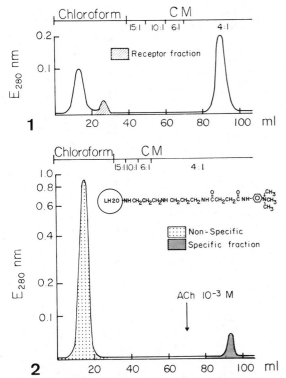

Figure 24–21. Separation of the cholinergic receptor from rat diaphragm. **1,** by conventional chromatography in Sephadex LH20; **2,** by affinity chromatography. Observe that the receptor protein fraction from **1** has been purified further (about fifteen times) in **2.** The specific receptor fraction appears only after a pulse of 10^{-3} M acetylcholine (*ACh*). The inset shows the chemical composition of the affinity column with the cholinergic end (p-Aminophenyl-trimethylammonium) linked to the gel by a 1.4 nm "arm" (3'-3'-iminobispropylamine). (From Barrantes, B., Arbilla, S., de Carlin, C, and De Robertis, E., 1975.)

which have a double innervation: excitatory (via glutamate) and inhibitory (via γ-aminobutyrate).[77]

After their extraction with the organic solvent, the proteins generally are separated by colum chromatography or by affinity chromatography. With the latter technique, the gel used contains an active group attached by a long arm. This group specifically binds the receptor protein and is then eluted by a pulse of the corresponding transmitter.

Figure 24–21 shows the separation of the cholinergic receptor of skeletal muscle in two steps. The small second peak of protein in Figure 24–21, *1* is the one having a high affinity for ACh and other cholinergic drugs. In Figure 24–21, *2* this protein has been further purified by affinity chromatography. Here, the protein peak, which appears after the ACh pulse, is the specific receptor protein fraction. With this procedure a total purification of about 15,000 times has been achieved in the rat diaphragm (Barrantes et al., *Biochim. Biophys. Res. Comm., 63*:194, 1975).

The Ionophoric Response

The previous considerations on the transmitter receptor interaction permit the belief that the transduction of energy at the synapse results from a conformational change of the receptor macromolecule, and that such a change is coupled with the translocation of ions across the membrane (Fig. 24–20). In the case of acetylcholine, these mechanisms can be expressed as shown at the bottom of the page. In this equation AChRc represents a reversible complex resulting from the binding of ACh to the receptor protein. In the second step, the ionophoric mechanism takes place, producing the opening of the channel for the translocation of the ions (AChRo). It is important to understand the essential mechanism of energy transduction which takes place at the synapse and that here there is a considerable amplification of the arriving electrical signal.

By the use of fine microelectrodes some of the best evidence of the functioning of receptors at the molecular level has been obtained in the myoneural junction.[78] It was mentioned before that in this preparation it is possible to record the spontaneous *miniature end-plate potentials* (*mepp*) which have an amplitude of about 0.1 mV and represent the discharge of single multimolecular quanta

$$\text{AChRc} \underset{K_{-1}}{\overset{K_1}{\rightleftharpoons}} \text{AChRc} \quad \underset{K_{-2}}{\overset{K_2}{\rightleftharpoons}} \text{AChRo} \qquad (1)$$

of ACh (Fig. 24–22). Katz and Miledi have observed that if a minimal but steady dose of acetylcholine is applied to the myoneural junction, there are minute fluctuations of the membrane potentials that are superimposed upon a small but steady depolarization of the membrane. The amplitude of these fluctuations is several hundred times smaller than the mepp. (For example, in Figure 24–22 the mepp is 700 μV and the fluctuation of the noise only about 0.2 μV). These elementary fluctuations have been interpreted as representing the opening of single channels by the interaction of the receptor with the transmitter. The current that is produced is equivalent to the translocation of about 5×10^4 univalent ions. The enormous amplification that has been produced at the chemical synapse is readily understandable; in fact, by the interaction of a few ACh molecules with a receptor, about 50,000 ions are translocated!

The ionophoric function of receptor proteins has been analyzed within a simpler system that uses artificial lipid membranes.

Minute amounts of cholinergic or adrenergic receptor protein were incorporated into these membranes, and then they were submitted to an application of the corresponding transmitter (i.e., ACh or noradrenalin) with a fine micropipette (see Fig. 7–8).[79, 80] In both cases, the interaction resulted in a conductance change of large magnitude. The pharmacologic studies conducted in the case of the cholinergic receptor from the electroplax demonstrated that the effect of ACh could be blocked by curare and other blocking agents.[81] Furthermore, when the membranes were observed under the electron microscope, they showed a more uneven planar structure at the height of the conductance response. They also showed dense spots of a maximum diameter of 2.0 nm in which the osmium tetroxide used as fixative was deposited[82] (Fig. 24–23). Such changes of the membrane are transient and disappear when the membrane regains the normal conductance. The dense spots have been interpreted as representing the opening of channels for the ionophoric response.

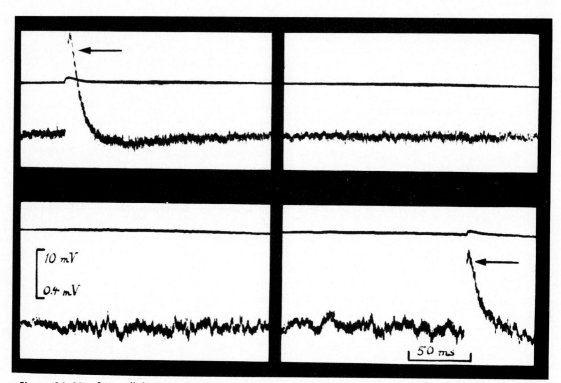

Figure 24–22.   Intracellular recording from an end-plate in frog sartorius muscle. In each block the upper trace is at a low amplification (scale, 10 mV), while the lower trace is at a much higher amplification (scale, 0.4 mV). In the two blocks of the bottom acetylcholine was applied; the result was the production of fine fluctuations in the membrane potential. The two upper blocks are controls. Two miniature end-plate potentials are indicated with arrows. (From Katz, B., and Miledi, R., *J. Physiol.* (London), *224*:665, 1972.)

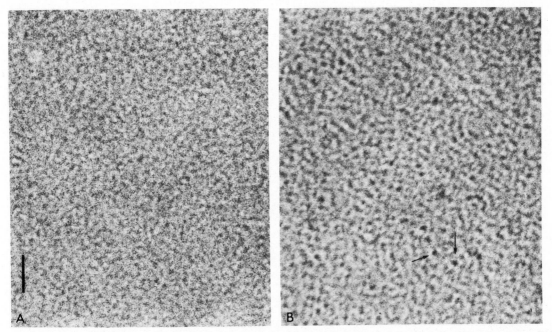

Figure 24–23. Electron micrographs of ultrathin lipid artificial membranes containing minute amounts of the cholinergic receptor protein from the electroplax. The membranes are observed in a planar view after fixation with glutaraldehyde and exposure to osmium vapor. The bar (in **A**) indicates 20 nm. **A,** control membrane showing a rather smooth and uniform texture; **B,** membrane "activated" with acetylcholine and showing a conformational change. The structure appears rougher and shows dense spots of 2.0 nm. Arrows point to some of these spots. (From Vásquez, C., Parisi, M., and De Robertis, E., *J. Membr. Biol.,* 6:353, 1971.)

Under certain conditions the membrane also showed fine fluctuations of the conductance, which could correspond to the opening of single channels. In fact, the conductance of such fluctuations is within the same order of magnitude as that found in the myoneural junction (Fig. 24–22). In the case of the adrenergic receptor it was found that the response to noradrenaline was stereospecific (i.e., found only with (−) noradrenaline) and was blocked by the specific adrenergic blocking agents (Fig. 24–24).

A Model of the Cholinergic Receptor

In 1971 De Robertis proposed the molecular model of the cholinergic receptor shown in Figure 24–25. This model is based

Figure 24–24. Records of current, in nanoamperes (nA), across artificial membranes containing an adrenergic receptor protein from the spleen capsule. The injection of (−) noradrenalin produces a transient increase in conductance, whose amplitude follows the concentration of the transmitter in the micropipette. In the presence of the adrenergic blocking agent, phentolamine, there is an inhibition of the response. S. injection of saline solution (see the experimental set-up in Fig. 7–8). (From Ochoa, E., Fiszer de Plazas, S., and De Robertis, E., *Mol. Pharmacol.,* 8:215, 1972.)

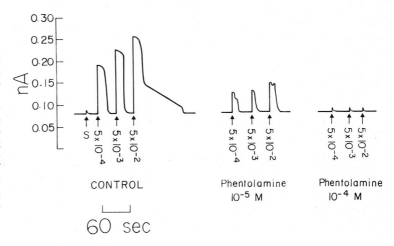

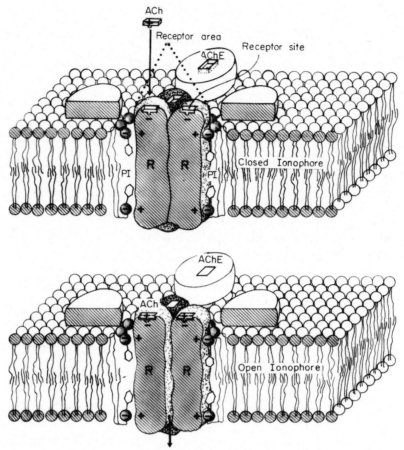

Figure 24–25. Tetrameric model of a cholinergic receptor area showing the receptor protein subunits traversing the lipid matrix (intrinsic protein). Each receptor subunit shows the site of binding for acetylcholine on the outer surface of the membrane; the four subunits in parallel constitute the ionophore. The presence of phosphatidyl-inositol (PI) attached to the receptor protein is indicated. The diagram also shows the presence of acetylcholinesterase (AChE) molecules that are peripheral to the membrane. **Upper part of figure,** the receptor site has not yet interacted and the ionophore is in the closed condition. **Lower part of figure,** the receptor site has been occupied by acetylcholine and has originated a conformational change resulting in the opening of the ionophore. (Modified from E. De Robertis, 1971.)

on the current concepts of membrane structure cited in Chapter 8 and especially on the fact that intrinsic proteins are intercalated within the lipid bilayer (see Fig. 8–8). In this model the receptor protein is shown to traverse the membrane having the specific binding site on the outer surface. The receptor macromolecules are shown to adopt an oligomeric (probably tetrameric) arrangement — a placement that, in the past, was postulated from purely pharmacologic considerations.

The four molecules in parallel constitute the ionophoric portion of the receptor, which is used as a channel for the translocation of ions across the membrane. In this model the binding of the ACh to the receptor is thought to produce a change in the net charge of each receptor molecule and in the degree of interaction between the monomeric units favoring the opening of the channel. Since the ACh-receptor interaction is reversible, it may easily go back toward the closed position once the ACh has been removed from the site of binding. Such a reversible effect would be facilitated by the fact that the receptor molecules are held in place by hydrophobic interactions within the framework of the membrane. There are several observations in favor of this model, particularly those made on the activated artificial membranes with the electron microscope, and others in which the isolated receptor proteins tend to form crystalline arrays when they interact with neuroactive drugs.[82]

Synapses are regions of contact between neurons, or between a neuron and a non-neuronal cell (i.e., a physiological receptor or effector) at which excitatory or inhibitory actions are transmitted in one direction. They may be *electrical,* in places in which there is a direct contact of membranes (tight junctions), and in cases like this the nerve impulse passes from one neuron to the other without delay. However, most synapses involve *chemical transmission* based on the release of specific transmitters such as acetylcholine (ACh), noradrenalin, dopamine and other biogenic amines, as well as amino acids such as glutamate, γ-aminobutyrate (GABA), and glycine.

The microphysiological study of synaptic transmission has demonstrated that the nerve impulse stops at the nerve ending, and that at the synapse a *synaptic potential* is generated in the postsynaptic cell. This potential may be *excitatory* (depolarization) or *inhibitory* (hyperpolarization); in both cases, it is graded and does not propagate. In some cases it may be observed that the same transmitter (i.e., ACh) may be excitatory or inhibitory (see Fig. 24–12); the different effect depends on the receptor protein with which the transmitter interacts. All the excitatory and inhibitory inputs acting on a neuron are added algebraically, and the neuron will fire new impulses in relation to these inputs as well as to its own spontaneous firing rhythm.

Morphologically, synapses may be *axodendritic, axosomatic,* or *axoaxonic.* With the electron microscope, *dendrodendritic, serial,* and *reciprocal* synapses may be distinguished by virtue of their presence on one side of synaptic vesicles and the differentiations of the synaptic membranes (i.e., presynaptic projections, intersynaptic filaments, and subsynaptic web).

The presence of *synaptic vesicles* is the main characteristic of synapses. Most of these structures are spherical, electron-translucent, and have a diameter of 40 to 50 nm and a membrane 4 to 5 nm thick. In the myoneural junction there are some 1000 vesicles per μm^3 and a total of 3×10^5; about 20 per cent of them are attached to the presynaptic membrane and are ready to discharge the transmitter. In many central and peripheral synapses there are *medium-sized* vesicles with a *dense core.* Some synapses, thought to be inhibitory, contain *flattened* or *elliptical vesicles.* In sympathetic axons and endings there are small *granulated vesicles;* the granule contains the transmitter, noradrenalin. Another type is the *complex or coated vesicle,* which has a shell of hexagonal material surrounding it. These vesicles are formed by endocytosis and may represent a mechanism of membrane recycling. Synaptic vesicles are related to the *quantal release* or transmitter, which produces the spontaneous *miniature end plate potentials* (*mepp*) in the myoneural junction. When the

SUMMARY:
Synaptic Transmission

nerve impulse arrives, 100 or more quanta are released, thereby producing the *synaptic or end plate potential.*

The relationship between vesicles and quantal release has been demonstrated by a variety of experiments. A direct demonstration that synaptic vesicles are the sites of storage of transmitter has been obtained by the isolation of the nerve endings and of the synaptic vesicles (Table 24–2).

In the central nervous system neurons produce different transmitters and have special biochemical mechanisms for their synthesis and catabolism. Neurons producing catecholamines or indolamines (i.e., serotonin) may be recognized by fluorescence. Recently, antibodies against specific enzymes have been used to identify a particular neuron (Fig. 24–13). In the *acetylcholine system* the synthesis is produced by the enzyme cholineacetyltransferase in the following reaction:

$$\text{choline} + \text{acetyl CoA} \rightarrow \text{Acetylcholine} \qquad (2)$$

After its release, acetylcholine is hydrolyzed by acetylcholinesterase, thereby releasing acetyl groups and choline which are again taken up by the nerve ending. The ACh content of vesicles may be very high. In *Torpedo,* its concentration is about 0.4 to 05M. Each quantum involves some 12,000 to 21,000 ACh molecules, which are packed within a vesicle. For the release, Ca^{2+} is indispensable, and the *synaptic delay* corresponds principally to the time taken by the actual secretion of the transmitter.

Receptor proteins are localized at the chemosensitive sites of the postsynaptic membrane. They are in low concentration, highly hydrophobic, and embedded in the membrane. They may be extracted with organic solvents or by the use of strong detergents, and can be purified by column chromatography and affinity chromatography (see Fig. 24–21). Receptor proteins from electroplax, skeletal muscle, brain, smooth muscle, and heart crustacean muscle have been isolated. These proteins are either cholinergic, adrenergic or related to the amino acids, glutamate, and GABA.

The essence of chemical transmission is in the interaction of the transmitter with the receptor. This interaction produces a conformational change in the protein by which channels for the translocation of ions are opened (ionophoric response). In the myoneural junction the application of a minimum dose of ACh produces fine fluctuations of the membrane potential. These are several hundred times smaller (i.e., 0.2 to 0.5 μV) than the *mepp* (700 to 1000 μV) and probably correspond to the opening of single channels. Each of these elementary fluctuations corresponds to the passage of some 50,000 univalent ions across the membrane.

The ionophoric response has been analyzed in ultrathin

artificial lipid membranes (see Fig. 7–8). By including minute amounts of cholinergic or adrenergic receptor protein in the membrane, investigators produced a response to the corresponding transmitter (i.e., ACh or noradrenalin). The conductance change was blocked by the corresponding blocking agent. In membranes activated with ACh, changes in planar fine structure were observed under the electron microscope with the appearance of dense spots, probably representing the opening of channels.

A molecular model of a cholinergic receptor has been proposed in which the receptor protein traverses the membrane. In this model the receptor is tetrameric, each subunit probably being able to bind ACh at the binding site, and the ionophore or channel is formed by the apposition of the four receptor subunits (Fig. 24–25).

REFERENCES

1. De Robertis, E., and Franchi, C. M. (1953) *J. Exp. Med., 98*:269.
2. Thornburg, W., and De Robertis, E. (1956) *J. Biophys. Biochem. Cytol., 2*:475.
3. De Robertis, E., and Schmitt, F. O. (1948) *J. Cell. Comp. Physiol., 31*:1.
4. Rodríguez Echandía, E. L., Piezzi, R. S., and Rodríguez, E. M. (1968) *Am. J. Anat., 122*:157.
5. Wuerker, R. B., and Palay, S. L. (1968) cited in *Neurosc. Res. Progr. Bull., 6*:125.
6. Weiss, P., and Hiscoe, H. B. (1948) *J. Exp. Zool., 107*:315.
7. Ramirez, G., Levitan, I. B., and Mushynski, W. E. (1972) *J. Biol. Chem., 247*:5382.
8. Gilbert, J. M. (1972) *J. Biol. Chem., 247*:6541.
9. Cuenod, M., and Schonbach, J. (1971) *J. Neurochem., 18*:809.
10. Fernandez, H. L., Burton, P. R., and Samson, F. E. (1971) *J. Cell Biol., 51*:176.
11. Droz, B., Koenig, H. L., and Di Giamberardino, L. (1973) *Brain Res., 60*:93.
12. Di Giamberardino, L., Bennett, G., Koenig, H. L., and Droz, B. (1973) *Brain Res., 60*:129.
13. Bennett, G., Di Giamberardino, L., Koenig, H. L., and Droz, B. (1973) *Brain Res., 60*:147.
14. Barondes, S. H. (1969) In: *Handbook of Neurochemistry,* Vol. 2, p. 435 (Lajtha, A., ed.) Plenum Publishing Corp., New York.
15. Giese, A. C. (1968) *Cell Physiology.* 3rd Ed. W. B. Saunders Company, Philadelphia.
16. Ruch, T. C. and Fulton, J. F. (1960) *Medical Physiology and Biophysics.* 18th Ed., Chap. 4. W. B. Saunders Company, Philadelphia.
17. Hodgkin, A. L., and Huxley, A. F. (1952) *Cold Spring Harbor Symp. Quant. Biol., 17*:43.
18. Tasaki, I. (1953) *Nervous Transmission.* Charles C Thomas, Springfield, Ill.
19. Davis, H. (1961) *Physiol. Rev., 41*:391.
20. Loewenstein, W. R. (1960) *Sci. Am., 203* (No. 4):98.
21. Cajal, S. R. (1934) *Trab. Inst. Cajal Invest. Biol.* (Madrid), *24*:1.
22. Eccles, J. C. (1957) *Physiology of Nerve Cells.* Johns Hopkins Press, Baltimore.
23. Arvanitaki, A. (1942) *J. Neurophysiol., 5*:108.
24. Furshpan, E. J., and Potter, D. D. (1957) *Nature* (London) *180*:342.
25. Ling, G., and Gerard, R. W. (1949) *J. Cell. Comp. Physiol., 34*:383.
26. Fatt, P., and Katz, B. (1951) *J. Physiol., 115*:320.
27. Nastuck, W. L. (1953) *J. Cell. Comp. Physiol., 42*:249.

28. Eccles, J. C. (1964) *The Physiology of Synapses.* Springer-Verlag, Berlin.
29. Tauc, L., and Gerschenfeld, H. M. (1960) *C. R. Acad. Sci. (Paris), 257*:3076.
30. De Robertis, E., and Bennett, H. S. (1955) *J. Biophys. Biochem. Cytol., 2*:307.
31. De Robertis, E., Pellegrino de Iraldi, A., Rodríguez de Lores Arnaiz, G., and Salganicoff, L. (1961) *Anat. Rec., 139*:220.
32. Bloom, F. E., and Aghajanian, G. K. (1966) *Science, 154*:1575.
33. Pfenninger, H., Sandri, C., Akert, K., and Eugster, C. H. (1969) *Brain Res., 12*:10.
34. Rall, W., and Shepherd, G. M. (1968) *J. Neurophysiol., 31*:884.
35. Dowling, J. E., and Boycott, B. B. (1966) *Proc. Roy. Soc. London [Biol.], 166*:80.
36. De Robertis, E. (1955) *Acta Neurol. Lat. Amer., 1*:1.
37. De Robertis, E. (1958) *Exp. Cell Res.,* Suppl. *5*:347.
38. Palay, S. L. (1958) *Exp. Cell Res.,* Suppl. *5*:275.
39. Birks, R., Huxley, H. E., and Katz, B. (1960) *J. Physiol.* (London), *150*:134.
40. Pellegrino de Iraldi, A., Farini-Duggan, H., and De Robertis, E. (1963) *Anat. Rec., 145*:561.
41. Uchizono, K. (1965) *Nature* (London), *207*:642.
42. De Robertis, E. (1968) In: *Structure and Function of Inhibitory Neuronal Mechanisms,* p. 511. (von Euler, C., et al., eds.) Pergamon Press, Oxford.
43. Kanaseki, T., and Kadota, K. (1969) *J. Cell Biol., 42*:202.
44. Heuser, J. E., Katz, B., and Miledi, R. (1971) *Proc. Roy. Soc. (London) Ser. B., 178*:407.
45. Heuser, J. E., and Miledi, R. (1971) *Proc. Roy. Soc. (London) Ser. B, 179*:247.
46. De Robertis, E., and Pellegrino de Iraldi, A. (1961) *Anat. Rec., 139*:298.
47. Birks, R. I., Katz, B., and Miledi, R. (1960) *J. Physiol., 150*:145.
48. De Robertis, E. (1959) *Int. Rev. Cytol., 8*:61.
49. Pellegrino de Iraldi, A., and De Robertis, E. (1963) *Int. J. Neuropharmacol., 2*:231.
50. De Robertis, E., and Vaz Ferreira, A. (1957) *J. Biophys. Biochem. Cytol., 3*:611.
51. Chen, I. I., and Lee, C. Y. (1970) *Arch. Abt. B. Tellpathol., 6*:318.
52. Clark, A. W., Hurlfut, W. P., and Mauro, A. (1972) *J. Cell Biol., 52*:1.
53. De Robertis, E., Pellegrino de Iraldi, A., Rodríguez de Lores Arnaiz, G., and Salganicoff, L. (1962) *J. Neurochem., 9*:23.
54. De Robertis, E., Rodríguez de Lores Arnaiz, G., and Pellegrino de Iraldi, A. (1962) *Nature,* (London) *194*:794.
55. De Robertis, E., Rodríguez de Lores Arnaiz, G., Salganicoff, L., Pellegrino de Iraldi, A., and Zieher, L. M. (1963) *J. Neurochem. 10*:225.
56. De Robertis, E., Fiszer, S., and Azcurra, J. M. (1967) *Brain Res., 4*:45.
57. De Robertis, E. (1967) *Science, 156*:907.
58. De Robertis, E., Rodríguez de Lores Arnaiz, G., Alberici, M., Sutherland, E. W., and Butcher, R. W. (1967) *J. Biol. Chem., 242*:3487.
59. Sutherland, E. W., Øye, I., and Butcher, R. W. (1965) *Rec. Progr. Hormone Res., 21*:632.
60. Falck, B., Hillarp, N. A., Thième, G., and Torp, A. (1962) *J. Histochem. Cytochem., 10*:348.
61. McCaman, R. E., Rodríguez de Lores Arnaiz, G., and De Robertis, E. (1965) *J. Neurochem., 12*:927.
62. Sheridan, M. N., Whittaker, V. P., and Israel, M. (1966) *Z. Zellforsch Mikroskop. Anat., 74*:291.
63. Musick, J., and Hubbard, J. I. (1972) *Nature* (London), *237*:279.
64. Jones, S. F., and Kwanbunbumpen, S. (1972) *J. Physiol.* (London), *207*:31.
65. Hodgkin, A. L., and Keynes, R. D. (1957) *J. Physiol.* (London), *138*:253.
66. Dodge, F. A., Jr., and Rahaminoff, R. (1967) *J. Physiol.* (London), *193*:419.

67. Katz, B., and Miledi, R. (1965) *Proc. Roy. Soc. (London) Ser. B,* *161*:483.
68. De Robertis, E., Fiszer, S., and Soto, E. F. (1967) *Science, 158*:928.
69. Changeux, J. P., Kasai, M., Huchet, M., and Meunier, J. C. (1970) *C. R. Acad. Sci. [D] (Paris), 270*:2864.
70. Miledi, R., Molinoff, P., and Potter, L. T. (1971) *Nature* (London), *229*:554.
71. La Torre, J. L., Lunt, G. S., and De Robertis, E. (1970) *Proc. Natl. Acad. Sci. U.S.A., 65*:716.
72. Lunt, G. G., Stefani, E., and De Robertis, E. (1971) *J. Neurochem., 18*:1545.
73. Ochoa, E. L. M., and De Robertis, E. (1973) *Biochim. Biophys. Acta, 295*:528.
74. De Robertis, E., and Fiszer de Plazas, S. (1969) *Life Sci.* [I] 8:1247.
75. Fiszer de Plazas, S., and De Robertis, E. (1972) *Biochim. Biophys. Acta, 266*:246.
76. Ochoa, E., Llorente de Carlin, C., and De Robertis, E. (1972) *Eur. J. Pharmacol., 18*:367.
77. Fiszer de Plazas, S., and De Robertis, E. (1973) *FEBS Letters, 33*:45.
78. Katz, B., and Miledi, R. (1972) *J. Physiol.* (London), *224*:665.
79. Parisi, M., Rivas, E., and De Robertis, E. (1971) *Science, 172*:56.
80. Ochoa, E., Fiszer de Plazas, S., and De Robertis, E. (1972) *Molec. Pharmacol., 8*:215.
81. Parisi, M., Reader, T. and De Robertis, E. (1972) *J. Gen. Physiol., 60*:454.
82. Vasquez, C., Parisi, M., and De Robertis, E. (1971) *J. Membr. Biol., 6*:353.

ADDITIONAL READING

Barondes, S. H. (1969) Axoplasmic transport. In: *Handbook of Neurochemistry* Vol. 2, p. 435. (Lajtha, A., ed.) Plenum Publishing Corporation, New York.
Bloom, F. E., Iversen, L. L., and Schmitt, F. O. (1970) Macromolecules in synaptic function. *Neurosci. Res. Prog.,* No. 4, 8:
Burnstock, N. (1972) Purinergic nerves. *Pharmacol. Rev., 24*:509.
Davson, H. (1959) *A Textbook of General Physiology.* 2nd Ed. Little, Brown and Co., Boston.
De Robertis, E. (1959) Submicroscopic morphology of the synapse. *Int. Rev. Cytol., 8*:61.
De Robertis, E. (1964) *Histophysiology of Synapses and Neurosecretion.* Pergamon Press, Oxford.
De Robertis, E. (1971) Molecular biology of synaptic receptors. *Science, 171*:963.
De Robertis, E. (1975) *Synaptic Receptors: Isolation and Molecular Biology,* pp. 1–387. Marcel Dekker, Inc., New York.
De Robertis, E., and Schacht, J. (1974) *Neurochemistry of Synaptic Receptors.* Raven Press, New York.
Droz, B., Koenig, H. L., and Di Giamberardino, L. (1973) Axonal migration of protein and glycoprotein. *Brain Res., 60*:93.
Eccles, J. C. (1957) *The Physiology of Nerve Cells.* Johns Hopkins Press, Baltimore.
Eccles, J. C. (1964) *The Physiology of Synapses.* Springer-Verlag, Berlin.
Euler, U. S. von (1961) Neurotransmission in adrenergic nervous system. *Harvey Lect.,* Ser. 55, p. 43.
Florey, E. (1961) Transmitter substances, *Ann. Rev. Physiol., 23*:501.
Grafstein, B. (1969) Axonal transport. In: *Advances in Biochemical Psychopharmacology,* Vol. 1, p. 11. (Costa, E., and Greengard, P., eds.) Raven Press, New York.
Hodgkin, A. L. (1958) Ionic movements and electrical activity in giant nerve fibers. *Proc. Roy. Soc. London,* Ser. B, *148*:1.
Hubbard, J. I. (1973) Microphysiology of vertebrate neuromuscular transmission. *Physiol. Rev., 53*:674.
Hydén, H. (1960) The neuron. In: *The Cell.* Vol. 4, p. 215 (Brachet, J., and Mirsky, A. E., eds.) Academic Press, Inc., New York.

Jaim-Etcheverry, J., and Zieher, L. M. (1971) Ultrastructural aspects of neurotransmitter storage in adrenergic nerves. In: *Advances in Cytopharmacology,* Vol. 1, p. 343, Raven Press, New York.

Katz, B. (1961) How cells communicate. *Sci. Am., 205* (No. 3):209.

Katz, B. (1966) *Nerve, Muscle, and Synapse.* McGraw-Hill Book Co., New York.

Loewenstein, W. R. (1960) Biological transducers. *Sci. Am., 203* (No. 2):99.

Ochs, S. (1972) Fast transport of materials in mammalian nerve fibers. *Science, 176*:252.

Quarton, G. C., et al, eds. (1967) *The Neurosciences.* The Rockefeller University Press, New York.

Rodriguez de Lores-Arnaiz, G., and De Robertis, E. (1973) Drugs affecting the synaptic components of the CNS. In: *Fundamentals of Cell Pharmacology,* p. 280, (Dikstein, S., ed.) Charles C Thomas, Pub., Springfield, Ill.

Ruch, T. C., and Fulton, J. F. (1960) *Medical Physiology and Biophysics.* 18th Ed. W. B. Saunders Company, Philadelphia.

twenty-five

CELL SECRETION

Secretion is one of the most common cellular functions. It may be defined as the process by which cells synthesize products that will be utilized by other cells or eliminated from the organism. Secretions are either (1) *external,* or *exocrine,* i.e., they are expelled into the outer environment or more frequently into natural cavities (e.g., the digestive or respiratory tract), or (2) *internal,* or *endocrine,* i.e., the secretions enter directly into the circulation to act on other tissues. Internal secretion is characteristic of the endocrine glands, such as the thyroid, parathyroid and adrenal glands, the hypophysis, and the islets of the pancreas. Typical exocrine secretion is that of the pancreatic acinus, the salivary glands, and the numerous small glands that are related to the digestive, respiratory, and genital tracts.

This basic cellular activity appears early in phylogeny. Sponges have mucus-secreting cells and cells that produce *spongin,* a collagen-like substance. Typical secretory cells that produce mucus and proteins are found in the gastrodermis of *Hydra,* a coelenterate, and in the epidermis of ctenophorans. Higher up on the phylogenetic scale secretory activity is manifest in cells of genital tissues and endocrines. In the evolution of organisms cell secretion participates not only as an adaptation to the environment, but also as an indispensable aid to reproduction of the species.[1]

Secretion is a complex function of the cell involving all the parts and organelles which have been discussed in previous chapters. The nucleus and the nucleolus, the ribosomes, the vacuolar system, including the endoplasmic reticulum and the Golgi complex, and also the mitochondria all participate, directly or indirectly, in the secretion process. Even the lysosomes may be involved in regulating the number of secretory granules or in disposing of any excess of membranes packing the granules.[2, 3]

The main interest of the cytologic study of secretion is in these coordinated series of events in which each part of the machinery of the cell is involved. It is advisable to review some of the fundamental functions presented in earlier chapters: (1) the production of the different RNA molecules by structural and other genes present in the DNA of the chromosomes, (2) the function of the nucleolus in the biogenesis of ribosomes, (3) the polyribosome as the site of protein synthesis. (4) the role of the endoplasmic reticulum in circulating proteins for export, and (5) the role of the Golgi complex in concentrating the secretion product and in providing a packing membrane. Thus, the study of secretion is a recapitulation of many chapters of cell biology.

The Secretory Cycle. Methods of Study

Secretion involves a continuous change that can be best interpreted by studying the cell throughout the different stages of cellular activity.

If fixed and stained secretory cells are studied under the microscope, the image obtained represents only a single stage of cell function. In cell secretion, more than in any other process, the *time factor* must be taken into account to interpret the results of cytomorphologic analysis.

In some secretory cells secretion is *continuous:* the secretion product is discharged as soon as it is elaborated. In these cells all the phases of the secretory process take place simultaneously. Under the microscope striking differences cannot be seen from one cell to another. This happens, for example, in some endocrine glands (e.g., thyroid, parathyroid, and adrenal cortex) and in the muciparous cells of the gastric epithelium.

In Chapter 9 the secretion of lipoproteins and glycoproteins by liver cells, as well as the release of antibodies by plasma cells were mentioned. In all these cases, the secretion product is not accumulated in special storage granules, and the release of secretory materials is more or less simultaneous with the synthesis and intracellular transport of these substances.

In other cells the secretory cycle is *discontinuous:* it is specially timed so that the synthesis and intracellular transport are followed by the accumulation of the secretion product in special storage granules which are finally released to the extracellular space.

The discontinuous type of activity is also called *rhythmic.* There are considerable differences in morphologic characteristics as well as in metabolism from one cell to another. Examples of rhythmic secretory cells are the goblet cells of the intestine and, to some extent, the pancreatic acini.

In some glands, even if the activity is continuous, the various cells may be in different stages of the secretory cycle. For example, the salivary glands of the rat and mouse are active continuously, whereas the individual acini show a rhythmic function. In such cases one may have to observe numerous sections through the gland to see the different stages. In other glands periods of almost complete inactivity may be followed by others of intense activity. To overcome these difficulties of observing the different stages of the secretory cycle, special stimuli can be used that rapidly modify the activity of the cells (which normally would be asynchronic or semisynchronic) and drive them in a given direction, thus establishing functional synchronization. If investigators wish to study the secretion of the exocrine pancreatic cells, they use a fasted animal whose pancreatic gland is in the resting state. The cells are stimulated by feeding the animals or by administering *pilocarpine,* which brings about the rapid excretion of the secretion products. In this way, the various phases of cellular activity are synchronized, and practically all the cells expel their contents and then recover gradually. Substances that induce release of secretory products are generally called *secretogogues.*

The cytologic study is carried out at various times after the application of the stimulus and can be done in a purely qualitative or quantitative way.

The methods for studying secretion are numerous; they involve not only the observation of living secretory cells for long periods of time or at different time intervals after fixation, but also cell fractionation methods to separate different parts of the secretory cell, the cannulation or fistulation of the excretory ducts of the gland to analyze the products that are eliminated after the application of the stimulus, and especially the use of radioactive precursors of the secretion. This last technique, when used on radioautographs at the light and electron microscopic levels, provides important information about the dynamics of the secretion process within the cell structure.

In studying the secretory cycle, fixation by freezing and drying (Chap. 6) is advantageous, since it stops the cellular processes rapidly and thus aids determination of the different stages. In addition, it permits one to observe, under the best conditions and without changes, the soluble products and protein secretion when present in high dilution. In the thyroid, this method demonstrates an intracellular colloid that is not readily observable by other methods and makes it possible to follow the different stages of its formation and excretion.[4, 5]

SUMMARY:
Secretion Methods of Study

Secretion is a general cellular function in which there is a coordinated series of intracellular events involving the nucleus and most of the cytoplasmic organelles. In the study of secretion the time factor is essential and must be taken into account to interpret the results of cytomorphologic analysis.

Secretion may be continuous when the secretion products are discharged without being stored. It is discontinuous when the synthesis and intracellular transport of the product is stored in special granules which are then released to the extracellular space.

Several physiologic or pharmacologic stimuli can induce the release of secretion *(secretogogues)* and can bring about synchronization of secretion in the various cells. Many of the qualitative and quantitative methods mentioned in this book (see Chapters 5 and 6) are used in the study of cell secretion.

MORPHOLOGY OF THE SECRETORY CYCLE

The secretory cycle has extremely variable cytologic expressions, but it is generally characterized by products, visible with the light microscope, which accumulate in the cell, and then are ultimately eliminated. These products may be dense and refractile granules, vacuoles, droplets, or other structures having a definite location in the cell and, at times, characteristic histochemical reactions.

The introduction of electron microscopy has helped to clarify the relationship between the fine structure of the cytoplasm and the secretion products. Since a detailed study of the submicroscopic morphology of secretion in the different types of glands is beyond the scope of this book (see histology textbooks),[6] consideration will now be given to a few examples in which a correlation between structure and function has been achieved by the use of ultrastructural and cytochemical methods.

Secretion in the Parotid Gland

The secretory cells of the parotid gland have a structural organization somewhat similar to that of the exocrine pancreas. These cells have various types of receptors on their surface by which they are able to react to nerve stimulation or to the action of neuroactive drugs acting as secretogogues. It is known that the parotid secretory cells have cholinergic receptors which can be stimulated with acetylcholine, pilocarpine, and other cholinergic drugs, and blocked by atropine. The activation of this mechanism causes the release of a watery secretion rich in potassium. This type of secretion can also be mediated by α-adrenergic receptors which are activated by certain adrenergic agents (i.e., epinephrine) and are blocked by α-adrenergic blocking agents (i.e., phentolamine). On the other hand, the secretion of the zymogen granules, which contain *amylase* (a carbohydrate-hydrolyzing enzyme), involves the activation of β-adrenergic receptors. In this case, the adrenergic effect is mediated by adenyl cyclase and cyclic AMP and can be blocked with β-adrenergic blocking agents (e.g., propranolol).

Figure 25–1 shows the sequence of events of the secretory cycle of the parotid gland of the rat after it has been synchronized by the injection of adrenergic drugs, acting on beta receptors. Between 20 and 50 minutes after the stimulus, the zymogen granules present in the apical region of the cell are discharged into the lumen in a sequential manner (i.e., Z_2 after Z_1, and so forth) by fusion with the cell membrane. As a consequence, the lumen of the acinus becomes wider and penetrates deeply into the cell until all the granules are discharged. At this time there is an excess of membrane material resulting from the fusion of the membranes of zymogen granules. These membranes are then reduced (i.e., at 120 to 300 min) by microvesiculation, with the formation of small apical vesicles. The phase of secretion release is correlated with the reduction in amylase content in the gland and with the decrease in number of zymogen granules (Fig. 25–2). The second phase of the cycle, which leads to restoration of secretory granules, is characterized by the appearance of the so-called condensing vacuoles in the Golgi region and of new zymogen granules in the apex of the cell. In this phase the restoration of cell secretion can also be correlated with the increase in amylase and the number of zymogen granules (Fig. 25-2).[7]

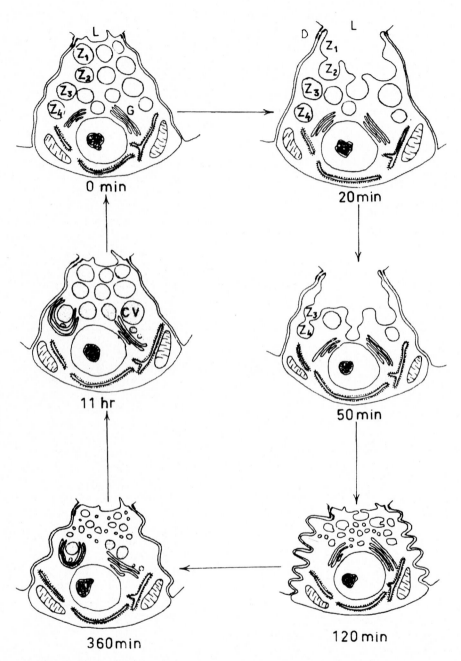

Figure 25–1. Diagram of the secretory cycle in a cell of the parotid gland after it received an adrenergic stimulus. The main cytoplasmic structures indicated. Z_1, Z_2, Z_3, and Z_4 represent zymogen granules that are in line to be released by exocytosis into the lumen (*L*). *G*, Golgi complex. *CV*, condensing vacuoles. (See the description in the text.) (From Amsterdam, A., Ohad, I., and Schramm, M., *J. Cell. Biol., 41*:753, 1969.)

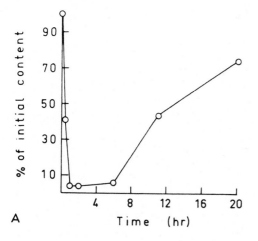

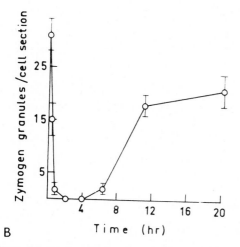

Figure 25–2. **A,** content of amylase as a function of time; and **B,** number of zymogen granules as a function of time, after the injection of the stimulus which induces cell secretion from the parotid gland. (From Amsterdam, A., Ohad, I., and Schramm, M., *J. Cell. Biol., 41*:753, 1969.)

The idea of a sequential release of the zymogen granules implies that the fusion of each granule takes place only after the membrane of a zymogen granule has already been incorporated into the lumen. Cyclic 3′-5′ AMP may be involved in this process of membrane modification as it is in other biological systems.[8] In fact, the secretion induced by β-adrenergic agonists can be obtained by using dibutyryl cyclic AMP directly. It has been observed that in the case of stimulation by β-adrenergic drugs, as well as by this cyclic nucleotide, a number of granules show a deformation with the appearance of a kind of pseudopod that is directed toward the apex of the cell and that

fuses with the cell membrane. This interesting effect has been reproduced in isolated zymogen granules in the presence of ATP.[9] An analogy with the effect of cyclic AMP in the case of ameboid movement in slime molds[10, 11] has been postulated. It was mentioned above that the parotid gland has both alpha and beta receptors for adrenergic agonists. Stimulation of the alpha receptors results in vacuolization of the cells with water and K^+ release. This may collaborate in the transport of the secreted protein from the lumen of the acini toward the gland ducts. The α-adrenergic response is not mediated by cyclic AMP but involves the use of energy and calcium ions.

Secretion in Pancreatic Cells

An example of a secretory cycle in which the secretion products are readily visible is that of the exocrine pancreatic cell, in which the cycle has been carefully studied. These cells belong to the group that produces serous or zymogenic secretion, so-called because they secrete a protein rich in enzymes.

In the resting state a pancreatic cell is typical in the polarization of its components (Fig. 9–5). The base of the cell is occupied by the nucleus, the basophilic substance containing ribonucleoproteins, and elongate mitochondria oriented in the apicobasal direction. The apical or excretory region is occupied by refractile granules with a high protein concentration. In the supranuclear zone and among the zymogen granules is a Golgi complex.

Under the electron microscope the great development of the endoplasmic reticulum with large cisternae oriented parallel to the cell axis is observed in the basal region. These cisternae are covered by numerous ribosomes attached to the membrane, while a smaller number of ribosomes are free in the cytoplasmic matrix. The supranuclear region contains the cisternae and vesicles of the Golgi complex with their characteristic lack of ribosomes. Some of the larger vesicles, called *condensing vacuoles,* contain a clear material or a more concentrated material, which by progressive condensation is transformed into the zymogen granules at the apex of the cell. Each one of these granules is bound by a membrane provided by the Golgi complex (Fig. 25–3, *A*).

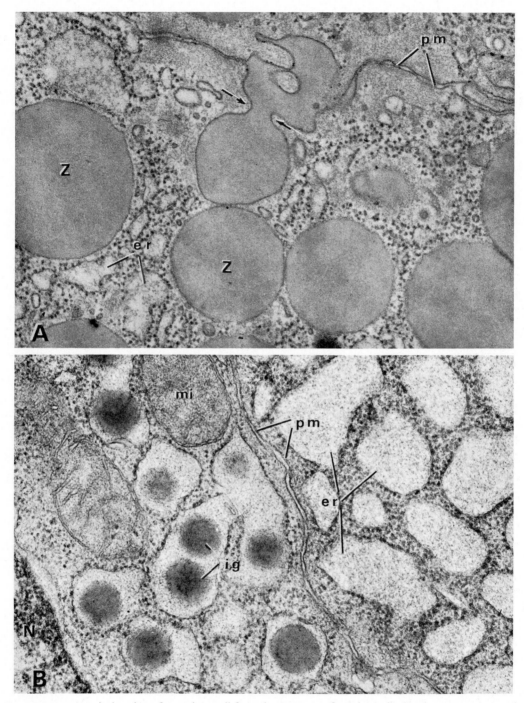

Figure 25–3. **A,** apical region of an acinar cell from the pancreas of a guinea pig showing zymogen granules (Z), one of which is being expelled into the lumen by exocytosis followed by membrane fusion; *er,* granular endoplasmic reticulum; *pm,* plasma membrane. × 30,000. (Courtesy of G. E. Palade.) **B,** the same as above, but from the basal portion showing the enlarged cisternae of the endoplasmic reticulum (*er*), some of which contain intracisternal granules (*ig*); *mi,* mitochondria; *N,* nucleus; *pm,* plasma membrane. × 30,000. (Courtesy of D. Zambrano.)

By using stereologic methods (see Chapter 5) investigators have obtained quantitative information about the relative percentage volume of different cellular structures. The rough endoplasmic reticulum occupies 60 per cent of the cytoplasm; the zymogen granules, 2.8 per cent; and the condensation vacuoles, 0.4 per cent.

Injection of the secretogogue, pilocarpine, brings about a liquefaction of the zymogen granules and the rapid expulsion of their contents.

Later, the cells elaborate more secretory granules, which accumulate at the apical pole; after several hours, the cells regain their original appearance. During this stage the Golgi apparatus hypertrophies and becomes intensely osmophilic.

The electron microscopic study of the secretory process of the pancreas has confirmed and extended the observation of a functional relationship between the endoplasmic reticulum and the Golgi complex in secretion. The material synthesized by the ribosomes may sometimes be observed within the cavities of the endoplasmic reticulum, forming the so-called intracisternal granules[12, 13] (Fig. 25–3, *B*). This material then passes into the Golgi complex and finally is concentrated and packed into the zymogen granules (Fig. 25–3, *A*).

Secretion in the Adrenal Medulla

Particularly favorable material for the study of the morphology of the secretory process are the cells of the adrenal medulla that produce and secrete catecholamines (epinephrine and norepinephrine). As shown by Plenick in 1902, catecholamines reduce osmium tetroxide intensely, and by this reaction they can be detected in minute amounts within the structure of the cell.

As shown in Figures 25–4 and 25–5, *B*, the first and smallest secretion droplets that appear are in the deepest region of the cytoplasm near the nuclear membrane. Some of the small vesicles belonging to the Golgi complex become filled with the dense material of the catecholamines. These droplets, always surrounded by the membrane, migrate toward the surface of the cell while increasing in size and density. As a result of this process of elaboration, the cytoplasm of the cell becomes filled with catechol-containing droplets about 160 mμ in diameter (Fig. 25–4).

The expulsion of the secretory material is mediated in this gland through the splanchnic nerves that innervate the cell by terminal endings filled with synaptic vesicles (see Chapter 24). These endings are cholinergic, which means that stimulation of the nerve releases acetylcholine, thus activating the release of catecholamines. It has been observed that an electrical stimulation that produces the maximum expulsion of the catechol secretion also increases the number of synaptic vesicles in the ending and the quantity of acetylcholine released.[14]

From the morphologic viewpoint, the mechanism of the actual expulsion of the secretory product into the intercellular spaces is of considerable interest. As is indicated in Figure 25–5, the catechol-containing droplets first become attached to the surface membrane. In a second stage, they increase in size and become less dense (swelling). In a final stage, the dense material is evacuated, leaving empty membranes. At the same time, new droplets are being actively formed in the Golgi region (Fig. 25–5).[15]

A somewhat similar mechanism for the secretion of synaptic vesicles at the synaptic endings was previously postulated. By this mechanism, acetylcholine, epinephrine, norepinephrine, or other active humoral agents may be synthesized by the cell, stored within a membrane, and then moved and discharged instantly at the surface membrane, when the appropriate stimulus is acting (see Chapter 24).

At variance with the catechol-containing droplets of the adrenal medulla, however, in which there is a true exocytosis (see below), the synaptic vesicles are used again and again after refilling with the transmitter at the nerve ending (see De Robertis, 1967). Exocytosis in the adrenal medulla has been confirmed by the technique of freeze-etching.[16]

The role of Ca^{2+} is of importance in the release of these neurohumors, and it has been demonstrated that catecholamines are released only in the presence of this cation.[17] It is possible that when the splanchnic nerves are stimulated, the release of acetylcholine by the nerve endings causes the entrance of Ca^{2+} into the adrenal cells and the simultaneous entrance of water into the secretory droplets attached to the plasma membrane. This results in swelling of the droplet and then fusion of the membranes.

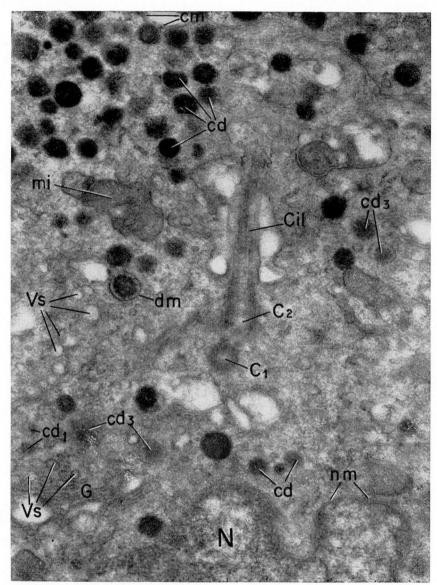

Figure 25–4. Electron micrograph of a chromaffin cell from the adrenal gland of the hamster. At the bottom, the nucleus (*N*) with the folded nuclear membrane (*nm*) can be seen. The supranuclear portion of the cytoplasm shows the Golgi complex (*G*), two centrioles (C_1, C_2) with a cilium (*Cil*) arising from one of them, several mitochondria (*mi*) and the catechol-containing droplets (*cd*). The smaller catechol droplets (cd_1, cd_3) appear in the Golgi zone. Within some of the small Golgi vesicles (*Vs*) the dense deposit is first observed. As the vesicles enlarge and the content increases (cd_2, cd_3), the clear space under the droplet membrane (*dm*) narrows. Completely formed catechol-containing droplets occupy the peripheral part of the cytoplasm near the cell membrane, (*cm*). × 51,500. (From E. De Robertis, and D. D. Sabatini.)

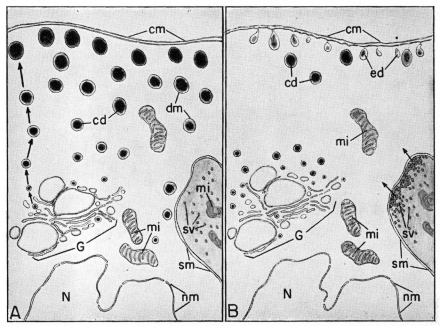

Figure 25–5. Diagrammatic interpretation of the mechanism of secretion in the chromaffin cell. **A,** cell in the resting stage, showing the storage of mature catechol droplets in the outer cytoplasm. Near the nucleus within the Golgi complex new secretion is being formed at a slow rate. At the right, a portion of a nerve terminal, showing the synaptic vesicles (*sv*) and mitochondria (*mi*); *cd*, catechol droplets; *cm*, cell membrane; *dm*, droplet membrane; *ed*, evacuated droplets; *G*, Golgi complex; *N,* nucleus; *nm*, nuclear membrane; *sm*, surface membrane. **B,** cell after strong electrical stimulation by way of the splanchnic nerve. Most of the catechol droplets have disappeared; the few that remain can be seen in different stages of excretion into the intercellular cleft. The Golgi complex is now forming new droplets at a higher rate. The nerve ending shows an increase of synaptic vesicles with accumulation at "active points" on the synaptic membrane. (From E. De Robertis and D. D. Sabatini.)

The morphology of the secretory cycle varies in the various cells described. The exocrine pancreatic cells are zymogenic and have a typical polarized structure. Release of secretion can be obtained with pilocarpine. In the parotid gland synchronization of secretion can be produced by adrenergic drugs acting on β-receptors. Within 20 to 50 minutes after the stimulus has been administered, zymogen granules are discharged in a sequential order by fusion of their membranes with the apical cell membrane. The excess membrane is taken up again, with the formation of apical microvesicles, and is re-utilized for the packing of new secretory products. The release phase is mediated by cyclic AMP. In the second phase of the cycle, condensing vacuoles appear in the Golgi complex, and there is restoration of zymogen granules. These changes are correlated with those in amylase content of the gland and in the number of zymogen granules. The adrenal medulla is a gland in which the secretory process can be followed histo-chemically. The cholinergic stimulation of this gland leads to rapid release of the catechol-containing droplets. In this mechanism, as in the release of other neurohumors, Ca^{2+} ions are indispensable.

SUMMARY:
Morphology of the
Secretory Cycle

CYTOCHEMISTRY OF THE SECRETORY CYCLE

Cytochemical Study

Radioautography at the electron microscopic level has provided important cytochemical information about the secretory process. The radiolabeled precursors may be given intravenously to a test animal or may be administered in vitro to slices of tissue maintained in saline solutions. Tritiated leucine (^{3}H-leucine) is one of the most frequently used precursors. At specific time intervals the tissues are fixed and processed for radioautography. The kinetics of the labeling of the secretory protein can be studied by the use of a *short pulse* followed by a so-called *chase*. For example, slices of pancreas, are exposed first to a leucine-free medium (i.e., a saline solution containing all amino acids except leucine) which produces depletion of the leucine-endogenous pool. After a short time the tissue is submitted to a medium containing ^{3}H-leucine for a few minutes, and this procedure is then followed by the chase, which consists of washing and incubation with a solution containing excess non-labeled leucine. In the fixed tissue, it is possible to observe that after a few minutes the isotope is localized in the endoplasmic reticulum of the basal region. Later, the newly synthesized protein passes into the Golgi complex. In this region apparently it becomes progressively concentrated into prozymogen granules or condensing vacuoles surrounded by a membrane. After a longer time, the label is found principally in the zymogen granules and in the lumen of the acinus. Figures 25–6 and 25–7 show electron microscopic radioautographs of the sequence of events in sections of pancreas previously pulse labeled with ^{3}H-leucine and followed by a chase at different time intervals.

Quantitative data may be obtained by measuring the number of radioautographic grains over the various cell components. Immediately after the pulse labeling there is a sudden increase in radioactivity in the rough endoplasmic reticulum with a tendency to decline rapidly. The radioactivity then becomes high in the Golgi complex, while it increases more slowly in both the immature and the mature secretory granules. Figure 25–8 readily shows that, in the case of the parotid gland, there is a wave-like movement of the pulse labeled secretory protein through the various intracellular compartments in the following order: rough endoplasmic reticulum → Golgi complex → immature granules (i.e., condensing vacuoles and prozymogen) → mature secretory (zymogen) granules.[18] In the rat pancreas the total life span of a zymogen granule has been estimated at 52.4 minutes.[19] From Figure 25–8 it may be concluded that although the processing of secretory proteins in the parotid is similar to that of the pancreas, the rate is slower and storage is more prolonged. In contrast to the secretory proteins, the synthesis of other proteins of the pancreas not used for export have a much longer life span.[19]

Biochemical Studies

The use of biochemical methods has greatly contributed to knowledge of the process of secretion. In the case of the pancreas, this study has been facilitated by the extensive information available on the physical, biochemical, chemical, and enzymatic properties of the pancreatic juice and by the use of suitable chromatographic procedures that permit the isolation of most of the proteins present in it.

The exocrine pancreas synthesizes and secretes a large variety of proteins: amylase, ribonuclease, chymotrypsinogen, trypsinogen, lipase, procarboxyl-peptidases A and B, and proelastase; other proenzymes or enzymes may be separated chromatographically, either from the tissue or from pancreatic juice. The various secreted proteins differ a great deal in their physical and chemical properties. For example, the molecular weight ranges from 70,000 to 6000 daltons, and the isoelectric point ranges from pH 11, in the most basic proteins, to pH 3, in the most acidic. The latter contain a sulfated material that may be demonstrated by the incorporation of radioactive sulfate ($^{35}SO_4$).[20] Some of the secretory proteins are glycoproteins.

The pancreas may be submitted to the cell fractionation methods described in Chapter 6. The cell fractions obtained are analyzed morphologically or followed by biochemical methods which involve the separation of the secretory proteins. The use of radiolabeled precursors, as in the case of radioautography, may be of considerable help in this study.

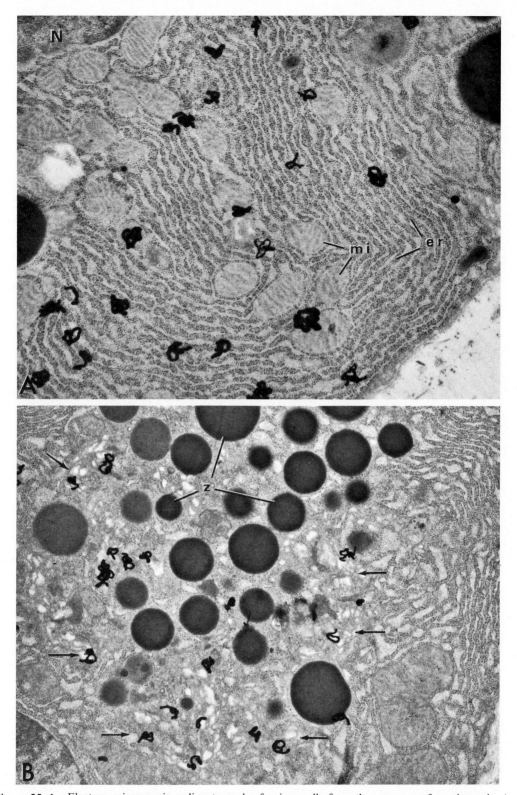

Figure 25–6. Electron microscopic radioautograph of acinar cells from the pancreas of a guinea pig. **A,** three minutes after pulse labeling with ^{3}H-leucine. The radioautographed grains are located almost exclusively on the granular endoplasmic reticulum (*er*); *mi,* mitochondria; *N,* nucleus. × 17,000. **B,** the same as above, but incubated for seven minutes after pulse labeling. The label is now in the region of the Golgi complex (arrows); *z,* zymogen granules. × 17,000. (Courtesy of J. D. Jamieson and G. E. Palade.)

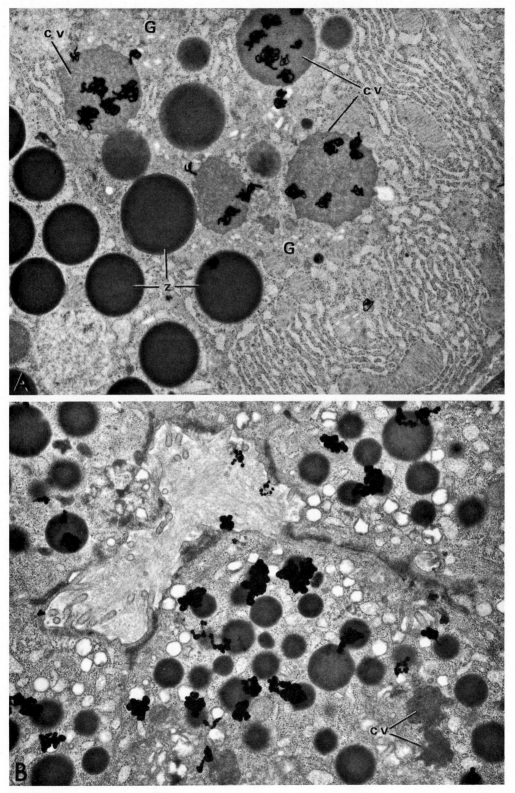

Figure 25–7. **A,** the same experiment as in Figure 25–6, but 37 minutes after pulse labeling. The label is now concentrated in the condensing vacuoles (*cv*) of the Golgi complex (*G*). The zymogen granules (*z*) are unlabeled. × 13,000. **B,** the same as above, but incubated for 117 minutes after pulse labeling. The radioautographed grains are now localized primarily over the zymogen granules, while the condensing vacuoles (*cv*) are devoid of label. Some grains are in the lumen of the acinus, indicating the secretion. × 13,000. (Courtesy of J. D. Jamieson and G. E. Palade.)

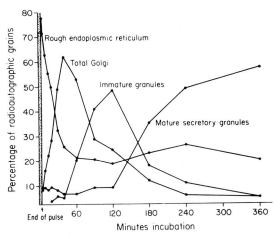

Figure 25–8. Diagram showing the radioactivity of ³H-leucine, measured as grains under the electron microscope, as a function of time. The curves indicate the variations that occur in the various cell compartments. (See the description in the text.) (From Castle, J. D., Jamieson, J. D., and Palade, G. E., *J. Cell. Biol., 53*:290, 1972.)

For example, a rather homogeneous and pure zymogen granule fraction has been isolated and studied under the electron microscope. This fraction has been found to contain about 94 per cent protein, 5 per cent phospholipid, and 1 per cent nucleic acid. At pH 8 the granules are solubilized, and a membrane fraction remains, representing the surface membranes that cover the zymogen granules within the cell. In the lysate of the granules the following enzymes can be isolated by column chromatography: trypsinogen, chymotrypsinogen A, ribonuclease, amylase, chymotrypsinogen B, procarboxypeptidase B, deoxyribonuclease, and procarboxypeptidase A. The great similarities in enzyme composition between the zymogen granules and the pancreatic juice, obtained by cannulation of the duct, is the most direct evidence that the granules contain the secretion products.

Sequence of the Secretory Process. Metabolic Requirements

Pancreatic secretion has been studied in slices that were pulse labeled with ³H-leucine for three minutes, followed by a chase. Cell fractionation carried out at different times permitted a kinetic analysis of the protein transport. Such a study confirmed that the first synthesis of the secretory proteins takes place in the rough endoplasmic reticulum. In fact, after three minutes incubation, this fraction (i.e., microsomes with ribosomes attached) contained more than twice the radioactivity than the smooth fraction, primarily represented by Golgi vesicles. After seven minutes the situation reversed completely, and the specific radioactivity in the Golgi vesicles was highest. This result provided direct evidence of the transport from the endoplasmic reticulum to the Golgi complex.

The involvement of the condensing vacuoles of the Golgi complex in secretion was analyzed in similar experiments in which the zymogen fraction contained a small population of condensing vacuoles that could be recognized morphologically. By radioautography with ³H-leucine it was demonstrated that these vesicles were strongly radioactive between 17 and 57 minutes, at which time the zymogen granules showed little or no labeling. This result confirms the idea that the proteins are transported into the large condensing vacuoles from the small vesicles of the Golgi region. The condensing vacuoles are converted into zymogen granules by the progressive filling and concentration of their content, and finally acquire their characteristic electron-opaque content. This conversion is not dependent on a supply of metabolic energy, since it continues after inhibition of glycolysis or respiration. Furthermore, the membranes of the zymogen granules do not contain a Na⁺-K⁺-ATPase that could participate in an active transport.[21]

The various steps in the secretion and discharge of zymogen granules have different energy requirements. The transport of the newly synthesized polypeptide chain from the ribosome into the cisternae of the endoplasmic reticulum does not require additional energy and seems to be controlled mainly by the structural relationship of the large ribosomal subunit with the membrane of the reticulum (see Chapter 19).[22–25] When the process of protein synthesis is inhibited by puromycin, the incomplete peptides formed are transported into the vacuolar cavity in the pancreas[24] and also in neurosecretory cells.[26] A similar study performed with another inhibitor of protein synthesis—cycloheximide— also demonstrated that the intracellular transport does not depend on the synthesis of secretory proteins and may continue even in the absence of such a synthesis.[27]

Transport from the endoplasmic reticulum to the condensing vacuole is insensitive to inhibitors of glycolysis (fluoride, iodoacetate), but it is blocked by inhibitors of cell respiration (nitrogen, cyanide, antimycin A) or inhibitors of oxidative phosphorylation (dinitrophenol, oligomycin). An important conclusion to be drawn from these studies is that at the periphery of the Golgi complex, in transitional elements between the endoplasmic reticulum and the small vesicles of the Golgi complex, there is a kind of *energy-dependent lock* in the transport. Such a lock may regulate the flow of secretory proteins — a flow which is slowed down by metabolic inhibitors.[28] In Figure 25–9 the lock would be situated somewhere in the region indicated by an arrow between *j* and the condensing vacuole.

The discharge of the zymogen granules by exocytosis requires energy. The release is strongly inhibited by compounds (e.g., antimycin A) or by conditions which interfere with ATP production by respiration. Presumably, part of this energy is related to the process of membrane fusion-fission taking place during exocytosis, but it is possible that energy could be consumed in the propulsion of the granule to the cell apex.

A pancreatic cell in which protein synthesis has been inhibited can go through an entire secretory cycle of transport, concentration, storage, and release. This fact suggests that within the period of time of one cycle (i.e., 60 to 90 min) the synthesis of new membrane proteins, serving as containers of the secretion, is not required. In other words, intracellular membranes have a much longer half-life and are probably re-utilized extensively during the secretory process. There are two sites in which membrane re-utilization and circulation occur: (1) between the rough endoplasmic reticulum and the Golgi complex, and (2) between the Golgi complex and the apical plasma membrane. In both sites there is an energy operated lock.[28] The membrane of the zymogen granules may be regarded as a vacuole that shuttles between the Golgi apparatus and the cell surface.

The mechanism of removal of excess membrane from the apical region of the cell is still unknown. It has been suggested that patches of membranes are invaginated from the surface as small vesicles that move back into the Golgi region, to be re-utilized in the packing of more secretion. This hypothesis is supported by the finding of an increase in the number and size of Golgi components after intense in vitro stimulation of the pancreas by carbamylcholine. This increase takes place even under conditions in which protein synthesis has been almost completely inhibited by cycloheximide.

Another interesting finding is that the in vitro stimulation of zymogen discharge is not accompanied by an increase in the rate of synthesis of secretory proteins. In both control and stimulated slices of pancreas, the secretory proteins always occur within membrane-bound compartments, and there is no evidence of their being transported freely through the cell matrix. Furthermore, in both conditions the rates of synthesis and discharge of secretory proteins are similar. The application of a secretory stimulus primarily affects the discharge of the secretory protein, but does not influence the rates of synthesis and intracellular transport.[29]

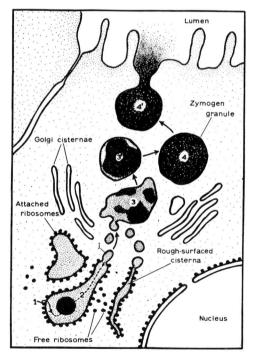

Figure 25–9. Diagram of the secretory process in the pancreatic acinus, showing the different stages described in the text. *1*, ribosomal stage; *2*, endoplasmic reticulum stage; *3,3'*, Golgi complex stage; *4*, zymogen stage; *4'*, release of zymogen into the lumen (intraluminal stage). (Courtesy of G. E. Palade.)

Processing of Secretory Proteins

The analysis just described (using radio-autography) does not provide information about the synthesis and transport of the individual proteins secreted by the pancreas. For example, it is not possible to determine whether each protein is synthesized and transported intracellularly by a different route, or if all of them are processed at the same time and by a similar mechanism. Recent work has provided evidence that there is a parallel processing of all the major proteins secreted by the pancreas. Cytochemical studies using immunological methods at the electron microscopic level have shown that not only each cell but also each zymogen granule contains several of the above-mentioned characteristic secretion proteins.[30] Another approach has been to incubate pancreatic lobules with amino acid precursors, and, at the same time, to stimulate the discharge by the use of the cholinergic agonist, carbamylcholine. The proteins secreted in the medium are submitted to microfractionation procedures involving isoelectric focusing, which separates the proteins according to their charge (isoelectric point). The use of gel electrophoresis separates the proteins according to size. The major proteins secreted in the medium could be identified by these two parameters and compared with similar proteins present in pancreatic juice. These exacting procedures for protein analysis have also been applied to fractions of zymogen granules isolated from the pancreas at different time intervals. Since the radioactivity reaches the condensing vacuoles at 15 minutes and is present in the zymogen granules generally after 30 minutes, it is possible — by proper timing — to follow the secretion of the various proteins in these two compartments of the cell.

The results obtained in subcellular fractions have led to the conclusion that the kinetics of the intracellular transport of all the major proteins is similar. They all originate in the rough endoplasmic reticulum and pass by way of the Golgi complex, the condensing vacuoles, and zymogen granules. When the gland is stimulated, the discharge of all these proteins is also simultaneous (see Tartakoff, 1974).

SUMMARY:
Cytochemistry of the
Secretory Cycle

Important information on the secretory cycle may be obtained by the use of radioactive precursors (e.g., ^{3}H-leucine). Frequently, a study of this type is carried out by pulse labeling, followed by a chase with an unlabeled precursor. Qualitative and quantitative data may be obtained on tissue sections by the observation and measurement of the radioautographic grains. The results obtained from the pancreas and parotid gland demonstrate that there is a wave-like transport of the secretory protein going through the following intracellular compartments: rough endoplasmic reticulum → Golgi complex → condensing vacuoles → zymogen granules. In the pancreas, the life span of a zymogen granule is about 50 minutes; in the parotid, it is longer.

The secretory process can be followed biochemically by using cell fractionation methods and studying the bulk secretory protein. Similar studies may be carried out on each individual protein. Through these techniques the sequence of transport from the rough microsomal fraction to the smooth microsomal fraction (Golgi complex), and then to the condensing vacuoles and zymogen granules was confirmed. The existence of special energy-dependent locks at the transitions, endoplasmic reticulum → Golgi complex and zymogen granule → apical membrane (i.e., exocytosis) was demonstrated. At these points the transport of secretion may be stopped by inhibitors of oxidative phosphorylation.

On the contrary the transport is not affected by inhibition of protein synthesis within the period of a secretory cycle. This indicates that the membranes serving as containers of the secretion have a much longer half-life. The results suggest that there are two sites (i.e., between endoplasmic reticulum and Golgi complex and between Golgi complex and apical plasma membrane) where there is re-utilization and circulation of membranes.

The application of a secretory stimulus affects the discharge of the secretory protein, but not the intracellular transport. The processing of the numerous proteins secreted by the pancreas is done in parallel. The proteins are all processed at the same time and by a similar mechanism.

Mechanisms of Secretion in the Pancreas

It is now possible to summarize all the available data in a coherent theory of protein secretion for the pancreas, which will probably apply for other protein secreting cells. The following sequential stages of secretion can be recognized (Fig. 25-9):

Ribosomal Stage. Proteins are synthesized in direct contact with polyribosomes present on the surface of the vacuolar system of the endoplasmic reticulum (Fig. 18–1). As indicated in Chapter 19, this is done by the interaction of messenger RNA (mRNA), which carries the genetic information from the DNA molecule contained in the chromosome, and the amino-acyl-transfer RNA complex, which attaches in the proper sequence, with subsequent polymerization of the amino acids. This first stage takes place in a matter of seconds or a few minutes. The disposition of the ribosomes on the membranes with the attachment of the 60S subunit (Chapter 18) probably facilitates the interaction with mRNA and also the rapid vectorial passage of the exportable protein into the endoplasmic reticulum system.

Endoplasmic Reticulum Stage. The newly synthesized proteins (i.e., enzymes) rapidly penetrate into the cisternae of the endoplasmic reticulum and migrate toward the apical zone of the cell. Sometimes this ma-

terial appears as small intercisternal granules, but in most cases it is a dilute solution of protein, in which some macromolecular material may be observed (Fig. 25–3, *B*).

Golgi Complex Stage. After a few minutes the secreted protein reaches the Golgi zone, probably by way of continuities with the endoplasmic reticulum, which may be permanent or, more probably, transient. In this region a twofold process may occur. The protein is first diluted and fills large vacuoles of the Golgi complex. Then it is progressively concentrated, forming *prozymogen granules* surrounded by a Golgi membrane.

Zymogen Stage. By progressive condensation, the enzymes migrate into the apical portion of the cell where they will be delivered by exocytosis accompanied by fusion of the surface membrane of the granule with the cell membrane at the luminal surface (Fig. 25–9, 4 to 4').

Intraluminal Stage. The enzymes progress slowly through the lumen of the acinus and the ducts and then are diluted by other secretions prior to entering the intestinal cavity.

Exocytosis and Secretion

The mechanism of exocytosis postulated for the secretion of catecholamines by the adrenal medulla has received interesting confirmation from biochemical studies. This

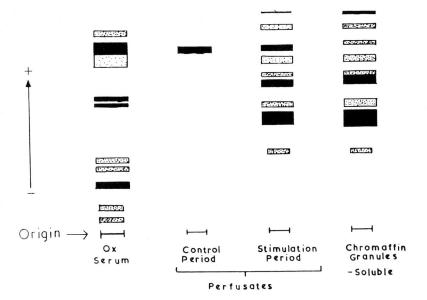

Figure 25–10. Starch-gel electrophoresis experiment demonstrating the secretion of several of the proteins contained in the chromaffin granules of the adrenal medulla. The perfusates were collected from bovine adrenals with and without stimulation by carbamylcholine. (From Smith, A. D., *in* Campbell, P. N. (Ed.), *The Interaction of Drugs on Subcellular Components of Animal Cells,* London, J. and A. Churchill Ltd., 1968, p. 239.)

mechanism implied that all the components present in the catechol-containing droplet (i.e., chromaffin granule) should be released at the same time.

Isolation of chromaffin granules by gradient centrifugation[31] has been achieved with a high degree of purity.[32] It was determined that the granules contained the following components (measured in per cent dry weight): protein (35.0%), lipids (22.0%), catecholamines (20.5%) and ATP (15.0%).

Most of the lipids, mainly phospholipids and cholesterol, are in the membrane, which can be separated by lysing the granules in hypotonic solutions. The membrane also contains some insoluble proteins and membrane-bound enzymes, particularly ATPase, dopamine β-hydroxylase and a special cytochrome (b-559). Within the granule there are several soluble proteins called *chromogranins*[33] (Fig. 25–10). It has been postulated that the amines are bound in a complex with the ATP and the proteins[34] and that for each molecule of chromogranin there are about 385

molecules of amine and 95 molecules of ATP.

Initial biochemical evidence for the mechanism of exocytosis was obtained from observations of the efferent venous blood of the adrenal gland, after stimulation, which showed adenine nucleotides in the same molar ratios with catecholamines as in the isolated granules.[35, 36] More direct evidence was provided by the detection of the soluble proteins of the granule also in the venous blood. Figure 25–10 shows that stimulation of the gland undergoing perfusion induces appearance of all the different chromogranins found in the chromaffin granules.[37] It has been demonstrated quantitatively that upon stimulation, together with the release of catecholamines, all the ATP and the chromogranins contained in the granule are secreted, while practically all the membrane-bound lipids and enzymes remain in the tissue. Such biochemical evidence fully supports the mechanism of exocytosis first proposed by De Robertis and Vaz Ferreira from electron microscopic observations.[14]

REFERENCES

1. Junqueira, L. C. V. (1967) In: *Secretory Mechanisms of Salivary Glands,* p. 286. (Schneyer, L. H., and Schneyer, C. A., eds.) Academic Press, Inc., New York.
2. Smith, R. E., and Farquhar, M. G. (1966) *J. Cell Biol., 31*:319.
3. Zambrano, D. (1969) *Z. Zellforsch., 93*:560.
4. De Robertis, E. (1940) *Anat. Rec., 78*:473.
5. De Robertis, E. (1941) *Anat. Rec., 80*:219.
6. Palay, S. L. (1958) In: *Frontiers in Cytology.* Yale University Press, New Haven.
7. Amsterdam, A., Ohad, I., and Schramm, M. (1969) *J. Cell Biol., 41*:753.
8. Robinson, G. A., Butcher, R. W., and Sutherland, E. W. (1967) *Ann. Acad. Sci. USA, 139*:703.
9. Schramm, M., Selinger, Z., Salomon Y., Eytan, E., and Batzri, S. (1972) *Nature (New Biol.), 240*:203.
10. Bonner, J. T., Barkley, D. S., Hall, E. M., Konijn, T. M., Mason, J. W., O'Keefe, G., and Wolfe, P. B. (1969) *Develop. Biol., 20*:72.
11. Pannbucker, R. G., and Bravard, L. J. (1972) *Science, 175*:1014.
12. Palade, G. E. (1956) *J. Biophys. Biochem. Cytol., 2*:417.
13. Siekevitz, P., and Palade, G. E. (1958) *J. Biophys. Biochem. Cytol., 4*:203.
14. De Robertis, E., and Vaz Ferreira, A. (1957) *J. Biophys. Biochem. Cytol., 3*:611.
15. De Robertis, E., and Sabatini, D. D. (1960) *Fed. Proc., 19*:70.
16. Smith, U., Smith, D. S., Winkler, H., et al. (1973) *Science, 179*:79.
17. Douglas, W. W. (1966) In: *Mechanisms of Release of Biogenic Amines* (Euler, U.S., von, et al. eds.) Pergamon Press, Inc., New York.
18. Castle, J. D., Jamieson, J. J., and Palade, G. E. (1972) *J. Cell Biol., 53*:290.
19. Warshawsky, H., Leblond, C. P., and Droz, B. (1963) *J. Cell Biol., 16*:1.
20. Berg, N. B., and Young, R. W. (1971) *J. Cell. Biol., 50*:469.
21. Jamieson, D., and Palade, G. E. (1971) *J. Cell Biol., 48*:503.
22. Redman, C. M., Siekevitz, P., and Palade, G. E. (1966) *J. Biol. Chem., 34*:597.
23. Redman, C. M. (1967) *J. Biol. Chem., 242*:761.
24. Redman, C. M., and Sabatini, D. D. (1966) *Proc. Natl. Acad. Sci. USA, 56*:608.
25. Sabatini, D. D., Tashiro, Y., and Palade, G. E. (1966) *J. Molec. Biol., 19*:503.
26. Zambrano, D., and De Robertis, E. (1967) *Z. Zellforsch., 76*:458.
27. Jamieson, J. D., and Palade, G. E. (1968) *J. Cell. Biol., 39*:580.
28. Schramm, M. (1967) *Ann. Rev. Biochem., 36*:307.
29. Jamieson, J. D., and Palade, G. E. (1971) *J. Cell Biol., 50*:135.
30. Kraehenbuhl, J. P., and Jamieson, J. D. (1972) *Proc. Natl. Acad. Sci. USA, 69*:1771.
31. Blaschko, H., Hagen, J. M., and Hagen, P. (1957) *J. Physiol., 139*:316.
32. Smith, A. D. (1968) In: *The Interaction of Drugs on Subcellular Components of Animal Cells,* p. 239 (Campbell, P. N., ed.) J. & A. Churchill Ltd., London.
33. Blaschko, H., Smith, A. D., Winkler, H., Van Den Bosch, H., and Van Deenen, L. L. M. (1967) *Biochem. J., 103*:30C–32C.
34. Hillarp, N. A. (1959) *Acta Physiol. Scand., 47*:271.
35. Douglas, W. W., and Poisner, A. M. (1966) *J. Physiol., 183*:236.
36. Banks, P. (1966) *Biochem. J., 101*:536.
37. Schneider, F. H., Smith, A. D., and Winkler, H. (1967) *Brit. J. Pharmacol., 31*:94.

ADDITIONAL READING

Ceccarelli, B., Meldolesi, J., and Clementi, F. (1974) Cytopharmacology of secretion. In: *Advances in Cytopharmacology:* Proceedings, Vol. 2, (Ceccarelli, B., Meldolesi, J., and Clementi, F., eds.) Raven Press, New York.

De Robertis, E., and Sabatini, D. D. (1960) Submicroscopic analysis of the secretory process in the adrenal medulla. *Fed. Proc., 19*:70.

Gabe, M., and Arvy, L. (1961) Gland cells. In: *The Cell,* Vol. 5, p. 1 (Brachet, J., and Mirsky, A. E., eds.) Academic Press, Inc., New York.

Jamieson, J. D. (1971) Role of the Golgi complex in the intracellular transport of secretory proteins. In: *Advances in Cytopharmacology: Proceedings,* Vol. 1, p. 181 (Clementi, F., and Ceccarelli, B., eds.) Raven Press, New York.

Junqueira, L. C., and Hirsch, G. C. (1956) Cell secretion: a study of pancreas and salivary glands. *Int. Rev. Cytol.,* 5:323.

Kurosomi, K. (1961) Electron microscopic analysis of the secretion mechanism. *Int. Rev. Cytol.,* 11:1.

Palay, S. L. (1958) The morphology of secretion. In: *Frontiers in Cytology.* (Palay, S. L., ed.) Yale University Press, New Haven.

Tartakoff, A. M., Greene, L. J., Jamieson, J. D., and Palade, G. E. (1974) Parallelism of processing of pancreatic proteins. In: *Advances in Cytopharmacology: Proceedings,* Vol. 2, p. 177 (Clementi, F., and Ceccarelli, B., eds.) Raven Press, New York.

INDEX